MODERN NONLINEAR OPTICS
Part 1

ADVANCES IN CHEMICAL PHYSICS

VOLUME LXXXV

EDITORIAL BOARD

MODERN NONLINEAR OPTICS
Part 1

Edited by

MYRON EVANS
Department of Physics
The University of North Carolina
Charlotte, North Carolina

STANISŁAW KIELICH
NonLinear Optics
Division
Institute of Physics
Adam Mickiewicz
University
Poznań, Poland

ADVANCES IN CHEMICAL PHYSICS
VOLUME LXXXV

Series Editors

ILYA PRIGOGINE
University of Brussels
Brussels, Belgium
and
University of Texas
Austin, Texas

STUART A. RICE
Department of Chemistry
and
The James Franck Institute
University of Chicago
Chicago, Illinois

AN INTERSCIENCE® PUBLICATION
JOHN WILEY & SONS, INC.
NEW YORK • CHICHESTER • BRISBANE • TORONTO • SINGAPORE

This text is printed on acid-free paper.

An Interscience® Publication

Library of Congress Catalog Number: 58-9935

ISBN 0-471-57548-8

Printed in the United States of America

10 9 8 7 6 5 4 3 2 1

CONTRIBUTORS TO VOLUME LXXXV
Part 1

WŁADYSŁAW ALEXIEWICZ, Institute of Physics, A. Mickiewicz University, Poznań, Poland

JIŘÍ BAJER, Department of Optics, Palacky University, Olomouc, Czechoslovakia

T. BANCEWICZ, Institute of Physics, A. Mickiewicz University, Poznań, Poland

J. BUCHERT, IUSL, City College of New York, New York

PAVEL CHMELA, Faculty of Mechanical Engineering, Technical University, Brno, Czechoslovakia

M. DĘBSKA-KOTŁOWSKA, Institute of Physics, A. Mickiewicz University, Poznań, Poland

Z. FICEK, Department of Physics, The University of Queensland, Queensland, Australia

ZDENĚK HRADIL, Department of Optics, Palacky University, Olomouc, Czechoslovakia

BOLESŁAWA KASPROWICZ-KIELICH, Institute of Physics, A. Mickiewicz University, Poznań, Poland

S. KIELICH, Institute of Physics, A. Mickiewicz University, Poznań, Poland

M. KOZIEROWSKI, Institute of Physics, A. Mickiewicz University, Poznań, Poland

J. R. LALANNE, CNRS Paul Pascal, Pessac, France

A. MIRANOWICZ, Institute of Physics, A. Mickiewicz University, Poznań, Poland

Z. OŻGO, Institute of Physics, A. Mickiewicz University, Poznań, Poland

JAN PEŘINA, Department of Optics, Palacky University, Olomouc, Czecholslovakia

VLASTA PEŘINOVÁ, Department of Optics, Palacky University, Olomouc, Czechoslovakia

K. PIĄTEK, Institute of Physics, A. Mickiewicz University, Poznań, Poland

GENEVIEVE RIVOIRE, POMA, University of Angers, France

ALEXANDER STANISLAW SHUMOVSKY, Joint Institute of Nuclear Research, Moscow, Russia

R. TANAŚ, Institute of Physics, A. Mickiewicz University, Poznań, Poland

R. ZAWODNY, Institute of Physics, A. Mickiewicz University, Poznań, Poland

INTRODUCTION

Few of us can any longer keep up with the flood of scientific literature, even in specialized subfields. Any attempt to do more and be broadly educated with respect to a large domain of science has the appearance of tilting at windmills. Yet the synthesis of ideas drawn from different subjects into new, powerful, general concepts is as valuable as ever, and the desire to remain educated persists in all scientists. This series, *Advances in Chemical Physics*, is devoted to helping the reader obtain general information about a wide variety of topics in chemical physics, a field which we interpret very broadly. Our intent is to have experts present comprehensive analyses of subjects of interest and to encourage the expression of individual points of view. We hope that this approach to the presentation of an overview of a subject will both stimulate new research and serve as a personalized learning text for beginners in a field.

ILYA PRIGOGINE
STUART A. RICE

PREFACE

Statistical molecular theories of electric, magnetic, and optical saturation phenomena developed by S. Kielich and A. Piekara in several papers in the late 1950s and 1960s clearly foreshadowed the developments of the next thirty years. In these volumes, we as guest editors have been honored by a positive response to our invitations from many of the most eminent contemporaries in the field of nonlinear optics. We have tried to give a comprehensive cross section of the state of the art of this subject. Volume 85 (Part 1) contains review articles by the Poznań School and associated laboratories, and volume 85 (Part 2 and Part 3) contains a selection of reviews contributed from many of the leading laboratories around the world. We thank the editors, Ilya Prigogine and Stuart A. Rice, for the opportunity to produce this topical issue.

The frequency with which the work of the Poznań School has been cited by others in these volumes is significant, especially considering the overwhelming societal difficulties that have faced Prof. Dr. Kielich and his School over the last forty years. Their work is notable for its unfailing rigor and accuracy of development and presentation, its accessibility to experimental testing, the systemic thoroughness of the subject matter, and the fact that it never seems to lag behind developments in the field. This achievement is all the more remarkable in the face of journal shortages and the lack of facilities that would be taken for granted in more fortunate centers of learning.

We hope that readers will agree that the contributors to these volumes have responded with readable and useful review material with which the state of nonlinear optics can be measured in the early 1990s. We believe that many of these articles have been prepared to an excellent standard. Nonlinear optics today is unrecognizably different from the same subject in the 1950s, when lasers were unheard of and linear physics ruled. In these three volumes we have been able to cover only a fraction of the enormous contemporary output in this field, and many of the best laboratories are not represented.

We hope that this topical issue will be seen as a sign of the ability of scientists all over the world to work together, despite the frailties of human society as a whole. In this respect special mention is due to Professor Mansel Davies of Criccieth in Wales, who was among the first in the West to recognize the significance of the output of the Poznań School.

Myron W. Evans

Charlotte, North Carolina
August 1993

CONTENTS

RELAXATION THEORY OF NONLINEAR PROCESSES IN THE SMOLUCHOWSKI ROTATIONAL DIFFUSION APPROXIMATION

WŁADYSŁAW ALEXIEWICZ AND BOLESŁAWA KASPROWICZ-KIELICH

Nonlinear Optics Department, Institute of Physics, Adam Mickiewicz University, Poznań, Poland

CONTENTS

This work was carried out within the framework of Project PB 2 0130 9101 of the State Committee for Scientific Research.

Modern Nonlinear Optics, Part 1, Edited by Myron Evans and Stanisław Kielich. Advances in Chemical Physics Series, Vol. LXXXV.
ISBN 0-471-57546-1

I. INTRODUCTION

It is our aim to give a review of the results achieved in the relaxation theory of nonlinear electro-optical effects in simple molecular liquids and dilute solutions of macromolecules in the approximation of equations of the Smoluchowski type for orientational molecular diffusion. Equations of this kind were first applied by Debye [1] in his classical theory of dielectric relaxation to elucidate the rotation of dipolar molecules in viscous media. With regard to the fundamental contribution of Smoluchowski's theory of Brownian motion [2–4] to the development of the molecular–kinetic theory of matter, we shall refer to the method used in the present paper as the Smoluchowski rotational diffusion approximation.

The classical theory of nonlinear electro-optical processes of molecular relaxation in intense electric fields of high and low frequency was first proposed by Kasprowicz-Kielich and coworkers [5, 6] and extended by the same authors at the Department of Nonlinear Optics in Poznań [7–14] to the more detailed description of the dispersional and absorptional behavior and time rise of the nonlinear electric polarization of the third order, induced in liquids by different reorienting fields.

The results of Refs. 5 and 6 are especially important. Owing to their high degree of generality, they cover a great variety of nonlinear optical effects [15–17] and can be extended to reorienting fields other than cosine-shaped field.

An extensive review of the present state of the theory concerning rotational Brownian motion has been given in the monograph of Evans et al. [18], the review by Coffey [19], and the monograph of Coffey et al. [20]. Smoluchowski's rotational diffusion equation, which provided the basis for the theory of optical birefringence proposed by Peterlin and Stuart [21] and Benoit [22] has also found applications in the theories of nonlinear depolarization of scattered light [23] and the optical Kerr effect [24] proposed by Kielich. It has also been applied in the theory of polarization of dipolar liquids by Novikov [25] and in that of dipolar but nonpolarizable liquids by Coffey and Paranjape [26].

We shall be considering a statistical assembly of mutually noninteracting, axially symmetric molecules, linearly and nonlinearly polarizable in an electric field and possessing a permanent dipole moment along their axis of symmetry. The potential energy of the molecule in the external reorienting field is then dependent on the polar angle ϑ only and, as shown by Morita [27], the rotational diffusion model leads to a scalar coefficient of rotational diffusion.

As shown by Perrin [28], if the potential energy of the molecule in the reorienting field $E(t)$ is moreover dependent on the azimuth angle ϕ, an asymmetry of the diffusion coefficients appears in the rotational diffusion equation leading to three different rotational relaxation times along the principal axes of symmetry of the ellipsoidal molecule [27–31]. It should be noted that the Smoluchowski equation in the form of Eq. 22 does not take into account molecular interactions, which play an essential role in liquids, and thus is applicable only to sufficiently dilute solutions of axial molecules and not excessively big macromolecules in spherical solvents.

The first attempt to include intermolecular interactions was made by Debye and Ramm [33], though the problem is by no means simple, as proved by the work of Frenkel [34]. The next step was taken by Budo [35], to whom is due the inclusion of intramolecular interactions between internally rotating dipolar assemblages in his treatment of dielectric relaxation. The theory was extended by Coffey et al. [36, 37] to comprise the influence of the moment of inertia. Quite recently, Morita and Watanabe [38] have shown that interactions of intramolecular rotating dipolar groups can lead to important divergences from the well-known Debye semicircle on the Cole–Cole graphs [39–41].

The equation of orientational diffusion has been used by Bloembergen and Lallemand [42] to explain the widening of the wings in induced Rayleigh and Raman scattering spectra from solutions of cyclohexane in carbon disulfide. The optical reorientation of molecules in liquids due to birefringence induced by the field of a light wave can also be applied in the synchronization if longitudinal laser modes [43]. Rotational molecular diffusion also plays an essential role in the depolarization of fluorescence from solutions of dipolar liquids as proved, for example, by the theory of Pierov [44]. Tolles [45] found a numerical solution of the Smoluchowski equation for a system of noninteracting nondipolar molecules showing that for high intensities of the electric field when the parameter of polarizability anisotropy reorientation takes values $q > 2$ (see Eq. (24)) the effective relaxation time becomes shorter.

Noteworthy is the so-called model of restricted rotational diffusion proposed by Warchol and Vaughan [46], who restricted the variability of the angle of rotation ϑ to the interval $(0 < \omega_0 < \pi/2)$ in the equation of rotational diffusion for noninteracting cylindrical molecules, for which $a_{33} - a_{11} > 0$. Their model was applied by Wang and Pecora [47] to calculate the time correlation functions of the spherical harmonics of the first and second order essential in dielectric relaxation, light scattering, and fluorescence depolarization. Kühle and Rapp [48], having recourse to the properties of coupled differential equations, solved the Smoluchowski equation for noninteracting, nonpolarizable dipolar molecules in a dc electric field with high jumpwise changes in field strength and showed that the averaged Legendre polynomials $\langle P_1(\cos\vartheta)\rangle$ and $\langle P_2(\cos\vartheta)\rangle$ are dependent on the effective relaxation times, which decrease with growing electric field strength.

Attention should be drawn to the extensive theoretical paper of Watanabe and Morita [49] on Kerr effect relaxation in strong electric fields of different shapes, which however, does not take into consideration the effect of nonlinear polarizability of the molecules. Their general method of solving the Smoluchowski equation (22) has recourse to the formalism of Laplace transforms.

Molecular light scattering (LS) is on its own a vast field of science where use is made of the equations of rotational and translational molecular diffusion in liquids. Extensive discussions of linear Rayleigh LS are to be found in the monographs of Berne and Pecora [50] and Chu [51]. Nonlinear LS has been dealt with in full detail by Kielich and his coworkers [52–55]. The earliest and by no means easy measurements of the spectral broadening of second-harmonic Rayleigh LS in different liquids due to rotational molecular diffusion were carried out by Maker [56]. His work has acted as a stimulus for further spectral theories of hyper-Raman [57] and hyper-Rayleigh [58–60] LS in liquids.

Smoluchowski-type equations modeling Brownian motion of molecules in the presence of external forces consistently play an important role in modern physics [19, 61]. Examples of their application are to be found in the theory of the spectral linewidth of nuclear magnetic resonance [62], in the theoretical description of incoherent scattering of slow neutrons in liquids [63, 64] and of current–voltage characteristics of the Josephson junction [65], in calculations of the influence of quantum noise on the mean frequency of ring laser gyroscopes [66], in the theory of harmonic frequency mixing by a Brownian particle in a cosine potential [67] and of the lineshape of single-mode semiconductor lasers [19], and in the description of the slowing down of electrons in gaseous moderators [19]. Further examples of their applications are given in Refs. 3, 19, and 61.

II. DISPERSION AND ABSORPTION OF MOLECULAR NONLINEAR POLARIZABILITIES IN THE APPROACH OF CLASSICAL ELECTRON THEORY

The classical theory of electron dispersion reaches back to Lorentz [68] and has been extended to nonlinear electron dispersion by Voigt [69]. Voigt's ideas concerning nonlinear distortion of molecular systems have stimulated numerous, repeatedly reviewed investigations [70–73].

Here, we discuss the classical results of Kasprowicz-Kielich and Kielich [6] for the effect of nonlinear electronic dispersion on the polarizability of molecules in liquids. To this aim, we consider a molecular system, composed of n electrons, the sth electron having the mass m, electric charge e, and radius vector $\mathbf{r}_s$. In the classical theory of Lorentz, the equation of motion of the electron is of the form [68]

$$m\frac{\mathrm{d}^2\mathbf{r}_s(t)}{\mathrm{d}t^2} + m\Gamma_s\frac{\mathrm{d}\mathbf{r}_s(t)}{\mathrm{d}t} = \mathbf{f}_s^{\mathrm{L}} + \mathbf{f}_s^{\mathrm{B}} \tag{1}$$

Quite generally, the electron is acted on by a Lorentz force

$$\mathbf{f}_s^{\mathrm{L}}(\mathbf{r}, t) = e\left[\mathbf{E}(\mathbf{r}, t) + \frac{1}{c}\frac{\mathrm{d}\mathbf{r}_s(t)}{\mathrm{d}t} \times \mathbf{H}(\mathbf{r}, t)\right] \tag{2}$$

with $\mathbf{E}(\mathbf{r}, t)$ and $\mathbf{H}(\mathbf{r}, t)$ denoting the electric and magnetic electromagnetic field vectors in the space–time point $(\mathbf{r}, t)$.

In accordance with Voigt's hypothesis [69], the electron is bound to the center of the microsystem by the anharmonic force

$$\begin{aligned} f_{si}^{\mathrm{B}}(\mathbf{r}, t) = &-\alpha_{ij}^{(s)}r_{si}(t) - \tfrac{1}{2}\beta_{ijk}^{(s)}r_{si}(t)r_{sk}(t) \\ &-\tfrac{1}{6}\gamma_{ijkl}^{(s)}r_{si}(t)r_{sk}(t)r_{sl}(t) - \cdots \end{aligned} \tag{3}$$

where the coefficients $\alpha_{ij}^{(s)}$ describe the harmonic vibrations and $\beta_{ijk}^{(s)}$ and $\gamma_{ijkl}^{(s)}$ the anharmonic vibrations of the sth electron.

In Eq. (2), Γ_s is a coefficient of the damping force acting on the sth electron. In the absence of the Lorentz force (2) if all damping forces are zero ($\Gamma = 0$), and in the absence of anharmonic force coefficients $\beta_{ijk}^{(s)}$, $\gamma_{ijkl}^{(s)}$, the electrons perform harmonic vibrations at the circular frequencies

$$\omega_m^2 = \frac{\alpha_{ij}^{(s)}}{m} \tag{4}$$

In the presence of a Lorentz force (2) with a time dependence of the form

$$E(t) = \sum_a E_a \exp(-\mathrm{i}\omega_a t) \tag{5}$$

the electron performs forced vibrations with the fundamental frequency ω_a and sum frequencies ω_{ab}, ω_{abc}:

$$\omega_{ab} = \omega_a + \omega_b \qquad \omega_{abc} = \omega_a + \omega_b + \omega_c \tag{6}$$

described by

$$\begin{aligned} r_{si}(t) = {} & \sum_a r_{si}(\omega_a)\exp(-\mathrm{i}\omega_a t) + \sum_{ab} r_{si}(\omega_{ab})\exp(-\mathrm{i}\omega_{ab} t) \\ & + \sum_{abc} r_{si}(\omega_{abc})\exp(-\mathrm{i}\omega_{abc} t) + \cdots \end{aligned} \tag{7}$$

The solution of the problem in Ref. 6 is restricted to the electric part of the Lorentz force (the magnetic field has been considered in Refs. 74 and 75). Hence, by Eqs. (1)–(7), the final result is

$$\begin{aligned} r_{si}(\omega_a) &= e\alpha_{ij}^{(s)} D_s(\omega_a) E_j(\omega_a) \\ r_{si}(\omega_{ab}) &= \tfrac{1}{2} e^2 \beta_{ijk}^{(s)} D_s(\omega_a) D_s(\omega_b) D_s(\omega_{ab}) E_j(\omega_a) E_k(\omega_b) \\ r_{si}(\omega_{abc}) &= \tfrac{1}{6}\{3e^2 \beta_{ijk}^{(s)} [D_s(\omega_a) D_s(\omega_{bc}) E_j(\omega_a) E_k(\omega_{bc}) \\ &\qquad + D_s(\omega_{ab}) D_s(\omega_c) E_j(\omega_{ab}) E_k(\omega_c)] \\ &\qquad + e^3 \gamma_{ijkl}^{(s)} D_s(\omega_a) D_s(\omega_b) D_s(\omega_c) E_j(\omega_a) E_k(\omega_b) E_l(\omega_c)\} \\ &\quad \times D_s(\omega_{abc}) \end{aligned} \tag{8}$$

In Eq. (8) the complex electron dispersion functions are of the form

$$\begin{aligned} D_s^{-1}(\omega_a) &= m(\omega_s^2 - \omega_a^2 - \mathrm{i}\omega_a \Gamma_s) \\ D_s^{-1}(\omega_{ab}) &= m[\omega_s^2 - (\omega_a + \omega_b)^2 - \mathrm{i}(\omega_a + \omega_b)\Gamma_s] \\ D_s^{-1}(\omega_{abc}) &= m[\omega_s^2 - (\omega_a + \omega_b + \omega_c)^2 - \mathrm{i}(\omega_a + \omega_b + \omega_c)\Gamma_s] \end{aligned} \tag{9}$$

Since, by definition, the total induced electric dipole moment of the molecule is

$$\mathbf{p}(t) = \sum_{s=1}^{n} e\mathbf{r}_s(t) \tag{10}$$

by using Eq. 7, we obtain

$$\begin{aligned}\mathbf{p}(t) = &\sum_{a} \mathbf{p}^{(1)}(\omega_a)\exp(-\mathrm{i}\omega_a t) + \sum_{ab} \mathbf{p}^{(2)}(\omega_{ab})\exp(-\mathrm{i}\omega_{ab} t)\\ &+ \sum_{abc} \mathbf{p}^{(3)}(\omega_{abc})\exp(-\mathrm{i}\omega_{abc} t) + \cdots\end{aligned} \tag{11}$$

where the dipole moment components of successive orders are

$$\begin{aligned}p_i^{(1)}(\omega_a) &= a_{ij}^{\omega_a}(\omega_a) E_j^0(\omega_a)\\ p_i^{(2)}(\omega_{ab}) &= \tfrac{1}{2} b_{ijk}^{\omega_{ab}}(\omega_a, \omega_b) E_j^0(\omega_a) E_k^0(\omega_b)\\ p_i^{(3)}(\omega_{abc}) &= \tfrac{1}{6} c_{ijkl}^{\omega_{abc}}(\omega_a, \omega_b, \omega_c) E_j^0(\omega_a) E_k^0(\omega_b) E_l^0(\omega_c)\end{aligned} \tag{12}$$

The tensors of the linear electron polarizability and nonlinear electron polarizabilities of the molecule are expressed as follows:

$$a_{ij}^{\omega_a}(\omega_a) = \sum_{s=1}^{n} e^2 D_s(\omega_a) \alpha_{ij}^{(s)} \tag{13}$$

$$b_{ijk}^{\omega_{ab}}(\omega_a, \omega_b) = \sum_{s=1}^{n} e^3 D_s(\omega_a) D_s(\omega_b) D_s(\omega_{ab}) \beta_{ijk}^{(s)} \tag{14}$$

$$\begin{aligned}c_{ijkl}^{\omega_{abc}}(\omega_a, \omega_b, \omega_c) = \frac{1}{2} \sum_{s=1}^{n} e^4 \Big[&3\beta_{iml}^{(s)} \beta_{mjk}^{(s)} D_s(\omega_{ab})\\ &+ 3\beta_{ijm}^{(s)} \beta_{mkl}^{(s)} D_s(\omega_{bc}) + 2\gamma_{ijkl}^{(s)}\Big]\\ &\times D_s(\omega_a) D_s(\omega_c) D_s(\omega_{abc})\end{aligned} \tag{15}$$

Once we know how these molecular polarizability tensors depend on high (optical) frequencies, we are in a position to determine the electron dispersion and absorption of the macroscopic polarizability tensors. In fact, high-frequency dispersion, which is due to electron processes practically independent of molecular correlations, can be written, for a medium of N molecules, to a very good approximation, in the form

$$
\begin{aligned}
A_{ij}^{\omega_a}(\Gamma, \omega_a) &= \sum_{p=1}^{N} a_{ij}^{\omega_a}(\Gamma_p, \omega_a) \\
B_{ijk}^{\omega_{ab}}(\Gamma, \omega_a, \omega_b) &= \sum_{p=1}^{N} b_{ijk}^{\omega_{ab}}(\Gamma_p, \omega_a, \omega_b) \\
C_{ijkl}^{\omega_{abc}}(\Gamma, \omega_a, \omega_b, \omega_c) &= \sum_{p=1}^{N} c_{ijkl}^{\omega_{abc}}(\Gamma_p, \omega_a, \omega_b, \omega_c)
\end{aligned}
\tag{16}
$$

On the assumption that all N molecules in V are of one species, the sum of Eq. (16) can simply be replaced by N. Thus, formulae (13) and (14) describe linear and nonlinear electron dispersion and absorption by resolution of the complex functions (9) and their products into real and imaginary parts.

III. THE RELAXATION FUNCTIONS WITHIN THE DEBYE–SMOLUCHOWSKI MODEL OF MOLECULAR ROTATIONAL DIFFUSION

Let us consider an isotropic dielectric of volume V consisting of a great number N of noninteracting molecules possessing a permanent dipole moment m directed along their symmetry 3 axis; components $a_{11} = a_{22} = a_{33}$ of linear electric polarizability; and nonzero components $b_{133} = b_{233}, b_{333}$ as well as $c_{1111}, c_{1133}, c_{3333}$ of their nonlinear polarizabilities.

The change in potential energy of the molecules in an external electric field E_Z applied to the dielectric along the laboratory Z axis can be expressed with sufficient accuracy in the form of the expansion

$$
\begin{aligned}
u(\vartheta, E_Z) = {} & -mE_Z P_1(\cos\vartheta) - \tfrac{1}{2}\left[a_0 + a_2 P_2(\cos\vartheta)\right] E_Z^2 \\
& - \tfrac{1}{6}\left[b_1 P_1(\cos\vartheta) + b_3 P_3(\cos\vartheta)\right] E_Z^3 \\
& - \tfrac{1}{24}\left[c_0 + c_2 P_2(\cos\vartheta) + c_4 P_4(\cos\vartheta)\right] E_Z^4
\end{aligned}
\tag{17}
$$

where we have introduced the irreducible components of linear and nonlinear polarizabilities equal, respectively, to [17]

$$
\begin{aligned}
a_0 &= \tfrac{1}{3}(a_{33} + 2a_{11}) \qquad a_2 = \tfrac{2}{3}(a_{33} - a_{11}) \\
b_1 &= \tfrac{3}{5}(b_{333} + 2b_{113}) \qquad b_3 = \tfrac{2}{5}(b_{333} - 3b_{113}) \\
c_0 &= \tfrac{1}{15}(8c_{1111} + 12c_{1133} + 3c_{3333}) \\
c_2 &= \tfrac{4}{21}(-4c_{1111} + 3c_{1133} + 3c_{3333}) \\
c_4 &= \tfrac{8}{35}(c_{1111} - 6c_{1133} + c_{3333})
\end{aligned}
\tag{18}
$$

For axially symmetric molecules, the energy (18) is dependent only on the polar angle ϑ between the symmetry axis of the molecule and the Z axis of the laboratory system of coordinates. The $P_i(\cos\vartheta)$ are Legendre polynomials.

An external electric field E_Z applied to the dielectric differs from the mean macroscopic field E_Z^m acting on the molecule. Assuming that the molecule is inside a spherical sample with isotropic electric permittivity ε and if this spherical sample is immersed in an isotropic continuous medium of electric permittivity ε_0, then classical electrostatics leads to

$$
E_Z = \frac{\varepsilon + 2\varepsilon_0}{3\varepsilon_0} E_Z^m = L E_Z^m \tag{19}
$$

To gain insight into the influence of molecular rotational diffusion on an arbitrary measured physical quantity $A[\Omega_t, E(t)]$ at a moment of time t subsequent to the application of the time-dependent electric field $E(t)$, we have recourse to a classical statistical averaging procedure defined as follows [3]:

$$
\langle A[\Omega_t, E(t)]\rangle = \frac{1}{V}\int A[\Omega_t, E(t)]\, f[\Omega_t; t, E(t)]\, d\Omega_t \tag{20}
$$

In Debye's model of dielectric relaxation [1], the time-dependent distribution function $f[\Omega_t; t, E(t)]$ can be determined from the Smoluchowski kinetic equation of diffusion [6, 28, 29]:

$$
\begin{aligned}
\frac{\partial f(\Omega_t, t)}{\partial t} = D_{ij}\Big\{ & \nabla_i \nabla_j f(\Omega_t, t) \\
& + \frac{1}{kT}\big[\nabla_i f(\Omega_t, t)\nabla_j u(\Omega_t, t) \\
& + f(\Omega_t, t)\nabla_i \nabla_j u(\Omega_t, t)\big]\Big\}
\end{aligned}
\tag{21}
$$

where D_{ij} is the second-rank symmetric diffusion tensor, ∇_i is the operator of spatial differentiation, k is Boltzmann's constant, and T is the absolute temperature. The function $f(\Omega_t, t)$ defines the probability of the molecule having the orientation Ω_t at time t, with Ω denoting the set of Euler angles.

In the present case, the Smoluchowski equation of rotational diffusion takes the form

$$\begin{aligned}\frac{1}{D}\frac{\partial f(\vartheta,t)}{\partial t} &= \frac{1}{\sin\vartheta}\frac{\partial}{\partial\vartheta}\left[\sin\vartheta\frac{\partial f(\vartheta,t)}{\partial\vartheta}\right]\\ &\quad+\frac{1}{kT}\left\{\frac{\partial u(\vartheta,E_Z)}{\partial\vartheta}\frac{\partial f(\vartheta,t)}{\partial\vartheta}\right.\\ &\quad\left.+\frac{f(\vartheta,t)}{\sin\vartheta}\frac{\partial}{\partial\vartheta}\left[\sin\vartheta\frac{\partial u(\vartheta,E_Z)}{\partial\vartheta}\right]\right\}\end{aligned} \tag{22}$$

$D = I\beta$ denotes the scalar coefficient of rotational diffusion of the molecule, β is the friction constant, and I is the moment of inertia of the molecule.

Equation (22) can be solved by the approximate method of classical statistical perturbation calculus [5–8]. We start by expanding the Maxwell–Boltzmann distribution function $f(\vartheta, E_Z)$ for a time-independent reorienting field E_Z:

$$\begin{aligned}f(\vartheta,E_Z) &= \frac{\exp[-u(\vartheta,E_Z)/kT]}{\int_0^\pi \exp[-u(\vartheta,E_Z)/kT]\sin\vartheta\, d\vartheta}\\ &= f_0 + f_1(\vartheta,E_Z) + f_2(\vartheta,E_Z) + f_3(\vartheta,E_Z) + \cdots\end{aligned} \tag{23}$$

in a power series in the molecular parameter of dipole moment reorientation p, the parameter of linear polarizability anisotropy q, and the parameters of nonlinear polarizability anisotropy s_1, s_3 defined as follows:

$$\begin{aligned}p &= \frac{mE_Z}{kT} \qquad q = \frac{|a_{33}-a_{11}|}{2kT}E_Z^2\\ s_1 &= \frac{(3b_{333}+2b_{113})}{5kT}E_Z^3 \qquad s_3 = \frac{2|b_{333}-b_{113}|}{5kT}E_Z^3\end{aligned} \tag{24}$$

The series expansion of (23), valid for $p, q, s_1, s_3 \ll 1$, leads to the successive components of the distribution function:

$$\begin{aligned} f_0 &= \tfrac{1}{2} \\ f_1(\vartheta, E_Z) &= \tfrac{1}{2} p P_1(\cos\vartheta) \pm \tfrac{1}{3} q P_2(\cos\vartheta) + \tfrac{1}{12} s_1 P_1(\cos\vartheta) \\ &\quad + \tfrac{1}{12} s_3 P_3(\cos\vartheta) \qquad (25) \\ f_2(\vartheta, E_Z) &= \tfrac{1}{6} p^2 P_2(\cos\vartheta) \pm \tfrac{1}{15} pq\left[P_1(\cos\vartheta) - 6P_3(\cos\vartheta)\right] \\ f_3(\vartheta, E_Z) &= \tfrac{1}{30} p^3\left[P_3(\cos\vartheta) - P_1(\cos\vartheta)\right] \end{aligned}$$

where the $P_i(\cos\vartheta)$ are Legendre polynomials. For a time-variable reorienting field

$$E_Z(t) = \sum_a E_a g_a(t) \qquad (26)$$

of the shape $g_a(t)$ we assume the statistical distribution function in the time-dependent form

$$\begin{aligned} f_1(\vartheta, t) &= \tfrac{1}{2} P_1(\cos\vartheta)\left[\sum_a p_a A_{11}^a(t) + \tfrac{1}{6}\sum_{abc} s_{1abc} B_{11}^{abc}(t)\right] \\ &\quad \pm \tfrac{1}{3} P_2(\cos\vartheta)\sum_{ab} q_{ab} A_{12}^{ab}(t) \qquad (27) \\ &\quad \pm \tfrac{1}{12} P_3(\cos\vartheta)\sum_{abc} s_{3abc} B_{13}^{abc}(t) \end{aligned}$$

$$\begin{aligned} f_2(\vartheta, t) &= \pm \tfrac{1}{15} P_1(\cos\vartheta)\sum_a p_a q_{bc} A_{21}^{abc}(t) \\ &\quad + \tfrac{1}{6} P_3(\cos\vartheta)\sum_{ab} p_a p_b A_{22}^{ab}(t) \qquad (28) \\ &\quad \pm \tfrac{2}{5} P_3(\cos\vartheta)\sum_{abc} p_a q_{bc} A_{23}^{abc}(t) \end{aligned}$$

$$\begin{aligned} f_3(\vartheta, t) &= \tfrac{1}{30}\sum_{ab} p_a p_b\left[P_3(\cos\vartheta) A_{33}^{ab}(t)\right. \\ &\quad \left. - P_1(\cos\vartheta) A_{31}^{ab}(t)\right] \qquad (29) \end{aligned}$$

with $p_a = mE_a/kT$, $q_{ab} = (|a_{33} - a_{11}|/2kT)E_a E_b$, and so on.

The unknown reorientation relaxation functions $A_{ij}^a(t)$, $A_{ij}^{ab}(t)$, $A_{ij}^{abc}(t)$, $B_{ij}^{abc}(t)$ introduced above are found on insertion of Eqs. (27)–(29) into the Smoluchowski equation (Eq. 22) and on having recourse to the orthogonality properties of Legendre polynomials:

$$\frac{1}{2}\int_0^{\pi} P_i(\cos\vartheta)P_j(\cos\vartheta)\sin\vartheta\, d\vartheta = \frac{\delta_{ij}}{2j+1} \tag{30}$$

where δ_{ij} is the Kronecker delta. On equating in (22) the terms with the same powers of the reorientation parameters, we get the following equations:

$$\tau_1 \dot{A}_{11}^{a} + A_{11}^{a} = g_a(t) \tag{31}$$

$$\tau_2 \dot{A}_{12}^{ab} + A_{12}^{ab} = g_a(t)g_b(t) \tag{32}$$

$$\tau_2 \dot{A}_{22}^{ab} + A_{22}^{ab} = A_{11}^{a} g_b(t) \tag{33}$$

$$\tau_1 \dot{A}_{21}^{abc} + A_{21}^{abc} = \tfrac{1}{2}\left[3A_{11}^{a} g_b(t) - A_{12}^{ab}\right] g_c(t) \tag{34}$$

$$\tau_1 \dot{A}_{31}^{abc} + A_{31}^{abc} = A_{22}^{ab} g_c(t) \tag{35}$$

$$\tau_1 \dot{B}_{11}^{abc} + B_{11}^{abc} = g_a(t)g_b(t)g_c(t) \tag{36}$$

where the dots denote time derivatives. Equations (31)–(36) with appropriate initial conditions permit the calculation of the respective shapes of the orientational relaxation rise functions for different shapes of the reorienting pulses $g_a(t)$. The τ_k are rotational relaxation times, related simply with the rotational diffusion coefficient D as follows:

$$\tau_k = [k(k+1)D]^{-1} = 2\tau_1[k(k+1)]^{-1} \tag{37}$$

It is worth noting that for $g_a(t) = g_b(t) = g_c(t) = 0$, the simplified equations (31)–(36) describe the shape of the polarization decay relaxation functions on switching off the electric reorienting fields.

IV. THIRD-ORDER ELECTRIC POLARIZATION IN LIQUIDS

On neglecting intermolecular interactions, the Z component of the total dipole moment of the liquid dielectric system $M_Z[E_Z(t)]$, induced by the time-variable field $E_Z(t)$, is given by the expansion [15, 17, 32]

$$\begin{aligned} \frac{1}{N} M_Z[E_Z(t)] &= mP_1(\cos\vartheta) + [a_0 + a_2P_2(\cos\vartheta)]E_Z(t) \\ &\quad + \tfrac{1}{2}[b_1P_1(\cos\vartheta) + b_3P_3(\cos\vartheta)]E_Z^2(t) \\ &\quad + \tfrac{1}{6}[c_0 + c_2P_2(\cos\vartheta) + c_4P_4(\cos\vartheta)]E_Z^3(t) + \cdots \end{aligned} \tag{38}$$

whereas the electric polarization of the system, acted on by the field $E_Z(t)$, is given by the statistical mean value

$$\begin{aligned}\langle P_Z(t)\rangle &= \langle P_Z[E_Z(t),t]\rangle_{\vartheta,t} \\ &= \frac{1}{V}\int_0^{\pi} M_Z[E_Z(t)]f[\vartheta, E_Z(t)]\sin\vartheta\, d\vartheta.\end{aligned} \tag{39}$$

The isotropically averaged electric polarization (38) splits into two components, proportional to the first and third power of $E_Z(t)$:

$$\langle P_Z(t)\rangle = \langle P_Z^{(1)}(t)\rangle + \langle P_Z^{(3)}(t)\rangle \tag{40}$$

By Eqs. (38), (39) and (27)–(30) these components take the form

$$\langle P_Z^{(1)}(t)\rangle = \frac{\rho m^2}{kT}\sum_a A_{11}^a(t)E_a \tag{41}$$

$$\begin{aligned}\langle P_Z^{(3)}(t)\rangle = \rho\sum_{abc}\{&\theta_1 A_{12}^{ab}(t)g_c(t) + \theta_2[A_{21}^{abc}(t) + A_{21}^{ab}(t)g_c(t)] \\ &-\theta_3 A_{31}^{abc}(t) \\ &+\theta_4[3A_{11}^a(t)g_b(t)g_c(t) + B_{11}^{abc}(t)] \\ &+\tfrac{1}{6}c_0 g_a(t)g_b(t)g_c(t)\}E_aE_bE_c\end{aligned} \tag{42}$$

with $\theta_1, \theta_2, \theta_3$ denoting the parameters of axially symmetric molecules.

$$\begin{aligned}\theta_1 &= \frac{2}{45kT}(a_{33}-a_{11})^2 \\ \theta_2 &= \frac{2m^2}{45k^2T^2}(a_{33}-a_{11}) \\ \theta_3 &= \frac{m^4}{45k^3T^3}\end{aligned} \tag{43}$$

intervening in the theory of dielectrics [17, 39–41].

In Eq. (42), the term in θ_1 is related with the process of Langevin reorientation of the polarizability ellipsoid of the molecule. That in θ_2 is the so-called Debye–Born term; it can take positive as well as negative values, and plays an essential role in Kerr's effect. The negative term involving θ_3, referred to as the Debye term, is related with reorientation of the permanent dipole moment of the molecule. In Eq. (42) a new term appears involving a parameter θ_4 related with the contribution from nonlinear molecular polarizability

$$\theta_4 = \frac{3m}{50kT}(b_{333} + 2b_{113}) \tag{44}$$

introduced by Piekara and Kielich [76, 77] in their theory of changes in the electric susceptibility of dielectrics in strong electric fields and denoted by them as Δ_2^{ee}.

Equations (41) and (42), which represent the essential results of this section, describe the electric polarization of liquids, induced by a reorienting electric field, as a function of time. Assuming that the dielectric experiences the action of three electric fields with frequencies $\omega_a, \omega_b, \omega_c$

$$\begin{aligned} E_Z(t) &= \tfrac{1}{2}\sum_a E(\omega_a)\exp(-\mathrm{i}\omega_a t) \\ &= \tfrac{1}{2}\sum_a L(\omega_a)E_m(\omega_a)\exp(-\mathrm{i}\omega_a t) \end{aligned} \tag{45}$$

where summation over a extends over positive as well as negative values of the ω_a, with $\omega_{-a} = -\omega_a$, its nonlinear electric polarization $\langle P_Z^{(3)}(\omega_{abc}, t)\rangle$ at the frequency $\omega_{abc} = \omega_a + \omega_b + \omega_c$ can be expressed by

$$\begin{aligned} \langle P_Z^{(3)}(\omega_{abc}, t)\rangle = \chi(-\omega_{abc}; \omega_a, \omega_b, \omega_c; t)E_m(\omega_a)E_m(\omega_b)E_m(\omega_c) \\ \times\exp(-\mathrm{i}\omega_{abc}t) \end{aligned} \tag{46}$$

In Eq. (46) the scalar electric susceptibility of order 3 of the isotropic medium is the sum of two parts:

$$\begin{aligned} \chi(-\omega_{abc}; \omega_a, \omega_b, \omega_c; t) = \chi(-\omega_{abc}; \omega_a, \omega_b, \omega_c) \\ + \delta\chi(-\omega_{abc}; \omega_a, \omega_b, \omega_c; t) \end{aligned} \tag{47}$$

the first part being independent of time and describing the steady-state behavior of the medium (a long time after switching on the fields (45)), and another, transient part, describing the rise in time of the polarization (46) after switching on the reorienting electric fields. The scalar susceptibility may also be written in the form [6]

$$\chi(-\omega_{abc};\omega_a,\omega_b,\omega_c) = \frac{1}{6V}L(\omega_{abc})L_m(\omega_a)L_m(\omega_b)L_m(\omega_c) \times\langle C(-\omega_{abc};\omega_a,\omega_b,\omega_c)\rangle \tag{48}$$

where $C(-\omega_{abc};\omega_a,\omega_b,\omega_c)$ denotes the scalar nonlinear polarizability of order 3 of the isotropic dielectric medium. Moreover, we have for the mean field, the relation

$$L(\omega) = \frac{\varepsilon(\omega) + 2\varepsilon_0}{3\varepsilon_0} \tag{49}$$

In our further discussion it will prove convenient to expand the nonlinear susceptibility in a power series in the inverse energy $(kT)^{-1}$, permitting an elegant separation of the various molecular contributions to the nonlinear electric polarization of order 3:

$$\begin{aligned}\chi(-\omega_{abc};\omega_a,\omega_b,\omega_c;t) = {}& \chi^{(0)}(-\omega_{abc};\omega_a,\omega_b,\omega_c;t)\\ &+\frac{1}{kT}\chi^{(1)}(-\omega_{abc};\omega_a,\omega_b,\omega_c;t)\\ &+\frac{1}{k^2T^2}\chi^{(2)}(-\omega_{abc};\omega_a,\omega_b,\omega_c;t)\\ &+\frac{1}{k^3T^3}\chi^{(3)}(-\omega_{abc};\omega_a,\omega_b,\omega_c;t)\end{aligned} \tag{50}$$

In the zeroth approximation, for the fields (45) we obtain

$$\chi^{(0)}(-\omega_{abc};\omega_a,\omega_b,\omega_c;t) = \tfrac{1}{6}c_0 \tag{51}$$

Thus, in the approximation of Smoluchowski rotational diffusion, nonlinear third-order polarizability contributes nothing to the process of rise in polarization, where the essential contributions come from fast electron processes of absorption and dispersion of the Lorentz type.

Further contributions to the susceptibility (50) are given by [8]

$$\begin{aligned}
\chi^{(1)}(-\omega_{abc};\omega_a,\omega_b,\omega_c;t) &= \underset{abc}{\mathbf{S}}\left[\tfrac{2}{45}\gamma(-\omega_{abc},\omega_a)\gamma(\omega_b,\omega_c)Q_{20}(\omega_{bc};t)\exp(-\mathrm{i}\omega_a t)\right. \\
&\quad + \tfrac{1}{10}m(-\omega_{abc})b(\omega_a,\omega_b,\omega_c)Q_{10}(\omega_{abc};t) \\
&\quad \left. + \tfrac{3}{10}b(-\omega_{abc},\omega_a,\omega_b)m(\omega_c)Q_{10}(\omega_c;t)\exp(-\mathrm{i}\omega_a t)\right]
\end{aligned} \tag{52}$$

$$\begin{aligned}
\chi^{(2)}(-\omega_{abc};\omega_a,\omega_b,\omega_c;t) &= \tfrac{2}{45}\underset{abc}{\mathbf{S}}\left(\tfrac{1}{2}m(-\omega_{abc})m(\omega_a)\gamma(\omega_b,\omega_c)\right. \\
&\quad \times\{3R_{10}(\omega_a)[Q_{10}(\omega_{abc};t)-Q_{11}(\omega_{bc};t)] \\
&\quad -R_{20}(\omega_{bc})[Q_{10}(\omega_{abc};t)-Q_{12}(\omega_a;t)]\} \\
&\quad +\gamma(-\omega_{abc},\omega_a)m(\omega_b)m(\omega_c)R_{10}(\omega_c) \\
&\quad \times[Q_{20}(\omega_{bc};t)-Q_{21}(\omega_b;t)] \\
&\quad \left.\times\exp(-\mathrm{i}\omega_a t)\right)
\end{aligned} \tag{53}$$

$$\begin{aligned}
\chi^{(3)}(-\omega_{abc};\omega_a,\omega_b,\omega_c;t) &= -\tfrac{1}{45}\underset{abc}{\mathbf{S}}\, m(-\omega_{abc})m(\omega_a)m(\omega_b)m(\omega_c) \\
&\quad R_{10}(\omega_c)\{R_{20}(\omega_{bc})[Q_{10}(\omega_{abc};t)-Q_{12}(\omega_a;t)] \\
&\quad -R_{21}(\omega_b)[Q_{11}(\omega_{ab};t)-Q_{12}(\omega_a;t)]\}
\end{aligned} \tag{54}$$

In Eqs. (52)–(54) we have introduced the time-dependent relaxational functions

$$Q_{km}(\omega_{abc};t) = R_{km}(\omega_{abc})\left[\exp\left(-\mathrm{i}\omega_{abc}t-\frac{t}{\tau_m}\right)-\exp\left(-\frac{t}{\tau_k}\right)\right] \tag{55}$$

where $\mathbf{S}_{abc}$ is the operator of symmetrization over all permutations of the frequencies $\omega_a, \omega_b, \omega_c$.

Equations (50)–(55) represent the chief result of our investigation. They describe the evolution in time of the process leading, in liquids, to a steady state of third-order polarization under the action of three, in general, time-variable electric fields (45).

By Eq. (47) each of the three terms, differing as to their kT-dependence, can be decomposed into a stationary and a transient component. The transient part consists (in addition to terms oscillating with the frequencies of the externally applied electric fields) of nonoscillating terms with exponential time growths, their dispersional properties determined by the factors

$$R_{km}(\omega_{abc}) = \left(1 - \frac{\tau_k}{\tau_m} - \mathrm{i}\omega_{abc}\tau_k\right)^{-1} \tag{56}$$

For $m = 0$, these generalized dispersional factors become the well-known Debye factors $R_k(\omega_{abc})$ defining dispersion and absorption of third-order polarization in the steady state, as discussed in Refs. 6 and 7:

$$R_k(\omega_{abc}) = (1 - \mathrm{i}\omega_{abc}\tau_k)^{-1} \tag{57}$$

Equations (52)–(54) are written in a way to stress the dependence of the electric properties of the molecule (its dipole moment m, anisotropy of polarizability $\gamma = \frac{3}{2}a_2 = a_{33} - a_{11}$, and mean electric hyperpolarizability $b = \frac{5}{9}b_1 = \frac{1}{3}(b_{333} + 3b_{113})$ on the frequencies of the external electric fields (5.6). The dependence in question is related to the well-known Voigt–Lorentz electron dispersion, which is apparent chiefly in the range of optical frequencies and is of essential importance in electro-optical effects involving Debye molecular reorientation in addition to electron dispersion due to the electric field of a laser light wave. The Debye reorientation can be due to difference frequencies between modes of laser light (the inverse difference frequencies being comparable to the rotational relaxation times τ_1, τ_2) or to modulation of the medium by an external slowly variable electric field. Equations (52)–(54) are of a form that makes it possible to analyze directly the time-evolution of the individual nonlinear electro-optical phenomena simply by specifying the frequencies of the electric fields applied.

These results provide a consistent description of the time-dependent variations in nonlinear relaxation phenomena in the approximation of Smoluchowski rotational diffusion, taking into account the processes of time-variable polarization that occur when the externally applied fields are switched on or off. This, obviously, is essentially important with regard to measurements of the dielectric and optical properties of the molecules by pulse techniques, when the switch-on time of the fields can be of the same order as the rotational relaxation times τ_1 and τ_2.

V. APPLICATIONS OF THE SMOLUCHOWSKI ROTATIONAL DIFFUSION APPROXIMATION IN THE RELAXATION THEORY OF NONLINEAR ELECTRO-OPTICAL PROCESSES

A. Phenomena Related to Nonlinear Electric Polarization in Liquids

Nonlinear dielectric phenomena, related to variations of the electric permittivity tensor proportional to the square of the electric field externally applied to a liquid, have been the subject of experimental and theoretical studies [15–17, 39–41, 76, 80–83]. The earliest and presumably best known is the nonlinear dielectric effect (NDE), first observed correctly by Herweg [84], Herweg and Pötsch [85], and Kautzsch [86]. NDE is to this day the subject of extensive studies [87]. Its molecular–statistical theory for isotropic media was proposed by Piekara and Kielich [76, 88] before the coming of lasers.

The NDE was later discovered by Piekara and Piekara [89] in solutions of nitrobenzene as positive (anomalous) dielectric saturation under the action of a strong dc electric field. At that time, Debye's theory [90] did not predict positive variations of permittivity. Piekara [91] interpreted them in terms of interactions leading to almost antiparallel reorientation of pairs of dipoles in the liquid. Afterward, NDE was also detected in other solutions of dielectric liquids [92]. The molecular–statistical theory of NDE successfully explaining the experimental results was proposed by Piekara and Kielich [76, 88]. Since then, NDE has become a valuable method for the study of the properties of liquids [93–96].

Kasprowicz-Kielich and Kielich [97] proposed a theory of the nonlinear changes of the electric permittivity tensor of dipolar solutions taking into account the influence of molecular redistribution as well as fluctuations of the molecular electric field, and gave an extensive analysis of the analogies between the changes in permittivity and other electro-optical phenomena. They proposed [5, 6] a now classical theory of nonlinear relaxational effects, related with third-order nonlinear electric polarization in liquids acted on by ac electric fields, stressing essentially the dependence of the electric permittivities on the frequencies of the reorienting fields resulting from the Smoluchowski rotational diffusion equation. Their results, comprising the frequency dependences of stationary nonlinear susceptibilities of liquids, bear on the following phenomena:

- Frequency tripling of the reorienting field vibrations
- Second-harmonic generation in an isotropic liquid in the presence of a dc electric field [15, 16, 98]
- The induction of optical birefringence
- The induction of the nonlinear dielectric effect.

Their method has been extended to other nonlinear effects, such as nonlinear changes in electric permittivity of liquids at mixed sum and difference frequencies, and optical rectification in liquids in the presence of a dc electric field [15, 16, 99], first observed in nitrobenzene by Ward and Guha [100]. Also noteworthy is the theory of oscillations of the polarization ellipsoid of light induced by a dc electric field proposed by Gadomski and Roman [101], closely related to the results of Ref. 100.

The influence of molecular rotational diffusion on NDE and on dc electric field-induced optical rectification has also been studied [8, 10, 11], leading to graphs of the frequency dependences of stationary nonlinear susceptibilities for different values of the parameter R (Eq. 63) derived from the Smoluchowski–Debye theory. These graphs serve to illustrate the behavior of nonlinear dispersion and absorption in liquids. Curves of the rise in time on switching on the reorienting field have been calculated for some of these effects.

If chemical reactions take place in a composite system, modifying the dipole moments of its components, we deal with a chemical mechanism of electric polarization determined by the process of chemical relaxation. The experimental method of NDE relaxation has been applied successfully to elucidate the kinetics of fast chemical reactions [102] since, as shown by Jadżyn and Hellemans [103], an interesting analogy exists between chemical relaxation on the one hand and phenomena of Smoluchowski orientational relaxation in strong electric fields on the other. Similarities between the stationary Kerr effect and stationary NDE in liquids have been studied by Hellemans and De Maeyer [104].

B. Dispersion and Absorption in Some Nonlinear Electro-optical Effects

We now consider certain results of Refs. 5–7 concerning the steady-state properties of the scalar nonlinear electric polarizability of order 3 of isotropic dielectric media occurring in Eq. (48). If the medium is acted on starting from $t = 0$ by dc as well as cosine electric fields (45) and after a period of time sufficiently long, $t \gg \tau_1$, for the steady state to set in throughout the medium, then Eq. (42) with Eqs. (52)–(55) leads to [7]

$$C(-\omega_{abc};\omega_a,\omega_b,\omega_c) = \theta_1\mathscr{R}_1(\omega_{abc}) + 2\theta_2\mathscr{R}_2(\omega_{abc}) - \theta_3\mathscr{R}_3(\omega_{abc}) + \tfrac{20}{9}\theta_4\mathscr{R}_4(\omega_{abc}) + \tfrac{1}{6}c_0 \tag{58}$$

Beside the molecular reorientation parameters given by Eqs. (43) and (44) there appear the following nonlinear relaxation functions:

$$\mathscr{R}_1(\omega_{abc}) = \tfrac{1}{6}\sum_{a,b} R_2(\omega_{ab}) \tag{59}$$

$$\mathscr{R}_2(\omega_{abc}) = \tfrac{1}{12}\sum_{a,b,c}\{R_1(\omega_a)R_2(\omega_{ab}) + \tfrac{1}{2}[3R_1(\omega_a) - R_2(\omega_{bc})]R_1(\omega_{abc})\} \tag{60}$$

$$\mathscr{R}_3(\omega_{abc}) = \tfrac{1}{6}\sum_{a,b,c} R_1(\omega_a)R_2(\omega_{ab})R_1(\omega_{abc}) \tag{61}$$

$$\mathscr{R}_4(\omega_{abc}) = \tfrac{1}{24}\sum_{a,b,c}[R_1(\omega_{abc}) + 3R_1(\omega_c)] \tag{62}$$

in the form of combinations of Debye factors $R_k(\omega_{abc})$, given by Eq. (57).

Equations (58)–(62) provide the basis for the theoretical analysis of various electro-optical effects, of which we shall consider the following:

1. Nonlinear dielectric effect, where $\omega_a = \omega$, $\omega_b = \omega_c = 0$
2. Frequency doubling in the presence of a dc electric field (dc electric field-induced second-harmonic generation DC-SHG), where $\omega_a = \omega_b = \omega$ and $\omega_c = 0$
3. The effect of nonlinear dielectric rectification in the presence of a dc electric field (DCIOR), where $\omega_a = \omega$, $\omega_b = -\omega$ and $\omega_c = 0$
4. Self-induced variations in the nonlinear polarizability, where $\omega_a = \omega_b = \omega$ and $\omega_c = -\omega$
5. The effect of third-harmonic generation at dielectric frequencies, where $\omega_a = \omega_b = \omega_c = \omega$.

It should be kept in mind that, in Refs. 6 and 7 the relaxation functions $\mathscr{R}_1(\omega_{abc})$, $\mathscr{R}_2(\omega_{abc})$, $\mathscr{R}_3(\omega_{abc})$, and $\mathscr{R}_4(\omega_{abc})$ are denoted by the symbols $B(\omega_{abc})$, $C(\omega_{abc})$, $D(\omega_{abc})$, and $A(\omega_{abc})$, respectively. The discussion simplifies considerably if in Eq. (58) one retains only the frequency dependence due to molecular reorientation. The dependence of the electric molecular properties on the frequencies causing linear and nonlinear electron dispersion, given in Eqs. (12)–(14), may be omitted in some cases. This occurs for low electric frequencies, because electron dispersion takes place at frequencies much higher than those at which the relaxation factors (57) play a role.

Henceforth, we shall neglect the term related to the nonlinear polarizability c_0 of the molecule, since it causes no reorientation in the present approximation and gives rise to a constant contribution to the polarizability $C(-\omega_{abc}; \omega_a, \omega_b, \omega_c)$. The relaxation functions $\mathscr{R}_1(\omega_{abc})$ are, in general, complex, the real and imaginary parts describing, respectively, dispersion and absorption of the process considered. In what follows, we shall give in analytical form the functions describing various temperature contributions to the nonlinear polarizability, for the selected processes, together with the respective graphs.

One notes that molecular reorientation effects in liquids are rather complicated, since the reorientation constants $\theta_1, \theta_2, \theta_3, \theta_4$ take very different values from one liquid to another. In general, we have $\theta_4 < \theta_1 < \theta_2 < \theta_3$ [39]. Here, then, are their values assessed to within the order of magnitude for dilute solutions of nitrobenzene in some solvents [77]. We have, in cgs units,

$$\theta_1 \cong 10^{-34} \qquad 2\theta_2 \cong 10^{-32} \qquad \theta_3 \cong 10^{-31}$$

with the negative dipolar term obviously predominant. In the case of nondipolar molecules with but weakly anisotropic linear polarizability, the essential role belongs to the term in the nonlinear electric polarizability θ_4, which can contribute as much as 50% [78].

With the dimensionless parameter R, defined as in Refs. 31 and 79,

$$R = \frac{2\theta_3}{\theta_2} = \frac{m^2}{(a_{33} - a_{11})kT} \tag{63}$$

and on neglecting the contributions to the total polarization from the nonlinear molecular polarizabilities b_{ijk} and c_{ijkl}, Eq. (58) takes the following form:

$$\begin{aligned} C(-\omega_{abc}; \omega_a, \omega_b, \omega_c) &= \theta_3\left[2R^{-2}\mathscr{R}_1(\omega_{abc}) + 4R^{-1}\mathscr{R}_2(\omega_{abc}) - \mathscr{R}_3(\omega_{abc})\right] \\ &= \theta_1\left[\mathscr{R}_1(\omega_{abc}) + 2R\mathscr{R}_2(\omega_{abc}) - \tfrac{1}{2}R^2\mathscr{R}_3(\omega_{abc})\right] \end{aligned} \tag{64}$$

which is well adapted to numerical computations. Equation (64) describes the steady-state nonlinear polarizability for dipolar anisotropically polarizable molecules. For nondipolar molecules, by Eq. (58) we have

$$C(-\omega_{abc};\omega_a,\omega_b,\omega_c) = \tfrac{1}{6}c_0 + \theta_1\mathscr{R}_1(\omega_{abc}) \tag{65}$$

Equation (64) permits the plotting of theoretical curves for dispersion and absorption by molecules for which the values of the parameter R are available. In Ref. 9 theoretical graphs of the real and imaginary polarizability components of Eq. (64) are given for the above considered effects in systems consisting of water molecules ($R = -57$), which behave like a typical dipolar liquid and of fibrinogen molecules ($R = 3$), at room temperature.

From Eq. (64) one easily notes that for the following two values of R, respectively,

$$R_{\pm} = \frac{2\mathscr{R}_2(\omega_{abc}) \pm \sqrt{4\mathscr{R}_2(\omega_{abc})^2 + 2\mathscr{R}_1(\omega_{abc})\mathscr{R}_3(\omega_{abc})}}{\mathscr{R}_3(\omega_{abc})} \tag{66}$$

the molecular polarizability and thus the respective nonlinear susceptibility $\chi(-\omega_{abc};\omega_a,\omega_b,\omega_c)$ vanish as a result of the mutual canceling out of the competing molecular contributions. Reference 8 draws attention to a particular case of Eq. (66) in the case of low-frequency NDE, when $\omega_a = \omega$, $\omega_b = \omega_c = 0$ and $\mathscr{R}_1(\omega_{abc}) = \mathscr{R}_2(\omega_{abc}) = \mathscr{R}_3(\omega_{abc}) = 1$. The values for $R_{\pm}$ for which $\chi(-\omega;\omega,0,0)$ vanishes are

$$R_{\pm} = 2 \pm 2\sqrt{6} \tag{67}$$

The last two formulae can be of some interest in describing the case of a dielectric still to be discovered, which, though exhibiting a nonzero transient component in its susceptibility,

$$\delta\chi(-\omega_{abc};\omega_a,\omega_b,\omega_c;t) \neq 0$$

has zero nonlinear steady-state susceptibility

$$\chi(-\omega_{abc};\omega_a,\omega_b,\omega_c) = 0 \tag{68}$$

in spite of the presence of reorienting electric fields.

1. Dispersion and Absorption of the Change in Nonlinear Electric Polarizability $C(-\omega;\omega,0,0)$ Due to an Intense dc Electric Field: Nonlinear Dielectric Effect

Here, we consider the so-called nonlinear dielectric effect, also referred to as the effect of dielectric saturation [6, 15, 39]. The effect consists of the induction in an isotropic liquid of changes in electric susceptibility proportional to the square of the strong reorienting electric field. NDE is apparent as a change in $\Delta\varepsilon \propto \chi(-\omega;\omega,0,0;t)$ due to the action of a strong dc external electric field on the system of molecules of a liquid, where $\Delta\varepsilon$ is measured with an optical probe field of frequency ω.

A detailed account of the steady-state and transient parts of the nonlinear susceptibility in NDE was given in Ref. 8. Here, we restrict ourselves to the results of Refs. 7 and 8 concerning the properties of NDE after the steady-state has been achieved. To this aim we put $\omega_a = \omega$ and $\omega_b = \omega_c = 0$ in Eqs. (58)–(62), leading to

$$\begin{aligned}
\mathscr{R}_1(\omega) &= \tfrac{1}{3}[1 + 2R_2(\omega)] \\
\mathscr{R}_2(\omega) &= \tfrac{1}{6}\{1 + R_2(\omega) + R_1(\omega)R_2(\omega) + [3 - R_2(\omega)]R_1(\omega) \\
&\qquad + \tfrac{1}{2}[3R_1(\omega) - 1]R_1(\omega)\} \qquad (69) \\
\mathscr{R}_3(\omega) &= \tfrac{1}{3}[1 + R_2(\omega) + R_1(\omega)R_2(\omega)]R_1(\omega) \\
\mathscr{R}_4(\omega) &= \tfrac{1}{2}[1 + R_1(\omega)]
\end{aligned}$$

Graphs of the real $\mathscr{R}_i'(\omega)$ and imaginary $\mathscr{R}_i''(\omega)$ parts of these functions are shown in Fig. 1 versus $\ln(\omega\tau_2)$. We see that the dispersion curves related to the molecular polarizabilities at high frequencies ($\omega\tau_2 \gg 1$) contribute to NDE contrary to the term in the third power of the permanent dipole moment proportional to $\mathscr{R}_3'(\omega)$, the reorientation of which cannot keep pace with the high-field frequency.

In NDE the nonlinear polarizability of order 3 for $\omega_a = \omega$ and $\omega_b = \omega_c = 0$ takes the form

$$\begin{aligned}
\theta_3^{-1}C(-\omega;\omega,0,0) = {}& \tfrac{2}{3}R^{-2}[1 + 2R_2(\omega)] \\
&+ \tfrac{2}{3}R^{-1}\{1 + R_2(\omega) + R_1(\omega)R_2(\omega) \\
&\qquad + [3 - R_2(\omega)]R_1(\omega) \qquad (70) \\
&\qquad + \tfrac{1}{2}[3R_1(\omega) - 1]R_1(\omega)\} \\
&- \tfrac{1}{3}R_1(\omega)[1 + R_2(\omega) + R_1(\omega)R_2(\omega)]
\end{aligned}$$

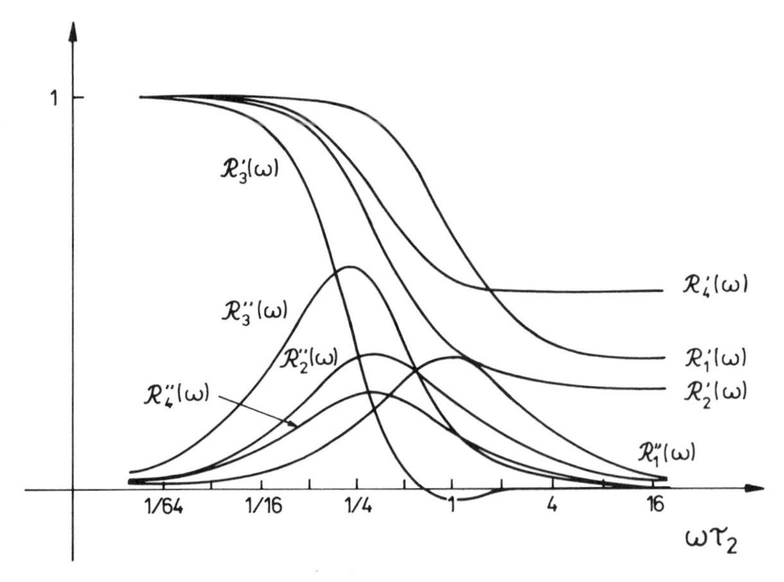

Figure 1. Temperature-dependent contributions to dispersion and absorption of the nonlinear polarizability $C(-\omega; \omega, 0, 0)$ in the nonlinear dielectric effect versus $\ln(\omega\tau_2)$.

In Fig. 2 the imaginary part $C''(-\omega; \omega, 0, 0)$ is shown versus the real part $C'(-\omega; \omega, 0, 0)$ for different values of the molecular parameter R (63). One sees that the curves of Fig. 2, which are counterparts of the well-known Cole–Cole graphs for linear dielectric relaxation [105], are essentially dependent on the parameter R and diverge markedly from the well-known semicircles.

In the particular case of nondipolar molecules, Eq. 70 gives

$$C(-\omega; \omega, 0, 0) = \frac{1}{3}\theta_1 \frac{3 - i\omega\tau_2}{1 - i\omega\tau_2} \tag{71}$$

On splitting the nonlinear polarizability (71) into real and imaginary parts

$$C = C' + iC''$$

we get the equation of the circumference:

$$C'^2 + C''^2 - \tfrac{4}{3}\theta_1 C' + \tfrac{1}{3}\theta_1^2 = 0 \tag{72}$$

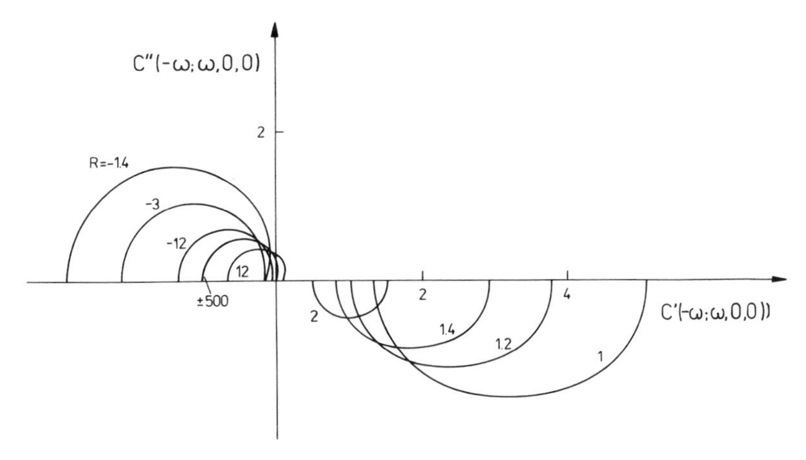

Figure 2. Cole–Cole graphs for NDE, for different values of the parameter R.

Thus, in coordinates (C', C'') the nonlinear Cole–Cole relation is given by the semicircle of radius equal to $\frac{1}{3}\theta_1$ centered at the point $(\frac{2}{3}\theta_1, 0)$ (since $C' > 0$, $C'' > 0$), as shown in Fig. 3.

Assuming $\omega\tau_2 \cong \infty$ for optical frequencies, we get

$$\mathscr{R}_1(\infty) = 2\mathscr{R}_2(\infty) = \tfrac{2}{3}\mathscr{R}_4(\infty) = \tfrac{1}{3} \qquad \mathscr{R}_3(\infty) = 0 \tag{73}$$

and the nonlinear polarizability, Eq. (58), in high-frequency NDE, $C(-\infty; \infty, 0, 0)$, is equal to

$$C(-\infty; \infty, 0, 0) = \tfrac{1}{6}c_0 + \tfrac{1}{3}\theta_1 + \tfrac{1}{3}\theta_2 + \tfrac{10}{9}\theta_4 \tag{74}$$

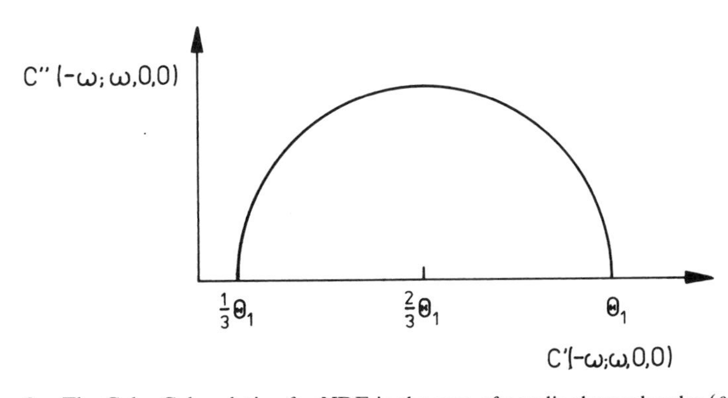

Figure 3. The Cole–Cole relation for NDE in the case of nondipolar molecules ($\theta_3 = 0$).

2. *Frequency Doubling in the Presence of a dc Electric Field*

Dc electric field-induced second-harmonic generation (DC-SHG) of laser light has been observed in dipolar and nondipolar gases [106]. As shown in Ref. 6, it would be of interest to investigate DC-SHG for low frequencies, at which the electric dipoles still keep pace with the field variation. Besides, as shown earlier [98], this effect is subject to considerable enhancement in the case of complete electric reorientation of molecules and macromolecules.

By insertion of $\omega_a = \omega_b = \omega$ and $\omega_c = 0$ in Eqs. (59)–(62) we obtain the following relations in DC-SHG:

$$\begin{aligned}
\mathscr{R}_1(2\omega) &= \tfrac{1}{3}[2R_2(\omega) + R_2(2\omega)] \\
\mathscr{R}_2(2\omega) &= \tfrac{1}{6}\{2R_1(\omega)R_2(\omega) + R_1(\omega)R_2(2\omega) \\
&\qquad + [3R_1(\omega) - R_2(\omega)]R_1(2\omega) \\
&\qquad + \tfrac{1}{2}[3 - R_2(2\omega)]R_1(2\omega)\} \\
\mathscr{R}_3(2\omega) &= \tfrac{1}{3}[R_1(\omega)R_2(2\omega) + R_1(\omega)R_2(\omega) + R_2(\omega)]R_1(2\omega) \\
\mathscr{R}_4(2\omega) &= \tfrac{1}{4}[1 + 2R_1(\omega) + R_1(2\omega)]
\end{aligned} \tag{75}$$

The respective dispersion and absorption curves are plotted in Fig. 4. The nonlinear relaxation functions $\mathscr{R}_2(2\omega)$ and $\mathscr{R}_3'(2\omega)$ specifically exhibit a change in sign of the dispersion and, on passing through a minimum tend to zero, thus making relaxational dispersion resemble resonance dispersion.

The nonlinear polarizability $C(-2\omega; \omega, \omega, 0)$ may be written, from Eqs. 59–62, in the form

$$\begin{aligned}
\theta_3^{-1}C(-2\omega; \omega, \omega, 0) &= \tfrac{2}{3}R^{-2}[2R_2(\omega) + R_2(2\omega)] \\
&\quad + \tfrac{2}{3}R^{-1}\{2R_1(\omega)R_2(\omega) \\
&\qquad + R_1(\omega)R_2(2\omega) + R_1(2\omega)[3R_1(\omega) - R_2(\omega)] \\
&\qquad + \tfrac{1}{2}R_1(2\omega)[3 - R_2(\omega)]\} \\
&\quad - \tfrac{1}{3}R_1(2\omega)[R_1(\omega) + R_2(\omega)R_2(\omega) + R_2(\omega)]
\end{aligned} \tag{76}$$

Figure 5 shows graphs $C''(-2\omega; \omega, \omega, 0)$ versus $C'(-2\omega; \omega, \omega, 0)$ calculated for different values of R. Equation (76) effectively describes low-frequency DC-SHG, whereas for SHG at appropriately high frequency

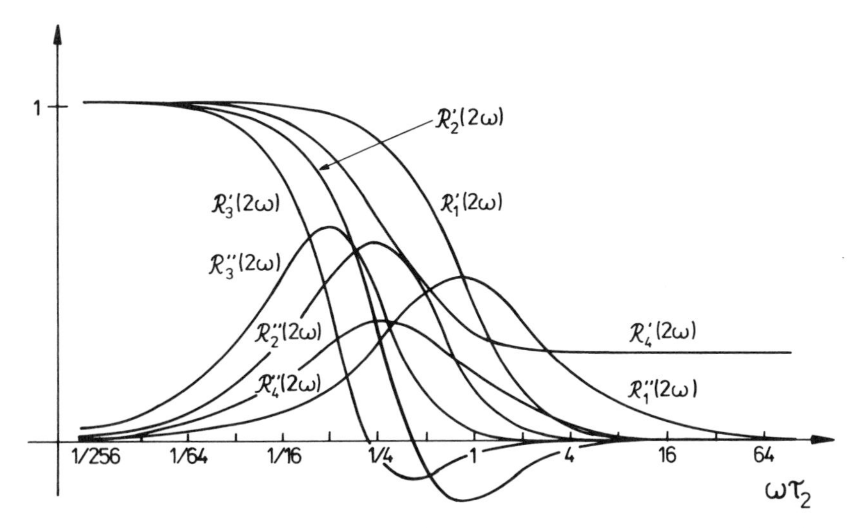

Figure 4. Temperature-dependent contributions to dispersion and absorption of the nonlinear polarizability $C(-2\omega; \omega, \omega, 0)$ describing dc second-harmonic generation versus $\ln(\omega\tau_2)$.

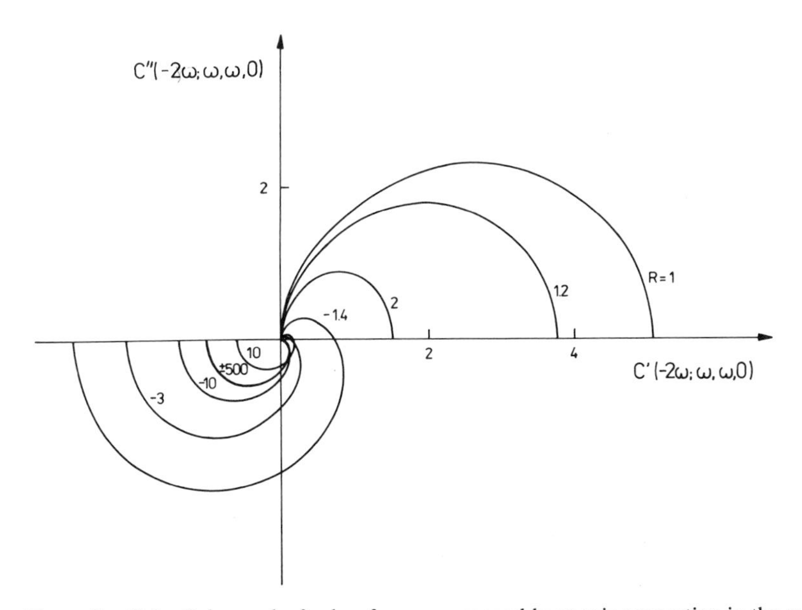

Figure 5. Cole–Cole graphs for low-frequency second-harmonic generation in the presence of a dc electric field for different values of the parameter R.

when $\omega\tau_2 \cong \infty$, since now only $\mathscr{R}_4(2\omega) = \frac{1}{4} \neq 0$, the following formula holds:

$$C(-\infty; \infty, \infty, 0) = \frac{1}{6}\left[c_0 + \frac{m(b_{333} + 2b_{113})}{5kT}\right] \tag{77}$$

Obviously, from Fig. 4, beside the nonlinear polarizability c_0 only the term in θ_4 related to the relaxation function $\mathscr{R}_4(2\omega)$ contributes.

3. *Nonlinear Polarizability $C(0; \omega, -\omega, 0)$: The Effect of Nonlinear Dielectric Rectification in the Presence of a Dc Electric Field*

The effect of electric field-induced optical rectification (DCIOR) was first observed, in nitrobenzene, by Ward and Guha [100]. By insertion in Eqs. (59)–(62) of $\omega_a = \omega$, $\omega_b = -\omega$ and $\omega_c = 0$ we arrive at the following relations:

$$\begin{aligned}
\mathscr{R}_1(0) &= \tfrac{1}{3}[1 + R_2(\omega) + R_2(-\omega)] \\
\mathscr{R}_2(0) &= \tfrac{1}{6}\{1 + R_1(\omega) + R_2(\omega) + R_1(-\omega)R_2(-\omega) \\
&\qquad + \tfrac{1}{2}[3R_1(\omega) + 3R_1(-\omega) - R_2(\omega) - R_2(-\omega)]\} \\
\mathscr{R}_3(0) &= \tfrac{1}{3}[R_1(\omega) + R_2(\omega) + R_1(-\omega)R_2(-\omega)] \\
\mathscr{R}_4(0) &= \tfrac{1}{4}[2 + R_1(-\omega) + R_1(-\omega)]
\end{aligned} \tag{78}$$

Curves illustrating the behavior of these contributions to the nonlinear polarizability $C(0; \omega, -\omega, 0)$ are shown in Fig. 6.

By Eqs. (59)–(62) the nonlinear polarizability $C(0; \omega, -\omega, 0)$ may be written in the form

$$\begin{aligned}
&\theta_3^{-1}C(-\omega; \omega, 0, 0) \\
&\quad = \tfrac{2}{3}R^{-2}[1 + R_2(\omega) + R_2(-\omega)] \\
&\qquad + \tfrac{2}{3}R^{-1}\{1 + R_1(-\omega)R_2(-\omega) + R_1(\omega) + R_2(\omega) \\
&\qquad\qquad + \tfrac{1}{2}[3R_1(\omega) + 3R_1(-\omega) - R_2(\omega) - R_2(-\omega)]\} \\
&\qquad - \tfrac{1}{3}[R_1(\omega) + R_2(\omega) + R_1(-\omega)R_2(-\omega)]
\end{aligned} \tag{79}$$

The imaginary part $C''(0; \omega, -\omega, 0)$ is plotted versus the real part $C'(0; \omega, -\omega, 0)$ for different values of R in Fig. 7; note the difference in scale of the coordinate axes. At optical frequencies the molecular polariz-

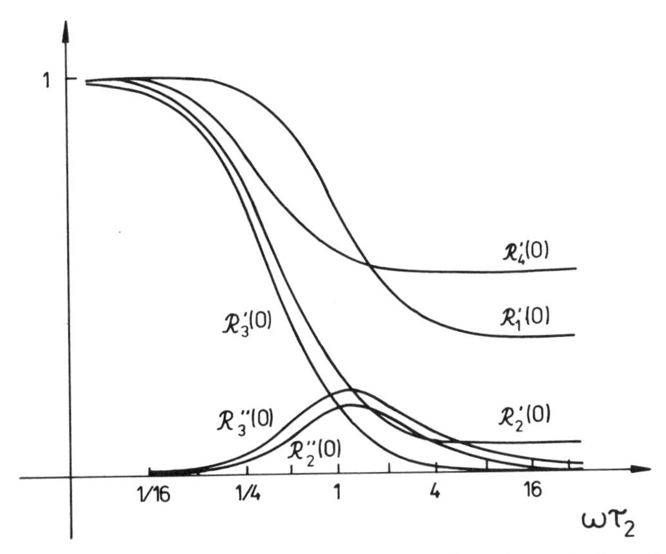

Figure 6. Temperature-dependent contributions to dispersion and absorption of the nonlinear polarizability $C(0; \omega, -\omega, 0)$ in the effect of nonlinear dielectric rectification versus $\ln(\omega\tau_2)$.

ability for DCIOR is found to be

$$C(0;\infty,-\infty,0) = \frac{1}{6}c_0 + \frac{1}{15kT}\left[\frac{2}{9}(a_{33}-a_{11})^2 + m(b_{333}+2b_{113})\right] + \frac{2m^2}{135k^2T^2}(a_{33}-a_{11}) \tag{80}$$

4. Dispersion and Absorption of Self-induced Variations in the Nonlinear Polarizability $C(-\omega; \omega, \omega, -\omega)$

The relaxational reorientation functions are now

$$\begin{aligned}
\mathscr{R}_1(\omega) &= \tfrac{1}{3}[2 + R_2(2\omega)] \\
\mathscr{R}_2(\omega) &= \tfrac{1}{6}\{R_1(\omega) + R_1(-\omega) \\
&\qquad + R_1(\omega)R_2(2\omega) + [3R_1(\omega) - 1]R_1(\omega) \\
&\qquad + \tfrac{1}{2}[3R_1(-\omega) + R_2(2\omega)]R_1(\omega)\} \\
\mathscr{R}_3(\omega) &= \tfrac{1}{3}[R_1(\omega) + R_1(-\omega) + R_1(\omega)R_2(2\omega)] \\
\mathscr{R}_4(\omega) &= \tfrac{1}{4}[3R_1(\omega) + R_1(-\omega)]
\end{aligned} \tag{81}$$

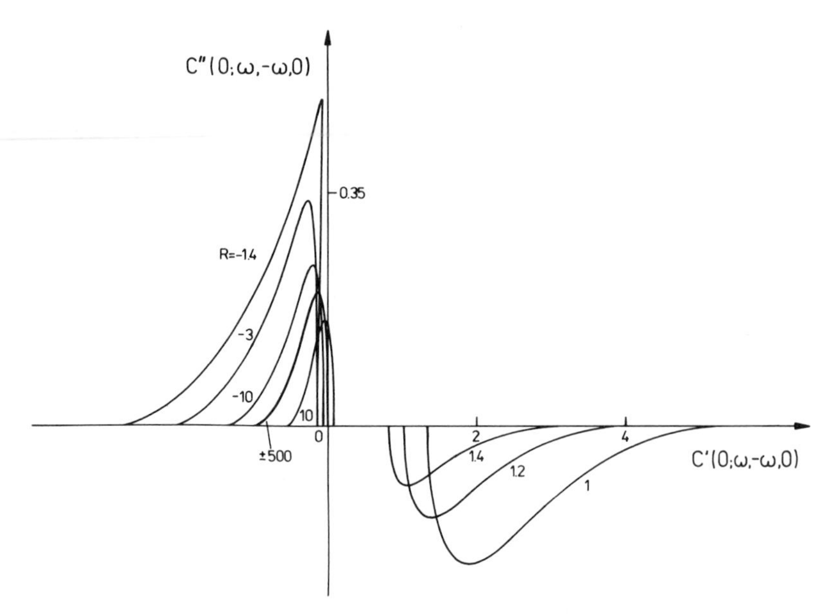

Figure 7. Cole–Cole graphs for nonlinear dielectric rectification in the presence of a dc electric field for different values of the parameter R.

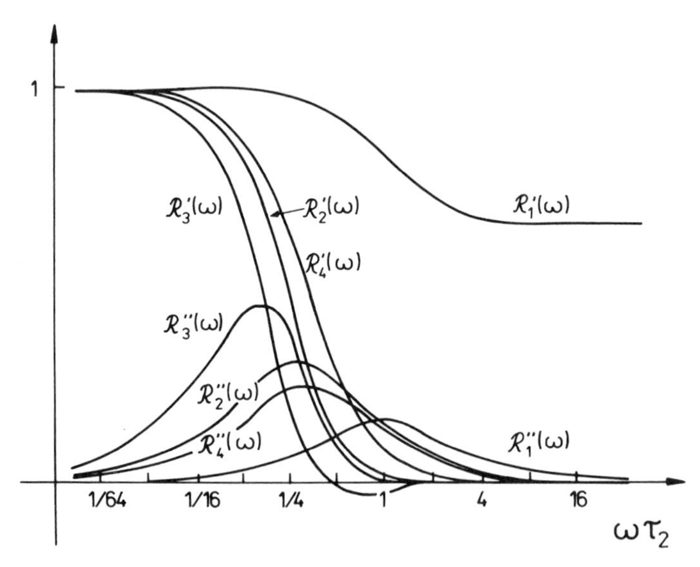

Figure 8. Temperature-dependent contributions to dispersion and absorption of the nonlinear polarizability $C(-\omega; \omega, \omega, -\omega)$ versus $\ln(\omega\tau_2)$.

Curves illustrating the behavior of these molecular contributions to the nonlinear polarizability $C(-\omega; \omega, \omega, -\omega)$ are shown in Fig. 8. Here, too, we find a change in sign for dispersion of the factor $\mathscr{R}_3'(\omega)$. In high-frequency fields, a contribution to the third-order susceptibility tensor arises from the factor related with the squared anisotropy in linear molecular polarizability. This contribution is proportional to the inverse temperature.

The nonlinear polarizability $C(-\omega; \omega, \omega, -\omega)$ may be written, with regard to the Eqs. (58)–(62) in the form

$$\begin{aligned}\theta_3^{-1}C(-\omega;\omega,\omega,\omega) = {} & \tfrac{2}{3}R^{-2}[2 + R_2(\omega)] \\ & + \tfrac{2}{3}R^{-1}\{R_1(\omega) + R_1(-\omega) + R_1(\omega)R_2(2\omega) \\ & \qquad + [3R_1(\omega) - 1]R_1(\omega) \\ & \qquad + \tfrac{1}{2}R_1(\omega)[3R_1(-\omega) - R_2(2\omega)]\} \\ & - \tfrac{1}{3}R_1(\omega)[R_1(\omega) + R_1(-\omega) + R_1(\omega)R_2(\omega)]\end{aligned} \tag{82}$$

The imaginary part $C''(-\omega; \omega, \omega, -\omega)$ is represented as a function of the real part $C'(-\omega; \omega, \omega, -\omega)$ for different values of R in Fig. 9. In the case

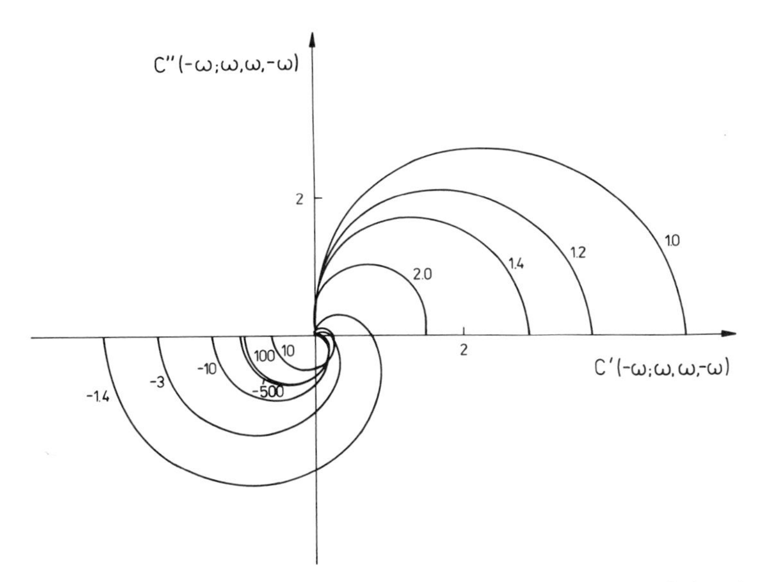

Figure 9. Cole–Cole graphs for the phenomenon of self-induced variations in the nonlinear polarizability $C(-\omega; \omega, \omega, -\omega)$ for different values of the parameter R.

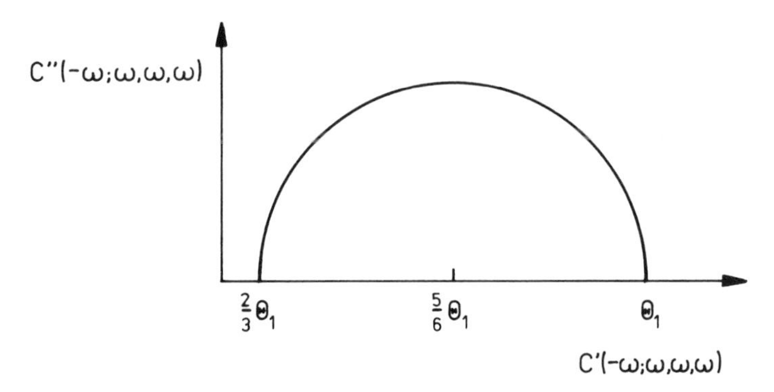

Figure 10. Cole–Cole graphs for self-induced variations in the nonlinear polarizability $C(-\omega; \omega, \omega, -\omega)$ for nondipolar molecules.

of nondipolar anisotropically polarizable molecules the nonlinear polarizability $C(-\omega; \omega, \omega, -\omega)$ is equal to

$$C(-\omega; \omega, \omega, \omega) = \frac{1}{3}\theta_1[2 + R_2(2\omega)] = \frac{1}{3}\theta_1 \frac{3 + 8\omega^2\tau_2^2 + 2i\omega\tau_2}{1 + 4\omega^2\tau_2^2} \quad (83)$$

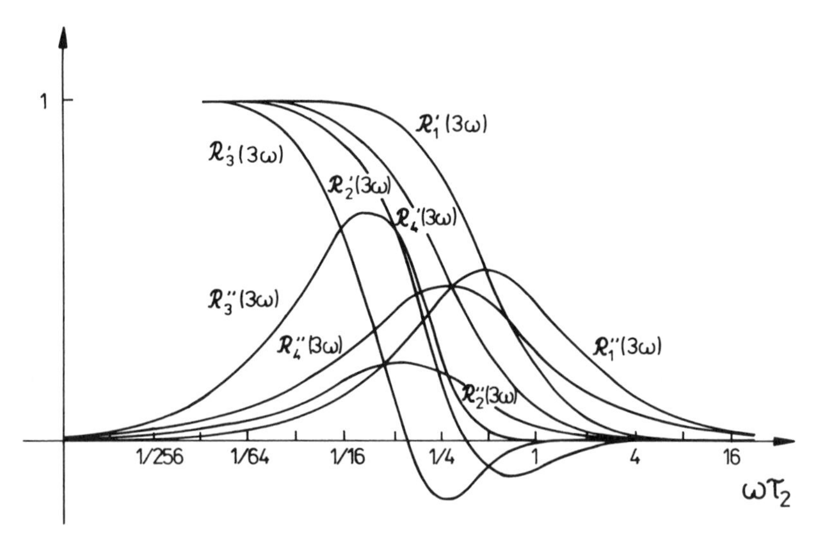

Figure 11. Temperature-dependent contributions to dispersion and absorption of the nonlinear polarizability $C(-3\omega; \omega, \omega, \omega)$ versus $\ln(\omega\tau_2)$.

On separating the polarizability (83) into real and imaginary parts and on eliminating the parameter $\omega\tau_2$, we get the equation of the circumference:

$$C'^2 + C''^2 - \tfrac{5}{3}\theta_1 C' + \tfrac{2}{3}\theta_1^2 = 0 \tag{84}$$

Accordingly the Cole–Cole graph for the nonlinear polarizability (83) is a semicircle with a radius of $\frac{1}{6}\theta_1$ and a center at $(\frac{5}{6}\theta_1, 0)$, as shown in Fig. 10. At optical frequencies, only $\mathscr{R}_1(-\infty) = \frac{2}{3}$ is nonzero, and Eq. (82) leads to

$$C(-\infty; \infty, \infty, -\infty) = \frac{1}{6}c_0 + \frac{4(a_{33} - a_{11})^2}{135kT} \tag{85}$$

5. *Dispersion and Absorption of the Nonlinear Polarizability $C(-3\omega; \omega, \omega, \omega)$ in the Process of Third-Harmonic Generation at Dielectric Frequencies*

The process, considered also in Refs. 7, 8, 13, and 14, is described by the relaxational functions (59)–(62) for $\omega_a = \omega_b = \omega_c = \omega$:

$$\begin{aligned}
\mathscr{R}_1(3\omega) &= R_2(2\omega) \\
\mathscr{R}_2(3\omega) &= \tfrac{1}{2}\{R_1(\omega)R_2(2\omega) \\
&\qquad + \tfrac{1}{2}[3R_1(\omega) - R_2(2\omega)]R_1(3\omega)\} \\
\mathscr{R}_3(3\omega) &= R_1(\omega)R_2(2\omega)R_1(3\omega) \\
\mathscr{R}_4(3\omega) &= \tfrac{1}{4}[3R_1(\omega) + R_1(3\omega)]
\end{aligned} \tag{86}$$

Curves illustrating the behavior of these contributions to the polarizability $C(-3\omega; \omega, \omega\ \omega)$ are shown in Fig. 11.

Here, as in the DC-SHG effect (Fig. 4), the dispersion curves of $\mathscr{R}_2'(3\omega)$ and $\mathscr{R}_3'(3\omega)$ exhibit a change in sign. However, at frequencies higher than those of the inverse relaxation times of molecular reorientation, the macroscopic polarizability of the medium is described solely by the contributions of the third-order nonlinear polarizability of the molecule, since all temperature-dependent contributions tend to zero (see Eq. (90)). Because of the nonlinearity of the relaxation functions, absorption still increases toward higher frequencies in some cases after the dispersion curve has fallen to zero.

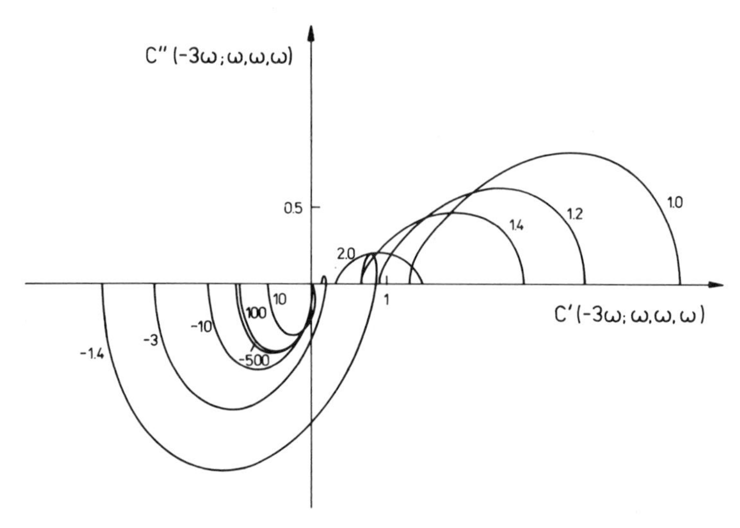

Figure 12. Cole–Cole graphs for the phenomenon of third-harmonic generation at dielectric frequencies for different values of the parameter R.

The nonlinear polarizability $C(-3\omega; \omega, \omega, \omega)$ may be written, using Eqs. (58)–(62), in the form

$$\begin{aligned}\theta_3^{-1} C(-3\omega; \omega, \omega, \omega) = {} & 2R^{-2} R_2(\omega) \\ & + \tfrac{1}{2} R^{-1} \{ R_1(\omega) R_2(2\omega) + \tfrac{1}{2} R_1(3\omega) \\ & \qquad \times [3R_1(\omega) - R_2(2\omega)] \} \\ & - R_1(\omega) R_2(2\omega) R_1(3\omega)\end{aligned} \tag{87}$$

The imaginary part $C''(-3\omega; \omega, \omega, \omega)$ is plotted as a function of the real part $C'(-3\omega; \omega, \omega, \omega)$ in Fig. 12 for different values of the parameter R. Note the difference in scale of the coordinate axes of Fig. 12. In the case of nondipolar molecules, Eq. (87) gives

$$C(-3\omega; \omega, \omega, \omega) = \theta_1 \frac{1 + i2\omega\tau_2}{1 + 4\omega^2\tau_2^2} \tag{88}$$

Taking the real and imaginary parts of Eq. (88) separately and eliminating

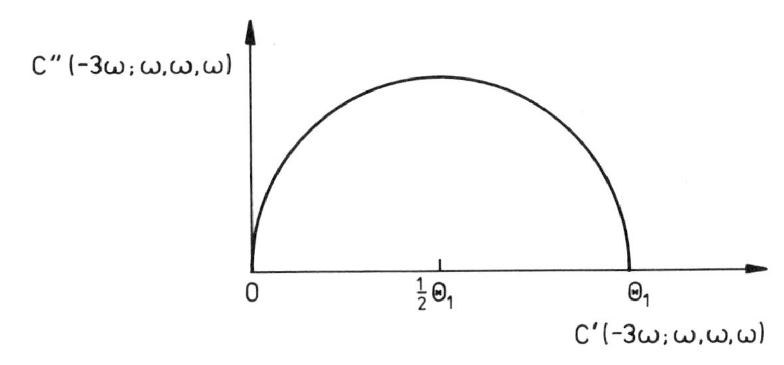

Figure 13. Cole–Cole plot for the phenomenon of third-harmonic generation at dielectric frequencies for nondipolar molecules.

the parameter $\omega\tau_2$, we obtain the equation of the circumference,

$$\left(C' - \tfrac{1}{2}\theta_1\right)^2 + C''^2 = \tfrac{1}{4}\theta_1^2 \tag{89}$$

with a radius of $\frac{1}{2}\theta_1$ and a center at $(\frac{1}{2}\theta_1, 0)$, as shown in Fig. 13. For the third-harmonic generation effect on optical frequencies one has simply

$$C(-3\omega; \omega, \omega, \omega) = \tfrac{1}{6}c_0 \tag{90}$$

C. Time Dependence of Nonlinear Dielectric Relaxation in Selected Electro-optical Phenomena in Liquids

We now proceed to the evolution in time of some of the nonlinear effects discussed in Section V.A. The description of the transients of the third-order electric susceptibility in the approach of the Smoluchowski rotational diffusion theory is founded on Eqs. (48)–(57). For these equations, we refer to Ref. 8, where the theory of the rise in nonlinear polarization, induced by harmonic electric fields, and the dependence of the susceptibility transients on the relaxation times τ_1 and τ_2 are studied. A more detailed discussion is devoted to the dynamics of NDE in slowly variable fields. The respective results have been extended by Buchert [9, 10] to NDE induced by rectangular and Gaussian electric pulses permitting the graphical representation of the theoretical susceptibility variations occurring in NDE for molecules with different values of the parameter R. The respective curves point, in the first place, to a change in sign of the susceptibility as the molecular system approaches the final stationary state.

Equations (52)–(57), which describe the dynamics of these nonlinear effects, are rather complicated. However, if the reorientation is caused by a single electric field, the results are much clearer.

At a first glance, most of these dispersion curves might be interpreted as functions, analytically described by simple Debye relaxation factors with a single relaxation time involving, according to the situation, doubled and tripled frequencies of the external field. However, with regard to the respective absorption curves, this interpretation is found to be unsound. In fact, absorption maxima exhibit relative heights both greater and less than 0.5, leading to Cole–Cole diagrams differing from regular semicircles, as should be the case on the interactionless model assumed here. Moreover, the inflection point of the dispersion curve often occurs at a different frequency from that of the absorption maximum. Also, due to nonlinearity of the relaxation functions, absorption in some cases still takes place at higher frequencies where dispersion is no longer nonzero.

1. *The Dynamics of the Nonlinear Dielectric Effect*

We consider NDE due to the application of a strong electric field at the moment of time $t = 0$. For a low-frequency measuring field, with regard to Eqs. (42)–(57), we obtain [8]

$$
\begin{aligned}
C(-\omega; \omega, 0, 0; t) &= \theta_3 \left\{ 2R^{-2}\left[1 - \exp\left(-\frac{t}{\tau_2}\right)\right] \right. \\
&\quad + 4R^{-1}\left[1 - \frac{9}{8}\exp\left(-\frac{t}{\tau_1}\right) - \frac{1}{8}\exp\left(-\frac{t}{\tau_2}\right)\right. \\
&\qquad \left. - \frac{3t}{4\tau_1}\exp\left(-\frac{t}{\tau_1}\right)\right] \\
&\quad - \left[1 - \frac{3}{4}\exp\left(-\frac{t}{\tau_1}\right) - \frac{1}{4}\exp\left(-\frac{t}{\tau_2}\right)\right. \\
&\qquad \left.\left. - \frac{3t}{2\tau_1}\exp\left(-\frac{t}{\tau_1}\right)\right]\right\}
\end{aligned} \tag{91}
$$

Hence, the nonlinear polarizability $C(-\omega, \omega, 0, 0, t)$ is found to grow from zero at $t = 0$ up to its steady-state value:

$$C(-\omega;\omega,0,0;t) = \theta_3 R^{-2}(2 + 4R - R^2) \tag{92}$$

The rise dynamics of NDE can be expressed in the form of the quotient [13]:

$$\begin{aligned}\eta_1(t) &= \frac{C(-\omega;\omega,0,0;t)}{C(-\omega;\omega,0,0;t \gg \tau_1)} \\ &= 1 + (2 + 4R - R^2) \\ &\quad \times \left[\frac{3}{4}R(R-6)\exp\left(-\frac{t}{\tau_1}\right)\right. \\ &\quad + \left(\frac{1}{4}R^2 + \frac{1}{2}R - 2\right)\exp\left(-\frac{t}{\tau_2}\right) \\ &\quad \left. + 4R\left(\frac{1}{2}R - 1\right)\frac{t}{\tau_1}\exp\left(-\frac{t}{\tau_1}\right)\right]\end{aligned} \tag{93}$$

indeterminate for $R_{\pm} = 2 \pm \sqrt{6}$. The rise dynamics are shown in Fig. 14

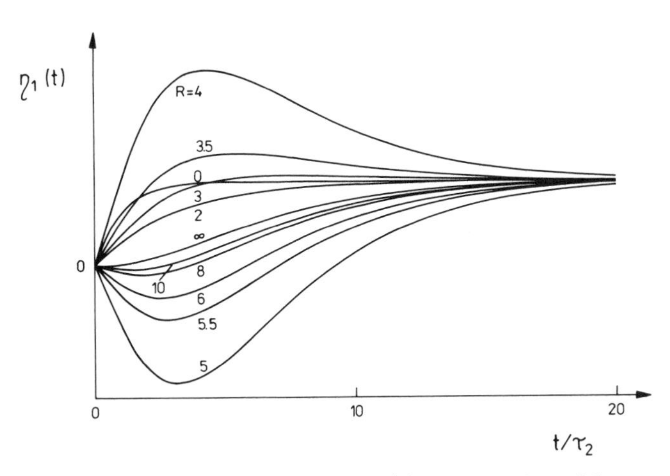

Figure 14. Evolution in time of the function $\eta_1(t)$ for some values of the parameter R.

versus $\omega\tau_2$ for different values of R. In a strongly dipolar liquid, when the term in θ_3 is predominant, we have

$$C(-\omega;\omega,0,0;t) = -\theta_3\left[1 - \frac{3}{4}\exp\left(-\frac{t}{\tau_1}\right) - \frac{1}{4}\exp\left(-\frac{t}{\tau_2}\right) - \frac{3t}{2\tau_1}\exp\left(-\frac{t}{\tau_1}\right)\right] \tag{94}$$

and the rise curve plotted in Ref. 8 is dependent on exponential terms with the relaxation times τ_1 and τ_2 as well as on the nonexponential term $t\exp(-t/\tau_1)$.

2. *The Dynamics of Dc Field-Induced Electric Rectification*

Considering reorientation due to a dc electric field, we obtain from Eqs. (42) and (52)–(57) the nonlinear polarizability in the following form:

$$\begin{aligned} C(0;\omega,-\omega,0;t) &= \theta_3\Bigg\{2R^{-2}\mathcal{R}_1(\omega_{abc})\left[1 - \exp\left(-\frac{t}{\tau_2}\right)\right] \\ &\quad + 4R^{-1}\mathcal{R}_1(\omega_{abc})\left[1 - \frac{9}{8}\exp\left(-\frac{t}{\tau_1}\right) + \frac{1}{8}\exp\left(-\frac{t}{\tau_2}\right) - \frac{3t}{4\tau_1}\exp\left(-\frac{t}{\tau_1}\right)\right] \\ &\quad - \mathcal{R}_3(\omega_{abc})\left[1 - \frac{3}{4}\exp\left(-\frac{t}{\tau_1}\right) - \frac{1}{4}\exp\left(-\frac{t}{\tau_2}\right) - \frac{3t}{2\tau_1}\exp\left(-\frac{t}{\tau_1}\right)\right]\Bigg\} \end{aligned} \tag{95}$$

In the case of dc-induced optical rectification (DCIOR), for $\omega_a = \infty$, $\omega_b = -\infty$, $\omega_c = 0$, Eqs. (59)–(62) give

$$\mathcal{R}_1(\infty) = 2\mathcal{R}_2(\infty) = \tfrac{1}{3} \qquad \mathcal{R}_3(\infty) = 0 \qquad \mathcal{R}_4(\infty) = \tfrac{1}{2}$$

and furthermore we have

$$C(0; \infty, -\infty, 0; t) = \frac{2\theta_3}{3R^2}\left[1 - \exp\left(-\frac{t}{\tau_2}\right)\right] + R\left[1 - \frac{9}{8}\exp\left(-\frac{t}{\tau_1}\right) + \frac{1}{8}\exp\left(-\frac{t}{\tau_2}\right) - \frac{3t}{4\tau_1}\exp\left(-\frac{t}{\tau_1}\right)\right] \quad (96)$$

showing that the DCIOR susceptibility grows from zero at time $t = 0$ to the stationary value

$$C(0; \infty, -\infty, 0; t \gg \tau_1) = \frac{2\theta_3}{3R^2}(1 + R) \quad (97)$$

On normalization the rise in time of the DCIOR effect takes the following form:

$$\begin{aligned}\eta_2(t) &= \frac{C(0; -\infty, \infty, 0; t)}{C(0; -\infty, \infty, 0; t \gg \tau_1)} \\ &= 1 + (1 + R)^{-1}\left[\frac{9}{8}R\exp\left(-\frac{t}{\tau_1}\right) + \left(1 - \frac{1}{8}R\right)\exp\left(-\frac{t}{\tau_2}\right) + \frac{3}{4}R\frac{t}{\tau}\exp\left(-\frac{t}{\tau_1}\right)\right] \qquad R \neq -1\end{aligned} \quad (98)$$

and is plotted in Fig. 15a, b as a function of the parameter R. The graphs of Fig. 15 show that the rise $\eta_2(t)$ is markedly dependent on R and, contrary to $\eta_1(t)$, takes only positive values. Obviously, $\eta_1(t > \tau_1) = \eta_2(t > \tau_1) = 1$. One readily notes that dc-induced low-frequency electric rectification, at $\mathscr{R}_1(0) = \mathscr{R}_2(0) = \mathscr{R}_3(0) = 1$, follows a course given by Eqs. (95) identical with that of $\eta_1(t)$, described by Eq. (93).

3. *The Dynamics of Optical Rectification Induced by Gaussian Electric Field Pulses*

We shall adduce the results of Ref. 10 concerning the theoretical description of the phenomenon of Gauss pulse-induced optical rectification (GIOR). It consists in the canceling out of the frequencies of oscillations

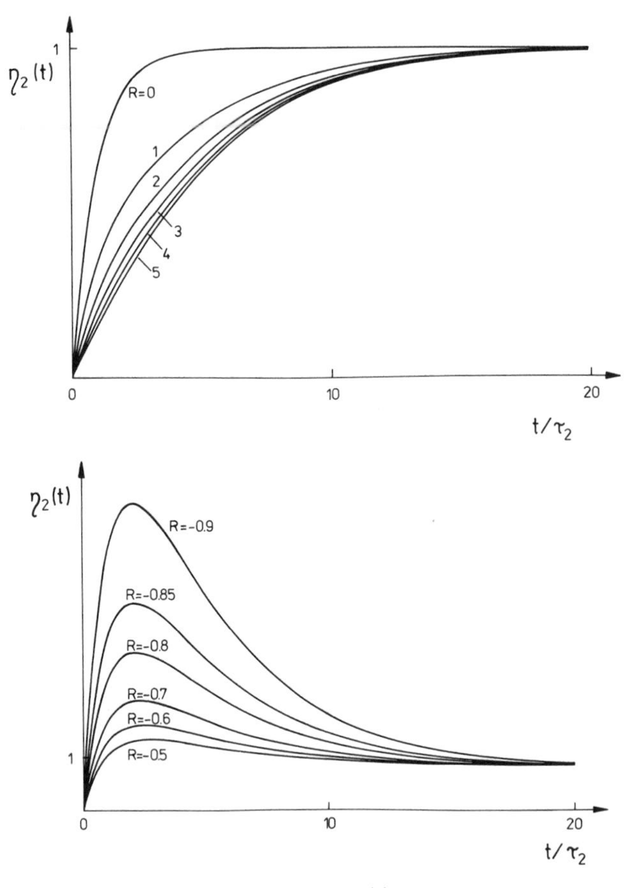

Figure 15. Evolution in time of the function $\eta_2(t)$, for some values of the parameter R.

of the optical fields:

$$g_a(t) = \exp(\mathrm{i}\omega t) \qquad g_b(t) = \exp(-\mathrm{i}\omega t) \tag{99}$$

under the action of the strong reorienting Gaussian pulse:

$$g_c(t) = -\exp\left[-4(\ln 2)\left(\frac{t^2}{T_{\mathrm{G}}}\right)\right] \tag{100}$$

where T_{G} is the pulse width at half-intensity. By Eqs. (42) and (50)–(57)

we obtain [10]

$$C_{\text{GIOR}}(0;\omega,-\omega,0;t) = \tfrac{1}{6}c_0 + \tfrac{1}{3}\theta_1\mathscr{G}_1(t) + \tfrac{1}{3}\theta_2\mathscr{G}_2(t) + \tfrac{10}{3}\theta_4\mathscr{G}_4(t) \tag{101}$$

where the time-dependent relaxation functions $G_i(t)$ are given by the following integrals:

$$\mathscr{G}_1(r) = \frac{1}{\tau_2}\exp\left(-\frac{t}{\tau_2}\right)\int_{-\infty}^{t} g_a(u)g_b(u)g_c(u)\exp\left(\frac{t}{\tau_2}\right)\mathrm{d}u \tag{102}$$

$$\mathscr{G}_2(t) = \frac{1}{2\tau_1\tau_2}\exp\left(-\frac{t}{\tau_1}\right) \times\left[3\int_{-\infty}^{t} g_a(u)g_b(u)\exp\left(\frac{u}{\tau_1}\right)\int_{-\infty}^{t} g_c(v)\exp\left(\frac{v}{\tau_1}\right)\mathrm{d}v\,\mathrm{d}u - \int_{-\infty}^{t} g_c(u)\exp\left(\frac{u}{\tau_1}-\frac{u}{\tau_2}\right)\int_{-\infty}^{t} g_a(v)g_b(v)\exp\left(\frac{v}{\tau_2}\right)\mathrm{d}v\,\mathrm{d}u\right] \tag{103}$$

$$\mathscr{G}_4(t) = \frac{1}{\tau_1}\exp\left(-\frac{t}{\tau_1}\right)\left[\frac{1}{2}g_a(t)g_b(t)\int_{-\infty}^{t} g_c(u)\exp\left(\frac{u}{\tau_1}\right)\mathrm{d}u + \int_{-\infty}^{t} g_a(u)g_b(u)g_c(u)\exp\left(\frac{u}{\tau_1}\right)\mathrm{d}u\right] \tag{104}$$

The shapes of the response functions reveal a competitive interplay of the various temperature contributions:

- Nonlinear mean second-order hyperpolarizability c_0; in the approximation of third-order polarizability, this term does not affect the rise in time since, here, the primary role belongs to electron rather than orientational processes.
- The two terms, θ_4 and θ_1 proportional to $(kT)^{-1}$, the one dependent on the product of the dipole moment m and mean first-order hyperpolarizability b of the molecule, and the other, which is positive, dependent on the squared anisotropy of polarizability γ^2.
- The term θ_2 containing the product $m^2\gamma$ with sign dependent on that of γ.

In optical work, i.e., if one external electric field frequency is of the order of 10^{14}–10^{15} Hz, no term with m and proportional to $(kT)^{-3}$ occurs, since the orientation of the dipole moment is unable to keep pace with the field frequency.

With regard to the complicated form of the relaxation functions (102)–(104), it is convenient to have recourse to numerical computations. As a typical dielectric medium we chose nitrobenzene, for which we assume the following values of the molecular parameters [107]:

$$m = 4.24 \times 10^{-18} \qquad [\text{esu cm}],$$
$$a_{33} - a_{11} = -7.64 \times 10^{-24} \qquad [\text{cm}^3]$$
$$R = 57 \qquad (T = 300\,\text{K})$$

Since the values of the mean hyperpolarizability b from various authors

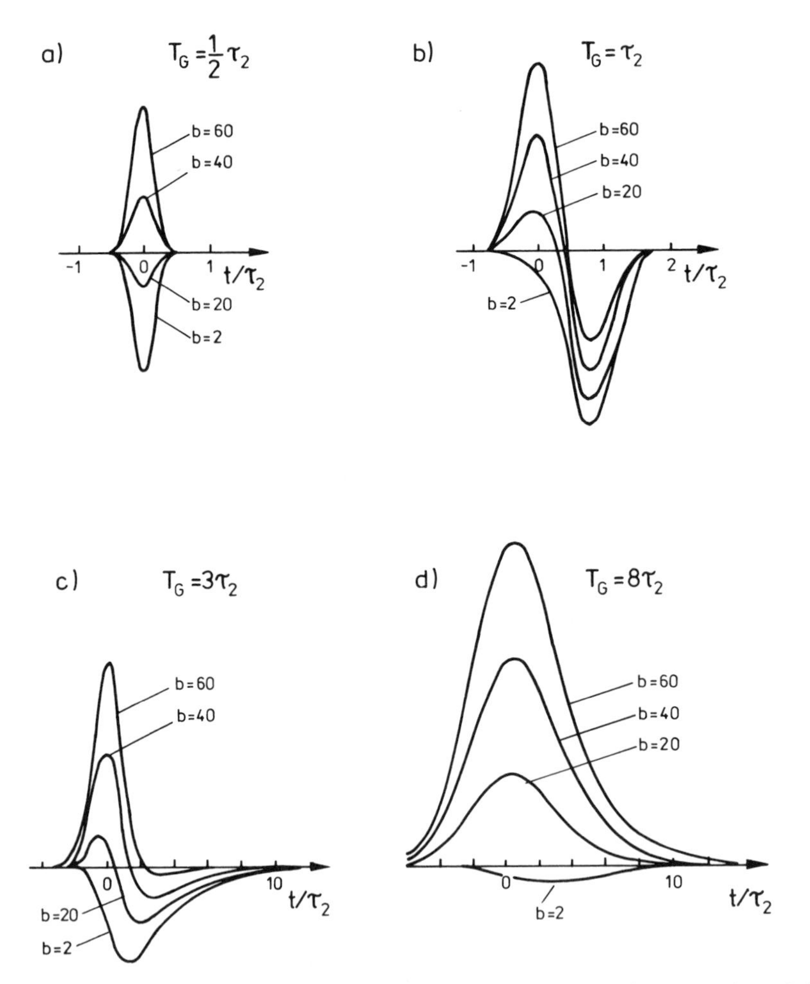

Figure 16. Time-dependence of the GIOR effect in nitrobenzene for some values of the mean hyperpolarizability b (in cgs units) and various Gaussian pulse half-widths.

diverge considerably, we plotted the time-dependent curves for various b values [108, 109].

Figure 16 shows the rise in time of the GIOR effect for various half-widths of the Gaussian pulse fields in terms of the rotational relaxation time τ_2 for nitrobenzene molecules. If the half-width is smaller than the relaxation time τ_2 (Fig. 16a), we obtain a response dependent on the molecular parameters similar to the Gaussian shape. If the values of b are small enough, the dominant role belongs to the term proportional to $(kT)^{-2}$ due to the negative anisotropy of polarizability of nitrobenzene molecules. For large values of b the system of molecules gives a positive response because of the predominance of the terms proportional to $(kT)^{-1}$. When the half-width of the pulse and the relaxation time τ_2 are of the same order of magnitude, the shape of the response (Fig. 16b, c) visualizes a competitive influence of different relaxation functions. Then the permittivity exhibits a change in sign. For large half-widths (Fig. 16d), positive terms predominate. Measurements of the time characteristics of the GIOR process on a time scale admitting of observation of the orientational molecular motions will permit the determination of the first-order hyperpolarizability of the molecule.

4. *The Influence of Molecular Rotational Diffusion on Gaussian Pulse-Induced Third-Order Electric Polarization in Liquids*

We now consider the dynamics of nonlinear electric polarization in a dielectric medium acted on by a Gaussian electric pulse field (100). A highly interesting analysis of the functions relevant to the description of cw autocorrelation measurements of picosecond pulses from a cw synchronously mode-locked laser has been given by Sala et al. [110]. If the field applied to the medium is that of Eq. (97), the nonlinear polarization of the medium (46) takes the form [111]

$$\langle P_Z^{(3)}(t)\rangle = \rho E_0^3\{\theta_1 A_{12}(t) + \theta_2[A_{21}(t) + A_{22}(t)g(t)] - \theta_3 A_{31}(t)\} \tag{105}$$

where the nonlinear relaxation functions $A_{ij}(t)$ are of the form

$$A_{ij}(t) = \frac{1}{\tau_j}\int_{-\infty}^{t} \exp\left[-4(\ln 2)^j\left(\frac{u}{T_G}\right)^j - \frac{t-u}{\tau_j}\right] du \qquad j = 1,2 \tag{106}$$

$$A_{22}(t) = \frac{1}{\tau_2}\int_{-\infty}^{t} A_{11}(u)\exp\left[-4(\ln 2)\left(\frac{u}{T_G}\right) - \frac{t-u}{\tau_2}\right] du \tag{107}$$

$$A_{31}(t) = \frac{1}{\tau_1}\int_{-\infty}^{t} A_{22}(u)\exp\left[-4(\ln 2)\left(\frac{u}{T}\right) - \frac{t-u}{\tau_1}\right] du \tag{108}$$

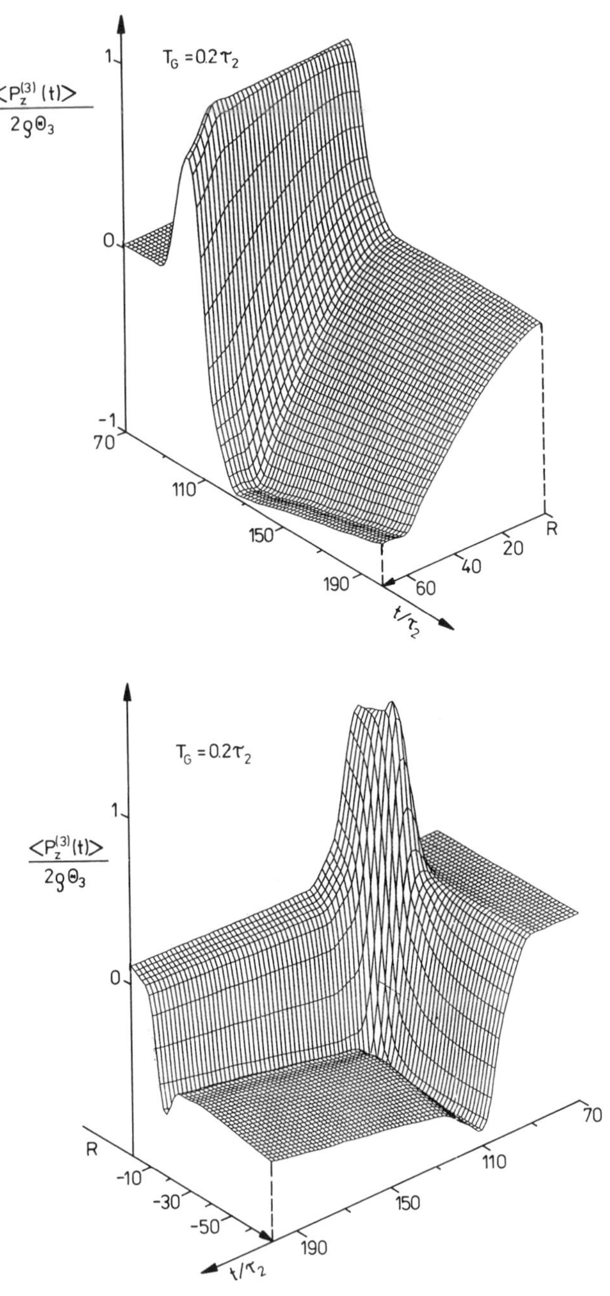

Figure 17. Time-evolution of the polarization $\langle P_Z^{(3)}(t) \rangle$ given by Eq. (105) for various Gaussian pulse half-widths and different values of the parameter R.

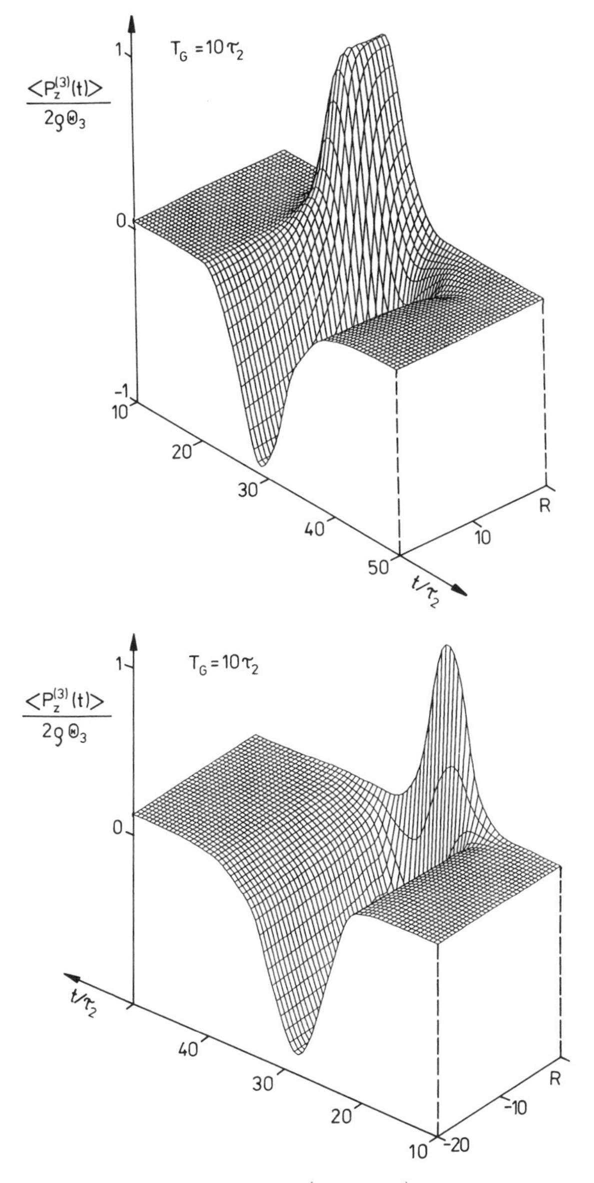

Figure 17. *(Continued).*

For the initial moment of time, when $t_0 = -\infty$, we have assumed $A_{ij}(t_0) = 0$.

Equations (105)–(108) permit the numerical computation of the time-evolution curves of third-order electric polarization for different widths of the reorienting pulse. Figure 17 shows the curves obtained for Gaussian pulses with half-widths $T_G = 0, 2\tau_2$ and $T_G = 10\tau_2$.

For better visualization, the values of the parameter R of Eq. (63) are plotted on the third coordinate axis of a three-dimensional relief. One immediately notes the differences in shape between the time evolutions of $\langle P_Z^{(3)}(t)\rangle$ for molecules with different values of R. For molecules with R close to zero we find a point of indeterminacy in Eq. (65); the polarization then becomes the strongest. The changes in sign of the pulse-induced polarization are due to competition between the different molecular contributions.

Predictably, the theoretical analysis of the evolution in time of nonlinear electro-optical phenomena can prove to be of essential importance to pulse laser technique. From this point of view, we draw attention to the theoretical calculations of the time evolution of optical birefringence caused by an ac electric field superimposed on a dc field [112–116] as well as to the results of Ref. 117.

D. The Process of Decay of Nonlinear Electric Polarization

The case is rather simple since the relaxation functions of decay describing the evolution in time on switching the reorienting pulses off at the moment of time t_0, namely $A_{21}(t > t_0)$, $A_{31}(t > t_0)$ and $B_{11}(t > t_0)$, are immediately obtained from Eqs. (34)–(36) for $g_a(t > t_0) = g_b(t > t_0) = g_c(t > t_0) = 0$. Thus, we have

$$A(t > t_0) = A(t_0)\exp\left(-\frac{t}{\tau_1}\right) \tag{109}$$

and the decay, with regard to Eqs. (45) and (109), is a simple exponential process running with the relaxation time τ_1:

$$\begin{aligned}\langle P_Z^{(3)}(t > t_0)\rangle = \rho \sum_{abc} E_a E_b E_c \\ \times\big[\theta_2 A_{21}^{abc}(t_0) - \theta_3 A_{31}^{abc}(t) \\ + \theta_4 B_{11}^{abc}(t_0)\big]\exp\left(-\frac{t}{\tau_1}\right)\end{aligned} \tag{110}$$

In Eq. (110) we have assumed that, at the moment t_0 when the field is switched off, the polarization $\langle P_Z^{(3)}(t > t_0)\rangle$ has already attained its steady state value, given by Eq. (58).

It is our hope that the preceding analysis may promote the idea that the investigation of time-evolution changes in nonlinear electro-optical phenomena in strong pulse fields will permit the effective separation of the various temperature-dependent contributions to the nonlinear third-order electric permittivity of dipolar as well as nondipolar molecules.

References

1. P. Debye, *Polare Molekeln*, Hirzel, Leipzig, 1929; *Polar Molecules*, Dover, New York, 1945.
2. M. V. Smoluchowski, *Ann. Phys.* **48**, 1103 (1915); in W. Natanson (Ed.), *Pisma Mariana Smoluchowskiego* (The works of Marian Smoluchowski), Vols. 1, 2, Chaps. 30, 32, Kraków, 1924–1928.
3. S. Chandrasekhar, *Rev. Mod. Phys.* **15**, 1 (1943).
4. E. N. Wax (Ed.), *Selected Papers on Noise and Stochastic Processes*, Dover, New York, 1954.
5. B. Kasprowicz-Kielich, S. Kielich, and J. R. Lalanne, in J. Lascombe (Ed.), *Molecular Motions in Liquids*, Reidel, Dordrecht, 1974, pp. 563–573.
6. B. Kasprowicz-Kielich and S. Kielich, *Adv. Mol. Relaxation Processes* **7**, 275 (1975).
7. J. Buchert, B. Kasprowicz-Kielich, and S. Kielich, *Adv. Mol. Relaxation Processes* **11**, 115 (1977).
8. W. Alexiewicz, J. Buchert, and S. Kielich, *Acta Phys. Pol.* **52A**, 445 (1977).
9. J. Buchert, in *Nonlinear Behaviour of Molecules, Atoms and Ions in Electric, Magnetic or Electromagnetic Fields*, Elsevier, Amsterdam, 1977, pp. 423–437.
10. W. Alexiewicz and J. Buchert, Proceedings of the III-rd Conference on Dielectric Materials, Measurements and Applications, Birmingham, 1979.
11. W. Alexiewicz, J. Buchert, and S. Kielich, in C. V. Shank, E. P. Ippen, and S. L. Shapiro (Eds.), *Picosecond Phenomena*, *Chemical Physics* **4**, Springer, Berlin, 1978, pp. 205–207.
12. W. Alexiewicz, *Mol. Phys.* **59**, 637 (1986).
13. W. Alexiewicz, *Acta Phys. Pol.* **A72**, 753 (1987).
14. W. Alexiewicz, *Physica* **A155**, 84 (1989).
15. S. Kielich, General molecular theory and electric field effects in isotropic dielectrics, in M. Davies (Ed.), *Dielectric and Related Molecular Processes*, Vol. 1, Chemical Society, London, 1972, pp. 192–387.
16. S. Kielich, Nonlinear electro-optics, in C. T. O'Konski (Ed.), *Molecular Electro-Optics*, Vol. 1, Dekker, New York, 1976, pp. 391–444.
17. S. Kielich, *Molekularna Optyka Nieliniowa*, PWN, Warsaw, 1977 (in Polish); *Nyelineynaya Molekularnaya Optika* (*Nonlinear Molecular Optics*), Nauka, Moscow 1981 (in Russian).
18. M. W. Evans, G. J. Evans, W. T. Coffey, and P. Grigolini, *Molecular Dynamics and Theory of Broad Band Spectroscopy*, Wiley, New York, 1982.

19. W. T. Coffey, in M. W. Evans (Ed.), *Dynamical Processes in Condensed Matter*, Wiley, New York, 1985, pp. 69–252.
20. W. T. Coffey, M. W. Evans, and P. Grigolini, *Molecular Diffusion and Spectra*, Wiley, New York, 1984.
21. A. Peterlin and H. Stuart, *Doppelbrechung Insbesondere Künstliche Doppelbrechung*, Akademische Verlagsgesellschaft, Leipzig, 1943.
22. H. Benoit, *Ann. Phys. (Paris)* **6**, 561 (1951); *J. Chem. Phys.* **49**, 517 (1952).
23. S. Kielich, *J. Phys.* **28**, 519 (1967); *Phys. Lett.* **25A**, 153 (1967); *Acta Phys. Pol.* **37A**, 447 (1970); **37A**, 719 (1970).
24. S. Kielich, *Acta Phys. Pol.* **30**, 683 (1966); **31**, 689 (1967).
25. M. A. Novikov, in *Nyelineynaya Optika*, Nauka, Novosibirsk, 1968, pp. 52–59.
26. W. T. Coffey and B. V. Paranjape, *Proc. R. Irish Acad.* **78A**, 17 (1979).
27. A. Morita, *J. Phys. D Appl. Phys.* **11**, L1, L9 (1978).
28. F. Perrin, *J. Phys. Radium* **5**, 497 (1934); **7**, 1 (1936).
29. A. Budo, E. Fisher, and S. Miyamoto, *Phys. Z.* **40**, 337 (1939).
30. E. Fisher and F. C. Frank, *Phys. Z.* **40**, 345 (1939).
31. A. Morita and H. Watanabe, *J. Chem. Phys.* **77**, 1193 (1982).
32. S. Kielich, *Acta Phys. Pol.* **17**, 239 (1958); **22**, 65 (1962); *Mol. Phys.* **6**, 49 (1963); *Proc. Phys. Soc.* **90**, 847 (1967).
33. P. Debye and W. Ramm, *Ann. Phys.* **28**, 28 (1937).
34. I. I. Frenkel, *Kineticheskaya Teoria Zhidkostey*, Nauka, Leningrad, 1975 (in Russian).
35. A. Budo, *Phys. Z.* **39**, 706 (1938); *J. Chem. Phys.* **17**, 686 (1949).
36. W. T. Coffey, *Mol. Phys.* **37**, 573 (1979).
37. W. T. Coffey, P. M. Corcoran, and M. W. Evans, *Proc. R. Soc. London Ser. A* **410**, 61 (1987).
38. A. Morita and H. Watanabe, *Chem. Phys. Lett.* **108**, 453 (1984); **112**, 319 (1984).
39. A. Chełkowski, *Dielectrics Physics*, PWN/Elsevier, Warsaw/Amsterdam, 1980.
40. C. J. F. Bötcher, *Theory of Electric Polarisation*, Vol. 1, Elsevier, Amsterdam, 1973.
41. C. J. F. Bötcher and P. Bordewijk, *Theory of Electric Polarisation*, Vol. 2, Elsevier, Amsterdam, 1978.
42. N. Bloembergen and P. Lallemand, *Phys. Rev. Lett.* **16**, 81 (1966); P. Lallemand, *Appl. Phys. Lett.* **8**, 276 (1966).
43. F. Kaczmarek, *Wstęp do Fizyki Laserów* (Introduction to Laser Physics), 2d ed., PWN, Warsaw, 1986 (in Polish).
44. A. A. Pierov, *Opt. Spektrosk.* **40**, 31 (1976).
45. W. M. Tolles, *J. Appl. Phys.* **46**, 991 (1975).
46. M. P. Warchol and W. E. Vaughan, *Adv. Mol. Relaxation Interactions Processes* **13**, 317 (1978).
47. C. C. Wang and R. Pecora, *J. Chem. Phys.* **72**, 5333 (1980).
48. T. Kühle and W. Rapp, *J. Mol. Struct.* **84**, 269 (1982).
49. H. Watanabe and A. Morita, *Adv. Chem. Phys.* **56**, 255–409 (1984).
50. B. J. Berne and R. Pecora, *Dynamic Light Scattering: With Application to Physics, Chemistry and Biology*, Wiley, New York, 1976.
51. B. Chu, *Laser Light Scattering*, Academic, New York, 1974.

52. S. Kielich, *Kvantovaya Elektronika* **4**, 2574 (1977).
53. S. Kielich, Multi-photon scattering molecular spectroscopy, *Progr. Opt.* **20**, 155 (1983); *Proc. Indian Acad. Sci. (Chem. Sci.)* **94**, 403 (1985).
54. S. Kielich, M. Kozierowski, Z. Ożgo, and R. Zawodny, *Acta Phys. Pol.* **A45**, 9 (1974).
55. M. Kozierowski, R. Tanaś, and S. Kielich, *Mol. Phys.* **31**, 629 (1976).
56. P. D. Maker, *Phys. Rev.* **A1**, 923 (1970).
57. W. Alexiewicz, T. Bancewicz, S. Kielich, and Z. Ożgo, *J. Raman Spectrosc.* **2**, 529 (1974).
58. T. Bancewicz and S. Kielich, *Mol. Phys.* **31**, 615 (1976).
59. W. Alexiewicz, Z. Ożgo, and S. Kielich, *Acta Phys. Pol.* **A48**, 243 (1975).
60. W. Alexiewicz, *Acta Phys. Pol.* **A47**, 657 (1975).
61. H. Risken, *The Fokker-Planck Equation: Methods of Solution and Applications*, Springer, Berlin, 1984.
62. A. Abragam, *The Principles of Nuclear Magnetism*, Oxford UP, London, 1961.
63. G. H. Vineyard, *Phys. Rev.* **110**, 999 (1958).
64. J. A. Janik, in M. Davies (Ed.), *Dielectric and Related Molecular Processes*, Vol. 3, Chemical Society, London, 1977, pp. 45–72.
65. G. Barone and A. Paterno, *Physics and Applications of the Josephson Effect*, Wiley, New York, 1982.
66. M. Buettiker, E. P. Harris, and R. Landauer, *Phys. Rev.* **B28**, 1268 (1983).
67. H. J. Breymayer, H. Risken, H. D. Vollmer, and W. Wonneberger, *Appl. Phys.* **B28**, 335 (1982).
68. H. A. Lorentz, *The Theory of Electrons*, Teubner, Leipzig, 1909.
69. W. Voigt, *Ann. Phys. Chem.* **69**, 297 (1899); *Ann. Phys.* **4**, 197 (1901); *Lehrb. Magnet-u. Electro-Optik*, Teubner, Leipzig, 1908.
70. R. J. W. Le Fèvre, in V. Gold (Ed.), *Advances in Physical Organic Chemistry*, Vol. 3, Academic, London, 1965, pp. 1–90.
71. S. Kielich, *Opto-Electronics* **2**, 125–151 (1970); *Ferroelectrics* **4**, 257 (1972).
72. P. Görlich and C. Hoffmann, *Feingerätetechnik* **19**, 49, 101 (1970); C. Hoffmann, *Feingerätetechnik* **20**, 544 (1971); **21**, 35 (1972).
73. A. D. Buckingham and B. J. Orr, *Quart. Rev.* **21**, 195 (1967); M. P. Bogard and B. J. Orr, *M. T. P. Review of Molecular Structure and Properties*, Butterworth, London, 1974.
74. S. A. Akhmanov and R. V. Khokhlov, *Problemy Nielinieynoy Optiki*, Akademiya Nauk USSR, Moscow, 1964 (in Russian).
75. B. Kasprowicz-Kielich, Thesis, Poznań, 1966.
76. A. Piekara and S. Kielich, *Acta Phys. Pol.* **17**, 209 (1958); *J. Chem. Phys.* **29**, 1297 (1958).
77. A. Piekara, *Phys. Z.* **38**, 671 (1937).
78. R. W. Hellwarth, A. Owyoung, and N. George, *Phys. Rev.* **A4**, 2342 (1971).
79. G. B. Thurston and D. I. Bowling, *J. Colloid Interface Sci.* **30**, 34 (1969).
80. A. Piekara, *Phys. Z.* **38**, 671 (1937).
81. S. Kielich and A. Piekara, *Acta Phys. Pol.* **18**, 439 (1959).
82. N. E. Hill, W. E. Vaughan, A. H. Price, and M. Davies, *Dielectric Properties and Molecular Behaviour*, Van Nostrand–Reinhold, London, 1969.

83. K. Sala, *Phys. Rev.* **A29**, 1944 (1984).
84. J. Herweg, *Z. Phys.* **3**, 36 (1920).
85. J. Herweg and W. Pötsch, *Z. Phys.* **8**, 1 (1922).
86. K. Kautzsch, *Phys. Z.* **29**, 105 (1928).
87. J. Małecki, *J. Chem. Soc. Faraday Trans. 2* **72**, 104 (1976).
88. A. Piekara and S. Kielich, *J. Phys. Rad.* **18**, 490 (1957).
89. A. Piekara and B. Piekara, *C. R. Acad. Sci. Paris* **203**, 852 (1936).
90. P. Debye, *Marx Hdb. Radiol.* **6**, 1933 (1925).
91. A. Piekara, *C. R. Acad. Sci. Paris* **204**, 1106 (1937); *Proc. R. Soc. London Ser. A* **172**, 360 (1939); *Acta Phys. Pol.* **10**, 37, 107 (1950).
92. A. Piekara and A. Chełkowski, *J. Chem. Phys.* **25**, 794 (1956); A. Piekara, A. Chełkowski, and S. Kielich, *Z. Phys. Chem. (Leipzig)* **206**, 375 (1957); *Arch. Sci.* **12**, 59 (1959).
93. M. Gregson, G. P. Jones, and M. Davies, *Trans. Faraday Soc.* **67**, 1630 (1975); B. L. Brown, G. P. Jones, and M. Davies, *J. Phys. D Appl. Phys.* **77**, 1192 (1974).
94. G. P. Jones, in M. Davies (Ed.), *Dielectric and Related Molecular Processes*, Vol. 2, Chemical Society, London, 1975, pp. 198–248.
95. L. Hellemans and L. De Maeyer, *J. Chem. Phys.* **63**, 3490 (1975).
96. L. Ruff, *Acta Chim. Hung.* **121**, 203 (1986).
97. B. Kasprowicz-Kielich and S. Kielich, *Acta Phys. Pol.* **A50**, 215 (1976).
98. S. Kielich, *IEEE J. Quantum Electron.* **QE-5**, 562 (1969); *Opto-Electronics* **2**, 5 (1970).
99. W. Gadomski, *J. Appl. Phys.* **54**, 1029 (1083).
100. J. F. Ward and J. K. Guha, *Appl. Phys. Lett.* **30**, 276 (1977).
101. W. Gadomski and M. Roman, *Opt. Commun.* **33**, 331 (1980).
102. G. Schwarz, *J. Phys. Chem.* **71**, 4021 (1957).
103. J. Jadżyn and L. Hellemans, *Acta Phys. Pol.* **A67**, 1093 (1985).
104. L. Hellemans and L. De Maeyer, *Chem. Phys. Lett.* **129**, 262 (1986).
105. K. S. Cole and R. H. Cole, *J. Chem. Phys.* **9**, 341 (1941).
106. G. Hauchecorne, F. Kerhervé, and G. Mayer, *J. Phys.* **32**, 47 (1971); R. S. Finn and J. F. Ward, *J. Chem. Phys.* **60**, 454 (1974).
107. S. Kielich, *Acta Phys. Pol.* **A37**, 447 (1970).
108. B. F. Levine and C. G. Bethea, *J. Chem. Phys.* **63**, 2666 (1975).
109. P. Bordewijk, *Chem. Phys. Lett.* **39**, 342 (1976).
110. K. L. Sala, G. A. Kenney-Wallace, and G. E. Hall, *IEEE J. Quantum Electron.* **QE-16**, 990 (1980).
111. W. Alexiewicz and H. Derdowska-Zimpel, *Physica* **A197**, **264** (1992).
112. W. Alexiewicz, *Mol. Phys.* **59**, 637 (1986).
113. J.-L. Déjardin and G. Debiais, *Physica* **A164**, 182 (1990).
114. J.-L. Déjardin, *J. Chem. Phys.* **95**, 576 (1991).
115. J.-L. Déjardin and G. Debiais, *J. Chem. Phys.* **95**, 2787 (1991); *Physica* **A175**, 407 (1991).
116. I. Teraoka and R. Hayakawa, *J. Chem. Phys.* **91**, 4920 (1989); **92**, 7653 (1990).
117. Y. H. Lee, D. Kim, S. H. Lee, and W. G. Jung, *J. Chem. Phys.* **91**, 5628 (1989).

SPECTRAL ANALYSIS OF LIGHT SCATTERED BY MONODISPERSE SOLUTIONS OF RIGID, ANISOTROPIC MACROMOLECULES IN A REORIENTING AC ELECTRIC FIELD

M. DĘBSKA-KOTŁOWSKA AND A. MIRANOWICZ

Nonlinear Optics Division, Institute of Physics, Adam Mickiewicz University, Poznań, Poland

CONTENTS

I. INTRODUCTION

Light scattering studies of molecular systems, acted on by an externally applied ordering factor (an ac or dc electric or magnetic field, an acoustic field) provide valuable information concerning the optical, electric, geometric, and dynamical properties of the molecules. For the sake of brevity, this chapter is essentially devoted to earlier results bearing on the autocorrelation function of light scattered by solutions of large macromolecules at free rotational diffusion as considered by Aragon and Pecora [1] in relation to our newer results obtained for rotational diffusion of macromolecules caused by an external ac electric field. Interesting results are also obtained from studies on the integral intensity of light scattered by solutions of macromolecules, reoriented by a dc or ac electric field or the strong electric field of laser light.

Modern Nonlinear Optics, Part 1, Edited by Myron Evans and Stanisław Kielich. Advances in Chemical Physics Series, Vol. LXXXV.
ISBN 0-471-57546-1

Work in this direction was initiated by Wippler [2, 3] from 1953 to 1957, who proposed a theory of light scattering on molecules reoriented in a dc electric field. In 1962, Wallach and Benoit [4] published results for benzyl poly-L-glutaminate (PBLG) with an interpretation based on Wippler's theory. Stoylov and coworkers [5–12] extended the theory, giving a first account of complete reorientation of the macromolecules into the direction of the external field. To Jennings et al. [13–19] are due experimental results for the relative variations in the scattered light intensity versus the square of the reorienting field strength.

Kielich [20–24] (for references see [25]) proposed a theory of light scattering by solutions of monodisperse macromolecules with linear dimensions much smaller than the wavelength of the incident light wave and reoriented under the action of a dc electric field, and/or the strong electric field of a laser beam. He moreover worked out a method for the determination of the sign and numerical value of the optical anisotropy of such macromolecules on the basis of electric saturation. Kielich's theory has since been extended to macromolecules with linear dimensions comparable to the light wavelength [26–28]; here, too, an interesting method has been obtained leading to the sign of the optical anisotropy of large macromolecules [29].

Laser technique has proved very efficient in this field of research [30]. The high power and short duration times of laser pulses permit the observation of complete reorientation in macromolecular systems. Experimental results concerning the relaxation times and the size of macromolecules have been derived from the studies of transient effects essentially due to Jennings and coworkers [31–33]. Optical methods for the study of macromolecule dynamics applying molecular light scattering have developed rapidly due to laser technique [34–38].

Here, we give new results achieved in the study of the polarized and depolarized components, the time autocorrelation function, and the amplitude of light, scattered by monodisperse solutions of large, rigid, anisotropic macromolecules (on the Rayleigh–Debye–Gans (R-D-G) [34] approximation) acted on by an external reorienting ac electric field. Formulae for the components are derived applying irreducible spherical tensor formalism to the Smoluchowski–Debye model of molecular diffusion in liquids [39, 40] generalized to comprise anisodiametric molecules, with special attention given to their anisotropy, induced by external field. We assume that the factor decisive for the final effect resides in rotational diffusion of the macromolecules in the external reorienting field **F**.

Our discussion is based on formulae obtained from the general expressions for the respective components of the time correlation function of the amplitudes of light scattered in a reorienting field of low intensity (approximation up to the square of the field strength). The formulae are expressed by Fourier transforms, permitting the determination of the spectral line shapes of the light scattered on the electrically reoriented macromolecules. We also discuss the influence of the size, shape, electric and optical properties, and reorientational molecular parameters on the shape of the spectral lines.

II. THEORY

We assume the monodisperse solution of macromolecules to fulfil the R-D-G approximation [34], that is, $2kL^W|n-1| \ll 1$, where $k = 2\pi/\lambda$ with λ the incident wavelength, and L^W the linear size of the macromolecule. The latter are assumed to be sufficiently dilute for us to neglect their mutual interaction as well as their interaction with the molecules of the solvent, assumed to be optically isotropic. The geometry of scattered light observation is shown in Fig. 1. We express the heterodyne autocorrelation function of the electric field amplitudes of the scattered light in the

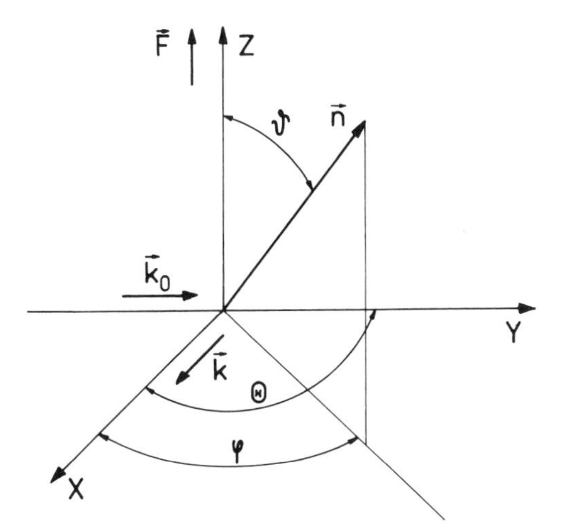

Figure 1. Geometry of observation of light scattering. The light wave, linearly polarized in the direction of vibration of the vector $\mathbf{E}(\omega)$, is incident in the direction of the laboratory Y axis. Scattering is observed along the X axis. We distinguish two possible directions of linear polarization of the scattered light wave: one vertical (perpendicular to the plane of observation) and the other horizontal (parallel thereto). Accordingly we have, respectively, the scattered intensities $I_{zz} = V$ (vertical component) and $I_{yy} = H$ (horizontal component). Likewise, in the incident light wave we distinguish a vertical and a horizontal direction of the polarization, which we denote by indices v and h at the intensity components of the scattered light wave.

form of the second-rank Cartesian tensor $I_{ij}^F(q', t)$ of the light intensity scattered by the macromolecular solution [38]. In laboratory coordinates X, Y, Z and for a moment of time t the tensor can be defined as follows [36, 38, 20]:

$$I_{ij}^F(q', t) = \rho' \left(\frac{\omega}{c}\right)^4 \left\langle \sum_p \sum_{p'} a_{ik}^p(t) a_{jl}^{p'*}(0) \times \exp\{\mathbf{iq}[\mathbf{r}_p(t) - \mathbf{r}_{p'}(0)]\} \right\rangle_{\Omega, F} I_{kl}^0(t) \tag{1}$$

where ρ' is the density of the macromolecules in the solution, $a_{ik}^p(t)$ is the second-rank Cartesian tensor of linear optical polarizability and $\mathbf{r}_p(t)$ is the position of the pth volume element (segment) of the macromolecule at the moment of time t, and $\mathbf{q} = \mathbf{k} - \mathbf{k}_0$ is the scattering vector ($\mathbf{k}_0$ is the wave vector of the incident wave and $\mathbf{k}$ that of the scattered wave observed); $\langle \cdots \rangle_{\Omega, F}$ stands for averaging over all possible orientations of the macromolecules, given by the set of Euler angles Ω in the presence of the external reorienting field $\mathbf{F}$; and $I_{kl}^0(t) = \frac{1}{2} E_{0k}(t) E_{0l}^*(0)$ is the intensity tensor of the light incident on the scattering medium at t. The electric field intensity of the incident light wave will be assumed as sufficiently weak to cause only linear polarization in the macromolecules. The local field of the latter will be neglected.

On expressing the vector $\mathbf{r}_p(t)$ as the sum [1]

$$\mathbf{r}_p(t) = \mathbf{R}(t) + \boldsymbol{\rho}_p(t)$$

where $\mathbf{R}(t)$ denotes the position of the center of mass of the macromolecule at t and $\boldsymbol{\rho}_p(t)$ the position of its pth segment at t with respect to its center of mass, and assuming the motion of the center of mass as statistically independent of the motions inside the macromolecule, we can separate the translational part of the tensor $I_{ij}^F(q', t)$ related to the motion of the center of mass from the other dynamical terms. The translational part is given by the well-known expression [36]

$$\langle \exp\{\mathbf{iq}'[\mathbf{R}(t) - \mathbf{R}(0)]\}\rangle = \exp(-q'^2 D_T t) \tag{2}$$

where D_T is the translational diffusion constant of the macromolecule. The phase constant occurring in Eq. 1 is expressed in terms of spherical

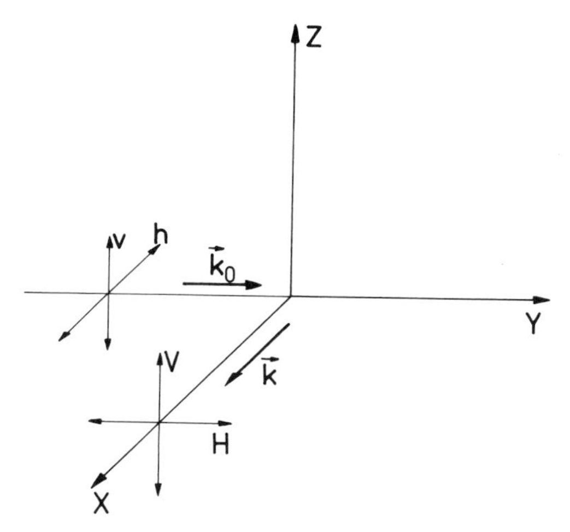

Figure 2. The axis of symmetry **n** of the macromolecule in the laboratory system of coordinates X, Y, Z; θ is the angle of observation of light scattering.

harmonics by the well-known formula [41]

$$\exp\left[i\mathbf{q}'\boldsymbol{\rho}_p(t)\right] = 4\pi \sum_{n,m} i^n j_n(q'\rho_p) Y_{nm}^*(\Omega_{q'}) Y_{nm}\left[\Omega_p(t)\right] \tag{3}$$

where $Y_{nm}(\Omega)$ is a spherical harmonic, and $j_n(q'\rho_p)$ is a spherical Bessel function of the order n. In the case of a rigid macromolecule, the length of the position vector of the segment is constant (time-independent); thus $|\boldsymbol{\rho}(t)| = \rho$, so that only the spherical harmonics are dependent on time t. $\Omega_p(t)$ is the set of Euler angles describing the orientation of the pth segment at the moment of time t, whereas $\Omega_{q'}$ describes the orientation of the scattering vector $\mathbf{q}'$. In accordance with the assumptions made, the angular coordinates of $\mathbf{q}'$ are (see Fig. 2)

$$\mathbf{q}' = 2k \sin\frac{\theta}{2}\left(\hat{x}\sin\frac{\theta}{2} - \hat{y}\cos\frac{\theta}{2}\right) \tag{4}$$

Hence

$$\Omega_{q'} = \left(\frac{\pi}{2}, \frac{\theta}{2} - \frac{\pi}{2}\right)$$

with $q' = 2k\sin(\theta/2)$ (θ is the angle at which we chose to observe the experiment of scattering). By Eq. (3) and with regard to Eq. (2), we rewrite

Eq. (1) in the form

$$\begin{aligned} I_{ij}^{F}(q', t) = A(4\pi)^2 \Big\langle \sum_{p,p'} \sum_{n,m} \sum_{n',m'} & i^n(-i)^{n'} \\ & \times j_n(q'\rho_p) j_n(q'\rho_{p'}) Y_{nm}^{*}(\Omega_{q'}) Y_{n'm'}(\Omega_{q'}) \\ & \times a_{ik}^{p}(t) a_{jl}^{p'*}(0) Y_{nm}[\Omega_p(t)] \\ & \times Y_{n'm'}^{*}[\Omega_{p'}(0)] \Big\rangle_{\Omega, F} I_{kl}^{0}(t) \end{aligned} \tag{5}$$

where

$$A = \rho' \left(\frac{\omega}{c} \right)^4 \exp\left(-q'^2 D_T t \right)$$

In our analysis of Eq. (5) we shall apply the method of irreducible spherical tensors [38, 41–43]. We express the Cartesian tensor $a_{ik}^{p}(t)$ by way of components of the spherical tensor $a_{LN_1}^{p}(t)$ as follows [42]:

$$a_{ik}^{p}(t) = \sum_{L, N_1} U_{LN_1}^{ik} a_{LN_1}^{p}(t) \tag{6}$$

with $L = 0, 2$. The values of the coefficient $U_{LN_1}^{ik}$ for the various components of the Cartesian tensor $a_{ik}^{p}(t)$ are given in Table 1 of Ref. 42.

The orientation Ω_p of the pth segment of the macromolecule was defined in the laboratory frame of reference, but we are concerned with the orientation of the macromolecule as a whole, i.e., the orientation of the macromolecular frame of reference with respect to the laboratory coordinate system. Accordingly, the substitutions to be made in (5) are

$$\begin{aligned} Y_{nm}[\Omega_p(t)] &= \sum_{r} D_{rm}^{n}(\Omega_t) Y_{nr}(\hat{\Omega}_p) \\ a_{LN_1}^{p}(t) &= \sum_{N_3} D_{N_3 N_1}^{L}(\Omega_t) a_{LN_3}(\hat{\Omega}_p) \end{aligned} \tag{7}$$

where $D_{rm}^{n}(\Omega_t)$ and $D_{N_3N_1}^{L}(\Omega_t)$ are Wigner rotation functions [1]. Ω_t describes the orientation of the macromolecular frame with respect to the laboratory system of reference at the moment of time t, and $\hat{\Omega}_p$ describes that of the pth segment in the macromolecular frame. With regard to the preceding expressions, Eq. (5) takes the form

$$\begin{aligned} I_{ij}^{F}(q',t) = A(4\pi)^2 \sum_{p,p'} \sum_{n,m} \sum_{n',m'} \sum_{L,L'} \sum_{N_1,N_2} \sum_{N_3,N_4} \sum_{r,r'} \\ \times i^n(-i)^{n'} U_{LN_1}^{ik} U_{L'N_2}^{il*} Y_{nm}^{*}(\Omega_{q'}) Y_{n'm'}(\Omega_{q'}) j_n(q'\rho_p) j_{n'}(q'\rho_{p'}) \\ \times Y_{nm}^{*}(\Omega_{q'}) Y_{n'm'}(\Omega_{q'}) a_{LN_3}(\hat{\Omega}_p) a_{L'N_4}^{*}(\hat{\Omega}_{p'}) Y_{nr}(\hat{\Omega}_p) \\ \times Y_{n'r'}^{*}(\hat{\Omega}_{p'}) \\ \times \left\langle D_{rm}^{n}(\Omega_t) D_{r'm'}^{n'*}(\Omega_0) D_{N_3N_1}^{L}(\Omega_t) D_{N_4N_2}^{L'*}(\Omega_0) \right\rangle_{\Omega,F} I_{kl}^{0}(t) \end{aligned} \tag{8}$$

where the factor

$$U_{LN_1}^{ik} U_{L'N_2}^{il*} Y_{nm}^{*}(\Omega_{q'}) Y_{n'm'}(\Omega_{q'}) \tag{8a}$$

accounts for the observation geometry applied in the experiment: the factor

$$\sum_{p} \sum_{p'} j_n(q'\rho_p) j_{n'}(q'\rho_{p'}) a_{LN_3}(\hat{\Omega}_p) a_{L'N_4}^{*}(\hat{\Omega}_{p'}) Y_{nr}(\hat{\Omega}_p) Y_{n'r'}^{*}(\hat{\Omega}_{p'}) \tag{8b}$$

accounts for the shape and symmetry of the macromolecule; and

$$\left\langle D_{rm}^{n}(\Omega_t) D_{r'm'}^{n'*}(\Omega_0) D_{N_3N_1}^{L}(\Omega_t) D_{N_4N_2}^{L'*}(\Omega_0) \right\rangle_{\Omega,F} \tag{8c}$$

accounts for the orientation of the macromolecule in the field **F**. The shape and symmetry factor (8b) can be replaced by the following analytical expression, derived by Aragon and Pecora [1]:

$$\begin{aligned} \sum_{p} \sum_{p'} j_n(q'\rho_p) j_{n'}(q'\rho_{p'}) a_{LN_3}(\hat{\Omega}_p) a_{L'N_4}^{*}(\hat{\Omega}_{p'}) Y_{nr}(\hat{\Omega}_p) Y_{n'r'}^{*}(\hat{\Omega}_{p'}) \\ = 3a\sqrt{(2n+1)(2n'+1)} \left[\delta_{L0}\delta_{L'0}\delta_{N_30}\delta_{N_40} d_n(x) d_{n'}(x) \right. \\ + \sqrt{2}\kappa\delta_{L0}\delta_{L'2}\delta_{N_30} d_n(x) b_{n'N_4}(x) + \sqrt{2}\kappa\delta_{L2}\delta_{L'0}\delta_{N_40} d_{n'}(x) b_{nN_3}(x) \\ \left. + 2\kappa^2\delta_{L2}\delta_{L'2} b_{nN_3}(x) b_{n'N_4}(x) \right] \end{aligned} \tag{9}$$

where

$$a = \frac{a_3 + 2a_1}{3} \qquad \kappa = \frac{a_3 - a_1}{2a_1 + a_3}$$

are, respectively, the mean optical polarizability and the anisotropy of optical polarizability of the macromolecule, whereas a_3 and a_1 are its optical polarizabilities along its axis of symmetry and along the symmetry axis perpendicular thereto. The expressions

$$\begin{aligned} d_n(x) &= \frac{1}{V}\left(\frac{4\pi}{2n+1}\right)^{1/2} \int_V \mathrm{d}^3\rho\, j_n(q'\rho) Y_{n0}(\hat{\Omega}) \\ b_{nN_3}(x) &= \frac{1}{V}\frac{4\pi}{\sqrt{5(2n+1)}} \int_V \mathrm{d}^3\rho\, j_n(q'\rho) Y_{2-N_3}(\hat{\Omega}) Y_{nN_3}(\hat{\Omega}) \end{aligned} \tag{10}$$

are the particle shape functions, with $x = q'L^W$ (L^W is the linear size characteristic for the macromolecule, and V its volume). The statistical averaging in the factor (9) is carried out as follows [44, 45]:

$$\begin{aligned} &\left\langle D^n_{rm}(\Omega_t) D^{n'*}_{r'm'}(\Omega_0) D^L_{N_3N_1}(\Omega_t) D^{L'*}_{N_4N_2}(\Omega_0)\right\rangle_{\Omega,F} \\ &\quad = \frac{1}{8\pi^2}\int_{\Omega_0}\int_{\Omega_t} \mathrm{d}\Omega_0\, \mathrm{d}\Omega_t D^{n'*}_{r'm'}(\Omega_0) D^{L'*}_{N_4N_2}(\Omega_0) f(\Omega_0, F) \\ &\quad\quad \times D^n_{rm}(\Omega_t) D^L_{N_3N_1}(\Omega_t) f(\Omega_0|\Omega_t, t, F) \end{aligned} \tag{11}$$

where $f(\Omega_0, F)$ is the stationary probability density distribution function in the external electric field $\mathbf{F}$, and $f(\Omega_0|\Omega_t, t, F)$ is the conditional probability density distribution function of orientation Ω_t of the macromolecule in the field $\mathbf{F}$ at time t, provided that its orientation was given by Ω_0 $t = 0$. It should be kept in mind that the external reorienting field $F(t)$

$$F(t) = \sum_a F(\omega_a)\exp(-\mathrm{i}\omega_a t)$$

oscillates along the Z axis of laboratory coordinates.

With the symmetry of the macromolecules assumed as above, the potential energy $u(\Omega_t, F)$ is, with accuracy to the square of the electric field strength [44, 45],

$$\begin{aligned} u(\Omega_t, F) = &-mD^1_{00}(\Omega_t)\sum_a F_Z(\omega_a)\mathrm{e}^{-\mathrm{i}\omega_a t} \\ &-\left[\tfrac{1}{3}\gamma D^2_{00}(\Omega_t) + \alpha D^0_{00}(\Omega_t)\right] \\ &\times \sum_{ab} F_Z(\omega_a) F_Z(\omega_b)\mathrm{e}^{-\mathrm{i}(\omega_a+\omega_b)t} \end{aligned} \tag{12}$$

where m is the permanent dipole electric moment of the macromolecule; $\alpha = \frac{1}{3}(\alpha_3 + 2\alpha_1)$ is its mean electric polarizability; and $\gamma = \alpha_3 - \alpha_1$ is the anisotropy of its linear electric polarizability (α_3 and α_1 are its electric polarizabilities in the direction of the axis of symmetry and in the direction perpendicular thereto, respectively).

The function $f(\Omega_0|\Omega_t, t, F)$ can be expanded in a series in Wigner functions [41, 44, 45]:

$$f(\Omega_0|\Omega_t, t, F) = \sum_{J,K,M,N} (2N+1) C_{KM}^{J,N}(t,F) D_{KM}^{N*}(\Omega_0) D_{KM}^{J}(\Omega_t) \tag{13}$$

where $C_{KM}^{J,N}(t, F)$ denotes the dynamical coefficients of Debye rotational diffusion. Now, by definition [46], the function $f(\Omega_0, F)$

$$f(\Omega_0, F) = \lim_{t\to\infty} f(\Omega_0|\Omega_t, t, F) \tag{14}$$

is expressed as follows [44, 45]:

$$f(\Omega_0, F) = \sum_I C_I(F) D_{00}^I(\Omega_0) \tag{15}$$

The expansion coefficients $C_I(F)$ are determined from the relation [41]

$$C_I = \frac{2I+1}{8\pi^2} \int_{\Omega_0} \mathrm{d}\Omega_0\, f(\Omega_0, F) D_{00}^I(\Omega_0) \tag{16}$$

on insertion into Eq. (16) of the equilibrium distribution function $f(\Omega_0, F)$, which for a strong time-independent field has the form [47, 48]

$$f(\Omega_0, F) = \frac{\exp(p\cos\vartheta \pm q\cos^2\vartheta)}{2\pi\int_0^\pi \exp(p\cos\vartheta \pm q\cos^2\vartheta)\sin\vartheta\,\mathrm{d}\vartheta} \tag{17}$$

where

$$p = \frac{mF}{k'T} \quad \text{and} \quad q = \frac{\alpha_3 - \alpha_1}{2k'T}F^2 \tag{18}$$

with, respectively, the successive dimensionless parameters of orientation for the permanent dipole moment m and the electric polarizability ellipsoid; ϑ is the angle between the symmetry axis $\mathbf{n}$ of the macromolecule and the field $\mathbf{F}$; k' is Boltzmann's constant; and T is the absolute temperature; we express the coefficients (16) as linear combinations of Langevin–Kielich functions $L_n(p,q)$ [20, 21, 47–50] (for references see [25]). For the initial coefficients $C_l(F)$ we have [44, 45]

$$\begin{aligned}
C_0 &= 1 \\
C_1 &= 3L_1(p,q) \\
C_2 &= \tfrac{5}{2}[3L_2(p,q) - 1] \\
C_3 &= \tfrac{7}{2}[5L_3(p,q) - 3L_1(p,q)] \\
C_4 &= \tfrac{9}{8}[35L_4(p,q) - 30L_2(p,q) + 3] \\
\cdots \quad &\cdots \; \cdots \; \cdots \; \cdots
\end{aligned} \tag{19}$$

The function $f(\Omega_t, F)$ describes a wide range of relaxational effects. It has been applied by Debye, in an approximation to the first power of the electric field, in his work on effects of dielectric relaxation [39]. In a form with accuracy to the square of the field it served Peterlin and Stuart [51] as well as Benoit [53] in their analyses of electric and optical birefringence in liquids [54]. Extended to the third power of the electric field, it has been applied to the description of dispersion and absorption in many nonlinear electro-optical effects [51, 52]. Methods of dielectric relaxation are currently in use in studies of rotational and translational motions of molecules, correlations between molecular dipoles, and the dynamics of the structure of liquids [53].

We restrict our considerations to macromolecules in the form of disks and rods. The shape functions (10) for rodlike macromolecules (prolate cylinders of height l much greater than their diameter $2a_D$, i.e., $l \gg 2a_D$, are of the form [1]

$$\begin{aligned}
d_n^P(x) &= \frac{1}{x}\int_0^x j_n(x')\,\mathrm{d}x' \\
b_{nN'}^P(x) &= \delta_{N'0}\, d_n^P(x)
\end{aligned} \tag{20}$$

whereas for disklike macromolecules (oblate cylinders, $l \ll 2a_D$) they are of the form

$$\begin{aligned}
d_n^D(x) &= (-1)^{n/2}\,\frac{(n-1)!!}{n!!}\,\frac{2}{x^2}\int_0^x j_n(x')\,x'\,\mathrm{d}x' \\
b_{nN'}^D(x) &= -\tfrac{1}{2}\delta_{N'0}\, d_n^D(x)
\end{aligned} \tag{21}$$

where $x^P = q'l/2$ (l is the length of the rod) and $x^D = q'a_D$ (a_D is the radius of the disk). Relations (20) and (21) hold for even n only. Making use of Eqs. (8), (13), (15), (20), and (21) we now derive an expression for the time-dependent autocorrelation function $I^F_{i,j}(q',t)$ valid both for rods and disks; in fact, by Eqs. (24) and (25), the conditions imposed on the values of n' and N are the same in both cases. Thus,

$$\begin{aligned}
I^F_{i,j}(q',t) &= A'\sum_I C_I(F)\Bigg\{\sum_{n'}\sum_{J,M}\sum_N(-1)^M i^J(-i)^{n'}\frac{2N+1}{2I+1}\frac{\sqrt{2n'+1}}{\sqrt{2J+1}} \\
&\times U^{ik}_{00}U^{jl*}_{00}Y^*_{J,-M}(\Omega_{q'})Y_{n',-M}(\Omega_{q'})C^{J,N}_{0M}(t,F)\begin{bmatrix} n' & N & I \\ 0 & 0 & 0\end{bmatrix} \\
&\times\begin{bmatrix} n' & N & I \\ -M & M & 0\end{bmatrix} d_J(x)d_{n'}(x) \\
&+\sqrt{2}\kappa\sum_{n',m'}\sum_{N_2}\sum_{J,M}\sum_G(-1)^M i^J(-i)^{n'}\frac{2N+1}{2I+1}\frac{\sqrt{2n'+1}}{\sqrt{2J+1}} \\
&\times U^{ik}_{00}U^{jl*}_{2N_2}Y^*_{J,-M}(\Omega_{q'})Y_{n'm}(\Omega_{q'})C^{J,N}_{0M}(t,F) \\
&\times\begin{bmatrix} n' & 2 & G \\ 0 & 0 & 0\end{bmatrix}\begin{bmatrix} n' & 2 & G \\ m', & N_2, & -M\end{bmatrix}\begin{bmatrix} G & N & I \\ 0 & 0 & 0\end{bmatrix} \\
&\times\begin{bmatrix} G & N & I \\ -M & M & 0\end{bmatrix} d_J(x)b_{n'0}(x) \\
&+\sqrt{2}\kappa\sum_{n'}\sum_{n,m}\sum_{N_1}\sum_N\sum_{J,M}(-1)^M i^n(-i)^{n'}\frac{2N+1}{2I+1}\frac{\sqrt{2n'+1}}{\sqrt{2n+1}} \\
&\times U^{ik}_{2N_1}U^{jl*}_{00}Y^*_{nm}(\Omega_{q'})Y_{n',-M}(\Omega_{q'})C^{J,N}_{0M}(t,F) \qquad (22) \\
&\times\begin{bmatrix} J & 2 & N \\ 0 & 0 & 0\end{bmatrix}\begin{bmatrix} J & 2 & n \\ M & N_1 & -m\end{bmatrix}\begin{bmatrix} n' & N & I \\ 0 & 0 & 0\end{bmatrix} \\
&\times\begin{bmatrix} n' & N & I \\ -M & M & 0\end{bmatrix} d_{n'}(x)b_{n0}(x) \\
&+2\kappa^2\sum_{n,m}\sum_{n',m'}\sum_{N_1,N_2}\sum_{J,M}\sum_G(-1)^M i^n(-i)^{n'}\frac{2N+1}{2I+1}\frac{\sqrt{2n'+1}}{\sqrt{2n+1}} \\
&\times U^{ik}_{2N_1}U^{jl*}_{2N_2}Y^*_{nm}(\Omega_{q'})Y_{n'm'}(\Omega_{q'})C^{J,N}_{0M}(t,F) \\
&\times\begin{bmatrix} J & 2 & n \\ 0 & 0 & 0\end{bmatrix}\begin{bmatrix} J & 2 & n \\ M & N_1 & -m\end{bmatrix}\begin{bmatrix} n' & 2 & G \\ 0 & 0 & 0\end{bmatrix}\begin{bmatrix} n' & 2 & G \\ m' & N_2 & -M\end{bmatrix} \\
&\times\begin{bmatrix} G & N & I \\ 0 & 0 & 0\end{bmatrix}\begin{bmatrix} G & N & I \\ -M & M & 0\end{bmatrix} b_{n0}(x)b_{n'0}(x)\Bigg\} I^0_{kl}(t)
\end{aligned}$$

with $A' = 96\pi^3 Aa^2$. For rods we have $b_{n0}^P(x) = d_n^P(x)$, and for disks $b_{n0}^D(x) = -\frac{1}{2}d_n^D(x)$.

As shown by Aragon and Pecora [1], in the case of macromolecules with the symmetry of a rotational ellipsoid and the shape of a disk or rod with $K = 0$ ($K = -2N'$, and $N' = 0$ (Eqs. (21) and (22)) at free diffusion, rotation about the symmetry axis of the macromolecule is invisible to the incident light beam. Accordingly, there is no diffusion coefficient $D_{\parallel}$, but only $D_{\perp}$. Similarly, if the reorienting field $\mathbf{F}$ is applied in the direction of the Z axis (Fig. 2), the equation of diffusion the solution of that will enable us to obtain the explicit form of the dynamical coefficient $C_{0M}^{J,N}(t, F)$ will involve only the coefficient $D_{\perp}$ of diffusion about the axis perpendicular to the symmetry axis of the macromolecule. Quite generally, the statistical distribution function $f(r_t, \Omega_t, t)$ determining the probability density for the macromolecule to assume the position $\mathbf{r}_t$ and orientation Ω_t at the moment of time t can be obtained for spherical top molecules from the Smoluchowski–Debye equation [39, 54]:

$$\frac{\partial}{\partial t} f(\mathbf{r}_t, \Omega_t, t) = D\nabla\{\nabla + \beta\nabla u(\mathbf{r}_t, \Omega_t)\} f(\mathbf{r}_t, \Omega_t, t) \tag{23}$$

where D is the isotropic diffusion coefficient; ∇ the spatial derivation operator; and $u(\mathbf{r}_t, \Omega_t)$ the Brownian potential energy of a particle in an external reorienting field. To find an analytical solution to the Smoluchowski–Debye equation one can assume in a first approximation that the molecular translational and rotational motions are statistically independent, so that [52]

$$f(\mathbf{r}_t, \Omega_t, t) = f(\mathbf{r}_t, t) f(\Omega_t, t) \tag{23a}$$

Having performed the preceding separation we obtain from (23) two independent equations for translational (Smoluchowski) and rotational (Debye) diffusion whence we determine the analytical forms of $f(\mathbf{r}_t, t)$ and $f(\Omega_t, t)$ [39, 54]. Accordingly, we get the dynamical coefficients $C_{0M}^{J,N}(t, F)$ by insertion of the expansion (13), with $D = D_{\perp}$, into Eq. (23).

Alexiewicz et al. [45] have solved (23) by applying the method of statistical perturbation calculus. They expanded the coefficients $C_{0M}^{J,N}(t, F)$ in a power series of $\beta = 1/(k'T)$ (the reciprocal of the thermal motion energy),

$$C_{0M}^{J,N}(t,F) = \sum_{n=0}^{\infty} \beta^{n(n)} C_{0M}^{J,N}(t,F) \tag{24}$$

and obtained

$$\begin{aligned} {}^{(n)}C_{0M}^{J,N}(t,F) = {} & \frac{1}{2} p \sum_{a} \sum_{J'=|J-1|}^{J+1} \frac{J(J+1) - J'(J'+1) + 2}{J(J+1)} \\ & \times \begin{bmatrix} J' & 1 & J \\ 0 & 0 & 0 \end{bmatrix} \begin{bmatrix} J' & 1 & J \\ M & 0 & M \end{bmatrix} \\ & \times \frac{1}{\tau_J} \mathrm{e}^{-t/\tau_J} \int_0^t e^{-\mathrm{i}\omega_a t + t/\tau_J} {}^{(n-1)}C_{0M}^{J',N}(t,F)\,\mathrm{d}t \\ & + \frac{q}{3} \sum_{a,b} \sum_{J'=|J-2|}^{J+2} \frac{J(J+1) - J'(J'+1) + 6}{J(J+1)} \\ & \times \begin{bmatrix} J' & 2 & J \\ 0 & 0 & 0 \end{bmatrix} \begin{bmatrix} J' & 2 & J \\ M & 0 & M \end{bmatrix} \\ & \times \frac{1}{\tau_J} \mathrm{e}^{-t/\tau_J} \int_0^t e^{-i(\omega_a + \omega_b)t + t/\tau_J} {}^{(n-1)}C_{0M}^{J',N}(t,F)\,\mathrm{d}t \end{aligned} \tag{25}$$

For free rotational diffusion $n = 0$ they found

$$^{(0)}C_{0M}^{J,N}(t) = \frac{1}{8\pi^2} \delta_{JN} \exp(-t/\tau_J) \tag{26}$$

where the τ_J are Debye rotational relaxation times, related as follows with the rotational diffusion coefficient $D_\perp$:

$$\tau_J = \frac{1}{J(J+1)D_\perp} \tag{26a}$$

By (22) and the data of Table I, in conformity with the experimental conditions adopted in Figs. 1 and 2, we finally arrive at the following expressions for the individual components of the time-autocorrelation function of light, scattered by solutions of rod- and disklike macromolecules reoriented by an external ac electric field **F**:

$$
\begin{aligned}
\frac{V_v^F(q',t)}{I_{zz}^0(t)} &= A' \sum_I C_I(F) \sum_N \sum_{J,M} (-1)^M C_{0M}^{J,N}(t,F) \\
&\times \Bigg\{ \frac{1}{3} \sum_{n'} i^J (-i)^{n'} \frac{2N+1}{2I+1} \frac{\sqrt{2n'+1}}{\sqrt{2J+1}} Y_{JM}(\Omega_{q'}) Y_{n'M}^*(\Omega_{q'}) \\
&\times \begin{bmatrix} n' & N & I \\ 0 & 0 & 0 \end{bmatrix} \begin{bmatrix} n' & N & I \\ -M & M & 0 \end{bmatrix} d_J(x) d_{n'}(x) \\
&+ \frac{2}{3}\kappa \sum_{n'} \sum_G i^J (-i)^{n'} \frac{2N+1}{2I+1} \frac{\sqrt{2n'+1}}{\sqrt{2J+1}} Y_{JM}(\Omega_q) Y_{n'M}^*(\Omega_{q'}) \\
&\times \begin{bmatrix} n' & 2 & G \\ 0 & 0 & 0 \end{bmatrix} \begin{bmatrix} n' & 2 & G \\ M & 0 & M \end{bmatrix} \begin{bmatrix} G & N & I \\ 0 & 0 & 0 \end{bmatrix} \\
&\times \begin{bmatrix} G & N & I \\ -M & M & 0 \end{bmatrix} d_J(x) b_{n'0}(x) \\
&+ \frac{2}{3}\kappa \sum_{n,n'} i^n (-i)^{n'} \frac{2N+1}{2I+1} \frac{\sqrt{2n'+1}}{\sqrt{2n+1}} Y_{nM}(\Omega_{q'}) Y_{n'M}^*(\Omega_{q'}) \\
&\times \begin{bmatrix} J & 2 & n \\ 0 & 0 & 0 \end{bmatrix} \begin{bmatrix} J & 2 & n \\ M & 0 & M \end{bmatrix} \begin{bmatrix} n' & N & I \\ 0 & 0 & 0 \end{bmatrix} \\
&\times \begin{bmatrix} n' & N & I \\ -M & M & 0 \end{bmatrix} d_n(x) b_{n'0}(x) \\
&+ \frac{4}{3}\kappa^2 \sum_{n,n'} \sum_G i^n (-i)^{n'} \frac{2N+1}{2I+1} \frac{\sqrt{2n'+1}}{\sqrt{2n+1}} \\
&\times Y_{nM}(\Omega_{q'}) Y_{n'M}^*(\Omega_{q'}) \\
&\times \begin{bmatrix} J & 2 & n \\ 0 & 0 & 0 \end{bmatrix} \begin{bmatrix} J & 2 & n \\ M & 0 & M \end{bmatrix} \begin{bmatrix} n' & 2 & G \\ 0 & 0 & 0 \end{bmatrix} \begin{bmatrix} n' & 2 & G \\ M & 0 & M \end{bmatrix} \\
&\times \begin{bmatrix} G & N & I \\ -M & 0 & M \end{bmatrix} \begin{bmatrix} G & N & I \\ 0 & 0 & 0 \end{bmatrix} b_{n0}(x) b_{n'0}(x) \Bigg\}
\end{aligned} \tag{27}
$$

$$
\begin{aligned}
\frac{H_v^F(q',t)}{I_{zz}^0(t)} &= \frac{V_h^F(q',t)}{I_{xx}^0(t)} \\
&= \kappa^2 A' \sum_I C_I(F) \sum_{J,M} \sum_N (-1)^M C_{0,M-1}^{J,N}(t,F) \sum_{n,n'} \sum_G \\
&\quad \times i^n(-i)^{n'} \frac{2N+1}{2I+1} \frac{\sqrt{2n'+1}}{\sqrt{2n+1}} Y_{nM}(\Omega_{q'}) Y_{n'M}^*(\Omega_{q'}) \\
&\quad \times \begin{bmatrix} J & 2 & n \\ 0 & 0 & 0 \end{bmatrix} \begin{bmatrix} J & 2 & n \\ M-1 & 1 & M \end{bmatrix} \begin{bmatrix} n & 2 & G \\ 0 & 0 & 0 \end{bmatrix} \\
&\quad \times \begin{bmatrix} n' & 2 & G \\ -M & 1 & 1-M \end{bmatrix} \\
&\quad \times \begin{bmatrix} G & N & I \\ 1-M & M-1 & 0 \end{bmatrix} \begin{bmatrix} G & N & I \\ 0 & 0 & 0 \end{bmatrix} b_{n0}(x) b_{n'0}(x)
\end{aligned}
\tag{28}
$$

$$
\begin{aligned}
\frac{H_h^F(q',t)}{I_{xx}^0(t)} &= \kappa^2 A' \sum_I C_I(F) \sum_{J,M} \sum_N \\
&\quad \times (-1)^M C_{0,M-2}^{J,N}(t,F) \sum_{n,n'} \sum_G i^n(-i)^{n'} \frac{2N+1}{2I+1} \frac{\sqrt{2n'+1}}{\sqrt{2n+1}} \\
&\quad \times \begin{bmatrix} J & 2 & n \\ 0 & 0 & 0 \end{bmatrix} \begin{bmatrix} J & 2 & n \\ M-2 & 2 & M \end{bmatrix} \begin{bmatrix} n' & 2 & G \\ 0 & 0 & 0 \end{bmatrix} \\
&\quad \times \begin{bmatrix} G & N & I \\ 2-M & M-2 & 0 \end{bmatrix} \begin{bmatrix} G & N & I \\ 0 & 0 & 0 \end{bmatrix} \\
&\quad \times \Bigg\{ Y_{nM}(\Omega_{q'}) Y_{n'M}^*(\Omega_{q'}) \begin{bmatrix} n' & 2 & n \\ -M & 2 & 2-M \end{bmatrix} \\
&\quad + Y_{nM}(\Omega_{q'}) Y_{n',M-4}^*(\Omega_{q'}) \\
&\quad \times \begin{bmatrix} n' & 2 & n \\ 4-M & -2 & 2-M \end{bmatrix} \Bigg\} b_{n0}(x) b_{n'0}(x)
\end{aligned}
\tag{29}
$$

TABLE I
Values of the Transformation Coefficients $U^{ik}_{LN_1}$

	LN_1					
ik	0 0	2 0	2 2	2 −2	2 1	2 −1
xx	$\frac{1}{\sqrt{3}}$	$-\frac{1}{\sqrt{6}}$	$\frac{1}{2}$	$\frac{1}{2}$	0	0
yy	$\frac{1}{\sqrt{3}}$	$-\frac{1}{\sqrt{6}}$	$-\frac{1}{2}$	$-\frac{1}{2}$	0	0
zz	$\frac{1}{\sqrt{3}}$	$\frac{2}{\sqrt{6}}$	0	0	0	0
xy	0	0	$-\frac{i}{2}$	$\frac{i}{2}$	0	0
yz	0	0	0	0	$\frac{i}{2}$	$\frac{i}{2}$
zx	0	0	0	0	$-\frac{1}{2}$	$\frac{1}{2}$

In (27)–(29) we have made use of the following relations [41]:

$$\begin{aligned}
&(1) \qquad \begin{bmatrix} a & b & c \\ 0 & 0 & 0 \end{bmatrix} \neq 0 \qquad \text{for even } d = a + b + c \\
&(2) \qquad \begin{bmatrix} a & b & c \\ \alpha & \beta & \gamma \end{bmatrix} = (-1)^{a+b+c} \begin{bmatrix} a & b & c \\ -\alpha, & -\beta, & -\gamma \end{bmatrix} \\
&(3) \qquad Y^{*}_{lm}(\theta', \phi) = (-1)^{m} Y^{*}_{l,-m}(\theta', \phi) \\
&(4) \qquad \Omega_{q'} = \left(\frac{\pi}{2}, -\frac{\pi}{4}\right) \qquad \text{from Eq. 4} \\
&(5) \qquad C^{J,N}_{0M}(t, F) = C^{J,N}_{0,-M}(t, F) \qquad \text{from Eq. 25}
\end{aligned} \tag{30}$$

In the absence of the reorienting field $\mathbf{F} = 0$, Eqs. (19) and (26) give

$$\begin{aligned}
I &= 0 \\
C_0 &= 1 \\
{}^{(0)}C^{J,N}_{0M}(t) &= \frac{1}{8\pi^2} \delta_{JN} \exp(-J(J+1)D_{\perp} t)
\end{aligned} \tag{31}$$

With the above conditions inserted into (27)–(29), we obtain

$$
\begin{aligned}
\frac{V_v(q',t)}{I_{zz}^0(t)}
&= \frac{A'}{8\pi^2}\sum_J (2J+1)\exp[-J(J+1)D_\perp t] \\
&\quad \times \Bigg\{ \frac{1}{3} d_J^2(x) + \frac{4}{3}\kappa \sum_n \sum_M \frac{(-1)^{J/2}}{\sqrt{2J+1}} \frac{(-1)^{n/2}}{\sqrt{2n+1}} \\
&\quad \times Y_{JM}(\Omega_{q'}) Y_{nM}^*(\Omega_{q'}) \begin{bmatrix} J & 2 & n \\ 0 & 0 & 0 \end{bmatrix} \\
&\quad \times \begin{bmatrix} J & 2 & n \\ M & 0 & M \end{bmatrix} d_J(x) b_{n0}(x) \\
&\quad + \frac{16\pi}{3}\kappa^2 \sum_{n,n'} \frac{(-1)^{n/2}}{\sqrt{2n+1}} \frac{(-1)^{n'/2}}{\sqrt{2n'+1}} \\
&\quad \times Y_{nM}(\Omega_{q'}) Y_{n'M}^*(\Omega_{q'}) \begin{bmatrix} J & 2 & n \\ 0 & 0 & 0 \end{bmatrix} \\
&\quad \times \begin{bmatrix} J & 2 & n' \\ 0 & 0 & 0 \end{bmatrix} \begin{bmatrix} J & 2 & n \\ M & 0 & M \end{bmatrix} \begin{bmatrix} J & 2 & n' \\ M & 0 & M \end{bmatrix} b_{n0}(x) b_{n'0}(x) \Bigg\}
\end{aligned}
\tag{32}
$$

$$
\begin{aligned}
\frac{H_v(q',t)}{I_{zz}^0(t)}
&= \frac{V_h(q',t)}{I_{xx}^0(t)} = \frac{A'\kappa^2}{8\pi^2} \sum_J (2J+1)\exp[-J(J+1)D_\perp t] \\
&\quad \times \sum_{n,n'} \sum_M \frac{(-1)^{n/2}}{\sqrt{2n+1}} \frac{(-1)^{n'/2}}{\sqrt{2n'+1}} Y_{nM}(\Omega_{q'}) Y_{n'M}^*(\Omega_{q'}) \\
&\quad \times \begin{bmatrix} J & 2 & n \\ 0 & 0 & 0 \end{bmatrix} \begin{bmatrix} J & 2 & n' \\ 0 & 0 & 0 \end{bmatrix} \begin{bmatrix} J & 2 & n \\ M-1 & 1 & M \end{bmatrix} \\
&\quad \times \begin{bmatrix} J & 2 & n' \\ M-1 & 1 & M \end{bmatrix} b_{n0}(x) b_{n'0}(x)
\end{aligned}
\tag{33}
$$

$$
\begin{aligned}
\frac{H_h(q',t)}{I^0_{xx}(t)}
&= \frac{A'\kappa^2}{8\pi^2}\sum_J (2J+1)\exp[-J(J+1)D_\perp t] \\
&\times \sum_{n,n'}\sum_M \frac{(-1)^{n/2}}{\sqrt{2n+1}}\frac{1}{\sqrt{2n'+1}}\begin{bmatrix} J & 2 & n \\ 0 & 0 & 0\end{bmatrix}\begin{bmatrix} J & 2 & n' \\ 0 & 0 & 0\end{bmatrix} \\
&\times \Big\{ Y_{nM}(\Omega_{q'})Y^*_{n'M}(\Omega_{q'})\begin{bmatrix} J & 2 & n \\ M-2 & 2 & M\end{bmatrix}\begin{bmatrix} J & 2 & n' \\ M-2 & 2 & M\end{bmatrix} \\
&\quad + Y_{nM}(\Omega_{q'})Y^*_{n',M-4}(\Omega_{q'})\begin{bmatrix} J & 2 & n \\ M-2 & 2 & M\end{bmatrix} \\
&\times \begin{bmatrix} J & 2 & n' \\ M-2 & -2 & M-4\end{bmatrix}\Big\} b_{n0}(x)b_{n'0}(x)
\end{aligned}
\tag{34}
$$

in agreement with the expressions derived by Aragon and Pecora [1] for free diffusion of rod-like macromolecules under identical conditions of observation of light scattering.

III. DISCUSSION AND CONCLUSIONS

Equations (27)–(29) enable us to obtain expressions for the components of the integral intensity of light scattered by solutions of small macromolecules ($L^W \ll \lambda$) in a dc electric field. The form of the successive approximations to the shape function $d_n(x)$ defined by Eqs. (20) and (21) permits the conclusion that, in the case of small macromolecules when $x \to 0$, we have $d_0(x) \to 1$ and all higher approximations tend to zero. Thus, for rods

$$
\begin{aligned}
d^P_0(x) &= \frac{1}{x}\int_0^x j_0(y)\,dy = \frac{1}{x}\int_0^x \frac{\sin y}{y}\,dy = \frac{\mathrm{Si}(x)}{x} \\
\lim_{x\to 0} d^P_0(x) &= 1 \\
d^P_2(x) &= \frac{1}{x}\int_0^x j_2(y)\,dy = \frac{-3\sin x}{2x^3} + \frac{3\cos x}{2x^2} + \frac{\mathrm{Si}(x)}{2x} \\
\lim_{x\to 0} d^P_2(x) &= 0
\end{aligned}
\tag{35}
$$

where $\mathrm{Si}(x)$ is the integral sine function. From (26) with condition (25) we have

$$ {}^{(n)}C_{kM}^{J,N}(0, F) = \frac{1}{8\pi^2} \tag{36} $$

Using Eqs. (27)–(29) and relations (19), and applying conditions (35) and (36) we can express as follows the respective components of the integral intensity of light scattered by solutions of small macromolecules in the presence of a dc electric field **F**:

$$ \begin{aligned} \frac{V_v^F}{I_{zz}^0} &= \rho\left(\frac{\omega}{c}\right)^4 a^2\left\{1 + \frac{4}{5}\kappa C_2(F) + \frac{4}{5}\kappa^2\left[1 + \frac{2}{7}C_2(F) + \frac{2}{7}C_4(F)\right]\right\} \\ \frac{H_v^F}{I_{zz}^0} &= \frac{V_h^F}{I_{xx}^0} = \frac{3}{5}\rho\left(\frac{\omega}{c}\right)^4 a^2\kappa^2\left[\frac{4}{21}C_4(F) - \frac{1}{7}C_2(F) - 1\right] \\ \frac{H_h^F}{I_{zz}^0} &= \frac{3}{5}\rho\left(\frac{\omega}{c}\right)^4 a^2\kappa^2\left[\frac{1}{21}C_4(F) - \frac{2}{7}C_2(F) + 1\right] \end{aligned} \tag{37} $$

The expressions are of the same form as the respective expressions of Kielich [20, 21] if the reorientation functions $C_2(F)$ and $C_4(F)$ are written in terms of Langevin–Kielich functions [20, 25, 42, 50].

With the form of the time autocorrelation function $I_{ij}^F(q', t)$ available, we are in a position to derive the expression for the spectral density $I_{ij}^F(q', \Delta\omega)$ of the scattered light intensity tensor. In fact, from the Wiener–Khintchine theorem we have [56]

$$ \begin{aligned} I_{ij}^F(q', \Delta\omega) &= \mathscr{F}_t I_{ij}^F(q', t) \\ &= \frac{1}{2\pi}\int_{-\infty}^{\infty} I_{ij}^F(q', t)\exp(i\Delta\omega t)\,\mathrm{d}t \end{aligned} \tag{38} $$

where $\mathscr{F}_t I_{ij}^F(q', t)$ is the Fourier transform $I_{ij}^F(q', t)$. With regard to Eqs. (27)–(29), the components of $I_{ij}^F(q', t)$ are dependent, among others, on the shape of the macromolecules as well as the reorientation parameters p and q related with the external reorienting field. Consequently, with regard to the theorem (38), the shape of the spectral lines will also be dependent on the factors stated above. The shape and size of the macromolecules are determined by the two functions $d_n(x)$ and $b_n(x)$, Eqs. (20) and (21). It will be remembered that the function $b_{n0}(x)$ differs from $d_n(x)$ only by a linear factor characterizing size. The forms of $d_0^2(x)$ and

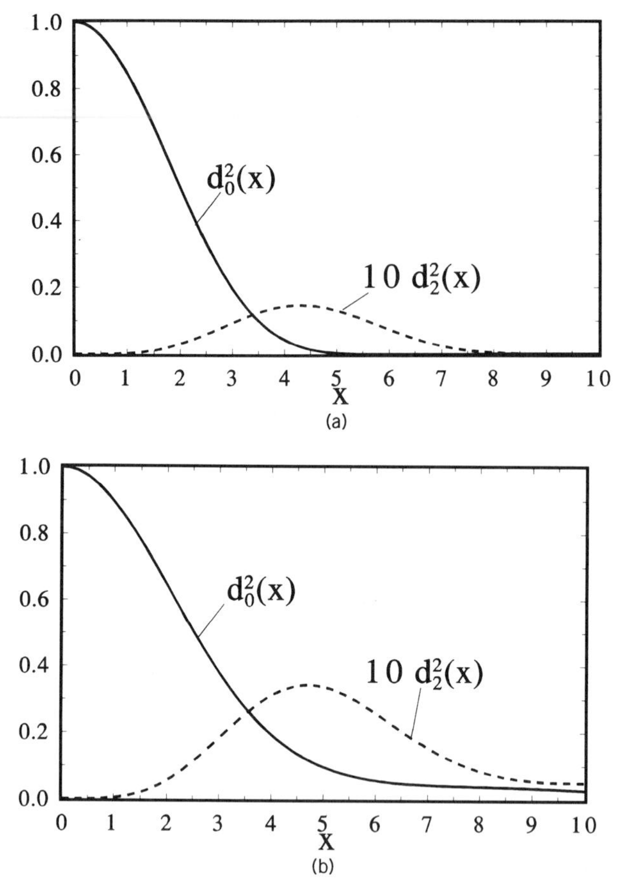

Figure 3. The particle shape functions d_0^2 (solid line) and d_2^2 (dashed line) for (a) disks, $x = q'a_D$, and (b) rods, $x = q'(l/2)$.

$d_2^2(x)$ for rods and disks are plotted in Fig. 3 where it is clearly seen that the function $d_0^2(x)$ takes markedly greater values than $d_2^2(x)$, particularly for x ranging from 0 to 3. In this range, and omitting higher approximations to $d_n(x)$, we get, for example, with $x = 2\sqrt{2}\pi(a_D/\lambda)$

$$\frac{a_D}{\lambda} = \frac{1}{10} \qquad x \cong 0.9 \qquad d_0^2(x) \gg d_2^2(x)$$

$$\frac{a_D}{\lambda} = \frac{1}{2} \qquad x \cong 4.4 \qquad d_0^2(x) \sim d_2^2(x)$$

$$\frac{a_D}{\lambda} = 1 \qquad x \cong 6.9 \qquad d_0^2(x) \ll d_2^2(x)$$

and likewise for rods; however, throughout the whole range of x values, $d_0^2(x)$ is markedly predominant.

We now discuss the influence of the size and the shape of the macromolecules as well as that of the reorientation parameters p and q on the spectral line shapes of the individual components of light scattered by solutions of rod- and disklike macromolecules, at reorientation in an external ac electric field of low intensity. In other words, we shall apply an approximation up to the square of the field. The dynamical coefficient (24) now becomes

$$^{(n)}C_{0M}^{J,N}(t,F) = {}^{(0)}C_{0M}^{J,N}(t,F) + {}^{(1)}C_{0M}^{J,N}(t,F) + {}^{(2)}C_{0M}^{J,N}(t,F)$$

where the successive terms are calculated with Eq. (26). However, the expansion coefficients $C_I(F)$ (19) are nonzero for $I = 0, 1, 2$ only. To determine the respective relations for $C_I(F)$ we have to expand the Langevin–Kielich functions [58] in a series in p and q leading to [59, 60]

$$\begin{aligned} C_0 &= 1 \\ C_1 &= pR_{10}(\omega_a) \\ C_2 &= \tfrac{2}{3}qR_{20}(\omega_{ab}) + \tfrac{1}{3}p^2R_{20}(\omega_{ab})R_{10}(\omega_a) \end{aligned} \tag{39}$$

where

$$\begin{aligned} R_{10}(\omega_a) &= \frac{1}{1 - i\omega_a\tau_1} \\ R_{20}(\omega_{ab}) &= \frac{1}{1 - i(\omega_a + \omega_b)\tau_2} \end{aligned} \tag{40}$$

The above two quantities are the Debye relaxation factors, with the Debye dipole relaxational time $\tau_1 = \tau_D$ and the reorientation time of a macromolecule induced by a strong electric field with $\tau_2 = \frac{1}{3}\tau_D$. The latter quantity is sometimes called the birefringence relaxation time [51].

With regard to the preceding relations, one can express as follows the Fourier transforms of the components $H_v^F(q', t)$, $H_h^F(q', t)$ and $V_v^F(q', t)$ for low-field **F** strengths and on approximating $d_n(x)$ and $b_n(x)$ to $n = 0, 2$ (for other approximations, $n = 4, 6, \ldots$ the shape function assumes very small values):

$$\frac{H_v^F(\Delta\omega)}{A''I_{zz}^0} = \kappa^2 \left\{ \frac{1}{20} d_0^2(x) r_2^{(0)}(\Delta\omega) + \frac{1}{28} d_0(x) d_2(x) r_2^{(0)}(\Delta\omega) + \frac{5}{392} d_2^2(x) \times \left[5 r_2^{(0)}(\Delta\omega) + 9 r_4^{(0)}(\Delta\omega)\right] + \frac{q}{21}\left[\frac{1}{20} d_0^2(x) A(\Delta\omega) + \frac{1}{28} d_0(x) d_2(x) B(\Delta\omega) + \frac{1}{56} d_2^2(x) C(\Delta\omega)\right] + \frac{p^2}{21}\left[\frac{1}{20} d_0^2(x) D(\Delta\omega) + d_0(x) d_2(x) G(\Delta\omega) + \frac{1}{7} d_2^2(x) H(\Delta\omega)\right]\right\} \tag{41}$$

$$\frac{H_h^F(\Delta\omega)}{A''I_{xx}^0} = \kappa^2 \left\{ \frac{1}{20} d_0^2(x) r_2^{(0)}(\Delta\omega) + \frac{5}{98} d_2^2(x) \left[\frac{1}{2} r_2^{(0)}(\Delta\omega) + 3 r_4^{(0)}(\Delta\omega)\right] - \frac{q}{21}\left[\frac{d_0^2(x)}{10} A(\Delta\omega) - \frac{d_0(x) d_2(x)}{7} K(\Delta\omega) + \frac{d_0^2(x)}{14} M(\Delta\omega)\right] - \frac{p^2}{21}\left[\frac{d_0^2(x)}{10} N(\Delta\omega) - \frac{d_0(x) d_2(x)}{14} O(\Delta\omega) + \frac{d_2^2(x)}{56} P(\Delta\omega)\right]\right\} \tag{42}$$

$$
\begin{aligned}
\frac{V_v^F(\Delta\omega)}{A'' I_{zz}^0} = {} & \frac{1}{3} d_0^2(x) \left[\frac{1}{4}\delta(\Delta\omega) + \frac{1}{4} r_2^{(0)}(\Delta\omega) + \frac{1}{5}\kappa^2 r_2^{(0)}(\Delta\omega) \right] \\
& + d_0(x) d_2(x) \left[\frac{1}{4}\kappa\left(\delta(\Delta\omega) + r_2^{(0)}(\Delta\omega)\right) + \frac{2}{21}\kappa^2 r_2^{(0)}(\Delta\omega) \right] \\
& + \frac{1}{3} d_2^2(x) \left[\frac{25}{16} r_2^{(0)}(\Delta\omega) - \frac{15}{28} r_2^{(0)}(\Delta\omega) \right. \\
& \qquad \left. + \kappa^2 \left(\frac{1}{4}\delta(\Delta\omega) + \frac{5}{49} r_2^{(0)}(\Delta\omega) \right) \right] \\
& + q \left[\frac{1}{45} d_0^2(x) \left(\kappa S(\Delta\omega) + \frac{2}{7}\kappa^2 A(\Delta\omega) \right) \right. \\
& \qquad \left. + \frac{1}{9} d_0(x) d_2(x) T(\Delta\omega) + \frac{1}{9} d_2^2(x) U(\Delta\omega) \right] \\
& + p^2 \left[\frac{1}{15} d_0^2(x) W(\Delta\omega) + d_0(x) d_2(x) Y(\Delta\omega) \right. \\
& \qquad \left. + d_2^2(x) Z(\Delta\omega) \right]
\end{aligned} \tag{43}
$$

where $d_0^P(x)$ and $d_2^P(x)$ are determined by Eq. (35), $x^P = \sqrt{2}\,\pi(l/\lambda)$, whereas for disks

$$
\begin{aligned}
d_0^D(x) &= \frac{1}{x^2}(\cos x - 1), \\
d_2^D(x) &= \frac{1}{x^2} - \frac{3\sin x}{2x^2} + \frac{\cos x}{2x^2} \\
x^D &= 2\sqrt{2}\,\pi \frac{a_D}{\lambda}
\end{aligned} \tag{44}
$$

and the constant A'' is equal to $A'/8\pi^3$. The expressions for $A(\Delta\omega), B(\Delta\omega), \ldots, Z(\Delta\omega)$ are somewhat cumbersome and, therefore, are given for rods in the appendix (A.1)–(A.17). They contain the functions $r_{JJ'}(\Delta\omega, \omega)$ and $r_J^{(0)}(\Delta\omega)$ of the form

$$
r_{JJ'}(\Delta\omega, \omega) = R_{JJ'}\left[L_{J'}(\Delta\omega + \omega) - L_J(\Delta\omega) \right] \tag{45}
$$

$$
r_J^{(0)}(\Delta\omega) = L_J(\Delta\omega) \tag{46}
$$

where

$$R_{JJ'}(\omega) = \frac{1}{1 - \tau_J/\tau_{J'} - i\omega\tau_J} \tag{47}$$

are generalized Debye relaxation factors. The $L_J(\Delta\omega)$ are Lorentzians [53–55] determining the Lorentz shape of the scattered light spectrum, with

$$L_J(\Delta\omega) = \frac{1}{\pi}\frac{\tau_J}{1 + (\Delta\omega\tau_J)^2} \tag{48}$$

whereas the

$$L_J(\Delta\omega + \omega_{ab}) = \frac{1}{\pi}\frac{\tau_J}{1 + (\Delta\omega + \omega_{ab})^2\tau_J^2} \tag{49}$$

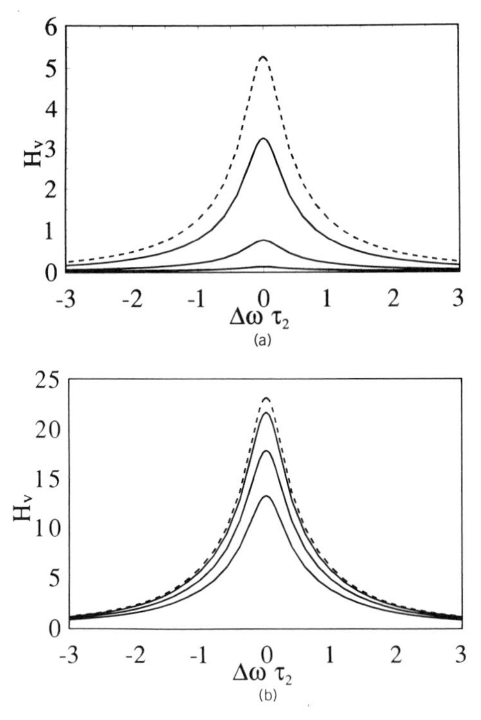

Figure 4. The spectral line shapes of the depolarized scattered component $H_v \equiv H_v^F(\Delta\omega)/A''I_{zz}^0$ for (a) disks at $p = 2$, $q = -1$, $\kappa = -0.1$, and (b) rods at $p = 2$, $q = 1$, $\kappa = 0.1$ for different values of the ratio $a_D/\lambda = l/\lambda = 0$ (dashed line), and 0.2, 0.4, 0.6 (solid lines).

are shape factors of the spectral lines, with maxima shifted by $\omega_{ab} = \omega_a + \omega_b$ with respect to the principal line, for which $\Delta\omega = 0$. For $J = 0$ these factors become

$$L_0(\Delta\omega) = \delta(\Delta\omega)$$
$$L_0(\Delta\omega, \omega_{ab}) = \delta(\Delta\omega + \omega_{ab}) \tag{50}$$

where $\delta(\Delta\omega)$ is Dirac's δ function.

For $J' = 0$, the generalized relaxation factors $R_{JJ'}(\omega)$ (47) become the well-known Debye relaxation factors (40) [52, 60] since the ratio of relaxation times in the Debye model of rotational diffusion has the form

$$\frac{\tau_J}{\tau_{J'}} = \frac{J'(J'+1)}{J(J+1)}$$

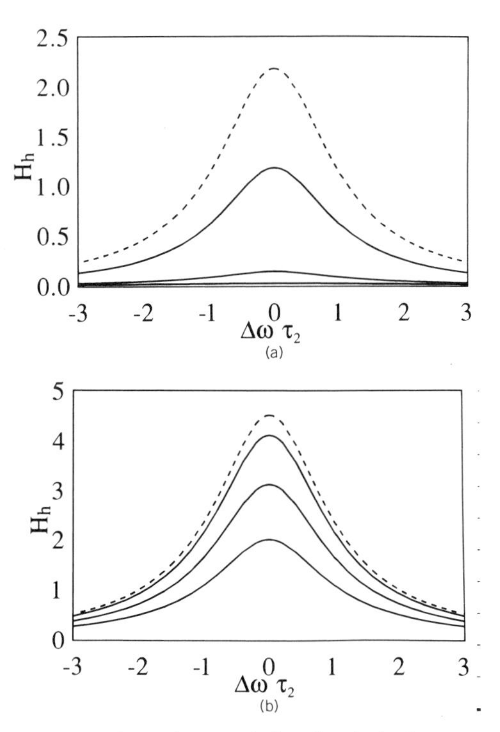

Figure 5. The spectral line shapes of the depolarized scattered component $H_h \equiv H_h^F(\Delta\omega)/A''I_{xx}^0$ for (a) disks at $p = 2$, $q = -1$, $\kappa = -0.1$, and (b) rods at $p = 2$, $q = 1$, $\kappa = 0.1$ for different values of the ratios $a_D/\lambda = l/\lambda = 0, 0.2, 0.4, 0.6$.

and for $J' = 0$ it is

$$\frac{\tau_J}{\tau_{J'}} = 0$$

The dispersion of these factors versus the reorienting field frequency, or difference frequencies (for instance, between the modes of laser oscillations), determines the amplitudes of the different Lorentzian factors composing the spectrum (41)–(43).

The relaxation factors $R_{JJ'}(\omega)$ become maximal at very low frequencies, when $\omega_a \ll 1/\tau_J$, that is, if the inverse relaxation times of the macromolecules greatly exceed the reorienting field frequency. For optical frequencies, $\omega_a \gg 1/\tau_J$, they tend to zero. The first terms of expressions (41)–(43), which do not involve the reorientation parameters p and q, describe the broadening caused by free rotational diffusion of the macro-

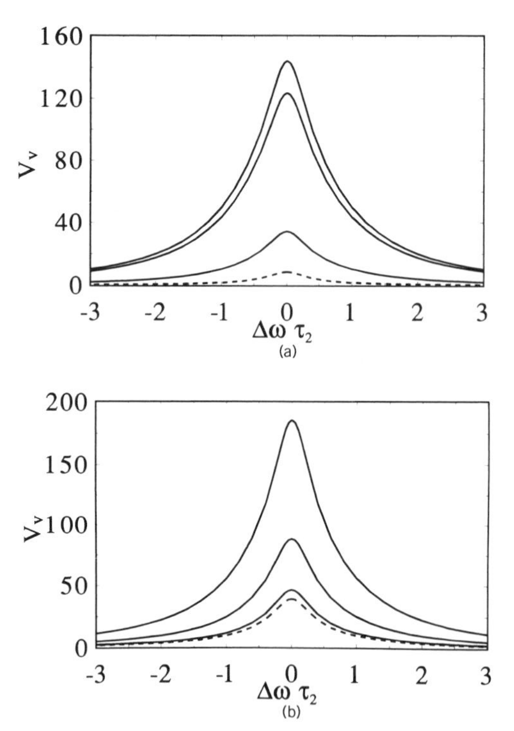

Figure 6. The spectral line shapes of the depolarized scattered component $V_v \equiv V_v^F(\Delta\omega)/A''I_{zz}^0$ for (a) disks at $p = 2$, $q = -1$, $\kappa = -0.1$, and (b) rods at $p = 2$, $q = 1$, $\kappa = 0.1$ for different values of the ratio $a_D/\lambda = l/\lambda = 0, 0.2, 0.4, 0.6$.

molecules in the spectral lines of light scattering. Whereas the superpositions of the functions $A(\Delta\omega), B(\Delta\omega), \ldots, Z(\Delta\omega)$ (Eqs. (A.1)–(A.17) in the appendix), which are also dependent on the optical anisotropy κ, determine the influence of macromolecular reorientation on the line broadening. Expressions (41)–(43) simplify considerably if the electric field **F** causing the macromolecules to reorient is assumed constant in time. In the time autocorrelation functions (27)–(29) the expressions of the form

$$R_{JJ}(\omega_{ab})\exp(-t/\tau_J)\left[\exp(-i\omega_{ab}t) - 1\right] \tag{51}$$

in the limit of $\omega_{ab} \to 0$ then become

$$\frac{t}{\tau_J}\exp\left(-\frac{t}{\tau_J}\right) \tag{52}$$

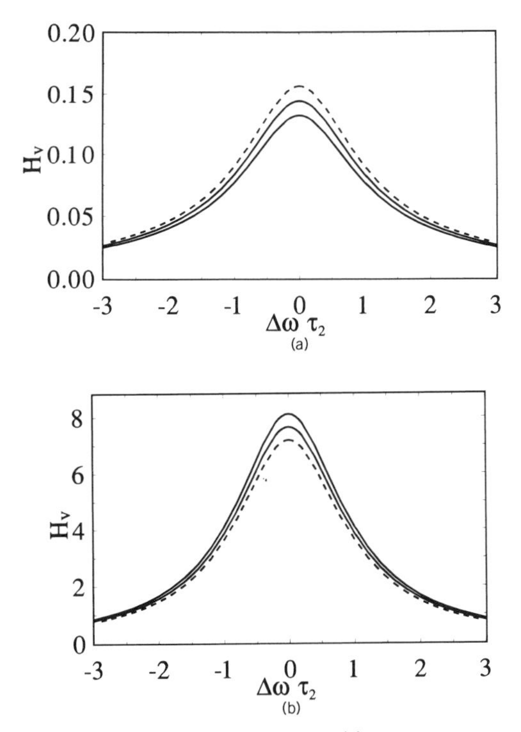

Figure 7. The influence of weak reorientation of (a) nondipolar disklike macromolecules at $q = 0, -0.5, -1$ and $\kappa = -0.1$, and (b) nondipolar rodlike macromolecules at $q = 0, 0.5, 1$ and $\kappa = 0.1$ on the depolarized component $H_v \equiv H_v^F(\Delta\omega)/A''I_{zz}^0$ of the spectrum of linear light scattering, for $a_D/\lambda = l/\lambda = 0.5$.

The Fourier transform of (52) gives a non-Lorentzian contribution to the spectral line shape, in the form [44, 45]

$$G_J(\Delta\omega) = \frac{\tau_J}{\pi} \frac{1 - (\Delta\omega\tau_J)^2}{\left(1 + \Delta\omega^2\tau_J^2\right)^2} \tag{53}$$

The factor (53) never occurs independently. It intervenes only as an additional component of the superposition of Lorentzian factors (Eq. 48). Figures 4–6 show the spectral line shapes for the depolarized $H_v^F(\Delta\omega)$ and polarized $H_h^F(\Delta\omega)$, $V_v^F(\Delta\omega)$ scattered light components from rod- and disklike molecules, for different values of the ratios l/λ and (a_D/λ) $[x^P = \sqrt{2}\pi(l/\lambda),\ x^D = 2\sqrt{2}\pi(a_D/\lambda)]$. The line intensities of the respective components decrease markedly for large macromolecules compared

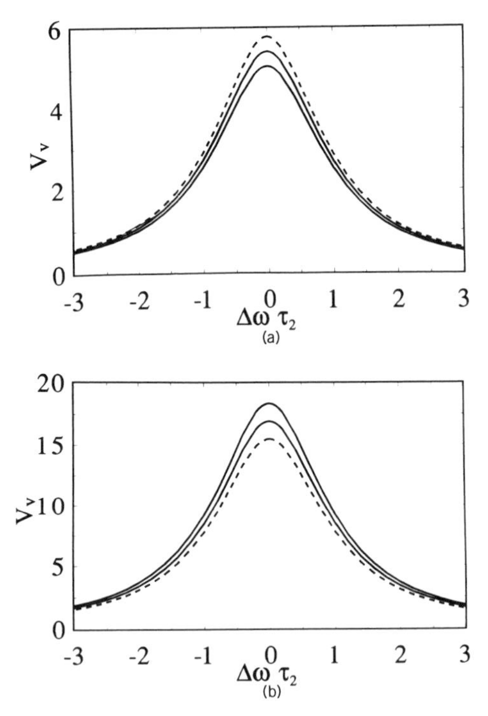

Figure 8. The influence of weak reorientation of (a) nondipolar disklike macromolecules at $q = 0, -0.5, -1$ and $\kappa = -0.1$, and (b) nondipolar rodlike macromolecules at $q = 0, 0.5, 1$ and $\kappa = 0.1$ on the depolarized component $V_v \equiv V_v^F(\Delta\omega)/A'' I_{zz}^0$ of the spectrum of linear light scattering, for $a_D/\lambda = l/\lambda = 0.1$.

with the intensities observed for small macromolecules ($l \ll \lambda$, $a_D \ll \lambda$). The greater the values of l/λ and a_D/λ, the weaker are the intensities. For example, in the case of disks (Fig. 5a), $H_h^F(\Delta\omega)$ has already decreased by 90% at $a_D/\lambda = 0.4$. The changes in intensity of $H_h^F(\Delta\omega)$, $H_h^F(\Delta\omega)$, and $V_v^F(\Delta\omega)$ (Figs. 4–6) for the same values of l/λ and a_D/λ are greater for disks than for rods. Similar studies have extended to the respective integral scattering components from solutions of large macromolecules in the presence of a dc electric field [26, 28]. Here, too, the intensities of the components decrease with growing size of the macromolecule as compared to the incident wavelength λ.

Figure 7 shows, for given values of l/λ ($l/\lambda = 0.5$) and for $p = 0$ and $q \neq 0$, that the line intensity for $H_v^F(\Delta\omega)$ at the center is greater than that of $H_v^F(\Delta\omega)$ at $p = 0$ and $q = 0$ for rods, but lower than for disks at

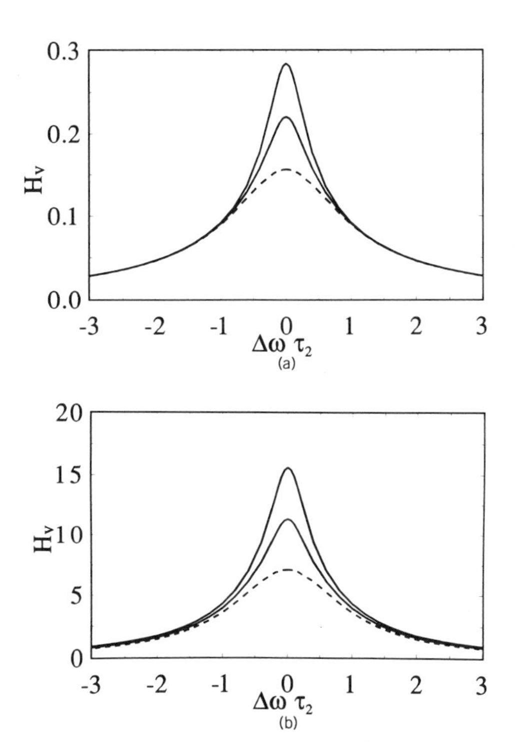

Figure 9. The influence of weak reorientation of (a) disklike macromolecules at $q = 0$ (dashed line) and $q = -0.5, -1$ (solid lines) for $\kappa = -0.1$, and (b) rodlike macromolecules at $q = 0$ (dashed line) and $q = 0.5, 1$ (solid lines) for $\kappa = 0.1$ on the depolarized component $H_v \equiv H_v^F(\Delta\omega)/A''I_{zz}^0$ of the spectrum of linear light scattering, for $a_D/\lambda = l/\lambda = 0.5$, $p = 2\sqrt{|q|}$.

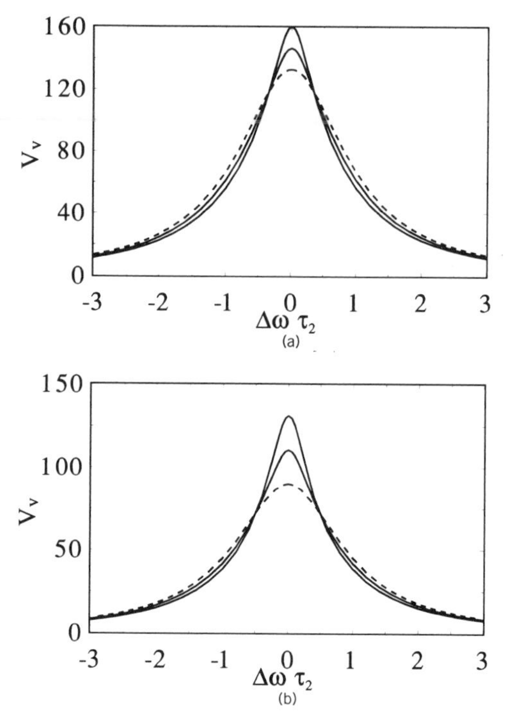

Figure 10. The influence of weak reorientation of (a) disklike macromolecules at $q = 0, -0.5, -1$ and $\kappa = -0.1$, and (b) rodlike macromolecules at $q = 0, 0.5, 1$ and $\kappa = 0.1$ on the depolarized component $V_v \equiv V_v^F(\Delta\omega)/A''J_{zz}^0$ of the spectrum of linear light scattering, for $a_D/\lambda = l/\lambda = 0.5$, $p = 2\sqrt{|q|}$.

$\Delta\omega = 0$. Thus, the line intensity measured at $\Delta\omega = 0$ can serve to determine the sign of the electric anisotropy of macromolecules as well as their shape. Similar conclusions have been reached by Alexiewicz et al. [44, 45] in their studies of small macromolecules.

The line intensity at the center of the component $V_v^F(\Delta\omega)$ for $p = 0$, $q \neq 0$ exceeds the intensity for $V_v^F(\Delta\omega)$ for $p = q = 0$ in the case of rods (Fig. 8b) but is less in that of disks (Fig. 8a). At $p \neq 0$ and $q \neq 0$ the intensity of $V_v^F(\Delta\omega)$ and $H_v^F(\Delta\omega)$ is greater than at $p = 0$, $q \neq 0$ both for rods and for disks (Figs. 9 and 10). Here, at $p \neq 0$ and small values of q reorientation of the permanent dipole moment, which is directed along the symmetry axis both in rods and disks, predominates.

The values of p and q exert a marked influence on the scattered intensity components. With growing q, the line intensities compared to those for $q = 0$, $p = 0$ increase in the case of rods ($q > 0$), but decrease for disks ($q < 0$). The expressions we derived for the components of the autocorrelation function of scattered light (32)–(34) are also valid for the rotational diffusion of macromolecules induced by the strong electric field of laser light (permanent dipole reorientation does not keep pace with the light oscillations).

In the present study we have not considered the effect of an electric field **F** on the translational motion of the macromolecules. However, at high anisotropy of the translational diffusion coefficients $(D_{\parallel}^T - D_{\perp}^T)/D^T$, the statistical distribution function contains terms that interrelate the translational and orientational motions, so that the two processes of diffusion cannot be dealt with as statistically independent [75–79]. As a consequence, in Eq. (1) for the heterodyne autocorrelation function, one has to take into consideration the translational effect due to the field **F** [80–83, 85]. As an example of macromolecules for which the influence of the mutual interdependence between the translational and reorientational motions on the spectrum is well apparent, we mention the case of the tobacco mosaic virus TMV, for which $(D_{\parallel}^T - D_{\perp}^T)/D^T = 0.4$ [84].

APPENDIX

In this appendix we give explicit expressions for functions occurring in Eqs. (41), (42), and (43) in the case of rods:

$$A(\Delta\omega) = \sum_{a,b} \left[2R_{20}(\omega_{ab})r_2^{(0)}(\Delta\omega) + r_{22}(\Delta\omega, \omega_{ab})\right] \tag{A.1}$$

$$\begin{aligned} B(\Delta\omega) = \sum_{a,b} \Big[&8R_{20}(\omega_{ab})r_2^{(0)}(\Delta\omega) + r_{22}(\Delta\omega, \omega_{ab}) \\ &+4r_{24}(\Delta\omega, \omega_{ab}) + 6R_{20}(\omega_{ab})r_4^{(0)}(\Delta\omega) \\ &+3r_{42}(\Delta\omega, \omega_{ab})\Big] \end{aligned} \tag{A.2}$$

$$\begin{aligned} C(\Delta\omega) = \sum_{a,b} \Big[&5R_{20}(\omega_{ab})r_2^{(0)}(\Delta\omega) + \tfrac{25}{7}r_{22}(\Delta\omega, \omega_{ab}) \\ &+ \tfrac{10}{7}r_{24}(\Delta\omega, \omega_{ab}) + \tfrac{15}{11}R_{20}(\omega_{ab})r_4^{(0)}(\Delta\omega) \\ &- \tfrac{15}{14}r_{42}(\Delta\omega, \omega_{ab}) + \tfrac{81}{154}r_{44}(\Delta\omega, \omega_{ab})\Big] \end{aligned} \tag{A.3}$$

$$D(\Delta\omega) = \sum_{a,b} \{R_{10}(\omega_a) R_{20}(\omega_{ab}) r_2^{(0)}(\Delta\omega)$$
$$+ \tfrac{21}{20} R_{12}(\omega_a)[r_{21}(\Delta\omega, \omega_b) - r_{22}(\Delta\omega, \omega_{ab})]$$
$$+ \tfrac{21}{20} R_{10}(\omega_a) r_{21}(\Delta\omega, \omega_b) - \tfrac{8}{5} R_{10}(\omega_a) r_{23}(\Delta\omega, \omega_b)$$
$$+ \tfrac{8}{15} R_{32}(\omega_a)[r_{23}(\Delta\omega, \omega_b) - r_{22}(\Delta\omega, \omega_{ab})]\} \quad \text{(A.4)}$$

$$G(\Delta\omega) = \sum_{a,b} \{\tfrac{1}{56} R_{10}(\omega_a) R_{20}(\omega_{ab})[8 r_2^{(0)}(\Delta\omega) + 6 r_4^{(0)}(\Delta\omega)]$$
$$+ \tfrac{3}{40} R_{10}(\omega_a) r_{21}(\Delta\omega, \omega_b)$$
$$- \tfrac{9}{70} R_{10}(\omega_a) r_{23}(\Delta\omega, \omega_b) \quad \text{(A.5)}$$
$$+ \tfrac{3}{56} R_{10}(\omega_a) r_{43}(\Delta\omega, \omega_b)$$
$$+ \tfrac{3}{80} R_{12}(\omega_a)[r_{21}(\Delta\omega, \omega_b) - r_{22}(\Delta\omega, \omega_{ab})]$$
$$+ \tfrac{2}{105} R_{32}(\omega_a)[r_{23}(\Delta\omega, \omega_b) - r_{22}(\Delta\omega, \omega_{ab})]$$
$$- \tfrac{1}{56} R_{32}(\omega_a)[r_{43}(\Delta\omega, \omega_b) - r_{42}(\Delta\omega, \omega_{ab})]$$
$$- \tfrac{1}{56} R_{34}(\omega_a)[r_{23}(\Delta\omega, \omega_b) - r_{24}(\Delta\omega, \omega_{ab})]\}$$

$$H(\Delta\omega) = \sum_{a,b} \{\tfrac{5}{16} R_{10}(\omega_a) R_{20}(\omega_{ab}) r_2^{(0)}(\Delta\omega)$$
$$+ \tfrac{15}{176} R_{10}(\omega_a) R_{20}(\omega_{ab}) r_4^{(0)}(\Delta\omega)$$
$$+ \tfrac{15}{16} R_{10}(\omega_a) r_{21}(\Delta\omega, \omega_b)$$
$$+ \tfrac{5}{8} R_{10}(\omega_a)[r_{23}(\Delta\omega, \omega_b) + r_{43}(\Delta\omega, \omega_b)]$$
$$- \tfrac{29}{44} R_{10}(\omega_a) r_{45}(\Delta\omega, \omega_b)$$
$$+ \tfrac{15}{32} R_{12}(\omega_a)[r_{21}(\Delta\omega, \omega_b) - r_{22}(\Delta\omega, \omega_{ab})] \quad \text{(A.6)}$$
$$+ \tfrac{5}{21} R_{32}(\omega_a)[r_{23}(\Delta\omega, \omega_b) - r_{22}(\Delta\omega, \omega_{ab})]$$
$$+ \tfrac{5}{224} R_{32}(\omega_a)[r_{43}(\Delta\omega, \omega_b) - r_{42}(\Delta\omega, \omega_{ab})]$$
$$+ \tfrac{155}{896} R_{34}(\omega_a)[r_{43}(\Delta\omega, \omega_b) - r_{44}(\Delta\omega, \omega_{ab})]$$
$$+ \tfrac{29}{220} R_{54}(\omega_a)[r_{45}(\Delta\omega, \omega_b) - r_{44}(\Delta\omega, \omega_{ab})]\}$$

$$K(\Delta\omega) = \sum_{a,b} [\tfrac{11}{4} R_{20}(\omega_{ab}) r_2^{(0)}(\Delta\omega) + r_{22}(\Delta\omega, \omega_{ab})$$
$$- \tfrac{1}{2} r_{24}(\Delta\omega, \omega_{ab}) + \tfrac{3}{4} R_{20}(\omega_{ab}) r_4^{(0)}(\Delta\omega) \quad \text{(A.7)}$$
$$+ \tfrac{3}{8} r_{42}(\Delta\omega, \omega_{ab})]$$

$$M(\Delta\omega) = \sum_{a,b} \left[\tfrac{5}{2}R_{20}(\omega_{ab})r_2^{(0)}(\Delta\omega) + \tfrac{1}{7}r_{22}(\Delta\omega, \omega_{ab}) + \tfrac{225}{22}R_{20}(\omega_{ab})r_4^{(0)}(\Delta\omega) + \tfrac{423}{308}r_{44}(\Delta\omega, \omega_{ab}) - \tfrac{5}{7}r_{24}(\Delta\omega, \omega_{ab}) + \tfrac{15}{28}r_{42}(\Delta\omega, \omega_{ab})\right] \tag{A.8}$$

$$N(\Delta\omega) = \sum_{a,b} \left\{R_{10}(\omega_a)R_{20}(\omega_{ab})r_2^{(0)}(\Delta\omega) + \tfrac{1}{2}R_{10}(\omega_a)r_{23}(\Delta\omega, \omega_b) + \tfrac{1}{6}R_{32}(\omega_a)\left[r_{22}(\Delta\omega, \omega_{ab}) - r_{23}(\Delta\omega, \omega_b)\right]\right\} \tag{A.9}$$

$$\begin{aligned} O(\Delta\omega) = \sum_{a,b} \Big\{ &\tfrac{11}{4}R_{10}(\omega_a)R_{20}(\omega_{ab})r_2^{(0)}(\Delta\omega) \\ &+ \tfrac{3}{4}R_{10}(\omega_a)R_{20}(\omega_{ab})r_4^{(0)}(\Delta\omega) \\ &+ \tfrac{1}{3}R_{32}(\omega_a)\left[r_{22}(\Delta\omega, \omega_{ab}) - r_{23}(\Delta\omega, \omega_b)\right] \\ &+ \tfrac{1}{2}R_{10}(\omega_a)\left[r_{23}(\Delta\omega, \omega_b) + \tfrac{3}{4}r_{43}(\Delta\omega, \omega_b)\right] \\ &+ \tfrac{1}{8}R_{34}(\omega_a)\left[r_{24}(\Delta\omega, \omega_{ab}) - r_{23}(\Delta\omega, \omega_b)\right] \\ &+ \tfrac{1}{8}R_{32}(\omega_a)\left[r_{42}(\Delta\omega, \omega_{ab}) - r_{43}(\Delta\omega, \omega_b)\right]\Big\} \end{aligned} \tag{A.10}$$

$$\begin{aligned} P(\Delta\omega) = \sum_{a,b} \Big\{ &5R_{10}(\omega_a)R_{20}(\omega_{ab})r_2^{(0)}(\Delta\omega) \\ &+ \tfrac{225}{11}R_{10}(\omega_a)R_{20}(\omega_{ab})r_4^{(0)}(\Delta\omega) \\ &+ \tfrac{10}{21}R_{32}(\omega_a)\left[r_{22}(\Delta\omega, \omega_{ab}) - r_{23}(\Delta\omega, \omega_b)\right] \\ &+ \tfrac{5}{14}R_{34}(\omega_a)\left[r_{24}(\Delta\omega, \omega_{ab}) - r_{23}(\Delta\omega, \omega_b)\right] \\ &+ \tfrac{5}{14}R_{32}(\omega_a)\left[r_{42}(\Delta\omega, \omega_{ab}) - r_{43}(\Delta\omega, \omega_b)\right] \\ &+ \tfrac{15}{56}R_{34}(\omega_a)\left[r_{44}(\Delta\omega, \omega_{ab}) - r_{43}(\Delta\omega, \omega_b)\right] \\ &+ R_{10}(\omega_a)r_{45}(\Delta\omega, \omega_b) + \tfrac{42}{11}R_{10}(\omega_a)r_{45}(\Delta\omega, \omega_b) \\ &+ \tfrac{42}{55}R_{54}(\omega_a)\left[r_{44}(\Delta\omega, \omega_{ab}) - r_{45}(\Delta\omega, \omega_b)\right]\Big\} \end{aligned} \tag{A.11}$$

$$S(\Delta\omega) = \sum_{a,b} \left\{\delta(\Delta\omega) + R_{20}(\omega_{ab})\left[\delta(\Delta\omega) + r_2^{(0)}(\Delta\omega)\right]\right\} \tag{A.12}$$

$$
\begin{aligned}
T(\Delta\omega) = \sum_{a,b} \Big\{ & \tfrac{1}{2} r_{20}(\Delta\omega, \omega_{ab}) \\
& + \tfrac{1}{2} R_{20}(\omega_{ab}) \delta(\Delta\omega) + \tfrac{1}{2} R_{20}(\omega_{ab}) r_2^{(0)}(\Delta\omega) \\
& + \tfrac{1}{7}\kappa \big[r_{20}(\Delta\omega, \omega_{ab}) + \tfrac{3}{2} r_{22}(\Delta\omega, \omega_{ab}) \\
& \quad + 2R_{20}(\omega_{ab})\delta(\Delta\omega) + 4R_{20}(\omega_{ab}) r_2^{(0)}(\Delta\omega) \big] \\
& + \kappa^2 \big[\tfrac{1}{5} r_{20}(\Delta\omega, \omega_{ab}) + \tfrac{4}{49} r_{22}(\Delta\omega, \omega_{ab}) \\
& \quad - \tfrac{24}{245} r_{24}(\Delta\omega, \omega_{ab}) + \tfrac{1}{5} R_{20}(\omega_{ab}) \delta(\Delta\omega) \\
& \quad + \tfrac{25}{49} R_{20}(\omega_{ab}) r_2^{(0)}(\Delta\omega) \big] \Big\}
\end{aligned} \tag{A.13}
$$

$$
\begin{aligned}
U(\Delta\omega) = \sum_{a,b} \Big\{ & -\tfrac{5}{56} r_{22}(\Delta\omega, \omega_{ab}) - \tfrac{5}{27} R_{20}(\omega_{ab}) r_2^{(0)}(\Delta\omega) \\
& + \tfrac{1}{2}\kappa \big[r_{20}(\Delta\omega, \omega_{ab}) \\
& \quad + \tfrac{45}{49} r_{22}(\Delta\omega, \omega_{ab}) - \tfrac{39}{49} r_{24}(\Delta\omega, \omega_{ab}) \\
& \quad + \tfrac{1}{2} R_{20}(\omega_{ab}) \delta(\Delta\omega) + \tfrac{395}{98} R_{20}(\omega_{ab}) r_2^{(0)}(\Delta\omega) \big] \\
& + \tfrac{1}{7}\kappa^2 \big[r_{20}(\Delta\omega, \omega_{ab}) \\
& \quad + \tfrac{10}{49} r_{22}(\Delta\omega, \omega_{ab}) - \tfrac{24}{49} r_{24}(\Delta\omega, \omega_{ab}) \\
& \quad + \tfrac{15}{7} R_{20}(\omega_{ab}) r_2^{(0)}(\Delta\omega) + R_{20}(\omega_{ab}) \delta(\Delta\omega) \big] \Big\}
\end{aligned} \tag{A.14}
$$

$$
\begin{aligned}
W(\Delta\omega) = \sum_{a,b} \Big(& \tfrac{1}{6}\kappa \big[R_{10}(\omega_a) R_{20}(\omega_{ab}) \\
& \quad \times \big(\delta(\Delta\omega) + r_2^{(0)}(\Delta\omega) \big) + R_{10}(\omega_a) r_{20}(\Delta\omega, \omega_{ab}) \big] \\
& + \tfrac{1}{5}\kappa^2 \Big\{ \tfrac{1}{3} R_{12}(\omega_a) \big[r_{21}(\Delta\omega, \omega_b) - r_{22}(\Delta\omega, \omega_{ab}) \big] \\
& \quad + R_{10}(\omega_a) \big[\tfrac{2}{3} r_{21}(\Delta\omega, \omega_b) - \tfrac{3}{7} r_{23}(\Delta\omega, \omega_b) \big] \\
& \quad + \tfrac{1}{7} R_{32}(\omega_a) \big[r_{23}(\Delta\omega, \omega_b) - r_{22}(\Delta\omega, \omega_{ab}) \big] \\
& \quad + \tfrac{10}{21} R_{10}(\omega_a) R_{20}(\omega_{ab}) r_2^{(0)}(\Delta\omega) \Big\} \Big)
\end{aligned} \tag{A.15}
$$

$$
\begin{aligned}
Y(\Delta\omega) = \sum_{a,b} \Big(& \tfrac{1}{36} \big\{ R_{10}(\omega_a) r_{20}(\Delta\omega, \omega_{ab}) \\
& + R_{10}(\omega_a) R_{20}(\omega_{ab}) \big[\delta(\Delta\omega) + r_2^{(0)}(\Delta\omega) \big] \big\} \\
& + \kappa \big\{ \tfrac{1}{126} R_{10}(\omega_a) r_{20}(\Delta\omega, \omega_{ab}) + \tfrac{1}{30} R_{10}(\omega_a) r_{21}(\Delta\omega, \omega_{ab}) \\
& - \tfrac{3}{140} R_{10}(\omega_a) r_{23}(\Delta\omega, \omega_b) + \tfrac{1}{63} R_{10}(\omega_a) R_{20}(\omega_{ab}) \delta(\Delta\omega) \\
& + \tfrac{2}{63} R_{10}(\omega_a) R_{20}(\omega_{ab}) r_2^{(0)}(\Delta\omega) + \tfrac{1}{60} R_{12}(\omega_a) r_{21}(\Delta\omega, \omega_b) \\
& - \tfrac{1}{60} R_{12}(\omega_a) r_{22}(\Delta\omega, \omega_b) - \tfrac{1}{140} R_{32}(\omega_a) r_{22}(\Delta\omega, \omega_{ab})
\end{aligned}
$$

$$
\begin{aligned}
&+ \tfrac{1}{140} R_{32}(\omega_a) r_{23}(\Delta\omega, \omega_{ab}) \\
&+ \tfrac{1}{60} R_{12}(\omega_a) [r_{21}(\Delta\omega, \omega_b) - r_{22}(\Delta\omega, \omega_b)] \\
&+ \tfrac{1}{140} R_{32}(\omega_a) [r_{23}(\Delta\omega, \omega_{ab}) - r_{22}(\Delta\omega, \omega_{ab})] \\
&+ \kappa^2 \Big[\tfrac{1}{90} R_{10}(\omega_a) r_{20}(\Delta\omega, \omega_{ab}) + \tfrac{4}{315} R_{10}(\omega_a) r_{21}(\Delta\omega, \omega_b) \\
&- \tfrac{2}{147} R_{10}(\omega_a) r_{23}(\Delta\omega, \omega_{ab}) + \tfrac{1}{90} R_{10}(\omega_a) R_{20}(\omega_{ab}) \delta(\Delta\omega) \\
&+ \tfrac{25}{882} R_{10}(\omega_a) R_{20}(\omega_{ab}) r_2^{(0)}(\Delta\omega) \\
&+ \tfrac{2}{315} R_{12}(\omega_a) [r_{21}(\Delta\omega, \omega_b) - r_{22}(\Delta\omega, \omega_{ab})] \\
&+ \tfrac{2}{735} R_{32}(\omega_a) [r_{23}(\Delta\omega, \omega_b) - r_{22}(\Delta\omega, \omega_{ab})] \\
&+ \tfrac{1}{735} R_{34}(\omega_a) [r_{24}(\Delta\omega, \omega_{ab}) - r_{23}(\Delta\omega, \omega_b)] \Big] \Big\} \Big)
\end{aligned} \tag{A.16}
$$

$$
\begin{aligned}
Z(\Delta\omega) = \sum_{a,b} \Big(&\tfrac{1}{36} \Big\{ R_{10}(\omega_a) r_{21}(\Delta\omega, \omega_{ab}) \\
&- \tfrac{33}{28} R_{10}(\omega_a) r_{23}(\Delta\omega, \omega_b) - \tfrac{5}{14} R_{10}(\omega_a) R_{20}(\omega_{ab}) r_2^{(0)}(\Delta\omega) \\
&+ \tfrac{1}{2} R_{12}(\omega_a) [r_{21}(\Delta\omega, \omega_b) - r_{22}(\Delta\omega, \omega_{ab})] \\
&+ \tfrac{11}{28} R_{32}(\omega_a) [r_{23}(\Delta\omega, \omega_b) - r_{22}(\Delta\omega, \omega_{ab})] \Big\} \\
&+ \kappa \Big\{ \tfrac{1}{72} R_{10}(\omega_a) r_{20}(\Delta\omega, \omega_{ab}) + \tfrac{1}{42} R_{10}(\omega_a) r_{21}(\Delta\omega, \omega_b) \\
&- \tfrac{1}{49} R_{10}(\omega_a) r_{23}(\Delta\omega, \omega_b) + \tfrac{1}{72} R_{10}(\omega_a) R_{20}(\omega_{ab}) \delta(\Delta\omega) \\
&+ \tfrac{395}{3528} R_{10}(\omega_a) R_{20}(\omega_{ab}) r_2^{(0)}(\Delta\omega) \\
&+ \tfrac{1}{84} R_{12}(\omega_a) [r_{21}(\Delta\omega, \omega_b) - r_{22}(\Delta\omega, \omega_{ab})] \\
&- \tfrac{1}{1764} R_{32}(\omega_a) [r_{23}(\Delta\omega, \omega_b) - r_{22}(\Delta\omega, \omega_{ab})] \\
&- \tfrac{13}{2352} R_{34}(\omega_a) [r_{23}(\Delta\omega, \omega_b) - r_{24}(\Delta\omega, \omega_{ab})] \Big\} \\
&+ \tfrac{1}{21} \kappa^2 \Big\{ \tfrac{1}{6} R_{10}(\omega_a) r_{20}(\Delta\omega, \omega_{ab}) + \tfrac{2}{21} R_{10}(\omega_a) r_{21}(\Delta\omega, \omega_b) \\
&- \tfrac{1}{7} R_{10}(\omega_a) r_{23}(\Delta\omega, \omega_b) + \tfrac{1}{6} R_{10}(\omega_a) R_{20}(\omega_{ab}) \delta(\Delta\omega) \\
&+ \tfrac{5}{14} R_{10}(\omega_a) R_{20}(\omega_{ab}) r_2^{(0)}(\Delta\omega) \\
&+ \tfrac{1}{21} R_{12}(\omega_a) [r_{21}(\Delta\omega, \omega_b) - r_{22}(\Delta\omega, \omega_{ab})] \\
&+ \tfrac{1}{49} R_{32}(\omega_a) [r_{22}(\Delta\omega, \omega_{ab}) - r_{23}(\Delta\omega, \omega_b)] \\
&+ \tfrac{1}{49} R_{34}(\omega_a) [r_{24}(\Delta\omega, \omega_{ab}) - r_{23}(\Delta\omega\ \omega_b)] \Big\} \Big)
\end{aligned} \tag{A.17}
$$

Functions $r_{JJ'}(\Delta\omega, \omega)$, $r_J^{(0)}(\Delta\omega)$, $R_{JJ'}(\omega)$, and $\delta(\Delta\omega)$ are defined by (45)–(47) and (50), respectively.

Acknowledgments

The authors express their indebtedness to Stanisław Kielich for his unceasing guidance and many helpful discussions. This work was carried out within the framework of Project PB 201309101 of the State Committee for Scientific Research.

References

1. S. R. Aragon and R. Pecora, *J. Chem. Phys.* **66**, 2506 (1977).
2. C. Wippler, *J. Chim. Phys.* **51**, 123 (1954).
3. C. Wippler, *J. Chim. Phys.* **53**, 316 (1956).
4. M. L. Wallach and H. Benoit, *J. Polym. Sci.* **57**, 41 (1962).
5. A. Scheludko, S. Stoylov, *Z. Kolloid; Z. Polym.* **199**, 36 (1964).
6. S. Stoylov, *Collection Czech. Chem. Commun.* **31**, 2866 (1966).
7. S. Stoylov, *J. Colloid Interface Sci.* **22**, 203 (1966).
8. S. Stoylov, *J. Polym. Sci.* **C16**, 2435 (1967).
9. S. Stoylov and S. Sokerov, *Eur. Polym. J.* **6**, 1125 (1970).
10. S. Stoylov and M. V. Stoimenova, *J. Colloid Interface Sci.* **40**, 154 (1972).
11. S. Stoylov and I. Petkanchin, *J. Colloid Interface Sci.* **40**, 159 (1972).
12. S. Sokerov and M. Stoimenova, *J. Colloid Interface Sci.* **46**, 94 (1974).
13. B. R. Jennings and H. Plummer, *J. Colloid Interface Sci.* **27**, 337 (1968).
14. B. R. Jennings, *Biopolymers* **9**, 1361 (1970).
15. B. R. Jennings, *Br. Polym. J.* **1**, 70 (1969).
16. J. Schweitzer and B. R. Jennings, *J. Phys. D: Appl. Phys.* **5**, 297 (1972).
17. V. J. Morris, P. J. Rudd, and B. R. Jennings, *J. Colloid Interface Sci.* **50**, 379 (1975).
18. J. C. Ravey and P. Mazeron, *J. Colloid Interface Sci.* **51**, 412 (1975).
19. B. R. Jennings, in S. Krause (Ed.), *Molecular Electro-optics*, Plenum, New York, 1981, p. 181.
20. S. Kielich, *Acta Phys. Pol.* **A37**, 447, 719 (1970).
21. S. Kielich, *Opt. Commun.* **1**, 345 (1970).
22. S. Kielich, *Acta Phys. Pol.* **23**, 321, 819 (1963).
23. S. Kielich, *Appl. Phys. Lett.* **13**, 371 (1968).
24. S. Kielich, *J. Colloid Interface Sci.* **27**, 432 (1968); **28**, 214 (1968).
25. S. Kielich, *Nonlinear Molecular Optics*, Nauka, Moscow, 1981.
26. M. Dębska-Kotłowska and S. Kielich, *J. Polym.* **17**, 1039 (1976).
27. M. Dębska-Kotłowska and S. Kielich, *J. Polym. Sci., Polym. Symp.* **61**, 101 (1977).
28. M. Dębska-Kotłowska, Thesis, Poznań, 1980.
29. M. V. Stoimenova and M. Dębska, *J. Colloid Interface Sci.* **52**, 265 (1975).
30. I. W. Parsons, R. L. Rowell, and R. S. Farinato, in B. R. Jennings (Ed.), *Electro-optics and Dielectrics of Macromolecules and Colloids*, Plenum, New York, 1979, p. 385.
31. B. R. Jennings and M. Bhanot, *Clay Minerals* **12**, 217 (1977).
32. V. J. Morris and B. R. Jennings, *J. Chem. Faraday Trans. 2* **71**, 1948 (1975).
33. B. R. Jennings and V. J. Morris, *J. Colloid Interface Sci.* **49**, 89 (1974).

34. H. C. Van de Hulst, *Light Scattering by Small Particles*, N.Y., Wiley; London, Chapman and Hall, 1957.
35. I. L. Fabelinski, *Molecular Scattering of Light*, Plenum, New York, 1968.
36. B. I. Berne and R. Pecora, *Dynamic Light Scattering with Applications to Chemistry Biology and Physics*, Wiley-Interscience, New York, 1976.
37. B. Chu, *Laser Light Scattering*, Academic, New York, 1974.
38. S. Kielich, *Proc. Indian Acad. Sci., Chem. Sci.* **94**, 403 (1985).
39. P. Debye, *Polare Molekeln*, Hirzel, Leipzig, 1929.
40. F. Perrin, *J. Phys. Radium* **5**, 497 (1934).
41. D. A. Varshalovich, A. N. Moskalyev, and V. K. Khersonskii, *Quantum Theory of Angular Momentum*, Nauka, Leningrad, 1975.
42. Z. Ożgo, *Multi-harmonic Molecular Light Scattering in the Treatment of Racach Algebra*, UAM, Poznań, 1975.
43. S. Kielich, *Progr. Opt.* **20**, 155 (1983).
44. W. Alexiewicz, J. Buchert, and S. Kielich, *Phys. Dielectrics Radiospectrosc.* **9**, 5 (1977).
45. W. Alexiewicz, J. Buchert, and S. Kielich, *Acta Phys. Pol.* **A52**, 442 (1977).
46. S. Chandrasekhar, *Rev. Mod. Phys.* **15**, 1 (1943).
47. S. Kielich, *IEEE J. Quantum Electron.* **QE-5**, 562 (1969).
48. S. Kielich, *Opto-Electr.* **2**, 5 (1970).
49. S. Kielich, *J. Colloid Interface Sci.* **33**, 142 (1970).
50. M. W. Evans, S. Woźniak, and G. Wagnière, *Physica* **B175**, 412 (1991).
51. A. Peterlin and H. Stuart, *Doppelbrechung, Insbesondere Künstliche Doppelbrechung*, Akad. Verlags, Leipzig, 1943.
52. W. A. Steele and R. J. Pecora, *Chem. Phys.* **42**, 1863, 1872 (1965).
53. H. Benoit, *Ann. Phys.* **6**, 561 (1951).
54. *The Works of Marian Smoluchowski* Vols. 1 and 2 (1927), Jagellonian UP, Cracow, 1924, 1927.
55. S. Kielich, *Acta Phys. Pol.* **30**, 683 (1966); **31**, 689 (1967).
56. M. C. Wang and G. E. Uhlenbeck, *Rev. Mod. Phys.* **17**, 323 (1945).
57. S. Kielich, *Phys. Dielectrics Radiospectrosc.* **5**, 183 (1972); **6**, 5, 41 (1972).
58. S. Kielich, *J. Colloid Interface Sci.* **34**, 228 (1970).
59. B. Kasprowicz-Kielich, S. Kielich, and J. R. Lalanne, in L. Lascombe (Ed.), *Molecular Motions in Liquids*, Reidel, Dordrecht, 1974, pp. 563–573.
60. B. Kasprowicz-Kielich and S. Kielich, *Adv. Mol. Relax. Processes* **7**, 275 (1975).
61. N. E. Hill, W. E. Vaughan, A. H. Price, and M. Davies, *Dielectric Properties and Molecular Behaviour*, Van Nostrand-Reinhold, London, 1969.
62. B. K. P. Scaife, in *Complex Permittivity*, English UP, London, 1971, pp. 2–51.
63. G. Wyllie, in M. Davies (Ed.), *Dielectric and Related Molecular Processes*, Chemical Society, London, 1971, pp. 21–64.
64. W. I. Orville-Thomas and R. Kosfeld, *Molecular Relaxation Studies*, Elsevier, Amsterdam, 1972.
65. J. E. Lascombe, in J. Lascombe (Ed.), *Molecular Motions in Liquids*, Reidel, Dordrecht, 1974.
66. G. Bondouris, *Riv. Nuovo Cimento* **1**, 1–56 (1969).

67. B. Keller and F. Kneubuhl, *Helv. Phys. Acta* **45**, 1127 (1972).
68. W. A. Steele, *Adv. Chem. Phys.* **34**, 1 (1976).
69. D. A. Pinnov, S. J. Candan, and T. A. Litovitz, *J. Chem. Phys.* **49**, 347 (1968).
70. F. J. Bartoli and T. A. Litovitz, *J. Chem. Phys.* **56**, 404, (1972).
71. P. D. Maker, *Phys. Rev.* **A1**, 923 (1970).
72. W. Alexiewicz, T. Bancewicz, Z. Ożgo, and S. Kielich, *J. Raman Spectrosc.* **2**, 529 (1974).
73. W. Alexiewicz, Z. Ożgo, and S. Kielich, *Acta Phys. Pol.* **A48**, 243 (1975).
74. W. Alexiewicz, *Acta Phys. Pol.* **A49**, 575 (1976).
75. H. Maeda and N. Saito, *Polym. J.* **4**, 309 (1973).
76. H. Maeda and N. Saito, *J. Phys. Soc. Jpn.* **27**, 989 (1969).
77. I. S. Hwang and H. Z. Cummins, *J. Chem. Phys.* **77**, 616 (1982).
78. M. W. Evans, *J. Chem. Phys.* **78**, 5403 (1983); **77**, 4632 (1982).
79. M. W. Evans, *Phys. Rev.* **30A**, 2062 (1984).
80. V. A. Storonkin, *L.G.Y. Vestnik*, No. **4**, 1983.
81. I. E. G. Lipson and W. H. Stockmayer, *J. Chem. Phys.* **89**, 5 (1988).
82. R. Piazza and V. Degiorgio, *Physica* **A182**, 576–592 (1992).
83. T. P. Burghardt, *J. Chem. Phys.* **78**, 5913 (1983).
84. P. Horn, Thesis, Strasbourg, 1954.
85. D. Tian and W. M. McClain, *J. Chem. Phys.* **91**, 4435 (1989); **90**, 4783, 6956 (1989).

HYPER-RAYLEIGH AND HYPER-RAMAN ROTATIONAL AND VIBRATIONAL SPECTROSCOPY

T. BANCEWICZ AND Z. OŻGO

Nonlinear Optics Division, Institute of Physics, Adam Mickiewicz University, Poznań, Poland

CONTENTS

I. INTRODUCTION

The study of light scattering is one of the most effective sources of information concerning the properties of matter. Among the numerous nonlinear optical phenomena caused by laser light of high intensity, scattering is foremost despite the difficulties that beset its observation. The nonlinear reaction of the medium to the incident light beam leads to the emergence of scattered higher harmonic components in addition to the component at ω, the incident frequency. These processes are referred to as multiharmonic scattering. If the scattering process is accompanied by quantum transitions in the molecular system, we deal with inelastic light

Modern Nonlinear Optics, Part 1, Edited by Myron Evans and Stanisław Kielich. Advances in Chemical Physics Series, Vol. LXXXV.
ISBN 0-471-57546-1

scattering. Components at frequencies $\omega_s = k\omega + \omega_{n'n}$ appear ($\omega_{n'n}$ is a transition frequency). Under certain conditions there can be strong coupling in the medium between the incident and scattered waves leading to qualitatively new effects such as generation of higher light harmonics and stimulated Raman effects. They exhibit rather high intensities and a well-determined direction of propagation of the generated wave and consequently were the first to be detected [1].

In the effects of spontaneous multiharmonic scattering the intensities are much weaker than in the stimulated case. Nonetheless, those are the elementary processes whereby the reaction of the medium to the incident light wave is apparent in the simplest manner.

The possibility of nonlinear scattering of light had been considered even before the coming of lasers [2]. Kielich [3–7] proposed a compact, deep-reaching theory of multiharmonic scattering processes, in both a classical and quantum treatment. The experimental detection of processes of this kind stimulated further work in the field. Terhune et al. [8] were the first to confirm the theory experimentally by their successful observations of scattered light with doubled frequency. They moreover observed inelastic second-harmonic scattering originating in transitions between vibrational levels of molecules.

To these studies is due the rapid development of the nonlinear spectroscopy of light scattering. Thus, hyper-Rayleigh scattering has been observed in solids [9–11] as well as in liquids and gases [12–19]. Inelastic second-harmonic scattering processes involving transitions between vibrational [20–23], rotational [24, 25], and electron [26] levels have been studied. Third-harmonic elastic and inelastic scattering in crystals [27] has also been the object of investigation.

Experiment and theory kept pace in this rapid progress [28–60]. Hyper-Rayleigh scattering in dense isotropic media, where intermolecular interactions and statistical fluctuations play an important role, has been analyzed in detail [34–39]. Work has also extended to third-harmonic scattering [40–42] and to the general properties of scattering processes of higher orders [43]. Much attention has been given to inelastic scattering related with vibrational and rotational transitions in molecules [44–53]. The selection rules governing nonlinear scattering effects have been studied in full detail. Studies of second-harmonic scattering under conditions of resonance have opened the way to the discovery of new phenomena involving nonlinear scattering [54–60]. In resonance scattering, the essential role belongs to the asymmetric part of the hyperpolarizability tensor.

We have proposed a compact theory of second- and third-harmonic scattering applying the method of irreducible spherical tensors. Scattering tensors are calculated, providing a complete description of the intensity and state of polarization of the scattered light for arbitrary directions of observation and determining the depolarization ratio and the reversal coefficient as well as their angular dependence. Throughout, we lay stress on the asymmetricity of the hyperpolarizability tensors. On the basis of the analytical expressions derived by us we calculate numerically the rotational structure of the hyper-Rayleigh line for gaseous carbon oxide CO and nitrogen oxide NO. We analyze the influence of the molecular fields on the process of second-harmonic scattering in liquids. We derive analytical expressions for the weight-one and weight-three parts of the effective hyperpolarizability tensor of a molecule immersed in a liquid. The feasibility of measuring the effective hyperpolarizability tensor is discussed from the experimental viewpoint. We also give a generalized and systematized treatment of our earlier results in the field of nonlinear light scattering [61–70].

II. MULTIHARMONIC SCATTERING

A. Second-Harmonic Scattering

An intense optical field

$$\mathbf{E}(t) = \mathbf{E}\cos\omega t$$

induces a transition dipole moment [7]:

$$\mathbf{m}(t)_{n'n} = \mathbf{m}^{\omega}(t)_{n'n} + \mathbf{m}^{2\omega}(t)_{n'n} + \mathbf{m}^{3\omega}(t)_{n'n} \tag{1}$$

in the molecule. Its amplitude can be expressed as follows in terms of the polarizability tensors

$$(m_i^{\omega})_{n'n} = (\alpha_{ij}^{\omega})_{n'n} E_j \tag{2a}$$

$$(m_i^{2\omega})_{n'n} = \tfrac{1}{4}(b_{ijk}^{2\omega})_{n'n} E_j E_k \tag{2b}$$

$$(m_i^{3\omega})_{n'n} = \tfrac{1}{24}(c_{ijkl}^{3\omega})_{n'n} E_j E_k E_l \tag{2c}$$

Within the approximation of polarizability theory we assume [71]

$$(a_{ij}^{\omega})_{n'n} = \langle R'V'|(a_{ij}^{\omega})_{ee}|RV\rangle \tag{3}$$

where $(a_{ij}^{\omega})_{ee}$ is the polarizability tensor operator, averaged with respect to the electronic ground state, whereas V, R represent the set of quantum numbers for the vibrational V and rotational R states of the molecule. Similar expressions hold for the polarizabilities of higher orders.

The properties of the scattered light are described by the scattering tensor

$$I_{ij}^{k\omega} = \frac{N_n \omega_s^4}{8\pi c^3 R_0^2} \left\langle \left(m_i^{k\omega}\right)_{n'n} \left(m_j^{k\omega}\right)_{n'n}^* \right\rangle_{\Omega'} \tag{4}$$

where N_n is the number of molecules in the initial state n; $\omega_s = k\omega + \omega_{n'n}$ is the scattered light frequency; R_0 the distance from the sample to the point of observation; and $\langle \ \rangle_{\Omega'}$ denotes averaging over all possible orientations of the molecules.

Let $\mathbf{n} \perp \mathbf{R}_0$ denote the direction of oscillation of the scattered light. Its intensity is then

$$I_n = I_{ij} n_i^* n_j \tag{5}$$

We now introduce three systems of reference:

$L_1(x, y, z)$, with the z axis pointing in the direction of light propagation

$L_2(X, Y, Z)$, with the Z axis in the direction of observation of the scattered light

$L_3(1, 2, 3)$, with axes rigidly fixed to the system of principal axes of the molecule.

The mutual orientation of these systems of reference is given by the Euler angles: L_1 with respect to L_2 by the angles Ω (α, β, γ) and L_3 with respect to L_1 by the angles Ω' $(\alpha', \beta', \gamma')$.

By calculating the scattering tensor in L_2 and performing a transformation to L_1 we obtain the angular distribution of the scattered light. We start by considering second-harmonic scattering. The scattering tensor in $L_2(X, Y, Z)$ has the form

$$I_{IJ}^{2\omega} = \frac{N_n \omega_s^4}{8\pi c^3 R_0^2} \left\langle \left(m_I^{2\omega}\right)_{n'n} \left(m_J^{2\omega}\right)_{n'n}^* \right\rangle_{\Omega'} \tag{6}$$

In what follows we shall be making use of spherical tensors, with phase chosen so that for the standard components $T_\lambda^{(l)}$ and contrastandard components $T_\lambda^{[l]}$ the following relation holds [72]:

$$T_\lambda^{[l]} = T_\lambda^{(l)*} = (-1)^{l-\lambda} T_{-\lambda}^{(l)} \tag{7}$$

For the vector **M** we have

$$M_A^{(1)} = \sum_I C_I^{1A} m_I$$
$$M_{\mp 1}^{(1)} = -\mathrm{i}/\sqrt{2}\,(m_x \mp \mathrm{i} m_y) \tag{8}$$
$$M_0^{(1)} = \mathrm{i} m_z$$

On rotation of the system of reference the components of a spherical tensor transform as

$$T_\lambda^{(l)} = \sum_\nu T_\nu^{(l)} D_{\nu\lambda}^{(l)}(\Omega) \tag{9}$$

With regard to the above and by applying the rules of multiplication of Wigner functions

$$D_{A_1\alpha_1}^{(1)}(\Omega) D_{A_2\alpha_2}^{(1)*}(\Omega) = \sum_f (-1)^{A_1-\alpha_1}(2f+1)\begin{pmatrix} 1 & 1 & f \\ -\alpha_1 & \alpha_2 & \phi_1 \end{pmatrix} \times \begin{pmatrix} 1 & 1 & f \\ -A_1 & A_2 & \phi_2 \end{pmatrix} D_{\phi_1\phi_2}^{(f)}(\Omega) \tag{10}$$

we get

$$I_{IJ}^{2\omega} = \frac{N_n \omega_s^4}{8\pi c^3 R_0^2} \left\langle \sum_{A_1, A_2, \alpha_1, \alpha_2, f} (-1)^{A_1-\alpha_1} C_I^{1A_1*} C_J^{1A_2} \left(M_{\alpha_1}^{(1)}\right)_{n'n}^* \times \left(M_{\alpha_2}^{(1)}\right)_{n'n} (2f+1) \begin{pmatrix} 1 & 1 & f \\ -\alpha_1 & \alpha_2 & \phi_1 \end{pmatrix} \times \begin{pmatrix} 1 & 1 & f \\ -A_1 & A_2 & \phi_2 \end{pmatrix} D_{\phi_1\phi_2}^{(f)}(\Omega) \right\rangle_{\Omega'} \tag{11}$$

where $\begin{pmatrix} a & b & c \\ \alpha & \beta & \gamma \end{pmatrix}$ stands for the 3-j Wigner coefficient.

The hyperpolarizability tensor $\beta_{ijk}^{2\omega}$ (2b) is symmetric in its last two indices. This has to be taken into account when transforming to irreducible form.

Quite generally, the symmetry properties of a Cartesian tensor of the nth rank are determined by the irreducible representations of the symmetry group of the tensor having the form of the direct product of the point group G and the group of permutations P_n. If the tensor is symmetric in all its indices, its symmetry group of index permutation is the symmetry group S_n of order $n!$; otherwise, it is a subgroup P_n of S_n which, in turn, can be represented in the form of direct and semidirect products of symmetry groups of lower orders. In turn, all point groups G are subgroups of the three-dimensional group of rotations K_h. A tensor of the rank n belongs to the representation $D(n)$ of the group K_h, which is the n-fold direct product of the vectorial representation D_u^1, taken n times. $D(n)$ is a reducible representation; its decomposition into a direct sum of representations $D_{g(u)}^1$ finally leads to the decompositionof the tensor into spherical tensors. The representations $D^1 \times [\lambda]$ are irreducible representations of the group $K_h \times S_n$, where $[\lambda]$ denotes irreducible representations of the group S_n in the form of a Young table. The results for tensors up to rank 4 are as follows [73]:

$$
\begin{aligned}
D(1) &= D_u^{(1)} \times [1] \\
D(2) &= D_g^{(0)} \times [2] + D_g^{(1)} \times [1^2] + D_g^{(2)} \times [2] \\
D(3) &= D_u^{(0)} \times [1^3] + D_u^{(1)} \times [3] + D_u^{(1)} \times [21] + D_u^{(2)} \times [21] \\
D(4) &= D_g^{(0)} \times [4] + D_g^{(0)} \times [2^2] + D_g^{(1)} \times [21^2] + D_g^{(1)} \times [31] \\
&\quad + D_g^{(2)} \times [4] + D_g^{(2)} \times [2^2] + D_g^{(2)} \times [31] + D_g^{(4)} \times [4]
\end{aligned}
\tag{12}
$$

For Cartesian tensors of the second rank A_{ij} we get three spherical tensors of different orders, $A^{(0)}$, $A^{(1)}$, $A^{(2)}$, each of which has a univocally determined permutational symmetry. Spherical tensors of the orders 0 and 2 are symmetric, whereas one of the order 1 is antisymmetric. This is by no means so for tensors of higher orders. Multiplication of spherical tensors of order 2 by a spherical tensor of order 1 representing a vector $V_v^{(1)}$ leads to the following tensors:

$$
B_\beta^{(a,b)} = \sum_{\alpha,\varphi} \begin{bmatrix} a & 1 & b \\ \alpha & \varphi & \beta \end{bmatrix} A_\alpha^{(a)} V_\varphi^{(1)} \tag{13}
$$

which have no well-defined permutation symmetry. In Eq. (13) $\begin{bmatrix} a & b & c \\ \alpha & \beta & \gamma \end{bmatrix}$ denotes the Clebsch–Gordan coefficient. Symmetrization can be achieved by having recourse to genealogical coefficients [14, 74]:

$$
B_\beta^{s(b)} = \sum_a \langle 2[\lambda']a, 1|3[\lambda]b\rangle B_\beta^{(a,b)} = \sum_a G_{sb}^a B_\beta^{(a,b)} \tag{14}
$$

The above results hold for the tensor of the highest permutational symmetry S_3. When imposing restrictions on the permutation of the indices we get tensors with the symmetries $S_1 \times S_2$ and $S_1 \times S_1 \times S_1$, corresponding to the partition of indices in the tensor $b_{i;jk}$ and $b_{i;j;k}$. Their properties result from the reduction $S_3 \Rightarrow S_1 \times S_2 \Rightarrow S_1 \times S_1 \times S_1$. The last reduction contributes nothing, since all the irreducible representations of the group $S_1 \times S_2$ are one dimensional. The representation [21] of S_3 reduces to two one-dimensional representations: $[1] \times [2]$ and $[1] \times [1^2]$ of the group $S_2 \times S_1$. They denote, respectively, tensors symmetric and antisymmetric in two of the indices. The remaining two representations [3] and $[1^3]$ are one dimensional; their reduction $[3] \Rightarrow [1] \times [2]$ and $[1^3] \Rightarrow [1] \times [1^2]$ does affect the properties of the tensors. The appurtenance of a tensor to the representation [3] or $[1^3]$ tells us whether it is symmetric or antisymmetric. To obtain a full description of the permutational properties of a spherical tensor we have to determine to which of the four representations it belongs. For convenience, we shall label them as follows: $s = 1$ to 4 ($1 = [3]$, $2 = [1] \times [2]$, $3 = [1] \times [1^2]$, $4 = [1^3]$).

The transition from components b_{ijk} to symmetrized components of spherical tensors is a unitary transformation, given by the formula

$$B_\beta^{s(b)} = \sum_{i,j,k} \langle sb\beta | ijk \rangle b_{ijk} \tag{15}$$

The coefficients of the preceding transformation depend on the type of the reference system in which the tensors b_{ijk} are determined. For the Cartesian case they are to be found in Ref. 14; for the circular case they are found in Refs. 74–76.

Proceeding along these lines one can decompose a tensor c_{ijkl} of rank 4 into spherical tensors $C^{s(c)}$. The coefficients of this transformation $\langle sc\gamma | ijkl \rangle$ are given in Ref. 74a, albeit only for tensors with complete permutational symmetry; ones symmetric in three indices only are given in Ref. 74b.

From the preceding analysis the tensor $b_{i;jk}^{2\omega}$, being symmetric in its last two indices, is representable by four spherical tensors $B_\beta^{s(b)}$: $B_\beta^{1(3)}$, $B_\beta^{1(1)}$, $B_\beta^{2(2)}$, and $B_\beta^{2(1)}$. If b_{ijk} can be dealt with as approximately symmetric in all its indices, only two of the spherical tensors, $B_\beta^{1(1)}$ and $B_\beta^{1(3)}$, will intervene.

To express the relation between the dipole moment and the hyperpolarizability tensor we have recourse to the equality of the scalar products:

$$\sum_\alpha M_\alpha^{(1)} E_\alpha^{[1]} = \frac{1}{4} \sum_{l,s,\lambda} B_\lambda^{s(l)} E_\lambda^{s[l]} = \frac{1}{4} \sum_{l,b,\lambda} B_\lambda^{(b,l)} E_\lambda^{[b,l]} \tag{16}$$

where $E_\lambda^{s(l)}$ represents the tensor $E_{ijk} = E_i E_j E_k$, and $B_\beta^{s(1)}$ represents the tensor b_{ijk}. Also,

$$E_\lambda^{[b,l]} = \sum_{\alpha,\beta} \begin{bmatrix} b & 1 & l \\ \beta & \alpha & \lambda \end{bmatrix} E_\beta^{[b]} E_\alpha^{[1]} \tag{17a}$$

$$E_\beta^{[b]} = \sum_{\gamma,\delta} \begin{bmatrix} 1 & 1 & b \\ \gamma & \delta & \beta \end{bmatrix} E_\gamma^{[1]} E_\delta^{[1]} \tag{17b}$$

From Eq. (16) it results that

$$M_\alpha^{(1)} = \frac{1}{4} \sum_{l,b,\lambda,\beta} (-1)^{l+b+\lambda+\beta} (2l+1)^{1/2} \begin{pmatrix} 1 & l & b \\ -\alpha & \lambda & \beta \end{pmatrix} B_\lambda^{(b,l)} E_\beta^{(b)} \tag{18}$$

From Eq. (18) we find that the dipole moment matrix elements occurring in Eq. (11) are determined by the matrix elements of $B_\beta^{(b,l)}$. We now proceed to determine the matrix elements of the hyperpolarizability spherical tensors

$$\left(B_\lambda^{(l)}\right)_{n'n} = \langle V'R' | B_\lambda^{(l)} | VR \rangle \tag{19}$$

The rotational wave functions of symmetric top molecules as well as linear and spherical ones with angular momenta J, the projections of which onto the axes z and z' are characterized, respectively, by the quantum number M and K, are given by way of the Wigner functions

$$|R\rangle = |JKM\rangle = i^{J-K} \left(\frac{2J+1}{8\pi^2} \right)^{1/2} D_{KM}^J(\Omega') \tag{20a}$$

The wave functions for asymmetric molecules can be expressed in the form of linear combinations

$$|J\tau M\rangle = \sum_K \langle J\tau M | JKM \rangle | JKM \rangle \tag{20b}$$

For simplicity, we shall restrict our considerations to symmetric top molecules. On transforming the spherical tensor from the laboratory system of coordinates $L_1(x, y, z)$ to molecular coordinates $L_3(1, 2, 3)$

$$B_\lambda^{(l)} = \sum_{\lambda'} \tilde{B}_{\lambda'}^{(l)} D_{\lambda'\lambda}^{(l)}(\Omega') \tag{21}$$

and invoking the Eckart–Wigner theorem [72] we obtain

$$\langle V'R'|B_{\lambda}^{(l)}|VR\rangle = (-1)^{J'-M'} i^{J+J'-K-K'}[(2J+1)(2J'+1)]^{1/2} \times \begin{pmatrix} J' & l & J \\ -K' & \lambda' & K \end{pmatrix} \begin{pmatrix} J' & l & J \\ -M' & \lambda & M \end{pmatrix} \langle V'|\tilde{B}_{\lambda}^{(l)}|V\rangle \tag{22}$$

On expanding $\tilde{B}_{\lambda}^{(l)}$ in a series of the normal coordinates Q ($\rho = 1, 2,$ $3n - 6$ or $3n - 5$ in the case of linear molecules) we have

$$\langle V'|\tilde{B}_{\lambda}^{(l)}|V\rangle = \langle V'|\left(\tilde{B}_{\lambda}^{(l)}\right)_0|V\rangle + \langle V'|\sum_{\rho}\left(\frac{\partial \tilde{B}_{\lambda}^{(l)}}{\partial Q_{\rho}}\right)_0 Q_{\rho}|V\rangle + \cdots \tag{23}$$

With regard to the properties of the vibrational wave functions, the first term of (23) can give scattering only if $v_{\rho} = v_{\rho}'$ for all ρ, whereas the second term can lead to scattering if $v_{\rho}' - v_{\rho} = \mp 1$ at all other $\Delta v_{\sigma} = 0$. The scattering related with the first term is referred to as rotational and that related with the second term as vibrational–rotational hyper-Raman scattering. Since the shifts due to rotation of the molecules are small compared with those due to vibrations, it is also customary to speak of the rotational structure of the hyper-Rayleigh spectrum and that of the hyper-Raman spectrum. To achieve a unique notation of the formulae describing the two scattering effects we introduce the following symbolism:

$$\tilde{B}_{\rho\lambda}^{(l)} = \left(\tilde{B}_{\lambda}^{(l)}\right)_0 \qquad \text{for } \rho = 0 \tag{24a}$$

$$\tilde{B}_{\rho\lambda}^{(l)} = \left(\frac{\partial \tilde{B}_{\lambda}}{\partial Q_{\rho}}\right)_0 \qquad \text{for } \rho = 1, 2, \ldots, 3n - 6 \tag{24b}$$

and for the matrix elements occurring in (23)

$$\begin{aligned} v_{\rho}^{+} &= \langle v_{\rho} + 1|Q_{\rho}|v_{\rho}\rangle = \frac{v_{\rho} + 1}{2\omega_{\rho}} \\ v_{\rho}^{-} &= \langle v_{\rho} - 1|Q_{\rho}|v_{\rho}\rangle = \frac{v_{\rho}}{2\omega_{\rho}} \\ v_{\rho}^{\mp} &= 1 \qquad \text{for } \rho = 0 \end{aligned} \tag{25}$$

Hence, for a vibration Q_{ρ} to be hyper-Raman active, the following two conditions must be fulfilled:

1. $$\Delta v_\rho = \mp 1, \quad \Delta v_\sigma = 0 \qquad \text{for } \rho \neq \sigma \tag{26a}$$

2. $$\tilde{B}^l_{\rho\lambda'} \neq 0 \tag{26b}$$

The selection rules for the rotational quantum numbers result directly from the properties of the $3j$ symbols. A rotational transition $JK \Rightarrow J'K'$ is permitted if

$$\Delta J = 0, \mp 1, \mp 2, \mp 3 \tag{27a}$$

$$\Delta K = \lambda' \tag{27b}$$

A set of rotational lines with the same ΔJ constitutes a branch. The branches are traditionally denoted by the capital letters N, O, P, Q, R, S, T, corresponding to ΔJ running from -3 to $+3$. The permitted values of λ' result from the vibrational selection rules $\tilde{B}^l_{\rho\lambda'} \neq 0$. Group theory enables us to determine when the condition $\tilde{B}^l_{\rho\lambda'} \neq 0$ is fulfilled. Once the decomposition of the tensor into spherical tensors is available, we are able to do this quite easily. Forming the appropriate combinations of the spherical tensor components, we find the basis of irreducible representations of the point groups. Denoting by $T^a_\alpha(l)$ the α vector of the basis of the irreducible representation Γ^a of the group G, composed of elements of the spherical tensor of lth order, we can write it as follows:

$$T^a_\alpha(l) = \sum_\lambda K^{a,l}_{\alpha,\lambda} T^{(l)}_\lambda \tag{28}$$

The literature provides exhaustive tables of the values of these coefficients. Leushin [77] has tabulated their values for all the crystallographical classes and $l = 0$– $\frac{17}{2}$. As an example we present in Table I the $T^a_\alpha(l)$ for the cubic point group O and $l = 0$–4.

Having available the matrices reducing Cartesian tensors of ranks 2, 3, and 4 and the values of $K^{a,l}_{\alpha,\lambda}$, the problem of decomposition of the polarizability tensors according to the point groups is solved. This, in turn, permits the determination of the selection rules. The normal coordinates Q_ρ also transform as the irreducible representations of the group G. For condition (26) to be fulfilled, the two quantities $\tilde{B}^{(l)}_\lambda$ and Q_ρ must belong to the same irreducible representation Γ^a of the symmetry group G of the molecule. Thus, the prescription as to which of the components of $\tilde{B}^{(l)}_\lambda$ belong to the individual representations Γ^a establishes simultaneously the vibrational and rotational selection rules. Whereas the rule $\Delta K = \lambda'$

TABLE I
Tensors $T_{\alpha}^{a}(l)$ for the point group O

$$A_1: \quad T_0^0, \sqrt{\frac{5}{24}}\,(T_4^4 + T_{-4}^4) + \sqrt{\frac{7}{12}}\,T_0^4$$

$$A_2: \quad \frac{i}{\sqrt{2}}(T_2^3 - T_{-2}^3)$$

$$E: \quad \left\{\frac{1}{\sqrt{2}}\left(T_2^2 + T_{-2}^2\right), T_0^2\right\},$$

$$\left\{\sqrt{\frac{7}{24}}\left(T_4^4 + T_{-4}^4\right) - \sqrt{\frac{5}{12}}\,T_0^4, \frac{1}{\sqrt{2}}\left(T_2^4 + T_{-2}^4\right)\right\}$$

$$T_1: \quad \{T_1^1, T_{-1}^1, T_0^1\}, \left\{\sqrt{\frac{3}{8}}\,T_1^3 + \sqrt{\frac{5}{8}}\,T_{-3}^3, \sqrt{\frac{5}{8}}\,T_3^3 + \sqrt{\frac{3}{8}}\,T_{-1}^3, -T_0^3\right\}$$

$$\left\{\sqrt{\frac{7}{8}}\,T_1^4 + \frac{1}{\sqrt{8}}T_{-3}^4, \frac{1}{\sqrt{2}}\left(-T_4^4 + T_{-4}^4\right), -\frac{1}{\sqrt{8}}T_3^4 - \frac{7}{\sqrt{8}}T_{-1}^4\right\}$$

$$T_2: \quad \left\{T_1^2, \frac{i}{\sqrt{2}}\left(T_2^2 - T_{-2}^2\right), T_{-1}^2\right\}$$

$$\left\{-\sqrt{\frac{5}{8}}\,T_1^3 + \sqrt{\frac{3}{8}}\,T_{-3}^3, \frac{i}{\sqrt{2}}\left(T_2^3 + T_{-2}^3\right),\right.$$
$$\left.-\sqrt{\frac{3}{8}}\,T_3^3 + \sqrt{\frac{5}{8}}\,T_{-}{-1}^3\right\}$$

$$\left\{-\frac{1}{\sqrt{8}}T_1^4 + \sqrt{\frac{7}{8}}\,T_{-3}^4, \frac{i}{\sqrt{2}}\left(T_2^4 - T_{-2}^4\right), \sqrt{\frac{7}{8}}\,T_3^4 - \frac{1}{\sqrt{8}}T_{-1}^4\right\}$$

Note. For the O_h group odd tensors such as $B_{\lambda}^{(l)}$ belong to the u type of irreducible representations, and even tensors such as C_{λ} to the g type.

results strictly from condition (26), the restrictions on ΔJ are not so rigorous. Nonetheless, the symmetry of the molecule and that of a normal vibration can jointly cause some branches to be inactive.

In the case of linear molecules, when $K = K' = 0$, and on the assumption that the tensor b_{ijk} is completely symmetric ($l = 1$ and 3), only transitions with $\Delta J = \mp 1$ and $\Delta J = \mp 3$ are permitted. This results from the circumstance that $3j$ symbols with zero arguments in their lower row differ from zero only if $J' + J + l$ is an even number. The assumption of asymmetricity in b_{ijk} leads to the activation of branches with $\Delta J = \mp 2$ as well. With regard to relations (18) and (22) and the fact that the averaging $\langle\ \rangle$ is equivalent to the operation $[1/(2J+1)]\Sigma_{M'M}$, we arrive at the ultimate form of the hyper-Raman scattering tensor (11):

$$I_{IJ} = \frac{N_{v_\rho JK}\left(v_\rho^{\mp}\right)^2 \omega_s^4}{128\pi c^3 R_0^2} \sum_{\substack{f, l, b_1, b_2 \\ A_1, A_2, \beta_1, \beta_2}} (-1)^{l+f+A_1+\beta_1} C_I^{1A_1*} C_J^{1A_2} \Pi_{fJ'}^2$$
$$\times \begin{pmatrix} J' & l & J \\ -K' & \lambda' & K \end{pmatrix}^2 \begin{pmatrix} 1 & 1 & f \\ -A_1 & A_2 & \phi_2 \end{pmatrix} \begin{pmatrix} b_1 & b_2 & f \\ -\beta_1 & \beta_2 & \phi_1 \end{pmatrix} \quad (29)$$
$$\times \begin{Bmatrix} 1 & 1 & \mathrm{f} \\ b_1 & b_2 & l \end{Bmatrix} D_{\phi_1\phi_2}^f(\Omega)$$
$$\times \tilde{B}_{\rho\lambda'}^{(b_1, l)} \tilde{B}_{\rho\lambda'}^{(b_2, l)*} E_{\beta_1}^{(b_1)} E_{\beta_2}^{(b_2)*}$$

where we have made use of the notation $\Pi_{ab\ldots f} = [(2a+1)(2b+1)\cdots(2f+1)]^{1/2}$; moreover, $\begin{Bmatrix} a & b & c \\ d & e & f \end{Bmatrix}$ stands for the Wigner $6-j$ symbol. The tensors $\tilde{B}_\lambda^{(b,l)}$ can be expressed in terms of the symmetrized tensors $\tilde{B}_\lambda^{s(l)}$ by the relation inverse to (14).

The process of hyper-Raman scattering, involving transitions from the initial state $v_\rho JK$ to the final state $v'_\rho J'K'$ is fed by only part of the molecules, $N_{v_\rho JK}$ in the initial state, as determined by the Boltzmann statistical distribution function. On carrying out summation over all possible rotational transitions we obtain the integral intensity of the vibrational bands. If harmonicity is taken into account, the contribution to the lines $2\omega \mp \omega_\rho$ is found to originate in all the transitions $v_\rho \Rightarrow v_\rho \mp 1$, causing the temperature dependence here to be the same as in linear Raman scattering, and is determined by the function $F_{\mp}(T)$:

$$F_+(T) = \frac{1}{1 - \exp(\hbar\omega_\rho/k_\mathrm{B}T)} \frac{\hbar}{2\omega_\rho} \quad (30a)$$

$$F_-(T) = \frac{1}{1 - \exp(-\hbar\omega_\rho/k_\mathrm{B}T)} \frac{\hbar}{2\omega_\rho} \quad (30b)$$

Finally, we obtain

$$I_{IJ}^{2\omega} = \frac{NF_{\mp}(T)\omega_s^4}{128\pi c^3 R_0^2} \sum_{\substack{f,l,b_1,b_2 \\ A_1,B_2,\beta_1,\beta_2}} (-1)^{l+f+A_1+\beta_1} C_I^{1A_1*} C_J^{1A_2}(2f+1)$$
$$\times \begin{pmatrix} 1 & 1 & f \\ -A_1 & A_2 & \phi_2 \end{pmatrix} \begin{pmatrix} b_1 & b_2 & f \\ -\beta_1 & \beta_2 & \phi_1 \end{pmatrix}$$
$$\times \begin{Bmatrix} 1 & 1 & f \\ b_1 & b_2 & l \end{Bmatrix} D_{\phi_1\phi_2}^{f}(\Omega) \tag{31}$$
$$\times \sum_{\lambda'} \left[\tilde{B}_{\rho\lambda'}^{(b_1,l)} \tilde{B}_{\rho\lambda'}^{(b_2,l)*} \right] E_{\beta_1}^{(b_1)} E_{\beta_2}^{(b_2)*}$$

The angular distribution of the scattered radiation is given by an appropriate linear combination of Wigner functions of $D_{\phi_1\phi_2}^{f}(\Omega)$ with $f = 0$ to 2. We shall write explicitly the results for several particular cases. Let us assume the scattered light to be observed in the plane defined by the directions z and Z, subtending the angle β mutually. For incident light horizontally polarized ($E_x \neq 0$, $E_y = 0$) the angles $\alpha = \gamma = 0$. The horizontal component of scattered radiation is expressed by $H_h^{2\omega} = I_{xx}^{2\omega}$ and the vertical component by $V_h^{2\omega} = I_{yy}^{2\omega}$. On performing the calculations implied by formula (31) we get the horizontal depolarization coefficient in the form:

$$D_h^{2\omega}(\beta) = \frac{H_h^{2\omega}}{V_h^{2\omega}} = \Big\{7|B^{1(1)}|^2 + 12|B^{1(3)}|^2 + 7(5|B^{2(1)}|^2 + 3|B^{2(2)}|^2$$
$$\times + \sqrt{5}\,|B^{1*2(1)}|^2)\Big\} \Big/ \Big\{7|B^{1(1)}|^2(1 + 8\cos^2\beta)$$
$$+ 6|B^{1(3)}|^2(2 + \cos^2\beta) \tag{32}$$
$$+ 7\big(5|B^{2(1)}|^2 + 3|B^{2(2)}|^2$$
$$+ \sqrt{5}\,|B^{1*2(1)}|^2\big)\sin^2\beta\Big\}$$

where we use the following notation

$$|B^{s(l)}|^2 = \sum_{\lambda'} |\tilde{B}_{\lambda'}^{(s(l)}|^2$$
$$|B^{s_1^* s_2(l)}|^2 = \sum_{\lambda'} \left[\tilde{B}_{\lambda'}^{s_1(l)*} \tilde{B}_{\lambda'}^{s_2(l)} + \tilde{B}_{\lambda'}^{s_1(l)} \tilde{B}_{\lambda'}^{s_2(l)*} \right]$$

The intensities $H_h^{2\omega}$ and $V_h^{2\omega}$ are, respectively, equal to the numerator and denominator of the above expression multiplied by a factor of

$$A_{\mp}^{2\omega} = \frac{\pi F_{\mp}(T)\omega_s^4 I_0^2}{630c^5 R_0^2}$$

with $I_0 = (c/8\pi)E_x^2$.

For incident light circularly right-polarized

$$\mathbf{E}_{+1} = \mathbf{e}_{+1}^{(1)} E \cos \omega t$$

the two intensity components, scattered with right circular polarization $R_{\mathrm{r}}^{2\omega} = I_{+1+1}^{2\omega}$ and left circular polarization $L_{\mathrm{r}}^{2\omega} = I_{-1-1}^{2\omega}$, are determined by the following reversal coefficients:

$$\begin{aligned} R^{2\omega}(\beta) = \frac{L_r^{2\omega}}{R_r^{2\omega}} = &\Big\{28|B^{1(1)}|^2\Phi_{-1}(\beta) + 6|B^{1(3)}|^2\Phi_{+2}(\beta) + \Big[35|B^{2(1)}|^2 \\ &\times - 14|B^{1*2(1)}|^2\Big]\Phi_{-1}(\beta) + 21|B^{2(2)}|^2\Phi_{-3}(\beta)\Big\}\Big/ \\ &\times\Big\{28|B^{1(1)}|^2\Phi_{+1}(\beta) + 6|B^{1(3)}|^2\Phi_{-2}(\beta) \\ &+\Big[35|B^{2(1)}|^2 - 14|B^{1*2(1)}|^2\Big]\Phi_{+1}(\beta) \\ &+21|B^{2(2)}|^2\Phi_{+3}(\beta)\Big\} \end{aligned} \tag{34}$$

with the angular distribution functions

$$\Phi_{\mp 1} = (1 \mp \cos\beta)^2 \tag{34a}$$

$$\Phi_{\mp 2} = 13 \mp 14\cos\beta + 3\cos^2\beta \tag{34b}$$

$$\Phi_{\mp 3} = 5 \mp 2\cos\beta - 3\cos^2\beta \tag{34c}$$

Above, we have made use of molecular parameters expressed in terms of irreducible spherical tensor components. Their dependence on the Cartesian hyperpolarizability tensor components (for clarity we omit the 2ω and ρ) is

$$|B^{1(1)}|^2 = \sum_{\lambda'} |\tilde{B}^{1(1)}_{\lambda'}|^2 = \frac{1}{15}(4b_{iij}b_{kkj} + 4b_{iij}b_{jkk} + b_{ijj}b_{ikk}) \quad (35a)$$

$$|B^{1(3)}|^2 = \sum_{\lambda'} |\tilde{B}^{1(3)}_{\lambda'}|^2 = \frac{1}{15}(5b_{ijk}b_{ijk} + 10b_{ijk}b_{jik} - 4b_{iij}b_{kkj} - 4b_{ijj}b_{kki} - b_{ijj}b_{ikk}) \quad (35b)$$

$$|B^{2(1)}|^2 = \sum_{\lambda'} |\tilde{B}^{2(1)}_{\lambda'}|^2 = \frac{1}{3}(b_{iij}b_{kkj} - 2b_{iij}b_{jkk} + b_{ijj}b_{ikk}) \quad (35c)$$

$$|B^{2(2)}|^2 = \sum_{\lambda'} |\tilde{B}^{2(2)}_{\lambda'}|^2 = \frac{1}{3}(2b_{ijk}b_{ijk} - 2b_{ijk}b_{jik} + 2b_{ijj}b_{kki} - b_{ijj}b_{ikk} - b_{iij}b_{kkj}) \quad (35d)$$

$$|B^{1*2(1)}|^2 = \sum_{\lambda'} \left(\tilde{B}^{1(1)*}_{\lambda'}\tilde{B}^{2(1)}_{\lambda'} + \tilde{B}^{1(1)}_{\lambda'}\tilde{B}^{2(1)*}_{\lambda'}\right) = \frac{2\sqrt{5}}{15}(b_{ijj}b_{ikk} + b_{iij}b_{jkk} - 2b_{iij}b_{kkj}) \quad (35e)$$

On the assumption that the tensor b_{ijk} is completely symmetric, the only two parameters remaining are

$$|B^{1(1)}|^2 = \tfrac{3}{5}b_{iij}b_{jkk} \quad (36a)$$

$$|B^{1(3)}|^2 = b_{ijk}b_{ijk} - \tfrac{3}{5}b_{iij}b_{jkk} \quad (36b)$$

The number of parameters describing a given hyper-Raman line depends on the symmetry of the molecule and the symmetry of the relevant vibration. For molecules of $O(O_h)$ symmetry it is seen from Table 1 that their vibrations depend on the following parameters:

$$A_1(A_{1u}) \Rightarrow \text{none}$$

$$A_2(A_{2u}) \Rightarrow |B^{1(3)}|^2$$

$$E(E_u) \Rightarrow |B^{2(2)}|^2$$

$$T_1(T_{1u}) \Rightarrow |B^{1(1)}|^2, |B^{1(3)}|^2, |B^{2(1)}|^2, |B^{1*2(1)}|^2$$

$$T_2(T_{2u}) \Rightarrow |B^{1(3)}|^2, |B^{2(2)}|^2$$

This means that vibrations A_1 are not hyper-Raman active and vibrations E are active only if the hyperpolarizability tensor is not completely symmetric and then the lines are fully depolarized with $D_h^{2\omega} = \infty$. For vibration A_2 the depolarization coefficients $D_h^{2\omega}(0) = \frac{2}{3}$, which is the maximum value when asymmetry of the tensor is not present. An interesting case occurs for T_2 vibrations, when two parameters are involved, one of which describes the asymmetry of the tensor. Hence,

$$D_h^{2\omega}(0) = \frac{2}{3} + \frac{7}{6}\frac{|B^{2(2)}|^2}{|B^{1(3)}|^2}$$

B. Third-Harmonic Scattering

The hyperpolarizability tensor $c_{i,jkl}$ intervening in third-harmonic scattering $3\omega + \omega_{n'n}$ is symmetric in its last three indices. On reduction to irreducible form, $C_\lambda^{s(l)}$, it is represented by three completely symmetric tensors $s = 1 = [4]$ and three asymmetric tensors $s = 2 = [1] \times [3]$:

$$c_{i,jkl} \Rightarrow C_\lambda^{1(4)} + C_\lambda^{1(2)} + C_\lambda^{1(0)} + C_\lambda^{2(3)} + C_\lambda^{2(2)} + C_\lambda^{2(1)} \tag{37}$$

We restrict our analysis of the scattered radiation to the simplest experimental geometries. Let incident light, polarized in the xz plane, propagate along the z axis. At observation in the direction of the y axis the components are $V_v^{3\omega} = I_{xx}^{3\omega}$ and $H_v^{3\omega} = I_{zz}^{3\omega}$; for direct observation they are $H_h^{3\omega} = I_{xx}^{3\omega} = V_v^{3\omega}$ and $V_h^{3\omega} = I_{yy}^{3\omega} = H_v^{3\omega}$. The intensity scattered with polarization parallel to an axis i of laboratory coordinates is given by the expression

$$I_{i=x,y,z}^{3\omega} = \frac{\pi^2}{9}\frac{N_n I_0^3 \omega_s^4}{c^6 R_0^2}\left\langle \left| \left(c_{i,xxx}^{3\omega}\right)_{n'n} \right|^2 \right\rangle_{\Omega'} \tag{38}$$

The polarizability tensors $c_{i,xxx}$ are given as follows in terms of spherical tensor components:

$$
\begin{aligned}
c^{3\omega}_{x,xxx} &= \frac{1}{4}\left(C^{1(4)}_{4} + C^{1(4)}_{-4}\right) - \frac{1}{2\sqrt{7}}\left(C^{1(4)}_{2} + C^{1(4)}_{-2}\right) \\
&\quad + \frac{3}{2\sqrt{70}}C^{1(4)}_{0} - \frac{3}{\sqrt{42}}\left(C^{1(2)}_{2} + C^{1(2)}_{-2}\right) + \frac{1}{\sqrt{7}}C^{1(2)}_{0} + \frac{1}{\sqrt{5}}C^{1(0)}_{0} \\
c^{3\omega}_{z,xxx} &= -\frac{1}{4\sqrt{2}}\left(C^{1(4)}_{3} - C^{1(4)}_{-3}\right) + \frac{3}{4\sqrt{14}}\left(C^{1(4)}_{1} - C^{1(4)}_{-1}\right) \\
&\quad + \frac{3}{2\sqrt{42}}\left(C^{1(2)}_{1} - C^{1(2)}_{-1}\right) - \frac{3}{4\sqrt{6}}\left(C^{2(3)}_{3} + C^{2(3)}_{-3}\right) \\
&\quad + \frac{1}{4\sqrt{10}}\left(C^{2(3)}_{1} + C^{2(3)}_{-1}\right) - \frac{3}{2\sqrt{15}}\left(C^{2(1)}_{1} + C^{2(1)}_{-1}\right) \\
&\quad - \frac{1}{2\sqrt{2}}\left(C^{2(2)}_{1} - C^{2(2)}_{-1}\right)
\end{aligned}
\tag{39b}
$$

On performing calculations similar to those for second-harmonic scattering we obtain the depolarization ratio of the vibrational bands $3\omega \mp \omega_\rho$ in the form

$$
\begin{aligned}
D^{3\omega} = \frac{I^{3\omega}_{zz}}{I^{3\omega}_{xx}} &= \left\{20|C^{1(4)}|^2 + 27|C^{1(2)}|^2 + 567|C^{2(2)}|^2\right. \\
&\quad \left. + 27\sqrt{21}\,|C^{1*2(2)}|^2 + 324|C^{2(3)}|^2 + 1134|C^{2(1)}|^2\right\} \Big/ \\
&\quad \times \left\{32|C^{1(4)}|^2 + 144|C^{1(2)}|^2 + 252|C^{1(0)}|^2\right\}
\end{aligned}
\tag{40}
$$

where

$$
|C^{s(l)}|^2 = \sum_{\lambda'} \left|\tilde{C}^{s(l)}_{\rho\lambda'}\right|^2 \tag{41a}
$$

$$
|C^{s_1*s_2(l)}|^2 = \sum_{\lambda'} \left(\tilde{C}^{s_1(l)*}_{\lambda'}\tilde{C}^{s_2(l)}_{\lambda'} + \tilde{C}^{s_1(l)}_{\lambda'}\tilde{C}^{s_2(l)*}_{\lambda'}\right) \tag{41b}
$$

The intensities $I^{3\omega}_{zz}$ and $I^{3\omega}_{xx}$ are, respectively, equal to the numerator and denominator of the above expression multiplied by a factor of

$$A^{3\omega} = \frac{\pi^2 N F_{\mp}(T) I_0^3 \omega_s^4}{11340 c^6 R_0^2} \tag{41c}$$

It is of interest to note that the polarized component $I_{xx}^{3\omega}$ is dependent only on the parameters imposed by the symmetrical part of the hyperpolarizability tensor. In general, seven molecular parameters are necessary to describe the scattering of linearly polarized light.

In the case of excitation with right-circularly polarized light one can distinguish two components, respectively right-circularly $I_{+1}^{3\omega}$ and left-circularly polarized $I_{-1}^{3\omega}$ in the scattered radiation at observation in the direction of propagation. They are given by the expression

$$I_{i=\mp 1}^{3\omega} = \frac{\pi^2}{9} \frac{N_n I_0^3 \omega_s^4}{c^6 R^2} \left\langle \left| \left(c_{i,-1-1-1}^{3\omega} \right)_{n'n} \right|^2 \right\rangle_{\Omega'} \tag{42}$$

The components of $c_{i,jkl}^{3\omega}$ in the circular system of reference are expressed as follows in terms of spherical tensor components:

$$c_{-1,-1-1-1}^{3\omega} = C_{-4}^{1(4)} \tag{43a}$$

$$c_{1,-1-1-1}^{3\omega} = \frac{1}{2\sqrt{7}} C_{-2}^{1(4)} - \frac{3}{\sqrt{42}} C_{-2}^{1(2)} + \frac{1}{2} C_{-2}^{2(3)} + \frac{1}{\sqrt{2}} C_{-2}^{2(2)} \tag{43b}$$

whence the reversal ratio $R^{3\omega}$ for the vibrational bands $3\omega \mp \omega_\rho$:

$$R^{3\omega} = \frac{I_{-1}^{3\omega}}{I_{+1}^{3\omega}} = 140 |C^{1(4)}|^2 \Big/ \Big\{ 5 |C^{1(4)}|^2 + 54 |C^{1(2)}|^2 + 126 |C^{2(2)}|^2 - 18\sqrt{21} \, |C^{1*2(2)}|^2 + 90 |C^{2(3)}|^2 \Big\} \tag{44}$$

Third-harmonic scattering of circularly polarized light is dependent on five molecular parameters. Two of them, $|C^{2(1)}|^2$ and $|C^{1(0)}|^2$, do not intervene in the reversal ratio. Hence, no univocal relation exists between the reversal ratio and the depolarization ratio, even in the approximation assuming complete permutational symmetry of the tensor c_{ijkl}.

The principle regarding the determination of the selection rules is similar to that for second-harmonic scattering processes. As an example, let us again consider the case of molecules having the symmetry $O(O_h)$.

For this case the nonzero parameters for the various types of vibrations are

$$
\begin{aligned}
A_1(A_{1g}) &\Rightarrow |C^{1(0)}|^2, |C^{1(4)}|^2 \\
A_2(A_{2g}) &\Rightarrow |C^{2(3)}|^2 \\
E(E_g) &\Rightarrow |C^{1(2)}|^2, |C^{2(2)}|^2, |C^{1*2(2)}|^2, |C^{1(4)}|^2 \\
T_1(T_{1g}) &\Rightarrow |C^{2(1)}|^2, |C^{2(3)}|^2, |C^{1(4)}|^2 \\
T_2(T_{2g}) &\Rightarrow |C^{1(2)}|^2, |C^{2(2)}|^2, |C^{1*2(2)}|^2, |C^{2(3)}|^2, |C^{1(4)}|^2
\end{aligned}
\tag{45}
$$

For the completely symmetric vibrations A_1, only two of the molecular parameters intervene and we have $D^{3\omega} < \frac{5}{8}$ and $R^{3\omega} = 28$. For vibrations of the type T_1, three parameters intervene, two of which determine the asymmetry of the tensor; if the latter is completely symmetric the ratios take the constant values $D_{T_1}^{3\omega} = \frac{5}{8}$ and $R_{T_1}^{3\omega} = 28$. The vibration A_2 is active only if the tensor $c_{i,jkl}$ is not completely symmetric.

C. Rotational Structure of Hyper-Rayleigh Scattering

The general expression (29) that we derived permits the numerical prediction of the rotational structure of the hyper-Rayleigh line. We assume that

1. Observation of polarized radiation is carried out at right angle.
2. The hyperpolarizability tensor $b_{ijk}^{2\omega}$ is totally symmetric.
3. The scattering is due to linear molecules.

The intensity of the rotational spectral line is then given by the expression [62]

$$
I^{2\omega} = N\frac{\pi(2\omega + \omega_{J'J})^4 I_0^2}{c^5 R_0^2}\,\frac{\exp(-E_J/k_B T)}{Z}(2J+1)(2J'+1)
$$

$$
\left\{\frac{1}{5}\begin{pmatrix} J' & 1 & J \\ 0 & 0 & 0\end{pmatrix}^2 |\tilde{B}_0^{1(1)}|^2 + \frac{2}{35}\begin{pmatrix} J' & 3 & J \\ 0 & 0 & 0\end{pmatrix}^2 |\tilde{B}_0^{1(3)}|^2\right\}\delta(\omega - \omega_{JJ'}) \tag{46}
$$

where $E_J = BJ(J+1)$; Z is the rotational partition function; and $\omega_{J'J} = B[J'(J'+1) - J(J+1)]$. Figures 1–3 show the theoretical Stokes-side spectral distribution of hyper-Rayleigh scattered radiation, respec-

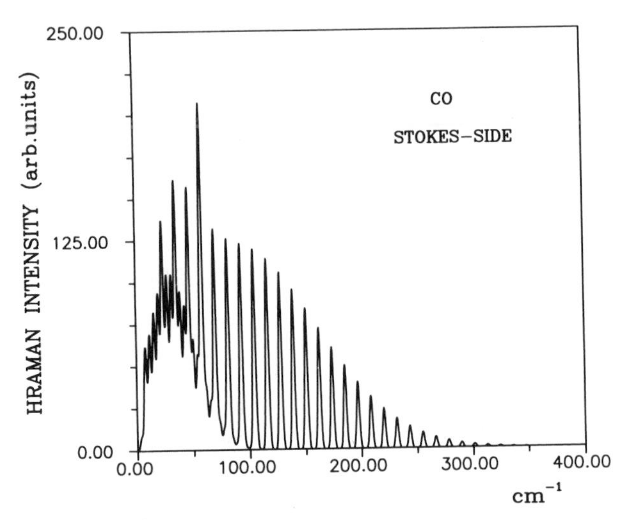

Figure 1. The theoretical rotational structure of hyper-Rayleigh scattering by CO. The Stokes-side spectrum is calculated with Hush and Williams' [78] values $b_{333}^{2\omega} = 19.73 \times 10^{-32}$ cm^5 and $b_{113}^{2\omega} = -17.10 \times 10^{-32}$ cm^5 of the hyperpolarizability tensor elements. In our computations we used $B = 1.931$ cm^{-1} and a Gaussian form of the apparatus function with full width at half-maximum (FWHM) 3 cm^{-1}.

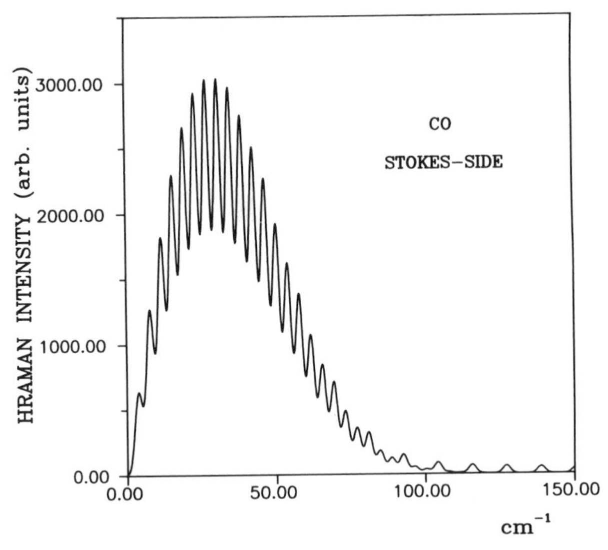

Figure 2. The theoretical rotational structure of hyper-Rayleigh scattering by CO. The Stokes-side spectrum is calculated with O'Hare and Hurst's [79] values $b_{333}^{2\omega} = -98.28 \times 10^{-32}$ cm^5 and $b_{113}^{2\omega} = -15.16 \times 10^{-32}$ cm^5 of the hyperpolarizability tensor elements ($B = 1.931$ cm^{-1}; FWHM = 3 cm^{-1}).

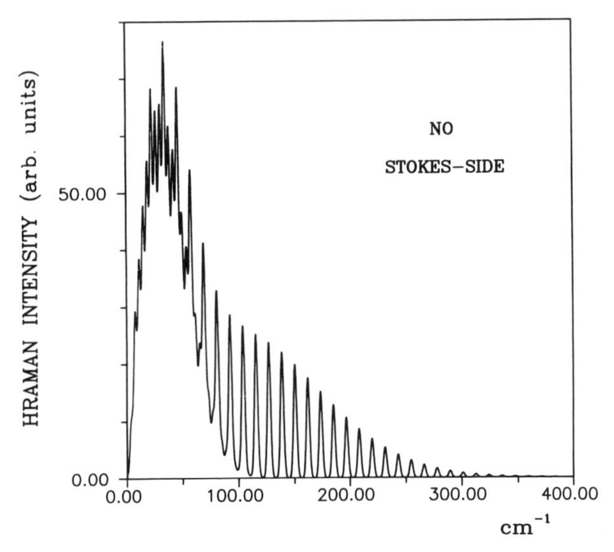

Figure 3. The theoretical rotational structure of hyper-Rayleigh scattering by NO. The Stokes-side spectrum is calculated with Hush and Williams' [78] values $b_{333}^{2\omega} = 22.39 \times 10^{-32}$ cm^5, $b_{113}^{2\omega} = -3.23 \times 10^{-32}$ cm^5 of the hyperpolarizability tensor elements ($B = 1.705$ cm^{-1}; FWHM = 3 cm^{-1}).

tively, for CO and two different sets of values of the hyperpolarizability tensor components, and for NO. To take into account the influence of the aperture slit on the spectrum we have broadened our theoretical stick spectrum (46) with a Gaussian function having a full width of 3 cm^{-1} at half-maximum.

III. INTERACTION-INDUCED CONTRIBUTIONS TO HYPER-RAYLEIGH AND HYPER-RAMAN SCATTERING

A. Influence of Interaction-Induced Contributions on the Spectral Shape and Integral Intensity of Scattered Light

Hitherto we considered hyper-Rayleigh and hyper-Raman scattering as coming from individual molecules of the scattering medium. However, in dense scattering systems intermolecular interactions (internal fields), as in the case of linear scattering [80], play an important role in nonlinear scattering processes. If we take into account the internal molecular fields existing in the scattering system, the collective hyperpolarizability tensor of the scattering system consists of two parts:

1. The sum of the individual molecular hyperpolarizabilities

$$ {}^{(0)}B^{2\omega}_{\alpha\beta\gamma} = \sum_i {}_{(i)}b^{2\omega}_{\alpha\beta\gamma} \tag{47} $$

2. Excess interaction-induced hyperpolarizability $\Delta B^{2\omega}_{\alpha\beta\gamma}$.

Then the total hyperpolarizability tensor of the system reads

$$ B^{2\omega}_{\alpha\beta\gamma} = {}^{(0)}B^{2\omega}_{\alpha\beta\gamma} + \Delta B^{2\omega}_{\alpha\beta\gamma} \tag{48} $$

where within the lowest order of the dipole-induced dipole approximation we have

$$ \Delta B^{2\omega}_{\alpha\beta\gamma} = \sum_{m=a,b,c} {}^{(m)}\Delta B^{2\omega}_{\alpha\beta\gamma} \tag{49} $$

with

$$ {}^{(a)}\Delta B^{2\omega}_{\alpha\beta\gamma} = \frac{1}{4} \sum_{i,j} {}_{(i)}b^{2\omega}_{\alpha\beta\varepsilon} T^{(ij)}_{\varepsilon\delta} \, {}_{(j)}a^{\omega}_{\delta\gamma} \tag{50a} $$

$$ {}^{(b)}\Delta B^{2\omega}_{\alpha\beta\gamma} = \frac{1}{4} \sum_{i,j} {}_{(i)}b^{2\omega}_{\alpha\varepsilon\gamma} T^{(ij)}_{\varepsilon\delta} \, {}_{(j)}a^{\omega}_{\delta\beta} \tag{50b} $$

$$ {}^{(c)}\Delta B^{2\omega}_{\alpha\beta\gamma} = \frac{1}{4} \sum_{i,j} {}_{(i)}a^{2\omega}_{\alpha\varepsilon} T^{(ij)}_{\varepsilon\delta} \, {}_{(j)}b^{2\omega}_{\delta\beta\gamma} \tag{50c} $$

where, along with the obvious meaning of $b^{2\omega}_{\alpha\beta\gamma}$ as a hyperpolarizability tensor, ${}_{(i)}a^{\omega}_{\alpha\beta}$ denotes the Rayleigh polarizability tensor of molecule i and $T^{(ij)}_{\varepsilon\delta}$ the dipole–dipole interaction tensor [7]:

$$ T_{\alpha\beta} = \frac{3\hat{r}_\alpha \hat{r}_\beta - \delta_{\alpha\beta}}{r^3} \tag{51} $$

The origin of Eqs. (50a) and (50b) is obvious [34]. Equation (50c) results from the following mechanism: Intense optical radiation induces, by way of the hyperpolarizability **b**, a dipole moment at frequency 2ω in molecule j, which in turn radiating at 2ω induces a dipole moment in molecule i, by way of the linear polarizability **a**. In general, tensors (50a) and (50b) are not symmetric to any permutation of their indexes, whereas tensor (50c) is symmetric in its last two indexes. We carry out our further calculations assuming that the one-molecule polarizabilities $a_{\alpha\beta}$ and $b^{2\omega}_{\alpha\beta\gamma}$ are fully symmetric in their indexes. Then the tensor ${}^{(a)}\Delta B^{2\omega}_{\alpha\beta\gamma}$ is symmetric in α and β, whereas ${}^{(b)}\Delta B^{2\omega}_{\alpha\beta\gamma}$ is symmetric in α and γ. The tensor ${}^{(c)}\Delta B^{2\omega}_{\alpha\beta\gamma}$ remains symmetric in its last two indexes. Then the collective hyperpolarizability tensor ${}^{(0)}B^{2\omega}_{\alpha\beta\gamma}$ can be decomposed into irreducible tensors of orders 1 and 3:

$$ {}^{(0)}B^{2\omega}_{\alpha\beta\gamma} = {}^{(0)}B^{(1)}_{\alpha\beta\gamma} + {}^{(0)}B^{(3)}_{\alpha\beta\gamma} \tag{52} $$

We restrict ourselves here to linear symmetry-centerless molecules. Then for the laboratory zzz component this decomposition has a relatively simple form, derived by Kielich [7]:

$$ {}^{(0)}B^{2\omega}_{zzz} = \frac{3}{5}\alpha^{2\omega}\sum_i D^1_{00}(\Omega_i) + \frac{2}{5}\beta^{2\omega}\sum_i D^3_{00}(\Omega_i) \tag{53} $$

where

$$ \alpha^{2\omega} = b^{2\omega}_{333} + 2b^{2\omega}_{113} \qquad \beta^{2\omega} = b^{2\omega}_{333} - 3b^{2\omega}_{113} \tag{53a} $$

with $b^{2\omega}_{123}$ denoting the Rayleigh hyperpolarizability tensor components in the frame of principal axes of the molecule; $D^l_{mn}(\Omega_i)$ is a Wigner rotation matrix and Ω_i stands for the set of Euler angles determining the orientation of the molecule i. To simplify our considerations we assume here a slightly different scattering geometry than in Section II. We suppose that the incident light is linearly polarized in the z direction and that z-polarized light is detected in conventional 90° scattering geometry.

From Eq. (53) we note that purely orientational hyper-Rayleigh scattering is governed by two collective orientational variables of first $Q^{(1)} = \Sigma_i D^1_{00}(\Omega_i)$ and third $Q^{(3)} = \Sigma_i D^3_{00}(\Omega_i)$ order. We can now formulate the following question: How does the interaction-induced part $\Delta B^{2\omega}_{\alpha\beta\gamma}$ of the total hyperpolarizability of the system influence the purely orientational hyper-Rayleigh scattering due to, respectively, $Q^{(1)}$ and $Q^{(3)}$? It can be easily answered with the help of the projection operator [80, 81]:

$$ P = \frac{\langle\ , A\rangle}{\langle A, A\rangle} A \tag{54} $$

which projects an arbitrary "vector" onto A. In our case we project $\Delta B^{2\omega}_{zzz}$ onto $Q^{(1)}$ and $Q^{(3)}$, respectively. Because the properties of a medium described by irreducible tensors of different ranks cannot be correlated in an isotropic fluid [80], we should project the irreducible first-rank term

$\Delta B_{zzz}^{(1)}$ of ΔB_{zzz} onto $Q^{(1)}$, and the third-rank term $\Delta B_{zzz}^{(3)}$ onto $Q^{(3)}$. Then we can write the irreducible first-rank term $B_{zzz}^{(1)}$ of the *total* hyperpolarizability tensor in the form (for the sake of clarity we omit the superscript 2ω)

$$B_{zzz}^{(1)} = \frac{3}{5}\alpha^{2\omega}Q^{(1)} + \frac{\langle \Delta B_{zzz}^{(1)}, Q^{(1)} \rangle}{\langle Q^{(1)}, Q^{(1)} \rangle} Q^{(1)} + \left\{ \Delta B_{zzz}^{(1)} - \frac{\langle \Delta B_{zzz}^{(1)}, Q^{(1)} \rangle}{\langle Q^{(1)}, Q^{(1)} \rangle} Q^{(1)} \right\} \tag{55a}$$

and for the third-rank term of B_{zzz} we obtain

$$B_{zzz}^{(3)} = \frac{2}{5}\beta^{2\omega}Q^{(3)} + \frac{\langle \Delta B_{zzz}^{(3)}, Q^{(3)} \rangle}{\langle Q^{(3)}, Q^{(3)} \rangle} Q^{(3)} + \left\{ \Delta B_{zzz}^{(3)} - \frac{\langle \Delta B_{zzz}^{(3)}, Q^{(3)} \rangle}{\langle Q^{(3)}, Q^{(3)} \rangle} Q^{(3)} \right\} \tag{55b}$$

We note that in Eq. (55a) the projection of $\Delta B_{zzz}^{(1)}$ onto $Q^{(1)}$ has the same time dependence as $Q^{(1)}$ so that we can define

$$\frac{3}{5}\Delta\alpha^{2\omega} = \frac{\langle \Delta B_{zzz}^{(1)}, Q^{(1)} \rangle}{\langle Q^{(1)}, Q^{(1)} \rangle} \tag{56}$$

and assume that $\Delta\alpha^{2\omega}$ behaves like an interaction-induced correction to the individual molecule's weight-1 hyperpolarizability parameter $\alpha^{2\omega}$. Similarly, from Eq. (55b) we have

$$\frac{2}{5}\Delta\beta^{2\omega} = \frac{\langle \Delta B_{zzz}^{(3)}, Q^{(3)} \rangle}{\langle Q^{(3)}, Q^{(3)} \rangle} \tag{57}$$

and $\Delta\beta^{2\omega}$ constitutes an interaction-induced correction to the individual molecule's weight-3 hyperpolarizability parameter $\beta^{2\omega}$. The remaining parts of Eqs. (55a) and (55b) placed between curly brackets are the purely interaction-induced parts of, respectively, $\Delta B_{zzz}^{(1)}$ if $l = 1$ and $\Delta B_{zzz}^{(3)}$ when $l = 3$:

$$^{(\mathrm{PII})}\Delta B_{zzz}^{(l)} = \Delta B_{zzz}^{(l)} - \frac{\langle \Delta B_{zzz}^{(l)}, Q^{(l)} \rangle}{\langle Q^{(l)}, Q^{(l)} \rangle} Q^{(l)} \tag{58}$$

where ${}^{(\mathrm{PII})}\Delta B_{zzz}^{(l)}$ are orthogonal to $Q^{(l)}$ at equal times. They are projections of $\Delta B_{zzz}^{(l)}$ into a subspace orthogonal to $Q^{(l)}$. The ${}^{(\mathrm{PII})}\Delta B_{zzz}^{(l)}$ are responsible for a new type of hyper-Rayleigh scattering, which, by analogy to linear scattering [80], we term purely interaction-induced hyper-Rayleigh scattering. Combining Eqs. (56)–(58) we can write Eqs. (55) in a new form:

$$B_{zzz}^{(1)}(t) = \frac{3}{5}(\alpha^{2\omega} + \Delta\alpha^{2\omega})Q^{(1)}(t) + {}^{(\mathrm{PII})}\Delta B_{zzz}^{(1)}(t) \tag{59a}$$

$$B_{zzz}^{(3)}(t) = \frac{2}{5}(\beta^{2\omega} + \Delta\beta^{2\omega})Q^{(3)}(t) + {}^{(\mathrm{PII})}\Delta B_{zzz}^{(3)}(t) \tag{59b}$$

Eq. (59a) introduces the interaction-induced renormalized effective weight-one hyperpolarizability parameter $\alpha_{\mathrm{eff}}^{2\omega} = \alpha^{2\omega} + \Delta\alpha^{2\omega}$ responsible for hyper-Rayleigh scattering due to $l = 1$ symmetry orientational fluctuations in dense fluids. From Eq. (59b) the effective weight-three hyperpolarizability parameter of linear molecules responsible for hyper-Rayleigh scattering due to $l = 3$ symmetry orientational fluctuations reads $\beta_{\mathrm{eff}}^{2\omega} = \beta^{2\omega} + \Delta\beta^{2\omega}$. The intensity as well as the spectral distribution of the hyper-Rayleigh scattered light are described by the autocorrelation function $C_{zz}(t)$ of the total hyperpolarizability tensor B_{zzz}:

$$C_{zz}(t) = \langle B_{zzz}(0)B_{zzz}(t)\rangle = C_{zz}^{(1)}(t) + C_{zz}^{(3)}(t) \tag{60}$$

where

$$\begin{aligned} C_{zz}^{(1)}(t) &= \left\langle B_{zzz}^{(1)}(0)B_{zzz}^{(1)}(t)\right\rangle \\ &= \tfrac{9}{25}\left(\alpha_{\mathrm{eff}}^{2\omega}\right)^2\left\langle Q^{(1)}(0)Q^{(1)}(t)\right\rangle + \left\langle {}^{(\mathrm{PII})}\Delta B_{zzz}^{(1)}(0)\,{}^{(\mathrm{PII})}\Delta B_{zzz}^{(1)}(t)\right\rangle \\ &\quad + \tfrac{3}{5}\alpha_{\mathrm{eff}}^{2\omega}\left\langle {}^{(\mathrm{PII})}\Delta B_{zzz}^{(1)}(0)Q^{(1)}(t) + {}^{(\mathrm{PII})}\Delta B_{zzz}^{(1)}(t)Q^{(1)}(0)\right\rangle \end{aligned} \tag{60a}$$

and

$$\begin{aligned} C_{zz}^{(3)}(t) &= \left\langle B_{zzz}^{(3)}(0)B_{zzz}^{(3)}(t)\right\rangle \\ &= \tfrac{4}{25}\left(\beta_{\mathrm{eff}}^{2\omega}\right)^2\left\langle Q^{(3)}(0)Q^{(3)}(t)\right\rangle + \left\langle {}^{(\mathrm{PII})}\Delta B_{zzz}^{(3)}(0)\,{}^{(\mathrm{PII})}\Delta B_{zzz}^{(3)}(t)\right\rangle \\ &\quad + \tfrac{2}{5}\beta_{\mathrm{eff}}^{2\omega}\left\langle {}^{(\mathrm{PII})}\Delta B_{zzz}^{(3)}(0)Q^{(3)}(t) + {}^{(\mathrm{PII})}\Delta B_{zzz}^{(3)}(t)Q^{(3)}(0)\right\rangle \end{aligned} \tag{60b}$$

The integrated hyper-Rayleigh intensity is proportional to $C_{zz}(t = 0)$.

From Eq. (60) for $t = 0$ we have

$$C_{zz}^{(1)}(t=0) = \langle B_{zzz}^{(1)}(0) B_{zzz}^{(1)}(0) \rangle = \tfrac{9}{75} N \left(\alpha_{\text{eff}}^{2\omega}\right)^2 (1 + J_1) + \left\langle \left({}^{(\text{PII})}\Delta B_{zzz}^{(1)} \right)^2 \right\rangle \tag{61a}$$

$$C_{zz}^{(3)}(t=0) = \langle B_{zzz}^{(3)}(0) B_{zzz}^{(3)}(0) \rangle = \tfrac{4}{175} N \left(\beta_{\text{eff}}^{2\omega}\right)^2 (1 + J_3) + \left\langle \left({}^{(\text{PII})}\Delta B_{zzz}^{(3)} \right)^2 \right\rangle \tag{61b}$$

where we make use of the relation

$$\langle Q^{(l)} Q^{(l)} \rangle = \left\langle \sum_{i=1}^{N} \sum_{j=1}^{N} D_{00}^{l}(\Omega_i) D_{00}^{l}(\Omega_j) \right\rangle = \frac{N}{2l+1}(1 + J_l) \tag{62}$$

with

$$J_l = \left\langle \sum_{j \neq i} P_l\left(\hat{k}_i \cdot \hat{k}_j\right) \right\rangle \tag{62$'$}$$

$\hat{k}$ is a unit vector along the main molecular axis of symmetry and $P_l(x)$ stands for the Legendre polynomial of lth order.

B. Analytical Expressions for $\Delta\alpha^{2\omega}$ and $\Delta\beta^{2\omega}$

We now derive formulas describing the interaction-induced changes $\Delta\alpha^{2\omega}$ and $\Delta\beta^{2\omega}$. Spherical coordinates are more convenient at this point. The transformation of the Cartesian component ΔB_{zzz} into spherical representation is straightforward:

$$\Delta B_{zzz}^{(1)} = -\mathrm{i}\left\{ \frac{\sqrt{3}}{3} B_0^{(0,1)} + \frac{2\sqrt{15}}{15} B_0^{(2,1)} \right\} \tag{63a}$$

$$\Delta B_{zzz}^{(3)} = \mathrm{i}\sqrt{\frac{2}{5}}\, B^{(2,3)} \tag{63b}$$

whereas Eq. (50a) can be transformed into spherical representation as

$$ {}^{(a)}\Delta B_{\alpha}^{(b,l)} = \sum_{i,j} {}^{(a)}_{(ij)}\Delta B_{\alpha}^{(b,l)} \tag{64} $$

where

$$ {}^{(a)}_{(ij)}\Delta B_{\alpha}^{(b,l)} = \frac{\sqrt{5}}{4} \sum_{g,h,f} (-)^{h+l+b} \Pi_{ghf} \begin{Bmatrix} g & h & l \\ 1 & b & 1 \end{Bmatrix} \begin{Bmatrix} 1 & 1 & 2 \\ 1 & h & f \end{Bmatrix} \times \left\{ {}_{(i)}\mathbf{b}^{(b,g)} \otimes \left[\mathbf{T}_{ij}^{(2)} \otimes {}_{(j)}\boldsymbol{\alpha}^{(f)} \right]^{(h)} \right\}_{\alpha}^{(l)} \tag{65} $$

and $\otimes$ stands for an irreducible tensor product.

From Eqs. (56), (57), and (63) we note that to calculate $\Delta\alpha^{2\omega}$ and $\Delta\beta^{2\omega}$ we should first calculate the correlation function of the type

$$ \left\langle Q^{(l)(a)}\Delta B_{0}^{(b,l)} \right\rangle = \left\langle \sum_{s} D_{00}^{l}(\Omega_s) \sum_{i,j} {}^{(a)}_{(ij)}\Delta B_{0}^{(b,l)} \right\rangle \tag{66} $$

The correlation function (66) consists of two binary terms with $s = i \neq j$ and $s = j \neq i$, as well as a ternary term when $i \neq j \neq s$. For the linear molecules considered here the essential part of the binary term of Eq. 66, when $s = i \neq j$, reads

$$ \left\langle \sum_{i,j} D_{00}^{l}(\Omega_i) \left\{ \mathbf{Y}^{(g)}(\Omega_i) \otimes \left[\mathbf{Y}^{(2)}(\Omega_{ij}) \otimes \mathbf{Y}^{(f)}(\Omega_j) \right]^{(h)} \right\}_{0}^{(l)} r_{ij}^{-3} \right\rangle \tag{67} $$

We now calculate the average value of Eq. (67) using for this purpose a pair correlation function $g\,(\mathbf{r}_{12}, \Omega_1, \Omega_2)$. Since Eq. (67) is written in the laboratory frame of reference, it is very convenient to expand the pair distribution function in that reference system as well. For linear molecules this expansion reads [83]

$$ g(\mathbf{r}_{12}, \Omega_1, \Omega_2) = \sum_{l_1,l_2,l} \sum_{m_1,m_2,m} g(l_1 l_2 l, r_{12}) \begin{bmatrix} l_1 & l_2 & l \\ m_1 & m_2 & m \end{bmatrix} \times Y_{m_1}^{(l_1)}(\Omega_1) Y_{m_2}^{(l_2)}(\Omega_2) Y_{m}^{(l)}(\Omega_{12})^{*} \tag{68} $$

Then, the mean value occurring in Eq. (67) is calculated as

$$\langle \cdots \rangle = \frac{N(N-1)}{(4\pi)^2 V} \iiint D^l_{00}(\Omega_1)\{\mathbf{Y}^{(g)}(\Omega_1)$$
$$\otimes [\mathbf{Y}^{(2)}(\Omega_{12}) \otimes \mathbf{Y}^{(f)}(\Omega_2)]^{(h)}\}^{(l)}_0 r_{12}^{-3} g(\mathbf{r}_{12}, \Omega_1, \Omega_2)\, d\Omega_1\, d\Omega_2\, d\mathbf{r}_{12} \tag{69}$$

Using the formula for the product of two spherical harmonics with the same arguments [82]:

$$Y^{(l)}_m(\Omega) Y^{(k)}_n(\Omega) = \sum_{L,M} \frac{\Pi_{lk}}{(4\pi)^{1/2}\Pi_L} \begin{bmatrix} l & k & L \\ 0 & 0 & 0 \end{bmatrix}$$
$$\times \begin{bmatrix} l & k & L \\ m & n & M \end{bmatrix} Y^{(L)}_M(\Omega) \tag{70}$$

and the orthogonality conditions of spherical harmonics for Eq. (69) we obtain the following relatively simple formula:

$$\langle \cdots \rangle = N\rho \sum_h (-)^{h+1} \frac{\Pi_{g2}}{\Pi_{hl}} \begin{bmatrix} l & g & h \\ 0 & 0 & 0 \end{bmatrix} H_3(h, f, 2) \tag{71}$$

where we have introduced the molecular parameter $H_3(a, b, c)$ based on the expansion coefficients of Eq. (68) and defined as

$$H_3(a, b, c) = 4\pi \int g(abc, r_{12}) r_{12}^{-n+2}\, dr_{12} \tag{72}$$

and $\rho = N/V$. A similar expression is obtained for the second binary term of Eq. (66) when $s = j \neq i$:

$$\left\langle \sum_{i,j} D^l_{00}(\Omega_j)\left\{\mathbf{Y}^{(g)}(\Omega_i) \otimes [\mathbf{Y}^{(2)}(\Omega_{ij}) \otimes \mathbf{Y}^{(f)}(\Omega_j)]^{(h)}\right\}^{(l)}_0 r_{ij}^{-3} \right\rangle$$
$$= N\rho \sum_j (-)^{g+l} \frac{\Pi_{fh2}}{\Pi_l} \begin{bmatrix} l & f & j \\ 0 & 0 & 0 \end{bmatrix} \begin{Bmatrix} f & j & l \\ g & h & 2 \end{Bmatrix} H_3(g, j, 2) \tag{73}$$

Finally, the binary terms of Eq. (66) read as follows:

when $s = i \neq j$

$$\left\langle \sum_{i,j} D^l_{00}(\Omega_i)^{(a)}_{(ij)} \Delta B_0^{(k,l)} \right\rangle$$

$$= \frac{N\rho\sqrt{30}}{4(4\pi)^{1/2}} \sum_{g,f,h} (-)^{k+h} \frac{\Pi_g}{\Pi_l} \begin{Bmatrix} g & h & l \\ 1 & k & 1 \end{Bmatrix} \tag{74}$$

$$\times \begin{Bmatrix} 2 & 1 & 1 \\ 1 & h & f \end{Bmatrix} \begin{bmatrix} g & h & l \\ 0 & 0 & 0 \end{bmatrix} H_3(h, f, 2) b_0^{(k,g)} \alpha_0^{(f)}$$

when $s = j \neq i$

$$\left\langle \sum_{i,j} D^l_{00}(\Omega_j)^{(a)}_{(ij)} \Delta B_0^{k,l)} \right\rangle = \frac{N\rho\sqrt{30}}{4(4\pi)^{1/2}} \sum_{g,f,j} (-)^{1+j} \frac{\Pi_f}{\Pi_l} \begin{Bmatrix} j & f & l \\ 1 & k & 1 \end{Bmatrix}$$

$$\times \begin{Bmatrix} 2 & 1 & 1 \\ k & g & j \end{Bmatrix} \begin{bmatrix} f & l & j \\ 0 & 0 & 0 \end{bmatrix} H_3(g, j, 2) b_0^{(k,g)} \alpha_0^{(f)} \tag{75}$$

The third term of Eq. (66) is the three-body term with $s \neq i \neq j$. We have to average this term with the triple correlation function $g^{(3)}$:

$$\left\langle \sum_s D^l_{00}(\Omega_s) \sum_{i,j} {}^{(a)}_{(ij)} \Delta B_0^{(k,l)} \right\rangle_{g^{(3)}(i,j,s)} \tag{76}$$

Analytically, this can be done only approximately. We follow here the approach proposed by Keyes and Ladanyi [85] in their theory of linear light scattering. The three-body correlation function $g^{(3)}(i, j, k)$ is written in the Kirkwood approximation as

$$g^{(3)}(i, j, s) = g^{(2)}(i, j) g^{(2)}(i, s) g^{(2)}(j, s) \tag{77}$$

Introducing the total correlation function $h(i, j)$ defined as

$$h(i, j) = g^{(2)}(i, j) - 1 \tag{78}$$

we obtain

$$g^{(3)}(i, j, s) = h(i, j)h(i, s) + h(i, j)h(j, s) + h(i, s)h(j, s) \times h(i, j)h(i, s)h(j, s) + h(i, j) + h(i, s) + h(j, s) + 1 \tag{79}$$

Obviously, the above transformation does not simplify the problem of calculating the three-body mean value of Eq. (76), but it suggests a possible approximate calculation. The mean value can be calculated analytically in a form suitable for numerical computations if we consider only the first two terms of Eq. (79). If we consider the last four terms of Eq. (79) the mean value (76) vanishes. However, the third and fourth terms of Eq. (79) lead to very complicated infinite series, practically useless for numerical estimations. Ladanyi and Keyes [85] refer to these terms as "irreducible connected" and propose to disregard them. Madden and Tildesley [86] also neglect them, arguing that the remaining part of Eq. (79) contains the lowest part of the h bonds. Here, we too shall apply this approximation. We then have

$$\left\langle \sum_{s} D^{l}_{00}(\Omega_s) \sum_{i,j} {}^{(a)}_{(ij)}\Delta B^{(k,l)}_{0} \right\rangle_{h(i,j)h(i,s)} = \left\langle \sum_{i,j} D^{l}_{00}(\Omega_i)^{(a)}_{(ij)}\Delta B^{(k,l)}_{0} \right\rangle J_l \tag{80}$$

$$\left\langle \sum_{s} D^{l}_{00}(\Omega_s) \sum_{i,j} {}^{(a)}_{(ij)}\Delta B^{(k,l)}_{0} \right\rangle_{h(i,j)h(j,s)} = \left\langle \sum_{i,j} D^{l}_{00}(\Omega_j)^{(a)}_{(ij)}\Delta B^{(k,l)}_{0} \right\rangle J_1 \tag{81}$$

Thus, finally taking into account in (66) the two-body terms of Eqs. (74) and (75) as well as the three-body contributions (80) and (81), we have

$$\begin{aligned}
\left\langle Q^{(l)(a)}\Delta B_0^{(k,l)}\right\rangle &= \left\langle \sum_s D^l_{00}(\Omega_s) \sum_{i,j} {}^{(a)}_{(ij)}\Delta B_0^{(k,l)}\right\rangle \\
&= \Big\{\langle \sum_{i,j} D^l_{00}(\Omega_i)^{(a)}_{(ij)}\Delta B_0^{k,l)}\rangle \\
&\quad + \langle \sum_{i,j} D^l_{00}(\Omega_j)^{(a)}_{(ij)}\Delta B_0^{(k,l)}\rangle\Big\}\ (1 + J_l)
\end{aligned} \tag{82}$$

Combining Eqs. (56), (74), (75), (80), and (81) and neglecting frequency dispersion we have for $\Delta\alpha^{2\omega}$

$$\begin{aligned}
\Delta\alpha^{2\omega} = \frac{3}{4}\rho\Big\{ &-\frac{11}{27}H_3(2,0,2)\alpha(b_{333} + 2b_{113}) \\
&-\frac{2}{9}H_3(2,0,2)\gamma(b_{333} + 2b_{113}) \\
&+\frac{11\sqrt{70}}{405}H_3(2,2,2)\gamma(b_{333} + 2b_{113}) \\
&-\frac{4}{45}H_3(2,0,2)\alpha(b_{333} - 3b_{113}) \\
&+\frac{2\sqrt{70}}{225}H_3(2,2,2)\gamma(b_{333} - 3b_{113}) \\
&-\frac{3\sqrt{30}}{45}H_3(1,1,2)\alpha(b_{333} + 2b_{113}) \\
&+\frac{4\sqrt{5}}{45}H_3(3,1,2)\alpha(b_{333} - 3b_{113}) \\
&-\frac{28\sqrt{30}}{675}H_3(1,1,2)\gamma(b_{333} + 2b_{113}) \\
&+\frac{8\sqrt{5}}{225}H_3(1,3,2)\gamma(b_{333} + 2b_{113}) \\
&-\frac{4\sqrt{105}}{1125}H_3(3,3,2)\gamma(b_{333} - 3b_{113})\Big\}
\end{aligned} \tag{83}$$

whereas for $\Delta\beta^{2\omega}$ we obtain

$$\begin{aligned}
\Delta\beta^{2\omega} = \frac{3}{4}\rho\Bigg\{ & -\frac{6}{35}H_3(2,0,2)\alpha(b_{333}+2b_{113}) \\
& -\frac{4}{35}H_3(2,0,2)\alpha(b_{333}-3b_{113}) - \frac{2}{21}H_3(0,2,2)\gamma(b_{333}-3b_{113}) \\
& +\frac{\sqrt{70}}{175}H_3(2,2,2)\gamma(b_{333}+2b_{113}) \\
& +\frac{\sqrt{70}}{175}H_3(2,2,2)\gamma(b_{333}-3b_{113}) \\
& +\frac{\sqrt{30}}{525}H_3(1,1,2)\gamma(b_{333}+2b_{113}) \\
& +\frac{2\sqrt{5}}{35}H_3(1,3,2)\alpha(b_{333}+2b_{113}) \\
& -\frac{2\sqrt{105}}{245}H_3(3,3,2)\alpha(b_{333}-3b_{113}) \\
& +\frac{4\sqrt{5}}{525}H_3(1,3,2)\gamma(b_{333}+2b_{113}) \\
& +\frac{6\sqrt{5}}{175}H_3(3,1,2)\gamma(b_{333}-3b_{113}) \\
& -\frac{8\sqrt{105}}{3675}H_3(3,3,2)\gamma(b_{333}-3b_{113})\Bigg\}
\end{aligned} \tag{84}$$

Where α denotes the trace of the linear palanizability tensor a divided over 3; $\alpha = \mathrm{Tr}(a)/3$; whereas γ denotes its amisotropy, $\gamma = a_{33} - a_{n}$. The laboratory frame expansion (68) of the pair correlation function offers the least cumbersome path leading to Eqs. (83) and (84). However, as a consequence of Eq. (68), the intermolecular parameters $H_3(a,b,c)$ are defined in the laboratory frame. The pair correlation function itself takes the simplest form if the molecular orientations are referred to the intermolecular frame, in which the z axis is parallel to the vector $\mathbf{r}_{12}$ connecting the centers of mass of the molecules. Then instead of Eq. (68) we have

$$g^{(2)}(r_{12},\Omega_1^{12},\Omega_2^{12}) = \sum_{l_1,l_2,m} g(l_1l_2m;r_{12})Y_m^{(l_1)}(\Omega_1^{12})Y_{-m}^{(l_2)}(\Omega_1^{12}) \tag{85}$$

The above spherical coefficients $g(l_1l_2m;r_{12})$ are usually calculated in theoretical [88] and computer simulation [87] modeling of $g^{(2)}$. It is then desirable to express our formulas for $\Delta\alpha^{2\omega}$ and $\Delta\beta^{2\omega}$ through the intermolecular frame coefficients $g(l_1l_2m; r_{12})$. Comparing (68) and (85) it is easy to find the relation between the coefficients of $g^{(2)}$ in both frames [83]:

$$g(l_1l_2l; r_{12}) = \sum_m \left(\frac{4\pi}{2l+1}\right)^{1/2} \begin{bmatrix} l_1 & l_2 & l \\ m & -m & 0 \end{bmatrix} g(l_1l_2m; r_{12}) \tag{86}$$

Let us define the following intermolecular parameter:

$$J_n(l_1l_2m) = 4\pi \int g(l_1l_2m; r_{12}) r_{12}^{-n+2} \, \mathrm{d}r_{12} \tag{87}$$

Then, by means of Eq. (86) we express the laboratory frame $H_3(a, b, c)$ coefficients of Eqs. (83) and (84) by way of the above intermolecular ones. We thus obtain

$$H_3(2,0,2) = \left(\frac{4\pi}{5}\right)^{1/2} J_3(2,0,0) \tag{88a}$$

$$H_3(2,2,2) = 2\left(\frac{4\pi}{70}\right)^{1/2} [2J_3(2,2,2) + J_3(2,2,1) - J_3(2,2,0)] \tag{88b}$$

$$H_3(1,1,2) = 2\left(\frac{4\pi}{30}\right)^{1/2} [J_3(1,1,0) + J_3(1,1,1)] \tag{88c}$$

$$H_3(3,1,2) = \left(\frac{4\pi}{35}\right)^{1/2} \left[-\sqrt{3}J_3(3,1,0) + 2\sqrt{2}J_3(3,1,1)\right] \tag{88d}$$

$$H_3(3,3,2) = \left(\frac{4\pi}{105}\right)^{1/2} [2J_3(3,3,0) - 3J_3(3,3,1) + 5J_3(3,3,3)] \tag{88e}$$

The first two intermolecular parameters (88a) and (88b) are analogous to the parameters τ_{20} and τ_{22} of linear scattering proposed by Keyes and Ladanyi [85]:

$$H_3(2,0,2) = \frac{1}{5}\sqrt{4\pi}\,\tau_{20} \qquad H_3(2,2,2) = \sqrt{4\pi}\left(\frac{2}{35}\right)^{1/2} \tau_{22} \tag{89}$$

The last three combinations (88c)–(88e) are novel intermolecular parameters, specific to hyper-Rayleigh scattering and reflecting the nonlinear nature of that scattering as well as the "dipolar" character of the scattering medium. When the spherical harmonic components of the pair correlation function (85) (or at least several first components active in Eq. (88)) are available either from computer simulation [87] or from theoretical calculations [88], Eqs. (83), (84), and (88) enable us to calculate, for the first time, the interaction-induced changes $\Delta\alpha^{2\omega}$ and $\Delta\beta^{2\omega}$ characterizing the effective hyperpolarizability of the scattering system.

Spectra due to, respectively, the weight-1 and weight-3 parts of the hyperpolarizability tensor can be determined separately as [14]

$$I^{(1)}(\Delta\omega) = I_{\parallel}^{2\omega}(\Delta\omega) - \tfrac{3}{2} I_{\perp}^{2\omega}(\Delta\omega) \tag{90a}$$

$$I^{(3)}(\Delta\omega) = I_{\perp}^{2\omega}(\Delta\omega) - \tfrac{1}{9} I_{\parallel}^{2\omega}(\Delta\omega) \tag{90b}$$

from measurements of the polarized $I_{\parallel}^{2\omega}(\Delta\omega)$ and depolarized $I_{\perp}^{2\omega}(\Delta\omega)$ components of the hyper-Rayleigh radiation. The first-order spectrum (90a) is described by the effective weight-1 part α_{eff} of the molecular hyperpolarizability as follows:

$$I^{(1)}(\Delta\omega) = A^{2\omega}\alpha_{\text{eff}}^{2} \frac{2\tau_c^{(1)}}{1 + \left(\Delta\omega\tau_c^{(1)}\right)^2}(1 + J_1) \tag{91a}$$

The third-order spectrum (90b) is determined by the effective weight-3 part β_{eff} of the molecular hyperpolarizability as

$$I^{(3)}(\Delta\omega) = A^{2\omega}\beta_{\text{eff}}^{2} \frac{2\tau_c^{(3)}}{1 + \left(\Delta\omega\tau_c^{(3)}\right)^2}(1 + J_3) \tag{91b}$$

where $\tau_c^{(n)}$ denotes the collective orientational relaxation time of rank n of the form [89]:

$$\tau_c^{(n)} = \tau_s^{(n)} \frac{(1 + J_n)}{(1 + g_n)} \tag{92}$$

where $\tau_s^{(n)}$ stands for the nth- order single-molecule orientational relaxation time, J_n for the static orientational correlation parameter (62′), and g_n describes collective dynamic orientational correlations [89] of the nth order. By means of Eqs. (90) and (91) we could measure, respectively, the weight-1 and weight-3 parts $\alpha_{\text{eff}}^2(1 + J_1)$ and $\beta_{\text{eff}}^2(1 + J_3)$ of the effective hyperpolarizability of a molecule surrounded by its high-density neighbors. However, this is no easy task. The total weight-1 hyper-Rayleigh spectrum

(60a) is a superposition of the orientational spectrum (91a), the purely interaction-induced spectrum, and the cross-orientational–purely interaction-induced spectrum. The orientation–fluctuation-induced spectrum should be extracted in some way from the experimental spectrum. In general, this is very difficult. However, if the single-molecule reorientational motion relaxes slowly in comparison with the interaction-induced light-scattering mechanisms, the orientational spectrum can be separated as a line relatively narrow in comparison with the spectrum resulting from the fluctuating interaction-induced hyperpolarizability. In this situation of time-scale separation between the reorientational and interaction-induced relaxations (when interaction-induced changes in $B_{\alpha\beta\gamma}$ decay much faster than orientational fluctuations) the cross-orientational–purely interaction-induced contribution to the experimental spectrum is negligible. The purely interaction-induced contribution to the final spectrum is determined as an exponent [90] with intensity and slope taken from the far wings of the experimental spectrum, which are believed to be almost interaction-induced in their origin. This procedure finally allows us to extract from the experimental spectra $I^{(1)}(\Delta\omega)$ and $I^{(3)}(\Delta\omega)$ their reorientational parts (91a) and (91b) and consequently study α_{eff} and β_{eff}.

Similar considerations should be performed for vibrational hyper-Raman scattering. In that case we have to start from the total Raman hyperpolarizability tensor of the system for a normal vibration S

$$(B_{\alpha\beta\gamma})_{\text{RAM}} = \sum_i \left(\frac{\partial_{(i)} b_{\alpha\beta\gamma}}{\partial Q} \right)_0 Q_i S + (\Delta B_{\alpha\beta\gamma})_{\text{RAM}} \tag{93}$$

instead of the total Rayleigh hyperpolarizability tensor (48). Further steps are analogous to the hyper-Rayleigh case.

IV. CONCLUSION

We have been applying throughout the method of Racah algebra in our analysis of the various aspects of multiharmonic light scattering. It permits the determination of the intensity and polarization state of the spectral lines for arbitrary directions of observation. It is applicable to classical as well as quantum-mechanical calculations; in both cases, the calculations proceed similarly.

Carrying out quantum calculations we have determined the intensities of various rotational and vibrational–rotational lines of second- and third-harmonic scattering. We have determined the depolarization ratio and reversal coefficient in their dependence on the scattering angles. We have also determined the selection rules for the permitted vibrational and rotational transitions in molecules according to the type of scattering, the symmetry of the molecule, and the state of polarization of the incident light wave.

On the basis of the formulae derived by us for the scattered light intensities we have calculated the rotational structure of the hyper-Rayleigh spectrum from gaseous CO, for two sets of the hyperpolarizability tensor components, and for NO. We discussed the influence of interaction induced effects on the hyper-Rayleigh and hyper-Raman scattered light and introduced the concept of the effective hyperpolarizability of a molecule in a liquid medium. We derived an analytical expression for the effective hyperpolarizability and discussed the experimental conditions for its measurement.

Acknowledgments

We express our gratitude to Stanisław Kielich for his guidance and many helpful and stimulating discussions throughout the years. This work was supported by Grant 2-0129-91-01 of the Polish Commission for Scientific Studies.

References

1. P. A. Franken, A. E. Hill, C. W. Peters, and G. Weinreich, *Phys. Rev. Lett.* **7**, 118 (1961).
2. J. Blaton, *Z. Phys.* **69**, 835 (1931).
3. S. Kielich, *Bull. Acad. Pol. Sci. Sér. Math. Astron. Phys.* **12**, 53 (1964).
4. S. Kielich, *Acta Phys. Pol.* **24**, 135 (1964).
5. S. Kielich, *Physica* **30**, 1717 (1964); *J. Phys.* **28**, 519 (1967).
6. S. Kielich, *Proc. Phys. Soc.* **86**, 709 (1965).
7. S. Kielich, *Nonlinear Molecular Optics*, Nauka, Moscow, 1981.
8. R. W. Terhune, P. D. Maker, and C. M. Savage, *Phys. Rev. Lett.* **14**, 681 (1965).
9. I. Freund, *Phys. Rev. Lett.* **21**, 1404 (1968).
10. G. Dolino, J. Lajzerowicz, and M. Vallade, *Solid State Commun.* **7**, 1005 (1969).
11. H. Vogt and H. Uwe, *Phys. Rev. B* **29**, 1030 (1984); H. Vogt, *Phys. Rev. B* **36**, 5001 (1987).
12. R. Bersohn, Y. H. Pao, and H. L. Frisch, *J. Chem. Phys.* **45**, 3184 (1966).
13. D. L. Weinberg, *J. Chem. Phys.* **47**, 1307 (1967).
14. P. D. Maker, *Phys. Rev.* **A1**, 923 (1970).
15. J. F. Verdieck, S. H. Peterson, M. Savage and P. D. Maker, *Chem. Phys. Lett.* **7**, 219 (1970).
16. S. Kielich, J. R. Lalanne, and F. B. Martin, *Phys. Rev. Lett.* **26**, 1295 (1971).
17. S. Kielich, J. R. Lalanne, and F. B. Martin, *Acta Phys. Pol.* **A41**, 479 (1972).
18. S. Kielich, J. R. Lalanne,and F. B. Martin, *J. Raman Spectrosc.* **1**, 119 (1973).
19. L. A. Nafie and W. L. Peticolas, *J. Chem. Phys.* **57**, 3145 (1972).
20. S. J. Cyvin, J. E. Rauch, and J. C. Decius, *J. Chem. Phys.* **43**, 4083 (1965).

21. Z. Ożgo, *Acta Phys. Pol.* **34**, 1087 (1968).
22. J. H. Christie and D. J. Lookwood, *J. Chem. Phys.* **54**, 1141 (1971).
23. V. L. Strizhevskii, *Kvantovaya Elektronika (Moskva)* **6**, 165 (1972).
24. P. D. Maker, in *Physics of Quantum Electronics*, McGraw-Hill, New York, 1966.
25. L. Stanton, *J. Raman Spectrosc.* **1**, 53 (1973).
26. S. Yatsiv, M. Rokni, and S. Barak, *IEEE J. Quant. Electron.* **4**, 900 (1968).
27. W. Yu and R. R. Alfano, *Phys. Rev.* **A11**, 188 (1975).
28. M. J. French and D. A. Long, in R. F. Barrow, D. A. Long and J. Sheridan (Eds.), *Specialist Periodical Report: Molecular Spectroscopy*, Vol. 4, Chemical Society, London, 1976.
29. D. A. Long, *Raman Spectroscopy*, McGraw-Hill, London, 1977.
30. M. J. French, in A. B. Harvey (Ed.), *Chemical Applications of Nonlinear Raman Spectroscopy*, Academic, New York, 1981.
31. R. Bonneville and D. S. Chemla, *Phys. Rev.* **A17**, 2046 (1978).
32. J. Jerphagnon, D. S. Chemla, and R. Bonneville, *Adv. Phys.* **27**, 609 (1978).
33. S. Kielich, *Prog. Opt.* **20**, 155 (1983).
34. S. Kielich, *Acta Phys. Pol.* **33**, 89 (1968).
35. S. Kielich, M. Kozierowski, and J. R. Lalanne, *J. Phys.* **34**, 1005 (1975).
36. M. Kozierowski and S. Kielich, *Acta Phys. Pol.* **A66**, 753 (1984).
37. S. Shin and M. Ishigame, *J. Chem. Phys.* **89**, 1892 (1988).
38. K. Clays and A. Persoons, *Phys. Rev. Lett.* **66**, 2980 (1991).
39. S. Kielich and M. Kozierowski, *Acta Phys. Pol.* **A38**, 271 (1970); **A66**, 753 (1974).
40. S. Kielich, *Acta Phys. Pol.* **30**, 393 (1966).
41. S. Kielich, *Chem. Phys. Lett.* **1**, 441 (1967).
42. S. Kielich, *Acta Phys. Pol.* **33**, 141 (1968).
43. S. Kielich and M. Kozierowski, *Opt. Commun.* **4**, 395 (1972).
44. E. Pascaud and G. Poussigue, *Can. J. Phys.* **56**, 1577 (1978).
45. D. L. Andrews and T. Thirunamachandran, *J. Chem. Phys.* **68**, 2941 (1978).
46. D. L. Andrews and M. J. Harlow, *Mol. Phys.* **49**, 937 (1983).
47. G. C. Nieman, *J. Chem. Phys.* **75**, 584 (1981).
48. R. A. Minard, G. E. Stedman, and A. G. McLellan, *J. Chem. Phys.* **78**, 5016 (1983); G. E. Stedman, *Adv. Phys.* **34**, 513 (1985).
49. (a) M. Kozierowski, Z. Ożgo, and S. Kielich, *J. Chem. Phys.* **84**, 5271 (1986); (b) M. Kozierowski, *Phys. Rev.* **A31**, 509 (1985).
50. A. V. Baranov, Y. S. Bobovich, and V. I. Petrov, *JETF* **88**, 741 (1985); *Uspiekhi Fizicheskikh Nauk* **160**, 35 (1990).
51. D. V. Murphy, K. U. Von Raben, R. K. Chang, and P. B. Dorain, *Chem. Phys. Lett.* **85**, 43 (1982).
52. T. Golab, J. R. Sprague, K. T. Carron, G. C. Schatz, and R. P. Van Duyne, *J. Chem. Phys.* **88**, 7942 (1988).
53. D. A. Long, M. J. French, T. J. Dines, and R. J. B. Hall, *J. Phys. Chem.* **88**, 547 (1984).
54. M. Kozierowski, Z. Ożgo, and R. Zawodny, *Mol. Phys.* **59**, 1227 (1986).
55. Yu. A. Ilinskii and W. D. Taranuckin, *Kvantovaya Elektronika (Moskva)* **1**, 1799 (1974).

56. (a) L. D. Ziegler, Y. C. Chung, and Y. P. Zhang, *J. Chem. Phys.* **87**, 4498 (1987); (b) Y. C. Chung and L. D. Ziegler, *J. Chem. Phys.* **89**, 4692 (1988); (c) L. D. Ziegler, *J. Raman Spectrosc.* **21**, 769 (1990).
57. V. N. Denisov, B. N. Mavrin, and V. B. Podobedov, *Phys. Rep.* **151**, 1 (1987).
58. Ch. C. Bonang and S. M. Cameron., *Chem. Phys. Lett.* **187**, 619 (1991); **192**, 303 (1992).
59. D. A. Long and L. Stanton, *Proc. R. Soc. London, Ser. A* **318**, 441 (1970).
60. S. Nie, L. A. Lipscomb, S. Feng, and N. T. Yu, *Chem. Phys. Lett.* **167**, 35 (1990).
61. S. Kielich and Z. Ożgo, *Opt. Commun.* **8**, 417 (1973).
62. T. Bancewicz, Z. Ożgo, and S. Kielich, *J. Raman Spectrosc.* **1**, 177 (1973).
63. T. Bancewicz, Z. Ożgo, and S. Kielich, *Phys. Lett.* **44A**, 407 (1973).
64. W. Alexiewicz, T. Bancewicz, S. Kielich, and Z. Ożgo, *J. Raman Spectrosc.* **2**, 529 (1974).
65. Z. Ożgo and S. Kielich, *Physica* **72**, 191 (1974).
66. T. Bancewicz, Z. Ożgo, and S. Kielich, *Acta Phys. Pol.* **A47**, 645 (1975).
67. T. Bancewicz and S. Kielich, *Mol. Phys.* **31**, 615 (1976).
68. T. Bancewicz, S. Kielich, and Z. Ożgo, in M. Balkanski (Ed.), *Light Scattering in Solids*, Flammarion, Paris, 1976.
69. T. Bancewicz and S. Kielich, *J. Raman Spectrosc.* **21**, 207 (1990).
70. S. Kielich and T. Bancewicz, *J. Raman Spectrosc.* **21**, 791 (1990).
71. G. Placzek, *Handb. Radiol.* **6**, 205 (1934).
72. A. R. Edmonds, *Angular Momentum in Quantum Mechanics*, Princeton UP, Princeton, NJ, 1957.
73. L. Boyle, *Int. J. Quant. Chem.* **3**, 231 (1969).
74. (a) Z. Ożgo and S. Kielich, *Physica* **81C**, 151 (1976); (b) K. Altman and G. J. Strey, *J. Raman Spectrosc.* **12**, 1, (1982).
75. Z. Ożgo, *Multi-Harmonic Molecular Light Scattering in the Approach of Racah Algebra*, Adam Mickiewicz University, Poznań, 1975 (in Polish).
76. Z. Ożgo and T. Bancewicz, to be published.
77. A. M. Leushin, *Tablitsy Funktsii Preobrazuyushchikhsia po Neprivodimym Predstavleniyam Kristallograficheskikh Tochechnikh Grupp*, (*Tables of Functions Transforming as Irreducible Representations of the Crystallographical Point Groups*), Moskva, 1968 (in Russian).
78. N. S. Hush and M. L. Williams, *Theor. Chim. Acta* **25**, 346 (1972).
79. J. M. O'Hare and R. P. Hurst, *J. Chem. Phys.* **46**, 2356 (1967).
80. D. Frenkel and J. P. McTague, *J. Chem. Phys.* **72**, 2801 (1980).
81. T. Bancewicz, S. Kielich, and W. A. Steele, *Mol. Phys.* **54**, 637 (1985).
82. D. A. Varshalovich, A. N. Moskalev, and V. K. Khersonskii, *Kvantovaya Teoria Uglovogo Momenta*, Nauka, Leningrad, 1975.
83. C. G. Gray and K. E. Gubbins, *Theory of Molecular Fluids*, Clarendon, Oxford, 1984.
84. B. Ladanyi and T. Keyes, *Mol. Phys.* **33**, 1063 (1977); **33**, 1247 (1977).
85. T. Keyes and B. Ladanyi, *Adv. Chem. Phys.* **56**, 411 (1984).
86. P. A. Madden and D. J. Tildesley, *Mol. Phys.* **55**, 969 (1985).
87. W. B. Streett and D. J. Tildesley, *Proc. R. Soc. London Ser. A* **355**, 239 (1977).
88. J. R. Sweet and W. A. Steele, *J. Chem. Phys.* **50**, 688 (1969).
89. B. J. Berne and R. Pecora, *Dynamic Light Scattering*, Wiley, New York, 1976.
90. T. I. Cox, M. R. Battaglia, and P. A. Madden, *Mol. Phys.* **38**, 1539 (1979).

POLARIZATION PROPERTIES OF HYPER-RAYLEIGH AND HYPER-RAMAN SCATTERINGS

M. KOZIEROWSKI

Nonlinear Optics Division, Institute of Physics, Adam Mickiewicz University, Poznań, Poland

CONTENTS

I. INTRODUCTION

Hyper-Rayleigh and hyper-Raman scatterings belong to nonlinear three-photon processes and consist of the annihilation of two photons ω of the incident light and the spontaneous emission of a third photon with the frequency ω_s fulfilling, with accuracy to the spectral width, the following relation: $\omega_s = 2\omega \pm \omega_{ij}$, where ω_{ij} is a frequency arising from the transition between the quantum vibration states i and j of the molecules. If $i = j$ we deal with hyper-Rayleigh (elastic second-harmonic) scattering, while for $i \neq j$ hyper-Raman (inelastic second-harmonic) scattering takes place. Three-photon light scattering was predicted in the early days of

Modern Nonlinear Optics, Part 1, Edited by Myron Evans and Stanisław Kielich. Advances in Chemical Physics Series, Vol. LXXXV.
ISBN 0-471-57546-1

quantum mechanics and its history has been widely presented by Altmann and Strey [1]. The invention of the laser and the progress achieved in detection technique has rendered possible the observation of this weak scattering and has also given a new impulse to its further discussion.

The first successful experiment in which hyper-Rayleigh and hyper-Raman scatterings were observed was carried out by Terhune et al. [2]. Weinberg [3] and Maker [4] studied, respectively, the temperature dependence and the spectral broadening of hyper-Rayleigh scattering. Kielich et al. [5] studied cooperative elastic second-harmonic light scattering by regions of short-range order of centrosymmetric noncentrally distributed molecules. The same process caused by angular fluctuations near the critical point was observed by Freund [6]. Of late, Dolino et al. [7] investigated this scattering by domains in ferroelectric crystals.

The theory of three-photon scattering for molecular gases was developed by Kielich [8] and Li [9] and then extended to dense media by Bersohn et al. [10], Kielich [11], and Strizhevsky and Klimenko [12]. Cyvin et al. [13] determined the vibrational selection rules for hyper-Raman scattering and showed for the fully symmetric scattering tensor that lines inactive in infrared and usual Raman can become active in the hyper-Raman effect; this was confirmed by Verdieck et al. [14]. In turn, Christie and Lockwood [15] determined the vibrational selection rules for the scattering tensor symmetric in its last two indices. In the course of time various theoretical [16–30] and experimental [31–37] aspects of the scattering in question were studied and the achievements have been reviewed in several books and articles [1, 38–40].

II. SCATTERED LIGHT INTENSITY TENSOR

As the basis for our considerations we take a macroscopic sample of a gaseous medium; i.e., we treat its molecules as statistically independent. Moreover, we assume the linear dimensions d of the molecules as much smaller than the incident wavelength λ ($\lambda \geq 20d$). This enables us to neglect interference effects from different parts of the same molecule and to discuss the problem in the electric-dipole approximation only.

The integral intensity tensor I_{ij} of the light scattered incoherently by N noninteracting randomly oriented molecules in the wave zone, i.e., at a distance R from the center of the scattering sample much greater than the dimensions of the scattering volume, is N times that for a single molecule:

$$I_{ij} = \frac{N}{4\pi R^2 c^3} \langle \ddot{m}_i \quad \ddot{m}_j^* \rangle_{\Omega, E, Q} \tag{1}$$

where c is the light velocity and $\ddot{m}_i$ denotes the second time-derivative of the ith component of the dipole moment induced in the molecule. The symbol $\langle \cdots \rangle_{\Omega, E, Q}$ stands for appropriate statistical averagings over molecular orientations Ω of the molecule-fixed frame (indices α, β, γ) with respect to the laboratory-fixed frame (indices i, j, k), molecular vibrations Q (for Raman-type scatterings), and amplitudes E of the incident field. The rate of nonlinear processes, contrary to that of linear ones, strongly depends on the statistical properties of the incident light. The asterisk denotes the complex conjugate.

In the classical description of scattering phenomena, a molecule in a strong electromagnetic field generally undergoes nonlinear polarization. This polarization is a source of new electromagnetic waves with frequencies that are in general multiples of the incident frequency and, in the case of Raman-type processes, with field amplitudes modulated by the vibrations of the molecules. Let the sample be irradiated by an intense plane wave of wave vector $\mathbf{k}$, the electric field of which at the point $\mathbf{r}$ is taken in the form

$$E_i(\mathbf{k}, t) = E_i(t)\exp[(-\mathrm{i}(\omega t - \mathbf{kr})] \tag{2}$$

The complex amplitude $E_i(t)$ fluctuates in general. For this reason, we have to perform the above-mentioned statistical averaging over the ensemble of the incident amplitudes which replaces time averaging for ergodic processes.

At sufficiently high intensity of the incident light, the electric dipole moment induced in an isolated molecule positioned at $\mathbf{r}$ has to be considered as nonlinear in $E_i(\mathbf{k}, t)$ and can be expanded in a series of harmonic contributions:

$$\begin{aligned} m_i(t) &= a_{ij}E_j(t)\exp[(-\mathrm{i}(\omega t - \mathbf{kr})] \\ &\quad + \tfrac{1}{2}b_{ijk}E_j(t)E_k(t)\exp[-2\mathrm{i}(\omega t - \mathbf{kr})] + \cdots \end{aligned} \tag{3}$$

The summation convention over repeated indices is applied throughout. a_{ij} is the tensor of linear molecular polarizability and b_{ijk} is the tensor of nonlinear polarizability of the first order. In general, both these tensors depend on vibrations of nuclei and can be expanded in a power series in normal coordinates Q_ν of vibration

$$\begin{aligned} a_{ij} &= a_{ij}(0) + \sum_\nu a_{ij}(\nu)Q_\nu + \cdots \\ b_{ijk} &= b_{ijk}(0) + \sum_\nu b_{ijk}(\nu)Q_\nu + \cdots \end{aligned} \tag{4}$$

Rayleigh scattering with the frequency ω is related to the tensor $a_{ij}(0)$, while hyper-Rayleigh scattering with the frequency 2ω is related to the tensor $b_{ijk}(0)$. In turn, Raman and hyper-Raman scatterings are related with the derivatives $a_{ij}(\nu)$ and $b_{ijk}(\nu)$, respectively. They comprise the Stokes $\omega_s = \omega - \omega_\nu$, $\omega_s = 2\omega - \omega_\nu$ and anti-Stokes $\omega_s = \omega + \omega_\nu$, $\omega_s = 2\omega + \omega_\nu$ frequencies, where ω_ν is the frequency corresponding to the vibration mode ν.

III. PERMUTATION SYMMETRY OF THE POLARIZABILITY TENSORS

Let us briefly recall now the main features of linear Raman and Rayleigh scatterings as related to the permutation symmetry of the polarizability tensors $a_{ij}(0)$ and $a_{ij}(\nu)$. In general, both these tensors are unsymmetrical in the indices i, j and can be broken down into two parts: a part a^{s}_{ij} symmetric with respect to permutation of i and j, and a part a^{ant}_{ij} antisymmetric to permutation of i and j. The symmetry properties of the tensors a_{ij} are the same under nonresonant and resonance conditions [41, 42] and the existence of their antisymmetric parts requires complex wave functions of the molecule; this is satisfied in the presence of magnetic perturbations. For molecules not immersed in an external magnetic field, magnetic perturbations arise from the degeneracy of the fundamental electronic level of the molecule and spin–orbit coupling. On introducing irreducible Cartesian sets of the tensor a_{ij}, we find that its symmetric part consists of two sets of weight 0 (isotropic) and weight 2 (anisotropic), having one and five components, respectively. The antisymmetric part is of weight 1 and has three components. Generally, one has

$$\begin{aligned} a_{ij} &= a^{\mathrm{s}}_{ij} + a^{\mathrm{ant}}_{i} \\ a^{\mathrm{s}}_{ij} &= a^{\mathrm{s}0}_{ij} + a^{\mathrm{s}2}_{ij} \\ a^{\mathrm{ant}}_{ij} &= a^{\mathrm{ant}1}_{ij} \end{aligned} \tag{5}$$

Only the tensors of the same weight mix under rotation in space. Therefore, the scattered intensity is simply the sum of isotropic, anisotropic, and antisymmetric contributions. In other words, there are no isotropic–anisotropic, isotropic–antisymmetric, and anisotropic–antisymmetric cross products. The linear electric-dipole Rayleigh and Raman scatterings, described by the antisymmetric molecular polarizability tensor, involve an inversed azimuth relative to the azimuth of the symmetric scattering [41, 42]. In the case of Rayleigh scattering the antisymmetric mechanism is always accompanied by an isotropic and/or anisotropic mechanism. The azimuth of the Rayleigh line can be inverted if, obviously, the antisymmetric contribution exceeds the symmetric one. This case is commonly referred to as anomalous polarization. In Raman scattering, except for the cases discussed above, we can moreover deal with another interesting situation. Namely, the antisymmetric part of the polarizability tensor is responsible for the emergence of new vibration lines in the scattered spectrum. These purely antisymmetric lines manifest the phenomenon of inverse polarization. No other polarization phenomena are expected in linear scattering by optically inactive molecules, i.e., in the pure electric-dipole approximation.

A third-rank tensor c_{ijk}, if it has no permutation symmetry, can be decomposed into a completely symmetric part with respect to permutation of all indices, an antisymmetric part related to the Levi-Civitá pseudotensor, and a remaining unsymmetrical part [1, 25, 26, 43]. The molecular hyperpolarizability scattering tensor b_{ijk} ($b_{ijk} = b_{ijk}(0)$ or $b_{ijk}(\nu)$ is by definition symmetric in its last two indices, since they are associated with the same incident frequency ω. Hence, b_{ijk} has no pure antisymmetric part related with the pseudotensor and can be decomposed into two parts only: a fully symmetric part b^{s}_{ijk} and a remaining unsymmetrical part b^{u}_{ijk} which, however, is still symmetric in j and k. One should stress that this unsymmetrical part, contrary to the antisymmetric tensor a^{ant}_{ij}, differs from zero even for molecules without internal magnetic structure, i.e., for diamagnetic molecules. Zhu [44] has shown that this tensor may be treated only approximately as completely symmetric for diamagnetic molecules if the incident ω and scattered ω_s frequencies are significantly less than the electronic transition frequency ω_e. In practice, at excitation at optical frequencies they are but slightly smaller than ω_e. Hence, the realistic description of the scatterings in question requires the use of a partly j, k symmetric tensor b_{ijk}. Index asymmetry in b_{ijk} leads to new selection rules for these scatterings [45]. Moreover, for paramagnetic molecules or in the case of resonance scattering, this tensor should also be considered as a complex tensor.

In general, the tensor b_{ijk} may be presented in the following form:

$$
\begin{aligned}
b_{ijk} &= b^{\mathrm{s}}_{ijk} + b^{\mathrm{u}}_{ijk} \\
b^{\mathrm{s}}_{ijk} &= \tfrac{1}{3}\left(b_{ijk} + b_{jki} + b_{kij}\right) \\
b^{\mathrm{u}}_{ijk} &= \tfrac{1}{3}\left(2b_{ijk} - b_{jki} - b_{kij}\right)
\end{aligned}
\tag{6}
$$

The symmetric tensor b^{s}_{ijk} can be further decomposed into two irreducible sets of weights 1 (b^{s1}_{ijk}) and 3 (b^{s3}_{ijk}) [1, 25, 26, 43, 46] with three and seven components, respectively:

$$
\begin{aligned}
b^{s}_{ijk} &= b^{s1}_{ijk} + b^{s3}_{ijk} \\
b^{s1}_{ijk} &= b^{s}_{i}\delta_{jk} + b^{s}_{j}\delta_{ik} + b^{s}_{k}\delta_{ij}
\end{aligned}
\tag{7}
$$

where b^{s}_{i} is obtained by contraction of b^{s}_{ijk} in two indices:

$$
b^{s}_{i} = \tfrac{1}{5} b^{s}_{ikk} = \tfrac{1}{15}(b_{ikk} + 2b_{kki}) \tag{8}
$$

and transforms like vectors under rotation in space. In turn, the tensor b^{s3}_{ijk} is simply the difference between b^{s}_{ijk} and b^{s1}_{ijk}.

The unsymmetrical part b^{u}_{ijk} can also be represented by two irreducible sets of weights 1 (b^{u1}_{ijk}) and 2 (b^{u2}_{ijk}), having three and five components, respectively [1, 25, 26, 43, 46]:

$$
\begin{aligned}
b^{u}_{ijk} &= b^{u1}_{ijk} + b^{u2}_{ijk} \\
b^{u1}_{ijk} &= 2b^{u}_{i}\delta_{jk} - b^{u}_{j}\delta_{ik} - b^{u}_{k}\delta_{ij}
\end{aligned}
\tag{9}
$$

where now

$$
b^{u}_{\mathrm{i}} = \tfrac{1}{4} b^{u}_{ikk} = \tfrac{1}{6}(b_{ikk} - b_{kki}) \tag{10}
$$

and, as previously, b^{u2}_{ijk} is obtained by subtraction of b^{u1}_{ijk} from b^{u}_{ijk}.

Hence, finally, we have

$$
b_{ijk} = b^{s1}_{ijk} + b^{u1}_{ijk} + b^{u2}_{ijk} + b^{s3}_{ijk} \tag{11}
$$

and by definition

$$
\begin{aligned}
&b^{s3}_{ijj} = b^{u2}_{ijj} = 0 \qquad b^{s1}_{ijj} = b^{s1}_{jji} \qquad b^{u1}_{ijj} = -2b^{u1}_{jji} \\
&b^{s1}_{ijj} b^{s1*}_{ikk} = \tfrac{5}{3}|b^{s1}_{ijk}|^{2} \qquad b^{u1}_{ijj} b^{u1*}_{ikk} = \tfrac{4}{3}|b^{u1}_{ijk}|^{2}
\end{aligned}
\tag{12}
$$

The irreducible set b^{u2}_{ijk} is, similarly to the antisymmetric tensor a^{ant1}_{ij}, responsible for the phenomenon of inverse polarization even in the case of hyper-Rayleigh scattering. Namely, molecules of the point symmetries D_4, D_5, and D_6 produce only pure hyper-Rayleigh unsymmetrical scattering. Since the tensor (11) contains two sets of the same weight mixing under rotation in space, one can, in general, expect cross symmetric–nonsymmetric contributions of the type $b^{s1}_{ijk} b^{u1*}_{ijk}$ and $b^{s1*}_{ijk} b^{u1}_{ijk}$ to the hyper-Rayleigh and hyper-Raman scattered intensities and, in principle, new polarization phenomena absent in linear scatterings can appear in nonlinear ones.

IV. GENERAL REMARKS ON THE POLARIZATION PROPERTIES OF LIGHT

Electromagnetic radiation interacting with matter experiences, in general, changes in the state of its polarization. The quality and magnitude of these changes depend on the kind of interaction and on the molecular and thermodynamical properties of the medium. The most consistent formulation of the subject in question is obtained by the use of Stokes parameters. If we assume the incident monochromatic (coherent) light to be propagating along the z axis, its Stokes parameters are [47]

$$\begin{aligned} s_0 &= E_x E_x^* + E_y E_y^* \\ s_1 &= E_x E_x^* - E_y E_y^* \\ s_2 &= E_x E_y^* + E_y E_x^* \\ s_3 &= \mathrm{i}(E_y E_x^* - E_x E_y^*) \end{aligned} \tag{13}$$

or, equivalently,

$$\begin{aligned} s_1 &= s_0 \cos 2\Phi \cos 2\Psi \\ s_2 &= s_0 \cos 2\Phi \sin 2\Psi \\ s_3 &= s_0 \sin 2\Phi \end{aligned} \tag{14}$$

where Ψ denotes the azimuth of the elliptic major axis relative to x $(0 \leq \Psi \leq \Pi)$, and Φ is the ellipticity $(-\Pi/4 \leq \Phi \leq \Pi/4)$. For linear polarization $\Phi = 0$, while for circular polarization $\Phi = \pm\Pi/4$. Only three of these parameters are independent for such light, since $s_0^2 = s_1^2 + s_2^2 + s_3^2$ which, in general, is not fulfilled for quasi-monochromatic light with fluctuating intensity. In linear optics, when describing processes caused by quasi-monochromatic light, we immediately use the parameters of Eq. (13) averaged over the ensemble of the incident field amplitudes. For nonlinear processes we have to use, in this case, unaveraged parameters (13) or (14). These processes depend on intensity fluctuations of the light causing them. But final results have to be averaged in the above-mentioned sense. If we use the form (14) of the Stokes parameters and if the incident light is unpolarized, then one has to perform additionally averaging over the angles Ψ and Φ.

We denote the Stokes parameters for the scattered light by capital letter S_i ($i = 0, 1, 2$, and 3.). The quantity $\sigma = (S_1^2 + S_2^2 + S_3^2)^{1/2}$ refers to its completely polarized portion I_p, while $S_0 - \sigma$ refers to its unpolarized portion I_u [47]. If the scattered light is observed along the Z axis, its Stokes parameters are

$$\begin{aligned} S_0 &= \frac{c}{4\pi}(I_{XX} + I_{YY}) \\ S_1 &= \frac{c}{4\pi}(I_{XX} - I_{YY}) \\ S_2 &= \frac{c}{4\pi}(I_{XY} + I_{YX}) \\ S_3 &= \frac{ci}{4\pi}(I_{YX} - I_{XY}) \end{aligned} \tag{15}$$

where the intensity tensor components are determined by Eq. (1).

One of the relative parameters characterizing the scattered light is given by the depolarization ratio ${}^{\mathrm{P}}D$:

$${}^{\mathrm{P}}D = \frac{{}^{\mathrm{P}}S_0 - {}^{\mathrm{P}}\sigma}{{}^{\mathrm{P}}S_0 + {}^{\mathrm{P}}\sigma} \tag{16}$$

where the superscript preceding D and the Stokes parameters will denote polarization of the incident light.

The polarized portion of the scattered light is described by the azimuth Ψ_s and the ellipticity Φ_s as [47]

$$\begin{aligned} \tan 2\Psi_s &= \frac{{}^{\mathrm{P}}S_2}{{}^{\mathrm{P}}S_1} \\ \sin 2\Phi_s &= \frac{{}^{\mathrm{P}}S_3}{{}^{\mathrm{P}}\sigma} \end{aligned} \tag{17}$$

When the incident radiation is circularly or elliptically polarized the scattered light is often characterized by the degree of circularity [2]:

$${}^{\pm}C = \frac{{}^{\pm}S_3}{{}^{\pm}S_0} \tag{18}$$

where the superscript "+" before the parameters stands for right-handedness of the incident radiation, and "−" stands for left-handedness, as viewed oppositely to the propagation direction (optical convention). One can further introduce the reversal ratio ${}^{\pm}R$, defined as the ratio of the scattered intensity transmitted by an optical system accepting the helicity contrary to the helicity of the incident light and the scattered intensity transmitted by the optical system accepting the same handedness as that of the incident radiation [48]. In terms of the Stokes parameters, the reversal ratio takes the form

$$ {}^{\pm}R = \frac{{}^{\pm}S_0 \mp {}^{\pm}S_3}{{}^{\pm}S_0 \pm {}^{\pm}S_3} \tag{19} $$

From Eqs. (18) and (19) we find

$$ {}^{\pm}R = \frac{1 \mp {}^{\pm}C}{1 \pm {}^{\pm}C} \tag{20} $$

Andrews and Harlow [49] have introduced a parameter p^{-1} representing the ratio of the scattered intensity with the same helicity as the incident light and the total scattered intensity. This ratio, in terms of the degree of circularity and the reversal ratio, may be written as follows:

$$ {}^{\pm}p^{-1} = \tfrac{1}{2}(1 \pm {}^{\pm}C) = \frac{1}{1 + {}^{\pm}R} \tag{21} $$

Andrews and Harlow [49] interpret $p^{-1} = \frac{1}{2}$ as signifying plane polarized light. This value of p^{-1} involves directly ${}^{\pm}C = 0$. Long [41] interprets ${}^{\pm}C = 0$ as meaning unpolarized light. In fact, ${}^{\pm}C = 0$ is not a decisive quantity and the two interpretations are really possible. Moreover, one can add a third interpretation according to which ${}^{\pm}C = 0$ signifies partly plane polarized light. The parameter ${}^{\pm}S_3$ is equal to the difference between the intensities transmitted by an optical system accepting oscillations with right-handed and with left-handed circular polarization. ${}^{\pm}S_3$ is equal to zero jut either for unpolarized, plane polarized, or partly plane polarized light. Hence, ${}^{\pm}C = 0$, $p^{-1} = \frac{1}{2}$, and ${}^{\pm}R = 1$ are not decisive quantities as for the determination of the polarization state of light.

V. STOKES PARAMETERS FOR THE SCATTERED LIGHT

Substituting the second term of Eq. (3) into Eq. (1) and performing unweighted averaging over molecular orientations, which involves averaging of the product of six directional cosines as first done by Kielich [50], one finds

$$\begin{aligned} I_{ij} = \frac{c}{4\pi} L\{ & A\delta_{ij}\langle E_k E_k^* E_l E_l^* \rangle_E + B\delta_{ij}\langle E_k E_k E_l^* E_l^* \rangle_E \\ & + C_1 \langle E_i^* E_j^* E_k E_k + E_i E_j E_k^* E_k^* \rangle_E \\ & + D_1 \langle E_i E_j^* E_k E_k^* \rangle_E + E_1 \langle E_i^* E_j E_k E_k^* \rangle_E \\ & + F \langle E_i^* E_j^* E_k E_k - E_i E_j E_k^* E_k^* \rangle_E \} \end{aligned} \tag{22}$$

where the six parameters A, B, C_1, D_1, E_1, and F are molecular rotational invariants comprising appropriate combinations of various products of the $b_{\alpha\beta\gamma}$ tensor components:

$$\begin{aligned} A &= 11 b_{\alpha\beta\gamma} b_{\alpha\beta\gamma}^* - 6 b_{\alpha\beta\gamma} b_{\beta\alpha\gamma}^* - 5 b_{\alpha\beta\beta} b_{\alpha\gamma\gamma}^* + 4 b_{\alpha\beta\beta} b_{\gamma\gamma\alpha}^* \\ &\quad + 4 b_{\beta\beta\alpha} b_{\alpha\gamma\gamma}^* - 6 b_{\beta\beta\alpha} b_{\gamma\gamma\alpha}^* \\ B &= -5 b_{\alpha\beta\gamma} b_{\alpha\beta\gamma}^* + 4 b_{\alpha\beta\gamma} b_{\beta\alpha\gamma}^* + 8 b_{\alpha\beta\beta} b_{\alpha\gamma\gamma}^* - 5 b_{\alpha\beta\beta} b_{\gamma\gamma\alpha}^* \\ &\quad - 5 b_{\beta\beta\alpha} b_{\alpha\gamma\gamma}^* + 4 b_{\beta\beta\alpha} b_{\gamma\gamma\alpha}^* \\ C_1 &= \tfrac{1}{2}(8 b_{\alpha\beta\gamma} b_{\alpha\beta\gamma}^* - 12 b_{\alpha\beta\gamma} b_{\beta\alpha\gamma}^* - 10 b_{\alpha\beta\beta} b_{\alpha\gamma\gamma}^* + 15 b_{\alpha\beta\beta} b_{\gamma\gamma\alpha}^* \\ &\qquad + 15 b_{\beta\beta\alpha} b_{\alpha\gamma\gamma}^* - 12 b_{\beta\beta\alpha} b_{\gamma\gamma\alpha}^*) \\ D_1 &= 2(-3 b_{\alpha\beta\gamma} b_{\alpha\beta\gamma}^* + b_{\alpha\beta\gamma} b_{\beta\alpha\gamma}^* + 2 b_{\alpha\beta\beta} b_{\alpha\gamma\gamma}^* - 3 b_{\alpha\beta\beta} b_{\gamma\gamma\alpha}^* \\ &\qquad - 3 b_{\beta\beta\alpha} b_{\alpha\gamma\gamma}^* + 8 b_{\beta\beta\alpha} b_{\gamma\gamma\alpha}^*) \\ E_1 &= 2(-3 b_{\alpha\beta\gamma} b_{\alpha\beta\gamma}^* + 8 b_{\alpha\beta\gamma} b_{\beta\alpha\gamma}^* + 2 b_{\alpha\beta\beta} b_{\alpha\gamma\gamma}^* - 3 b_{\alpha\beta\beta} b_{\gamma\gamma\alpha}^* \\ &\qquad - 3 b_{\beta\beta\alpha} b_{\alpha\gamma\gamma}^* + b_{\beta\beta\alpha} b_{\gamma\gamma\alpha}^*) \\ F &= \frac{7i}{2}(b_{\alpha\beta\beta}^* b_{\gamma\gamma\alpha} - b_{\alpha\beta\beta} b_{\gamma\gamma\alpha}^*) \end{aligned} \tag{23}$$

Obviously, $b_{\alpha\beta\gamma} = b_{\alpha\beta\gamma}(0)$ for hyper-Rayleigh and $b_{\alpha\beta\gamma} = b_{\alpha\beta\gamma}(\nu)$ for hyper-Raman scattering. The form of the constant factor for hyper-Raman scattering reads $L = N(2\omega \pm \omega_\nu)^4 \langle Q_\nu^2 \rangle_Q H_\pm / 420 R^2 c^4$, where the coefficient $H_\pm$ ensures the correct ratio of Stokes-scattered $(-)$ and anti-Stokes-scattered $(+)$ intensities resulting from the principle of detailed equilibrium at scattering and stating that $H_+ = \exp(-\hbar\omega_\nu/kT)H_-$. For hyper-Rayleigh scattering one has to put $\omega_\nu = 0$ and to omit $\langle Q_\nu^2\rangle_Q$ and $H_\pm$.

If the scattered light is observed in the $YZ = yz$ plane along the Z axis at an angle ϑ relative to z, then from Eqs. (15) and (22) after some algebra we get

$$\begin{aligned}
{}^{\mathrm{P}}S_0(\vartheta) = \tfrac{1}{2}L\{&4A\langle s_0^2\rangle + 2B[\langle s_0^2\rangle + \langle s_1^2\rangle + \langle s_2^2\rangle - \langle s_3^2\rangle] \\
&+C_1[\langle s_0^2\rangle + \langle s_1^2\rangle + \langle s_2^2\rangle - \langle s_3^2\rangle \\
&\quad +2\langle s_0 s_1\rangle + (\langle s_0^2\rangle + \langle s_1^2\rangle + \langle s_2^2\rangle - \langle s_3^2\rangle \\
&\qquad -2\langle s_0 s_1\rangle)\cos^2\vartheta] \\
&+2E[\langle s_0^2\rangle + \langle s_0 s_1\rangle + (\langle s_0^2\rangle - \langle s_0 s_1\rangle)\cos^2\vartheta] \\
&+2F\langle s_2 s_3\rangle \sin^2\vartheta\} \\
{}^{\mathrm{P}}S_1(\vartheta) = \tfrac{1}{2}L\{&2E[\langle s_0^2\rangle + \langle s_0 s_1\rangle - (\langle s_0^2\rangle - \langle s_0 s_1\rangle)\cos^2\vartheta] \\
&+C_1[\langle s_0^2\rangle + \langle s_1^2\rangle + \langle s_2^2\rangle - \langle s_3^2\rangle + 2\langle s_0 s_1\rangle \\
&\quad -(\langle s_0^2\rangle + \langle s_1^2\rangle + \langle s_2^2\rangle - \langle s_3^2\rangle - 2\langle s_0 s_1\rangle)\cos^2\vartheta] \\
&+2F\langle s_2 s_3\rangle(1 + \cos^2\vartheta)\} \qquad (24) \\
{}^{\mathrm{P}}S_2(\vartheta) = L&(D\langle s_0 s_2\rangle - 2F\langle s_1 s_3\rangle)\cos\vartheta \\
{}^{\mathrm{P}}S_3(\vartheta) = L&G\langle s_0 s_3\rangle\cos\vartheta
\end{aligned}$$

where $D = D_1 + E_1 + 2C_1$, $2E = D_1 + E_1$, and $G = D_1 - E_1$. In the above equations we deal with the seven molecular invariants A–G; in fact, three of them are mutually dependent: $D = 2C_1 + 2E$. All of them intervene only in the case of elliptical or partly elliptical polarization of the incident light; then the products $s_1 s_3$ and $s_2 s_3$ differ from zero. For other polarization states of the incident light their number reduces by one since the terms containing the parameter F vanish.

Moreover, from these general forms of the Stokes parameters it is possible to read some polarization properties of three-photon scattered light. Namely, for perpendicular observation ($\vartheta = \Pi/2$ or $3\Pi/2$), we have ${}^{\mathrm{P}}S_2(\Pi/2) = {}^{\mathrm{P}}S_3(\Pi/2) = 0$. This means that the polarized portion of the scattered intensity will be linearly polarized irrespective of the polarization state of the incident light. ${}^{\mathrm{P}}S_2(\Pi/2) = 0$ corresponds to the azimuths $\Psi_s = 0$ or Π and $\Psi_s = \Pi/2$, which signify that the electric field of this polarized portion may oscillate along the X or Y axis according to the sign of ${}^{P}S_1(\Pi/2)$.

Making use of Eqs. (14) instead of Eqs. (24), we have

$$\begin{aligned}
{}^{P}S_0(\vartheta) &= \tfrac{1}{2}\Gamma\{4A + [4B + C(1 + \cos^2\vartheta)]\cos^2 2\Phi \\
&\quad + D\cos 2\Phi \cos 2\Psi \sin^2\vartheta \\
&\quad + 2E(1 + \cos^2\vartheta) + F\sin 4\Phi \sin 2\Psi \sin^2\vartheta\} \\
{}^{P}S_1(\vartheta) &= \tfrac{1}{2}\Gamma[(C\cos^2 2\Phi + 2E)\sin^2\vartheta \\
&\quad + (D\cos 2\Phi\cos 2\Psi + F\sin 4\Phi \sin 2\Psi)(1 + \cos^2\vartheta)] \\
{}^{P}S_2(\vartheta) &= \Gamma(D\cos 2\Phi \sin 2\Psi - F\sin 4\Phi\cos 2\Psi)\cos\vartheta \qquad (25) \\
{}^{P}S_3(\vartheta) &= \Gamma G \sin 2\Phi \cos\vartheta
\end{aligned}$$

where $\Gamma = L\langle s_0\rangle^2 g^{(2)}$ and $\langle s_0\rangle$ is the mean total intensity of the incident light having the second-order coherence degree $g^{(2)} = \langle s_0^2\rangle/\langle s_0\rangle^2$. The molecular invariant $C = 2C_1$ and, obviously, $D = C + 2E$. Note that chaotic (i.e., with Gaussian distribution of amplitudes) completely polarized light is scattered twice as effectively as coherent light, as was first pointed out by Shen [51]. This is so because $g^{(2)} = 2$ for polarized chaotic radiation, whereas $g^{(2)} = 1$ for coherent radiation.

With respect to relations (12), the molecular rotational invariants may be presented as follows:

$$\begin{aligned}
A &= 5\mathrm{B}_3^{\mathrm{s}} + 14\mathrm{B}_2^{\mathrm{u}} \\
B &= \tfrac{7}{3}\mathrm{B}_1^{\mathrm{s}} - \mathrm{B}_3^{\mathrm{s}} + \tfrac{35}{3}\mathrm{B}_1^{\mathrm{u}} - 7\mathrm{B}_2^{\mathrm{u}} + 7\,{}^{1}\mathrm{B}_{11}^{\mathrm{su}} \\
C &= \tfrac{28}{3}\mathrm{B}_1^{\mathrm{s}} - 4\mathrm{B}_3^{\mathrm{s}} - \tfrac{70}{3}\mathrm{B}_1^{\mathrm{u}} + 14\mathrm{B}_2^{\mathrm{u}} + 7\,{}^{1}\mathrm{B}_{11}^{\mathrm{su}} \\
D &= \tfrac{56}{3}\mathrm{B}_1^{\mathrm{s}} + 2\mathrm{B}_3^{\mathrm{s}} - \tfrac{35}{3}\mathrm{B}_1^{\mathrm{u}} - 7\mathrm{B}_2^{\mathrm{u}} - 7\,{}^{1}\mathrm{B}_{11}^{\mathrm{su}} \qquad (26) \\
E &= \tfrac{14}{3}\mathrm{B}_1^{\mathrm{s}} + 3\mathrm{B}_3^{\mathrm{s}} + \tfrac{35}{6}\mathrm{B}_1^{\mathrm{u}} - \tfrac{21}{2}\mathrm{B}_2^{\mathrm{u}} - 7\,{}^{1}\mathrm{B}_{11}^{\mathrm{su}} \\
F &= \tfrac{21}{2}\,{}^{2}\mathrm{B}_{11}^{\mathrm{su}} \\
G &= \tfrac{28}{3}\mathrm{B}_1^{\mathrm{s}} - 14\mathrm{B}_3^{\mathrm{s}} + \tfrac{35}{3}\mathrm{B}_1^{\mathrm{u}} + 7\mathrm{B}_2^{\mathrm{u}} - 14\,{}^{1}\mathrm{B}_{11}^{\mathrm{su}}
\end{aligned}$$

where

$$\begin{aligned}
&\mathrm{B}_1^{\mathrm{s}} = |b_{\alpha\beta\gamma}^{\mathrm{s}1}|^2 \qquad \mathrm{B}_1^{\mathrm{u}} = |b_{\alpha\beta\gamma}^{\mathrm{u}1}|^2 \qquad \mathrm{B}_3^{\mathrm{s}} = |b_{\alpha\beta\gamma}^{\mathrm{s}3}|^2 \qquad \mathrm{B}_2^{\mathrm{u}} = |b_{\alpha\beta\gamma}^{\mathrm{u}2}|^2 \\
&{}^{1}\mathrm{B}_{11}^{\mathrm{su}} = \tfrac{1}{2}\left(b_{\alpha\beta\beta}^{\mathrm{s}1} b_{\alpha\gamma\gamma}^{\mathrm{u}1*} + b_{\alpha\beta\beta}^{\mathrm{s}1*} b_{\alpha\gamma\gamma}^{\mathrm{u}1}\right) \qquad (27) \\
&{}^{2}B_{11}^{\mathrm{su}} = \frac{\mathrm{i}}{2}\left(b_{\alpha\beta\beta}^{\mathrm{s}1} b_{\alpha\gamma\gamma}^{\mathrm{u}1*} - b_{\alpha\beta\beta}^{\mathrm{s}1*} b_{\alpha\gamma\gamma}^{\mathrm{u}1}\right)
\end{aligned}$$

Obviously, the number of molecular parameters can be reduced with regard to molecular and vibrational symmetries. In particular, both scatterings can be described by one parameter B_2^{u} only. It is precisely pure unsymmetric scattering that is represented by this invariant.

VI. DIFFERENTIAL SCATTERING OF ELLIPTICALLY POLARIZED LIGHT AND ROTATION OF THE AZIMUTH

To start with, let us discuss three-photon scattering of elliptically polarized light as offering new polarization effects compared to linear electric-dipole Rayleigh and Raman scatterings. The elliptical total intensity differential ratio is, by definition [52]

$$\Delta = \left({}^{E+}S_0 - {}^{E-}S_0\right) / \left({}^{E+}S_0 + {}^{E-}S_0\right) \tag{28}$$

Inserting the parameter ${}^{e}E^{\pm}S_0(\vartheta)$ of Eqs. (25) into the above definition, one gets [53]

$$\Delta(\vartheta) = \frac{F|\sin 4\Phi|\sin 2\Psi \sin^2\vartheta}{4A + \left[4B + C(1+\cos^2\vartheta)\right]\cos^2 2\Phi + D\cos 2\Phi \cos 2\Psi \sin^2\vartheta + 2E(1+\cos^2\vartheta)} \tag{29}$$

It is readily seen that, as to the angle ϑ, electric-dipole elliptical differential scattering (EDEID) of the total intensity vanishes only for forward ($\vartheta = 0$) and backward ($\vartheta = \Pi$) directions of observation. However, EDEID can still occur for these directions in the intensity components perpendicular (X) and parallel (Y) to the plane of observation. Hence, it is convenient to introduce the following differential ratios [52]:

$$\begin{aligned} \Delta_X &= \frac{{}^{E+}S_0 + {}^{E+}S_1 - \left({}^{E-}S_0 + {}^{E-}S_1\right)}{{}^{E+}S_0 + {}^{E+}S_1 + {}^{E-}S_0 + {}^{E-}S_1} \\ \Delta_Y &= \frac{{}^{E+}S_0 - {}^{E+}S_1 - \left({}^{E-}S_0 + {}^{E-}S_1\right)}{{}^{E+}S_0 - {}^{E+}S_1 + {}^{E-}S_0 - {}^{E-}S_1} \end{aligned} \tag{30}$$

By Eqs. (25) these components are [54]

$$\begin{aligned} \Delta_X &= \frac{F|\sin 4\Phi|\sin 2\Psi}{2A + (2B + C)\cos^2 2\Phi + D\cos 2\Phi\cos 2\Psi + 2E} \\ \Delta_Y(\vartheta) &= \frac{-F|\sin 4\Phi|\sin 2\Psi\cos^2\vartheta}{2A + (2B + C\cos^2\vartheta)\cos^2 2\Phi - D\cos 2\Phi\cos 2\Psi\cos^2\vartheta + 2E\cos^2\vartheta} \end{aligned} \tag{31}$$

The component Δ_X does not depend on the scattering angle at all and hence can be nonzero for forward and backward scattering as well. In turn, the component Δ_Y is dependent on ϑ and tends to zero for perpendicular scattering. On adding the numerators and denominators in the above equations we get the elliptical total intensity differential ratio (29). EDEID tends to zero without any exception if the axes of the incident ellipse coincide with the axes x and y, i.e., if either of them lies in the observation plane.

In turn, on insertion of the appropriate Stokes parameters (Eqs. (25)) into the first relation (17), we get for the azimuth of the scattered light

$$\tan 2^{\mathrm{E}\pm}\Psi_s(\vartheta) = \frac{2(D\cos 2\Phi \sin 2\Psi \mp F|\sin 4\Phi|\cos 2\Psi)\cos\vartheta}{(C\cos^2 2\Phi + 2E)\sin^2\vartheta + (D\cos 2\Phi\cos 2\Psi \pm F|\sin 4\Phi|\sin 2\Psi)(1+\cos^2\vartheta)} \tag{32}$$

which, for forward scattering, transforms to [55]

$$\tan 2^{\mathrm{E}\pm}\Psi_s(0) = \tan 2\Psi \frac{D \mp 2F\cot 2\Psi|\sin 2\Phi|}{D \pm 2F\tan 2\Psi|\sin 2\Phi|} \tag{33}$$

The above formulas point to rotation of the scattered polarization ellipse relative to that of the incident light. The sense of rotation depends on the incident handedness and on the sign of the invariant F, which, in fact, is decisive for the existence of EDEID and rotation of the azimuth.

In Eq. (32) the phenomenon of rotation of the polarization ellipse is partly masked by the purely geometrical rotation related to the changes in magnitude of the projections of the elliptic axes onto the plane of observation with varying angle ϑ. This additional rotation can be eliminated by an appropriate choice of the incident azimuth: $\Psi = 0$ or Π (the elliptic major axis is then vertical (V) to the observation plane), and $\Psi = \Pi/2$ (the elliptic major axis then lies in the observation plane (H)). Then, from Eq. (32) one finds that

$$\begin{aligned} \tan 2^{\mathrm{EV}\pm}\Psi_s(\vartheta) &= \frac{\mp 2F|\sin 4\Phi|\cos\vartheta}{(C\cos^2 2\Phi + 2E)\sin^2\vartheta + D\cos 2\Phi(1+\cos^2\vartheta)} \\ \tan 2^{\mathrm{EH}\pm}\Psi_s(\vartheta) &= \frac{\pm 2F|\sin 4\Phi|\cos\vartheta}{(C\cos^2 2\Phi + 2E)\sin^2\vartheta - D\cos 2\Phi(1+\cos^2\vartheta)} \end{aligned} \tag{34}$$

For perpendicular scattering, as said and as is also evident directly from Eqs. (32) and (34), the scattered ellipse transforms into a straight line (${}^{\mathrm{E}\pm}S_2(\Pi/2) = 0$) and, depending on the sign of the parameter ${}^{\mathrm{E}\pm}S_1(\Pi/2)$, the scattered azimuth becomes either parallel or perpendicular to the observation plane.

It is obvious that in the symmetric approximation or for pure unsymmetric scattering the cross symmetric–unsymmetric parameter F vanishes and for forward scattering Eq. (33) then yields

$$\tan 2^{\mathrm{E}\pm}\Psi_s(0) = \tan 2\Psi \tag{35}$$

In the symmetric approximation the invariant $D = 56\mathrm{B}_1^{\mathrm{s}}/3 + 2\mathrm{B}_3^{\mathrm{s}}$ is positive. Thus, the only acceptable solution of Eq. (35) is $\Psi_s = \Psi$, which means retention of the direction of the incident azimuth [16]. In turn, for the purely unsymmetric scattering the invariant $D = -7\mathrm{B}_2^{\mathrm{u}}$ is negative and in this case the other solution of the above equation is acceptable, namely that implying a change of the scattered azimuth by $\Pi/2$ [56].

Let us emphasize once more that rotation of the azimuth occurs only in the case of elliptically or partly elliptically polarized incident light. For linearly polarized incident light, $\Phi = 0$ and, for instance, the last term in the numerator and in the denominator of Eq. (33) vanishes and this equation again transforms to Eq. (35).

VII. QUANTUM-MECHANICAL FORM OF THE HYPER-RAYLEIGH SCATTERING TENSOR

The quantum-mechanical definition of the hyperpolarizability tensor $b_{\alpha\beta\gamma}(0) = b_{\alpha\beta\gamma}(2\omega)$ can be found in numerous papers [8, 54, 57–62]. However, only Refs. 8, 54, and 60–62 take into account the widths of the excited levels. According to Kielich [8, 61], this tensor may be written as follows:

$$b_{\alpha\beta\gamma}(2\omega) = \frac{1}{4\hbar^2}\rho_{kk}\sum_{l,r}\left[A(+\omega)_{k,l,r} + A(-\omega)^*_{k,l,r}\right] \tag{36}$$

where

$$\begin{aligned} A(\pm\omega)_{k,l,r} = {} & \frac{P_{\alpha\beta\gamma} + P_{\alpha\gamma\beta}}{(\mp 2\omega + \omega_{lk} - \mathrm{i}\Gamma_l)(\mp\omega + \omega_{rk} - \mathrm{i}\Gamma_r)} \\ & + \frac{P_{\beta\alpha\gamma} + P_{\gamma\alpha\beta}}{(\pm\omega + \omega_{lk} + \mathrm{i}\Gamma_l)(\mp\omega - \omega_{rk} - \mathrm{i}\Gamma_r)} \\ & + \frac{P_{\beta\gamma\alpha} + P_{\gamma\beta\alpha}}{(\pm\omega + \omega_{lk} + \mathrm{i}\Gamma_l)(\pm 2\omega + \omega_{rk} + \mathrm{i}\Gamma_r)} \end{aligned} \tag{37}$$

and the matrix elements P are

$$P_{\alpha\beta\gamma} = \langle k|m_\alpha|l\rangle\langle l|m_\beta|r\rangle\langle r|m_\gamma|k\rangle \tag{38}$$

The quantity ρ_{kk} is the mean quantum value of the unperturbed density matrix in the ground state $|k\rangle$; $\omega_{lk} = \omega_l - \omega_k$ is the transition frequency between states $|k\rangle$ and $|l\rangle$ with the lifetime γ_l^{-1} ($\Gamma_l = \gamma_l/2$); and m_α is the component ($\alpha = 1, 2, 3$) of the electric-dipole moment operator. From the above equations it is obvious that $b_{\alpha\beta\gamma}(0)$ is a polar tensor symmetric in the last two indices. If the wave functions of the electronic stationary states are complex, this tensor may be presented in the following general form [54]:

$$b_{\alpha\beta\gamma}(0) = \alpha_{\alpha\beta\gamma} + \mathrm{i}\beta_{\alpha\beta\gamma} + \mathrm{i}(\gamma_{\alpha\beta\gamma} + \mathrm{i}\delta_{\alpha\beta\gamma}) \tag{39}$$

where the four tensors $\alpha_{\alpha\beta\gamma}$–$\delta_{\alpha\beta\gamma}$ are real and their form is as follows:

$$\begin{aligned}
\alpha_{\alpha\beta\gamma} = \frac{1}{2\hbar^2}\rho_{kk}\sum_{l,r}\big[&G_{21}(\omega^2,k,l,r)\mathrm{Re}(P_{\alpha\beta\gamma}+P_{\alpha\gamma\beta})\\
&+G_{11}(\omega^2,k,l,r)\mathrm{Re}(P_{\beta\alpha\gamma}+P_{\gamma\alpha\beta})\\
&+G_{12}(\omega^2,k,l,r)\mathrm{Re}(P_{\beta\alpha\gamma}+P_{\gamma\beta\alpha})\big]\\
\beta_{\alpha\beta\gamma} = \frac{\omega}{2\hbar^2}\rho_{kk}\sum_{l,r}\big[&H_{21}(\omega^2,k,l,r)\mathrm{Im}(P_{\alpha\beta\gamma}+P_{\alpha\gamma\beta})\\
&+H_{11}(\omega^2,k,l,r)\mathrm{Im}(P_{\beta\alpha\gamma}+P_{\gamma\alpha\beta})\\
&-H_{12}(\omega^2,k,l,r)\mathrm{Im}(P_{\beta\alpha\gamma}+P_{\gamma\beta\alpha})\big]\\
\gamma_{\alpha\beta\gamma} = \frac{\omega}{2\hbar^2}\rho_{kk}\sum_{l,r}\big[&K_{21}(\omega^2,k,l,r)\mathrm{Re}(P_{\alpha\beta\gamma}+P_{\alpha\gamma\beta})\\
&+K_{11}(\omega^2,k,l,r)\mathrm{Re}(P_{\beta\alpha\gamma}+P_{\gamma\alpha\beta})\\
&+K_{12}(\omega^2,k,l,r)\mathrm{Re}(P_{\beta\alpha\gamma}+P_{\gamma\beta\alpha})\big]\\
\delta_{\alpha\beta\gamma} = \frac{1}{2\hbar^2}\rho_{kk}\sum_{l,r}\big[&L_{21}(\omega^2,k,l,r)\mathrm{Im}(P_{\alpha\beta\gamma}+P_{\alpha\gamma\beta})\\
&+L_{11}(\omega^2,k,l,r)\mathrm{Im}(P_{\beta\alpha\gamma}+P_{\gamma\alpha\beta})\\
&-L_{12}(\omega^2,k,l,r)\mathrm{Im}(P_{\beta\alpha\gamma}+P_{\gamma\beta\alpha})\big]
\end{aligned} \tag{40}$$

Above we have used the following notation:

$$G_{21}(\omega^2,k,l,r) = \frac{\omega_{lk}\omega_{rk}\big[f_2^+(lk)f_1^+(rk) - 8\omega^2\Gamma_l\Gamma_r\big]}{g_2(lk)g_1(rk)} + \frac{2\omega^2 f_2^-(lk)f_1^-(rk) - h_2(lk)h_1(rk)\Gamma_l\Gamma_r}{g_2(lk)g_1(rk)}$$

$$
\begin{aligned}
G_{11}(\omega^2, k, l, r) &= \frac{\omega_{lk}\omega_{rk}[f_1^+(lk)f_1^+(rk) - 4\omega^2\Gamma_l\Gamma_r]}{g_1(lk)g_1(rk)} \\
&\quad + \frac{-\omega^2 f_1^-(lk)f_1^-(rk) + h_1(lk)h_1(rk)\Gamma_l\Gamma_r}{g_1(lk)g_1(rk)} \\
H_{21}(\omega^2, k, l, r) &= \frac{\omega_{lk}[f_2^+(lk)f_1^-(rk) - 4h_1(rk)\Gamma_l\Gamma_r]}{g_2(lk)g_1(rk)} \\
&\quad + \frac{2\omega_{rk}[f_2^-(lk)f_1^+(rk) - h_2(lk)\Gamma_l\Gamma_r]}{g_2(lk)g_1(rk)} \\
H_{11}(\omega^2, k, l, r) &= \frac{\omega_{lk}[f_1^+(lk)f_1^-(rk) - 2h_1(rk)\Gamma_l\Gamma_r]}{g_1(lk)g_1(rk)} \\
&\quad + \frac{-\omega_{rk}[f_1^-(lk)f_1^+(rk) - 2h_1(lk)\Gamma_l\Gamma_r]}{g_1(lk)g_1(rk)} \\
K_{21}(\omega^2, k, l, r) &= \frac{\Gamma_l[h_2(lk)f_1^-(rk) + 4\omega_{lk}\omega_{rk}f_1^+(rk)]}{g_2(lk)g_1(rk)} \\
&\quad + \frac{2\Gamma_r[f_2^-(lk)h_1(rk) + \omega_{lk}\omega_{rk}f_2^+(lk)]}{g_2(lk)g_1(rk)} \\
K_{11}(\omega^2, k, l, r) &= \frac{\Gamma_l[-h_1(lk)f_1^-(rk) + 2\omega_{lk}\omega_{rk}f_1^+(rk)]}{g_1(lk)g_1(rk)} \\
&\quad + \frac{\Gamma_r[-f_1^-(lk)h_1(rk) + 2\omega_{lk}\omega_{rk}f_1^+(lk)]}{g_1(lk)g_1(rk)} \\
L_{21}(\omega^2, k, l, r) &= \frac{\Gamma_l[\omega_{rk}h_2(lk)f_1^+(rk) + 4\omega^2\omega_{lk}f_1^-(rk)]}{g_2(lk)g_1(rk)} \\
&\quad + \frac{\Gamma_r[\omega_{lk}f_2^+(lk)h_1(rk) + 4\omega^2\omega_{rk}f_2^-(lk)]}{g_2(lk)g_1(rk)} \\
L_{11}(\omega^2, k, l, r) &= \frac{\Gamma_l[-\omega_{rk}h_1(lk)f_1^+(rk) + 2\omega^2\omega_{lk}f_1^-(rk)]}{g_1(lk)g_1(rk)} \\
&\quad + \frac{\Gamma_r[-\omega_{lk}f_1^+(lk)h_1(rk) + 2\omega^2\omega_{rk}f_1^-(lk)]}{g_1(lk)g_1(rk)}
\end{aligned}
\tag{41}
$$

where

$$g_n(lk) = \left[-(n\omega)^2 + \omega_{lk}^2\right]^2 + 2\left[(n\omega)^2 + \omega_{lk}^2\right]\Gamma_l^2 + \Gamma_l^4$$
$$f_n^{\pm}(lk) = -(n\omega)^2 + \omega_{lk}^2 \pm \Gamma_l^2 \tag{42}$$
$$h_n(lk) = (n\omega)^2 + \omega_{lk}^2 + \Gamma_l^2$$

The frequency-dispersional parameters G_{12}, H_{12}, K_{12}, and L_{12} can be obtained from G_{21}, H_{21}, K_{21}, and L_{21}, respectively, on interchanging l and r. The parameters K and L are proportional to Γ_l and Γ_r. Hence, the tensors $\gamma_{\alpha\beta\gamma}$ and $\delta_{\alpha\beta\gamma}$ exist only in the case of nonnegligible level widths. This makes sense if the incident frequency ω or the scattered frequency 2ω approaches an atomic transition frequency. The tensors $\alpha_{\alpha\beta\gamma}$ and $\delta_{\alpha\beta\gamma}$ are even functions of the frequency ω, while the two others are odd functions of ω. The tensors $\beta_{\alpha\beta\gamma}$ and $\delta_{\alpha\beta\gamma}$ vanish in the case of real wave functions. Moreover, one can check that the tensors $\alpha_{\alpha\beta\gamma}$ and $\gamma_{\alpha\beta\gamma}$ are invariant under time inversion and are i tensors, whereas $\beta_{\alpha\beta\gamma}$ and $\delta_{\alpha\beta\gamma}$ are antisymmetric with respect to time inversion and are c tensors. All the above tensors have been tabulated by us for the 102 magnetic point groups in Ref. 54.

Each of the tensors (40) can be decomposed into symmetric and unsymmetric parts similarly to the way it was done in Eqs. (6)–(12). In particular, the unsymmetric parts of the i tensors $\alpha_{\alpha\beta\gamma}$ and $\gamma_{\alpha\beta\gamma}$ are nonzero; asymmetry of the hyperpolarizability tensor does not require complex molecular wave functions.

The molecular rotational invariant F (26), proportional to ${}^2B_{11}^{\mathrm{su}}$ (27), would contain among others products of the i tensor $\alpha_{\alpha\beta\gamma}$ or $\gamma_{\alpha\beta\gamma}$ and the c tensor $\beta_{\alpha\beta\gamma}$ or $\delta_{\alpha\beta\gamma}$. Such products differ from zero for a single molecule with internal magnetic structure. The scattering medium at thermal equilibrium contains equal numbers of molecular time-inverse counterparts with c tensors of opposite signs [63]. The quantity NF is simply equal to $\sum_{i=1}^{N} F_i$. The above-mentioned products can affect F_i but not F. Therefore, we have generally

$$F = \tfrac{21}{2}\left(\alpha_{\alpha\beta\beta}^{\mathrm{s1}}\gamma_{\alpha\gamma\gamma}^{\mathrm{u1}} - \alpha_{\alpha\beta\beta}^{\mathrm{u1}}\gamma_{\alpha\gamma\gamma}^{\mathrm{s1}} + \beta_{\alpha\beta\beta}^{\mathrm{s1}}\delta_{\alpha\gamma\gamma}^{\mathrm{u1}} - \beta_{\alpha\beta\beta}^{\mathrm{u1}}\delta_{\alpha\gamma\gamma}^{\mathrm{s1}}\right) \tag{43}$$

The first two terms of the above equation are the only terms for diamagnetic molecules.

The tensors $\gamma_{\alpha\beta\gamma}$ and $\delta_{\alpha\beta\gamma}$ are related to resonance conditions; hence, the polarization phenomena discussed in Section VI for hyper-Rayleigh scattering are the resonance phenomena. The resonance conditions also give rise to a considerable increase in the scattered intensity by up to several orders. In the case of resonant hyper-Rayleigh scattering by optically active molecules, an enhancement of the magnetic-dipole and electric-quadrupole contributions will also occur. But elliptical differential scattering due to these cross mechanisms should remain several orders smaller than that due to the pure electric-dipole transitions.

In a similar way one can decompose the hyper-Raman hyperpolarizability tensor. For this scattering EDEID and rotation of the polarization ellipse will also take place under the resonance conditions.

VIII. SYMMETRIC–UNSYMMETRIC SCATTERING OF LINEARLY AND CIRCULARLY POLARIZED AND NATURAL LIGHT

Let us assume linear (L) polarization ($\Phi = 0$) of the incoming light vertical (V) to the observation plane ($\Psi = 0$). Then from Eqs. (25) we get

$$\begin{aligned} {}^{LV}S_0(\vartheta) &= \Gamma(2A + 2B + D) \\ {}^{LV}S_1(\vartheta) &= \Gamma D \\ {}^{LV}S_2(\vartheta) &= {}^{LV}S_3(\vartheta) = 0 \end{aligned} \tag{44}$$

irrespective of the scattering angle. Moreover, $\Phi_s = 0$ for all ϑ. In this case the process is described by the three molecular parameters A, B, and D, which, however, depend on five of the invariants (26). The scattered light is, in general, partly linearly polarized: ${}^{LV}S_0 \neq {}^{LV}S_1$.

Inserting the parameters A, B, and D from Eqs. (26) one finds [46]

$$\begin{aligned} {}^{LV}S_0 &= \tfrac{1}{3}\Gamma\left(70\mathrm{B}_1^{\mathrm{s}} + 30\mathrm{B}_3^{\mathrm{s}} + 35\mathrm{B}_1^{\mathrm{u}} + 21\mathrm{B}_2^{\mathrm{u}} + 21\,{}^1\mathrm{B}_{11}^{\mathrm{su}}\right) \\ {}^{LV}S_1 &= \tfrac{1}{3}\Gamma\left(56\mathrm{B}_1^{\mathrm{s}} + 6\mathrm{B}_3^{\mathrm{s}} - 35\mathrm{B}_1^{\mathrm{u}} - 21\mathrm{B}_2^{\mathrm{u}} - 21\,{}^1\mathrm{B}_{11}^{\mathrm{su}}\right) \end{aligned} \tag{45}$$

The light scattered by a great number of noninteracting statistically independent molecules is incoherent. The elementary scattering processes from different molecules are uncorrelated. However, the magnitude of these elementary processes is not in general a simple additive sum of symmetric and unsymmetric contributions owing to the presence of the cross term ${}^1\mathrm{B}_{11}^{\mathrm{su}}$. On the other hand, this as it were interference term does not influence the scattered azimuth. It is worth noting that the scattered intensity component X is purely symmetric:

$$^{\mathrm{LV}}S_0 + {}^{\mathrm{LV}}S_1 = 6\Gamma(7\mathrm{B}_1^{\mathrm{s}} + 2\mathrm{B}_3^{\mathrm{s}}) \tag{46}$$

The magnitude of the component Y is the additive sum of the symmetric and unsymmetric contributions resulting from the symmetric and unsymmetric irreducible tensors of weights 3 and 2, respectively, and of a remaining nonadditive part due to the symmetric and unsymmetric components of the irreducible tensors of weight 1:

$$^{\mathrm{LV}}S_0 - {}^{\mathrm{LV}}S_1 = 2\Gamma(4\mathrm{B}_3^{\mathrm{s}} + 7\mathrm{B}_2^{\mathrm{u}}) + \tfrac{14}{3}\Gamma\left(\mathrm{B}_1^{\mathrm{s}} + 5\mathrm{B}_1^{\mathrm{u}} + 3\,{}^1\mathrm{B}_{11}^{\mathrm{su}}\right) \tag{47}$$

The cross term ${}^1\mathrm{B}_{11}^{\mathrm{su}}$ may be positive for certain molecules and kinds of vibrations but negative for some others, leading to a constructive or destructive influence on the scattered intensity component Y. In principle, the destructive influence can totally extinguish the second right-hand nonnegative term of the above equation. Making use of relations (12) we can present this term in the following form:

$$\begin{aligned} \alpha &= \tfrac{14}{3}\Gamma\left(\mathrm{B}_1^{\mathrm{s}} + 5\mathrm{B}_1^{\mathrm{u}} + 3\,{}^1\mathrm{B}_{11}^{\mathrm{su}}\right) \\ &= \tfrac{7}{10}\Gamma\left(2b_{\alpha\beta\beta}^{\mathrm{s}1} + 5b_{\alpha\beta\beta}^{\mathrm{u}1}\right)\left(2b_{\alpha\gamma\gamma}^{\mathrm{s}1*} + 5b_{\alpha\gamma\gamma}^{\mathrm{u}1*}\right) \end{aligned} \tag{48}$$

This term can be zero for $2b_{\alpha\beta\beta}^{\mathrm{s}1} = -5b_{\alpha\beta\beta}^{\mathrm{u}1}$. Should, additionally, the invariants $\mathrm{B}_3^{\mathrm{s}}$ and $\mathrm{B}_2^{\mathrm{u}}$ then be zero, the scattered light would be fully linearly polarized in the direction X and would arise from $\mathrm{B}_1^{\mathrm{s}}$ only.

Had we defined the "depolarization ratio" in the usual way, i.e., as the ratio of the scattered intensity component Y and the component X [1, 11, 13], we would have

$$^{\mathrm{LV}}\rho = \frac{^{\mathrm{LV}}I_{YY}}{^{\mathrm{LV}}I_{XX}} = \frac{^{\mathrm{LV}}S_0 - {}^{\mathrm{LV}}S_1}{^{\mathrm{LV}}S_0 + {}^{\mathrm{LV}}S_1} \tag{49}$$

whence

$$^{\mathrm{LV}}\rho = \frac{8\mathrm{B}_3^{\mathrm{s}} + 14\mathrm{B}_2^{\mathrm{u}} + \alpha}{42\mathrm{B}_1^{\mathrm{s}} + 12\mathrm{B}_3^{\mathrm{s}}} \tag{50}$$

Such a depolarization ratio ranges within $0 \leq {}^{\mathrm{LV}}\rho \leq \infty$ [1], where 0 corresponds to $\mathrm{B}_3^{\mathrm{s}} = \mathrm{B}_2^{\mathrm{u}} = 0$ and to the negative invariant ${}^1\mathrm{B}_{11}^{\mathrm{su}}$ extinguishing the term (48) ($\alpha = 0$), whereas the infinity corresponds to pure unsymmetric scattering: $\mathrm{B}_2^{\mathrm{u}} \neq 0$, $\mathrm{B}_1^{\mathrm{s}} = \mathrm{B}_3^{\mathrm{s}} = \mathrm{B}_1^{\mathrm{u}} = {}^1\mathrm{B}_{11}^{\mathrm{su}} = 0$. This type of scattering will be considered in detail in Chapter 5. Both these limit values of the depolarization ratio ${}^{\mathrm{LV}}\rho$ represent fully linearly polarized light with azimuth perpendicular or parallel to the observation plane, respectively.

On the other hand, the depolarization ratio defined in Eq. (16) in this case takes the following form:

$$ {}^{\mathrm{LV}}D = \frac{{}^{\mathrm{LV}}S_0 - |{}^{\mathrm{LV}}S_1|}{{}^{\mathrm{LV}}S_0 + |{}^{\mathrm{LV}}S_1|} \tag{51} $$

and its values lie within the interval $0 \leq {}^{\mathrm{LV}}D \leq 1$.

The equality ${}^{\mathrm{LV}}S_0 = |{}^{\mathrm{LV}}S_1|$, pointing to completely polarized light, ${}^{\mathrm{LV}}D = 0$, occurs either for pure unsymmetric scattering or for molecules whose tensor components cause vanishing of α (48) and simultaneously $\mathrm{B}_3^{\mathrm{s}} = \mathrm{B}_2^{\mathrm{u}} = 0$. Obviously, to this value of ${}^{\mathrm{LV}}D$ correspond two different values of the quantity ${}^{\mathrm{LV}}\rho$ (0 or ∞). In turn, if ${}^{\mathrm{LV}}S_1 = 0$, which could take place for $63\mathrm{B}_1^{\mathrm{s}} + 6\mathrm{B}_3^{\mathrm{s}} = 3\alpha/2 + 21\mathrm{B}_2^{\mathrm{u}}$, the scattered light would be unpolarized, ${}^{\mathrm{LV}}D = 1$. It is an open question whether this condition is realistic. In the symmetric approximation the greatest possible value of ${}^{\mathrm{LV}}D$ amounts to $\frac{2}{3}$ and corresponds to tetrahedral molecules.

For the scattering of circularly (C) polarized light ($\Phi = \pm\Pi/4$), we have by Eqs. (25)

$$ \begin{aligned} {}^{\mathrm{C}\pm}S_0(\vartheta) &= \Gamma\left[2A + E(1 + \cos^2\vartheta)\right] \\ {}^{\mathrm{C}\pm}S_1(\vartheta) &= \Gamma E \sin^2\vartheta \\ {}^{\mathrm{C}\pm}S_3(\vartheta) &= \pm\Gamma G \cos\vartheta \\ {}^{\mathrm{C}\pm}S_2(\vartheta) &= 0 \end{aligned} \tag{52} $$

As previously, the process depends on three molecular parameters (this time on A, E, and G), which, obviously, are expressed by the same five of the invariants related to the irreducible tensors (26). The polarized portion of the scattered light is, in general, for the angles ϑ different from 0, $\Pi/2$, Π, and $3\Pi/2$ elliptically polarized.

With respect to relations (26) one finds that for forward scattering

$$ \begin{aligned} {}^{\mathrm{C}\pm}S_0(0) &= \tfrac{1}{3}\Gamma\left(28\mathrm{B}_1^{\mathrm{s}} + 48\mathrm{B}_3^{\mathrm{s}} + 35\mathrm{B}_1^{\mathrm{u}} + 21\mathrm{B}_2^{\mathrm{u}} - 42\,{}^1\mathrm{B}_{11}^{\mathrm{su}}\right) \\ {}^{\mathrm{C}\pm}S_3(0) &= \pm\tfrac{7}{3}\Gamma\left(4\mathrm{B}_1^{\mathrm{s}} - 6\mathrm{B}_3^{\mathrm{s}} + 5\mathrm{B}_1^{\mathrm{u}} + 3\mathrm{B}_2^{\mathrm{u}} - 6\,{}^1\mathrm{B}_{11}^{\mathrm{su}}\right) \end{aligned} \tag{53} $$

The reversal ratio (19) takes the form

$$ {}^{C\pm}R(0) = \frac{45B_3^s}{28B_1^s + 3B_3^s + 35B_1^u + 21B_2^u - 42\,{}^1B_{11}^{su}} \tag{54} $$

The denominator of the above reversal ratio can be written as the following sum of the nonnegative invariants:

$$ 3B_3^s + 21B_2^u + \beta $$

$$ \beta = 7\left(4B_1^s + 5B_1^u - 6\,{}^1B_{11}^{su}\right) = \tfrac{21}{20}\left(4b_{\alpha\beta\beta}^{s1} - 5b_{\alpha\beta\beta}^{u1}\right)\left(4b_{\alpha\gamma\gamma}^{s1*} - 5b_{\alpha\gamma\gamma}^{u1*}\right) \tag{55} $$

The destructive influence of ${}^1B_{11}^{su}$ now occurs for its positive values, i.e., when for scattering of linearly polarized light ${}^1B_{11}^{su}$ acts constructively, and reaches a maximum ($\beta = 0$) for $4b_{\alpha\beta\beta}^{s1} = 5b_{\alpha\beta\beta}^{u1}$. The numerator is proportional to B_3^s. Hence, the maximal value of the reversal ratio (54) will then be reached if additionally $B_2^u = 0$ and will amount to 15. The same upper limit is obtained in the symmetric approximation at $B_1^s = 0$ (tetrahedral molecules). In general, the reversal ratio varies within $0 \leq {}^{C\pm}R(0) \leq 15$. Its lower limit is either for molecules with $B_3^s = 0$ or for pure unsymmetric scattering. In this case the light scattered forward remains fully circularly polarized. ${}^{C\pm}R(0) < 1$ denotes preserved handedness of the incident light, whereas for ${}^{C\pm}R(0) > 1$ we deal with reversed helicity. In both cases the scattered light is partly circularly polarized. Finally, ${}^{C\pm}R(0) = 1$ here means unpolarized light. From the definition of the ellipticity (17) and the form of the parameter ${}^{C\pm}S_3(0)$ it is evident that helicity is retained for $28B_1^s + 35B_1^u + 21B_2^u > 42B_3^s + 42\,{}^1B_{11}^{su}$ and reversed for $28B_1^s + 35B_1^u + 21B_2^u < 42B_3^s + 42\,{}^1B_{11}^{su}$.

For the completeness of our discussion, let us mention scattering of natural light (N). In the early stages of the investigation of three-photon scattering of natural light, its description was borrowed bodily from linear scattering processes. Thus, natural light was treated as a superposition of two modes with perpendicular polarizations and equal constant amplitudes, albeit with phases fluctuacting independently, which led to Strizhevsky and Klimenko's result [12], or was dealt with as one wave with constant amplitude but with fluctuating polarization direction. Already Strizhevsky [64] proved this approach to be inadequate for the correct description of nonlinear scattering of natural light, and the matter was discussed by some authors [1, 25, 65].

One can define natural light in three ways: (1) as a superposition of two orthogonal linearly polarized waves with independently fluctuating Gaussian amplitudes of equal mean intensities; (2) as a superposition of two contrary-handed circularly polarized waves with equal mean independently fluctuating intensities; and (3) as a superposition of two elliptically polarized waves with azimuths differing by $\Pi/2$ and equal mean independently fluctuating intensities. For all three definitions

$$\begin{aligned}\langle s_0^2\rangle &= 3\langle s_1^2\rangle = 3\langle s_2^2\rangle = 3\langle s_3^2\rangle = \tfrac{3}{2}\langle s_0\rangle^2\\ \langle s_0 s_1\rangle &= \langle s_0 s_2\rangle = \langle s_0 s_3\rangle = \langle s_1 s_3\rangle = \langle s_2 s_3\rangle = 0\end{aligned} \tag{56}$$

On insertion of the above relations into Eqs. (24) we arrive at

$$\begin{aligned}{}^{\mathrm{N}}S_0(\vartheta) &= \tfrac{1}{2}\Gamma_{\mathrm{N}}\left[6A + 4B + (D+E)(1+\cos^2\vartheta)\right]\\ {}^{\mathrm{N}}S_1(\vartheta) &= \tfrac{1}{2}\Gamma_{\mathrm{N}}(D+E)\sin^2\vartheta\end{aligned} \tag{57}$$

where Γ_{N} is equal to Γ with $g^{(2)} = \frac{3}{2}$. It is readily seen that the light scattered forward or backward is unpolarized (the parameter ${}^{\mathrm{N}}S_1$ then vanishes).

In particular, for perpendicular scattering due to Eqs. (26), one finds

$$\begin{aligned}{}^{\mathrm{N}}S_0(\Pi/2) &= \tfrac{1}{12}\Gamma_{\mathrm{N}}\left(196\mathrm{B}_1^{\mathrm{s}} + 186\mathrm{B}_3^{\mathrm{s}} + 245\mathrm{B}_1^{\mathrm{u}} + 231\mathrm{B}_2^{\mathrm{u}} + 84\,{}^1\mathrm{B}_{11}^{\mathrm{su}}\right)\\ {}^{\mathrm{N}}S_1(\Pi/2) &= \tfrac{1}{12}\Gamma_{\mathrm{N}}\left(140\mathrm{B}_1^{\mathrm{s}} + 30\mathrm{B}_3^{\mathrm{s}} - 35\mathrm{B}_1^{\mathrm{u}} - 105\mathrm{B}_2^{\mathrm{u}} - 84\,{}^1\mathrm{B}_{11}^{\mathrm{su}}\right)\end{aligned} \tag{58}$$

The anomalous depolarization ratio calculated from the definition (49)

$${}^{\mathrm{N}}\rho\left(\frac{\Pi}{2}\right) = \frac{26\mathrm{B}_3^{\mathrm{s}} + 56\mathrm{B}_2^{\mathrm{u}} + 2\alpha}{56\mathrm{B}_1^{\mathrm{s}} + 36\mathrm{B}_3^{\mathrm{s}} + 35\mathrm{B}_1^{\mathrm{u}} + 21\mathrm{B}_2^{\mathrm{u}}} \tag{59}$$

can range within the limits

$$0 \leq {}^{\mathrm{N}}\rho\left(\frac{\Pi}{2}\right) \leq \frac{8}{3} \tag{60}$$

where the upper limit corresponds to pure unsymmetric scattering [1, 46], while the lower one corresponds to $\alpha = 0$ (48) and $\mathrm{B}_3^{\mathrm{s}} = \mathrm{B}_2^{\mathrm{u}} = 0$. In the latter case the X component of the scattered intensity depends on one rotational invariant, $\mathrm{B}_1^{\mathrm{s}}$. In turn, the depolarization ratio (16), having here the simplified form (51), could reach its maximal value 1 if $140\mathrm{B}_1^{\mathrm{s}} + 30\mathrm{B}_3^{\mathrm{s}} = 35\mathrm{B}_1^{\mathrm{u}} + 105\mathrm{B}_2^{\mathrm{u}} + 84\,{}^1\mathrm{B}_{11}^{\mathrm{su}}$.

In the symmetric approximation both depolarization ratios have the same form:

$$ {}^{\mathrm{N}}D^{\mathrm{s}}\left(\frac{\Pi}{2}\right) = {}^{\mathrm{N}}\rho^{\mathrm{s}}\left(\frac{\Pi}{2}\right) = \frac{14\mathrm{B}_1^{\mathrm{s}} + 39\mathrm{B}_3^{\mathrm{s}}}{84\mathrm{B}_1^{\mathrm{s}} + 54\mathrm{B}_3^{\mathrm{s}}} \tag{61} $$

ranging within the interval $[\frac{1}{6}, \frac{13}{18}]$ in accordance with Refs. 1 and 46, but differing from Ref. 64 as to the upper limit.

IX. UNSYMMETRIC SCATTERING

This scattering is described by one molecular rotational invariant only, i.e., by the parameter $\mathrm{B}_2^{\mathrm{u}}$. As mentioned, hyper-Rayleigh unsymmetric light scattering is produced by molecules of the point symmetries D_3, D_4, and D_6. In turn, in the case of hyper-Raman scattering all vibrational modes related to the unsymmetric tensor $b_{\alpha\beta\gamma}^{\mathrm{u}2}$ are inactive in IR and fall into three classes with regard to their activity in ordinary Raman scattering. Vibrations responsible for pure unsymmetric hyper-Raman scattering and their Raman activity are listed in Table I [56].

Except for the mode E of the cubic groups, the only nonzero components of the tensor $b_{\alpha\beta\gamma}^{\mathrm{u}2}$ are

$$ b_{123}^{\mathrm{u}2} = b_{132}^{\mathrm{u}2} = -b_{213}^{\mathrm{u}2} = -b_{231}^{\mathrm{u}2} \tag{62} $$

and it is easily seen that this tensor is moreover antisymmetric in the indices 1 and 2. With respect to the above relations, the rotational

TABLE I
Vibrational Modes Responsible for Pure Unsymmetric Hyper-Raman Scattering and Their Raman Activity

	Modes/Raman Activity		
Point Groups	Symmetric	Antisymmetric	Inactive
D_4, D_5, D_6	A_1		
D_{2d}	B_1		
T, T_d, O	E		
C_{4v}, C_{5v}, C_6		A_2	
$C_{\infty v}$		Σ^-	
D_{3h}, D_{5h}			A''
D_{4h}, D_{6h}, D			A_{1u}
D_{4d}, D_{6d}			B_1
$D_{\infty h}$			Σ'_{u}
T_h, O_h			E_{u}

invariant B_2^u equals

$$B_2^u = 4|b_{123}^{u2}|^2 \tag{63}$$

In the case of the cubic groups, the vibrational mode E is double-degenerate and described by the following nonvanishing tensor components:

$$\begin{aligned} b_{123}^{u2}(1) &= b_{132}^{u2}(1) = b_{213}^{u2}(1) = b_{231}^{u2}(1) = -\tfrac{1}{2}b_{321}^{u2}(1) = -\tfrac{1}{2}b_{312}^{u2}(1) \\ b_{123}^{u2}(2) &= -b_{213}^{u2}(2) = \sqrt{3}\,b_{123}^{u2}(1) \end{aligned} \tag{64}$$

where the symbols 1 and 2 in parentheses refer to the first and second component of the degenerate vibrational mode E. Now, the only component (2) is antisymmetric in the indices 1 and 2. The invariant B_2^u for this mode reads

$$B_2^u = 18|b_{123}^{u2}(1)|^2 \tag{65}$$

From Eqs. (25) for unsymmetric scattering we get

$$\begin{aligned} {}^{P}S_0^u(\vartheta) &= \Gamma_B\left[2 + (1 + 2\sin^2 2\Phi - \cos 2\Phi \cos 2\Psi)\sin^2\vartheta\right] \\ {}^{P}S_1^u(\vartheta) &= -\Gamma_B\left[(1 + 2\sin^2 2\Phi)\sin^2\vartheta \right. \\ &\qquad \left. + \cos 2\Phi \cos 2\Psi(1 + \cos^2\vartheta)\right] \\ {}^{P}S_2^u(\vartheta) &= -2\Gamma_B \cos 2\Phi \sin 2\Psi \cos\vartheta \\ {}^{P}S_3^u(\vartheta) &= 2\Gamma_B \sin 2\Phi \cos\vartheta \end{aligned} \tag{66}$$

For the sake of brevity we have introduced the notation $\Gamma_B = 7\Gamma B_2^u/2$.

From the above equations it arises that for polarized incident light and $\vartheta = 0$ we have ${}^{P}S_0^u(0) = [S_1^u(0) + {}^{P}S_2^u(0) + {}^{P}S_3^u(0)]^{1/2}$, meaning that light unsymmetrically scattered forward remains fully polarized. In turn, for natural incident light, with respect to $\langle\cos 2\Phi\rangle_E = \langle\sin 2\Phi\rangle_E = \langle\cos 2\Psi\rangle_E = \langle\sin 2\Phi\rangle_E = 0$, only ${}^{P}S_0^u(0)$ differs from zero. This signifies unpolarized scattered light.

Let us now discuss unsymmetric scattering for some chosen polarizations of the incident light. To start with, we chose elliptical polarization and $\Psi = 0$. So, the elliptic major axis is normal to the observation plane.

The Stokes parameter ${}^{\mathrm{EV}}S_2^{\mathrm{u}}(\vartheta) = 0$ for all scattering angles. The remaining ones (66) then read

$$
\begin{aligned}
{}^{\mathrm{EV}\pm}S_0^{\mathrm{u}}(\vartheta) &= \Gamma_{\mathrm{B}}\left[2 + (1 + 2\sin^2 2\Phi - \cos 2\Phi)\sin^2 \vartheta\right] \\
{}^{\mathrm{EV}\pm}S_1^{\mathrm{u}}(\vartheta) &= -\Gamma_{\mathrm{B}}\left[(1 + 2\sin^2 2\Phi)\sin^2 \vartheta + \cos 2\Phi(1 + \cos^2 \vartheta)\right] \\
{}^{\mathrm{EV}\pm}S_3^{\mathrm{u}}(\vartheta) &= \pm 2\Gamma_{\mathrm{B}}|\sin 2\Phi|\cos \vartheta
\end{aligned}
\tag{67}
$$

Except for right-angled scattering, the polarized portion of the scattered light is elliptically polarized and its ellipticity depends on the scattering angle. Since ${}^{\mathrm{EV}}S_1^{\mathrm{u}}(\vartheta)$ is always negative, the scattered azimuth is horizontally oriented irrespective of ϑ. For forward scattering the reversal ratio (19)

$$
{}^{\mathrm{EV}\pm}R^{\mathrm{u}}(0) = \frac{1 - |\sin 2\Phi|}{1 + |\sin 2\Phi|} \tag{68}
$$

is identical with that of the incident light and the scattered light is fully elliptically polarized. In turn, for perpendicular scattering the parameter ${}^{\mathrm{EV}\pm}S_3^{\mathrm{u}}(\Pi/2)$ vanishes (as it should) and the depolarization ratio (16) takes the form

$$
{}^{\mathrm{EV}\pm}D^{\mathrm{u}}(\Pi/2) = \frac{1 - \cos 2\Phi}{2(1 + \sin^2 2\Phi)} \tag{69}
$$

dependent on the incident ellipticity and ranges within $0 < {}^{\mathrm{EV}\pm}D^{\mathrm{u}}(\Pi/2) < \frac{1}{4}$. Obviously, the polarized portion is polarized linearly.

Let us now assume the incident azimuth as $\Psi = \Pi/2$. In this case the Stokes parameter ${}^{\mathrm{EH}\pm}S_2^{\mathrm{u}}(\vartheta) = 0$ for all scattering angles, and for the remaining parameters from Eqs. (66) one gets

$$
\begin{aligned}
{}^{\mathrm{EH}\pm}S_0^{\mathrm{u}}(\vartheta) &= \Gamma_{\mathrm{B}}\left[2 + (1 + 2\sin^2 2\Phi + \cos 2\Phi)\sin^2 \vartheta\right] \\
{}^{\mathrm{EH}\pm}S_1^{\mathrm{u}}(\vartheta) &= -\Gamma_{\mathrm{B}}\left[(1 + 2\sin^2 2\Phi)\sin^2 \vartheta - \cos 2\Phi(1 + \cos^2 \vartheta)\right] \\
{}^{\mathrm{EH}\pm}S_3^{\mathrm{u}}(\vartheta) &= \pm 2\Gamma_{\mathrm{B}}|\sin 2\Phi|\cos \vartheta
\end{aligned}
\tag{70}
$$

It is easily seen that the parameter ${}^{\mathrm{EH}\pm}S_1^{\mathrm{u}}(\vartheta)$ can now change its sign periodically; therefore, for some angles ϑ; ${}^{\mathrm{EH}\pm}S_1^{\mathrm{u}}(\vartheta) = 0$. For these angles the parameter ${}^{\mathrm{EH}\pm}S_3^{\mathrm{u}}(\vartheta) \neq 0$, which together with ${}^{\mathrm{EH}\pm}S_2^{\mathrm{u}}(\vartheta) = 0$ means that the polarized portion of the scattered light is then polarized circularly. These angles depend on the incident ellipticity and are determined by the relation

$$\sin^2 \vartheta = \frac{2\cos 2\Phi}{1 + 2\sin^2 2\Phi + \cos 2\Phi} \tag{71}$$

which has the four solutions ϑ_C, $\Pi \pm \vartheta_C$, and $2\Pi - \vartheta_C$, where $0 < \vartheta_C < \Pi/2$. ${}^{EH\pm}S_1^u$ is positive for angles $\Pi - \vartheta_C < \vartheta < \Pi + \vartheta_C$ and $2\Pi - \vartheta_C < \vartheta < \vartheta_C$, signifying an azimuth inverted with respect to the azimuths of the light scattered in the symmetric approximation and the incident light. For $\vartheta_C < \vartheta < \Pi - \vartheta_C$ and $\Pi + \vartheta_C < \vartheta < 2\Pi - \vartheta_C$, ${}^{EH\pm}S_1^u$ is negative and the scattered azimuth is coplanar with the incident azimuth but still normal to that of the symmetric approximation. The reversal ratio (20) reads

$${}^{EH\pm}R^u(\vartheta) = \frac{2 + (1 + 2\sin^2 2\Phi + \cos 2\Phi)\sin^2 \vartheta - 2|\sin 2\Phi|\cos \vartheta}{2 + (1 + 2\sin^2 2\Phi + \cos 2\Phi)\sin^2 \vartheta + 2|\sin 2\Phi|\cos \vartheta} \tag{72}$$

and for forward scattering reduces to that of Eq. (68), as could be suspected. For the two scattering angles for which ${}^{EH\pm}S_1^u = 0$ and $\cos \vartheta > 0$, the above reversal ratio is identical to the depolarization ratio. In turn, for perpendicular scattering

$${}^{EH\pm}D^u\left(\frac{\Pi}{2}\right) = \frac{1 + \cos 2\Phi}{2(1 + \sin^2 2\Phi)} \tag{73}$$

ranging within $\frac{1}{4}{}^{EH} < {}^{\pm}D^u(\Pi/2) < 1$.

We now consider unsymmetric scattering of the linearly polarized incident light ($\Phi = 0$) and let its azimuth be vertical to the observation plane. Hence, from Eqs. (67) or (45) at $B_1^s = B_3^s = B_1^u = {}^1B_{11}^{su} = 0$ we find

$$\begin{aligned} {}^{LV}S_0^u(\vartheta) &= 2\Gamma_B \\ {}^{LV}S_1^u(\vartheta) &= -{}^{LV}S_0^u(\vartheta) \end{aligned} \tag{74}$$

irrespective of the scattering angle. The above equations signify that the scattered light is completely polarized linearly along the direction Y, which in fact means inverse polarization (with respect to the light scattered in the symmetric approximation [46] and, in this case, to the incident light as well) and ${}^{LV}D^u = 0$ for all scattering angles.

For incident light linearly polarized in the observation plane Eqs. (70) give

$$
\begin{aligned}
{}^{\mathrm{LH}}S_0^{\mathrm{u}}(\vartheta) &= 2\Gamma_{\mathrm{B}}(1 + \sin^2\vartheta) \\
{}^{\mathrm{LH}}S_1^{\mathrm{u}}(\vartheta) &= 2\Gamma_{\mathrm{B}}\cos^2\vartheta \\
{}^{\mathrm{LH}}S_2^{\mathrm{u}}(\vartheta) &= {}^{\mathrm{LH}}S_3^{\mathrm{u}}(\vartheta) = 0
\end{aligned}
\tag{75}
$$

The polarized portion of the scattered light is for the majority of scattering angles polarized linearly with the azimuth perpendicular to the observation plane and the depolarization ratio equaling

$$
{}^{\mathrm{LH}}D^{\mathrm{u}}(\vartheta) = \sin^2\vartheta \tag{76}
$$

As in the symmetric approximation, for right-angled scattering, the polarized portion of the scattered intensity vanishes: ${}^{\mathrm{LH}}S_1^{\mathrm{u}}(\Pi/2) = 0$; the scattered light becomes unpolarized.

We now consider circular polarization of the incident radiation. In this case we have to take $\Phi = \pm\Pi/4$. From Eqs. (66) or (52) at $\mathrm{B}_1^{\mathrm{s}} = \mathrm{B}_3^{\mathrm{s}} = \mathrm{B}_1^{\mathrm{u}} = {}^{1}\mathrm{B}_{11}^{\mathrm{su}} = 0$ we arrive at

$$
\begin{aligned}
{}^{\mathrm{C}\pm}S_0^{\mathrm{u}}(\vartheta) &= \Gamma_{\mathrm{B}}(2 + 3\sin^2\vartheta) \\
{}^{\mathrm{C}\pm}S_1^{\mathrm{u}}(\vartheta) &= -3\Gamma_{\mathrm{B}}\sin^2\vartheta \\
{}^{\mathrm{C}\pm}S_2^{\mathrm{u}}(\vartheta) &= 0 \\
{}^{\mathrm{C}\pm}S_3^{\mathrm{u}}(\vartheta) &= \pm 2\Gamma_{\mathrm{B}}\cos\vartheta
\end{aligned}
\tag{77}
$$

In particular, for forward scattering ${}^{\mathrm{C}\pm}S_1^{\mathrm{u}}(0) = 0$, and at ${}^{\mathrm{C}+}S_3^{\mathrm{u}}(\vartheta) > 0$ and ${}^{\mathrm{C}-}S_3^{\mathrm{u}}(\vartheta) < 0$ the depolarization ratio (16) plays simply the same role as the reversal ratio (19) and both amount to

$$
{}^{\mathrm{C}\pm}D^{\mathrm{u}}(0) = {}^{\mathrm{C}\pm}R^{\mathrm{u}}(0) = 0 \tag{78}
$$

pointing to completely circularly polarized radiation with preserved incident helicity. In turn, for perpendicular scattering ${}^{\mathrm{C}\pm}D^{\mathrm{u}}(\Pi/2) = \frac{1}{4}$ and ${}^{\mathrm{C}\pm}R^{\mathrm{u}}(\Pi/2) = 1$ (${}^{\mathrm{C}\pm}S_3^{\mathrm{u}}(\Pi/2) = 0$), meaning that the scattered light is partly linearly polarized. Since ${}^{\mathrm{C}\pm}S_1^{\mathrm{u}}$ is negative, the azimuth of the polarized portion is oriented horizontally.

In general,

$$^{C\pm}R^{u}(\vartheta) = \frac{5 - 2\cos\vartheta - 3\cos^2\vartheta}{5 + 2\cos\vartheta - 3\cos^2\vartheta} \tag{79}$$

and for angles ϑ different from those discussed above the scattered light is partly elliptically polarized.

For unsymmetric scattering of natural light, from Eq. (57) one finds

$$\begin{aligned} {}^{N}S_0^{u}(\vartheta) &= \tfrac{1}{3}\Gamma_B(11 - 5\sin^2\vartheta) \\ {}^{N}S_1^{u}(\vartheta) &= -\tfrac{5}{3}\Gamma_B\sin^2\vartheta \end{aligned} \tag{80}$$

The depolarization ratio (16) is equal to

$$^{N}D(\vartheta) = \frac{3}{8 - 5\cos^2\vartheta} \tag{81}$$

where $^{N}D(0) = 1$, meaning unpolarized light, and $^{N}D(\Pi/2) = \frac{3}{8}$, being obviously the inverse of the upper limit of Eq. (60), and denoting partly linearly polarized light with azimuth parallel to the observation plane.

The observation of the purely unsymmetric scattering (lines) is particularly important in the case of modes that are both IR and Raman inactive. One should emphasize that each of the relative polarization parameters characterizing pure unsymmetric scattering takes the same value for all molecules.

Both scatterings in question are still intensely studied both theoretically [66, 67] and experimentally, particularly the resonance hyper-Raman process [68–71]. Recent developments in this field are reviewed in Refs. 72 and 73.

References

1. K. Altmann and G. Strey, *J. Raman Spectrosc.* **12**, 1 (1982).
2. R. W. Terhune, P. D. Maker, and C. M. Savage, *Phys. Rev. Lett.* **14**, 681 (1965).
3. D. L. Weinberg, *J. Chem. Phys.* **47**, 1307 (1967).
4. P. D. Maker, *Phys. Rev.* **A1**, 923 (1970).
5. S. Kielich, J. R. Lalanne, and F. B. Martin, *Phys. Rev. Lett.* **26**, 1295 (1971).
6. L. Freund, *Phys. Rev. Lett.* **19**, 1288 (1968).
7. G. Dolino, J. Lajzerowicz, and M. Vallade, *Phys. Rev.* **B2**, 2194 (1970).
8. S. Kielich, *Physica* **30**, 1717 (1964); *Acta Phys. Pol.* **26**, 135 (1964); *J. Phys. (France)* **28**, 519 (1967).
9. Y. Y. Li, *Acta Phys. Sinica* **20**, 164 (1965).

10. R. Bersohn, Y. H. Pao, and H. L. Frisch, *J. Chem. Phys.* **45**, 3184 (1966).
11. S. Kielich, *Acta Phys. Pol.* **33**, 89 (1968); *IEEE J. Quant. Electron.* **QE-4**, 744 (1968).
12. V. L. Strizhevsky and V. M. Klimenko, *JETF* **53**, 244 (1967).
13. S. J. Cyvin, J. E. Rauch, and J. C. Decius, *J. Chem. Phys.* **44**, 4083 (1965).
14. J. F. Verdieck, S. H. Peterson, C. M. Savage, and P. D. Maker, *Chem. Phys. Lett.* **7**, 219 (1970).
15. J. H. Christie and D. J. Lockwood, *J. Chem. Phys.* **54**, 1141 (1971).
16. L. Stanton, *Mol. Phys.* **23**, 120 (1972).
17. D. A. Long and L. Stanton, *Mol. Phys.* **24**, 57 (1972).
18. T. Bancewicz, Z. Ożgo, and S. Kielich, *J. Raman Spectrosc.* **1**, 417 (1974).
19. Z. Ożgo and S. Kielich, *Physica* **72**, 191 (1974).
20. W. Alexiewicz, T. Bancewicz, S. Kielich, and Z. Ożgo, *J. Raman Spectrosc.* **2**, 59 (1974).
21. S. Kielich and M. Kozierowsky, *Acta Phys. Pol.* **A45**, 231 (1974).
22. W. Alexiewicz, *Acta Phys. Pol.* **A48**, 243 (1975).
23. S. Kielich, M. Kozierowski, and Z. Ożgo, *Chem. Phys. Lett.* **48**, 491 (1977).
24. K. Altmann and G. Strey, *Z. Naturforsch.* **A32**, 307 (1977).
25. D. L. Andrews and T. Thirunamachandran, *Opt. Commun.* **22**, 312 (1977); *J. Chem. Phys.* **68**, 2941 (1977).
26. R. Bonneville and D. S. Chemla, *Phys. Rev.* **A17**, 2046 (1978).
27. R. A. Minard, G. E. Stedman, and A. G. McLellan, *J. Chem. Phys.* **78**, 5016 (1983).
28. C. D. Churcher, *Mol. Phys.* **46**, 621 (1982).
29. D. L. Andrews and M. J. Harlow, *Mol. Phys.* **49**, 937 (1983).
30. D. V. Murphy, K. U. von Raben, R. K. Chang, and P. B. Dorain, *Chem. Phys. Lett.* **85**, 43 (1982).
31. V. I. Ostrovskii, I. Ya. Ogurtsov, and I. B. Bersuker, *Mol. Phys.* **48**, 13 (1983).
32. D. A. Long, M. J. French, T. J. Dines, and R. J. B. Hall, *J. Phys. Chem.* **88**, 547 (1984).
33. V. I. Petrov, *Opt. Spektrosk.* **59**, 469 (1985); A. V. Baranov, Y. S. Bobovich, and V. I. Petrov, *JETF* **88**, 741 (1985).
34. L. D. Ziegler, Y. C. Chung, and Y. P. Zhang, *J. Chem. Phys.*, **87**, 4498 (1987).
35. J. T. Golab, J. R. Sprague, K. T. Carron, G. C. Schatz, and R. P. van Duyne, *J. Chem. Phys.* **88**, 547 (1988).
36. S. Shin and M. Ishigame, *J. Chem. Phys.* **89**, 1892 (1988).
37. J. P., Neddersen, S. A. Mounter, J. M. Bostick, and C. K. Johnson, *J. Chem. Phys.* **90**, 4719 (1989).
38. M. J. French and D. A. Long, *Mol. Spectrosc.* **4**, 225 (1976).
39. M. J. French, in A. B. Harvey (Ed.), *Chemical Applications of Nonlinear Raman Spectroscopy*, Academic, New York, 1981, Chapter 6.
40. S. Kielich, *Prog. Opt.* **20**, 155 (1983).
41. D. A. Long, *Raman Spectroscopy*, McGraw-Hill, New York, 1977.
42. L. D. Barron, *Molecular Light Scattering and Optical Activity*, Cambridge UP, 1982.
43. S. Kielich, *Nonlinear Molecular Optics*, izd. Nauka, 1981 (in Russian).
44. Z. H. Zhu, *Acta Phys. Sinica* **21**, 1587 (1965).
45. J. H. Christie and D. J. Lockwood, *J. Chem. Phys.* **54**, 1141 (1971).

46. M. Kozierowski and S. Kielich, *Acta Phys. Pol.* **A66**, 753 (1984).
47. M. Born and E. Wolf, *Principles of Optics*, Pergamon, New York, 1964.
48. G. Placzek, *Marx Handbuch der Radiologie* **6**, 205 (1934); V. B. Berestetskii, E. M. Lifshitz, and L. P. Pitayevskii, *Relativistic Quantum Theory*, Pergamon, New York, 1971.
49. D. L. Andrews and M. J. Harlow, *Mol. Phys.* **49**, 937 (1983).
50. S. Kielich, *Acta Phys. Pol.* **20**, 433 (1961).
51. Y. Shen, Phys. Rev. **155**, 921 (1967).
52. L. D. Barron and A. D. Buckingham, *Mol. Phys.* **20**, 1111 (1971).
53. M. Kozierowski, *Phys. Rev.* **A31**, 509 (1985).
54. M. Kozierowski, Z. Ożgo, and R. Zawodny, *Mol. Phys.* **59**, 1227 (1986).
55. M. Kozierowski, *Mol. Phys.* **54**, 197 (1985).
56. M. Kozierowski, Z. Ożgo, and S. Kielich, *J. Chem. Phys.* **84**, 5271 (1986).
57. P. S. Armstrong, N. Blombergen, N. Ducuing, and P. S. Pershan, *Phys. Rev.* **127**, 1918 (1962).
58. P. S. Pershan, *Phys. Rev.* **130**, 919 (1963).
59. P. A. Franken and J. E. Ward, *Rev. Mod. Phys.* **35**, 23 (1963).
60. P. N. Butcher and T. P. McLean, *Proc. Phys. Soc.* **81**, 219 (1963); **83**, 579 (1964).
61. S. Kielich, *Proc. Phys. Soc.* **86**, 709 (1965); *Physica* **32**, 385 (1966); *Acta Phys. Pol.* **29**, 875 (1966); **30**, 393 (1966).
62. D. A. Long and L. Stanton, *Proc. R. Soc. London Ser. A* **318**, 441 (1970).
63. A. D. Buckingham, C. Graham, and R. E. Raab, *Chem. Phys. Lett.* **8**, 622 (1971).
64. V. L. Strizhevsky, *Kvantovaya Elektronika* **6**, 395 (1972).
65. L. Wołejko, M. Kozierowski, and S. Kielich, *Conf. Abstracts, EKON'78, ed UAM*, Poznań, 1978, Vol. B, p. 120.
66. S. Kielich and T. Bancewicz, *J. Raman Spectrosc.* **21**, 791 (1990).
67. D. L. Andrews and N. B. Blake, *Phys. Rev.* **A41**, 4550 (1990).
68. L. D. Ziegler and J. L. Roebber, *Chem. Phys. Lett.* **136**, 377 (1987).
69. Y. C. Chung and L. D. Ziegler, *J. Chem. Phys.* **89**, 4692 (1988).
70. S. Nie, L. A. Lipscomb, S. Feng, and N. T. Yu, *Chem. Phys. Lett.* **167**, 35 (1990).
71. C. C. Bonang and S. M. Cameron, *Chem. Phys. Lett.* **187**, 619 (1991).
72. A. V. Baranov, Ya.S. Bobovich, and V. I. Petrov, *UPhN* **160**, 35 (1990).
73. L. D. Ziegler, *J. Raman Spectrosc.* **21**, 769 (1990).

FAST MOLECULAR REORIENTATION IN LIQUID CRYSTALS PROBED BY NONLINEAR OPTICS

J. R. LALANNE

CNRS Paul Pascal, Pessac, France

J. BUCHERT

IUSL City of College of New York, New York, New York

S. KIELICH

Nonlinear Optics Department, Adam Mickiewicz University, Poznań, Poland

CONTENTS

Modern Nonlinear Optics, Part 1, Edited by Myron Evans and Stanisław Kielich. Advances in Chemical Physics Series, Vol. LXXXV.
ISBN 0-471-57546-1

I. INTRODUCTION

Liquid crystals are generally characterized by a long-range correlation between molecules that respond cooperatively to external perturbations, such as electric, magnetic flow, or optical field [1–6]. These cooperative effects give rise to large nonlinear effects generally due to optical field induced reorientation [7]. In nematics, for instance, light-induced reorientation of the director results in effects that can be used in many optical applications of these phases [8]. These torque-induced optical nonlinearities have been intensively investigated recently [9] because of the importance of their technological applications. Both the static and dynamic properties of these field-induced reorientations are found to obey the Leslie theory of director rotation [10]. Associated with cooperative rotations under the influence of the mean field created by long-range intermolecular interactions, the corresponding slow relaxation times range from microseconds to seconds [11]. Such a wide time range can be studied by excitation of the medium with short laser pulses generally obtained from the well-known mode-locking technique.

But these very short pulses can also induced pseudo-individual orientational fluctuations of molecules, as first observed fifteen years ago, with characteristic times in the picosecond range [12–15]. These fast reorientations are mainly connected with the so-called local order in the liquid crystal and have also been thoroughly studied by techniques such as dielectric spectroscopy (DS) [16], far-infrared spectroscopy (FIR) [16], quasi-elastic neutron spectroscopy (QNS) [17–20], electron spin relaxation (ESR) [21], or nuclear magnetic resonance (NMR) [22].

Particular attention should be given to depolarized Rayleigh scattering (DRS) [23] and to the three nonlinear optics (NLO) techniques, namely the optical Kerr effect (OKE) [24–26], induced transient grating (ITG) [27], and degenerate four-wave mixing (DFWM) [28]. All of these techniques have been used, not only to study isotropic (I) phases, but also to explore fast rotational dynamics in oriented liquid-crystal samples in the nematic (N) or smectic (A) (Sm-A) phases.

We provide here a review of these investigations. Section II pertains to general considerations about the third-order interaction between an optical field and matter. We successively discuss the comparison between electronic and nuclear contributions, the definition of the first-order polarizability anisotropy, the link between NLO techniques and DRS, and the relation between the various relaxation times involved in the previously cited techniques. Section III is related to the three NLO techniques used for studying fast reorientations in liquid crystals, while Section IV describes the structures of the liquid crystal phases, the compounds studied, the sample preparation, and the control of the optical quality of the alignments. Section V describes the three NLO investigations of I, N and Sm-A phases and reports the obtained results. A general conclusion and the implications of this work are presented in Section VI.

II. GENERAL CONSIDERATIONS

A. Third-Order Interaction Between a Laser Field $E(\mathbf{r}, t)$ and Matter (Assumed to Be Macroscopically Isotropic)

From a purely phenomenological point of view, the third-order electric polarization, induced at r and t, takes the form

$$P_i^3(\mathbf{r}, t) = \chi_{ijkl}^{(3)} E_j(\mathbf{r}, t) E_k(\mathbf{r}, t) E_l(\mathbf{r}, t) \tag{1}$$

with an electric field given by the classical Fourier form:

$$\check{E}(\mathbf{r}, t) = \sum_a \check{E}(\omega_a) \exp[\mathrm{i}(\mathbf{k}_a \cdot \mathbf{r} - \omega_a t)] \tag{2}$$

The a summation is here extended both over the frequencies and the wave vectors of the components. $P_i^3(\mathbf{r}, t)$ can be written [29] as

$$\begin{aligned} P_i^3(\omega_\Sigma) = \sum_{a,b,c} & E_j^{\omega_a} \exp\left[\mathrm{i}(\mathbf{k}_a + \mathbf{k}_b + \mathbf{k}_c) \cdot \mathbf{r} - \omega_\Sigma t\right] \\ & \times \Big[\chi_{\text{iso}}^{(3)}(-\omega_\Sigma, \omega_a, \omega_b, \omega_c) \delta_{ij} E_k^{\omega_b} E_k^{\omega_c} \\ & + \chi_{\text{ani}}^{(3)}(-\omega_\Sigma, \omega_a, \omega_b, \omega_c) \left(\tfrac{3}{2} E_i^{\omega_b} E_j^{\omega_c} \right. \\ & \left. + \tfrac{3}{2} E_j^{\omega_b} E_i^{\omega_c} - \delta_{ij} E_k^{\omega_b} E_k^{\omega_c}\right) \Big] \end{aligned} \tag{3}$$

where $\omega_a = 2\pi c/\lambda_a$, $\omega_b = 2\pi c/\lambda_b$, $\omega_c = 2\pi c/\lambda_c$ are the frequencies of the fields corresponding to the related wavelengths λ; c is the speed of light; and ω_Σ denotes the frequencies sum $(\omega_a + \omega_b + \omega_c)$. $\chi_{\text{iso}}^{(3)}$ and $\chi_{\text{ani}}^{(3)}$ respectively describe the isotropic and anisotropic parts of the optical nonlinearities. In fact, they are statistical parameters, functions of the temperature which can be written, by use of perturbation theory, in the form

$$\chi^{(3)}(T) = \sum_n \chi_{(n)}^{(3)} T^{(-n)} \tag{4}$$

The first two orders $n = 0$ and $n = 1$ are of great interest within the framework of this work:

1. *Zero-order approximation.* In this case $\chi_{(0)}^{(3)}$ is temperature independent and describes pure electronic phenomena, classically described about 100 years ago by Voigt [30], and later by Born [31] and Buckingham [24]. If one assumes that the molecular third-order tensor $C_{\alpha\beta\gamma\delta}$ is symmetric in the indexes $\alpha, \beta, \gamma, \delta$, one finds

$$\chi_{(0)\text{iso}}^{(3)} = \tfrac{5}{2} \chi_{(0)\text{ani}}^{(3)} = \tfrac{1}{18} \rho C_{\alpha\alpha\beta\beta} \tag{5}$$

   where ρ is the number density of molecules. This will be the case for all the liquid crystal molecules studied, because our choice of laser optical frequencies is very far from the electronic absorption ones. This effect can be viewed as an induced deformation of the electron clouds by the field. The associated times are expected to be in the 10^{-15}-s range.
2. *First-order approximations.* A complete treatment can be deduced from Debye's famous work [32]. In our case, limited to orientational phenomena induced by laser fields, we can neglect the interaction of permanent electric dipoles with the optical field. A very simple explanation was proposed by Mayer and Gires [25] in the 1960s: The

time-averaged orientational torque vanishes because of the incapability of the permanent dipoles to "follow" the optical field. In this case, we find

$$\chi^{(3)}_{(1)\mathrm{iso}}(-\omega_\Sigma, \omega_a, \omega_b, \omega_c) = 0 \tag{6}$$

$$\chi^{(3)}_{(1)\mathrm{ani}}(-\omega_\Sigma, \omega_a, \omega_b, \omega_c) = \frac{2\rho}{45k} \frac{\gamma(-\omega_\Sigma, \omega_a)\gamma(\omega_b, \omega_c)}{1 - \mathrm{i}(\omega_b + \omega_c)\tau_2} \tag{7}$$

where k is the Boltzmann constant and γ is the anisotropy of the first-order polarizability α, which will be defined further on. τ_2 is connected with the Debye time τ_D by $\tau_2 = \tau_\mathrm{D}/3$. For ordinary molecules, τ_D is about 10^{-11} s, leading to values of τ_2 in the picosecond range. Extended theory [33], which is not reported here, shows that time τ_2 also characterizes rise and decay processes in the medium. The second and third approximations contribute nothing in our case, and we immediately understand why picosecond laser pulses are mainly qualified to probe, by analysis in real time, pseudo-individual molecular reorientations, which exhibit τ_2 values in the same range of time of the pulse duration. However, the responses of liquid crystals to picosecond optical pulses integrate the electronic contributions and it is important to know their magnitude.

B. Comparison Between the Electronic and Orientational Contributions

For usual molecules, such as CS_2, electronic deformation is about 10^{-36} esu [34], while γ^2 is about 10^{-46} cm^6 [35], leading to a value of the ratio $\chi^{(3)}_{(0)\mathrm{iso}}/\chi^{(3)}_{(0)\mathrm{ani}}$ about 5×10^{-4} at room temperature. Such values can be obtained for a large number of organic molecules, with a framework not very different from that of liquid crystals molecules [34].

We have measured [36] γ^2 in 4-cyano-4-*n*-octylbiphenyl (8CB) in cyclohexane diluted solution [36] and found $\gamma^2 \neq 9 \times 10^{-46}$ cm^6 from an experiment of depolarized Rayleigh scattering performed at a wavelength of 510 nm, i.e., at approximately the same frequency as those used in nonlinear optics (532 nm). This value is about nine times larger than the corresponding one for CS_2.

Recent measurements [37] of the nonlinear part of the refractive index in 5CB (4-cyano-4′-*n*-pentylcyanobiphenyl), performed by the so-called Z-scan technique [38], using 33-ps pulses, gives values of the nonlinear parts of the refractive index (at 532 nm) $n_{2\parallel} = 1.04 \times 10^{-11}$ esu and $n_{2\perp} = 0.69 \times 10^{-11}$ esu (the subscripts $\parallel$ and $\perp$ respectively refer to light polarized parallel and perpendicular to the director of the oriented sample at 24° C, i.e., in the nematic phase). The values are smaller than that of CS_2, which is about 1.1×10^{-11} esu [34], thus confirming results obtained some years ago [39, 40].

Because the nonlinear part n_2 is directly proportional to the electronic third polarizability C, it can be easily calculated that C is about 10^{-36} esu for 5CB in the nematic state. This experimental value is about three orders of magnitude smaller than the result of a recent calculation on the cyanobiphenyl series, using Parr, Pariser and Pople or Huckel approximations [41], which predicts the electronic contribution of the mean third-order polarizability to be about 10^{-33} esu. Such high polarizabilities are, in fact, measured only for large molecules having high conjugation [42] in their central framework and appear to be unrealistic in the case of liquid crystals. Recent measurements performed in a series of naphtyl-core liquid crystals also lead to values of $\chi^{(3)}$ largely inferior to the ones of CS_2. Electronic contributions are neglected in studies using pulses of some tens of picoseconds.

C. Definition of the First-Order Polarizability Anisotropy γ

γ^2 is defined by $\gamma^2 = \frac{3}{2}[\mathrm{Tr}(\alpha\alpha) - \frac{1}{3}(\mathrm{Tr}\,\alpha)^2]$, where α is the first-order polarizability and Tr defines the trace of the second-order tensor $\boldsymbol{\alpha}$. When applied in the principal frame of the molecule, this last relation leads to [43]

$$\gamma^2 = \frac{1}{2}\left[(\alpha_{\parallel} - \alpha_{\perp,1})^2 + (\alpha_{\parallel} - \alpha_{\perp,2})^2 + (\alpha_{\perp,1} - \alpha_{\perp,2})^2\right] \qquad (8)$$

where $\alpha_{\parallel}$ is the principal polarizability along the long axis of the molecule, and $\alpha_{\perp,1}$ and $\alpha_{\perp,2}$ the ones in the planes perpendicular to this axis. It is often admitted that $\alpha_{\parallel} \gg \alpha_{\perp,1} = \alpha_{\perp,2}$ then the molecule is assumed to be optically rodlike. Such an assumption is always highly unrealistic: By assuming that the C_8H_{17} chain of 8 CB has an isotropic first-order polarizability in the plane perpendicular to the long molecular axis, we have been able to calculate [36] $\gamma_{\perp}^2 = (\alpha_{\perp,1} - \alpha_{\perp,2})^2$ for 8 CB and found the very large value 441×10^{-48} cm^6 (which is still four times larger than that of $\gamma^2 = (\alpha_{\parallel} - \alpha_{\perp})^2$ in CS_2). This point appears to be important, because it explains why an optical polarized field can produce molecular orientational effects in oriented liquid crystal samples where the director of the phase is parallel to the wave vector of the light (see Section V). All the effects described in this paper should not exist in oriented N and *Sm-A* phases composed of rod like molecules.

The parameter γ^2, which has been intensively studied for nematogens —where it determines the birefringence of the nematic directly connected to display performance—cannot be considered here as an intrinsic molecular parameter. Strongly connected with the *local* orientational order [44] in the fluid and with the shape of the cavity used for the evaluation of the so-called local field $\mathscr{L}$ [45], it should be considered as an effective parameter and should be written $\mathscr{L}\,\gamma_{\text{eff}}^2$. Molecular theories [46, 47] of $\mathscr{L}$ suggest an expression of the form $\mathscr{L} = [(n^2 + 2)/3]^4$, where n is the refractive index. The expression of γ_{eff}^2 is difficult to write in the general case of asymmetric molecules. Flygare et al. [48] proposed an expression that can be written

$$\gamma_{\text{eff}}^2 = N_{11}\,\Delta\alpha_1^2 + 2N_{22}\,\Delta\alpha_2^2 + 2\sqrt{2}\,N_{12}\,\Delta\alpha_1\,\Delta\alpha_2 \tag{9}$$

where $\Delta\alpha_1$ and $\Delta\alpha_2$ are polarizability anisotropies that can be expressed in terms of the components of the so-called principal first-order polarization tensor of the molecule:

$$\begin{aligned} \Delta\alpha_1 &= \tfrac{1}{2}(2\alpha_\perp - \alpha_{\parallel,1} - \alpha_{\perp,2}) \\ \Delta\alpha_2 &= \tfrac{1}{2}(\alpha_{\perp,1} - \alpha_{\perp,2}) \end{aligned} \tag{10}$$

The first term of Eq. (9) is the most important and can be written

$$N_{11}\,\Delta\alpha_1^2 = \gamma^2\big[1 + (N-1)\big\langle D_{00}^{(2)}(1)\,D_{00}^{(2)}(2)\big\rangle\big] \tag{11}$$

where N is the number of molecules in the scattering volume and $D_{00}^{(2)}(i)$ are the second-order Wigner rotation functions [49] of the Euler angles that describe the orientation of molecule i expressed in the laboratory coordinate system. The brackets indicate the space average. The $D_{km}^{(l)}(\boldsymbol{\Omega})$ Wigner functions are introduced in the generalized Van Hove correlation function [50] $G(\mathbf{r}, \boldsymbol{\Omega}, t)$, which can be expanded in terms of a complete set of Wigner functions ($\boldsymbol{\Omega}$ here is the orientational parameter). In the case of rodlike molecules $\Delta\alpha_1 = \alpha_\parallel - \alpha_\perp = \gamma$; $\Delta\alpha_2 = 0$ and Eq. 12 leads to the well-known expression [51]

$$\gamma_{\text{eff}}^2 = \gamma^2(1 + J_{\text{A}}) \tag{12}$$

with

$$J_{\text{A}} = (N-1)\big\langle \tfrac{3}{2}\cos^2\theta_{ij} - \tfrac{1}{2}\big\rangle \tag{13}$$

where θ_{ij} is the angle between the different C_∞ axes of the molecules.

TABLE I
Relative Values of $\chi_{(1)}^{(3)}$ and γ_{eff}^2 Versus Temperature for Simple Molecules Used in the Central Framework of Liquid Crystals

Compound	T (K) (± 0.1)	d (g/cm^3) (± 0.001)	n (632.8 nm) ($\pm 10^{-4}$)	n (1060 nm) ($\pm 10^{-4}$)	$\frac{\chi_{(1)}^{(3)}(T)}{\chi_{(1)}^{(3)}(T_0)}$	$\frac{\gamma_{\text{eff}}^2(T)}{\gamma_{\text{eff}}^2(T_0)}$
	294	0.872	1.4969	1.4902	1	1
Benzene	328.5	0.841	1.4775	1.4708	0.94	1.12
	342	0.827	1.4699	1.4629	0.97	1.24
	294	0.867	1.4933	1.4812	1	1
Toluene	324	0.838	1.4775	1.4662	0.87	1.01
	353	0.817	1.4628	1.4519	0.78	1.04
	294	1.203	1.5473	1.5277	1	1
	314.2	1.182	1.5384	1.5203	0.90	0.99
Nitrobenzene	328.5	1.166	1.5321	1.5151	0.84	0.98
	343	1.151	1.5257	1.5098	0.75	0.94
	357	1.136	1.5197	1.5047	0.69	0.92

Note. The nonlinear technique used is OKE. The reference temperature is 294 K. The wavelengths used are $\lambda_a = 632.8$ nm, $\lambda_b = \lambda_c = 1060$ nm (see Eq. 4). d is the density.

In the general case, the parameters N_{22} and N_{21} are difficult to calculate. The variations of the orientational local order with temperature are also barely predictable. However, previous work [51] performed on simple molecules shows that the local angular correlations between anisotropic molecules can be rather important. In the case of nitrobenzene, we have found $\gamma_{\text{eff}}^2/\gamma^2 = 2.2 \pm 0.1$ (this value can be compared to the corresponding one (2.8 ± 0.3) obtained by Alms et al. [44]). These correlations are in general temperature dependent. We illustrate this point by reporting results obtained on three anisotropic molecules basically used in the central framework in liquid crystals [52]. The measured values of useful physical parameters are listed in Table I.

These results are reported in Fig. 1 and are different for the three liquids studied. For toluene the decrease of $\chi_{(1)}^{(3)}$ with temperature (22%, from 293 to 353 K) is exactly as predicted by the $1/T$ temperature dependence (Fig. 1a). γ_{eff}^2 appears to be temperature independent (atropic behavior) (Fig. 1b). For benzene and nitrobenzene the $1/T$ law is not fulfilled (Fig. 1a) leading to γ_{eff}^2 respectively increasing and decreasing functions of the temperature (Fig. 1b). These two behaviors are respectively denoted diatropism and paratropism [51], and can be important. For nitrobenzene, the OKE signal, which varies as the squared value of $\chi_{(1)}^{(3)}$, is divided by a factor about two from 294 to 357 K. In this case, the short-range molecular organization (paratropic tendency to a parallel

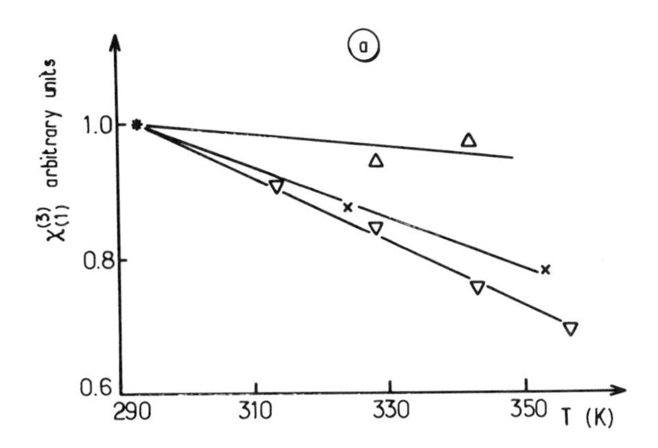

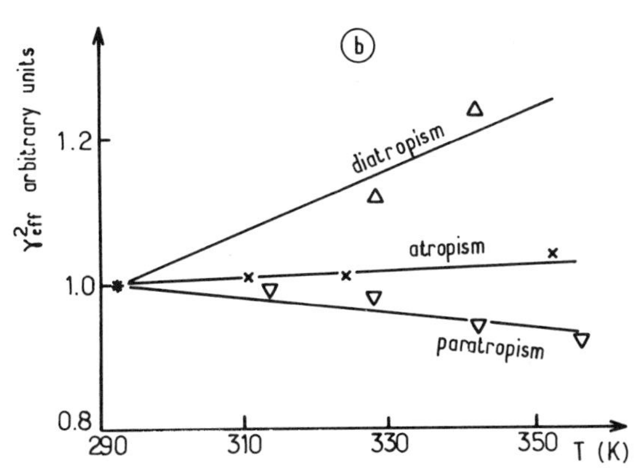

Figure 1. Variations of (a) $\chi^{(3)}_{(1)}$ and (b) γ^2_{eff} versus temperature. The values are referred to the corresponding one at $T = 294$ K; Δ, benzene; ∇, nitrobenzene; X, toluene.

organization) is progressively destroyed when the temperature rises and the effective optical anisotropy decreases. The comparison of these preliminary results to corresponding results obtained in liquid crystals should thus be fruitful.

D. The Link Between Nonlinear Optical Properties Described by $\chi^{(3)}_{(1)\text{ani}}$ and Those Probed by Depolarized Rayleigh Scattering (DRS)

The intensity of the depolarized scattered light from liquids composed of optically anisotropic molecules can be used to measure polarizability anisotropies and angular pair correlations between particles in dense fluids and solutions [35]. Starting with vertically (V) polarized incident light (frequency ω_R), the intensity of the depolarized light I_{VH} (electric field in a horizontal (H) plane) scattered by spatially isotropic fluids appears to be basically proportional to $\gamma^2(-\omega_R, \omega_R)$. This parameter includes a summation (over the molecules in the scattering volume) of the second-order spherical harmonic, which describes the orientation of the molecules and is then directly connected to the short-range orientational order of the medium. Then, if as previously assumed, the electronic contribution can be neglected in the induced third-order polarization, nonlinear optical techniques measuring $\chi^{(3)}_{(1)\mathrm{ani}}$ (Eq. (8)) and depolarized Rayleigh scattering lead to the same parameter γ^2, directly connected to the local molecular orientational order of the fluid. The proof of such a conclusion can be found from experimental work concerning strongly anisotropic simple liquids [53].

Table II reports our contribution in this domain. All the measurements relate to benzene. Comparison between the two columns on the right in Table II shows that, taking into account the experimental errors, there is rather good agreement between the two kinds of results, when they are corrected for the dispersion [54] due to the experimental use of different frequencies. This agreement is important because it allows us to directly compare results obtained in liquid crystals by NLO techniques and by Rayleigh scattering. The latter has often been used recently in liquid crystal research.

Moreover, Rayleigh light-scattering techniques can be used not only to determine the degree of orientational order in liquids and to provide

TABLE II
Comparison Between Nonlinear Optical Techniques and DRS results

	$\chi^{(3)}_{(1)\mathrm{ani}}/\chi^{(3)}_{(1)\mathrm{ani,\,benz}}$			$\dfrac{\chi^{(3)}_{(1)\mathrm{ani}}}{\chi^{(3)}_{(1)\mathrm{ani,\,benz}}}$	$i_{\mathrm{HV}}/i_{\mathrm{HV,\,benz}}$
Liquid	$\lambda_a = 441.6$ nm $\lambda_b = \lambda_c =$ 694.3 nm	$\lambda_a = 632.8$ nm $\lambda_b = \lambda_c =$ 1064 nm	$\lambda_a = 530$ nm $\lambda_b = \lambda_c =$ 1064 nm	$\lambda_a = \lambda_b = \lambda_c =$ 546.1 nm	$\lambda_R =$ 546.1 nm
Benzene	1	1	1	1	1
Toluene	1.3 ± 0.3	1.4 ± 0.3		1.3 ± 0.3	1.31 ± 0.01
Mesitylene	1.6 ± 0.3	1.5 ± 0.3		1.5 ± 0.3	1.58 ± 0.01
Nitrobenzene		5.3 ± 0.8		5.6 ± 0.8	6.18 ± 0.06
Carbon disulfide	7.1 ± 1.3	6.6 ± 1.3	6.9 ± 1.3	7.2 ± 1.3	7.59 ± 0.07

Note. The dispersion with wavelength is performed according to Ref. 54. $T = 293$ K. The nonlinear technique used here is OKE. The wavelengths $\lambda_a, \lambda_b, \lambda_c$ refer to Eq. 4.

information on the spatial extent of these correlations, but also to study how this local order affects the spectrum of the scattered light. It is not the aim of this paper to give a detailed presentation of depolarized scattering of light. It can be shown that the normalized Rayleigh spectrum leads to $g_2(\Delta_\omega)$, i.e., the Fourier transform of the $l = 2$ spherical component of the spatially averaged orientation-dependent pair correlation function included in the expression of γ^2_{eff}. The simplest model of isotropic orientational diffusion predicts pure Lorentzian line shapes deriving from a purely exponentially decaying correlation function.

The associated relaxation time [29] is $\tau_2 = \tau_D/3$ where $\tau_D = 4\pi a^3 \eta / KT$. In Debye's expression η is a shear viscosity and a the mean radius of the molecule, assumed to be spherical. In the case of asymmetric molecules, $4\pi a^3$ can be replaced by kV, where V is the volume of the particle and k an empirical shape parameter that increases as the molecule becomes more elongated [46]. The main point here is that DRS exhibits the same characteristic time τ_2 as the one included in the denominator of the $\chi^{(3)}_{1\,\text{ani}}$ expression (7). Then, direct comparison is allowed between the NLO investigations, performed in real time, and the studies of the frequency spectrum of DRS, generally recorded by spectrometric or interferometric methods. This method is used in this paper.

The relaxation time τ_2 is also an effective one, denoted $\tau_{2,\text{eff}}$ which can be calculated from the spectral half-width at half-height (HWHH) $\Delta\omega_{1/2}$ using the relation deduced from the Fourier transform:

$$\tau_{2,\text{eff}} = \frac{1}{2\pi\,\Delta\omega_{1/2}} \tag{14}$$

The exact link between $\tau_{2,\text{eff}}$ and τ_2 is difficult to establish. It is generally assumed [46], in agreement with some experimental observations [55], that $\tau_{2,\text{eff}} = (1 + J_A)\tau_2$ for symmetric top molecules, τ_2 being, as already noted, a linear function of shear viscosity.

E. Relation Between Various Relaxation Times

The orientational motions of molecules in liquids and liquid crystals also contribute to many other physical effects. It will be interesting to compare the results obtained by these techniques to those obtained by NLO and DRS. For instance, these orientational motions give an important contribution to the broadening of nuclear magnetic resonance (NMR), far-infrared absorption (FIR), Raman scattering (RS), quasi-elastic neutron spectroscopy (QNS), and electron spin relaxation (ESR) lines. In all of these cases, comparisons between various components of the spatially averaged function $G(\mathbf{\Omega}, t)$ already introduced will be possible and confrontation between the various characteristic times τ_l (l is here the order of the corresponding Wigner function) are generally worthwhile. In the case of isotropic orientational diffusion of free particles, the various τ_l are connected by the relation

$$\tau_l = 2\tau_D/l(l+1) \qquad \text{where } l \neq 0 \tag{15}$$

An important case corresponds to $l = 1$. $\tau_1 = \tau_\text{D}$ is then the dielectric decay time, obtained from dielectric relaxation (DR), and measured via the frequency dependence of the dielectric losses in the medium. In fact, τ_DR is not directly equal to τ_1, because of the response field of the surroundings to the permanent dipole $\boldsymbol{\mu}$ of the molecule, which can be important in liquid crystals [56]. Complete analysis given by Cole [57] leads to the relation

$$\tau_\text{DR} = \left(\frac{3\varepsilon_0}{2\varepsilon_0 + \varepsilon_\infty}\right)\tau_1 \tag{16}$$

where ε_0 and ε_∞ are the dielectric constants taken, respectively, for $\omega = 0$ and $\omega \to \infty$ (optical régime). For 8CB, $\varepsilon_0 \approx 6.0$ and $\varepsilon_\infty = n^2 \approx 2.4$ lead to $\tau_\text{DR} \neq 1.3\tau_1$. One should also note that $\tau_1 \sim 3\tau_2$. Relaxation rates $\tau_2 >$ 1 ps correspond to dielectric frequencies up to 250 GHz, which are presently impossible to obtain and to use correctly.

The reader should not forget that the Brownian mechanism of orientational relaxation, which has been assumed in the previous analysis, has been found to be inappropriate in many instances, especially when short-range correlations exist. Many alternative jump models have been proposed, without any convincing proof of their adequacy to a true description of the real rotations in the liquids.

III. NONLINEAR OPTICAL TECHNIQUES USED FOR THE STUDY OF FAST MOLECULAR RELAXATIONS IN LIQUID CRYSTALS

We restrict our presentation to techniques directly related to the third-order polarization $\chi^{(3)}_{(1)}$. However, we must mention the well-known technique of induced dichroism, developed in 1969 to study the effects of hydrogen bonding on orientational relaxation of laser dyes, such as rhodamine 6G [58, 59] and DODCI [60] and which is, at this time, used in our laboratory for the study of reorientation of dyes included in *Sm-A* liquid crystal films. A dichroism is induced by electronic excitation of molecules, for instance, aligned along the picrosecond pump linear polarization. The orientational distributional distribution of molecules in the ground state becomes anisotropic and the return to disorder can be investigated with the use of an absorbed auxiliary probe wave. We directly measure the sum $(1/\tau_R + 1/\tau_F)$, where τ_R and τ_F are respectively the orientational relaxation time and the characteristic time of decay of the excited species to the ground state. When $\tau_F \gg \tau_R$, direct investigation of molecular rotation is possible without any other measurement. Let us now present the principle of the three main techniques of NLO that have been used in the picosecond range.

A. Optical Kerr Effect

The principle of the optical Kerr effect (OKE) is given in Fig. 2. It is described by putting $\omega_A = \omega_{pr}$; $\omega_b = -\omega_c = \omega_p$; $\omega_\Sigma = \omega_{pr}$ in Eq. 8. In this case we obtain

$$\chi^{(3)}_{(1)\mathrm{OKE}} = \frac{\rho}{45kT}\gamma(-\omega_{pr}, \omega_{pr})\gamma(\omega_{pu}, -\omega_{pu}) \tag{17}$$

where ω_{pr} and ω_{pu} are respectively the frequencies of the probe and pump waves.

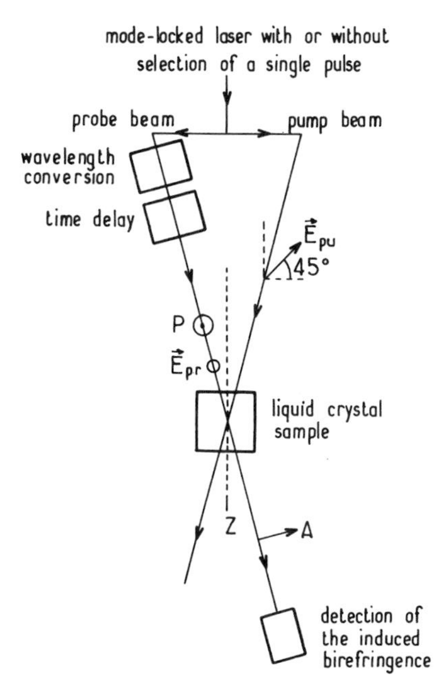

Figure 2. Principle of the OKE experiment; P and A are linear polarizations respectively perpendicular and parallel to the plane of the sheet. $\mathbf{E}_{pu}$ is at 45° from this plane.

By using the relation

$$\Delta\varepsilon_{r,ij} - \delta_{ij} = \frac{1}{\varepsilon_0}\frac{P^{(3)}_{i(1)}}{E_j} \tag{18}$$

we can calculate the following expression of the refractive index increment $\Delta n(\omega_{\mathrm{pr}})$, induced by the field $|E_{\mathrm{pu}}|$ of the pump, and analyzed by the probe:

$$\Delta n(\omega_{\mathrm{pr}}) = \frac{\chi^{(3)}_{(1)\mathrm{OKE}}|E_{\mathrm{pu}}|^2}{30\varepsilon_0 nkT} = \lambda_{\mathrm{pr}} B(\omega_{\mathrm{pr}}, \omega_{\mathrm{pu}})|E_{\mathrm{pu}}|^2 \tag{19}$$

where B is a temperature-dependent parameter describing OKE.

In the calculation of the transmission of the analyzing wave, convolution products between $\Delta n(t)$ and the rectangular function of amplitude

$$\frac{n_{\mathrm{pr}} n_{\mathrm{pu}} z}{c}\left(\frac{1}{n_{\mathrm{pu}}} \pm \frac{1}{n_{\mathrm{pr}}}\right)$$

are used [61]. z is the spatial coordinate of the traveling waves in the cell and c the speed of light. n_{pr} and n_{pu} are the mean values of the refractive indexes of the probe and pump waves; $(-)$ is used when the two waves are propagating in the same direction, and $(+)$ in the opposite case. Such a product is not adequate for thin films of liquid crystals. But for cells having some centimeters of thickness (such as the ones frequently used for the study of isotropic phases of liquid crystals) and optical waves of opposite directions, the width of the convolution product can reach values of about 100 ps, one or two orders of magnitude larger than the theoretical resolution of such an optical gate. In picosecond investigation, the probe and pump pulses must always be pointed in the same direction to avoid such a loss of time resolution. The pump light can be:

1. A high-repetition rate system, for instance, the wave given by passively or actively (or both) mode-locked dye or YAG lasers. The latter gives 25-ps pulses, which can be shortened to the picosecond range by time compression in monomode fibers. The use of cavity dumpers is often necessary to avoid accumulation effects due to the high frequency (about 100 MHz) of the delivered pulses.
2. A single pulse selected from a mode-locked laser train given by Nd/glass or Nd/YAG lasers. Techniques for achieving pulse selection have now reached an acceptable degree of reliability. Frequencies between 1 Hz and 1 kHz can be used.

The probe light can be:

1. A cw laser wave (He–Ne, He–Cd, or Ar and Kr lasers)
2. Second-harmonic pulses, delayed in time.

B. Induced Transient Grating

The induced transient grating (ITG) principle is illustrated in Fig. 3. Samples are exposed to two picosecond excitation pump pulses, having the same wavelength but crossed linear polarizations. The two optical fields cannot interfere and thus do not give the usual intensity spatial modulation, which is known to generate the acoustic stationary waves that have recently been intensively studied. However, they add vectorially and create periodic variations of the orientation of molecules, giving rise to a phase modulation. Such an effect acts as a diffraction grating for a variably delayed probe. The changes in diffracted probe intensities as a function of time reflect the orientational processes within the sample. Generally, the detected signal comes from both the real and imaginary parts of the modulated complex refractive index. But, the effects due to the imaginary part are only important—as an induced dichroism—when both the pump and probe wavelengths lie within the absorption bands of the compounds, or when the laser intensity is high enough to induce two-photon absorption. In this case, the decay of the excited species can become important and can hide the orientational part of the signal. It is possible, however, to separate the different contributions by using variable polarizations of the waves [62]. With liquid crystal, wavelengths less than 400 nm must, for this reason, be systematically avoided. The polarization grating is located on equidistant planes, perpendicular to the vector:

$$\mathbf{K} = \mathbf{K}_{\mathrm{pu},1} - \mathbf{K}_{\mathrm{pu},2} \tag{20}$$

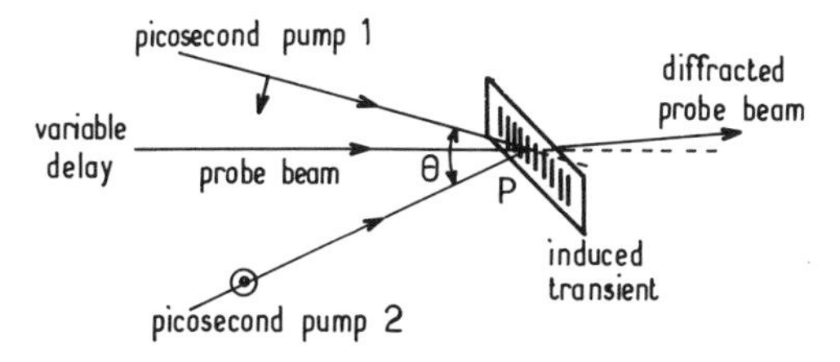

Figure 3. Principle of the ITG used for the measurement of molecular rotations. The two pump waves have crossed polarizations (horizontal and vertical). The polarization of the probe wave can be undetermined.

and separated by the spatial period:

$$\Lambda = 2\pi/|\mathbf{K}| = \lambda_{\text{pu}}/2\sin(\theta/2) \tag{21}$$

where θ is the angle $(\mathbf{k}_{\text{pu},2}, \mathbf{k}_{\text{pu},2})$ and λ_{pu} the wavelength of the pump.

The distinction between thick and thin gratings, not too strongly modulated, can be made with the aid of the parameter Q defined as

$$Q = \frac{2\pi\lambda d}{n\Lambda^2} \tag{22}$$

where d is the thickness of the material and n its refractive index. With the typical values $\lambda = 0.53\ \mu\text{m}$, $d_{\text{min}} = 250\ \mu\text{m}$, $n \simeq 1.5$, and $\Lambda \simeq 5\ \mu\text{m}$, we find $Q \simeq 20$. We are in the case of thick grating. From a holographic treatment of [63], we know that only zero and first diffraction orders are present. As a result, high diffraction efficiency can be achieved. In this case, the diffraction efficiency η is given by

$$\eta = \frac{\sin^2\left(\nu\sqrt{1 + (\Delta k_z/2\nu)^2}\right)}{1 + (\Delta k_z/2\nu)^2} \tag{23}$$

with

$$\nu = \pi\,\Delta n\,d/\lambda\cos\theta$$

$$\Delta k_z = \frac{d}{\Lambda\cos\theta}(2\pi\sin\theta - \pi\lambda/n\Lambda)$$

Δn is the amplitude of the refractive index modulation and θ the angle between the diffracted beam and the normal to the grating.

If the Bragg condition is fulfilled, $\Delta k_z = 0$ and

$$\eta = \sin^2\frac{\pi\,\Delta n\,d}{\lambda\cos\theta_B} \tag{24}$$

The maximum, ideally $\eta = 1$, is achieved for

$$\Delta n\,d = \frac{\lambda}{2}\cos\theta_B$$

with $|k_i| = 2\pi n_i/\lambda_i$ and θ_{pr}, θ_{d}, $\theta_{\text{pu},1}$ and $\theta_{\text{pu},2}$, respectively, the angles with the normal of the probe, diffracted, pump 1, and pump 2 waves is fulfilled; i.e., $\Delta k_z = 0$, the diffraction efficiency is highest. The maximum, ideally $\nu = 1$, is achieved for $\nu_d = \pi/2$. The intensity of the signal recorded versus time is basically related to $\chi^{(3)}_{(1)\mathrm{ani}}(-\omega_{\mathrm{pr}}; \omega_{\mathrm{pr}}; \omega_{\mathrm{pu}}; -\omega_{\mathrm{pu}})$ given by Eq. 7 and proportional to $\Delta n^2(\omega_{\mathrm{pr}})$ given by Eq. 20. The characteristic time is also τ_2.

C. Degenerate Four-Wave Mixing

The principle of degenerate four-wave mixing (DFWM) is illustrated in Fig. 4. In the DFWM, the sample is irradiated by two oppositely propagating pump waves (frequency ω_{L}; wave vectors $\mathbf{k}_{\mathrm{pu},1} = -\mathbf{k}_{\mathrm{pu},2}$) and by a probe (frequency ω_{L}; wave vector $\mathbf{k}_{\mathrm{pr}}$) with a linear polarization perpendicular to that of the pumps. The angle $(\mathbf{k}_{\mathrm{pu},1}, \mathbf{k}_{\mathrm{pr}})$ is about a few degrees. DFWM describes a process in which three input waves of the same frequency, but different propagation directions, mix and generate a fourth wave at the same frequency ω_{L}. The time dependence of the process, induced with picosecond pulses, follows the decay of the pertinent material excitations, providing a way to study their dynamic behaviors. It is described by $\chi^{(3)}_{(1)\mathrm{ani}}(-\omega_{\mathrm{L}}; \omega_{\mathrm{L}}; \omega_{\mathrm{L}}; -\omega_{\mathrm{L}})$ and directly linked to $\gamma^2(-\omega_{\mathrm{L}}; \omega_{\mathrm{L}})$. We list here some of the most important features of this

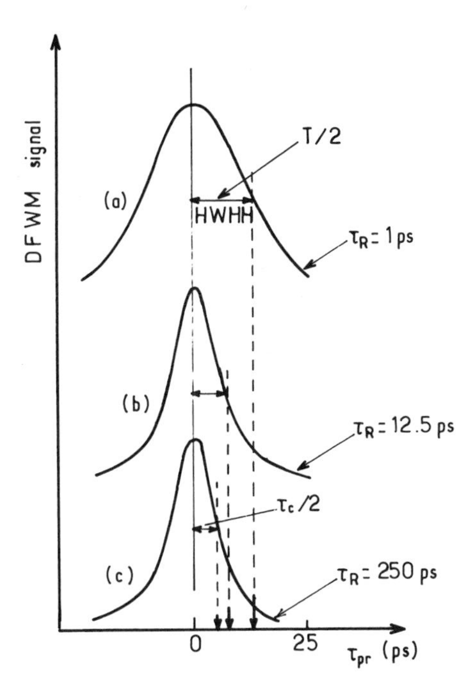

Figure 4. Computer calculation of the DFWM signal as a function of the probe time delay τ_{pr} for three values of the reorientation time τ_{R}. The pulse duration τ and the coherence time τ_{c} are held fixed at respectively 25 and 12.5 ps. (a) $\tau_{\mathrm{R}} \ll \tau$; (c) $\tau_{\mathrm{R}} \gg \tau$; (b) intermediate value of τ_{R}.

effect, which have numerous applications:

- From a phase-matching consideration, it can be shown that the fourth wave has a conjugate phase of the probe and propagates in its reverse direction [28].
- DFWM is often described in terms of real-time holography [63]. As in ITG, two of the three incoming waves produce gratings. As already stated, any gratings built between pump 1 and pump 2 do not markedly contribute to the backward wave because of the phase mismatch in this direction. Pump 1 (respectively pump 2) and the probe produce polarization gratings of periods Λ_1 and Λ_2, respectively, given by

$$\begin{aligned} \Lambda_1 &= \lambda/2\sin(\theta/2) \qquad \text{(pump 1–probe)} \\ \Lambda_2 &= \lambda/2\cos(\theta/2) \qquad \text{(pump 2–probe)} \end{aligned} \tag{25}$$

  The angle θ being small (a few degrees), we have $\Lambda_1 \gg \Lambda_2$. The DFWM signal is mainly due to the diffraction of pump 1 (respectively 2) by the phase grating produced by pump 2 (respectively 1) and probe, two waves having crossed polarizations. The analogy with ITG is obvious.
- $\gamma^2(-\omega_L, \omega_L)$ obtained with the DFWM technique does not suffer from the dispersion with frequencies, which affects both OKE and ITG and which can be important. Thus, DFWM and DRS lead to single-frequency determinations and can easily be performed at the same wavelength. This advantage should not be neglected.
- The most important advantages of DFWM lie in the fact that, besides the already mentioned conservation of linear momentum $\hbar|\mathbf{k}|$ of the photons, the energy ($\hbar\omega$) is also conserved, and the incident probe field is amplified simultaneously to the emission of intense backward waves. The sample behaves as a conjugate mirror with increased reflectivity [64]. This point is crucial because it allows the study of very thin samples (down to 50 μm) of well-oriented liquid crystal molecules, which cannot be studied by OKE, for which no coherent energy transfer occurs, and which need the use of thicker samples, if one want to avoid laser pulses that are too intense.
- As regards wave vector and phase, DFWM behaves like an ideal effect. For polarization, the situation appears to be more complicated [65]. In fact, the polarization of the backward wave should be deduced from the general laws of conservation of the angular momentum in a fixed quantization frame. It depends on the isotropic or anisotropic

character of the illuminated material. In the choosen geometry, the directions of polarization of the pumps and of the probe define the normal axes of the experiment, which remains unchanged during the nonlinear interaction in an isotropic medium. The result is that possible optical activity of samples does not play an important part in DFWM, while it can lead to a permanent, chirality-dependent, additional transparency in OKE.

A difficulty occurs in the NLO techniques when the duration τ of the laser pulse (HWHH) is of the same order of magnitude as $\tau_R = \tau_2$, the relaxation time of the species. The usual solid-state lasers exhibit nonlinear optical behavior of their active material, leading to phase modulation of the pulses. Phase-modulated light pulses [66] are characterized by the coherence time $\tau_c \approx 1/\Delta\nu$, where $\Delta\nu$ is the width of the pulse spectrum. For the different kinds of lasers used in this work, phase modulation is mainly due to the nonlinear refractive index of the glass or YAG rods and is known to be moderate (i.e., $0.5 \leq \tau_c \leq \tau$, where τ is the pulse duration; $\Delta\nu$ is about 50 GHz). In this case, the pulse amplitude varies with time:

$$E(t) \approx \varepsilon(t)\exp\left[ia\varepsilon^2(t)\right]$$

where $a \approx 2\tau/\tau_c$ measures the phase modulation, and $\varepsilon(t)$ is given by

$$\varepsilon(t) = \exp\left[-2\ln 2(t/\tau)^2\right] \tag{26}$$

When the reorientation time τ_R is in the same range of values as τ_c, which is the case in this work, the shape of the phase modulation is essential and the HWHH of the recorded signal as a function of the probe time delay τ_{pr} is largely dependent on the ratio τ_R/τ.

Figure 5 gives computer simulations [67] of the recorded signals of ITG and DFWM, when there is no delay between the two pumps, in the three important cases $\tau_R = 250$ ps, $\tau_R = 12.5$ ps, and $\tau_2 = 1$ ps. The value of τ has been fixed at 25 ps (YAG laser pulse mean duration). The difference between the HWHH values are obvious: They allow one to estimate the relaxation time of nonlinear media. Such a method has already been used for the last few years and can be considered a useful supplement to the conventional pump–probe techniques, when the relaxation time to be measured is closed to the technical resolution limit. Results have been reported concerning crystalline silicon and liquids. Some of us have also studied oriented liquid crystals [67], but precise results can be obtained only in the two limiting cases $\tau_c \approx \tau \ll \tau_R$ and $\tau_c \approx \tau \gg \tau_R$. Moreover, the results depend on the knowledge of the magnitude of the phase

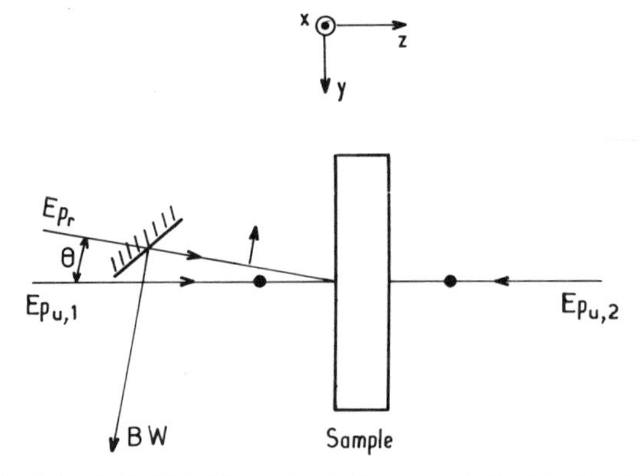

Figure 5. Principle of DFWM. The probe is linearly polarized in the plane of the figure. The two pumps have linear polarization perpendicular to the plane of the figure. BW, backward wave; $\theta \approx 7°$.

modulation. We have preferred to avoid such difficulties in our DFWM experiments.

IV. THE LIQUID CRYSTALS AND THEIR PREPARATION

A. Structure of the Phases

An illustration of the molecular structure of isotropic (I), nematic (N) and smectic A and C (*Sm-A*; *Sm-C*) is given in Fig. 6. I phases are often considered as ordinary liquids. In fact, they exhibit rather complicated behavior. In N phases, there exists an orientational order only of the long molecular axis. This privileged direction is referred to by the unit vector **n**, called the director of the phase. In *Sm-A* phases, an additional translational order leads to a layer structure characterized by the unit vector **N** perpendicular to the layers. For *Sm-A* phases at rest, **n** and **N** are parallel to the Oz axis of the laboratory frame. Such a vision is highly idealized and fluctuations of both **n** and **N** occur in the two phases, due to all thermal, mechanical, and acoustical fluctuations in the samples. By lowering the temperature, one successively induces $I \rightarrow N$; $N \rightarrow$ *Sm-A*; *Sm-A* $\rightarrow$ *Sm-C* phase transitions. Some phases can be lacking in the classical sequence. By increasing the temperature, the opposite sequence is observed. Moreover, pretransitionnal behavior [1] exists in each phase. Near the phase-transition temperature, fluctuations develop in the high-temperature phase, giving rise to the local instantaneous order, which prefigures the order existing in the phase stable at lower temperature.

When chiral molecules are involved, the N phase is replaced by a cholesteric one, N^*, and original properties, due to a new helical order, are observed in the I^*, $Sm\text{-}A^*$, and $Sm\text{-}C^*$ phases. The most important of these new properties is ferroelectricity.

B. The Compounds Studied

The nine compounds studied along with some of their principal physical properties are listed in Table III. Compounds 1–3 do not exhibit any stable thermotropic liquid crystal phases above their melting points. They can be considered as pseudo-nematogens in the temperature domain experimentally available. Compound 8 (MBBA) has a low chemical stability and its correct study remains difficult. All the other compounds can be studied without marked difficulty. Compounds 4, 5, 8, and 9 were purchased and used without further purification. The Merck mixture 5 has a compensated helical pitch in the N^* phase that diverges close to the $Sm\text{-}A$ phase transition.

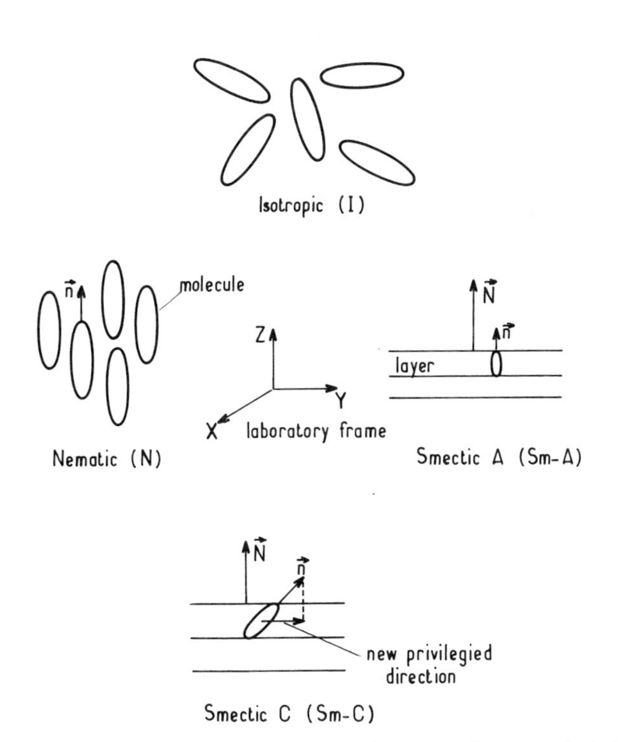

Figure 6. The structures of idealized isotropic (I), nematic (N), smectic-A ($Sm\text{-}A$), and smectic-C ($Sm\text{-}C$) phases at rest.

TABLE III
The Studied Compounds and Some of Their Physical Properties

No.	Compound	Origin	Molecular Formula	Chirality	Ferro-electricity	T (°C) $S_C \rightleftharpoons S_A$ or $S_C^* \rightleftharpoons S_A$	$S_A \rightleftharpoons N$ or $S_A \rightleftharpoons N^*$	$N \rightleftharpoons I$ or $N^* \rightleftharpoons I$
1	MBA (*p*-methyl-benzilidene-*p*′,*n*′-butylaniline)	Home-synthesized	$H_3C-C_6H_4-CH=N-C_6H_4-C_4H_9$	No	No			
2	MPA (*p*-methyl-benzilidene-*p*′,*n*-propylaniline)	Home-synthesized	$H_3C-C_6H_4-CH=N-C_6H_4-C_3H_7$	No	No			
3	MMA (*p*-methyl-benzilidene-*p*′-methylaniline)	Home-synthesized	$H_3C-C_6H_4-CH=N-C_6H_4-CH_3$	No	No			
4	8CB (*p*,*n*-octyl-*p*′-cyanobiphenyl)	BDH	$CN-C_6H_4-C_6H_4-C_8H_{17}$	No	No		33.5	40.8

5	ZLI 3488	Merck	Mixture unknown	Yes	Yes	61.0	66.0	85
6	4-*n*-Heptyl-oxyphenyl	Home-synthesized	$C_{10}H_{21}O-C_6H_4-C(=O)-O-C_6H_4-OC_7H_{15}$	No	No	83.2	87.1	90.6
7	4′-*n*-Decyloxy-benzoate-4-[2-methyl-butanoyloxy-phenyl]-4-decyl-oxythiobenzoate (racemic mixture)	Home-synthesized	$C_{10}H_{21}O-C_6H_4-C(=O)-S-C_6H_4-OC(=O)-\overset{*}{C}H(CH_3)-CH_2-CH_3$	Achiral by com-pensation	No	72.7	74.6	76
8	MBBA (*p*-methoxy-benzilidene-*p*′,*n*-methylaniline)	BDH	$H_3C-O-C_6H_4-CH=N-C_6H_4-C_4H_9$	No	No			42.5
9	MBPB (*p*-methoxy-benzoate-*p*′,*n*-pentylbenzene)	Thomson	$H_3C-O-C_6H_4-C(=O)-O-C_6H_4-C_5H_{11}$	No	No			42.4

C. Sample Preparation

From a general point of view, we must note that all the compounds studied were filtered through a pore size of 0.1 μm to remove dust particles. All were warmed up to their clearing points to eliminate all gas bubbles, which would exist in high-viscosity phases (*Sm-A*, for instance). All the cells were glass cells, always used without any adhesive or elastomer, and were housed in temperature-controlled ovens with a nominal temperature stability of 0.01°C. In the studies of OKE in isotropic phases (I), cells in the range of the centimeter range were used. For oriented phases, two thicknesses, 200 and 500 μm, were used. The most difficult problem is obtaining samples of high optical quality with a good alignment of particles. The reported work concerns homeotropic alignment in the nematic (N) and smectic A (*Sm-A*) phases. In such an alignment, the director $\mathbf{n}$ of the phase (along the Oz axis; see Fig. 6) is perpendicular to the faces of the cells (xOy plane). This was obtained by coating the inner faces of the windows of the cell with octadecyl triethoxysilane in propanol-2 solution. A partial polymerization of the product was performed by heating the coating at 120°C for 1 h. The cell was filled with the compound in the I phase. Then, the temperature was slowly decreased to the *Sm-A* phase in a few hours. The alignment can be greatly improved with static magnetic inductions perpendicular to the surfaces. Two magnets of about 3000g and 10 000g were used. The orientation of molecules occurs in the N phase. When it does not exist, orientation can be obtained at the $I \rightleftharpoons$ *Sm-A* phase transition by using the highest induction. For the Merck mixture 5, alignment can be obtained only at the $N^* \rightleftharpoons$ *Sm-A* phase transition, where the pitch diverges. One must note that there is no helical pitch in the classical *Sm-A** phases of our studies, except in the vicinity of the *Sm-A** $\rightleftharpoons$ *Sm-C** phase transition, where, as previously mentioned, pretransitional order can occur (the *Sm-C** phase has a helical structure).

D. Control of the Optical Quality of the Samples

The *Sm-C* and *Sm-C** phases of the compounds in this experiment are opaque and cannot be studied. All the other phases are transparent and look like glass. They must be free from dust and bubbles when observed with a magnifying lens (I phases). For N and *Sm-A* phases, two additional steps were performed: First, the alignment was readily verified by observing the characteristic dark cross in the conoscopic pattern under a polarizing microscope using focused light. Then, the extinction ratio was measured with parallel light, between crossed polarizers both normal to the wave vector of the light and to the director of the phase. Typical values of about 1/3000 were obtained for *Sm-A* phases. Samples were rejected if the ratio was higher than 1/1000. For *Sm-A* phases, we also verified that optical activity does not play an important part in the DFWM experiment. For 500-μm thickness, rotations less than some minutes of the arc were measured, far from the $Sm\text{-}A \rightleftharpoons Sm\text{-}C^*$ phase transition, in agreement with other investigations [68, 69].

The optical quality of the sample governs its behavior in the applied high-power light pulses. The laser intensity is always adjusted, using glass filters, to values where laser-induced dielectric breakdown within the sample occurs very infrequently. For all the liquid crystals studied, the refuse produced by a single breakdown would reduce the dielectric strength of the sample enough to necessitate its change. Moreover, when it accidentally occurs, visible damage is observed on the sample or on the windows of the cell. Typically, the laser has to be attenuated to less than 0.1 MW to prevent breakdown. In OKE experiments, we do not focus the pump and probe waves. In DFWM experiments, lenses of long focal lengths (some tens of centimeters) are used, not to focus the waves, but to slightly reduce their diameters.

V. EXPERIMENTS AND RESULTS

A. OKE Investigations

As previously noted, the laser-induced refractive index increment Δn_{OKE} is proportional to the OKE constant B (Eq. (19)), which basically varies as the length l of the sample. The OKE signal varies as l^2 and it is difficult to detect reliable OKE signals with samples of some hundreds of micrometers such as the ones containing oriented *N* or *Sm-A* phases. Thus, the reported OKE investigations concern only *I* phases.

1. *Studies of Slow Responses to Picosecond Pulses [12]*

a. Methods and Materials. The experimental setup is shown in Fig. 7. The pump wave was provided by a home-made passively mode-locked Nd^{3+}/glass laser. The dye used was the Eastman Kodak 9740 solution housed in a cell, 100 μm thick, placed inside an afocal optical (diverging lens-concave mirror $R = 5$ m) system constituting the rear reflector of the cavity. This allowed us to reduce the energy density inside the dye solution and thus to increase its lifetime. The cavity round-trip was about 6 ns. Pulse selection was performed by a home-made electronic shutter driven by the mode-locked oscillator. An amplifier (gain about 30) increased the

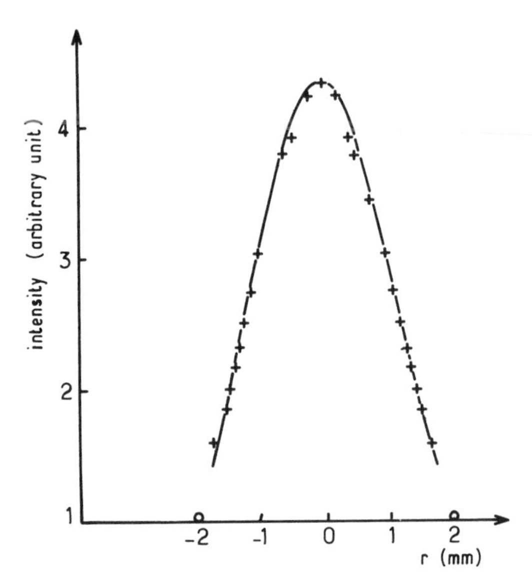

Figure 7. Experimental setup. P_1, P_3, Glan prism; P_2, Wollaston prism; D_1, CSF CPA 1443 photocell; D_2, Hewlett-Packard 5082-4207 photodiode; PM_1, La Radiotechnique 150 CVD photomultiplier; PM_2, La Radiotechnique XP 2018 photomultiplier; L, beamsplitters; the pump laser is a single pulse Nd^{3+}/glass laser; the cw probing laser can be He–Ne or He–Cd Spectra Physics models 125 or 185.

energy of a single pulse up to a few millijoules. Of course, the transverse structure of the inducing wave must be highly reproductible. The amplitude of the OKE signal appears to be proportional to

$$\left[\int_x \int_y \left[E_{pu}^2(x, y) E_{2pr}(x, y)\, dx\, dy \right]^2,\right.$$

where (x, y) are the coordinates in a plane perpendicular to the common direction of the wavevectors of both pump and probe. Usually, $E_{\mathrm{pr}}(x, y)$ corresponds to the TEM_{00} mode of a gas laser and has a good time range stability. Thus, a simple condition for keeping the value of the last expression constant is to use a pump laser oscillating on the TEM_{00} mode. The mode selection was obtained by inserting a pinhole with 2.0-mm diameter into the laser cavity. Figure 8 shows the transverse structure of the 1064-nm pump laser recorded near the sample. As a consequence of Fresnel diffraction, the structure was expected to become nearly Gaussian at some distance from the laser. The fit of the experimental results by a

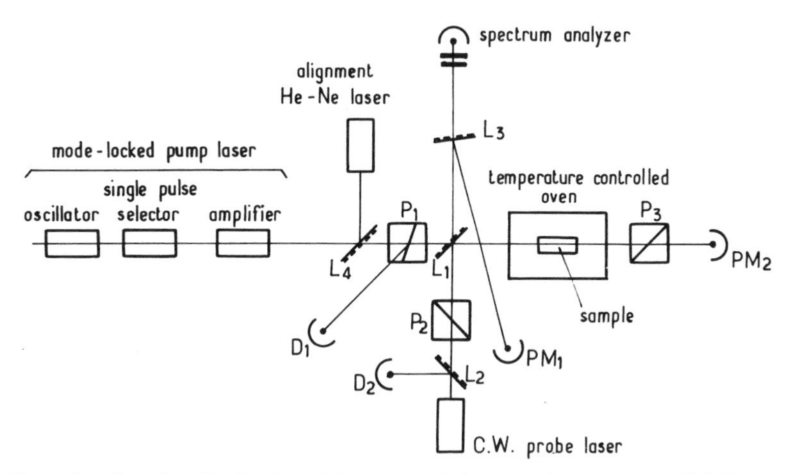

Figure 8. Gaussian distribution of the energy of the pump laser versus radial distance to the axis. The detection photodiode was placed near the sample. +, experimental points; −, fit of experimental results to $\exp(-r^2/a^2)$. a was kept free during the fit.

Gaussian distribution seems to be good. The divergence is near $\Delta\theta \neq 6 \times 10^{-4}$ rad, corresponding to the theoretical value of the far-field divergence of a fundamental mode of 560-μm beam waist ($\gamma_0 = \lambda_{pu}/\pi\,\Delta\theta$).

The natural complement to the study of transverse structure of the pump wave is the study of its time duration. We measured τ_{pu} with a two-photon fluorescence method performed on the single pulse and found $\tau_{pu} = (7 \pm 1)$ ps. The pump wave was linearly polarized by the Brewster cut of the active rod and by the Glan prism P_1 at 45° to the vertical. The reflected part of the pump was sampled by a fast photocell which triggered the oscilloscope. The energy of the pump pulse was measured by a photomultiplier coupled to a photon-coupling device.

For probe waves we used both single-mode He–Ne ($\lambda_{pr} = 632$ nm) and He–Cd ($\lambda_{pr} = 441.6$ nm) lasers (mean powers of a few milliwatts). The intensity of the probe during the pump action in the sample was measured by means of a pin photodiode coupled with a storage oscilloscope. The probe was linearly and vertically polarized by a Wollaston prism, carefully crossed with the Glan prism placed after the cell.

Detection of the OKE signal was performed by a fast photomultiplier. We measured its impulse response by using pulses of less than 1-ns duration delivered by a flash generator. The response is then quasi-symmetrical with no trailing tail. The rise time and the FWHH were, respectively, 1.5 and 2.4 ns. The photomultiplier was directly connected to a 500-MHz digitizer.

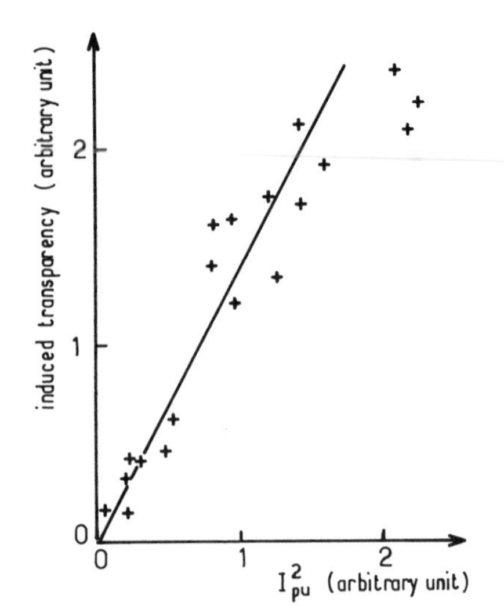

Figure 9. Variations of the induced transparency τ versus I_{pu}^2. $+$, experimental points; $-$, linear dependence for weak values of I_{pu}^2; compound: MBPB at 45.00 ± 0.01°C; $\lambda_{\text{pr}} =$ 632.8 nm; length of the cell: 4 cm.

Figure 9 shows the variations of the OKE signal (which is proportional to $\sin^2(\Delta\psi/2)$, where $\Delta\psi$ is the phase change in the cell) versus $I_{\text{pu}}^2 \propto E_{\text{pu}}^4$. The variation of the phase is easily related to Δn_{OKE} and B by

$$\Delta\psi = \frac{2\pi\,\Delta n_{\text{OKE}}}{\lambda_{\text{pr}}} = 2\pi l B E_{\text{pu}}^2 \tag{27}$$

where l is the length of the cell. We see that, for the highest values of I_{pu}, the signal measured is no longer proportional to I_{pu}^2. We may assume that, in this case, $\sin^2 \Delta\psi/2$ can no longer be approximated by $\Delta\psi^2/4$. In fact, the observed divergence can also be attributed to a self-focusing effect [70]. We shall not discuss here the numerous and complex mechanisms responsible for this effect. As is well known, in the picosecond range and in the case of liquids with large values of OKE constants, molecular reorientation plays the main role [71]. Correlatively, parasitic nonlinear effects (Raman and Brillouin stimulated effects, for instance) are generated in the sample and can dissipate an important part of the pump energy. We do not try to estimate the relative importance of these two contributions and we make an effort to perform the experiment with the lowest intensity of the pump, for which the OKE signal is always linear in I_{pu}^2.

b. Results. Compounds 8 and 9 were studied. In Fig. 10 we show an example of an OKE signal recorded versus time at fixed temperature for

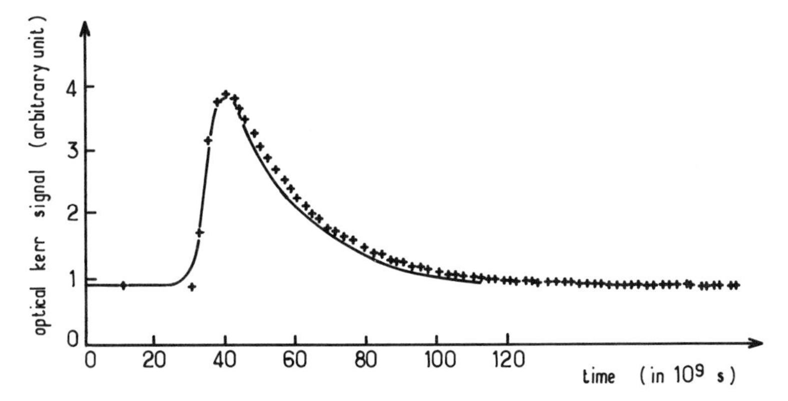

Figure 10. Variations of OKE signal versus time. $\tau_{pu} = 7 \pm 1$ ps; $\lambda_{pr} = 632.8$ nm; compound: MBPB at (50.00 ± 0.02)°C; length of the cell: 30 cm; –, experimental oscilloscope trace; $\cdots$, fit of results to Eq. 42.

the compound MBPB in its isotropic phase. The rise time is in the nanosecond range, i.e., of the same order as the rise time of the photomultiplier that records the signal. The relaxation of the birefringence is in the 10^{-8}-s range. A logarithmic plot (not reported here) is in favor of a pure exponential relaxation with a characteristic time τ_R. Obviously, such a relaxation time is two or three orders of magnitude larger than the one expected from pure molecular reorientation. The Landau–de Gennes theory [72, 73] of the isotropic-nematic phase transition can be successfully used for the description of the observed relaxation. As we have noted, the nematic phase of a liquid crystal is composed of anisotropic particles having their long axis nearly parallel to the director **n** and their centers of mass randomly distributed. Macroscopically, the medium is uniaxial, but the molecular permanent electrical dipoles are rather uncoupled and the symmetry of the phase is quadrupolar, with a microscopic order parameter describing the orientational correlations between the long axes given by

$$Q_{ij} = \tfrac{1}{2} S(n_i n_j - \delta_{ij}) \tag{28}$$

Here n_i is the component of the local optical axis **n** along the direction **i** of the laboratory frame, S is a scalar order parameter, and Q_{ij} is a traceless second-order tensor. If θ denotes the angle between **n** and the direction of the long axis, then S can be written in the form

$$S = \tfrac{1}{2}\langle 3\cos^2\theta - 1\rangle$$

Beyond T_{NI}, the value of S decreases to zero discontinuously and the sample becomes macroscopically isotropic (I phase). In the I phase, when T decreases toward T_{NI}, pretransitional nematic order develops, leading to fluctuations characterized by a lifetime τ and a geometrical extension measured by the correlation length ξ.

Let us now briefly recall the de Gennes theory. The free energy per unit volume F is expanded as a power function of the order parameter (the elastic terms have been omitted):

$$F = F_0 + \tfrac{1}{2}AQ_{ij}Q_{ij} - \tfrac{1}{3}BQ_{ij}Q_{jk}Q_{ki} + \tfrac{1}{4}CQ_{ij}Q_{jk}Q_{kl}Q_{li} - \tfrac{1}{2}\chi_{ij}E_iE_j + \cdots \tag{29}$$

where E is a field and χ_{ij} is the corresponding susceptibility. Q_{ij} is a new order parameter, defined from macroscopic data only and such that

$$Q_{ij} = \frac{3}{2\,\Delta\chi}\left(\chi_{ij} - \frac{1}{3}\chi_{kk}\delta_{ij}\right) \tag{30}$$

Here $\Delta\chi = (\chi_{\parallel} - \chi_{\perp})$, where $\chi_{\parallel}$ and $\chi_{\perp}$ refer to a perfectly aligned medium with $S = 1$. The coefficient A is assumed to vary with temperature:

$$A = a(T - T_c)^{\alpha} \qquad \alpha > 0 \tag{31}$$

The NI transition occurs at T_{NI} slightly above T_c, given by

$$A = a(T_{NI} - T_c)^{\alpha} = 2B^2/27C \tag{32}$$

At T_{NI}, S jumps from 0 to $2B/9C$. According to the thermodynamics of irreversible processes, we can define a viscosity η by

$$\eta\dot{Q}_{ij} = -\frac{\partial F}{\partial Q_{ij}} \tag{33}$$

We have neglected the coupling between Q_{ij} and the velocity gradients by assuming the absence of flow constraint in the sample.

Taking into account only the first term in the right-hand part of Eq. (29), we can write

$$\eta\dot{\mathrm{Q}}_{ij} \neq -\mathrm{AQ}_{ij} + \tfrac{1}{2}\,\Delta\chi_{\omega_{\mathrm{pu}}}\left(E_iE_j - \tfrac{1}{3}E^2\delta_{ij}\right) \tag{34}$$

where $\Delta\chi(\omega_{pu})$ is related to the optical Kerr constant $B(\omega_{pu}, \omega_{pr})$ and to the refractive index $n(\omega_{pr})$ at frequency ω_{pr} and wavelength λ_{pr} of the probe wave by

$$B(\omega_{pr}, \omega_{pu} = \frac{\Delta\chi(\omega_{pr})\Delta\chi(\omega_{pu})}{4\varepsilon n(\omega_{pr})\lambda_{pr} A} \tag{35}$$

with $A = a(T - T_c)^{\alpha}$ (Eq. 32). The OKE constant varies as A^{-1}, i.e., as $1/(T - T_c)^{\alpha}$.

The theoretical prediction of the value of α requires a microscopic description of the system, allowing the calculation of the angular correlation parameter J_A (Eq. (13)). The link with the phenomenological theory is then ensured by the microscopic expression of the OKE constant:

$$B(\omega_{pr}, \omega_{pu}) \propto \gamma(-\omega_{pr}, \omega_{pr})\gamma(-\omega_{pu}, \omega_{pu})(1 + J_A) \tag{36}$$

Let us recall here that mean-field theory [74] leads to $(1 + J_A) \propto 1/(T - T_c)$, that is, $\alpha = 1$. It is now possible, from Eq. (35), to calculate the phase variation $\Delta\varphi$ induced at time t and length l. We find

$$\Delta\varphi = \frac{2\pi}{\lambda_{pr}} \int_0^l \Delta n(z, t)\, dz \tag{37}$$

with

$$\Delta n(z, t) = B(\omega_{pr}, \omega_{pu})\lambda_{pr} \int_{-\infty}^{t} E_{pu}^2(z, t') f(t - t')\, dt' \tag{38}$$

$f(t)$ is given by integration of Eq. (34) and appears as a pure exponential relaxation which can be written for positive times:

$$f_1(t) = \frac{1}{\tau_R} \exp\left(\frac{-t}{\tau_R}\right) \tag{39}$$

where $\tau_R = \eta/a(T - T_c)^{\alpha}$.

Because the orientational relaxation times involved here are at least of the order of several nanoseconds, one may neglect the group mismatch [75] between pump and probe. Moreover, the time dependent of E_{pu}^2 will be approximated by a Dirac pulse. Thus,

$$E_{pu}^2(z, t') = I_{pu} e^{-\alpha_{pu} z} \delta(t')$$

where $I_{pu} = \int_{-\infty}^{+\infty} E_{pu}^2 \, dt$ is the energy of the pump pulse. Equation (38) becomes

$$\Delta n(z,t) = \frac{B(\omega_{pr}, \omega_{pu}) \lambda_{pu} I_{pu} e^{-\alpha_{pu} z}}{\tau_R} Y(t) \exp\left(\frac{-t}{\tau_R}\right) \tag{40}$$

Then, putting this result in Eq. (37), we find

$$\Delta\varphi = \frac{2\pi B(\omega_{pr}, \omega_{pu}) I_{pu}\left[1 - \exp(-\alpha_{pl} l)\right]}{\alpha_{pu}\tau_R} Y(t) \exp\left(\frac{-t}{\tau_R}\right) \tag{41}$$

where $Y(t) = 1$ for $t > 0$ and $Y(t) = 0$ for $t < 0$.

The OKE signal is proportional to

$$\mathscr{T} = \exp(-\alpha_{pr} l) \sin^2 \frac{\Delta\varphi}{2} \approx \frac{1}{4} \exp(-\alpha_{pr} l) \Delta\varphi^2$$

where α_{pr} is the absorption rate of the probe assumed to be polarization independent. $\mathscr{T}$ will be called induced transparency. The OKE signal decreases when τ_R increases as $1/\tau_R^2$. We must observe a pure exponential decay with a time constant $\tau_R/2$. Figure 11 shows the experimental results [76] obtained with MBPB—taking into account the response time

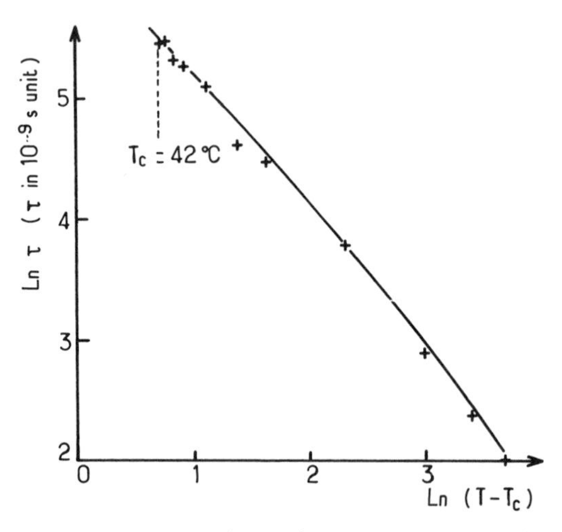

Figure 11. Variations of τ_R with $\ln(T - T_c)$. Compound: MBPB; length of the cell: 30 cm; τ_{pu}: 7 ± 1 ps; $\lambda_{pr} = 632.8$ nm; +, experimental values; −; fit to Eq. 42.

TABLE IV
Absorption Rate of MBPB at the Three Wavelengths Used

	$\lambda = 632.8$ nm										
T (°C)	42	42.2	42.4	42.6	42.8	43	44	45	50	57	65
α (m^{-1})	4.38	4.08	3.45	2.93	2.70	2.68	1.74	1.44	1.40	1.11	0.14
	$\lambda = 441.6$ nm										
T (°C)	42.10	42.31	42.72	43.12	44.20	45.06	47.15	49.11	51.19	53.16	57.11
α (m^{-1})	32.10	24.0	14.78	12.08	7.58	7.58	3.40	3.56	3.20	2.40	2.02
	$\lambda = 1060$ nm										
T (°C)	42	42.2	42.4	42.6	42.8	43	44	45	50	57	62
α (m^{-1})	3.48	3.48	3.48	3.48	3.17	3.17	2.86	2.25	1.65	1.36	0.79

of the photomultiplier used for detection—and its fit to Eq. (41). Agreement is good and confirms the single relaxation expected. This has also been observed for all other temperatures checked. The relaxation times vary from about 7 ns (at 38° above the $I \rightleftharpoons N$ phase transition) to 230 ns at the phase transition temperature. In these experiments, a 632.8-nm probe wavelength was used. Quite the same results have been observed by probing at 441.6 nm, i.e., for a wave with an 8 times greater absorption coefficient. Table IV gives the measured absorption coefficients at the three wavelengths used. Their variations with temperature appear to be important.

The identity of the observed behaviors beings the proof that the probe wave does not induce any appreciable thermal effects in the long cells (many centimeters) used here. A quantitative comparison with de Gennes behavior is presented in Fig. 12. The results have been fit to the relation

$$\tau_R = \eta / a(T - T_c) \tag{42}$$

by assuming a temperature dependence of the viscosity coefficient η of the type $\eta = \eta_0 \exp(T_0/T)$ (where η_0 and T_0 are constants) and putting $\alpha = 1$, as predicted by mean-field theory. The agreement is good, leading to values of $T_0 \neq 1475$ K and $T_c \neq 40.0$°C. The same results were obtained with MBBA. They confirm older results obtained with Q-switched laser pulses in the range of about 10 ns, a duration that cannot be neglected when compared to the relaxation time of the sample.

Let us now discuss the variations of the parameter B with temperature. Equation (36) shows that $B^{-1}(\omega_{pr}, \omega_{pu})$ varies as $(T - T_c)^{\alpha}$. We have shown that $\alpha \neq 1$. Then $B^{-1}(\omega_{pr}, \omega_{pu})$ must exhibit a linear variation with

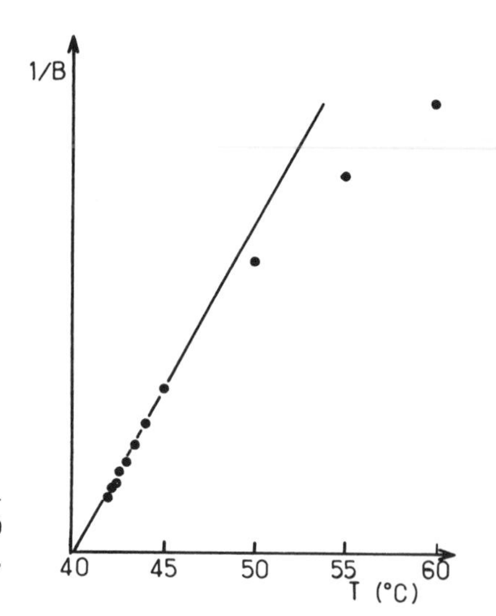

Figure 12. Variations of B^{-1} versus T. Compound: MBPB; length of the cell: 4 cm; $\tau_{pu} \neq 30$ ns; $\lambda_{pr} = 441.6$ nm; ●, experimental points; −, linear dependence in the vicinity of T_{NI}.

temperature. This linear behavior was first observed by Shen et al. [77, 78] about twenty years ago. During the same period we observed, for many compounds, nonlinear thermal behavior [74]. Figure 13 shows results obtained with MBPB, and nanosecond pulse excitation, where what we observe is an increasing departure from the linear behavior for high temperatures, i.e., far from the $N \rightleftharpoons I$ phase transition. Such an anomaly was also reported in some Cotton–Mouton and Kerr effects investigations [80, 81]. Our investigations led to the conclusion that a $(T - T_c)^{-1}$-type law is obeyed only if $(T - T_c)$ is less than about 10°. At higher temperatures, the decrease of B is slower than predicted by the mean-field law. An explanation can be found if we keep in mind that, till now, the theoretical background chosen has been that of the $I \rightleftharpoons N$ phase transition.

We have assumed that the fluctuations of the order parameter, characteristic of the phase transition are the only relevant mechanism describing the response of the system to the optical field. Such a simplification appears to be relevant in the vicinity of the transition temperature, where these fluctuations are large enough to hide any other noncritical mechanism. When the temperature increases, the fluctuations of the macroscopic order parameter and, consequently, the nematogenic character of the liquid, decrease drastically. Thus, we may expect its behavior to approach that of ordinary liquids; that is, a supplementary noncritical contribution should appear at high temperatures. In Fig. 14, B is plotted vs. $(T - T_c)^{-1}$, with the value of T_c taken from the previously described relaxation

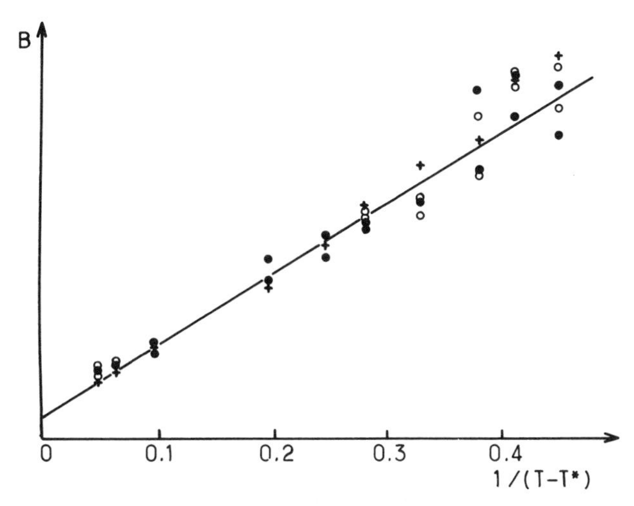

Figure 13. Variations of B versus $1/T - T^*$. Compound: MBPB; length of the cell: 4 cm; $\tau_{pu} \neq 30$ ns; $\lambda_{pr} = 632.8$ nm; •, +, 0: experimental points corresponding to three different methods of evaluation of B (see Ref. 76); −, fit to Eq. 44. During the fit, the value $T_c = 40.0°C$ (obtained from independent relaxation data) was held fixed.

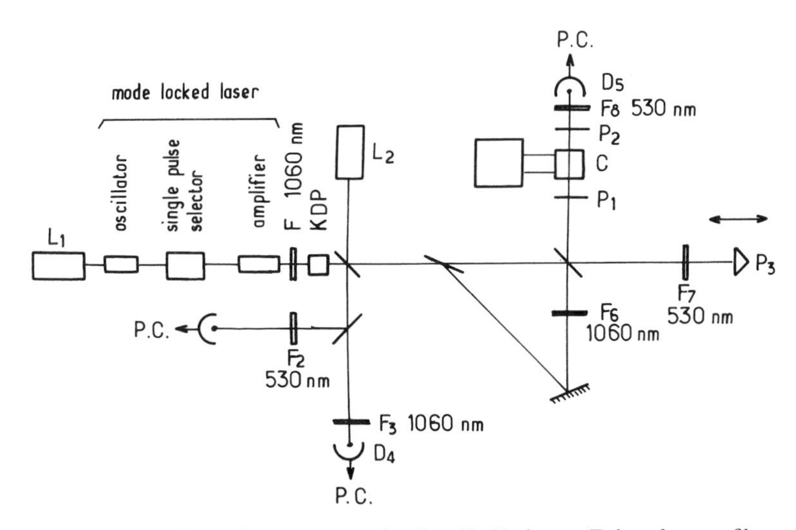

Figure 14. Experimental arrangement. L_1, L_2; He-Ne lasers; F, interference filters; P_1, P_2, HN22 polaroid; P_3, prism; D, photomultipliers associated to photon-counting devices; C, sample in its oven.

studies. The graph seems effectively well fitted by the function

$$B = B_1 + B_2 = \frac{b}{(T - T_c)} + \frac{b'}{T} \tag{43}$$

where B_1 is the contribution related to de Gennes' theory in the mean-field approximation, while B_2 appears to be a smoothly varying contribution. This additional contribution must also be included in a correct treatment of a depolarized Rayleigh scattering experiment performed in MBBA [76]. Of course, the additional contribution is expected to relax much more quickly than the "cooperative" one of the order parameter. It is quite impossible to detect it in the OKE relaxation presented in Fig. 11. We therefore decided to use a more appropriate approach in our OKE investigations.

2. *Studies of Fast Responses in OKE Investigations [14, 15]*

a. Methods and Materials. The experimental setup is shown in Fig. 14. The OKE is induced by the $\lambda_{pu} = 1060$ nm, $\tau_{pu} = (7 \pm 1)$-ps pulses given by the Nd^{3+}/glass laser previously described, with a linear polarization at 45° to the optical axes, respectively horizontal and vertical of the polarizers P_1 and P_2. Green pulses at 530 nm were generated by second-harmonic generation (SHG) in a KDP crystal, slightly mistuned to reduce the conversion and giving linear horizontal polarization. The green pulses were used to probe the induced birefringence created by the infrared pump pulses in the sample. We measured the duration τ_{pr} of these pulses using classical techniques of NLO [15] and found $\tau_{pr} = (5 \pm 1)$ ps.

The probe pulses can be continuously delayed in time with respect to the pump pulses, up to a value of 5 ns and with an uncertainty not larger than 1 ps, by varying the position of the prism PR. At each fixed position of PR, photomultipliers associated with photon counting techniques simultaneously measured the energies of the pump, the probe, and the OKE pulses. We controlled both the dependence of the signal with I_{pu}^2 and the linear transmission of the sample with I_{pr} (absence of nonlinear effects on the probe).

As we mentioned, the pump and probe pulses propagated in the same direction to avoid a loss of time resolution during the investigations. An example of the recorded signal is given in Fig. 15. As expected from previous discussion, fast relaxation can be observed, associated with slower relaxation.

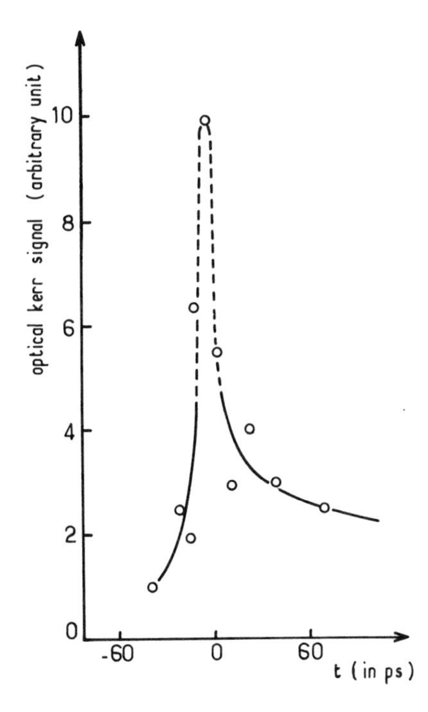

Figure 15. Example of OKE signal obtained with MPBP; (50.00 ± 0.00)°C.

Let us assume the presence of two relaxations, fast (F) and slow (S). The response function must now be written

$$f_2(t) = \frac{C_S}{\tau_{R,S}} \exp\left(-\frac{t}{\tau_{R,S}}\right) + \frac{C_F}{\tau_{R,F}} \exp\left(-\frac{t}{\tau_{R,F}}\right) \tag{44}$$

The transient birefringence induced at time t and coordinate z along the axis of the sample is given by

$$\Delta n(z,t) = B\lambda_{pr} \int_{-\infty}^{t} f_2(t-t') E_{pu}^2(z,t')\, dt' \tag{45}$$

By assuming Gaussian time distributions for the pulses

$$\begin{aligned} I_{pu}(z,t) &= I_{pu,O} \exp\left[-\frac{(t - n_{pu} z/c)^2}{\tau_{pu}^2}\right] \\ I_{pt}(z,t) &= I_{pr,O} \exp\left[-\frac{(t - n_{pr} z/c)^2}{\tau_{pr}^2}\right] \end{aligned} \tag{46}$$

where $n_{pu}, \alpha_{pu}, n_{pr}, \alpha_{pr}$ are pump and probe refractive indexes (n) and absorption rates (α).

One finds

$$\Delta n(z,t) = B\lambda_{pr} I_{pu,O} \exp(-\alpha_{pu} z) \sum_{K=S,F} (C_K/\tau_{R,K}) \exp\left(\tau_{pu}^2/4\tau_{R,K}^2\right) \times \exp\left[-\frac{(t - n_{pu} z/c)}{\tau_{R,K}}\right] \times \left\{1 + \text{erf}\left[\frac{\left(t - z n_{pu}/c - \tau_{pu}^2/2\tau_{R,K}\right)}{\tau_{pu}}\right]\right\} \tag{47}$$

and the phase variation $\Delta\varphi$ reads

$$\begin{aligned}\Delta\varphi(l,t) &= \frac{2\pi}{\lambda_{pr}} \int_0^l \Delta n(z, t + n_{pr} z/c)\, dz \\ &= 2\pi B \int_0^l \exp(-\alpha_{pu} z) \sum_{K=S,F} (C_K/\tau_{R,K}) \\ &\quad \times \exp\left(\frac{\tau_{pu}^2}{4\tau_{R,K}^2}\right) \exp\left[-\frac{(t + \beta z)}{\tau_{R,K}}\right] \\ &\quad \times \left\{1 + \text{erf}\left[\frac{\left(t + \beta z - \tau_{pu}^2/2\tau_{R,K}\right)}{\tau_{pu}}\right]\right\}\end{aligned} \tag{48}$$

where $\beta = (n_{pr} - n_{pu})/c$. Thus, we can obtain the OKE signal (S), with a delay Δt_{pr} between pump and probe pulses, by calculating the convolution product

$$S(\Delta t) = \exp(-\alpha_{pr} l) \int_{-\infty}^{+\infty} I_{pr}(\Delta t_{pr} - t') \sin^2 \frac{\Delta\varphi(t')}{2}\, dt' \tag{49}$$

b. Results. We report here the results concerning MBBA and MMA, MPA, and MBA, which have no nematic stable phases in the temperature range studied. The calculation of $S(\Delta t)$ assumes the knowledge of both refractive indexes (at λ_{pr} and λ_{pu}) and absorption rates at the same wavelengths. We measured the transmission of the studied compounds, at each temperature, for the pump and probe wavelengths. We found $\alpha_{pr} \ll \alpha_{pu}$, except in the vicinity of the phase transition, where the absorption is mainly due to scattering. The obtained value of α_{pu}, at 47°C and for MBBA, is $16.4 \pm 0.8\ m^{-1}$.

The refractive index of MBBA at λ_{pr} is deduced from earlier results [82]; $n(\lambda_{pu})$ is calculated by applying classical dispersion laws [15]. We found, always at 47°C, $n_{pr} = 1.624 \pm 0.001$ and $n_{pu} = 1.425 \pm 0.004$. Thus, we calculated $S(\Delta t)$ for different values of $\tau_{R,F}$, keeping $\tau_{R,S}$ fixed at the value 100 ns, deduced from an earlier experiment at 47.00 ± 0.01°C performed in our laboratory. We then measured the FWHH time and its variations with $\tau_{R,F}$. Results are reported in Fig. 16. This curve can be used to directly determine an approximate value of the relaxation time of the fast component from recorded OKE signals. Results are shown in Fig. 17. MBBA exhibits fast relaxations with τ_{RF} lying in the 5-ps range, against a flat background which is obviously connected with the slow response ($\tau_{R,S} = 100$ ns). The ratio C_F/C_S has been found equal to $(1.7 \pm 0.3) \times 10^{-3}$ at 47°C. Of course, it is impossible to go beyond this purely qualitative conclusion. MBPB (Fig. 18) exhibits a similar behavior, with perhaps a more marked tendency to many relaxations in the picosecond range. The three other compounds, which have no nematic phases, show only the fast relaxations with comparable HWHH times. These reorientation times are in agreement with the results of Rayleigh scattering investigations of Amer et al. [13] who reported bandwidths measured in the wings of the spectrum in the 3.5- to 5.5-cm^{-1} range. (A reorientational time of 5 ps corresponds to a bandwidth of about 1 cm^{-1}). The variations in the intensity of the fast OKE signal and its FWHH time with

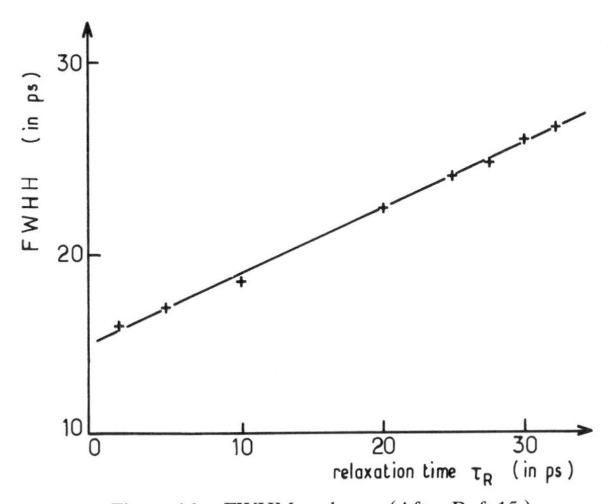

Figure 16. FWHM vs. ln τ_R. (After Ref. 15.)

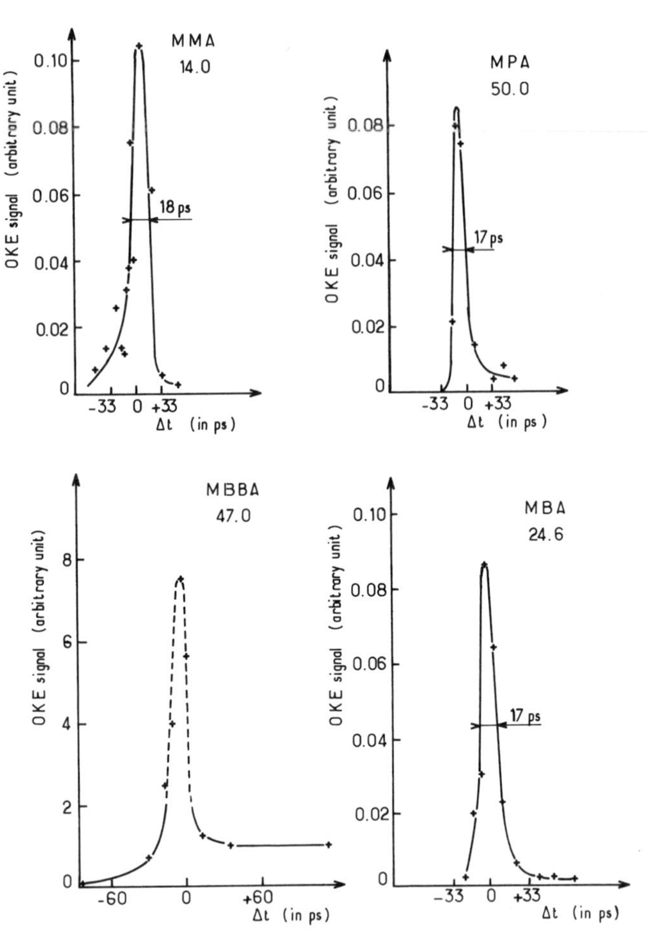

Figure 17. OKE signals versus pump–probe delay Δt. MMA: 94.00 ± 0.01°C: MPA: 50.00 ± 0.01°C; MBA: 24.60 ± 0.01°C; MBBA: 47.00 ± 0.01°C; $\tau_{pu} = 7 \pm 1$ ps; $\tau_{pr} = 5 \pm 1$ ps; length of the cell: 4 cm. (After Ref. 15.)

temperature are reported in Fig. 19. As expected, both appear to be quite temperature independent.

OKE measurements performed in the isotropic phases of nematogens confirm the adequacy of both the de Gennes theory and the mean-field approximation to describe the response of such phases when submitted to intense optical fields. But, from both picosecond OKE and depolarized Rayleigh scattering investigations, it appears that fast contributions with no critical temperature dependence must be taken into account. These

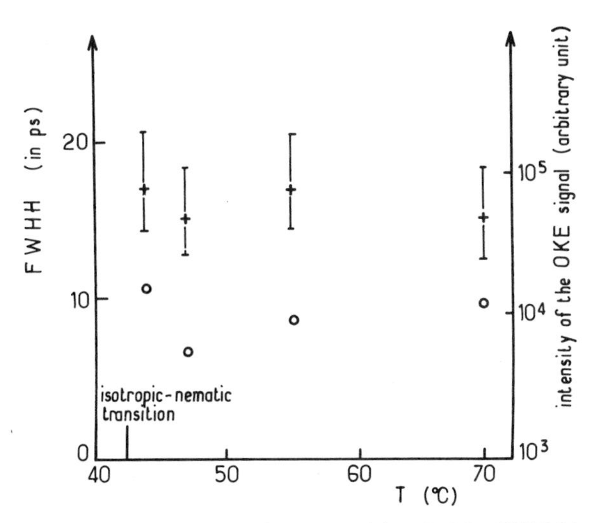

Figure 18. Variations of the intensity (0, right scale) and of the FWHM (+, left scale) of MBBA vs. temperature; $\tau_{pu} = 7 \pm 1$ ps; $\tau_{pr} = 5 \pm 1$ ps; length of the cell: 4 cm.

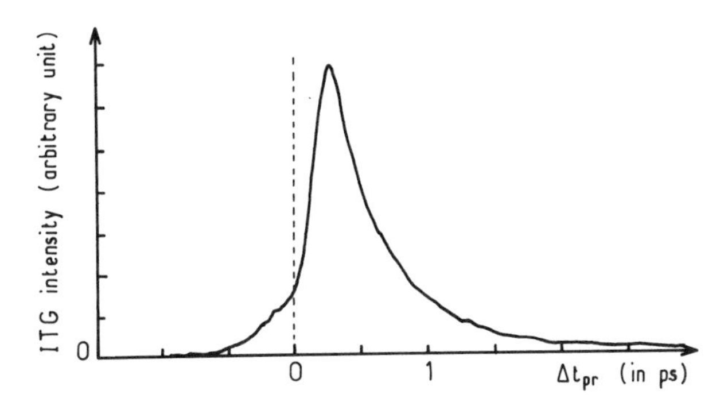

Figure 19. ITG signal (pure nuclear origin) versus probe delay. Compound: 5CB (*I* phase); $t = (42.0 \pm 0.2)$°C; $\lambda_{pu} = \lambda_{pr} = 665$ nm; $\tau_{pu} = \tau_p r \simeq 200$ fs; thickness of the cell: 1 mm. (Reprinted from F. W. Deeg and M. D. Fayer, *J. Chem. Phys.* **91**, 2269 (1989), Fig. 8d, p. 2277.)

fast reorientations can probably be interpreted by involving local order fluctuations in the orientations of the correlated molecules in the phase. Of course, this work can only lead to an approximate value of the decay time (about 5 ps) of the "mean" fast relaxation. The two nematogens exhibit identical behavior, virtually temperature independent. We shall now see how these results have been recently confirmed and their accuracy exhanced by using subpicosecond ITG investigations.

B. ITG Investigations

ITA investigations concern all the *I*, *N*, and *Sm-A* phases. The last two phases have been studied in homeotropic alignment. The previous description of the OKE results of the *I* phases suggested the presence of many relaxations from the picosecond up to the microsecond range. We will see that such a complexity is also one of the main characteristics of these two much-ordered phases.

1. Isotropic Phases

We report here investigations of *p-n*-pentyl-*p*′-cyanobiphenyl (5CB) and *p*-cyanobenzylidene-*p*-octyloxyaniline (CBOA) by Eyring and Fayer (EF) [83] and by Deeg, Fayer, et al. (DF) [84, 85]. EF used pump ($\lambda_{pu} = 1064$ nm) and probe ($\lambda_{pr} = 530$ nm) pulses of the duration $\tau_{pu} = \tau_{pr} = 100$ ps. The pulse frequency was 400 Hz. The energy of the pumps was about 15 μJ (meanpower on the 280-μm-thick sample: 14 mW). In the crossed polarization geometry, they observed results in very good agreement with those of the OKE. Slow relaxation ($\tau_{R,S} = 60$ ns in 5CB at 42°C) was observed and a fast component, nonexponential and composed of a range of decay times from 2 ns down to the duration of the pulses, was evidenced. This fast relaxation is associated with individual rotational reorientation in "pockets" of different characteristics.

Five years later, DF developed a powerful and elegant technique using 50-μJ, 200 to 300-fs pulses, and a 1-kHz repetition rate at wavelengths of 575 and 665 nm, focused down to 120-μm spot sizes (50 GW/cm^2). By varying both the angles between the pumps and detected signal polarizations, they were able to separate many contributions and, especially, to isolate pure nuclear reorientation. Figure 19 shows such a relaxation obtained with 5CB sample, 1 mm thick. While reorientational relaxation seems not to be hydrodynamical (it cannot be described in terms of the Debye model and displays complex dynamics up to 200 ps), the "mean" relaxation time appears to be about 1 ps. This is exactly what can be predicted from rotational hydrodynamics of species with a moment of inertia of about 10^{-13} g/cm, associated to viscosity of about 10^{-1} P, a typical value in *I* phases. A model describing such complex relaxations has been proposed. Values less than 1 ps are probably associated with intramolecular rotations of atom groups in the molecule. Moreover, the observed behavior is weakly temperature dependent down to the $I \rightleftharpoons N$ phase transition. This beautiful experiment brings quantitative confirmation of the pioneering OKE work, performed fifteen years ago with less highly evolved techniques.

2. *Nematic Phases*

Under laser pulses from cw to a nanosecond regime, the third-order nonlinearities are mainly due to director-axis reorientation and thermal or densities changes. These slow responses, which may be very useful in optical applications of liquid crystals, have been intensely studied lately [5, 87–90]. They also appear in picosecond-induced gratings as recently demonstrated by Eichler and Macdonald (EM) [91], and Khoo et al. [90]. EM used 80-ps pulses, at $\lambda_{pu} = 532$ nm, with crossed polarizations and a cw probing wave at $\lambda_{pr} = 488$ nm. The energy of the pumps was less than 1 mJ. Slow effects with rise and decay times respectively some tens of nanoseconds and milliseconds were observed. The rise time is explained by the pump-induced director deformation of the phase, associated with laser induced flow, while the relaxation is governed by the visco-elastic properties of the sample. These slow effects generally occur when the pump electric field exhibits a component in the direction of the director $\mathbf{n}$ of the phase. In the EM experiment $(\mathbf{k}_{pu}, \mathbf{n}) = 22.5°$ and the nematic barrier [92] was overcome. Khoo used 66-ps pulses at $\lambda_{pr} = 532$ nm.

If the wave vectors of the pumps are almost parallel to the director, these slow responses strongly decrease in the case of orthogonal polarizations of the pumps [91]. EF observed that the slow component of the Kerr relaxation disappears. This was interpretated in terms of onset of the

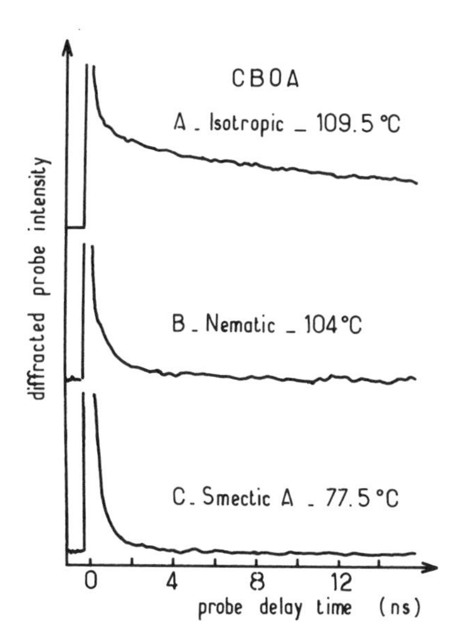

Figure 20. ITG signal versus probe delay. Compound: CBOA; $\lambda_{pu} = 1064$ nm; $\lambda_{pr} = 532$ nm; $\tau_{pu} = \tau_{pr} \simeq 100$ ps; thickness of the cell: 280 μm. (Reprinted from G. Eyring and M. D. Fayer, *J. Chem. Phys.* **81**, 4314 (1984) Fig. 4, p. 4318.)

namatic barrier, the optical torques not being sufficient to produce the collective slow rotation of the molecules. But the fast relaxations remain measurable, despite a strong reduction of their amplitude (by a factor of 10), as shown in Fig. 20. We see in the next section that we have confirmed this reduction from the I to the N phase with our DFWM experiments.

3. *Smectic-A Phases*

As far as we know, the only work remaining is that of EF (Fig. 20). As in the N phase, the slow component does not appear. The fast relaxations are slightly reduced in amplitude (by a factor of 3) and the FWHH time approaches the same order of magnitude as the response time of the detection, showing a more pronounced evolution toward pure individual molecular relaxations.

C. DFWM Investigations [95, 96]

Before describing our experimental setup, let us mention here the pioneering work of Yariv et al. [93], who used the large and nonresonant third-order susceptibility of the I phases to perform DFWM investigation, and that of Madden et al. [94], who used 25-ns pulses at 1064 nm to study I samples of 1 cm thickness by using the same technique. Our experimental setup is described in Fig. 21. We used a simultaneously Q-switched and mode-locked home-made YAG laser. The mode-locking mode was performed by using an acousto-optic mode locker coupled to passive dye locking. A single pulse (λ = 1060 nm: $\tau \simeq$ 25 ps) was selected from the first half of the pulse train by a Pockels single-pulse selector at a 1-Hz frequency rate. After amplification by a double passing YAG amplifier, a second-harmonic generator produced single green (λ = 530 nm) pulses of about 1 mJ energy. An expanding telescope was used to increase the diameter of the selected TEM_{00} mode up to about 3.5 mm. These pulses were monitored by a fast photodiode coupled to a fast digitizer that triggered the signal acquisition of a computer that drove the laser, the mechanical drivers, the photon counting devices, and performed the mathematical treatment of the results. The two pump and probe waves were separated by several beamsplitters and sent on the 250- or 500-μm samples after a reduction of their diameter to about 1.5 mm. The intensities of the two pumps and the probe were respectively about 200 and 100 μJ when they reached the sample (500 MW/cm^2). The sample was placed inside an oven where the temperature was regulated to ± 0.01 K. The probe and the first pump can be delayed by mechanical drivers. The distance length between the sample and the photon-counting device was about 3 m. This distance, together with the pinhole (diameter about 5 mm) placed in front of the detector, largely reduced the spurious signals coming

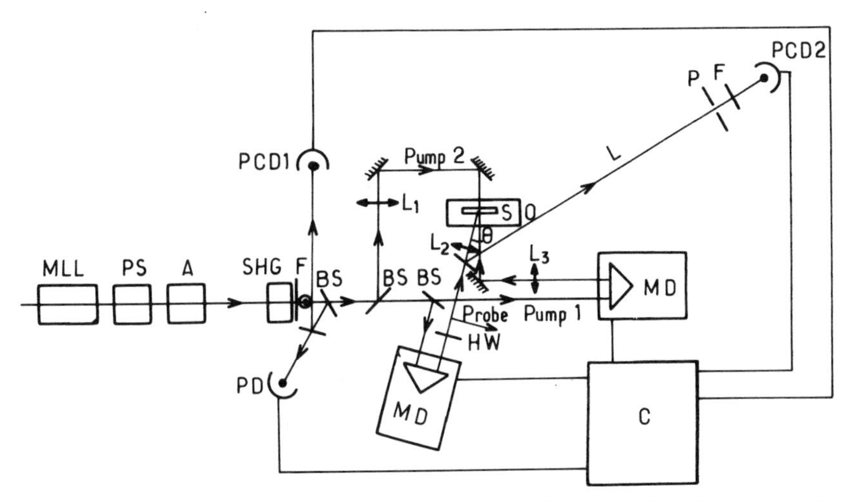

Figure 21. DFWM experimental setup. MLL, mode-locked laser; PS, pulse selector; A, double-scan amplifier; SHG, second-harmonic generator; F, interference filters; $\lambda = 530$ nm; BS, beamsplitters; HW, half-wave plate, $\lambda = 530$ nm; S, sample; O, oven; PD, fast photodiode; PCD, photon counting device; C, computer; MD, mechanical delays; P, pinhole, $\phi \simeq 5$ mm; $L \simeq 3$ m; $\theta \simeq 7°$; L_1, lens $f = 50$ cm; L_2, lens $f = 16$ cm; L_3, lens $f = 40$ cm. (After Ref. 96.)

from all the components of the setup, thus affording the detection of the DFWM signal on a very weak background. At each laser pulse, the laser wave intensity was monitored, allowing the necessary check of the signal dependence with the third power of the laser intensity. Because of this power law, the laser intensity stability must be carefully controlled. An accurate adjustment of the oscillator cavity length, as required by the active mode-locking, yielded a ratio of about 0.05 of the standard deviation and the signal. Spurious laser pulses, which can sometimes occur, were rejected by the computer. The linear transmission of the material was carefully controlled for the laser pump before investigation.

The third-order polarization $P^{(3)}(t)$ can be written (in the hypothesis of two relaxations)

$$P^{(3)}(t) = E_{\text{pu},2}(t)\int_{-\infty}^{t} f_2(t-t')E_{\text{pr}}(t'-\Delta t_{\text{pr}})E_{\text{pu},1}(t'-\Delta t_{\text{pu}})\,\mathrm{d}t' \quad (50)$$

where Δt_{pr} and $\Delta t_{\text{pu},1}$ are the delays of the probe and pump 1 with respect to pump 2, chosen as the time reference. The intensity of the backward wave is given by

$$I_{\mathrm{BW}} \propto \int_{-\infty}^{+\infty} |P^{(3)}(t)|^2 \, \mathrm{d}t \tag{51}$$

The polarization of the probe and the pump are crossed. Thus, it appears that I_{BW} is basically a function of the relaxation constant (*s*) τ_{R} involved in $f_2(t)$, Δt_{pu}, and Δt_{pr}. It will be denoted by $I_{\mathrm{BW}}(\tau_{\mathrm{R}}, \Delta t_{\mathrm{pu}}, \Delta t_{\mathrm{pr}})$.

The natural method for measuring τ_{R} (method 1) is to plot $I_{\mathrm{BW}}(\tau_{\mathrm{R}}, \Delta t_{\mathrm{pu}}, 0)$ versus Δt_{pu}. Figure 22 reports a computer simulation of the variations of $I_{\mathrm{BW}}(\tau_{\mathrm{R}}, \Delta t_{\mathrm{pu}}, 0)$ versus Δt_{pu}, in the simplified case where self-modulation is absent ($a = 0$ in Eq. (27)). The duration τ of the pulses was held fixed at 25 ps. As expected, the signal duration increases with τ_{R}. But it is possible to determine accurately τ_{R} from the measurement of FWHH only if $\tau_{\mathrm{R}} \gg \tau$. Unfortunately, in this work $\tau \approx \tau_{\mathrm{R}}$ and the measurements of τ_{R} with this method cannot be accurately performed. However, it will be used later.

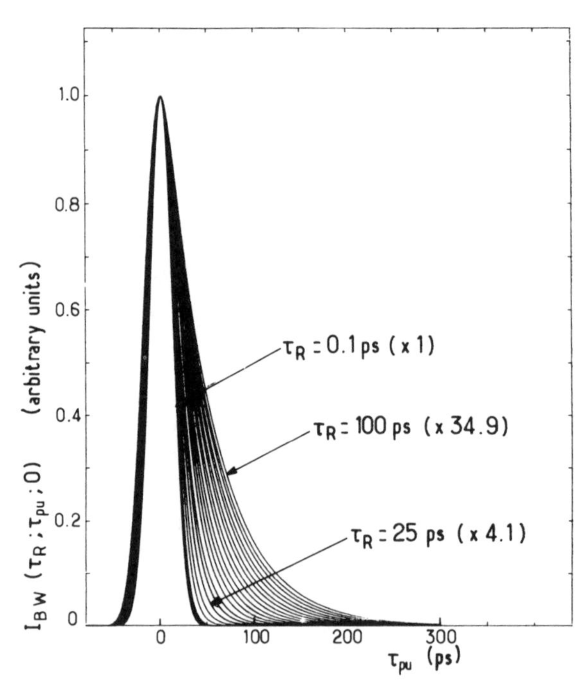

Figure 22. $I_{\mathrm{BW}}(\tau_{\mathrm{R}}; \Delta t_{\mathrm{pu}}; 0)$ vs. Δt_{pu} for same values of τ_{R}; the amplitude is normalized to 1. (After Ref. 96.)

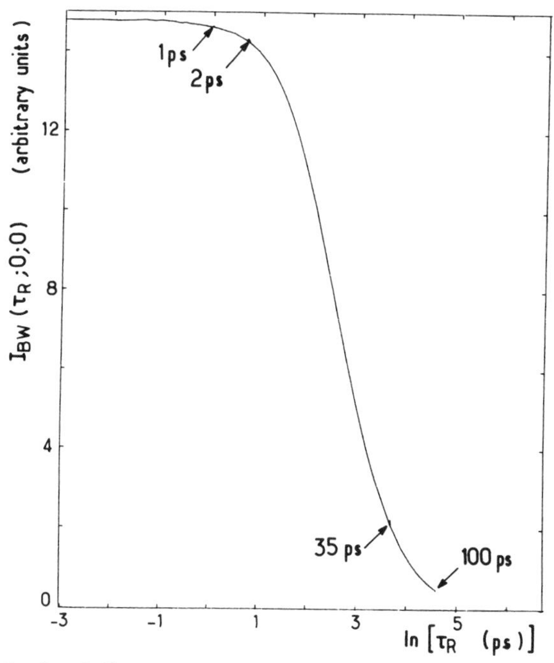

Figure 23. $I_{BW}(\tau_R; 0; 0)$ vs. τ_R; $\tau = 25$ ps; $a = 0$. The amplitude of the response function f_2 is assumed to be constant. (After ref. 96.)

For $\Delta t_{pr} = \Delta t_{pu,1} = 0$, the three pulses reach the sample at the same time. Thus, Eq. (51) becomes

$$P^{(3)}(t) = E_{pu,2}(t) \int_{-\infty}^{+t} f_2(t - t') E_{pr}^*(t') E_{pu,1}(t')\, dt' \tag{52}$$

In this case

$$E_{pr}^*(t') E_{pu,1}(t') = |\varepsilon(t')|^2 = \exp\left[-4 \ln 2 (t'/\tau)^2\right] \tag{53}$$

and $I_{BW}(t) \approx |P^{(3)}(t)|^2$ becomes phase-modulation independent.

A computer simulation (Fig. 23) shows that, in this case, the amplitude $I_{BW}(\tau, 0, 0)$ strongly decreases when τ_R increases in the pertinent time domain (1–100 ps). Then, the second method, i.e., plotting the variations of $I_{BW}(\tau_R, 0, \Delta t_{pr})$ versus Δt_{pr} and taking the maximum value $I_{BW}(\tau_R, 0, 0)$, can be used for evaluating τ_R by computer simulation. This

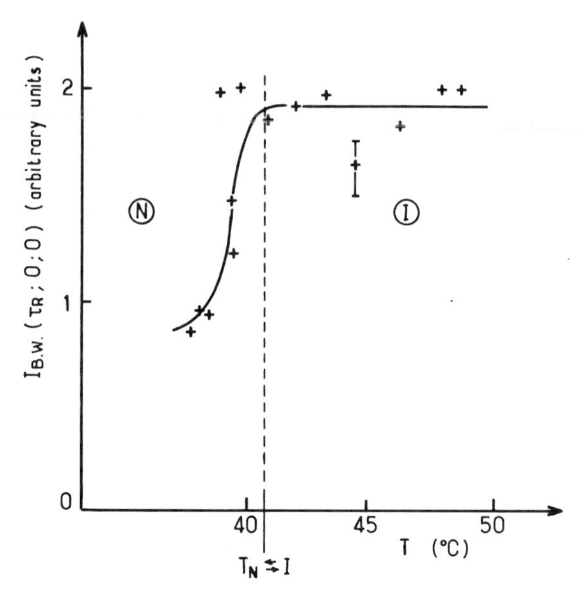

Figure 24. $I_{BW}(\tau_R; 0; 0)$ vs. temperature. Compound: 8CB; thickness of the sample: 200 μm; $\lambda_{pu} = \lambda_{pr} = 532$ nm; $\lambda_{pu} = \lambda_{pr} \simeq 25$ ps.

second method obviously will reveal only fast reorientations in the range of some tens of picoseconds.

Let us now report our main results.

1. *Isotropic and Nematic Phases*

For 8CB in the I phase, $I_{BW}(\tau_R, 0, 0)$ appears to be temperature independent down to the $I \rightleftharpoons N$ phase transition, where the signal suddenly decreases by a factor of about 2.5. The decrease occurs because of (1) the reduction of the effective anisotropy γ^2_{eff}, connected with the increased hindrance of the rotations about the short molecular axes in the homeotropic N phase (in I phases, the highest first-order polarizability along the long molecular axis plays an important role that is forbidden in the N phase) and (2) the sudden increase of the shear viscosity at the $I \rightarrow N$ transition [97]. Figure 24 illustrates this important reduction. No such change appears at the $N \rightleftharpoons Sm\text{-}A$ phase transition.

2. *Smectic A Homeotropic Phases*

In the picosecond range, the DFWM signal is mainly induced by the reorientations about the long molecular axis, which, as already mentioned in the introduction, are associated to biaxiality, and to ferroelectricity in

*Sm-C** phases (which are basically biaxial phases, as shown in Fig. 6). They are associated with the second-rank tensor $\Delta\varepsilon$ describing the fluctuations of the dielectric constant and, more especially, to $\Delta\varepsilon_{yx}$ for the homeotropic *Sm-A* phase oriented according to Fig. 6. These fluctuations can be called in-plane fluctuations, by reference to the layer planes. They appear to be basically connected to the already introduced restricted effective optical anisotropy $\gamma^2_{\perp,\mathrm{eff}} = (\alpha_{\perp,1,\mathrm{eff}} - \alpha_{\perp,2,\mathrm{eff}})^2$. Then, keeping the orientational situation described in Fig. 6, the corresponding response of *Sm-A* phases to picosecond pulses, driving optical fields linearly polarizated in the *XY* plane, is of the fast type and can be written

$$f_3(t) \propto \rho\mathscr{L}\frac{\gamma^2_{\perp,\mathrm{eff}}}{kT\,\tau_{\mathrm{R,F}}}\exp\left(t - \frac{t}{\tau_{\mathrm{R,F}}}\right) = \frac{C'_{\mathrm{F}}}{\tau_{\mathrm{R,F}}}\exp\left(\frac{-t}{\tau_{\mathrm{R,F}}}\right) \tag{54}$$

All the parameters of this equation have already been defined. In fact, the situation appears to be more complicated. We have recently shown [98], both theoretically and experimentally, that because of the fluctuations of both **N** and **n** with respect to the laboratory axis *Oz* (see Fig. 6), there appears an additional contribution to $\Delta\varepsilon_{yx}$ leading to an additional term $\chi^{(3)}_{\mathrm{ani,S}}$ which can be written, far from the *Sm-A* ⇌ *N* phase transition,

$$\chi^{(3)}_{\mathrm{ani,S}} \propto \Delta\varepsilon_{\mathrm{a}}(kT)\left[\frac{A_1}{\sqrt{\tilde{B}\delta k_i^3}} + \frac{A_2}{\sqrt{Dk_i^3}}\right]^{1/2} \tag{55}$$

where $\Delta\varepsilon_{\mathrm{a}}$ denotes the anisotropy $(\varepsilon_{\parallel} - \varepsilon_{\perp})$ of the dielectric constant at optical frequencies, A_1 and $A_2 \simeq A_1/10$ are constants; $\tilde{B}$ and D respectively describe layer compressibility and reorientation of the director [1]; k_i and δk_i are generic notations for the Franck constants of the phase [99]. These contributions diverge at the *Sm-A* ⇌ *N* phase transition. The associated relaxation times are of the slow type, but have not been measured at this time. The whole response function is then of the type

$$f_4(t) = \frac{C'_{\mathrm{F}}}{\tau_{\mathrm{R,F}}}\exp\left(\frac{-t}{\tau_{\mathrm{R,F}}}\right) + \frac{C'_{\mathrm{S}}}{\tau_{\mathrm{R,F}}}\exp\left(\frac{-t}{\tau_{\mathrm{R,S}}}\right) \tag{56}$$

which looks like Eq. (44), but with the difference that $C'_{\mathrm{F}} \propto \gamma^2_{\perp,\mathrm{eff}}$, while $C_{\mathrm{F}} \propto \gamma^2_{\mathrm{eff}}$. Moreover, the temperature dependences of C_{S} and C'_{S} differ strongly.

The DFWM signal appears to be very small. We have chosen CS_2 as a reference, because of its importance as a nonlinear standard. We have obtained a mean ratio for the signals given by CS_2 and 8CB about 18 ± 2. 8CB was studied at 31.5°C, i.e., in the *Sm-A* phase. One must evaluate the ratio $(C_{F,CS_2}/C'_{8CB})^2$ by using the second method already cited. In this case,

$$I_{BW}(\tau_{R,F}, 0, \Delta t_{pr}) = \frac{C'^2_F}{\tau^2_{R,F}} \int_{-\infty}^{+\infty} \varepsilon^2(t) \times \left[\int_{\infty}^{+\infty} \exp - [(t - t')/\tau_{R,F}] \varepsilon(t' - \Delta t_{pr}) \varepsilon(t') \, dt' \right]^2 dt \tag{57}$$

With the numerical values $\rho_{CS_2} = 10^{22}$ cm^{-3} [100], $\rho_{8CB} = 2.0 \times 10^{21}$ cm^{-3} [101], $n_{CS_2} = 1.673$ [100], $n_{8CB} = 1.56$ [102], $\gamma^2_{CS_2} = 117 \times 10^{-48}$ cm^6 [35], $\gamma^2_{\perp,8CB} = 441 \times 10^{-48}$ cm^3 [103], and taking $\mathscr{L} = (n^2 + \frac{2}{3})^4$ [94], we find a ratio of about 2.8, i.e., 6.4 times smaller than the one measured. Therefore, the slowing down factor, due to a larger τ_{RF} for 8CB as compared to CS_2, is about 6.4. If one refers to Fig. 23 and to the 2-ps reorientational time of CS_2, one finds for 8CB at 31.5°C in the *Sm-A* phase, $\tau_{R,F} \simeq 35$ ps. We checked this value using the first method; the results are given in Fig. 25. As expected, in the 28–34°C temperature range, we find a quite linear variation of ln $I_{BW}(\tau_{R,F}, \Delta t_{pu} 0)$ versus Δt_{pu} leading to a FWHH of about 30 ps. Of course, this linear behavior, observed over only one decade, does not prevent slower relaxations from occurring for larger values of Δt_{pu}. But, in this case, the signal-to-noise ratio strongly decreases and we cannot perform accurate determinations of the slower rates. The fastest one recorded can be interpreted as the time describing rotational librations of individual molecules about their long axis, modified by short-range coupling to neighboring particles. In the frequently used models of rotational relaxation in dense media composed of rodlike molecules [56, 104, 105], these fast components are associated with reorientation of the rod within the cylinder formed by its nearest neighbors. These models also include slower components generated by the rotational diffusion of the cylinder (coupled rotations of some tens of particles) with associated times in the range of some hundreds of picoseconds. These times have been obtained by FIR [19] as well as by time domain spectroscopy and significantly slow down upon entering the *Sm-C* phase (from 342 ps in *Sm-A* to 517 ps in *Sm-C* [20]).

We have not detected any decrease of I_{BW} down to the *Sm-A* ⇌ *Sm-C* phase transition [95, 96] for compound 6, which is achiral. However, for compound 5, a ferroelectric chiral mixture, we observe an important decrease of $I_{BW}(\tau_{R,F}, 0, 0)$ in the last degree before the transition. Figure 26 reports the results for compounds 5 and 6. Our observations suggest the

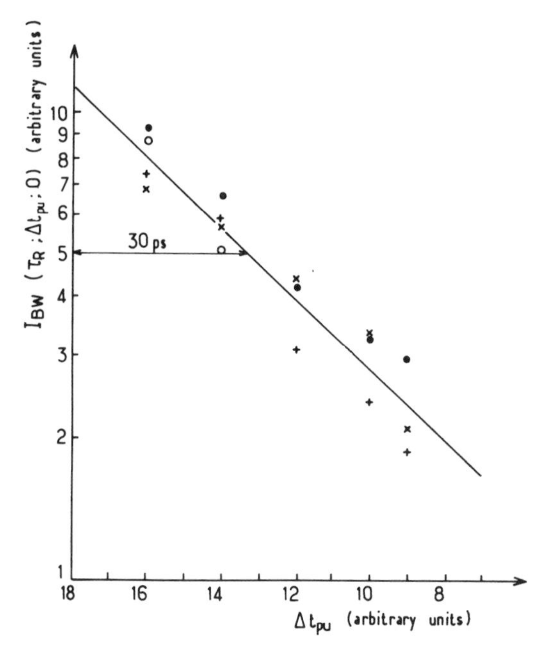

Figure 25. $I_{BW}(\tau_R; \Delta t_{pu}; 0)$ vs. Δt_{pu}. Compound: 8CB in the *Sm-A* phase; thickness of the sample: 200 μm; $\lambda_{pu} = \lambda_{pr} = 532$ nm; $\tau_{pu} = \tau_{pr} \simeq 25$ ps. (After Ref. 96.)

presence of a pretransitional *Sm-C** effect in the *Sm-A** phase, in the vicinity of the phase transition, an idea supported by recent studies of ferroelectric liquid crystals dynamics, showing helical fluctuations far under the *Sm-C** $\rightleftharpoons$ *Sm-A** transition, and by X-ray investigations showing *Sm-C** density modulations inside a 1-K temperature range across the transition [106]. In spite of the fact that we have not proved that C'_F does not decrease in the vicinity of the *Sm-C** phase, the observed effect appears to be more probably connected to a two- or threefold increase of $\tau_{R,F}$ in the last degree of the *Sm-A** phase, in agreement with the idea of a slowing down of the rotation about the long molecular axis in ferroelectric phases [107]. This means that the isotropic rotation in the *Sm-A** phase is largely modified near the ferroelectric phase, the short molecular axes spending more and more time in some privileged directions. In this case, $\tau_{R,F}$, which gives only an effective macroscopic view of the rotational relaxation, increases.

The theory of the effect [96] leads to an increase of the effective time given by

$$\tau_{R,F,eff} = \left[1/\tau_{R,F} - B/(T - T_c)\right]^{-1} \tag{58}$$

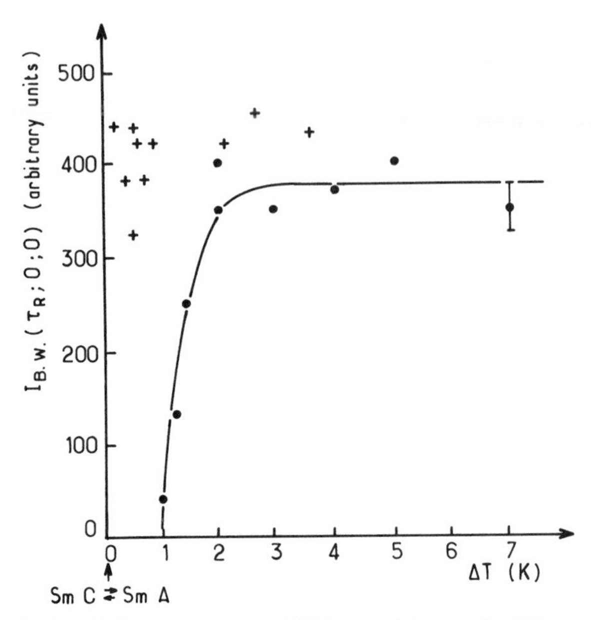

Figure 26. $I_{BW}(\tau_R; 0; 0)$ vs. temperature. Thickness of the sample: 500 μm; $\lambda_{pu} = \lambda_{pr} =$ 532 nm; $\tau_{pu} = \tau_{pr} \simeq$ 25 ps; +, compound 6 of Table 3; •, compound 5 of Table 3.

where B is a constant and the fit of our experimental results by Eq. (58) leads to $\tau_{R,F} \simeq 20$ ps; $B \simeq 4.4 \times 10^{-2}$ ps^{-1} K.

We must note here that an alternative interpretation of our work has been proposed [108] and a recent OKE investigation of the same mixture has not confirmed our observation [109]. However, it must be pointed out that this OKE investigation was performed with an inducing wave of high power (many watts), strongly focused in the sample (beam-waist of 80 μm), at a wavelength (1.053 μm) where liquid crystals absorb, and mode-locked at a very high repetition rate (100 MHz). These strong differences between the two experiments can perhaps explain the different results obtained. For instance, the OKE signal for 8CB, at zero delay between pump and probe, is reported to be three times larger than that from CS_2, while our DFWM signal from 8CB was 18 times smaller than the corresponding one from CS_2. A ratio larger than 1 cannot be explained from numerical evaluation of the contribution due to pure molecular reorientation about the long axis. As reported before, the calculated amplitude ratio is about 0.36, and the difference between reorientational times (8CB: some tens of ps; CS_2: 2 ps) still diminishes the value of the ratio.

Moreover two recent DS experiments performed on pure compounds, near 1 GHz, give opposite results. The first (110, 111) does not show any variation of the relaxation frequency from 10 K above the *Sm-A* $\rightleftharpoons$ I^* transition, all across the *Sm-C** phase down to the crystal, in contrast to previous results [112]. The second is interpreted as invoking a component that slows down near the ferroelectric phase [113]. But, none of the DS experiments seem to probe pure individual reorientation.

Finally, let us point out that Eq. (58) is deduced from a model involving a coupling (B constant) between laser-induced biaxiality and fluctuating permanent polarization in the cybotactic ferroelectric groups appearing in the *Sm-A** phase. Then, the role played by the molecular transversal permanent dipoles, which are responsible for the ferroelectricity, is determinant. Compound 7, which is a racemic mixture without permanent transversal polarization, does not give any coupling ($B = 0$) and, as expected, the corresponding I_{BW} does not reveal any noticeable decrease down to the *Sm-C* phase [96].

Thus, we have presented results of DFWM on *I*, *N*, and *Sm-A* phases consisting of chiral, achiral, or racemic species. In *I* phases, a time-unresolved temperature-independent response has been detected. It slows down at the $I \rightleftharpoons N$ phase transition in agreement with the observations resulting from ITG investigations. Signals monitored in homeotropic *N* and *Sm-A* phases reveal complex relaxations, the main signal having probably been due to individual molecular reorientations about the long molecular axis, affected by local order. In chiral compound, I_{BW} falls near the ferroelectric phase, probably in connection with a shift of the mean relaxation time toward longer times. This effect is interpreted as being due to a coupling between the fluctuating permanent polarization and the transient biaxial mode, induced by the laser field in the cybotactic groups of molecules [96]. Achiral and racemic compounds do not present this pretransitional behavior, which seems directly connected to the vicinity of the ferroelectric *Sm-C** phase.

VI. GENERAL CONCLUSIONS AND PROSPECTS

We have tried to show the interest of nonlinear optics in studies of the fast reorientations in liquid crystals. Actually, only the isotropic phase seems to be correctly investigated. In addition to the de Gennes contribution, *I* phases exhibit a distribution of orientational relaxation times, ranging from some hundreds of femtoseconds up to some hundreds of picoseconds with a mean relaxation time of a few picoseconds. The ITG technique [85, 86] has given the most accurate results, in agreement with older OKE investigations. Such results are in good agreement with the pioneering work, both theoretical [114] and experimental (Rayleigh scattering) of Shen et al. [13]. The exact origin of this new reorientational dynamics remains an open question. Reorientation times less than some picoseconds can be due to intramolecular dynamics. Times of some tens or hundreds of picoseconds probably arise from orientational fluctuations of individual molecules modified by coupling with neighboring particles. The decay rates can strongly change—by orders of magnitude—according to the duration of the laser pulses used. Thus, subpicosecond pulses are needed, as in the Fayer work, with the additional conditions (1) that the electronic contribution can be accurately separated from the nuclear one, and (2) that the extremely high values of the optical fields delivered by these very short pulses do not lead to undesirable effects in the compounds and thus to erroneous interpretations.

The homeotropic *N* and *Sm-A* phases cannot be correctly oriented in samples of more than 500-μm thickness (except in free-surface geometry, which has recently allowed us to obtain 2.5-mm *Sm-A** phases of good optical quality). Thus, OKE and ITG induced by subpicosecond pulses generally lead to poor signal-to-noise ratios. DFWM appears to be a more useful technique, because of its high efficiency. It has allowed us to study the orientational in-plane fluctuations in the *Sm-A* phases, thus opening the investigations of pretransitional behavior in relation with the ferroelectric order exhibited by *Sm-C** phases. But the separation of the different processes involved appears to be rather difficult in these anisotropic phases. Here, also, the use of subpicosecond pulses might be useful to solve the still open question concerning the molecular origin of ferroelectricity in liquid crystals. Finally, let us note that the nonlinear techniques already mentioned could be applied to the study of the recently discovered *Sm-O** [115] and $Sm\text{-}C_{A^*}$ [116] phases, the exact nature of which are still unknown, and which appear to be optically equivalent to biaxial *Sm-A* phases.

References

1. P. G. De Gennes, *The Physics of Liquid Crystals*, Clarendon, Oxford, UK, 1974.
2. R. M. Herman and R. J. Serinko, *Phys. Rev. A* **19**, 1757 (1979).
3. A. S. Zolot'Ko, V. F. Kitaeva, N. Kroo, N. N. Sobolev, and L. Chillag, *Pis'ma Zh. Eksp. Teor. Fiz.* **32**, 170 (1980) (*JETP Lett.* **32**, 158 (1980)).
4. See, for example, S. M. Arakelian, and Y. S. Chilingarian, *Non Linear Optics of Liquid Crystals*, Nauka, Moscow, 1984; I. C. Khoo, *Prog. Opt.* **26**, 105 (1988); E. M. Averyanow and M. A. Osipov, *Usp. Fiz. Nauk* **160**, 89 (1990).
5. S. D. Durbin, S. M. Avakelian, and Y. R. Shen, *Phys. Rev. Lett.* **47**, 1411 (1981).
6. Mi-Mee Cheung, S. D. Durbin, and Y. R. Shen, *Opt. Lett.* **8**, 39 (1983).
7. H. Hsiung, L. P. Shi, and Y. R. Shen, *Phys. Rev. A* **30**, 1453 (1984).
8. I. C. Khoo and Y. R. Shen, *Opt. Eng.* **24**, 579 (1985).
9. H. J. Eichler and R. Macdonald, in R. C. Sze and F. J. Duarte (Eds.), *Proceedings of the International Conference on Lasers 88*, STS Press, McLean, VA, 1989.

10. See S. Chandrasekhar, *Liquid Crystals*, Cambridge UP, Cambridge, UK, 1977, Chapter 3.
11. H. J. Eichler and R. Macdonald, *Phys. Rev. Lett.* **67**, 2666 (1991).
12. J. R. Lalanne, *Phys. Lett.* **51A**, 74 (1975).
13. N. M. Amer, Y. S. Lin, and Y. R. Shen, *Solid State Commun.* **16**, 1157 (1975).
14. J. R. Lalanne, B. Martin, B. Pouligny, and S. Kielich, *Opt. Commun.* **19**, 440 (1976).
15. J. R. Lalanne, B. Martin, B. Pouligny, and S. Kielich, *Mol. Cryst. Liq. Cryst.* **42**, 153 (1977).
16. J. A. Janik, M. Godlewska, T. Grochulski, A. Kocot, E. Sciesinska, J. Sciesinski, and W. Witko, *Mol. Cryst. Liq. Cryst.* **98**, 67 (1983).
17. F. Volino, A. J. Dianoux, and A. Heideman, *J. Phys. Lett.* **40**, 583 (1979).
18. A. J. Leadbetter, J. M. Richardson, and J. C. Frost, *J. Phys. Colloq.* **40**, 125 (1979).
19. J. A. Janik, J. Krawczyk, J. M. Janik, and K. Otnes, *J. Phys. Colloq.* **40**, 169 (1979).
20. J. Chrusciel and W. Zajac, *Liq. Cryst.* **10**, 419 (1991).
21. A. Nayeem and J. H. Freed, *J. Phys. Chem.* **93**, 6539 (1989).
22. J. W. Emslev and C. A. Veracini, *NMR in Liquid Crystals*, Reidel, Dordrecth, 1985.
23. See, for example, B. J. Berne and R. Pecora, *Dynamic Light Scattering*, Wiley, New York, 1976; M. W. Evans, *Simulation and Symmetry in Molecular Diffusion and Spectroscopy*, *Adv. Chem. Phys.* **81**, 361 (1992) (I. Prigogine and S. A. Rice, (Eds.), Wiley, New York).
24. A. D. Buckingham, *Proc. Phys. Soc.* **B69**, 344 (1956).
25. G. Mayer and F. Gires, *C. R. Acad. Sci. Paris* **258**, 2039 (1961).
26. S. Kielich, *Acta Phys. Pol.* **30**, 683 (1966); **31**, 689 (1967).
27. H. J. Eichler, Special Issue on Dynamic Gratings and Four Waves Mixing, *IEEE J. Quantum Electron.* **QE-22**, 1194 (1986).
28. R. W. Hellwarth, *J. Opt. Soc. Am.* **67**, 1 (1977).
29. B. Kasprowicz-Kielich, S. Kielich, and J. R. Lalanne, in J. Lascombe (Ed.), *Molecular Motions in Liquids*, Reidel, Dordrecht, 1974, p. 563.
30. W. Voigt, *Magneto-und Electro-Optik*, Teubner, Leipzig, 1908.
31. M. Born, *Optik*, Springer, Berlin, 1933.
32. P. Debye, *Polare Molekeln*, Leipzig, 1929.
33. W. Alexiewicz, J. Buchert, and S. Kielich, *Acta Phs. Pol.* **A52**, 445 (1977).
34. B. F. Levine and C. G. Bethea, *J. Chem. Phys.* **63**, 2666 (1975).
35. See, for example, J. R. Lalanne, *J. Phys.* **30**, 643 (1969); S. Kielich, *Proc. Indian Acad. Sci. (Chem. Sci.)* **94**, 403 (1985).
36. J. R. Lalanne, B. Lemaire, J. Rouch, C. Vaucamps, and A. Proutiere, *J. Chem. Phys.* **73**, 1927 (1980).
37. P. Papffy-Muhoray, H. J. Yuan, L. Li, M. A. Lee, J. R. Desalvo, T. H. Wei, M. Sheik-Bahae, D. J. Hagan, and E. W. Van Stryland, *Mol. Cryst. Liq. Cryst.* **207**, 291 (1991).
38. M. J. Soileau, T. H. Wei, M. Sheik-Bahae, D. J. Hagan, M. Sence, and E. W. Van Sryland, *Mol. Cryst. Liq. Cryst.* **207**, 97 (1991).
39. M. J. Soileau, E. W. Van Stryland, S. Guha, E. J. Sharp, G. L. Wood, and J. L. W. Pohlmann, *Mol. Cryst. Liq. Cryt.* **143**, 139 (1987).

40. M. J. Soileau, S. Guha, W. E. Williams, E. W. Van Stryland, and H. Vanherzeele, *Mol. Cryst. Liq. Cryst.* **127**, 321 (1985).
41. R. Risser, D. W. Allender, M. A. Lee, and K. E. Schmidt, *Mol. Cryst. Liq. Cryst.* **179**, 335 (1990).
42. J. P. Hermann, D. Ricard, and J. Ducuing, *Appl. Phys. Lett.* **23**, 178 (1973).
43. J. R. Lalanne and P. Bothorel, *J. Chim. Phys.* **11**, 1538 (1966).
44. G. R. Alms, T. D. Gierke, and W. H. Flygare, *J. Chem. Phys.* **61**, 4083 (1974).
45. A. K. Burnham, G. R. Alms, and W. H. Flygare, *J. Chem. Phys.* **62**, 3289 (1975).
46. D. Kivelson and P. A. Madden, *Annu. Rev. Phys. Chem.* **31**, 523 (1980).
47. T. Keyes and B. M. Ladanyi, *Mol. Phys.* **34**, 765 (1977).
48. S. J. Bertucci, A. K. Burnham, G. R. Alms, and W. H. Flygare, *J. Chem. Phys.* **66**, 605 (1977), T. D. Gierke, *J. Chem. Phys.* **65**, 3873 (1976).
49. M. E. Rose, *Elementary Theory of Angular Momentum*, Wiley, New York, 1967.
50. L. Van Hove, *Phys. Rev.* **95**, 249 (1954).
51. See, for example, B. Kasprowicz and S. Kielich, *Acta Phys. Pol.* **33**, 495 (1968); S. Kielich, *J. Colloid Interface Sci.* **33**, 142 (1970); S. Kielich, J. R. Lalanne, and F. B. Martin, *J. Phys.* **33**, C1-191 (1972); M. W. Evans, S. Wozniak, and G. Wagniere, *Physica B* **175**, 412 (1991). In Eq. 13, the negative, zero, and positive values of J_A successively describe the diatropic, atropic, and paratropic behaviors, according to the terminology introduced by J. A. Prins and W. Prins, *Physica* **23**, 253 (1957).
52. F. B. Martin and J. R. Lalanne, *Opt. Commun.* **2**, 219 (1970).
53. F. B. Martin and J. R. Lalanne, *Phys. Rev. A* **4**, 1275 (1971).
54. J. R. Lalanne, F. B. Martin, and P. Bothorel, *J. Colloid Interface Sci.* **39**, 601 (1972).
55. D. R. Bauer, G. R. Alms, J. I. Brauman, and R. Pecora, *J. Chem. Phys.* **61**, 2255 (1974).
56. P. D. Maker, *Phys. Rev. A* **1**, 923 (1970).
57. R. H. Cole, *J. Chem. Phys.* **42**, 637 (1965).
58. K. B. Eisenthal and K. H. Drexhage, *J. Chem. Phys.* **51**, 5720 (1969).
59. T. J. Chuang and K. B. Eisenthal, *Chem. Phys. Lett.* **11**, 368 (1971).
60. C. V. Shank and D. H. Auston, *Phys. Rev. Lett.* **34**, 479 (1975).
61. J. R. Lalanne and R. Lefebvre, *J. Chim. Phys.* **73**, 337 (1976).
62. J. Etchepare, G. Grillon, J. P. Chambaret, G. Harmoniaux, and A. Orszag, *Opt. Commun.* **63**, 329 (1987).
63. A. Yariv, *Opt. Commun.* **25**, 23 (1978).
64. A. Yariv, *IEEE J. Quantum. Electron.* **QE-14**, 650 (1978).
65. D. Bloch and M. Ducloy, *J. Phys. B* **14**, 1471 (1981).
66. M. A. Vasil'eva, J. Vischakas, V. Kabelka, and A. V. Masalov, *Opt. Commun.* **53**, 412 (1985).
67. M. Kaczmarek, J. M. Buchert, S. Kielich, and J. R. Lalanne, *J. Mod. Opt.* **38**, 193 (1991).
68. E. I. Demikhov, V. K. Dolganov, and V. M. Filev, *Pis'ma Zh. Eksp. Theor. Fiz.* **37**, 305 (1983) (*JETP Lett.* **37**, 361 (1983)).
69. P. J. Collings (private communication).
70. N. Bloembergen, *Non Linear Optics*, Benjamin, New York, 1965, p. 147.
71. J. Reintjes and R. L. Carman, *Phys. Rev. Lett.* **28**, 1697 (1972).

72. P. G. De Gennes, *Phys. Lett.* **30A**, 454 (1969).
73. P. G. De Gennes, *Mol. Cryst. Liq. Cryst.* **12**, 193 (1971).
74. W. Maier and A. Saupe, *Z. Natürforsch.* **A14**, 882 (1959); **15**, 287 (1960).
75. W. S. Struve, *Opt. Commun.* **21**, 215 (1977).
76. B. Pouligny, E. Sein, and J. R. Lalanne, *Phys. Rev. A* **21**, 1528 (1980).
77. G. K. L. Wong and Y. R. Shen, *Phys. Rev. A* **10**, 1277 (1974).
78. E. G. Hanson, Y. R. Shen, and G. K. L. Wong, *Phys. Rev. A* **14**, 1281 (1976).
79. J. Prost and J. R. Lalanne, *Phys. Rev. A* **8**, 2090 (1970).
80. J. C. Filippini and Y. Poggi, *J. Phys. Lett.* **37**, 17 (1976).
81. M. Schadt and W. Helfrich, *Mol. Cryst. Liq. Cryst.* **17**, 355 (1972).
82. M. Brunet-Germain, *C. R. Acad. Sci. Paris* **271**, 1075 (1970).
83. G. Eyring and M. D. Fayer, *J. Chem. Phys.* **81**, 4314 (1984).
84. F. W. Deeg, J. J. Stankus, S. K. Greenfield, V. J. Newell, and M. D. Fayer, *J. Chem. Phys.* **90**, 6893 (1989).
85. F. W. Deeg and M. D. Fayer, *J. Chem. Phys.* **91**, 2269 (1989).
86. F. W. Deeg, S. R. Greenfield, J. J. Stankus, V. J. Newell, and M. D. Fayer, *J. Chem. Phys.* **93**, 3503 (1990).
87. I. C. Khoo, T. H. Liu, and P. Y. Yan, *J. Opt. Soc. Am.* **B4**, 115 (1987).
88. I. C. Khoo and R. Normandin, *Opt. Lett.* 9, 285 (1984); *IEEE J. Quantum Electron.* **QE-21**, 329 (1985).
89. I. C. Khoo, R. Michael, and P. Y. Yan, *IEEE J. Quantum Electron.* **QE-23**, 267 (1987).
90. I. C. Khoo, R. G. Lindquist, R. R. Michael, R. J. Mansfield, and P. Lopresti, *J. Appl. Phys.* **69**, 3853 (1991).
91. H. J. Eichler and R. Macdonald, *Phys. Rev. Lett.* **67**, 2666 (1991).
92. C. Druon and J. M. Waerenier, *Mol. Cryst. Liq. Cryst.* **98**, 201 (1983).
93. D. Fekete, J. A. Yeung, and A. Yariv, *Opt. Lett.* **5**, 51 (1980).
94. P. A. Madden, F. C. Saunders, and A. M. Scott, *IEEE J. Quantum Electron.* **QE-22**, 1287 (1986).
95. J. R. Lalanne, J. Buchert, C. Destrade, H. T. Nguyen, and J. P. Marcerou, *Phys. Rev. Lett.* **62**, 3046 (1989).
96. J. R. Lalanne, C. Destrade, H. T. Nguyen, and J. P. Marcerou, *Phys. Rev. A* **44**, 6632 (1991).
97. P. Martinoty, S. Candau, and F. Debeauvais, *Phys. Rev. Lett.* **27**, 1123 (1981).
98. O. Mondain-Monval, J. R. Lalanne, and J. P. Marcerou, to be published.
99. F. C. Frank, *Discuss. Faraday Soc.* **25**, 19 (1958).
100. *Handbook of Chemistry and Physics*, 64th ed., R. C. Weast (Ed.), CRC Press, Boca Raton, FL, 1984.
101. D. A. Dunmut and W. H. Miller, *J. Phys.* **40**, 471 (1978).
102. J. D. Bunning, D. A. Crellin, and T. E. Faber, *Liq. Cryst.* **1**, 37 (1986).
103. Value taken from Ref. 36, with the assumption that the C_8H_{17} chain has an isotropic first-order polarizability in the plane perpendicular to the long molecular axis.
104. K. M. Zero and R. Pecora, *Macromolecules* **15**, 87 (1982).
105. V. I. Gajduk and Y. P. Kalmykov, *J. Chem. Soc. Faraday Trans.* **2**, 929 (1981).

106. C. W. Garland, private communication.
107. R. B. Meyer, *Mol. Cryst. Liq. Cryst.* **40**, 33 (1977).
108. H. R. Brand and H. Pleiner, *Phys. Rev. Lett.* **64**, 1309 (1990).
109. E. Freyss, private communication.
110. K. Kremer, S. U. Vallerien, H. Kapitza, R. Zentel and E. W. Fisher, *Phys. Rev. A* **42**, 3667 (1990).
111. F. Kremer, A. Schönfeld, S. U. Vallerien, A. Hofmann, and N. Schwenk, *Ferroelectrics* **121**, 13 (1991).
112. L. Benguigui, *J. Phys.* **43**, 915 (1982).
113. R. Nozaki, T. K. Bose, and J. Thoen, *Ferroelectrics* **121**, 1 (1991).
114. C. Flytzanis and Y. R. Shen, *Phys. Rev. Lett.* **33**, 14 (1974).
115. Y. Galerne and L. Liebert, *Phys. Rev. Lett.* **64**, 906 (1990); **66**, 2891 (1991).
116. K. Hirakao, A. Taguichi, Y. Ouchi, H. Takezoe, and A. Fukuda, *Jpn. J. Appl. Phys.* **29**, L103 (1990).

NONLINEAR PROPAGATION OF LASER LIGHT OF DIFFERENT POLARIZATIONS

GENEVIEVE RIVOIRE

Laboratoire des Propriétés Optiques des Matériaux et Applications, Université d'Angers, Angers, France

CONTENTS

Modern Nonlinear Optics, Part 1, Edited by Myron Evans and Stanisław Kielich. Advances in Chemical Physics Series, Vol. LXXXV.
ISBN 0-471-57546-1

I. INTRODUCTION

This text about light polarization in nonlinear optics pays a tribute to Professor S. Kielich. Indeed, he was one of the very first, through the lectures he gave in 1971 at Bordeaux University, to teach us the tensor notation that enables a synthetic presentation of the problems encountered in nonlinear optics, specifically those related to nonlinear molecular optics. This is the only means to consider globally the influence of the polarization state of the waves propagating in a nonlinear medium on the phenomena involved. It can be said that right from the discovery of the laser, Kielich has given a clear view of the linear and nonlinear interactions between matter and radiations without restricting himself to the simplest models. The latter consist of, for example, studying only the linearly polarized waves or leaving out certain phenomena, such as birefringence or optical activity, which can significantly influence the development of nonlinear effects. He certainly was the first, as well, to state that among the new optical effects observed when using power field lasers in the 1960s and 1970s, some effects were not symptoms of the optical properties of matter in electric dipolar approximation but originated in multipolar electric and magnetic polarizations. In addition, he was the first to consider intermolecular correlations in nonlinearities.

How is it possible to discuss polarization state in nonlinear optics without repeating the whole of nonlinear optics? We choose a main thread that all in all is very simple: What becomes of the polarization state of a wave propagating in a medium in which it excites nonlinearities? The study is carried out first for a wave traveling alone, then for a wave in the presence of another wave (two-wave mixing), and then for a wave in the presence of two other waves (called four-wave mixing because the presence of three incident waves gives rise to a fourth wave). A model for the general case of N waves is also discussed.

The quest for the answer to the apparently simple question posed above leads to different results that depend on the linear optical properties of the medium—particularly connected with its symmetry—and on the nonlinear phenomena present. Studying these results leads us to three types of conclusions:

- Knowledge of the polarization eigenstates, that is, the polarization states that are not altered as the wave travels through a nonlinear medium
- Measurement of the nonlinear susceptibility tensor components involved in these experiments
- Applications.

We shall restrict the field of study to third-order nonlinearities and devote ourselves to macroscopically isotropic media.

II. PROPAGATION OF ONE WAVE IN A NONLINEAR MEDIUM

We first consider isotropic media, such as transparent liquids, without linear absorption and without linear and nonlinear optical activity.

A. Experiments

When an intense light wave propagates through an isotropic nonlinear medium, the following behavior is observed [1–4]. A circularly polarized incident wave retains its polarization state. An elliptically polarized beam undergoes a rotation of its axes and moreover an ellipticity change if nonlinear absorption (NLA) is present. The polarization changes remain simple when the power per unit surface is below the thresholds of self-induced phenomena, such as self-focusing and stimulated scatterings. Within these limits, both the rotation $\Delta\theta$ of the ellipse's axes, and the parameter ρ connected to the ellipticity e by the expression $\rho = \ln[(1 + e)/(1 - e)]$ increase quasilinearly with the incident intensity I_0. The measured rotation $\Delta\theta$ can be large: It reaches 10° or more after a few centimeters of propagation in a liquid such as CS_2, C_6H_6, or $C_6H_5NO_2$ excited by nano- or picosecond laser pulses in the 10- to 100-MW/cm^2 irradiance range.

Figure 1 gives some examples. In nanosecond excitation the rules $\theta \approx I_0$ and $\rho =$ constant observed in the absence of NLA are altered and an ellipticity change appears when the stimulated Raman scattering threshold is reached (Fig. 1a–c). In picosecond excitation, for liquids with a large Kerr constant, instabilities of the polarization parameters $\Delta\theta$ and e appear far below the Raman threshold (Fig. 1d–f).

B. Equations: Third-Order Nonlinear Susceptibility Tensor

The propagation equation of the electric field Fourier component at frequency ω_n in the nonlinear medium, in the absence of linear absorption and optical activity, is

$$\Delta \mathbf{E}(\omega_n) + k_n^2 \mathbf{E}(\omega_n) + \frac{4\pi k_n^2}{n_n^2}\left(\mathbf{P}^{\mathrm{NL}} + \frac{1}{k_n^2}\,\mathbf{grad\,div\,P}^{\mathrm{NL}}\right) = 0 \quad (1)$$

where the real electric field $\mathscr{E}(\omega_n)$ and polarization $\mathscr{P}^{\mathrm{NL}}(\omega_n)$ have been

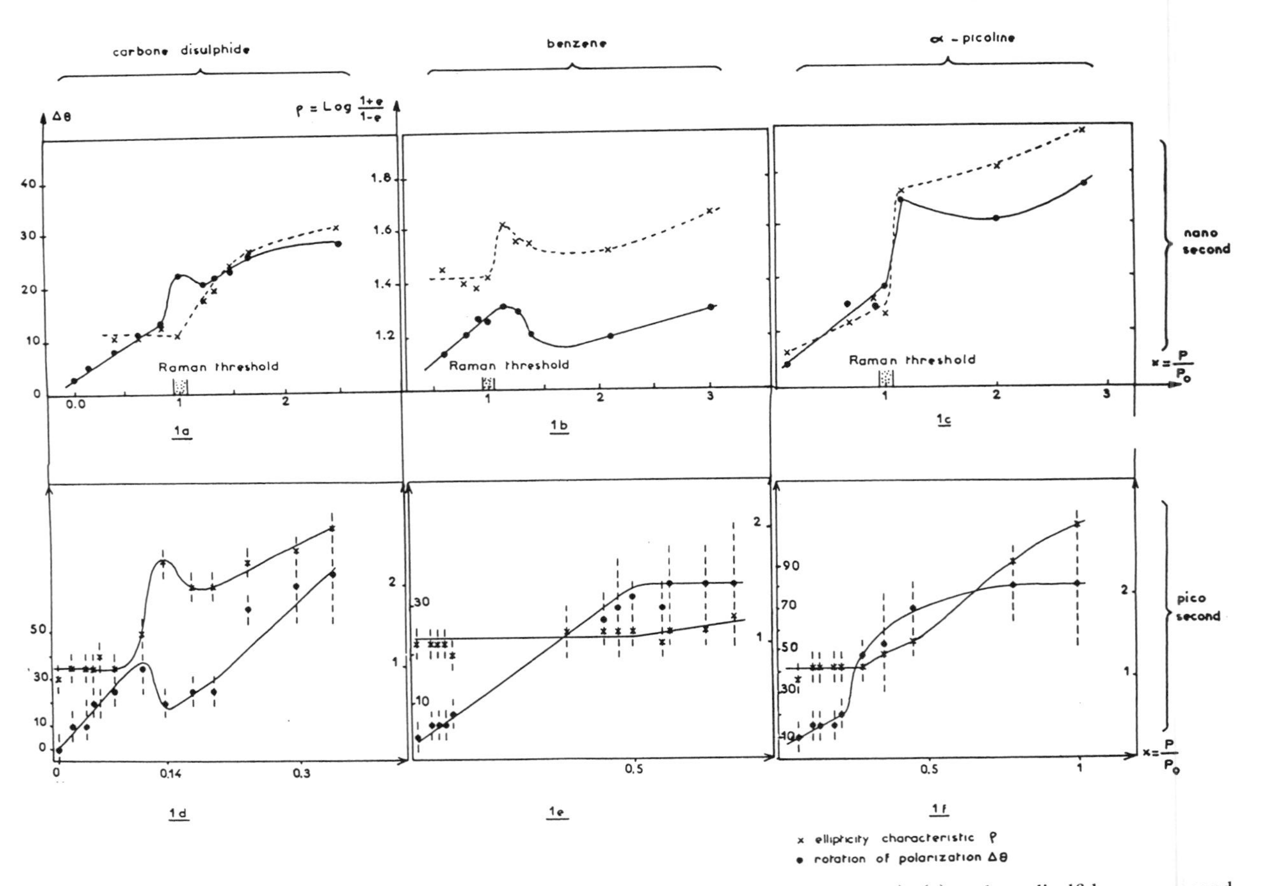

Figure 1. Evolution of the polarization state of one wave in different liquids ($l = 16$ cm): (a) carbon disulfide, nanosecond excitation, $P_0 = 7.5$ MW/cm^2; (b) benzene, nanosecond excitation, $P_0 = 27$ MW/cm^2; (c) α-picoline, nanosecond excitation, $P_0 = 30$ MW/cm^2; (d) carbon disulfide, picosecond excitation, $P_0 = 900$ MW/cm^2; (e) benzene, picosecond excitation, $P_0 = 1000$ MW/cm^2; (f) α-picoline, picosecond excitation, $P_0 = 1000$ MW/cm^2. $\Delta\theta$, rotation of polarization ellipse; $\rho = \ln[(1 + e)/(1 - e)]$; P/P_0, relative power ($P/P_0 = 1$ at the stimulated Raman threshold).

written in the form

$$\begin{aligned}\mathscr{E}(\omega_n) &= \tfrac{1}{2}\left[\mathbf{E}(\omega_n)e^{i\omega_n t} + cc\right]\\ \mathscr{P}(\omega_n) &= \tfrac{1}{2}\left[\mathbf{P}^{NL}(\omega_n)e^{i\omega_n t} + cc\right]\end{aligned} \tag{2}$$

n_n is the refraction index for ω_n, and $k_n = n_n\omega_n/c$. In the hypothesis of slowly varying envelopes, the term in $1/k_n^2$ can be reglected, and Eq. (1) gives

$$\Delta\mathbf{E}(\omega_n) + k^2\mathbf{E}(\omega_n) + \frac{4\pi k_n^2}{n_n^2}\mathbf{P}^{NL}(\omega_n) = 0 \tag{3}$$

The resolution of Eq. 3 enables the description of the polarization state evolution of the wave at frequency ω_n.

The third-order nonlinear polarization $\mathbf{P}^3(\omega_n)$ is generally described [5] by

$$P_i^3(\omega_n;\mathbf{k}_n) = D\chi_{ijkl}^3 E_j(\omega_1,\mathbf{k}_1)E_k(\omega_2,\mathbf{k}_2)E_t(\omega_3,\mathbf{k}_3) \tag{4}$$

where the medium characterized by susceptibility tensor χ_{ijkl}^3 is illuminated by three electric fields having frequencies $\omega_1, \omega_2, \omega_3$ and wave vectors $\mathbf{k}_1, \mathbf{k}_2, \mathbf{k}_3$, with

$$\begin{aligned}\omega_n &= \omega_1 + \omega_2 + \omega_3\\ \mathbf{k}_n &= \mathbf{k}_1 + \mathbf{k}_2 + \mathbf{k}_3\end{aligned} \tag{5}$$

The indices i, j, k, l stand for spatial coordinates (x, y, z). D is an integer depending on the number of possible permutations of the electric fields in relation (4) ($D = 1$ when $\omega_1 = \omega_2 = \omega_3$, $D = 3$ when $\omega_1 = \omega_2 \neq \omega_3$, and $D = 6$ when $\omega_1 \neq \omega_2$, $\omega_2 \neq \omega_3$, $\omega_1 \neq \omega_3$).

We notice that the factorization implied by relation (4) assumes that tensor χ^3 obeys some symmetry rules. But this is not always the case, and we prefer the following definition [6]:

$$P_i^3(\omega_n,\mathbf{k}_n) = 6\chi_{ijkl}^S E_j(\omega_1,\mathbf{k}_1)E_k(\omega_2,\mathbf{k}_2)E_l(\omega_3,\mathbf{k}_3) \tag{6}$$

where symmetrized susceptibility tensor χ^S is defined by

$$\begin{aligned}6\chi_{ijkl}^S = {}& \chi_{ijkl}^3(\omega_1 + \omega_2 + \omega_2) + \chi_{ijlk}^3(\omega_1 + \omega_3 + \omega_2)\\ &+ \chi_{ikjl}^3(\omega_2 + \omega_1 + \omega_3) + \chi_{iklj}^3(\omega_2 + \omega_3 + \omega_1)\\ &+ \chi_{iljk}^3(\omega_3 + \omega_1 + \omega_2) + \chi_{ilkj}^3(\omega_3 + \omega_2 + \omega_1)\end{aligned} \tag{7}$$

With this definition, the order in which the three electric fields are written in expression (6) is fundamental and must always remain the same.

When only one wave is present, Eq. (6) becomes

$$P_i^3(\omega) = 6\chi_{ijkl}^{S}(\omega + \omega - \omega)E_j(\omega)E_k(\omega)E_l^*(\omega) \tag{8}$$

This expression can be inserted in propagation Eq. (3). We use

$$\begin{aligned} \mathbf{E} &= \mathbf{A}(\mathbf{r}, z)e^{ik_z z} \\ \mathbf{P}^{NL} &= \mathbf{P}(\mathbf{r}, z)e^{ik_z z} \end{aligned} \tag{9}$$

where k_z is the component of wave vector $\mathbf{k}$ in the propagation direction $\mathbf{z}$. Equation (3) yields

$$\nabla_t \mathbf{A} + 2ik_z \frac{\partial \mathbf{A}}{\partial z}(k_z^2 - k^2)\mathbf{A} + \frac{4\pi k^2}{n}\mathbf{P} = 0 \tag{10}$$

∇_t is the operator $\partial^2/\partial x^2 + \partial^2/\partial y^2$. In the plane-wave approximation, below the self-focusing threshold, wave vector $\mathbf{k}$ has a very large component on $\mathbf{z}$ and $k_z \cong k$. Relation (10) is simplified as

$$2ik\frac{\partial \mathbf{A}}{\partial z} + \frac{4\pi k^2}{n}\mathbf{P} = 0 \tag{11}$$

For an isotropic (or cubic) medium, the general formula (7) yields

$$\begin{aligned} \mathbf{P} = 6\Big\{\big[\chi_{xxyy}^{S}(\omega + \omega - \omega) + \chi_{xyxy}^{S}(\omega + \omega - \omega)\big](\mathbf{A} \cdot \mathbf{A}^*) \cdot \mathbf{A} \\ + \chi_{xyyx}^{S}(\omega + \omega - \omega)(\mathbf{A} \cdot \mathbf{A})\mathbf{A}^*\Big\} \end{aligned} \tag{12}$$

Equations (11) and (12) lead to:

$$\frac{\partial \mathbf{A}}{\partial z} = iH\left[a(\mathbf{A} \cdot \mathbf{A}^*)\mathbf{A} + b(\mathbf{A} \cdot \mathbf{A})\mathbf{A}^*\right] \tag{13}$$

where

$$\begin{aligned} H &= \frac{24\pi^2}{n\lambda_S} \\ a &= \chi_{xxyy}^{S}(\omega + \omega - \omega) + \chi_{xyxy}^{S}(\omega + \omega - \omega) \\ b &= \chi_{xyyx}^{S}(\omega + \omega - \omega) \end{aligned} \tag{14}$$

Notice that $a + b = \chi_{xxxx}^{S}(\omega + \omega - \omega)$.

When an elliptically polarized wave travels through the nonlinear medium, the description of its polarization state—defined by its ellipticity e and the angle θ of its axes with respect to $\mathbf{0}_x$—is deduced from [13]

$$\Delta\theta(z) = \theta(z) - \theta(0) = H\,\mathrm{Re}(b)\int_0^z \frac{2e(z)}{1+e^2(z)} I(z)\,\mathrm{d}z \quad (15a)$$

$$\frac{1-e(z)}{1+e(z)} = \frac{1-e(0)}{1+e(0)} \exp\left[-\mathrm{Im}(b)H\int_0^z \frac{2e}{1+e^2} I(z)\,\mathrm{d}z\right] \quad (15b)$$

$$\frac{\partial I(z)}{\partial z} = -2H\left\{\mathrm{Im}(a) + \mathrm{Im}(b)\left[\frac{1-e^2(z)}{1+e^2(z)}\right]^2\right\} I^2(z)\,\mathrm{d}z \quad (15c)$$

where $I(z) = \mathbf{A}\cdot\mathbf{A}^*$ is the intensity. The polarization eigenstates, i.e., the states that retain their polarization characteristics in the medium, can be deduced from Eq. (15).

C. Polarization Eigenstates

A circularly polarized wave $[e(0) = 1]$ is the only wave to maintain its polarization in the medium, even in the presence of nonlinear absorption (this absorption implies $\mathrm{Im}(b) \neq 0$ in Eq. (15b)). A linearly polarized wave ($e(0) = 0$ or ∞) also keeps its polarization state in the absence of NLA. In the absence of NLA, the ellipticity of an elliptically polarized wave is constant. Its axes undergo a rotation $\Delta\theta$ given by

$$\Delta\theta = H\,\mathrm{Re}(b)\frac{2e}{1+e^2} I_0 l \quad (16)$$

The values of the susceptibility tensor components a and b can be deduced from the measurements of $\Delta\theta$, e, and I (Eqs. (15) and (16)).

D. Determination of the Susceptibility Tensor Components by Means of Measurements Using Single Polarized Beam

1. Ellipsometry Measurements

The component $\mathrm{Re}[\chi^{\mathrm{S}}_{xyyx}(\omega + \omega - \omega)]$ of the susceptibility tensor can be deduced from the measurement of the polarization rotation $\Delta\theta$ of an elliptically polarized beam [14]. However, if $\mathrm{Im}(\chi^{\mathrm{S}}_{xyyx}) \neq 0$, the slope of the curve $\Delta\theta = f(I_0)$ is not simply proportional to $\mathrm{Re}(\chi^{\mathrm{S}}_{xyyx})$. The numerical resolution of Eq. 15, illustrated by Fig. 2, shows that both the intensity decrease and the ellipticity change undergone by the wave in the presence of NLA influence the rotation [4]: $\Delta\theta$ is significantly reduced when the ratio $\mathrm{Im}(\chi^{\mathrm{S}}_{xyyx})/\mathrm{Re}(\chi^{\mathrm{S}}_{xyyx})$ is increased. Thus, all the interpretations of

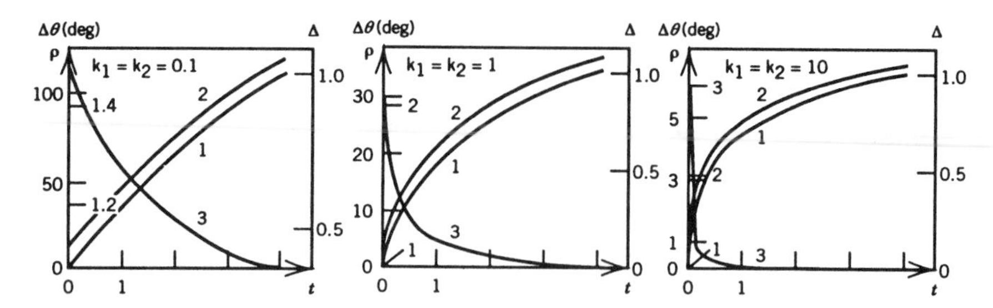

Figure 2. Evolution of the polarization state in presence of nonlinear absorption. 1, rotation $\Delta\theta$ of polarization ellipse; 2, $\rho = \ln[(1+e)/(1-e)]$; 3, relative output intensity $\Delta = I/I_0$. Calculated results for an input ellipticity $e_0 = 0.5$ and for different values of the characteristics $k_1 = \mathrm{Im}(\chi^{\mathrm{S}}_{xyyx})/\mathrm{Re}(\chi^{\mathrm{S}}_{xyyx})$ and $k_2 = \mathrm{Im}(\chi^{\mathrm{S}}_{xxyy} + \chi^{\mathrm{S}}_{xyxy})/\mathrm{Re}(\chi^{\mathrm{S}}_{xyyx})$.

rotation ellipse measurements made in the presence of a component $\mathrm{Im}(\chi^{\mathrm{S}}_{xyyx}) \neq 0$ have to be made carefully. Let us consider for instance the case of carbon disulfide, which is considered a reference liquid in nonlinear optics. In picosecond excitation, we have observed much below the stimulated Raman scattering threshold Rayleigh wing scattering and two-photon absorption. These effects result in $\mathrm{Im}(\chi^{\mathrm{S}}_{xyyx}) \neq 0$ and can explain a slope in the curves $\Delta\theta = f[I_0]$ smaller in picosecond excitation (Fig. 1d) than in nanosecond excitation (Fig. 1a).

In Table IIIa we gather the experimental results for the relative values of the components χ^{S}_{ijkl} of CS_2 obtained by different techniques. The value obtained by nanosecond ellipsometry for χ^{S}_{xyyx} (Fig 1a) is presented and will be compared to calculated values in Section IV.

2. *Refraction Index Measurements*

When $\mathrm{Im}(b) = 0$, the refraction index change Δn is given by

$$\Delta n = n - n_0 = \frac{12\pi}{n_0}\left[a + b\left(\frac{1-e^2}{1+e^2}\right)^2\right]|\mathbf{E}|^2 = K|\mathbf{E}^2| \qquad (17)$$

Δn depends on the polarization state of the wave. From the refraction index measurements of one wave propagating through a nonlinear medium, the two components $a = \chi^{\mathrm{S}}_{xxyy}(\omega+\omega-\omega) + \chi^{\mathrm{S}}_{xyxy}(\omega+\omega-\omega)$ and $b = \chi^{\mathrm{S}}_{xyyx}(\omega+\omega-\omega)$ can be deduced.

Several methods have been developed to measure the refraction index change. The earlier ones use self-focusing threshold measurements. Others, more recent and more accurate, use exciting powers lower than this threshold. We give a short description of the different methods.

a. Self-focusing Threshold Method (SFT). The influence of the polarization state on the refractive index change appears particularly in the self-focusing phenomena. The self-focusing threshold P_t is in first approximation given by [7]

$$P_t = \left(\frac{nr_0^2}{4l^2}\right)\left(\frac{1}{K}\right) \approx \frac{1}{a + b[(1 - e^2)/(1 + e^2)]} \tag{18}$$

where r_0 is the beam radius. For a linearly polarized wave,

$$P_{tl} \approx \frac{1}{\chi^S_{xxxx}(\omega + \omega - \omega)}$$

and for a circularly polarized wave,

$$P_{tc} \approx \frac{1}{\chi^S_{xxyy}(\omega + \omega - \omega) + \chi^S_{xyxy}(\omega + \omega - \omega)}$$

Several authors, considering that SFT and stimulated Raman threshold (SRT) are the same, deduced the susceptibility components a and b from the Raman threshold measurements [8–10]. Some results appear in curves $P_t = f(e)$ in Fig. 3 [11]. Relation (18) works well for some materials ($CHCl_3, C_6H_6$) but not so well for other ($CS_2, C_6H_5NO_2$). The values deduced from these measurements for the ratio $(\chi^S_{xxyy} + \chi^S_{xyxy})/\chi^S_{xxxx}$ in CS_2 are presented and discussed in Table IIIa and in Section IV.

All the conclusions derived from these measurements are based on several hypotheses:

- Identity between SFT and SRT
- SFT given by relation (18)
- $\mathrm{Im}(\chi_{ijkl}) = 0$ at the SRT.

We have studied the influence of self-focusing (SF) and stimulated Raman scattering (SRS) on the polarization state of the exciting wave [11]. Theory and experiments lead to the following conclusions:

- A circularly polarized input wave remains circularly polarized during SF even in the presence of TPA.
- In the absence of both TPA and stimulated scatterings, an elliptically polarized wave conserves its ellipticity during SF, while the axes rotation $\Delta\theta$ is significantly increased by SF. A linearly polarized wave keeps its polarization state.
- The ellipticity of an exciting wave is modified by SRS.

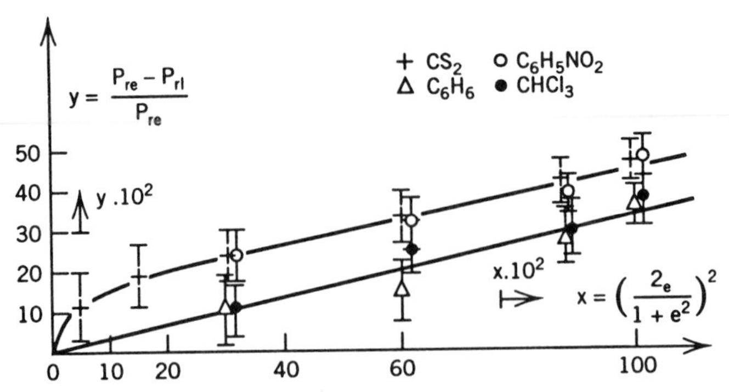

Figure 3. Measured stimulated Raman scattering threshold P_{te} as a function of the ellipticity e of the exciting wave

$$x = \left(\frac{2e}{1+e^2}\right)^2$$

$$y = \frac{P_{\mathrm{te}} - P_{\mathrm{tl}}}{P_{\mathrm{te}}} \left[y \approx \frac{\chi_{xyyx}}{\chi_{xxxx}} \left(\frac{2e}{1+e^2}\right)^2 \right]$$

These results confirm that it is better to choose methods using exciting powers lower than SFT and SRT to measure the susceptibility tensor components. This is particularly true in picosecond excitation where nonlinear absorption and instabilities occur much below the SRS Fig. 1d–f.

Recently, two new methods have been proposed for refraction index measurements, which are presented briefly below.

b. Z-scan Method [12, 13]. In this method, the transmittance of a sample is measured through a finite aperture in the far field as the sample is moved along the propagation path z of a focused Gaussian beam (Fig. 4). When the scan is started far away from the focus, the beam irradiance is small in the medium, and nonlinearities remain small. When the sample is near the focus, the irradiance and thus the nonlinearities increase, leading to self-lensing. According to the sign of the refractive index change, the self-focusing or self-defocusing respectively results in increase or decrease of the transmittance. The comparison between the calculated and measured transmittance allows the determination of the real and imaginary parts of the nonlinear susceptibility tensor.

Actually, the authors [12, 13] have used their technique without reference to the influence of the polarization state of the beam. In this way,

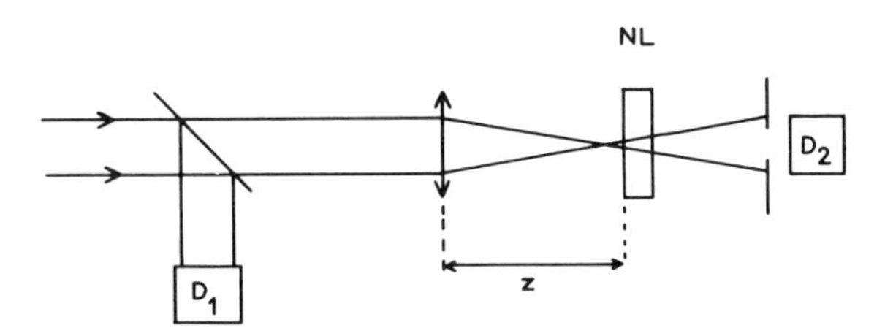

Figure 4. The Z-scan apparatus: The ratio D2/D1 is measured as a function of the nonlinear material position z.

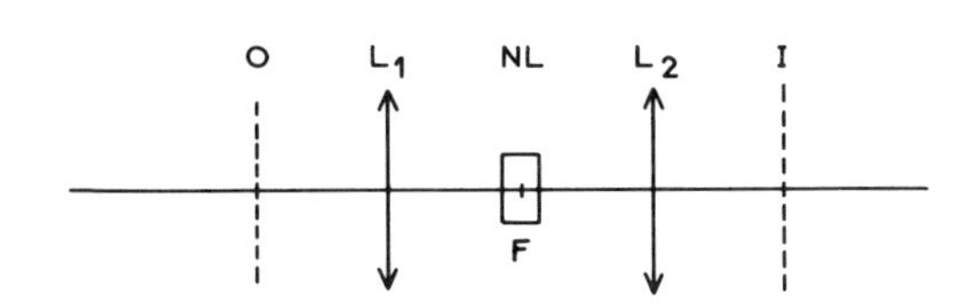

Figure 5. The nonlinear imaging setup: L_1, L_2, lenses; O, object; NL, nonlinear material; I, image of O, filtered by NL.

they have measured χ_{xxxx} with linearly polarized excitation for different materials, and especially CS_2 at different wavelengths. The method, extended to circularly and elliptically polarized beams, could lead to the determination of χ_{xyyx} and $\chi_{xxyy} + \chi_{xyxy}$.

c. Nonlinear Imaging Method. We have shown that the image of an object can be modified in a Zernike setup where the usual absorbing and (or) dephasing screen is replaced by a nonlinear material (Fig. 5) [14, 15]. From the comparison between the image obtained at low irradiance (called the linear image) and the image obtained at high irradiance (called the nonlinear image), nonlinear material characteristics can be deduced [15].

Actually, the measurements are based either on the contrast inversion in the image of a matrix made of 0 and 1 or on the appearance of lateral diffracted structures in the image of slits. The results show that the method is a good candidate for the determination of third-order nonlinearities in liquids [14] and photorefractive crystals at different polarizations [15, 16] (see a result for CS_2 in Table IIIa).

3. *Conclusions*

All the one-wave methods described are able to give, in isotropic materials, the values of $\chi^{S}_{xyyx}(\omega + \omega - \omega)$, $\chi^{S}_{xxyy}(\omega + \omega - \omega) + \chi^{Sn}_{xyxy}(\omega + \omega +$

ω), and their sum $\chi^{\mathrm{S}}_{xxxx}(\omega + \omega - \omega)$. It is known that the ratios between the components of the susceptibility tensor depend on the physical phenomena creating the nonlinearities. Two main local effects contribute to the nonlinearities in isotropic materials: The first is due to molecular movements (rotations, translations, vibrations), and the second is connected to electronic deformations. These two contributions, called nuclear and electronic, will be referred to as χ^{n} and χ^{e}:

$$\chi^{\mathrm{S}}_{ijkl} = \chi^{\mathrm{Sn}}_{ijkl} + \chi^{\mathrm{Se}}_{ijkl} \tag{19}$$

According to Hellwarth [17], we write

$$\chi^{\mathrm{n}}_{ijkl}(\omega_1 + \omega_2 + \omega_3) = d_{ijkl}(\omega_2 + \omega_3)$$

with

$$\begin{aligned} d_{ijkl}(\omega_2 + \omega_3 = 0) &= -\frac{B_0}{3}\delta_{ij}\delta_{kl} + \frac{B_0}{2}(\delta_{ik}\delta_{jl} + \delta_{il}\delta_{jk}) \\ d_{ijkl}(\omega_2 + \omega_3 \neq 0) &= 0 \end{aligned} \tag{20}$$

This means that the molecular movements due to the simultaneous presence of two waves at frequencies ω_2 and ω_3 exist only if $\omega_2 + \omega_3 = 0$ (exactly $\omega_2 + \omega_3 \ll 1/\tau$, where τ is the relaxation time of the movements considered).

The tensor χ^{n}_{ijkl} does not verify the intrinsic symmetry permutation property [6]. When only one wave is present ($\omega_1 = \omega_2 = \omega, \omega_3 = -\omega$), the calculation of $\chi^{\mathrm{Sn}}_{ijkl}$ made from relation (6) yields

$$\begin{aligned} 6\chi^{\mathrm{Sn}}_{xxxx}(\omega + \omega - \omega) &= \frac{8B_0}{3} \\ 6\chi^{\mathrm{Sn}}_{xxyy}(\omega + \omega - \omega) &= 6\chi^{\mathrm{Sn}}_{xyxy}(\omega + \omega - \omega) = \frac{B_0}{3} \\ 6\chi^{\mathrm{Sn}}_{xyyx} &= 2B_0 \end{aligned} \tag{21}$$

The nonresonant electronic tensor obeys the relations

$$\chi^{\mathrm{e}}_{ijkl}(\omega_1 + \omega_2 + \omega_3) = \sigma_{ijkl} = \tfrac{1}{6}\sigma(\delta_{ij}\delta_{kl} + \delta_{ik}\delta_{jl} + \delta_{il}\delta_{jk}) \tag{22}$$

which verifies the intrinsic symmetry permutation property, and thus we have

$$
\begin{aligned}
6\chi^{\mathrm{Se}}_{xxxx} &= 3\sigma \\
6\chi^{\mathrm{Se}}_{xxyy} &= 6\chi^{\mathrm{Se}}_{xyyx} = 6\chi^{\mathrm{Se}}_{xyxy} = \sigma
\end{aligned}
\tag{23}
$$

B_0 and σ characterize respectively the nuclear and electronic contributions.

The expected values for components a and b obtained with the one-wave techniques described in Section I.B are given by

$$
\begin{aligned}
a(\omega+\omega-\omega) &= \chi^{\mathrm{S}}_{xxyy}(\omega+\omega-\omega) \\
+\,\chi^{\mathrm{S}}_{xyxy}(\omega+\omega-\omega) &= \frac{1}{6}\left(\frac{2B_0}{3}+2\sigma\right) \\
b(\omega+\omega-\omega) &= \chi^{\mathrm{S}}_{xyyx} = \tfrac{1}{6}(2B_0+\sigma) \\
a(\omega+\omega-\omega)+b(\omega+\omega-\omega) &= \chi^{\mathrm{S}}_{xxxx}(\omega+\omega-\omega) = \frac{1}{6}\left(\frac{8B_0}{3}+3\sigma\right)
\end{aligned}
\tag{24}
$$

and thus

$$
\frac{\chi^{\mathrm{S}}_{xyyx}(\omega+\omega-\omega)}{\chi^{\mathrm{S}}_{xxxx}(\omega+\omega-\omega)} = \frac{6+3\sigma/B_0}{8+9\sigma/B_0}
\tag{25}
$$

This ratio is therefore expected to vary between 0.75 and 0.33 according to the relative importance of the nuclear and electronic contributions. All the values measured by the SRT in nanosecond excitation are in the 0.5 to 0.33 range [8, 10, 18] (see Section I.D.2.a). However, we cannot conclude that the electronic effects play the dominant role. At first the presence of nonlocal effects and especially electrostriction and heating can contribute to the nonlinearities [18, 19]. The measurements made at SFT and SRT are subject to many systematic errors, as analyzed above.

Finally, we stress that the absolute measurements need the calibration of the exciting power per unit surface in the medium. When two components, such as χ_{xyyx} and χ_{xxxx}, are measured in different experiments (for instance, ellipsometry for χ_{xyyx} and refraction index measurements for χ_{xxxx}), large errors are made on their ratio. Therefore, the determination of the relative parts of nuclear and molecular effects τ and B_0 in formulas such as Eq. (25) becomes impossible.

Nonlinear susceptibility measurements performed with several wave mixing, as described hereafter, will allow more reliable results than most of the one-wave methods described above. Before their study, we give some information about single-wave propagation in media with linear absorption and optical activity.

E. Influence of Linear Absorption Optical Activity

Several materials used in nonlinear optics applications feature linear absorption, birefringence, or optical activity. Moreover, they can show a nonlinear optical activity. Several authors have considered these effects. A synthesis devoted to nonlinear optical activity has been done by Zheludev [21]. A complete description of the polarization can be written in the following way [22]:

$$P_i = \varepsilon_{ij}E_j + \gamma_{ijk}\nabla_k E_j + \chi_{ijkl}E_jE_kE_l + \Gamma_{ijklm}E_jE_k\nabla_m E_l + \cdots \tag{26}$$

where γ_{ijk} and Γ_{ijklm} describe respectively the linear and nonlinear optical activities, connected to spatial dispersion [23].

We illustrate the results obtained when Eq. (26) is inserted in the propagation equation for the typical crystals $Bi_{12}SiO_{20}$ (BSO) and $Bi_{12}GeO_{20}$ (BGO) used in dynamic holography. They present a high linear optical activity and large linear and nonlinear absorptions (a birefringence is created by a static field—here we describe crystals without static field). The propagation equation is

$$\frac{\partial \mathbf{A}}{\partial z} = -\frac{\alpha}{2}\mathbf{A} + \mathrm{i}[G]\mathbf{A} + \mathrm{i}H\chi^{\mathrm{S}}\mathbf{AAA}^* \tag{27}$$

where α is the linear absorption coefficient, $[G]$ is the antisymmetric tensor describing linear optical activity, and χ^{S} is the third-order susceptibility tensor. χ^{S} has non-nul components with a large imaginary part:

$$\begin{gathered}\chi_{xxxx} = \chi_{yyyy} = a_1 \qquad \chi_{xxyy} = a_2 \qquad \chi_{yyxx} = a_3 \\ \chi_{xyxy} = a_4 \qquad \chi_{yxyx} = a_5 \qquad \chi_{xyyx} = a_6 \qquad \chi_{yxxy} = a_7\end{gathered} \tag{28}$$

Equation (27) leads to

$$\begin{aligned}\frac{\partial Ax}{\partial z} &= -\frac{\alpha}{2}A_x - gA_y \\ &\quad + \mathrm{i}H\left[a_1A_xA_xA_x^* + (a_2 + a_4)A_xA_yA_y^* + a_6A_yA_yA_x^*\right] \\ \frac{\partial Ay}{\partial z} &= -\frac{\alpha}{2}A_y \pm gA_x \\ &\quad + \mathrm{i}H\left[a_1A_yA_yA_y^* + (a_3 + a_5)A_yA_xA_x^* + a_7A_xA_xA_y^*\right]\end{aligned} \tag{29}$$

From Eq. (29) we deduce that only circularly polarized waves are eigenpolarization states, if the condition $a_2 + a_4 - a_6 = a_3 + a_5 - a_7$ is verified (the experiment confirms this result in BSO and BGO crystals).

Right- and left-handed circularly polarized waves undergo the same attenuation, described by the change in their intensities I_+ and I_-:

$$\frac{\partial I_\pm}{\partial z} = (-\alpha + HI_\pm \operatorname{Im} P)I_\pm \tag{30}$$

Due to the optical activity, their phases φ_+ and φ_- are different:

$$\frac{\partial \varphi_\pm}{\partial z} = \pm g + \frac{H}{2}\operatorname{Re}(p)I_\pm \tag{31}$$

where

$$P = \tfrac{1}{4}\left[3a_1 + \tfrac{5}{2}(a_2 + a_3 + a_4 + a_5) - \tfrac{3}{2}(a_6 + a_7)\right]$$

The values of Re(P) and Im(P) can be deduced from measurements performed on one circularly polarized wave propagating in a crystal [24, 25]. However, measurements on noncircularly polarized waves are necessary to obtain other components of the susceptibility tensor. The corresponding equations deduced from Eq. (29) are complicated and are not presented here [26].

F. Applications

Many applications exist of the polarization state of one wave propagating through a nonlinear medium. An important one is the optical Kerr effect, whereby the intensity of the polarization components of a light wave can be self-modulated as it travels through a nonlinear medium. Applications to fibers (polarization switching; also called all-optical switching) have been proposed for fused silica and glass fibers, and recently for liquid CH_5NO_2 core fibers. In the latter, the large nonlinearity allows self-switching of one polarization at peak powers as low as 1 W in a 10-cm-long fiber [27].

III. PROPAGATION OF TWO WAVES

Interaction between a nonlinear medium and two waves can lead to various phenomena. Energy exchanges can occur between the two light beams via the material when the intensity grating built by the two interfering input waves and the refraction index grating registered in the material are phase shifted, owing to the physical mechanisms involved in the registration. In the usual Kerr-like mediums, in the absence of scatterings and of nonlocal effects, no energy exchange is possible without higher-order processes or moving grating techniques.

Here, we describe the polarization properties mostly in the absence of energy coupling between the waves. In these experiments, the directions, frequencies, intensities, polarization states, and temporal delays of the two waves can be chosen. One kind of experiment has been particularly studied: That where an intense pump wave creates a nonlinear perturbation and influences the propagation of a signal wave of low intensity. From the measurements of the polarization changes of the signal wave, components of the susceptibility tensor χ^{S} can be deduced. The influence of the wavelength of the waves, and of their temporal delay can be studied. We present the propagation equations, the results actually obtained in χ^{3} measurements, and some applications.

A. Equations: Polarization Eigenstates

The evolution of the electric fields of N waves propagating through a nonlinear medium can be described with the definitions and methods used in Section I.B. The electric field Fourier component A_j of the jth wave (frequency ω_j) is decomposed into its two right- and left-handed circular components A_{j+} and $A_{j\text{-}}$. We write $\mathscr{A}_{j\pm} = A_{j\pm}(r, z)e^{ikj\varnothing j\pm(r,z)}$. From the propagation equation, we deduce the following equations [28] for waves having the same propagation directions and in the absence of self-focusing:

$$\frac{\partial I_{j\pm}}{\partial z} = \mp \frac{4\pi k_j}{\varepsilon_j} \sum_{l=1}^{N} a'' I_l I_{j\pm}$$

$$+ (b''_{lj} \cos \Delta_{lj} + b'_{lj} \sin \Delta_{lj})(I_{ll+} I_{l-} I_{j+} I_{j-})^{1/2} \tag{32a}$$

$$\frac{\partial \varnothing}{\partial z} = \frac{2\pi}{\varepsilon_j} \sum_{l=1}^{N} \left[a'_{lj} I_l (b'_{lj} \cos \Delta_{lj} \mp b''_{lj} \sin \Delta_{lj})(I_{l+} I_{l-} l_{j+}/I_{j\pm})^{1/2} \right] \tag{32b}$$

where

$$I_{j\pm} = A_{j\pm} A^{+}_{j\pm}$$

$$I_j = I_{j+} + I_{j-}$$

$$a_{lj} = a'_{lj} + ja''_{lj} = \chi^{S}_{xxyy}(\omega_j + \omega_l - \omega_l) + \chi^{S}_{xyxy}(\omega_j + \omega_l - \omega_l)$$

$$b_{lj} = b'_{lj} + b''_{lj} = \chi^{S}_{xyyx}(\omega_j + \omega_l - \omega_l) + \chi^{S}_{xyxy}(\omega_j + \omega_l - \omega_l) \tag{33}$$

$$\Delta_{lj} = k_l(\varnothing_{l+} - \varnothing_{l-}) - k_j(\varnothing_{j+} - \varnothing_{j-})$$

In the absence of nonlinear absorption ($a'' = b'' = 0$), Eqs. (32) yield

$$\frac{\partial I_{j+}}{\partial z} + \frac{\partial I_{j-}}{\partial z} = 0 \tag{34a}$$

$$\sum_{j=1}^{N} \frac{\varepsilon_j}{k_j} \frac{\partial I_{j\pm}}{\partial z} = 0 \tag{34b}$$

Equation (34a) shows that the total power of each wave remains constant, while Eq. (34b) means that the sum of the intensities of the right (and left) circularly polarized components of the N waves is constant. We use the physical parameters

$$\begin{aligned} u_j^2 &= \frac{I_{j+}}{I_{j-}}\left(\frac{1-e_j}{1+e_j}\right) \\ \theta_j &= \tfrac{1}{2}k_j(\varnothing_{j+} - \varnothing_{j-}) \end{aligned} \tag{35}$$

where l_j and θ_j are respectively the ellipticity and the angular position of the axes of the polarization ellipse of the jth wave. Equation (32) reduces to

$$\frac{1}{u_j}\frac{\partial u_j}{\partial z} = -\frac{2\pi}{\varepsilon_j}\sum_{l-1}^{N} k_j b'_{lj} \sin \Delta_{lj} \frac{u_l}{u_j}\frac{1-u_j^2}{1+u_l^2} I_l \tag{36}$$

The polarization eigenstates are deduced from Eq. (36): If the incident beam comprises a mixture of parallel or orthogonal vibration directions and circularly polarized waves, no change is observed in the polarization states (in the absence of NLA).

B. Propagation of a Signal Wave in the Presence of Pump Wave

The polarization of the pump wave obeys the rule described in Section I. For the signal wave, in a nonabsorbing medium, Eqs. (32) yield

$$\frac{\partial u_{\rm s}}{\partial z} = -\frac{H}{2} b'_{\rm sp} \sin \theta_{\rm sp} u_{\rm p} \frac{1-u_{\rm s}^2}{1+u_{\rm p}^2} I_{\rm p}$$

$$\frac{\partial \theta_{\rm s}}{\partial z} = H b'_{\rm sp} \frac{u_{\rm p}}{u_{\rm s}} \frac{1-u_{\rm s}^2}{1+u_{\rm p}} \cos \theta_{\rm sp} I_{\rm p}$$

with $\theta_{sp} = \theta_s - \theta_p$, and

$$\frac{\partial \theta_p}{\partial z} = Hb'_{pp} \frac{1 - u_p^2}{1 + u_p^2} I_p$$
$$n_s = n_0 + \frac{2\pi}{n_{0s}} \left(a'_{sp} + b'_{sp} u_p^2 u_s^2 \cos \theta_{sp} \right) \tag{37}$$

where the values $j = \mathrm{s}$ and $l = \mathrm{p}$ stand respectively for signal and pump beams.

Even in the absence of NLA, the ellipticity of the signal wave can change due to the presence of the pump wave. In the presence of NLA, more complex phenomena appear. For instance, saturable absorption in gases can induce an optical activity without circular dichroism or a dischroism without optical activity [29].

Coefficients χ^{S}_{ijkl} depend on the frequencies ω_s and ω_p [29]. Table I gives their calculated values in the presence of nuclear and electronic effects, particularly in the two cases $\omega_s = \omega_p$ and $\omega_s \neq \omega_p[(\omega_p - \omega_s)\tau \gg 1]$.

1. *Polarization State of the Probe Wave: Experimental Results*

Experimental studies concerning the polarization state of a signal wave in the presence of a pump beam began in 1969. They have shown that in

TABLE I
Contribution of Nuclear Effects (B_0) and Electronic Effects (σ) in the Third-Order Susceptibility Tensor Components in Two-Wave Mixing Experiments

	$\omega_s - \omega_p$	
χ^{S}_{ijkl}	$\omega_s - \omega_p = 0$	$\omega_s - \omega_p \gg \frac{1}{\tau}$
$6\chi^{S}_{xxxx}$	$\frac{8B_0}{3} + 3\sigma$	$\frac{4B_0}{3} + 3\sigma$
$6\chi^{S}_{xxyy}$	$\frac{B_0}{3} + \sigma$	$-\frac{2B_0}{3} + \sigma$
$6\chi^{S}_{xyxy}$	$\frac{B_0}{3} + \sigma$	$B_0 + \sigma$
$6\chi^{S}_{xyyx} = b$	$2B_0 + \sigma$	$B_0 + \sigma$
$6\chi^{S}_{xxyy} + 6\chi^{S}_{xyxy} = a$	$\frac{2B_0}{3} + 2\sigma$	$\frac{B_0}{3} + 2\sigma$

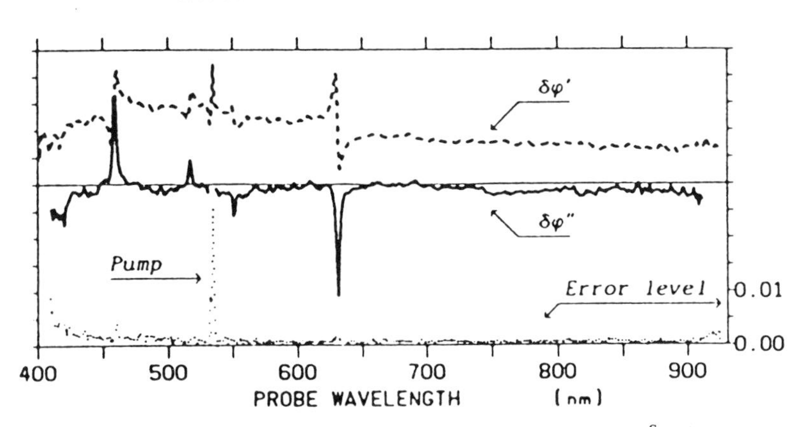

Figure 6. Real ($\Delta\varphi'$) and imaginary ($\Delta\varphi''$) parts of the component $\chi^{S}_{xxyy}(\omega_s + \omega_p - \omega_p) + \chi^{S}_{xyxy}(\omega_s + \omega_p - \omega_p)$ in tetramethylsilane excited with a picosecond pulse laser (λ = 532 nm, I_0 = 250 MW/cm^2). Two Raman lines are observed at 594 and 2884 cm^{-1}. (From Ref. 31.)

liquids such as those in Fig. 1, the frequency difference $\omega_s - \omega_p$ influences the results, according to Table I [30].

An effective method was recently proposed [31]: The incident pump and probe beams are linearly polarized, with a 45° angle between their electric fields. The analyzer makes a small angle with the probe field and the intensity is measured as a function of this angle. The real and imaginary parts of χ^{S}_{xxxx} and $(\chi^{S}_{xxyy} + \chi^{S}_{xyyx})$ can be deduced from the results. The frequency of the signal beam ω_s is varied, while ω_p remains constant. Figure 6 shows the results obtained for the variations of the component $a_{sp} = \chi^{S}_{xxyy}(\omega_s + \omega_p - \omega_p) + \chi^{S}_{xyxy}(\omega_s + \omega_p - \omega_p)$ with the frequency difference $\omega_s - \omega_p$. It reveals the influence of the Raman lines (Stokes and anti-Stokes sides) on the real and imaginary parts of a_{sp}, and is a good illustration of the Kramers–Kronig relations.

Using the same types of techniques, different authors [32, 33] have studied the temporal behavior of χ^{S}_{ijkl}. For instance, Lotshaw et al. [33] used 65-fs pump pulses to induce a birefringence read by a probe beam with an adjustable delay. They showed the existence of several nuclear phenomena (see Section III.B).

2. *Refraction Index of the Signal Wave*

The first refraction index measurements have been performed by interferometry techniques [34, 35]. Two probe waves interfere after their travel through the nonlinear material. The change in the interference pattern caused by the superposition of a pump beam to one of the probe beams is measured. The real and imaginary parts of χ^{S}_{ijkl} are deduced respectively from the displacement of the fringes, and from their visibility change. When the pump and signal beams have the same linear polarization, the measured component is $\chi^{S}_{xxxx}(\omega_s + \omega_p - \omega_p)$, while it is $[\chi^{S}_{xxyy}(\omega_s + \omega_p - \omega_p) + \chi^{S}_{xyxy}(\omega_s + \omega_p - \omega_p)]$ when the pump is circularly polarized.

The Z-scan method described in Section I.D.2.b has been used to measure the nonlinearities excited on the probe wave by a pump wave at a different wavelength [36]. The components $[\chi^{S}_{xxxx}(\omega_s + \omega_p - \omega_p)]$ and $\chi_{xxyy}(\omega_s + \omega_p - \omega_p)$ were measured for pump and probe beams polarized respectively parallelly and orthogonally. For carbon disulfide, the measurements in the picosecond range give

$$\frac{\chi^{S}_{xxyy}(\omega_s + \omega_p - \omega_p)}{\chi^{S}_{xxxx}(\omega_s + \omega_p - \omega_p)} = -0.31$$

This value, compared to the calculated value of $-(2B_0 + 3\sigma)/(4B_0 + 9\sigma)$ (Table I) yields a ratio $\sigma/B_0 = 0.13$, indicating that the electronic contribution is near 15% of the total nonlinear effects, in the coefficient $\chi^{S}_{xxxx}(\omega + \omega - \omega)$, in agreement with other measurements described below (Section II.B).

C. Propagation of Two Intense Waves

When an intense wave travels through a nonlinear medium disrupted by a second intense wave, it undergoes a polarization state change due to self- and cross-induced effects. The case of two intense waves propagating in opposite directions has been studied because it plays an important role in phase conjugation experiments [26].

Recently, Altshuller et al. [37] have studied the interaction of two counter propagating elliptically polarized waves in isotropic nonlinear media. They have shown rotation and ellipticity changes of the polarization ellipses even in the absence of NLA. They propose an application to a modulator, illustrated in Fig. 7: The transmittance of the return wave

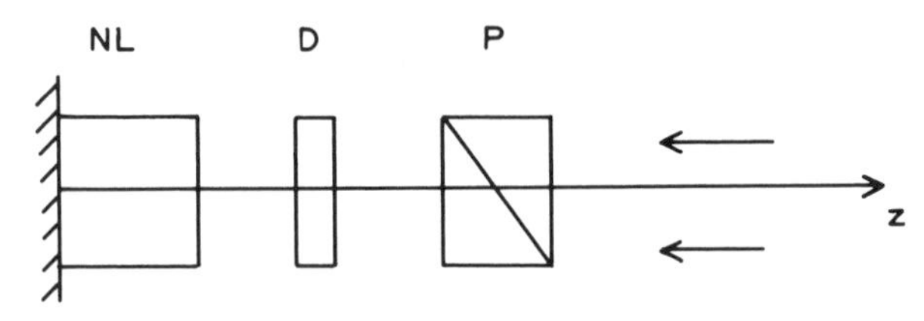

Figure 7. Schematic diagram of an optical modulator: P, polarizer; D, optical delay; NL, nonlinear element. (From Ref. 37.)

shows a saturation when the incident intensity increases, and is a multiple-valued function of the light intensity. This modulator, more sensitive than those using only one-wave propagation (Section I.F), could work in monomode optical fibers.

D. Conclusions

The optical Kerr effect induced on a probe beam by a pump beam depends largely on the polarization states of both beams. It provides a good measurement method for the susceptibility tensor components. However, it has been little used. The systematic measurements undertaken from 1965 to 1970 [30] revealed several fuzzy results, partly due to technical difficulties. New methods like ellipsometry [31], Z-scan [36], or nonlinear imaging [16] can now be more widely used to bring information on the physical effects involved in the χ^{S}_{ijkl} components.

IV. PROPAGATION OF THREE INCIDENT WAVES: FOUR-WAVE MIXING

A lot of studies have been carried out, since the discovery of phase conjugation in 1971 [38–42], on wave mixing experiments and particularly on four-wave mixing (FWM). In the classical degenerate four-wave mixing (DFWM) setup, three incident waves with the same frequency have their wave vectors $\mathbf{k}_1$, $\mathbf{k}_2$, and $\mathbf{k}_3$ verifying the relation $\mathbf{k}_1 = -\mathbf{k}_2$. A fourth wave, phase conjugated with the third one, is created, with a wave vector $\mathbf{k}_4 = -\mathbf{k}_3$ (Fig. 8). The incident waves can also have different frequencies (nondegenerate FWM).

FWM can be analyzed with the language of dynamic holography. Two of the three incident waves interfere and build an illumination grating that creates a refraction index grating or (and) absorption grating: Both are recorded in the medium. The third wave is diffracted on the recorded grating(s), and a fourth wave is thus created. Several physical phenomena contribute to the recording of the refraction index grating: Molecular movements, electronic deformations, electrostriction, heating, and

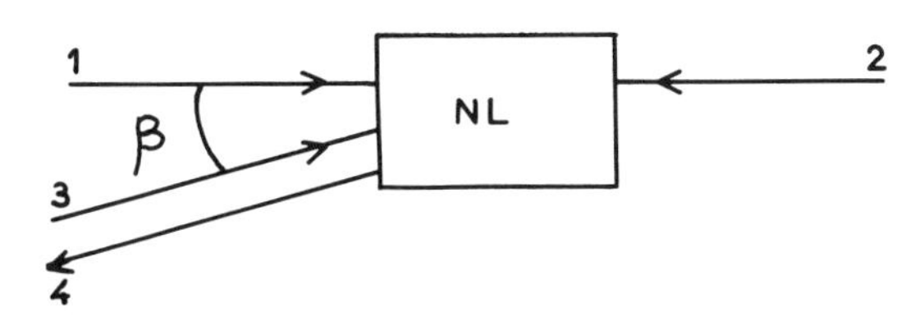

Figure 8. Four-wave mixing theoretical setup.

electro-optics effects, the latter being particularly studied in photorefractive materials.

The dynamic holographic language, or grating language, is mostly used for nonlocal, and rather slow effects. For local effects and especially for those that occur in Kerr-like media, the language of nonlinear susceptibilities is more convenient. For instance, it is known in DFWM, that, although the grating R_{12} built by the two opposite waves 1 and 2 (Fig. 8) is inactive and cannot diffract wave 3, a fourth wave is observed in Kerr-like mediums when the other two gratings, R_{13} and R_{23}, are made inefficient by the choice of wave 3 orthogonally polarized with respect to the waves 1 and 2.

Arrangements of the incident wave vectors [43] other than the arrangement described (Fig. 8) have been used by some authors. Thomazeau et al. [44] and Etchepare et al. [45] chose either a Raman Nath or nonplanar Bragg configuration, with three incident forward waves. All the arrangements can be used to measure the components χ^{S}_{ijkl} with a correct choice of the polarization states of the three incident waves. Over the last decade the FWM method (Fig. 8) has produced more publications than all the other methods using one or two waves for the nonlinear susceptibility measurements. Its results are described below.

A. Propagation Equations and Susceptibility Tensor in Four-Wave Mixing

The propagation equations for the four waves are written with the method described in Section I. In the absence of linear absorption, of linear and nonlinear optical activities, supposing that waves 1 and 2 are high-intensity pump waves, while wave 3 is a low-intensity signal wave $[I_1 \approx I_2 \gg I_3]$ and that the angle β is small (Fig. 8), the propagation equations are

$$\frac{\partial A_{1i}}{\partial z} = \mathrm{i}H\left[\chi^{S}_{ijkl}(\omega_1 + \omega_1 - \omega_1)A_{i1}A_{j1}A^{*}_{k1} + \chi^{S}_{ijkl}(\omega_1 + \omega_2 - \omega_2)A_{i1}A_{j2}A^{*}_{k2}\right] \tag{38a}$$

$$\frac{\partial A_{2i}}{\partial z} = \mathrm{i}H\left[\chi^{S}_{ijkl}(\omega_2 + \omega_1 - \omega_1)A_{i2}A_{j1}A^{*}_{k1} + \chi^{S}_{ijkl}(\omega_2 + \omega_2 - \omega_2)A_{i2}A_{j2}A^{*}_{k2}\right] \tag{38b}$$

$$\frac{\partial A_{3i}}{\partial z} = \mathrm{i}H\chi^{S}_{ijkl}(\omega_1 + \omega_2 - \omega_3)A_{j1}A_{k2}A^{*}_{l3} \tag{38c}$$

$$\frac{\partial A_{4i}}{\partial z} = -\,\mathrm{i}H\chi^{S}_{ijkl}(\omega_1 + \omega_2 - \omega_3)A_{j1}A_{k2}A^{*}_{l4} \tag{38d}$$

If FWM is degenerated ($\omega_1 = \omega_2 = \omega$), the tensor used in Eqs. 38 is the same as that calculated for one wave by formula 21 and written in column 1 of Table I. If $\omega_1 = \omega_3 \neq \omega_2$, the second column of Table I is used. When waves 1, 2, and 3 are polarized linearly in directions $\mathbf{0}_x$ and $\mathbf{0}_y$ or circularly, Eqs. 38 are easily solved. For instance, if waves 1, 2, and 3 have linear polarizations states on $\mathbf{0}_x$ or $\mathbf{0}_y$, respectively described by the indices α, β, and γ, then Eqs. (38) lead to

$$I_{4\delta} = I_{3\gamma} tg^2\left(\left|\chi^{\mathrm{S}}_{\delta\alpha\beta\gamma}\right|\sqrt{I_{1\alpha} I_{2\beta}}\, l\right) \tag{39}$$

where $I_{1\alpha}, I_{2\beta}, I_{3\gamma}$ are the input intensities of the three incident waves, and $I_{4\delta}$ the intensity of the fourth wave linearly polarized along $\gamma(\alpha, \beta, \gamma, \delta \equiv x \text{ or } y)$.

When the efficiency $R = I_{4\delta}/I_{3\gamma}$ remains much smaller than 1, then Eq. (39) gives

$$R_{\delta\alpha\beta\gamma} = \left|\chi^{\mathrm{S}}_{\delta\alpha\beta\gamma}\right|^2 I_{1\alpha} I_{2\beta} l^2 \tag{40}$$

Table II shows the expected values of $R_{\delta\alpha\beta\gamma}$ according to the polarization

TABLE II
Nonlinear Susceptibility Tensor Components Involved in Degenerate and Nondegenerate Four-Wave Mixing

Polarization of the Incident Waves			Polarization of Wave 4	Calculated Value	Calculated Value of
α	β	γ	δ	of $\chi^{\mathrm{S}}_{\delta\alpha\beta\gamma}$ in DFWM	$\chi^{\mathrm{S}}_{\delta\alpha\beta\gamma}$ for $\omega_1 = \omega_3 \neq \omega_2$
x	x	x	x		
or				$6\chi^{\mathrm{S}}_{xxxx} = \dfrac{8B_0}{3} + 3\sigma$	$\dfrac{4B_0}{3} + 3\sigma$
y	y	y	y		
x	x	y	y		
or				$6\chi_{xyyx} = 2B_0 + \sigma$	$B_0 + \sigma$
y	y	x	x		
x	y	x	y		
or				$6\chi_{xyxy} = \dfrac{B_0}{3} + \sigma$	$B_0 + \sigma$
y	x	y	x		
y	x	x	y		
or				$6\chi_{xxyy} = \dfrac{B_0}{3} + \sigma$	$-\dfrac{2B_0}{3} + 3\sigma$
x	y	y	x		

states of beams 1, 2, 3, for a Kerr-like medium. The values of $\chi^{S}_{\delta\alpha\beta\gamma}$ can be deduced from the measurements of $R_{\delta\alpha\beta\gamma}$. Let us examine the experimental results.

B. Measurements of the Susceptibility Tensor Components

It is surprising to discover that very few authors have measured more than two of the four components χ^{S}_{xxxx}, χ^{S}_{xxyy}, χ^{S}_{xyxy}, and χ^{S}_{xyyx}, which can, however, be obtained easily in the same setup (Fig. 8) by turning the polarizer and analyzer plates on the four waves. For instance, for the isotropic liquid carbon disulfide, often considered as a reference material, to our knowledge no publication gives these four components. We have recently carried out this measurement [6]. The values obtained are shown in Table IIIa. They are compared with values obtained in other kinds of experiments. A synthesis is proposed in the conclusion.

TABLE IIIa
Values of the Third-Order Susceptibility Tensor Elements of CS_2 Measured Using Various Methods

		One-Wave Methods		Nonlinear	Two-Wave Methods	Four-Wave Mixing
	Ellipsometry	Raman Thresholds		Imaging	Z scan	Degenerate
	ns	ns	ps	ps	ps	ps
Coefficient	1 = 2.4 cm	1 = 2.4 cm	1 = 2.4 cm	1 = 1 mm	1 = 1 mm	1 = 1 mm
$\dfrac{\chi_{xyyx}}{\chi_{xxxx}}$	0.75[(3)a] 0.70					0.67[(6)] 0.64[(58)] 0.70
$\dfrac{\chi^{S}_{xxyy}}{\chi^{S}_{xxxx}}$					−0.31[(36)] −0.33	0.14[(6)] 0.15
$\dfrac{\chi^{S}_{xyxy}}{\chi^{S}_{xxxx}}$						0.14[(6)] 0.15
$\dfrac{\chi^{S}_{xxyy}+\chi^{S}_{xyxy}}{\chi^{S}_{xxxx}}$		0.5[(8)(10)(11)] 0.30	0.63[(11)] 0.30	0.29[(59)] 0.30	0.28[(6)] 0.30	

Note. Values in the boxes are the calculated ones at $\lambda = 532$ nm (see Table IIIb).

[a]The absolute value of χ^{S}_{xyyx} measured by ellipsometry is related to the absolute value of χ^{S}_{xxxx} measured by another technique.

In most studies, pump waves 1 and 2 have the same linear polarization, while the probe beam polarization is either parallel or perpendicular to the pump polarization [20, 46]. In this way, only the coefficients χ^{S}_{xxxx} and χ^{S}_{xyyx} are measured. We notice here that in DFWM, the calculated ratio $\chi^{Sn}_{xyyx}/\chi^{Sn}_{xxxx}$ is equal to the ratio of the nonsymmetrized components $\chi^{n}_{xyyx}/\chi^{n}_{xxxx} = \frac{3}{4}$. But for the other components we observe $\chi^{sn}_{xxyy}/\chi^{Sn}_{xxxx} = \chi^{Sn}_{xyxy}/\chi^{Sn}_{xxxx} = \frac{1}{8}$, while $\chi^{n}_{xxyy}/\chi^{n}_{xxxx} = -\frac{1}{2}$ and $\chi^{n}_{xyxy}/\chi^{n}_{xxxx} = \frac{3}{4}$. If the discrepancy between $\chi^{Sn}_{ijkl}(\omega + \omega - \omega)$ and $\chi^{n}_{ijkl}(\omega + \omega - \omega)$ is left out, measurements of R_{xyxy} and R_{xxyy} in DFWM experiments performed in materials such as carbon disulfide, where the molecular reorientation is the dominant third-order process, lead to results that do not coincide with the models.

On the contrary, in nondegenerate FWM, the symmetrized and nonsymmetrized components have the same ratios. Echepare et al. [45] have measured them in femtosecond excitation. They have separated for the first time the nuclear and electronic components of χ in CS_2. The electronic contribution in $\chi^{3}_{xxxx}(\omega_s + \omega_p - \omega_p)$ is 11%. The nuclear contribution comprises two parts, with very different relaxation times (these times are measured by delaying the reading wave with respect to the writing ones). The first part, due to reorientation of individual molecules, has a relaxation time $\tau_R = 1.5$ ps. For the second part, due to collective nuclear processes, the relaxation time is $\tau_R = 0.1$ ps. The relative contributions of the individual and collective effects are respectively 61 and 28%.

Many authors have used the FWM technique to measure the susceptibility tensor of organic materials, and particularly the component χ_{xxxx}. Several studies use polarization discrimination to measure different components of χ, in order to identify the physical phenomena involved [46–52]. For the electrostrictive effects and the acousto thermic effects always present in absorbing mediums, the grating language is generally used. Because grating R_{12} is inefficient, the thermal component χ^{St}_{xyyx} is expected to be zero, while the components χ^{St}_{xyxy} and χ^{St}_{xxyy}, respectively associated with gratings R_{13} and R_{23}, are nonzero ($\chi_{xyyx} = 0$ is also obtained by other physical reasonings) [61]. Thus, in soluble polydiacetylene excited by 180-ps pulses (532 nm) Dennis and Blau [49] find $\chi^{St}_{xyxy}|\chi^{St}_{xxxx} = 0.71$, $\chi^{St}_{xxyy}|\chi^{St}_{xxxx} = 0.36$, and $\chi^{St}_{xyyx}|\chi^{St}_{xxxx} < 0.07$.

Generally speaking, a very low measured ratio $\chi^{S}_{xyyx}/\chi^{S}_{xxxx}$ is interpreted in terms of nonlocal effects. Durations in the nanosecond range are necessary for the development of electrostrictive and thermal effects. In all the measurements performed with very short pulses (femto- and picoseconds), these effects play no role, while they become important in nanosecond excitation (for one or several waves).

In the presence of absorptions and optical activity (linear and nonlinear) formula (40) is no longer valid. Calculations have been performed by various authors. For instance, in BSO and BGO crystals, the large NLA and linear optical activity influence strongly the polarization and efficiency of FWM [24, 25].

C. Polarization Wavefront Conjugation

The phase distorsions of signal wave 3 in FWM can be accompanied by polarization distortions, for instance if the wave has gone through a birefringent medium. The compensation of these polarization distorsions as well as that of the phase distorsions in FWM needs the creation of a fourth wave displaying a Jones vector complex conjugate of the Jones vector of the third wave. This problem has been studied theoretically [54, 55] and experimentally [56, 57].

V. CONCLUSION

After a summary of the main features concerning polarization eigenstates, we gather the results of third-order nonlinear susceptibility components measurements obtained with different techniques using one or several waves, degenerated or not, and compare them with calculations.

A. Polarization Eigenstates

Polarization eigenstates of one wave propagating through a nonlinear medium depend not only on optical linear and nonlinear properties of the medium, but on the possible presence of other waves too. In the absence of absorptions and scatterings, eigenstates are easily calculated and the results are in keeping with the experiments. The following features appear in the absence of birefringence and optical activity:

- A single wave keeps its polarization state when it is polarized linearly or circularly (with NLA, only circular polarization states are eigenstates).
- A probe wave in the presence of a pump wave keeps its polarization state if the pump wave is circularly polarized. If the pump is linearly polarized, only probe waves polarized linearly, either parallelly or orthogonally to the pump, keep their polarization state.
- Generally speaking, several intense waves displaying the same propagation direction maintain their polarization states if they consist of

waves polarized either circularly or linearly with one common polarization direction, or two perpendicular ones.

B. Third-Order Nonlinear Susceptibility Components

In Table IIIa are values of the components of χ^{S}_{ijkl} deduced from polarization states and refraction index measurements of one wave propagating either alone or in the presence of other waves, in liquid carbon disulfide (a material often considered a reference). In Table IIIb, relative values of these components are calculated for both cases $\omega_1 = \omega_2 = -\omega_3 = \omega$ (propagation of one wave, or of several waves in degenerate mixings) and $\omega_1 = \omega_2 \neq -\omega_3$ (nondegenerate wave mixings). Calculations are performed with the hypothesis that only molecular and electronic effects occur, to the exclusion of acoustothermal effects. The latter are negligible for excitations in the visible and near-infrared range [36]. Molecular individual orientation effects have a relaxation time of 1.5 ps, while collective molecular effects last about 100 ps [45, 60]. These two kinds of effects occur totally in nanosecond and picosecond excitations, but not in femtosecond excitation. All measurements displayed in Table IIIa are performed at $\lambda = 532$ nm, in nano- or picosecond excitation. To our knowledge, only a few experiments leading to the determination of all the components χ^{S}_{ijkl} have been made, as can be seen in Table IIIa. All the results displayed in Table IIIa, to the exclusion of those deduced from Raman threshold measurements, are in agreement with the calculations. This confirms that it is necessary to use the susceptibility symmetrized tensor, especially for molecular orientation phenomena.

The one-, two-, or four-wave polarization method is a powerful tool to measure the susceptibility tensor components, and thus to separate the different physical effects that make up the susceptibility. For instance, this method has been able to ascribe the fast effects observed in CS_2 and some other liquids [45, 62] to molecular phenomena (fast effects are those having a relaxation time in the order of 100 fs). It is also able to separate the contributions of electronic, molecular, thermal, and acoustothermal effects in new organic materials.

We have not tackled the topic of quantum effects, which affect the polarization states when the quantum description of the field is used. Kielich, Tanas, and Gantsog [63, 64] have shown that purely quantum effects arise during the propagation of light in a Kerr medium. Concerning the polarization state of an incident elliptically polarized beam, the main quantum effects theoretically discovered are the appearance of an unpolarized component of the field, a modification of the rotation angle of the polarization ellipse, and an ellipticity change in the absence of nonlinear

TABLE IIIb

Relative values of the Third-Order Susceptibility Tensor Elements in Four-Wave Mixing, Either Degenerate ($\omega_1 = \omega_2 = -\omega_3 = \omega$) or Nondegenerate ($\omega_1 = \omega_2 \neq -\omega_3$)

	Calculated Value for	Numerical Value for $\omega_1 = \omega_2 = -\omega_3 = \omega$			Calculated Value for	Numerical Value for $\omega_1 = \omega_2 \neq -\omega_3$		
	$\omega_1 = \omega_2 = -\omega_3 = \omega$	$\frac{\sigma}{B_0} = 0$	$\frac{\sigma}{B_0} = 0.11$	$\frac{B_0}{\sigma} = 0$	$\omega_1 = \omega_2 \neq -\omega_3$	$\frac{\sigma}{B_0} = 0$	$\frac{\sigma}{B_0} = 0.11$	$\frac{B_0}{\sigma} = 0$
$\frac{\chi^S_{xyyx}}{\chi^S_{xxxx}}$	$\frac{6B_0 + 3\sigma}{8B_0 + 9\sigma}$	0.75	0.70	0.33	$\frac{3(B_0 + \sigma)}{4B_0 + 9\sigma}$	0.75	0.66	0.33
$\frac{\chi^S_{xxyy}}{\chi^S_{xxxx}}$	$\frac{B_0 + 3\sigma}{8B_0 + 9\sigma}$	0.125	0.15	0.33	$\frac{-2B_0 + 3\sigma}{4B_0 + 9\sigma}$	-0.5	-0.33	$+0.33$
$\frac{\chi^S_{xyxy}}{\chi^S_{xxxx}}$	$\frac{B_0 + 3\sigma}{8B_0 + 9\sigma}$	0.125	0.15	0.33	$\frac{3(B_0 + \sigma)}{4B_0 + 9\sigma}$	0.75	0.66	0.33
$\frac{\chi^S_{xxyy} + \chi^S_{xyxy}}{\chi^S_{xxxx}}$	$\frac{2(B_0 + 3\sigma)}{8B_0 + 9\sigma}$	0.25	0.30	0.66	$\frac{B_0 + 6\sigma}{4B_0 + 9\sigma}$	0.25	0.33	0.66

Note. Numerical values are calculated for CS_2, with a ratio of the electronic to nuclear contributions $\tau/B_0 = 0.11$. On both sides of the column referred to as CS_2, values are listed for pure molecular effects ($\tau = 0$) and pure electronic effects ($B_0 = 0$).

absorption. These quantum effects are expected to be small in most real physical situations, but could be observed at high laser intensities.

Acknowledgments

This paper is supported by several studies carried out in the Laboratory Poma (Angers), and particularly those performed by P. X. Nguyen and J. P. Bourdin, I thank them for their contribution, as I thank R. Chevalier, J. P. Lecoq, D. Lamaury, and J. Marolleau for their technical help.

References

1. P. D. Maker, R. W. Terhune, and C. M. Savage, *Phys. Rev. Lett.* **12**, 507 (1964).
2. Y. Le Duff and R. Dupeyrat, *C.R. Acad. Sci. Paris* **268**, 1346 (1969).
3. P. X. Nguyen and G. Rivoire, *Opt. Acta* **25**, 233 (1978).
4. P. X. Nguyen, J. L. Ferrier, J. Gazengel, and G. Rivoire, *Opt. Commun.* **46**, 329 (1983).
5. P. D. Maker and R. W. Terhume, *Phys. Rev.* **137**, A801 (1965).
6. J. P. Bourdin, P. X. Nguyen, G. Rivoire, and J. M. Nunzi, to be published.
7. P. L. Kelley, *Phys. Rev. Lett.* **15** 1005 (1965).
8. C. C. Wang, *Phys. Rev.* **152** 149 (1966); **173**, 908 (1968).
9. R. Y. Chiao, E. Garmire, and CH. Townes, *Phys. Rev. Lett.* **13**, 479 (1964).
10. D. H. Close, C. R. Giuliano, R. W. Hellwarth, L. D. Hess, F. J. McClung, and W. G. Wagner, *IEEE* **2**, 9 (1966).
11. G. Rivoire, C. Desblancs, J. L. Ferrier, J. Gazengel, and P. X. Nguyen, *Opt. Quantum. Electron.* **15**, 209 (1982).
12. M. Sheik-Bahae, A. A. Said, and E. W. Van Stryland, *Opt. Lett.* **14**, 955 (1989).
13. M. Sheik-Bahae, A. A. Said, T. H. Wei, D. J. Hagan, and E. W. Van Stryland, *IEEE J. Quantum Electron.* **QE-26**, 760 (1990).
14. P. X. Nguyen, J. L. Ferrier, J. Gazengel, G. Rivoire, G. L. Brekhovskhikh, A. D. Kudriavtseva, A. I. Sokolovskaia, and N. Tcherniega, *Opt. Commun.* **68**, 244 (1988).
15. J. L. Ferrier, J. Gazengel, P. X. Nguyen, J. J. Zhang, and G. Rivoire, *Opt. Commun.* **76**, 13 (1990).
16. G. Boudebs, P. X. Nguyen, J. Gazengel, J. P. Lecoq, and G. Rivoire, *Opto 91 Paris*, ESI Publications, Boca Raton, FL, 1991, p. 80.
17. R. W. Hellwarth, *Prog. Quantum Electron.* **5**, 1 (1977).
18. P. X. Nguyen, Thesis, University of Angers, 1976.
19. K. A. Nelson, R. J. D. Wayne Muller, D. R. Lutz, and M. D. Fayer, *J. Appl. Phys.* **33**, 1444 (1982).
20. C. K. Wu, P. Agostini, G. Petite, and F. Fabre, *Opt. Lett.* **8**, 67 (1983).
21. N. I. Zheludev, *XIII Int. Conf. on Coherent and NL Optics*, Minsk, Sep. 1987.
22. S. A. Akhmanov and V. I. Zharukov, *Zh. ETF Pisma* **6**, 644 (1967).
23. S. Kielich, Lectures, University of Bordeaux, 1971; Podstanry Optyki Nieliniowej, Poznan, 1972 (Universystet Im. A. Mickiewicza).
24. M. Sylla, P. X. Nguyen, D. Rouede, and G. Rivoire, *J. Appl. Phys.* **71**, 11 (1992).

25. M. Sylla, D. Rouede, R. Chevalier, P. X. Nguyen, and G. Rivoire, *Opt. Commun.* 89 (1992).
26. M. Sylla, Thesis, University of Angers, 1991.
27. Kashyap, *Opt. Lett.* **17**, 405 (1992).
28. P. X. Nguyen, V. P. Nayyar, and G. Rivoire, *Opt. Quantum Electron.* **13**, 95 (1981).
29. A. A. Kurbatov and T. Y. Popova, *Opt. Spectrosc.* **45**, 679 (1978).
30. M. Paillette, *Ann. Phys.* **4**, 671 (1969).
31. N. Pfeffer, P. Charra, and J. M. Nunzi, *Opt. Lett.* **16**, 1987 (1991).
32. B. I. Greene and R. C. Farrow, *Chem. Phys. Lett.* **98**, 273 (1983).
33. W. T. Lotshaw, D. McMarrow, C. Kalpouzos, G. Kenney-Wallace, *Chem. Phys. Lett.* **136**, 323 (1987).
34. A. Owyoung, *Opt. Commun.* **16**, 266 (1976).
35. P. X. Nguyen, J. L. Ferrier, J. Gazengel, and G. Rivoire, *Opt. Commun.* **51**, 433 (1984).
36. M. Sheik-Bahae, J. Wang, R. DeSalvo, D. J. Hagan, and E. W. Van Stryland, *Opt. Lett.* **17**, 258 (1992).
37. G. B. Altshuller, V. B. Karassev, S. A. Kozlov, and L. I. Pavlov, *J. Mod. Opt.* **35**, 727 (1988).
38. B. I. Stepanov, E. V. Ivakin, and A. S. Rubanov, *Sov. Phys. Dokl.* **16**, 46 (1971).
39. R. W. Hellwarth, *J. Opt. Soc. Am.* **67**, 1 (1977).
40. A. Yariv, *IEEE J. Quantum Electron.* **QE-14**, 651 (1978).
41. R. Fisher, *Optical Phase Conjugation*, Academic, New York, 1983.
42. B. Y. Zeldovich, N. F. Pilipetski, and V. V. Schkunov, *Principles of Phase Conjugation*, Springer, New York, 1985.
43. H. J. Eichler, *Opt. Acta* **4**, 631 (1977).
44. I. Thomaseau, J. Etchepare, G. Grillon, G. Hamoniaux, and A. Orszag, *Opt. Commun.* **35**, 442 (1985).
45. J. Etchepare, G. Grillon, J. P. Chambaret, G. Hamoniaux, and A. Orszag, *Opt. Commun.* **63**, 329 (1987).
46. D. J. McGraw, A. E. Siegman, G. M. Walraff, and R. D. Muller, *Appl. Phys. Lett.* **54**, 1713 (1989).
47. C. Maloney, H. Byne, W. M. Dennis, and W. Blau, *Chem. Phys.* **121**, 21 (1988).
48. J. M. Nunzi, J. L. Ferrier, and R. Chevalier, *Nonlinear Optical Effect in Organic Polymers*, Kluwer Academic, Norwell, MA, 1989, p. 365.
49. W. M. Dennis and W. Blau, *Opt. Commun.* **57**, 371 (1986).
50. B. K. Rhee, W. E. Bron, and J. Kuhl, *Phys. Rev. B* **30**, 7358 (1984).
51. A. E. Neeves and M. H. Birnboim, *Opt. Soc. Am.* **5**, 701 (1988).
52. W. M. Dennis, W. Blau, and D. J. Bradley, *Appl. Phys. Lett.* **48**, 200 (1986).
53. J. M. Nunzi, Thesis, Paris, 1990.
54. B. Y. Zeldovich, F. V. Bunkin, and D. V. Vlasov, *Sov. J. Quantum Electron.* **9**, 379 (1979).
55. L. A. Bol'shov, F. v. Bunkin, and D. V. Vlasov, *Sov. J. Quantum Electron.* **10**, 1197 (1980).
56. G. Martin, L. Lam, and R. Hellwarth, *Opt. Lett.* **5**, 185 (1980).
57. L. Nikolova, K. Stoyanova, T. Todorore, and V. Tatanenko, *Opt. Commun.* **64**, 75 (1987).

58. C. K. Wu, P. Agostini, G. Petite, and F. Fabre, *Opt. Lett.* **8**, 67 (1983).
59. G. Boudebs and M. Chis, to be published.
60. S. Ruhman, L. R. Williams, A. G. Joly, B. Kohler, and K. A. Nelson, *J. Phys. Chem.* **91**, 2237 (1987).
61. R. W. Boyd, *Nonlinear Optics*, Academic, New York, 1992, p. 331.
62. J. Etchepare, G. Grillon, G. Hamaniaux, A. Antonetti, and A. Orszag, *Rev. Phys. Appl.* **22**, 1749 (1987).
63. R. Tanas, and S. Kielich, *J. Mod. Opt.* **37**, 1935 (1990).
64. R. Tanas and T. Gantsog, *Opt. Commun.* **87**, 369 (1992).

SELF-ORGANIZED NONLINEAR OPTICAL PHENOMENA IN OPTICAL FIBERS

PAVEL CHMELA

Faculty of Mechanical Engineering, Technical University, Brno, Czech Republic

CONTENTS

Modern Nonlinear Optics, Part 1, Edited by Myron Evans and Stanisław Kielich. Advances in Chemical Physics Series, Vol. LXXXV.
ISBN 0-471-57546-1

Dedicated to the 67th birthday of Professor Stanisław Kielich.

I. INTRODUCTION

Optical fibers are usually made of isotropic centrosymmetric materials (doped silica glass), for which the lowest-order nonlinear susceptibility $\chi^{(2)}$ should be zero (see, e.g., [1–4]). That is why they had been believed not to be able to exhibit second-order nonlinear optical phenomena such as second-harmonic generation (SHG), sum-frequency generation (SFG), or difference-frequency generation (DFG). Third-order nonlinearities were studied mostly in optical fibers only up to the end of the 1970s (see, e.g., [5–7]). Weak SHG and SFG, being near the limit of visibility, were observed in optical fibers at the beginning of the 1980s only [8–10].

In the fall of 1985 two young physicists, U. Österberg and W. Margulis, at the Royal Institute of Technology in Stockholm observed unusually efficient SHG, as high as 3%, in 1-m-long germanium-phosphorus-doped optical fibers, when the fibers were subjected to illumination by mode-locked, Q-switched Nd:YAG laser radiation for more than 10 h [11]. In the next experiment [12] the efficiency of SHG was raised to above 5%. The prepared fibers exhibited a memory effect, namely, they were able to generate SH radiation immediately after the fundamental infrared radiation being launched again, and this capability lasted for a few weeks or months.

The optical preparation of fibers for an efficient SHG was considerably accelerated by the method proposed by Stolen and Tom [13], who launched a weak SH seeding radiation simultaneously with the strong fundamental pump into the fiber. The conditioning time was reduced to about 5 min in this way.

The record, a 13% SH conversion efficiency, was achieved by Farries [14].

The ingenious experiments by Österberg and Margulis [11, 12] and Stolen and Tom [13] have stimulated a wide research of self-organized nonlinear optical phenomena in optical fibers (for review, see [15–17, 66]). The main aims are, on the one hand, to develop cheap fiber-optical nonlinear elements that could replace the traditional nonlinear crystals, and on the other hand, to ascertain the transmission changes, which appear in the prepared fibers as an attendant effect, in order to avoid the contingent induced losses in optical communication channels [18–20].

Another aspect represents the self-organized Bragg grating formation [21–23] that might find very important practical applications in optical communication systems and fiber sensors [21, 24, 25].

A lot of ingenious experiments were performed (see, e.g., [27–54]) and many brilliant models have been suggested (see, e.g., [13, 16, 27, 31, 34, 39, 41, 42, 45, 55–67]) to explain the self-organized nonlinear optical effects in fibers, but the basic physics of these phenomena remains a puzzle.

Because the research of microscopic mechanisms that are responsible for the formation of a new structure in the fiber core, which enables an efficient SHG, contingently SFG or DFG, is in the early stages, it is difficult to advance a consistent microscopic theory of self-organized nonlinear optical phenomena at the present time. Therefore, we shall base our next treatment on the phenomenological description only. Simple microscopic models will be mentioned if necessary merely to complete the phenomenological theory.

II. SOME BASIC EXPERIMENTS

A. Self-Seeding Second-Harmonic Generation

In the early experiments on efficient SHG in optical fibers by Österberg and Margulis [11, 12] the germanium-phosphorus-doped fibers were simply prepared by illuminating with intense infrared radiation alone (see Fig. 1). The fibers used were manufactured by the conventional chemical-vapor-deposition technique. Germanium was used as dopant for the formation of

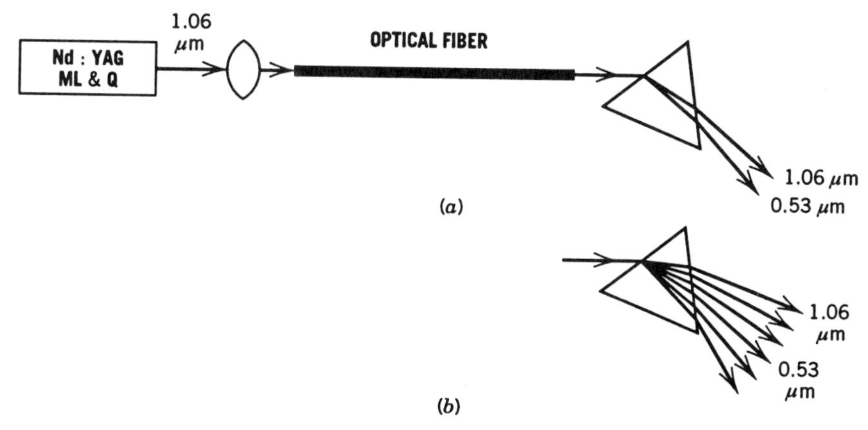

Figure 1. (*a*) Scheme of experimental arrangement for self-preparation of short (0.5-m) optical fibers for efficient SHG. (*b*) In somewhat longer (5- to 10-m) conditioned fibers bright visible spectrum manifesting 15 discrete peaks occurs [12].

the proper refractive-index profile and both the fiber core and the fiber cladding were additionally doped by 0.5 wt% phosphorus. The fibers possessed a core diameter between 8 and 10 μm, the cutoff wave-length ranged between 0.95 and 1.38 μm, and the respective core-cladding refractive-index difference Δn was between 0.0030 and 0.0045.

The preparation of fibers for an efficient SHG was performed using the radiation of Nd:YAG laser, operating in the cw mode-locked and Q-switched regime, producing 100- to 130-ps pulses at 1.064 μm, typically lasting for 450 ns, with the repetition rate 1250 Hz. The peak power was about 70 kW. The infrared laser radiation was focused into the fiber with 30–50% gain so that the peak power propagating in the fiber core was as high as 20 kW. In all experiments the infrared pump propagated in the fundamental LP_{01} mode.

The times necessary for preparing fibers were typically hours. After the initial slow stage, the effectiveness of SHG grew nearly exponentially with time until saturation. In the first experiment [11] saturation occurred after approximately 12 h. The rate of fiber preparation depends strongly on the pumping input power and it was found that for a peak power below 5 kW the fibers cannot be prepared for efficient SHG even if illuminated for a few days [12].

The prepared fibers exhibited a memory effect, namely, they were able to generate SH radiation immediately after being illuminated again by fundamental pumping radiation. The optically induced changes in the prepared fibers were nearly permanent. The capability of SHG dropped slowly when the fibers were not used for a few weeks, but SHG recovered its original value after a few minutes of illumination.

The maximum conversion efficiency slightly exceeded 5%. The generated green SH radiation at 0.532 μm propagated mostly in the LP_{11} mode, but other modes, including LP_{01}, were sometimes observed [12]. The pulse duration of SH light was about one-half of the pump pulse width [11].

The effective region of SHG was investigated by both cutting the fiber from the input and exit sides and measuring the side SH light that was emitted by the fiber [12]. It was found that the first few centimeters of the input side of prepared fiber plays the dominant role in stimulating the SHG process, and the total effective fiber length does to exceed 40 cm.

B. Second-Harmonic Generation with External Seeding

The long buildup time for efficient SHG can be considerably shortened when launching a weak SH seeding radiation simultaneously with the strong pump into the fiber under preparation (Fig. 2), as shown by Stolen and Tom [13]. The germanium-phosphorus-doped fiber was drawn to be a single mode at both 1.064 and 0.532 μm. The core diameter was 4.1 μm

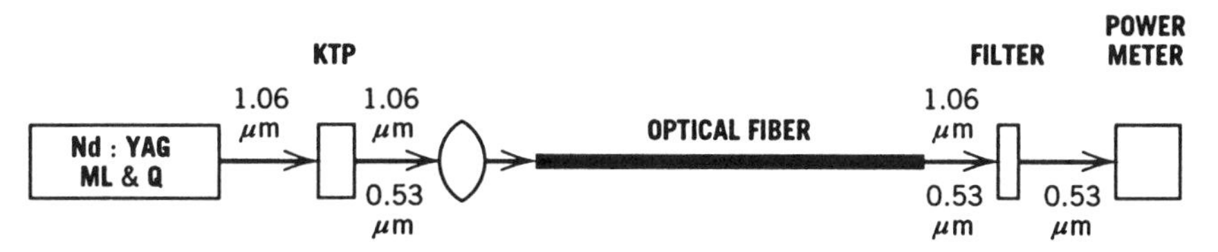

Figure 2. Scheme of experimental arrangement for conditioning optical fibers to efficient SHG with external seeding [13].

and the core–cladding difference Δn was 0.0029. The effective areas were 3.29×10^{-7} cm^2 for the fundamental mode and 8.65×10^{-8} cm^2 for the SH mode. The fiber length was 55 cm.

The fiber was prepared using the fundamental radiation at 1.064 μm of a 76-MHz mode-locked, Q-switched Nd:YAG laser, producing 1.2-ns pulses in a 480-ns envelope with a peak power of 3.5 kW, and the frequency doubled light at 0.532 μm of the peak power 10 W. Both radiations were linearly polarized and then coupled into the fiber. After only 5 min of illumination, saturation occurred and the fiber was able to generate weak SH radiation with infrared pump input alone. No SHG was observed after 12 h of fiber preparation with the pump alone (without the external seeding). The SH conversion efficiency attained was 0.03%, which is rather low. However, it can be considerably raised by increasing both the pump and the SH seed power (see, e.g., [43, 48]).

C. Sum-Frequency and Difference-Frequency Generation

The SHG is the most frequently observed self-organized nonlinear optical phenomenon in optical fibers. However, other second-order nonlinear optical processes, such as SFG and DFG, are occasionally reported. The SFG and DFG occur mostly in the fibers that were originally prepared for efficient SHG, but special structures for efficient SFG and DFG can also be formed in optical fibers.

An accessory generation of bright visible spectrum, manifesting 15 discrete peaks of various intensities, in 5- to 10-m-long fibers, which were conditioned to efficient SHG with a strong pump at 1.064 μm, was already reported by Österberg and Margulis in their early paper [12]. Unfortunately, no details were mentioned and also no further communication concerning this very interesting phenomenon has been published. It seems very likely that the appearance of a bright visible spectrum is a consequence of the parametric generation from quantum noise and the subsequent sum- and difference-frequency generations at a self-written, rather complicated photoinduced structure in the far distant regions of fiber.

In another experiment [68] the germanium-phosphorus-doped, 50-cm-long fiber had been prepared for efficient SHG with the fundamental radiation at 1.064 μm from a Q-switched and mode-locked Nd:YAG laser. Afterward, the Raman-broadened infrared pump, exhibiting the Stokes peak at 1.49 μm and the anti-Stokes peak at 0.99 μm, along with the dominant central peak at 1.064 μm, was launched into the prepared fiber. In the generated spectrum three peaks were promoted at 0.574, 0.532, and 0.496 μm, corresponding to the frequency-doubled light of infrared Stokes, central, and anti-Stokes peaks, respectively. In addition, two other lower peaks were observed. The first, at 0.512 μm, is evidently a consequence of SFG with the infrared central peak at 1.064 μm and the anti-Stokes peak at 0.99 μm subfrequency radiations. As for the second peak, at 0.544 μm, its wavelength is somewhat longer than the expected wavelength of SF light generated by mixing the central peak at 1.064 μm and the Stokes peak at 1.149 μm, which should be 0.552 μm. This wavelength shift has not been explained yet. The SHG prepared with the Stokes and the anti-Stokes pump, as well as the SFG obtained by mixing various infrared wavelengths in the fiber, which had been prepared with the fundamental pump at 1.064 μm, were rather surprising, because the frequency-doubled light bandwidth of the central peak alone was only 3 Å.

The first SFG of ultraviolet light owing to the mixing of infrared and green light in optical fiber was observed by Poumellec et al. [69]. The fiber used was doped by germanium and fluorine in the core and by phosphorus and fluorine in the cladding. The core-cladding refractive index difference Δn was 0.02. The core profile was elliptical with a rate of 1.43, so that the pump light was guided in the fundamental mode only, while the SH light could propagate in three modes. About 10–20 mW average infrared power at 1.064 μm from a Q-switched, mode-locked Nd:YAG laser was coupled simultaneously with 0.4–0.8 mW average frequency-doubled green power at 0.532 μm into the fiber. The fundamental light was polarized linearly, while the SH light was polarized circularly. Tree peaks at 0.736, 0.715, and 0.355 μm appeared in the output light fluorescence spectrum after a few minutes of illumination. The first peaks result from spontaneous four-wave mixing caused by the interaction of SH light at 0.532 μm and fluorescent emissions at 0.432 and 0.416 μm. The third peak, at 0.355 μm, corresponds to the third harmonic of the pump. However, it vanished when the input SH beam was removed. Since third-harmonic generation cannot be very effective in the fiber because of the phase mismatch, the origin of the ultraviolet peak at 0.355 μm is attributed to the self-organized SFG.

D. Second-Harmonic Generation in Fiber Preform

The efficient self-organized SHG had been reported to occur in drawn fibers only. A question arose as to whether this phenomenon could also be observed in the preform of which the fiber is drawn. This question was answered in the affirmative by Lawandy and Selker [47], who managed to prepare a small piece of fiber preform for SHG by the seeding method. A 2-cm-long piece of germanium-phosphorus-doped fiber preform was used as a sample. The preform was manufactured by MO CVD methods, containing 3 wt% Ge and 0.5 wt% P in the core. The core diameter was 1.5 mm.

The fiber preform was prepared using a 76-MHz Q-switched and mode-locked Nd:YAG laser producing 110-ps pulses at 1.064 μm of the peak power of several megawatts. A part of the laser radiation was frequency-doubled in KTP crystal producing 90-ps SH pulses at 0.532 μm of the peak power in 100-kW range. The linearly polarized fundamental and SH pulses were focused into the center of the preform core in such a way that the focal spot diameter was about 90 μm.

It has been found that the fiber preform can be prepared for SHG with infrared peak intensities above 10 GW/cm^2 after about 30 min of illumination. When the infrared radiation was subsequently launched into prepared sample, a visible green light was observed. The conversion efficiency attained was 2×10^{-7}, which is rather small. However, only a 0.5-mm preform length was prepared at most. Compared with conventionally drawn fibers, the optically induced effective quadratic susceptibility in the preform was about one-half greater than that in the fiber, which indicates that the drawing process lowers the achievable SHG efficiency.

Attempts to prepare the fiber preform by the self-seeding procedure (without any external SH input radiation) failed.

E. Preparation of Optical Fibers for Effective Second-Harmonic Generation by Poling Technique

A special method of preparing optical fibers for an effective SHG by applying a strong electric dc field across the fiber core was proposed and demonstrated by Fermann et al. [40, 41]. D-shaped fibers with a built-in electrode [70] that allows the application of very large transverse dc electric fields were used. The fiber used in [41] possessed an elliptical core with an aspect ratio of 1.7 and a core area of about 5 μm^2. It was doped by germanium with a concentration of 18 mol%. The respective cutoff wavelengths were 1.12 and 1.7 μm. The length of the fiber was typically 30 cm.

In the first stage of the experiment [41] a strong transverse dc electric field of 140 V/μm was applied in the direction of the major axis of the fiber core and, simultaneously, 40 mW of argon-laser light at 0.488 μm was launched with equal power along both axes into the fiber core. Preparation of the fiber lasted for 10 min and then both the argon-laser light and the poling field were switched off.

A tunable Raman-shifted dye pulse laser was used as the source of infrared radiation from 1.04 to 1.09 μm for recording the SH conversion efficiency of the prepared fiber as a function of pump wavelength. The SHG was solely observed for the pump and SH polarizations parallel to the dc poling field. The fundamental infrared radiation was scanned from 1.04 to 1.09 μm and launched in the LP_{01} mode into the fiber. The SH radiation generated was observed in four modes. The highest conversion efficiencies were obtained for the LP_{11} SH mode at 1.046- and 1.051-μm pump wavelengths and for the LP_{31} SH mode at 1.062- and 1.067-μm pump wavelengths. A maximum efficiency of 1% was obtained with a peak pump power of 150 W at 1.051 μm for the LP_{11} SH mode.

The preparation of fibers for SHG by poling with a periodic dc electric field of the order of 1 V/μm and one exciting mode of either infrared radiation at 1.064 μm or argon-laser radiation at 0.514 μm was demonstrated by Kashyap [112]. The SH conversion efficiency of about 4×10^{-6}% was relatively very low.

F. Enhancement of Second-Harmonic Conversion Efficiency by Irradiation and Heating

In the experiments described above fibers were prepared for efficient SHG (SFG or DFG) at room temperature and without any external irradiation. However, external conditions such as irradiation and heating can considerably affect the fiber preparation process.

Farries et al. [71] attained the highest SH conversion efficiency of the order of 10% by exposing fibers to green/blue light before the preparation process.

An enhancement of SH conversion efficiency by irradiating fibers with gamma radiation was reported by Anoikin et al. [52]. Several fibers were irradiated using a ^{60}Co source with a dose of 10^6 rad at the rate of 400 rad/s. Simultaneously a strong infrared pump at 1.064 μm from a Q-switched Nd:YAG laser of the peak power of 1 kW and a frequency-doubled SH seed at 0.532 μm of the peak power of 300 W were launched into the fibers. The preparation process lasted for approximately 1 h. The greatest increase in SH conversion efficiency by a factor of 20 was observed in germanium-doped fibers, while the lowest enhancement by a factor of 2 occurred in natural silica fibers. The gamma-ray-induced conversion efficiency dropped slowly to about one-tenth after 40 days.

Margulis et al. [54] prepared fibers in a "bath" of ultraviolet radiation. A 15-cm-long germanium-doped fiber was placed inside a capillary glass tube. Approximately 70% of Nd:YAG laser radiation was twice frequency doubled and the fourth harmonic at 0.266 μm of the power of 0.3 mW was directed into the capillary containing the fiber. The fiber cladding was transparent enough to enable leaking of the ultraviolet radiation into the fiber core. The rest of about 30% of the laser radiation at a fundamental wavelength of 1.064 μm and an SH wavelength of 0.532 μm was launched into the fiber and used for preparation and reading. It was found that the simultaneous illumination of fiber with ultraviolet radiation and fundamental and SH seed accelerates the preparation process up to two orders. The preparation started off exponentially and became linear in time after a few seconds. The saturation occurred in a few minutes. It was also observed that the SH light generated exceeded the SH seed level as much as 30 times before saturation.

The effect of heating was studied by Krol et al. [72]. Germanium-doped glass fibers were heated in a hydrogen atmosphere at 250°C for 18 h or at 1000° C for 1 h, and then slowly cooled to room temperature. After this treatment the fibers were prepared for SHG when launching the pump at 1.064 μm from a Nd:YAG laser and frequency doubled SH seed at 0.532 μm. It was found that the thermal treatment considerably increases the aptitude of fiber preparation for SHG. The SH conversion efficiency was about 20 times higher in the fiber that had been heated to 250°C and it was enhanced up to two orders in fiber heated to 1000°C compared to the thermally unconditioned fibers.

G. Erasure Experiments

The prepared fibers preserve their SHG ability for weeks or months [12] and some fibers do not lose their conversion efficiency even after many years [73]. However, the SHG ability can be erased when fibers are exposed to strong radiation or when they are heated to extreme temperatures.

Fermann [74] observed that when annealing the prepared fibers their conversion efficiency remains approximately constant up to 100°C, but goes down to about 10% at a temperature of 400°C.

Ouelette et al. [31] reported that germanium-doped prepared fibers can be optically erased by injection of SH light alone (without the fundamental). Erasure also occurred when the prepared fibers were irradiated with intense green or blue light from a cw argon laser. The average bleaching power of green (SH) light at 0.532 μm from a Q-switched, mode-locked Nd:YAG laser was about 3 mW and that of green light at 0.514 μm or blue light at 0.488 μm from a cw argon laser was about 600 mW. The bleaching ran about two times faster with irradiation at 0.488 μm than at 0.514 μm. It was also observed that the erasure rate increases with increasing bleaching power. The typical bleaching times were tens of minutes. The erased fibers can recover their original conversion efficiency if they are reseeded with the fundamental and SH light.

The erasure of prepared fibers by strong SH radiation alone was also observed by Krotkus and Margulis [29].

In a later paper Hibino et al. [75] reported that the high-germanium-doped prepared fibers can even be erased by the strong fundamental infrared radiation.

The optical erasure by SH green radiation alone and by infrared radiation alone, as well as the thermal erasure at temperatures from 65°C to above 100°C were also reported in prepared samples of semiconductor-doped glass by Lawandy and McDonald [65].

III. INTRINSIC GENERATION OF INITIAL SECOND-HARMONIC SEED

One of the most puzzling questions is how the initial SH radiation can be generated in centrosymmetric optical fibers by the fundamental pump alone without any external seed [8–12].

A. Electric-Quadrupole Interaction Model

Most physicists have believed that the initial self-seeded SHG occurs due to electric-quadrupole interactions [76, 77]. The lowest-order nonlinear polarization allowed, which can be responsible for SHG in a centrosymmetric medium, is electric-quadrupole polarization:

$$\mathbf{P}_{2\omega}^{(2)\mathrm{quadr}} = \boldsymbol{\chi}_{(2\omega=\omega+\omega)}^{(2)\mathrm{quadr}} \vdots \mathbf{E}_{\omega} \nabla \mathbf{E}_{\omega} \tag{1}$$

where the fourth-rank tensor $\boldsymbol{\chi}^{(2)\mathrm{quadr}}$ describes the electric-quadrupole-response nonlinear susceptibility. The SHG due to the quadrupole interactions has a very low efficiency and can be observed only in anisotropic media, where the electric field intensity and electric field induction vectors are not parallel [80].

The problem of SHG in optical fibers owing to the quadrupole second-order polarization has been studied by Terhune and Weinberger [77]. They have shown that the core–cladding interface effects contribute dominantly to the SHG in optical fibers and that the efficiency of SHG depends considerably upon the overlap integral of interacting radiation modes [81], which is related to the core–cladding refractive index difference Δn. They have also found the selection rule for the radiation modes that can produce an efficient quadrupole second-order nonlinear interaction. Using the $\mathrm{HE}_{\nu n}$ representation [81], the following relationship must be true for a nonzero azimuthal Φ overlap integral [77]:

$$\nu^{\omega} \pm \nu^{\omega} + \nu^{2\omega} = 0 \tag{2}$$

where ν^{ω} and $\nu^{2\omega}$ are first indexes of the fundamental and SH modes, respectively. For example, considering the most usual case, when the fundamental radiation propagates in the LP_{01} (HE_{11}) mode possessing an even symmetry, the electric-quadrupole interaction could generate SH light only in the asymmetric modes LP_{11}(HE_{21}, TE_{01}, and TM_{01}), LP_{31} (EH_{21}), etc.

Compared with the early experiments by Sasaki and Ohmori [9, 10], the quadrupole interaction model [77] predicts a SH conversion efficiency of seven orders less than that observed. However, the observed instantaneous self-seeded SH conversion efficiencies of 10^{-11}–10^{-10} [43, 55] are in good agreement with those predicted by Terhune and Weinberger [77].

As for the modal distribution of SH radiation generated in the intrinsic self-seeded SHG with pump in the LP_{01} mode, the SH output occurs initially in the LP_{11} mode [12, 13, 53], which has the right symmetry; but later it can switch to either the LP_{01} [12, 53] or the LP_{02} (see quotation 29 in Ref. 77) modes, which are forbidden for the electric-quadrupole interaction. It is interesting to note that at high infrared pump power the SH light stays in the LP_{11} mode for the whole preparation process [53].

Other models for exotic higher-order electric-dipole polarizations have also been proposed.

B. Four-Wave Mixing Model

A four-wave mixing with the Stokes wave at zero frequency and the anti-Stokes wave at 2ω was considered by Stolen [15]. From the phenomenological point of view such interaction can be described by third-order nonlinear electric-dipole polarizations [4]:

$$\begin{aligned}\mathbf{P}_0^{(3)} = {} & \boldsymbol{\chi}^{(3)}_{(0=0+\omega-\omega)} \vdots \mathbf{E}_0 \mathbf{E}_\omega \mathbf{E}_\omega^* \\ & + \boldsymbol{\chi}^{(3)}_{(0=\omega+\omega-2\omega)} \vdots \mathbf{E}_\omega \mathbf{E}_\omega \mathbf{E}_{2\omega}^*\end{aligned} \tag{3a}$$

and

$$\begin{aligned}\mathbf{P}_{2\omega}^{(3)} = {} & \boldsymbol{\chi}^{(3)}_{(2\omega=2\omega+\omega-\omega)} \vdots \mathbf{E}_{2\omega} \mathbf{E}_\omega \mathbf{E}_\omega^* \\ & + \boldsymbol{\chi}^{(3)}_{(2\omega=\omega+\omega-0)} \vdots \mathbf{E}_\omega \mathbf{E}_\omega \mathbf{E}_0^*\end{aligned} \tag{3b}$$

It is evident from Eqs. (3) that the SHG cannot start with initial zero intensities of Stokes (dc) and anti-Strokes (SH) waves. The author of this model assumes the occurrence of excited defects in the fiber with a random distribution of polarizations. This random field can act as a nonpropagating Stokes wave. The only components of the random field that are expected to grow up are those that have the right direction and the period equal to the quasi-phase-matching period for effective SHG. The outlined model has not been elaborated into greater details

C. Second-Harmonic Generation From Quantum Noise Owing to Fifth-Order Nonlinearity

In another model [59] the fifth-order electric-dipole polarization,

$$\mathbf{P}^{(5)}_{2\omega} = \boldsymbol{\chi}^{(5)}_{(2\omega=\omega+\omega+\omega+\omega-2\omega)} \vdots \mathbf{E}_{\omega}\mathbf{E}_{\omega}\mathbf{E}_{\omega}\mathbf{E}_{\omega}\mathbf{E}^{*}_{2\omega} \tag{4}$$

was considered, which is the lowest-order nonlinear dipole polarization that is allowed for the interaction of two waves at ω and 2ω in a centrosymmetric medium. Classical theory dictates that the SHG process cannot start with an initial SH intensity $\mathbf{E}_{2\omega} = 0$. However, the SHG can be initiated due to the quantum noise in a similar way to the parametric generation from quantum noise [82, 126–128].

In the quantum picture such nonlinear optical interaction can be described by the time-dependent interaction Hamiltonian

$$\hat{H}_{\text{int}}(t) = -\tfrac{1}{2}\hbar\, g\, \hat{a}^{+2}_{2\omega}(t)\hat{a}^{4}_{\omega}(t) + \text{h.c.} \tag{5}$$

where $\hat{a}_j$ and $\hat{a}_j^{+}$ $(j = \omega, 2\omega)$ label the annihilation and creation operators relative to the jth mode, $\hbar$ is Planck's constant divided by 2π, and g is the coupling constant.

The appurtenant equation of motion in a Dirac representation was solved and the average time necessary for generating the first SH photon was calculated to be [59]

$$t_{\text{phot average}} \approx \left(\frac{\varepsilon_0}{\mu_0}\right)^{3/2} \frac{n_{\text{average}}}{4\omega c \chi^{(5)}_{\text{eff}} I^{2}_{\omega\ \text{average}}} \tag{6}$$

where ε_0 is the electric permittivity, μ_0 is the magnetic permeability of vacuum in SI units, c is the velocity of light in vacuum, n_{average} is the average index of refraction, $I_{\omega\ \text{average}}$ is the average light intensity of the pump, and $\chi_{\mathrm{eff}}^{(5)}$ is the effective fifth-order susceptibility. It was rather difficult to estimate the time necessary for the buildup the SHG process because of great uncertainty in the value of $\chi_{\mathrm{eff}}^{(5)}$, and also because it is difficult to transform the results of the above simple quantum model to the real case of the interaction of light pulses in optical fiber. The advantages of the above model can be seen in the fact that it does not restrict the nonlinear coupling of various fundamental and SH fiber modes, and it is compatible with the phenomenological models of $\chi^{(2)}$ grating formation [13, 64].

Finally, note that the failure of attempts to prepare a fiber-preform sample for SHG by self-seeding procedure [47] indicates that the core–cladding interface contributions are of vital importance for initiating SHG in the self-preparing fibers, which supports the quadrupole interaction model. However, other effects, like anisotropic refractive-index changes in the fiber core [83, 84] may play an important role as well. At any rate, the question of initial self-seeding SHG remains open.

IV. FORMATION OF $\chi^{(2)}$ GRATING

Optical fibers, which have not been previously subjected to SHG (SFG or DFG) preparation process, cannot manifest the effective second-order nonlinear optical phenomena for the following reasons. First, they are made of amorphous (centrosymmetric) material whose macroscopic electric-dipole response quadratic susceptibility $\chi^{(2)}$ equals zero. Second, the phase matching of fundamental (subfrequency) and SH (sum-frequency) waves cannot be generally fulfilled because of the dispersion of refractive indices.

It is obvious that a proper permanent or quasi-permanent $\chi^{(2)}$ structure must be formed in the fiber core in the course of the preparation process. The new optically written $\chi^{(2)}$ structure must be periodical along the fiber so that the phase shift of interacting waves owing to the refractive-index difference can be balanced. This method of synchronization is usually called quasi-phase matching [85].

It is hard to believe that any considerable structural changes, such as the formation of a new crystalline structure, could happen at room temperatures. Since no SHG was seen in fibers with pure silica cores, the formation of $\chi^{(2)}$ grating has to be attributed to the behavior of dopant defects in the fiber core.

There is now fair agreement that the optically induced second-order nonlinearities in fibers are associated with the excitation of dopant defects and the alignment of created elementary dipoles due to the nondegenerate optical rectification, which causes the deprivation of centrosymmetric structure of the fiber core.

A. Excitation of Dopant Defects

As shown by Tsai et al. [38] the SHG conversion efficiency is dominantly related to the concentration of germanium in the fiber core. Therefore, we shall focus our attention on the excitation of germanium defects only.

Two kinds of centers can be created in doped glass. Positively charged centers are created by releasing an electron from the wrong Ge-Si or Ge-Ge bonds due to the sufficient excitation energy amount. Negatively charged centers arise when the released electron is taken into a Ge-related trap.

A question arises as to how we can know that there are defects in the prepared glass. If the defects are paramagnetic they can be detected by a technique called electron spin resonance (ESR) (see, e.g., [38, 67, 86–89]). The defects that are not paramagnetic can be identified only by spectroscopical methods (see, e.g., [16, 19, 90–92, 94]). A good way to characterize the optically induced defects is to measure the ESR and the absorption spectrum before and after irradiation and to find new bands corresponding to the creation of new defects.

A typical positively charged defect, which can be identified in germanium-doped optical fibers, is a positively charged point defect comprising an O vacancy at a Ge site, the so-called GeE' center [16, 38, 67, 95], also referred as a hole trap [96]. Assuming a neutrally charged net, there must be for each GeE' center a trapped electron somewhere, and long-range fields associated with these defect centers must arise if the centroid of the trapped holes (GeE' centers) is different from the centroid of the trapped electrons. Thus, the GeE' centers are paramagnetic and, therefore, detectable by ESR [38].

Two negatively charged dopant defects, Ge(1) and Ge(2) centers, are assumed to play an important role. Referring to Tsai and Griscom [67], the Ge(1) center is one electron trapped at a four coordinated Ge ion substitutional for an Si ion in the SiO_2 network with no Ge ions in next-nearest-neighbor positions. The structure of the Ge(2) center is similar to that of the Ge(1) center, but there must be one Ge ion as a next-nearest neighbor. Both Ge(1) and Ge(2) centers are less paramagnetic and they are not easily detectable by ESR.

Other paramagnetic and diamagnetic defect centers can also occur in germanium-doped fibers [67]. Note that there is not a univocal opinion about the structure of defect centers (e.g., compare with [16, 93]). The absorption band of the GeE' center has not been identified experimentally, but it is probably situated in the deep ultraviolet region [67, 97]. The absorption bands of the Ge(1) and Ge(2) centers were observed at 0.280 and 0.214 μm, respectively [98].

The correlation of GeE' centers in germanium-phosphorus-doped fibers with optically induced SH conversion efficiency was investigated by Tsai et al. [38]. They reported a considerable enhancement of GeE' center concentration in prepared fibers, but no phosphorus-associated centers were observed. The fractional increase in GeE' center concentration was 3.4 in germanium-phosphorus-doped fiber and only 0.96 and 0.20 in two different germanium-doped fibers. A correlation was found between the SH conversion efficiency and GeE' center concentration enhancement. The observed SH conversion efficiency was 5% in the germanium-phosphorus-doped fiber and 1.5 and 0.5 in two different germanium-doped fiber samples. On the other hand, the concentration of Ge(1) and Ge(2) centers was estimated to be orders of magnitude less than that of GeE' centers [67]. The last estimation, however, is in contradiction with the observation of light-induced absorption enhancement of germanosilicate fibers by Poyntz-Wright et al. [19], who attributed the increase of absorption predominantly to the Ge(1) center formation. An exchange of released electrons among GeE' centers and Ge(1) and Ge(2) centers was considered and the balance between positively and negatively charged center concentrations was assumed to occur at equilibrium.

There are two general questions here: Are the electrons that are released from the GeE' centers free or are they trapped by the Ge(1), Ge(2) or other centers? Is the interaction of light with dopant defects local or is it a long-space-ranging process? Another important question is, How can the defect centers be excited in the course of SHG preparation process? It has been found that energy of about 4.4 eV is sufficient to excite the dopant defects in germanium-doped glass [91, 97, 99, 100]. Since a new structure of $\chi^{(2)}$ can be induced by infrared or visible light, it seems highly probable that higher-order nonlinear absorption processes are responsible for the excitation of dopant defects in optical fibers. Considering the interaction of a strong infrared pump at 1.064 μm and relatively weak green SH seed at 0.532 μm, the following interactions are plausible: two-photon absorption of two green photons, three-photon absorption of one green and two infrared photons, and four-photon absorption of four infrared photons. The n-photon absorption is necessarily a $(2n - 1)$-order nonlinear process; thus, the highest-order absorption of four infrared photons represents an exotic seventh-order nonlinear interaction, which might explain the extremely long buildup times of the self-seeding preparation process [11, 12]. On the other hand, when preparing fibers by the poling technique [40, 41], the typical times of preparation are considerably shorter, though the light intensities are rather small, which can explain the fact that dopant-defect excitation is due to the two-photon absorption of green/blue light (third-order process).

B. The Simple Optical-Rectification Model

The first phenomenological model of $\chi^{(2)}$ grating formation in optical fibers was outlined by Stolen and Tom [13]. They proposed that an internal electrostatic field is produced in the fiber owing to the nondegenerate third-order optical rectification,

$$\mathbf{P}_{\text{dc}}^{(3)\,rect} = \tfrac{1}{2}\boldsymbol{\chi}_{(0=\omega+\omega-2\omega)}^{(3)\,\text{rect}} \vdots (\mathbf{E}_\omega \mathbf{E}_\omega \mathbf{E}_{2\omega}^* + \mathbf{E}_\omega^* \mathbf{E}_\omega^* \mathbf{E}_{2\omega}) \tag{7}$$

where $\boldsymbol{\chi}_{(0=\omega+\omega-2\omega)}^{(3)\,\text{rect}}$ is the real fourth-rank susceptibility tensor possessing 21 nonvanishing components and 2 independent components in an isotropic medium [104]. When considering the interaction of two monochromatic radiations at ω and 2ω,

$$\mathbf{E}_\omega(x, y, z, t) = \mathbf{e}_\omega A_\omega(x, y)\exp[i(\omega t - \beta_\omega z)] \tag{8a}$$

$$\mathbf{E}_{2\omega}(x, y, z, t) = \mathbf{e}_{2\omega} A_{2\omega}(x, y,)\exp[i(2\omega t - \beta_{2\omega} z)] \tag{8b}$$

(where $\mathbf{e}_\omega, \mathbf{e}_{2\omega}$ are the polarization unit vectors and $\beta_\omega, \beta_{2\omega}$ are the propagation constants), it is easy to see that the generated third-order nonlinear polarization at zero frequency is spatially periodic along the z axis (fiber axis) as follows:

$$\begin{aligned}\mathbf{P}_{\text{dc}}^{(3)\,\text{rect}}(x, y, z) = \boldsymbol{\chi}_{(0=\omega+\omega-2\omega)}^{(3)\,\text{rect}} &\vdots \mathbf{e}_\omega \mathbf{e}_\omega \mathbf{e}_{2\omega} |A_\omega(x, y)|^2 |A_{2\omega}(x, y)| \\ &\times \cos(\Delta\beta z - \Delta\varphi)\end{aligned} \tag{9}$$

where $\Delta\beta = \beta_{2\omega} - 2\beta_\omega$ represents the phase mismatch and $\Delta\varphi = \varphi_{2\omega} - 2\varphi_\omega$ is the relative phase shift between the interacting waves.

As shown by Franken and Ward [101] and Kielich [102–105] the applied electrostatic field $\mathbf{E}_{\text{dc}}$ deprives the centrosymmetric structure of isotropic medium, and SHG can occur due to the third-order nonlinear polarization

$$\mathbf{P}_{2\omega}^{(3)} = \boldsymbol{\chi}_{(2\omega=\omega+\omega+0)}^{(3)} \vdots \mathbf{E}_\omega \mathbf{E}_\omega \mathbf{E}_{\text{dc}} \tag{10}$$

It is easy to see that the electrically induced effective quadratic susceptibility is then given by

$$\chi_{\text{eff}}^{(2)} = \mathbf{e}_{2\omega} \cdot \boldsymbol{\chi}_{(2\omega=\omega+\omega+0)}^{(3)} \vdots \mathbf{e}_\omega \mathbf{e}_\omega \mathbf{E}_{\text{dc}} \tag{11}$$

Stolen and Tom [13] simply assumed that the induced distribution of electric charges breaks the inversion symmetry and that the written structure of $\chi^{(2)}$ is proportional to the third-order nonlinear dc polarization:

$$\chi_{\text{eff}}^{(2)} = \gamma P_{\text{dc}}^{(3)\,\text{rect}} \tag{12}$$

Unfortunately, the mathematical description of the problem given in Ref. 13 is not quite correct. Namely, the $\chi^{(2)}$ writing process was not separated from the SHG reading process, which means that the description by Stolen and Tom corresponds to the SHG via fifth-order nonlinearity and, thus, the expected SHG efficiency should be extremely small [64]. Moreover, the first terms on the right sides of Eqs. 3 in Ref. 13 are not suitable because they stand for nonlinear polarizations at the wrong frequencies.

Nevertheless, the basic idea of Stolen and Tom's model—that the self-organized $\chi^{(2)}$ grating is written by the optically induced spatially periodical electrostatic field—was an ingenious one and has been accepted as a plausible phenomenological explanation of self-organized nonlinear optical effects in fibers.

Many attempts were made to verify Stolen and Tom's model. Mizrahi et al. [35] prepared optical fibers in a strong external transverse dc electric field, $E_{\text{dc}} \approx 6$ kV/cm, and in the absence of external field. The excited dipoles in the fiber core were expected to become aligned with the direction of the applied field, which should inhibit the formation of $\chi^{(2)}$ grating. Surprisingly, no significant effect of the external electric field on the preparation process was observed. On the contrary, the SH conversion efficiency in the fibers that had been prepared in the presence of the strong external dc electric field was slightly enhanced compared with that of those fibers prepared in a zero external field. Moreover, a transverse dc electric field being again three orders of magnitude larger enhanced considerably the formation of self-written $\chi^{(2)}$ grating [106].

Note that the results of Mizrahi's [35] and Li's [106] experiments are apparently in contradiction with the model of SHG fiber preparation by poling proposed by Fermann et al. [40, 41], who believed that the light-excited dipoles are aligned due to the strong external poling field in the direction perpendicular to the fiber axis.

It was also argued that the internal optically induced dc electric field, which was estimated to be of the order of 1 V/cm [34], is too weak to overcome the thermal fluctuations and to generate a large permanent $\chi^{(2)}$ structure [34, 66, 107]. However, the numerical value of $\chi_{\text{eff}}^{(3)\,\text{rect}}$ was substituted with that of Kerr third-order nonlinear susceptibility for fused silica when estimating the magnitude of internal induced permanent dc electric field in the fiber, which might lead to considerable errors.

In another experiment, Mizrahi et al. [45] studied the polarization properties of photoinduced self-organized SHG in a germanium-doped optical fiber. They have shown quite generally that the photoinduced $\chi^{(2)}$ grating is consistent with the deprivation of centrosymmetric structure in the fiber core owing to the internal transverse dc electric field which is periodic along the fiber axis. The magnitude of the internal dc field was estimated to be 4×10^4 V/cm.

The first direct measurement of an optically induced dc electric field of the order of 10^4 V/cm in SHG-conditioned fiber was performed by Kamal et al. [108], who observed changes in the polarization state of a probe laser beam passing through an electro-optic crystal held in close proximity to the fiber. The spatially periodic structure in the core of fiber conditioned to SHG, fitting with the expected period of $\chi^{(2)}$ grating, was identified by Kamal et al. [49] in a Raman scattered light spectrum, which was emitted perpendicular to the fiber axis.

Summarizing the above experimental results, we can see that, on the one hand, there is both direct [108] and indirect [45] evidence that the formation of $\chi^{(2)}$ grating in self-preparing fibers is a consequence of optically induced permanent internal electric dc field. On the other hand, the formation of permanent electric-dipole polarization in the fiber core cannot be considerably affected by a strong external transverse electric dc field [35, 106]. Now, how can we work out this puzzling affair?

A proper modus vivendi can be found by considering longitudinal field components. As suspected by Poyntz-Wright et al. [19], permanent electric dipoles can be formed of excited Ge centers within strings of interconnected dopant defects. It seems likely that the strings of interconnected dopant defects are dominantly aligned with the fiber axis owing to the layer structure of dopant distribution and the drawing process. Thus, the doped fiber cannot be considered as isotropic medium with regard to the $\chi^{(2)}$ formation process, but it possesses the rotational symmetry with respect to the fiber axis. Considering such an anisotropic dopant distribution, $\chi^{(2)}$ can be shaped by the formation of an inhomogeneous structure of permanent dipoles that are dominantly oriented along the fiber axis [109, 110]. Although the elementary dipoles are aligned with the fiber axis, the macroscopic dc polarization and the field intensity possess considerable transverse components because of inhomogeneous dipole distributions.

The above conception is strongly supported by a recent model of Sceats and Poole [111], who proposed that longitudinal field components, being about 13% of the transverse field strength, play the dominant role in the $\chi^{(2)}$ grating formation process. A mechanism of stress relief was proposed to be generally based on the writing $\chi^{(2)}$ structure in the fiber core. The experimental data proved to be in good agreement with the proposed model [111].

C. Rigorous Phenomenological Theory for Interaction of Monochromatic Radiation Modes

In a more realistic picture of $\chi^{(2)}$ grating formation in doped optical fibers at the interaction of monochromatic radiation modes, the following facts must be taken into account:

1. The formation of the $\chi^{(2)}$ structure is evidently connected with the excitation of dopant defects, which is likely due to the higher-order photon absorptions. Thus, the process of writing $\chi^{(2)}$ is strongly nonlinear with respect to the transmitted light intensities.
2. The doped drawn fiber cannot be considered an isotropic medium with regard to the $\chi^{(2)}$ formation process, but it possesses rotational symmetry with respect to the fiber axis and mirror symmetry with respect to the plane perpendicular to the fiber axis. The inhomogeneous dopant distribution across the fiber core must be taken into account as well.
3. The longitudinal field components cannot be ignored; on the contrary, they are assumed to play an important role in the $\chi^{(2)}$ grating formation process.
4. The limited fiber volume must be considered when calculating the light-induced dc electric field intensity appurtenant to the third-order nonlinear optical rectification polarization.

In the next section a new phenomenological model is proposed that enables us to describe the $\chi^{(2)}$ grating formation in a common way for both the degenerate nonlinear optical interaction (one fundamental and one SH waves) and the nondegenerate nonlinear optical interaction (two subfrequency waves and one sum-frequency wave), as well as for the poling technique.

1. *Degenerate Two-Wave Interaction*

Considering the interaction of the fundamental mode $\mathbf{E}_\omega(t, R)$ and the SH mode $\mathbf{E}_{2\omega}(t, R)$ in a doped optical fiber, the formation of permanent dc polarization occurs due to the third-order cumulative nonlinear optical rectification that can be described as follows:

$$\mathbf{P}_{\mathrm{dc}}^{(3)}(t,R) = \frac{1}{2}\int_0^t \Big[\boldsymbol{\chi}^{(3)\,\mathrm{resp}}_{(0=\omega+\omega-2\omega)}\big(I_\omega(t-\tau,R), I_{2\omega}(t-\tau,R), x, y, \tau\big) \\ \vdots \mathbf{E}_\omega(t-\tau,R)\mathbf{E}_\omega(t-\tau,R) \\ \times \mathbf{E}_{2\omega}^*(t-\tau,R) + \text{c.c.}\Big]\, d\tau \tag{13}$$

where

$$\chi^{(3)\,\mathrm{resp}}_{(0=\omega+\omega-2\omega)}\big(I_\omega(t-\tau,R), I_{2\omega}(t-\tau,R), x, y, \tau\big)$$

is the fourth-rank tensor describing the medium response and

$$I_j = \frac{n_j}{2}\left(\frac{\varepsilon_0}{\mu_0}\right)^{1/2} E_j E_j^* \qquad (j = \omega, 2\omega)$$

denotes the light intensities of interacting radiations. The dependence of $\chi^{(3)\,\mathrm{resp}}_{(0=\omega+\omega-2\omega)}$ upon the light intensities I_ω and $I_{2\omega}$ involves the excitation of dopant defects, and x, y coordinates stand for the inhomogeneous distribution of dopants across the fiber. The fourth-rank tensor $\boldsymbol{\chi}^{(3)\,\mathrm{resp}}_{(0=\omega+\omega-2\omega)}$ possesses rotational symmetry with respect to the z axis (fiber axis) and mirror symmetry with respect to the x, y, plane.

The response function $\boldsymbol{\chi}^{(3)\,\mathrm{resp}}_{(0=\omega+\omega-2\omega)}(\tau)$ is assumed to be a very slow time-varying function. All the components of $\boldsymbol{\chi}^{(3)\,\mathrm{resp}}_{(0=\omega+\omega-2\omega)}(\tau)$ equal zero for negative τ because of the causality principle,

$$\chi^{(3)\,\mathrm{resp}}_{(0=\omega+\omega-2\omega)jmpq}(\tau < 0) = 0,$$

and for times greater than the saturation time t_{sat},

$$\chi^{(3)\,\mathrm{resp}}_{(0=\omega+\omega-2\omega)jmpq}(\tau > t_{\mathrm{sat}}) = 0$$

It is highly probable that drastic changes in dopant distribution happen during the fiber drawing process and that relief stress occurs in the drawn fiber. Thus, it is reasonable to assume that the absolute values of $\boldsymbol{\chi}^{(3)\,\mathrm{resp}}_{(0=\omega+\omega=2\omega)}$ components $\chi^{(3)\,\mathrm{resp}}_{xzzx}$, $\chi^{(3)\,\mathrm{resp}}_{xxzz}$, $\chi^{(3)\,\mathrm{resp}}_{zxzx}$, $\chi^{(3)\,\mathrm{resp}}_{zxxz}$, and $\chi^{(3)\,\mathrm{resp}}_{zzzz}$ can be considerably greater in comparison with $\chi^{(3)\,\mathrm{resp}}_{xxxx}$, which may explain the important role of longitudinal field components in the $\chi^{(2)}$ grating formation process, even though they are very small compared to the

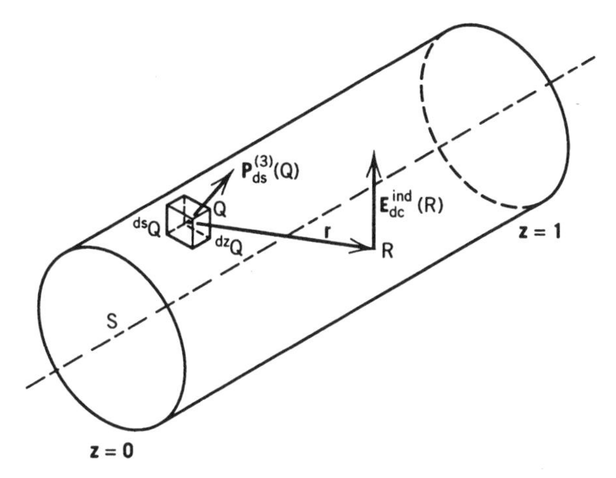

Figure 3. Calculation of induced dc electric field intensity owing to the third-order optical rectification in an optical fiber.

transversal ones. Consequently, the optically induced dc polarization $\mathbf{P}_{dc}^{(3)}(t, R)$ can possess both the transversal and the longitudinal components.

The appurtenant dc field intensity at a fiber point R is calculated as the sum of contributions of all the elementary dipoles in the fiber volume [113] (see Fig. 3):

$$\mathbf{E}_{dc}^{ind}(t, R) = \frac{1}{4\pi\varepsilon_0}\int_0^l \mathrm{d}z_Q \iint_{(S)} \mathrm{d}S_Q \frac{1}{r^3}\left\{\frac{3\left[\mathbf{P}_{dc}^{(3)}(t, Q)\cdot\mathbf{r}\right]\mathbf{r}}{r^2} - \mathbf{P}_{dc}^{(3)}(t, Q)\right\} \tag{14}$$

where $\mathbf{r}$ represents the distance vector from an elementary dipole point Q to the field point R, l is the fiber length, and S is the cross section of the fiber.

Taking into consideration the nonuniform distribution of induced permanent dc polarization, owing to the inhomogeneous dislocation of dopants across the fiber core and the nonuniform distribution of radiation in the fiber modes, especially the azimuthal dependence of their longitudinal components, it is possible to see from Eq. (14) that, generally, the longitudinal components of $\mathbf{P}_{dc}^{(3)}$ can contribute to building up the transversal components of $\mathbf{E}_{dc}^{ind}$ and vice versa. At any rate, the spatial distribution of the induced internal dc field $\mathbf{E}_{dc}^{ind}(R)$ is rather complicated and it is not possible to describe it analytically without knowing the component structure of the response tensor $\boldsymbol{\chi}^{(3)\,\text{resp}}_{(0=\omega+\omega-2\omega)}$.

The induced internal dc field $\mathbf{E}^{\text{ind}}_{\text{dc}}$ causes a deprivation of centrosymmetric structure in the fiber core [101–104] and, consequently, the SH radiation can be generated due to the third-order nonlinear dipole polarization:

$$\mathbf{P}^{(3)}_{2\omega}(t,R) = \boldsymbol{\chi}^{(3)}_{(2\omega=0+\omega+\omega)} \vdots \mathbf{E}^{\text{ind}}_{\text{dc}}(t,R)\mathbf{E}_{\omega}(t,R)\mathbf{E}_{\omega}(t,R) \tag{15}$$

Hence, the time- and space-dependent second-order susceptibility tensor is defined in the following way:

$$\boldsymbol{\chi}^{(2)\,\text{ind}}_{(2\omega=\omega+\omega)}(t,R) = \boldsymbol{\chi}^{(3)}_{(2\omega=0+\omega+\omega)} \cdot \mathbf{E}^{\text{ind}}_{\text{dc}}(t,R) \tag{16}$$

Note that the induced second-order susceptibility $\boldsymbol{\chi}^{(2)\,\text{ind}}_{(2\omega=\omega+\omega)}(t,R)$ behaves as any usual quadratic susceptibility of transparent media; namely, it is real and it satisfies Kleinman's permutation symmetry [see, e.g., 45].

The induced second-order susceptibility enables the generation of new SH radiation in the fiber, which further stimulates writing the $\chi^{(2)}$ structure in the fiber. Moreover, the Kerr nonlinearities cause so-called self-phase and cross-phase modulation, which affect considerably the process of $\chi^{(2)}$ formation with intense radiation.

To describe the dynamics of the $\chi^{(2)}$ grating formation we shall consider the interaction of fundamental and SH fiber modes [81]:

$$\mathbf{E}_{\omega}(t,R) = \left[\mathbf{e}_x F_{T\omega}(x,y) + \mathbf{e}_z F_{L\omega}(x,y)\right] A_{\omega}(t,z) \exp\left[\mathrm{i}(\omega t - \beta_{\omega} z)\right] \tag{17a}$$

$$\begin{aligned} \mathbf{E}_{2\omega}(t,R) = {} & \left[\mathbf{e}_x F_{T2\omega}(x,y) + \mathbf{e}_z F_{L2\omega}(x,y)\right] A_{2\omega}(t,z) \\ & \times \exp\left[\mathrm{i}(2\omega t - \beta_{2\omega} z)\right] \end{aligned} \tag{17b}$$

where $\mathbf{e}_x$ and $\mathbf{e}_z$ are unit vectors in the perpendicular (x) and longitudinal (z) directions, respectively (the fiber is assumed to be polarization preserving); $\beta_{\omega}, \beta_{2\omega}$ are propagation constants; and $F_{\text{T}\omega}(x,y)$, $F_{\text{T}2\omega}(x,y)$, and $F_{L\omega}(x,y)$, $F_{\text{L}2\omega}(x,y)$ are dimensionless functions describing the transverse spatial profiles of transverse and longitudinal field components, respectively, which satisfy the normalization condition

$$W_j = \frac{n_j}{2}\left(\frac{\varepsilon_0}{\mu_0}\right)^{1/2} |A_j|^2 \iint_{(S)} \left[F^2_{\text{T}j}(x,y) + F^2_{\text{L}j}(x,y)\right] \mathrm{d}x\,\mathrm{d}y \qquad j = \omega, 2\omega \tag{18a}$$

where W_j $(j = \omega, 2\omega)$ is the total power of respective radiation mode propagating in the fiber and n_j $(j = \omega, 2\omega)$ is the effective index of refraction.

Because the longitudinal field strength is very small compared to the transverse one [81], the second integral on the right side of Eq. (18a) can be ignored and it holds approximately

$$W_j \approx \frac{n_j}{2}\left(\frac{\varepsilon_0}{\mu_0}\right)^{1/2} |A_j|^2 \iint_{(S)} F_{\mathrm{T}j}^2(x, y)\, \mathrm{d}x\, \mathrm{d}y \qquad j = \omega, 2\omega \quad (18\mathrm{b})$$

The transverse and longitudinal components of induced dc electric field will be considered as well:

$$\mathbf{E}_{\mathrm{dc}}^{\mathrm{ind}}(t, R) = \mathbf{e}_x E_{\mathrm{dc\,T}}^{\mathrm{ind}}(t, x, y, z) + \mathbf{e}_z E_{\mathrm{dc\,L}}^{\mathrm{ind}}(t, x, y, z) \qquad (19)$$

The evolution of complex field amplitudes can be described by means of two coupled first-order partial differential equations:

$$\frac{\partial A_\omega(t, z)}{\partial z} + \frac{1}{v_{g\omega}} \frac{\partial A_\omega(t, z)}{\partial t}$$
$$= -\mathrm{i}\left(\frac{\mu_0}{\varepsilon_0}\right)^{1/2} \frac{\omega}{2n_\omega}\Big\{2\chi_{\omega\,\mathrm{eff}}^{(2)}(t, z) A_{2\omega}(t, z) A_\omega^*(t, z) \exp(-\mathrm{i}\,\Delta\beta z)$$
$$+ \Big[\chi_{\omega\,\mathrm{deg\,eff}}^{\mathrm{Kerr}} |A_\omega(t, z)|^2 + \chi_{\omega\,\mathrm{nondeg\,eff}}^{\mathrm{Kerr}} |A_{2\omega}(t, z)|^2\Big] A_\omega(t, z)\Big\}$$
$$(20\mathrm{a})$$

and

$$\frac{\partial A_{2\omega}(t, z)}{\partial z} + \frac{1}{v_{g2\omega}} \frac{\partial A_{2\omega}(t, z)}{\partial t}$$
$$= -\mathrm{i}\left(\frac{\mu_0}{\varepsilon_0}\right)^{1/2} \frac{\omega}{n_{2\omega}}\Big\{\chi_{2\omega\,\mathrm{eff}}^{(2)}(t, z) A_\omega^2(t, z) \exp(\mathrm{i}\,\Delta\beta z) \qquad (20\mathrm{b})$$
$$+ \Big[\chi_{2\omega\,\mathrm{deg\,eff}}^{\mathrm{Kerr}} |A_{2\omega}(t, z)|^2 + \chi_{2\omega\,\mathrm{nondeg\,eff}}^{\mathrm{Kerr}} |A_\omega(t, z)|^2\Big] A_{2\omega}(t, z)\Big\}$$

where n_ω and $n_{2\omega}$ are the effective indexes of refraction, $v_{g\omega}$ and $v_{g2\omega}$ label the effective group velocities of interacting radiation modes, $\Delta\beta =$

$\beta_{2\omega} - 2\beta_\omega$ represents the phase mismatch, and the effective nonlinear susceptibilities are defined as follows:

- Second-order effective susceptibilities

$$\chi^{(2)}_{\omega\,\mathrm{eff}}(t,z) = \frac{1}{O_\omega}\left\{\chi^{(3)}_{\alpha\alpha\alpha\alpha} O^{\mathrm{TTTT}}_{\omega,2\omega}(t,z) + \chi^{(3)}_{\alpha\alpha\beta\beta} O^{\mathrm{TLTL}}_{\omega,2\omega}(t,z) + \chi^{(3)}_{\alpha\beta\beta\alpha}\left[O^{\mathrm{TTLL}}_{\omega,2\omega}(t,z) + O^{\mathrm{TLLT}}_{\omega,2\omega}(t,z)\right]\right\} \tag{21a}$$

and

$$\chi^{(2)}_{2\omega\,\mathrm{eff}}(t,z) = \frac{1}{O_{2\omega}}\left[\chi^{(3)}_{\alpha\alpha\alpha\alpha} O^{\mathrm{TTTT}}_{\omega,2\omega}(t,z) + 2\chi^{(3)}_{\alpha\alpha\beta\beta} O^{\mathrm{TLTL}}_{\omega,2\omega}(t,z) + \chi^{(3)}_{\alpha\beta\beta\alpha} O^{\mathrm{LLTT}}_{\omega,2\omega}(t,z)\right] \tag{21b}$$

where $\chi^{(3)}_{jklm}$ are the components of fourth-rank tensor $\boldsymbol{\chi}^{(3)}_{(2\omega=\omega+\omega+0)}$, which possesses 21 nonvanishing components and two independent components, $\chi^{(3)}_{\alpha\beta\alpha\beta} = \chi^{(3)}_{\alpha\alpha\beta\beta}$ and $\chi^{(3)}_{\alpha\beta\beta\alpha}(\alpha, \beta = x, y, z$ and $\alpha \neq \beta)$, and it holds that and $\chi^{(3)}_{\alpha\alpha\alpha\alpha} = 2\chi^{(3)}_{\alpha\alpha\beta\beta} + \chi^{(3)}_{\alpha\beta\beta\alpha}$ [2, 102, 105, 114, 115]. Note that the third-order susceptibility tensors

$$\boldsymbol{\chi}^{(3)}_{(2\omega=\omega+\omega+0)} \quad \text{and} \quad \boldsymbol{\chi}^{(3)}_{(\omega=2\omega-\omega+0)}$$

satisfy Kleinman's permutation symmetry [105, 114, 115]:

$$\chi^{(3)}_{jklm}(2\omega = \omega + \omega + 0) = \tfrac{1}{2}\chi^{(3)}_{kjlm}(\omega = 2\omega - \omega + 0) \tag{22}$$

- Third-order effective Kerr susceptibilities

$$\chi^{\mathrm{Kerr}}_{\omega\,\mathrm{deg\,eff}} = \frac{O^{\mathrm{Kerr}}_{\omega,\omega}}{O_\omega}\chi^{(3)}_{\alpha\alpha\alpha\alpha}(\omega = \omega - \omega + \omega) \tag{23a}$$

$$\chi^{\mathrm{Kerr}}_{2\omega\,\mathrm{deg\,eff}} = \frac{O^{\mathrm{Kerr}}_{2\omega,2\omega}}{O_{2\omega}}\chi^{(3)}_{\alpha\alpha\alpha\alpha}(2\omega = 2\omega - 2\omega + 2\omega) \tag{23b}$$

$$\chi^{\mathrm{Kerr}}_{\omega\,\mathrm{nondeg\,eff}} = \frac{O^{\mathrm{Kerr}}_{\omega,2\omega}}{O_\omega}\chi^{(3)}_{\alpha\alpha\alpha\alpha}(\omega = 2\omega - 2\omega + \omega) \tag{23c}$$

$$\chi^{\mathrm{Kerr}}_{2\omega\,\mathrm{nondeg\,eff}} = \frac{O^{\mathrm{Kerr}}_{\omega,2\omega}}{O_{2\omega}}\chi^{(3)}_{\alpha\alpha\alpha\alpha}(2\omega = \omega - \omega + 2\omega) \tag{23d}$$

The Kerr third-order susceptibilities satisfy Kleinman's permutation symmetry [105, 114, 115]:

$$\chi^{(3)}_{jklm}(\omega = 2\omega - 2\omega + \omega) = \chi^{(3)}_{kjml}(2\omega = \omega - \omega + 2\omega) \tag{24}$$

and for negligible frequency dispersion the following approximate relations are valid [114]:

$$\begin{aligned}\chi^{(3)}_{jklm}(\omega = \omega - \omega + \omega) &\approx \chi^{(3)}_{jklm}(2\omega = 2\omega - 2\omega + 2\omega) \\ &\approx \tfrac{1}{2}\chi^{(3)}_{jklm}(\omega = 2\omega - 2\omega + \omega) \\ &\approx \tfrac{1}{2}\chi^{(3)}_{jklm}(2\omega = \omega - \omega + 2\omega)\end{aligned} \tag{25}$$

- Overlap integrals

$$O_{\omega} = \iint_{(S)} F^2_{\mathrm{T}\omega}(x, y)\,\mathrm{d}x\,\mathrm{d}y \tag{26a}$$

$$O_{2\omega} = \iint_{(S)} F^2_{\mathrm{T}2\omega}(x, y)\,\mathrm{d}x\,\mathrm{d}y \tag{26b}$$

$$O^{\mathrm{TTTT}}_{\omega,2\omega}(t, z) = \iint_{(S)} F^2_{\mathrm{T}\omega}(x, y) F_{\mathrm{T}2\omega}(x, y) \times E^{\mathrm{ind}}_{\mathrm{dc\,T}}(t, x, y, z)\,\mathrm{d}x, \mathrm{d}y \tag{27a}$$

$$O^{\mathrm{TLTL}}_{\omega,2\omega}(t, z) = \iint_{(S)} F_{T\omega}(x, y) F_{\mathrm{L}\omega}(x, y) F_{\mathrm{T}\,2\omega}(x, y) \times E^{\mathrm{ind}}_{\mathrm{dc\,L}}(t, x, y, z)\,\mathrm{d}x\,\mathrm{d}y \tag{27b}$$

$$O^{\mathrm{TTLL}}_{\omega,2\omega}(t, z) = \iint_{(S)} F^2_{\mathrm{T}\omega}(x, y) F_{\mathrm{L}\,2\omega}(x, y) \times E^{\mathrm{ind}}_{\mathrm{dc\,L}}(t, x, y, z)\,\mathrm{d}x\,\mathrm{d}y \tag{27c}$$

$$O^{\mathrm{TLLT}}_{\omega,2\omega}(t, z) = \iint_{(S)} F_{T\omega}(x, y) F_{\mathrm{L}\omega}(x, y) F_{\mathrm{L}\,2\omega}(x, y) \times E^{\mathrm{ind}}_{\mathrm{dc\,T}}(t, x, y, z)\,\mathrm{d}x\,\mathrm{d}y \tag{27d}$$

$$O^{\mathrm{LLTT}}_{\omega,2\omega}(t, z) = \iint_{(S)} F^2_{\mathrm{L}\omega}(x, y) F_{\mathrm{T}\,2\omega}(x, y) \times E^{\mathrm{ind}}_{\mathrm{dc\,T}}(t, x, y, z)\,\mathrm{d}x\,\mathrm{d}y \tag{27e}$$

$$O^{\mathrm{Kerr}}_{\omega,\omega} = \iint_{(S)} F^4_{\mathrm{T}\omega}(x, y)\,\mathrm{d}x\,\mathrm{d}y \tag{28a}$$

$$O^{\mathrm{Kerr}}_{2\omega,2\omega} = \iint_{(S)} F^4_{\mathrm{T}\,2\omega}(x, y)\,\mathrm{d}x\,\mathrm{d}y \tag{28b}$$

$$O^{\mathrm{Kerr}}_{\omega,2\omega} = \iint_{(S)} F^2_{\mathrm{T}\,\omega}(x, y) F^2_{\mathrm{T}\,2\omega}(x, y)\,\mathrm{d}x\,\mathrm{d}y \tag{28c}$$

The contributions of longitudinal field components to the Kerr nonlinear terms on the right sides of Eqs. (20) were ignored.

Equations (13), (14), (16), (17), (19), (20), (21), (23), and 26–28 fully describe the process of forming $\chi^{(2)}$ grating at the interaction of monochromatic fundamental and SH radiation modes in a doped fiber from the phenomenological point of view. However, these equations cannot be solved analytically without knowing the explicit form of the response tensor function $\boldsymbol{\chi}^{(3)\text{resp}}_{(0=\omega+\omega-2\omega)}(I_\omega(t-\tau,R), I_{2\omega}(t-\tau,R), x, y, \tau)$. Even if the response function were known, the solution of the problem would be rather exacting; thus, an analytical solution in a closed form seems to be impossible.

It can be deduced from Eqs. (13), (14), (16), (17), (20), (21) and (27) that, except for the short fiber edges, a quasi-periodical effective $\chi^{(2)}$ grating is formed in the saturation state:

$$\chi^{(2)}_{\text{eff sat}}(z) = \chi^{(2)}_0(z) f[\Delta\beta z - \psi(z)] \tag{29}$$

where $\chi^{(2)}_0(z)$ and $\psi(z)$ are slowly varying amplitude and phase, respectively, and $f(\zeta)$ is a periodical alternating function with the period 2π (a modified sine or cosine function).

For a rather simplified case, when ignoring the changes of pump and SH writing intensities and assuming that the SH radiation generated in the fiber is relatively weak so that it cannot affect considerably the preparation process, the response tensor can be treated as a simple time and transverse coordinate complex function:

$$\begin{aligned} \boldsymbol{\chi}^{(3)\text{resp}}_{(0=\omega+\omega-2\omega)}(\tau, x, y) &= \text{Re}\left[\boldsymbol{\chi}^{(3)\text{resp}}_{(0=\omega+\omega-2\omega)}(\tau, x, y)\right] \\ &\quad + \text{i}\,\text{Im}\left[\boldsymbol{\chi}^{(3)\text{resp}}_{(0=\omega+\omega-2\omega)}(\tau, x, y)\right] \end{aligned} \tag{30}$$

The induced dc polarization can then be calculated using Eqs. (13) and (17) to be

$$\begin{aligned} \mathbf{P}^{(3)}_{\text{dc}}(t, R) &= \mathbf{P}^{(3)}_{\text{dc Re}}(t, x, y)\cos(\Delta\beta z - \Delta\varphi) \\ &\quad - \mathbf{P}^{(3)}_{\text{dc Im}}(t, x, y)\sin(\Delta\beta z - \Delta\varphi) \end{aligned} \tag{31}$$

where

$$
\begin{aligned}
&\mathbf{P}^{(3)}_{\mathrm{dc}{\mathrm{Re} \atop \mathrm{Im}}}(t, x, y) \\
&\quad = |A_\omega|^2 |A_{2\omega}| \int_0^t \Bigg(\mathbf{e}_x \Bigg\{ {\mathrm{Re} \atop \mathrm{Im}} \left[\chi^{(3)\mathrm{resp}}_{xxxx}(\tau, x, y) \right] F^2_{\mathrm{T}\omega}(x, y) F_{\mathrm{T}2\omega}(x, y) \\
&\quad + 2 {\mathrm{Re} \atop \mathrm{Im}} \left[\chi^{(3)\mathrm{resp}}_{xxzz}(\tau, x, y) \right] F_{\mathrm{T}\omega}(x, y) F_{\mathrm{L}\omega}(x, y) F_{\mathrm{L}2\omega}(x, y) \\
&\quad + {\mathrm{Re} \atop \mathrm{Im}} \left[\chi^{(3)\mathrm{resp}}_{xzzx}(\tau, x, y) \right] F^2_{\mathrm{L}\omega}(x, y) F_{\mathrm{T}2\omega}(x, y) \Bigg\} \\
&\quad + \mathbf{e}_z \Bigg\{ {\mathrm{Re} \atop \mathrm{Im}} \left[\chi^{(3)\mathrm{resp}}_{zzzz}(\tau, x, y) \right] F^2_{\mathrm{L}\omega}(x, y) F_{\mathrm{L}2\omega}(x, y) \\
&\qquad + 2 {\mathrm{Re} \atop \mathrm{Im}} \left[\chi^{(3)\mathrm{resp}}_{zxzx}(\tau, x, y) \right] F_{\mathrm{T}\omega}(x, y) F_{\mathrm{L}\omega}(x, y) F_{\mathrm{T}2\omega}(x, y) \\
&\qquad + {\mathrm{Re} \atop \mathrm{Im}} \left[\chi^{(3)\mathrm{resp}}_{zxxz}(\tau, x, y) \right] F^2_{\mathrm{T}\omega}(x, y) F_{\mathrm{L}2\omega}(x, y) \Bigg\} \Bigg) \mathrm{d}\tau
\end{aligned}
\tag{32}
$$

and $\Delta\varphi = \varphi_{2\omega} - 2\varphi_\omega$ is the relative phase shift between the pump and SH seeding radiation.

Two third-order dc polarization gratings, which are mutually shifted by $\pi/2$, are induced in the fiber under preparation. The cosine grating is connected with the real part of response tensor $\mathrm{Re}[\boldsymbol{\chi}^{(3)\mathrm{resp}}_{(0=\omega+\omega-2\omega)}(\tau, x, y)]$, while the sine grating is related to the imaginary response tensor part $\mathrm{Im}[\boldsymbol{\chi}^{(3)\mathrm{resp}}_{(0=\omega+\omega-2\omega)}(\tau, x, y)]$.

The spatial period of induced dc electric field is very large in comparison with the diameter of fiber core (see, e.g., [13, 49]). Hence, the induced internal dc electric field in the fiber must possess the same spatial periodicity, but it has the opposite sign:

$$
\begin{aligned}
\mathbf{E}^{\mathrm{ind}}_{\mathrm{dc}}(t, R) = &-\mathbf{E}^{\mathrm{ind}}_{\mathrm{dc\,Re}}(t, x, y) \cos(\Delta\beta z - \Delta\varphi) \\
&+ \mathbf{E}^{\mathrm{ind}}_{\mathrm{dc\,Im}}(t, x, y) \sin(\Delta\beta z - \Delta\varphi)
\end{aligned}
\tag{33}
$$

Strictly, $\mathbf{E}^{\mathrm{ind}}_{\mathrm{dc\,Re}}(t, x, y)$ and $\mathbf{E}^{\mathrm{ind}}_{\mathrm{dc\,Im}}(t, x, y)$ are to be calculated by means of Eqs. (14), (31), and (32). However, for practical purposes it is sufficient to assume that the maxima of $E^{\mathrm{ind}}_{\mathrm{dc}}$ gratings coincide with the minima of $P^{(3)}_{\mathrm{dc}}$ gratings, or, in other words, that the $E^{\mathrm{ind}}_{\mathrm{dc}}$ gratings are shifted by π with respect to the $P^{(3)}_{\mathrm{dc}}$ gratings, respectively. As a consequence, two $\chi^{(2)}$

gratings are induced in the fiber:

$$\chi^{(2)}_{2\omega\text{eff}}(t,z) = \chi^{(2)}_{0\,\text{Re}}(t)\cos(\Delta\beta z - \Delta\varphi) + \chi^{(2)}_{0\,\text{Im}}(t)\sin(\Delta\beta z - \Delta\varphi) \tag{34}$$

where

$$\chi^{(2)}_{0\,{\text{Re} \atop \text{Im}}}(t) = \frac{1}{O_{2\omega}}\left\{\chi^{(3)}_{\alpha\alpha\alpha\alpha}\,{\text{Re} \atop \text{Im}}\left[O^{\text{TTTT}}_{\omega,2\omega}(t)\right] + 2\chi^{(3)}_{\alpha\alpha\beta\beta}\,{\text{Re} \atop \text{Im}}\left[O^{\text{TLTL}}_{\omega,2\omega}(t)\right] + \chi^{(3)}_{\alpha\beta\beta\alpha}\,{\text{Re} \atop \text{Im}}\left[O^{\text{LLTT}}_{\omega,2\omega}(t)\right]\right\} \tag{35}$$

and the overlap integrals $O^{\text{TTTT}}_{\omega,2\omega}$, $O^{\text{TLTL}}_{\omega,2\omega}$, and $O^{\text{LLTT}}_{\omega,2\omega}$ are to be calculated by means of Eqs. (27a), (27b), and (27e), respectively, when substituting the correspondent components of $\mathbf{E}^{\text{ind}}_{\text{dc Re}}(t,x,y)$ or $\mathbf{E}^{\text{ind}}_{\text{dc Im}}(t,x,y)$. We stress that the cosine $\chi^{(2)}$ grating is connected with the real part of the response tensor $\boldsymbol{\chi}^{(3)\text{resp}}_{(0=\omega+\omega-2\omega)}$ and the sine $\chi^{(2)}$ grating is related to the imaginary response tensor part $\boldsymbol{\chi}^{(3)\text{resp}}_{(0=\omega+\omega-2\omega)}$.

Because the functions

$$Re\left[\boldsymbol{\chi}^{(3)resp}_{(0=\omega+\omega-2\omega)}(\tau,x,y)\right] \text{ and } \text{Im}\left[\boldsymbol{\chi}^{(3)resp}_{(0=\omega+\omega-2\omega)}(\tau,x,y)\right]$$

can have a completely different time behavior, we can expect that the sine or cosine $\chi^{(2)}$ grating might dominate in particular stages of the preparation process and a period of competition between both these gratings is also presumed.

2. *Nondegenerate Three-Wave Interaction*

The formation of second-order susceptibility at the interaction of one fundamental and one SH wave represents the special case of a more general case, wherein the permanent $\chi^{(2)}$ structure is written by the interaction of three radiation modes at ω_1, ω_2, and ω_3 with

$$\omega_3 = \omega_1 + \omega_2 \tag{36}$$

Considering the interaction of two subfrequency radiation modes $\mathbf{E}_1(t,R)$, $\mathbf{E}_2(t,R)$ and one sum-frequency radiation mode $\mathbf{E}_3(t,R)$, the formation of permanent dc polarization can be described, from the point of view of phenomenological theory, as follows:

$$\begin{aligned}\mathbf{P}_{\text{dc}}^{(3)}(t,R) = \frac{1}{2}\int_0^t \Big[\boldsymbol{\chi}^{(3)\text{resp}}_{(0=\omega_1+\omega_2-\omega_3)}\big(I_1(t-\tau,R), I_2(t-\tau,R), \\ I_3(t-\tau,R), x, y, \tau\big) \\ \vdots \mathbf{E}_1(t-\tau,R)\mathbf{E}_2(t-\tau,R)\mathbf{E}_3^*(t-\tau,R) + \text{c.c.}\Big]\, d\tau\end{aligned} \tag{37}$$

The complex response tensor function

$$\boldsymbol{\chi}^{(3)\text{resp}}_{(0=\omega_1+\omega_2-\omega_3)}\big(I_1(t-\tau,R), I_2(t-\tau,R), I_3(t-\tau,R), x, y, \tau\big)$$

is assumed to have a similar behavior to that in the degenerate case. The appurtenant internal induced dc electric field intensity in the fiber is calculated by means of Eq. (14).

The induced dc field, $\mathbf{E}_{\text{dc}}^{\text{ind}}$, breaks the centrosymmetric structure in the fiber core, which enables the generation of sum- or difference-frequency generation owing to the third-order nonlinear polarizations

$$\mathbf{P}_3^{(3)}(t,R) = \boldsymbol{\chi}^{(3)}_{(\omega_3=0+\omega_1+\omega_2)} \vdots \mathbf{E}_{\text{dc}}^{\text{ind}}(t,R)\mathbf{E}_1(t,R)\mathbf{E}_2(t,R) \tag{38a}$$

$$\mathbf{P}_2^{(3)}(t,R) = \boldsymbol{\chi}^{(3)}_{(\omega_2=0+\omega_3-\omega_1)} \vdots \mathbf{E}_{\text{dc}}^{\text{ind}}(t,R)\mathbf{E}_3(t,R)\mathbf{E}_1^*(t,R) \tag{38b}$$

$$\mathbf{P}_1^{(3)}(t,R) = \boldsymbol{\chi}^{(3)}_{(\omega_1=0+\omega_3-\omega_2)} \vdots \mathbf{E}_{\text{dc}}^{\text{ind}}(t,R)\mathbf{E}_3(t,R)\mathbf{E}_2^*(t,R) \tag{38c}$$

The time- and space-dependent second-order susceptibility tensors can be defined in the following way:

$$\begin{aligned}\boldsymbol{\chi}^{(2)}_{(\omega_j=\omega_k\pm\omega_l)}(t,R) = \boldsymbol{\chi}^{(3)}_{(\omega_j=0+\omega_k\pm\omega_l)} \cdot \mathbf{E}_{\text{dc}}^{\text{ind}}(t,R) \\ j,k,l = 1,2,3 \\ j \neq k \neq l\end{aligned} \tag{39}$$

Note that all the components of induced $\chi^{(2)}$ are real and they satisfy Kleinman's permutation symmetry

$$\begin{aligned}\chi^{(2)}_{(\omega_1=\omega_3-\omega_2)jkl}(t,R) &= \chi^{(2)}_{(\omega_2=\omega_3-\omega_1)lkj}(t,R) \\ &= \chi^{(2)}_{(\omega_3=\omega_1+\omega_2)klj}(t,R)\end{aligned} \tag{40}$$

As for the dynamics of $\chi^{(2)}$ grating formation, three first-order partial differential equations, similar to Eqs. (20), can be formulated to describe the evolution of complex field amplitudes in the course of the $\chi^{(2)}$ writing process. However, these equations are not introduced here.

Considering the nondepleted writing radiation modes

$$\mathbf{E}_1(t,R) = [\mathbf{e}_{\mathrm{T}1}F_{\mathrm{T}1}(x,y) + \mathbf{e}_z F_{\mathrm{L}1}(x,y)]A_1 \exp[\mathrm{i}(\omega_1 t - \beta_1 z)] \quad (41a)$$

$$\mathbf{E}_2(t,R) = [\mathbf{e}_{\mathrm{T}2}F_{\mathrm{T}2}(x,y) + \mathbf{e}_z F_{\mathrm{L}2}(x,y)]A_2 \exp[\mathrm{i}(\omega_2 t - \beta_2 z)] \quad (41b)$$

$$\mathbf{E}_3(t,R) = [\mathbf{e}_{\mathrm{T}3}F_{\mathrm{T}3}(x,y) + \mathbf{e}_z F_{\mathrm{L}3}(x,y)]A_3 \exp[\mathrm{i}(\omega_3 t - \beta_3 z)] \quad (41c)$$

where $\mathbf{e}_{\mathrm{T}j}$ $(j = 1, 2, 3)$ are polarization unit vectors in the transverse direction, and ignoring the contributions of radiation generated in the fiber, we can see from Eqs. (37), (14), (39), and (41), using a similar procedure as in the above degenerate case, that two effective $\chi^{(2)}$ gratings are induced in the fiber:

$$\begin{aligned}\chi^{(2)}_{\omega_j\,\mathrm{eff}}(t,z) &= -\chi^{(2)}_{j0\,\mathrm{Re}}(t)\cos(\Delta\beta z - \Delta\varphi) \\ &\quad + \chi^{(2)}_{j0\,\mathrm{Im}}(t)\sin(\Delta\beta z - \Delta\varphi) \qquad j=1,2,3\end{aligned} \quad (42)$$

where $\Delta\beta = \beta_3 - \beta_2 - \beta_1$ represents the phase mismatch and $\Delta\varphi = \varphi_3 - \varphi_2 - \varphi_1$ is the relative phase shift of writing modes. Similarly as in the previous case, the real part of the response tensor function, $\mathrm{Re}[\boldsymbol{\chi}^{(3)\mathrm{resp}}_{(0=\omega_1+\omega_2-\omega_3)}(x,y,\tau)]$, gives rise to the cosine $\chi^{(2)}$ grating, while the imaginary part $\mathrm{Im}[\boldsymbol{\chi}^{(3)\mathrm{resp}}_{(0=\omega_1+\omega_2-\omega_3)}(x,y,\tau)]$ produces the sine $\chi^{(2)}$ grating.

3. *Interaction of Radiation with an External DC Poling Field*

Early in the experiment on preparing fibers for effective SHG using the poling technique by Fermann et al. [41], the exciting light was launched into the fiber in two modes with different propagation constants:

$$\begin{aligned}\mathbf{E}^{(\mathrm{I})}_{\mathrm{ex}}(t,R) &= [\mathbf{e}^{(\mathrm{I})}_{\mathrm{T}}F^{(\mathrm{I})}_{\mathrm{T\,ex}}(x,y) + \mathbf{e}_z F^{(\mathrm{I})}_{\mathrm{L\,ex}}(x,y)]A^{(\mathrm{I})}_{\mathrm{ex}} \\ &\quad \times \exp[\mathrm{i}(\omega_{\mathrm{ex}}t - \beta^{(\mathrm{I})}_{\mathrm{ex}}z)]\end{aligned} \quad (43a)$$

$$\begin{aligned}\mathbf{E}^{(\mathrm{II})}_{\mathrm{ex}}(t,R) &= [\mathbf{e}^{(\mathrm{II})}_{\mathrm{T}}F^{(\mathrm{II})}_{\mathrm{T\,ex}}(x,y) + \mathbf{e}_z F^{(\mathrm{II})}_{\mathrm{L\,ex}}(x,y)]A^{(\mathrm{II})}_{\mathrm{ex}} \\ &\quad \times \exp[\mathrm{i}(\omega_{\mathrm{ex}}t - \beta^{(\mathrm{II})}_{\mathrm{ex}}z)]\end{aligned} \quad (43b)$$

(the longitudinal field components were not considered in the original model presented in [41]), and simultaneously, a strong transverse uniform dc electric field $\mathbf{E}^{\mathrm{pol}}_{\mathrm{dc}}$ was applied across the fiber core. Fermann et al. [41] believed that the defect centers were excited by blue exciting light and the created dipoles were then aligned owing to the external poling field, which results in the formation of $\chi^{(2)}$ structure in the fiber core. The periodical structure of the written $\chi^{(2)}$ with grating constant $G = \beta^{(\mathrm{I})}_{\mathrm{ex}} - \beta^{(\mathrm{II})}_{\mathrm{ex}}$ was attributed to the interference pattern of exciting radiation modes in the overlap regions.

In another experiment Kashyap [112] conditioned optical fibers to a strong periodical dc poling field possessing both transverse and longitudinal components, being periodic along the z axis, which can be approximately described as

$$\mathbf{E}_{\text{dc}}^{\text{pol}}(x, y, z) = \mathbf{e}_{\text{T}} E_{\text{dcT}}(x, y)\cos(Gz) + \mathbf{e}_z E_{\text{dcL}}(x, y)\sin(Gz) \quad (44)$$

(for more precise field distribution see Refs. 116 and 117) and simultaneously launching the exciting radiation at either 1.064 or 0.514 μm into the fiber in one single mode. By adjusting the right period of the poling field so that $G = \beta_{2\omega} - 2\beta_{\omega}$, the fiber could be prepared for the $\text{LP}_{01}^{\omega} \rightarrow \text{LP}_{01}^{2\omega}$ quasi-phase-matched SHG after a period of few hours. Note that additional nonlinear processes, such as electric-field-induced SHG and normally and self-seeding SHG, occurred during the preparation process. The poled nonlinearity was found to be opposite in sign to the electric-field-induced instantaneous nonlinearity.

It seems very likely that the microscopic mechanisms responsible for creating the permanent $\chi^{(2)}$ structure in fibers with both the above poling techniques [41, 112] and the conventional (purely optical) procedure [11, 12, 13] must be of the same nature. Consequently, the induced permanent dc polarization written by the poling technique should be treated as a time-dependent cumulative medium response as follows:

$$\mathbf{P}_{\text{dc}}^{(3)\text{pol}}(t, R) = \tfrac{1}{2}\left[\boldsymbol{\chi}_{\text{integral}}^{(3)\text{resp}}(I_{\text{ex}}(R), x, y, t)\,\vdots\,\mathbf{E}_{\text{ex}}(R)\mathbf{E}_{\text{ex}}^{*}(R)\mathbf{E}_{\text{dc}}^{\text{pol}}(R) + \text{c.c.}\right] \quad (45)$$

where

$$\boldsymbol{\chi}_{\text{integral}}^{(3)\text{resp}}(I_{\text{ex}}(R), x, y, t) = \int_0^t \boldsymbol{\chi}_{(0=\omega_{\text{ex}}-\omega_{\text{ex}}+0)}^{(3)\text{resp}}(I_{\text{ex}}(R), x, y, \tau)\,d\tau$$

and

$$I_{\text{ex}}(R) = \frac{n_{\text{average}}}{2}\left(\frac{\varepsilon_0}{\mu_0}\right)^{1/2} E_{\text{ex}}(R) E_{\text{ex}}^{*}(R)$$

denotes the local intensity of exciting radiation. The rest of the procedure is similar to that in the previous cases.

It follows from Eqs. (14), (16), (43), and (45) and Eq. (44), respectively, that a periodical effective $\chi^{(2)}$ structure is formed in the fibers being prepared by the both considered poling methods [41, 112]:

$$\chi_{\text{eff}}^{(2)}(z) = \chi_0^{(2)} g(Gz - \psi) \tag{46}$$

where $g(\zeta)$ is a periodical function with the period 2π.

Considering the fiber preparation with the uniform transverse poling field $\mathbf{E}_{\text{dc}}^{\text{pol}}$ and the two different exciting radiation modes being described by Eqs. (43) [41], the periodicity of created $\chi^{(2)}$ grating is a consequence of the interference pattern of exciting radiation modes in the overlap regions. Neglecting the small contributions of longitudinal field components, it follows from Eqs. (43) for exciting light intensity distribution that

$$I_{\text{ex}}(x, y, z) \approx \frac{n_{\text{average}}}{2}\left(\frac{\varepsilon_0}{\mu_0}\right)^{1/2}\Big[F_{\text{Tex}}^{(\text{I})^2}(x, y)|A_{\text{ex}}^{(\text{I})}|^2 + F_{\text{Tex}}^{(\text{II})^2}(x, y)|A_{\text{ex}}^{(\text{II})}|^2$$
$$+ 2F_{\text{Tex}}^{(\text{I})}(x, y)F_{\text{Tex}}^{(\text{II})}(x, y)|A_{\text{ex}}^{(\text{I})}||A_{\text{ex}}^{(\text{II})}|\cos(Gz - \Delta\varphi_{\text{ex}})\Big] \tag{47}$$

where $G = \beta_{\text{ex}}^{(\text{I})} - \beta_{\text{ex}}^{(\text{II})}$ and $\Delta\varphi_{\text{ex}} = \varphi_{\text{ex}}^{(\text{I})} - \varphi_{\text{ex}}^{(\text{II})}$ are the relative phase shifts of the exciting radiation modes. However, the written $\chi^{(2)}$ grating does not simply copy the interference maxima and minima of exciting light intensity. The situation is rather more complicated. Namely, the periodicity of the $\chi^{(2)}$ structure has two sources. On the one hand, the response function must strongly depend upon the exciting light intensity $\boldsymbol{\chi}_{\text{integral}}^{(3)\text{resp}}(I_{\text{ex}}(x, y, z), x, y, t)$, meaning that $\boldsymbol{\chi}_{\text{integral}}^{(3)\text{resp}}$ has considerable value only in those localities in which I_{ex} is large. On the other hand, the periodicity of $\boldsymbol{\chi}^{(2)}$ arises from the driving mechanism of nonlinear interaction itself, as can be seen from Eqs. (43) and (45).

By considering two exciting radiation modes to be polarized parallel with the poling field, the induced dc polarization can be calculated using Eqs. (43) and (45):

$$\mathbf{P}_{\text{dc}}^{(3)\text{pol}}(t, R) = \mathbf{P}_{\text{dc Re}}^{(3)}(t, R)\cos(\Delta\beta z - \Delta\varphi_{\text{ex}})$$
$$+ \mathbf{P}_{\text{dc Im}}^{(3)}(t, R)\sin(\Delta\beta z - \Delta\varphi_{\text{ex}}) \tag{48}$$

where $\Delta\beta = \beta_{\text{ex}}^{(\text{I})} - \beta_{\text{ex}}^{(\text{II})}$ and $\Delta\varphi_{\text{ex}} = \varphi_{\text{ex}}^{(\text{I})} - \varphi_{\text{ex}}^{(\text{II})}$ and

$$\begin{aligned}
\mathbf{P}_{\text{dc}\,\substack{\text{Re}\\ \text{Im}}}^{(3)}(t,R) = \Bigg(\mathbf{e}_x\Bigg\{ & \begin{matrix}\text{Re}\\ \text{Im}\end{matrix}\left[\chi_{xxxx}^{(3)\text{resp}}(I_{\text{ex}}(R),x,y,t)\right] F_{\text{Tex}}^{(\text{I})}(x,y)F_{\text{Tex}}^{(\text{II})}(x,y) \\
& + \begin{matrix}\text{Re}\\ \text{Im}\end{matrix}\left[\chi_{xzzx}^{(3)\text{resp}}(I_{\text{ex}}(R),x,y,t)\right] F_{\text{Lex}}^{(\text{I})}(x,y)F_{\text{Lex}}^{(\text{II})}(x,y)\Bigg\} \\
& + \mathbf{e}_z \begin{matrix}\text{Re}\\ \text{Im}\end{matrix}\left[\chi_{zxzx}^{(3)\text{resp}}(I_{\text{ex}}(R),x,y,t)\right] \\
& \times \Big[F_{\text{Tex}}^{(\text{I})}(x,y)F_{\text{Lex}}^{(\text{II})}(x,y) \\
& \qquad + F_{\text{Lex}}^{(\text{I})}(x,y)F_{\text{Tex}}^{(\text{II})}(x,y)\Big]\Bigg)\left|A_{\text{ex}}^{(\text{I})}\right|\left|A_{\text{ex}}^{(\text{II})}\right| E_{\text{dc}}^{\text{pol}}
\end{aligned} \tag{49}$$

where $\chi_{jklm}^{(3)\text{resp}}(I_{\text{ex}}(R), x, y, t)$ are the components of the response tensor function $\boldsymbol{\chi}_{\text{integral}}^{(3)\text{resp}}(I_{\text{ex}}(R), x, y, t)$.

As a consequence, two $\chi^{(2)}$ gratings mutually shifted by $\pi/2$ are induced in the fiber:

$$\begin{aligned}
\chi_{\text{eff}}^{(2)}(t,z) = & -\chi_{0\,\text{Re}}^{(2)}\big(t, I_{\text{ex average}}(z)\big)\cos(\Delta\beta z - \Delta\varphi_{\text{ex}}) \\
& + \chi_{0\,\text{Im}}^{(2)}\big(t, I_{\text{ex average}}(z)\big)\sin(\Delta\beta z - \Delta\varphi_{\text{ex}})
\end{aligned} \tag{50}$$

where $I_{\text{ex average}}(z)$ is some average of exciting light intensity over the cross section of the fiber. The grating amplitudes $\chi_{0\,\text{Re}}^{(2)}$ and $\chi_{0\,\text{Im}}^{(2)}$ are calculated in a similar way as in Section IV.C.1. Similarly to the previous case, the cosine $\chi^{(2)}$ grating is connected with the real part of the response tensor function $\text{Re}[\boldsymbol{\chi}_{\text{integral}}^{(3)\text{resp}}(I_{\text{ex}}(R), x, y, t)]$ and the sine $\chi^{(2)}$ grating is related to the imaginary response tensor part $\text{Im}[\boldsymbol{\chi}_{\text{integral}}^{(3)\text{resp}}(I_{\text{ex}}(R), x, y, t)]$.

However, the induced gratings do not represent regular cosine and sine gratings because their amplitudes are functions of exciting light intensity. As $I_{\text{ex average}}(z)$ reaches its minimum for $\cos(\Delta\beta z - \Delta\varphi_{\text{ex}}) = -1$, the positive parts of the cosine $\chi^{(2)}$ grating are negligibly small and only negative parts are outstanding. On the other hand, the positive and negative parts of the sine $\chi^{(2)}$ grating are symmetric, even though the sine grating does not obey the regular sine function shape.

The uniform (nonperiodic) dc polarizations were not considered when calculating $\mathbf{P}_{\text{dc}}^{(3)\text{pol}}$, because they cannot contribute significantly to the SHG process.

In the case of preparing fibers with the spatial periodic dc poling field and the exciting radiation propagating in one single fiber mode [112], the periodicity of $\chi^{(2)}$ issues evidently from the periodicity of the applied poling field. The response function depends upon the transverse profile of exciting light intensity only, $\boldsymbol{\chi}^{(3)\mathrm{resp}}_{\mathrm{integral}}(I_{\mathrm{ex}}(x, y), x, y, t)$.

Considering the simplified case wherein the polarization direction of exciting radiation is parallel with the transverse direction of the poling field, i.e.,

$$\mathbf{E}_{\mathrm{ex}}(t, R) = [\mathbf{e}_x F_{\mathrm{Tex}}(x, y) + \mathbf{e}_z F_{\mathrm{Lex}}(x, y)] A_{\mathrm{ex}} \exp[\mathrm{i}(\omega_{\mathrm{ex}} t + \beta_{\mathrm{ex}} z)] \tag{51}$$

and

$$\mathbf{E}^{\mathrm{pol}}_{\mathrm{dc}}(R) = \mathbf{e}_x E_{\mathrm{dcT}}(x, y)\cos(Gz) + \mathbf{e}_z E_{\mathrm{dcL}}(x, y)\sin(Gz) \tag{52}$$

the induced dc polarization can then be easily calculated using Eqs. (45), (51), and (52) to be

$$\mathbf{P}^{(3)\mathrm{pol}}_{\mathrm{dc}}(t, R) = \mathbf{P}^{(3)}_{\mathrm{C}}(t, x, y)\cos(Gz) + \mathbf{P}^{(3)}_{\mathrm{S}}(t, x, y)\sin(Gz), \tag{53}$$

where

$$\begin{aligned}\mathbf{P}^{(3)}_{\mathrm{C}}(t, x, y) = \{\mathbf{e}_x[\chi^{(3)\mathrm{resp}}_{xxxx}(t, x, y) F^2_{\mathrm{Tex}}(x, y) \\ + \chi^{(3)\mathrm{resp}}_{xzzx}(t, x, y) F^2_{\mathrm{Lex}}(x, y)] \\ + 2\mathbf{e}_z \chi^{(3)\mathrm{resp}}_{zzxx}(t, x, y) F_{\mathrm{Tex}}(x, y) F_{\mathrm{Lex}}(x, y)\} E_{\mathrm{dcT}}(x, y) |A_{\mathrm{ex}}|^2\end{aligned} \tag{54a}$$

and

$$\begin{aligned}\mathbf{P}^{(3)}_{\mathrm{S}}(t, x, y) = \{2\mathbf{e}_x \chi^{(3)\mathrm{resp}}_{xxzz}(t, x, y) F_{\mathrm{Tex}}(x, y) F_{\mathrm{Lex}}(x, y) \\ + \mathbf{e}_z[\chi^{(3)\mathrm{resp}}_{zxxz}(t, x, y) F^2_{\mathrm{Tex}}(x, y) \\ + \chi^{(3)\mathrm{resp}}_{zzzz}(t, x, y) F^2_{\mathrm{Lex}}(x, y)]\} E_{\mathrm{dcL}}(x, y) |A_{\mathrm{ex}}|^2\end{aligned} \tag{54b}$$

and $\chi^{(3)\mathrm{resp}}_{jklm}(t, x, y)$ are the components of the response tensor function $\boldsymbol{\chi}^{(3)\mathrm{resp}}_{\mathrm{integral}}(I_{\mathrm{ex}}(x, y), x, y, t)$. As a consequence, two $\chi^{(2)}$ gratings are induced in the fiber:

$$\chi^{(2)}_{\mathrm{eff}}(t, z) = -\chi^{(2)}_{0\mathrm{C}}(t)\cos(Gz) - \chi^{(2)}_{0\mathrm{S}}(t)\sin(Gz) \tag{55}$$

where the grating amplitudes $\chi^{(2)}_{0\mathrm{C}}$ and $\chi^{(2)}_{0\mathrm{S}}$ are calculated in a similar way as in Section IV.C.1. Unlike the purely optical procedure, the cosine and sine $\chi^{(2)}$ gratings are related to the transverse and longitudinal components of the applied periodical poling field, respectively, in this case.

It is interesting to note that the only real response function $\boldsymbol{\chi}^{(3)\mathrm{resp}}_{(0=\omega_{\mathrm{ex}}-\omega_{\mathrm{ex}}+0)}(I_{\mathrm{ex}}(R), x, y, \tau)$ has a physical meaning when describing the $\chi^{(2)}$ grating formation with the periodic dc poling field. The imaginary part of $\boldsymbol{\chi}^{(3)\mathrm{resp}}_{(0=\omega_{\mathrm{ex}}-\omega_{\mathrm{ex}}+0)}(I_{\mathrm{ex}}(R), x, y, \tau)$ could have a physical meaning only when describing the spatially periodic poling field in the complex form,

$$\mathbf{E}^{\mathrm{pol}}_{\mathrm{dc}}(R) = \mathbf{e}_x E_{\mathrm{dcT}}(x, y)\exp(\mathrm{i}Gz) + \mathbf{e}_z E_{\mathrm{dcL}}(x, y)\exp\left(\mathrm{i}Gz - \mathrm{i}\frac{\pi}{2}\right)$$

The contributions of transverse and longitudinal field components play a crucial role in determining the component structure of the response tensor function $\boldsymbol{\chi}^{(3)\mathrm{resp}}(t, R)$. Therefore, other modifications of Kashyap's experiment [112] with various distributions of the poling field and facilities of measuring the phase of generated SH radiation would be advisable, which might provide a better understanding of the mechanisms behind the fiber preparation process.

As for the microscopis mechanisms behind the $\chi^{(2)}$ grating formation, the model of the photovoltaic effect has received considerable attention [57, 60–62, 131–134]. This model is based on the idea that an atom subjected to the fundamental and SH fields exhibits an anisotropic photoelectron emission. The direction of electron emission depends on the relative phase shift between the fundamental and SH fields. The induced coherent photovoltaic current depends strongly on the ratio of the transverse size of the light beam r_0 and the grating period $l_{\mathrm{per}} = 2\pi/\Delta\beta$, r_0/l_{per} [132]. The model was tested by Dianov et al. [132], who prepared optical fibers with fundamental pumping and SH seeding radiation propagating both in one direction and in opposite directions. It was predicted that the SHG efficiency should be considerably greater when the pump and SH seed propagate in the same direction than when the radiation is propagated in opposite directions. The experimental results obtained proved a very good agreement with the photovoltaic model of the $\chi^{(2)}$ grating formation. We are of the opinion that the proposed photovoltaic model is fully consistent with the above outlined phenomenological model of the $\chi^{(2)}$ grating formation.

V. QUADRATIC NONLINEAR PHENOMENA AT PERIODICAL $\boldsymbol{\chi}^{(2)}$ GRATINGS

A medium with induced permanent second-order susceptibility behaves like any other natural quadratic medium, except that the spatial distribution of $\boldsymbol{\chi}^{(2)}(x, y, z)$ must be taken into consideration when formulating the appurtenant differential equations describing the evolution of complex field amplitudes. In the following calculations we shall not consider the dynamics of second-order susceptibility formation, but we shall assume the presence of the regular periodical $\chi^{(2)}$ grating in the fiber.

The effective quadratic susceptibilities can be defined in a similar way as in Eqs. (21). However, it is necessary to distinguish between the writing field and the reading field. Generally, the relation between the induced dc electric polarization and the appurtenant internal dc electric field intensity is given by Eq. (14). Thus, it is difficult to express the correspondent overlap integrals of Eqs. (27) in terms of transverse distribution functions of writing and reading radiation modes.

As a rough approximation Kashyap et al. [118] assumed that the spatial distribution of the internal dc electric field is almost the same as that of the induced dc polarization (see also [119, 120]). Considering the writing $\boldsymbol{\chi}^{(2)}$ in a purely optical way with one fundamental and one SH radiation mode (Eqs. (17)), it follows that

$$E_{\mathrm{dc}p}^{\mathrm{ind}}(x,y,z) \sim |A_\omega|_{\mathrm{write}}^2 |A_{2\omega}|_{\mathrm{write}} F_{l\omega}(x,y)_{\mathrm{write}} F_{m\omega}(x,y)_{\mathrm{write}} F_{n2\omega}(x,y)_{\mathrm{write}} \times \frac{\cos}{\sin}(\Delta\beta z - \Delta\varphi_{\mathrm{write}}) \qquad p,l,m,n = \mathrm{T,L} \tag{56}$$

where $F_{l\omega}(x,y)_{\mathrm{write}}$, $F_{m\omega}(x,y)_{\mathrm{write}}$, and $F_{n2\omega}(x,y)_{\mathrm{write}}$ describe the spatial profiles of either transverse or longitudinal writing radiation modes. The overlap integrals of Eqs. (27) can then be expressed as

$$O^{\substack{lmn\\pqr}}(z) \sim |A_\omega|_{\mathrm{write}}^2 |A_{2\omega}|_{\mathrm{write}} I_{\mathrm{total}}^{\substack{lmn\\pqr}} \frac{\cos}{\sin}(\Delta\beta z - \Delta\varphi_{\mathrm{write}}) \tag{57}$$

where

$$I_{\mathrm{total}}^{\substack{lmn\\pqr}} = \iint_{(S)} F_{p\omega}(x,y)_{\mathrm{read}} F_{q\omega}(x,y)_{\mathrm{read}} F_{r2\omega}(x,y)_{\mathrm{read}} \times F_{l\omega}(x,y)_{\mathrm{write}} F_{m\omega}(x,y)_{\mathrm{write}} F_{n2\omega}(x,y)_{\mathrm{write}}\,\mathrm{d}x\,\mathrm{d}y \qquad l,m,n,p,q,r = \mathrm{T,L} \tag{58}$$

The distribution functions can be described as the product of radial and azimuthal distribution functions [118]

$$F_{pj}(x,y) = \xi_{pj}(r)\cos(\nu_j\Phi) \qquad p = \mathrm{T,L} \qquad j = \omega, 2\omega \tag{59}$$

where both transverse and longitudinal field components of one fiber mode are assumed to have the same azimuthal distribution [81].

The overlap integral $I_{\text{total}}^{\substack{lmn\\pqr}}$ can be split into radial and azimuthal parts [118], so that

$$I_{\text{total}}^{\substack{lmn\\pqr}} = I_r^{\substack{lmn\\pqr}} I_\Phi \tag{60}$$

where the radial integrals

$$\begin{aligned} I_r^{\substack{lmn\\pqr}} = \int_0^a & \xi_{p\omega}(r)_{\text{read}} \xi_{q\omega}(r)_{\text{read}} \xi_{r2\omega}(r)_{\text{read}} \\ & \times \xi_{l\omega}(r)_{\text{write}} \xi_{m\omega}(r)_{\text{write}} \xi_{n2\omega}(r)_{\text{write}} r \, \mathrm{d}r \end{aligned} \tag{61}$$

describe the efficiency of reading operation (SHG) and the azimuthal integral

$$\begin{aligned} I_\Phi = \int_0^{2\pi} & \cos^2(\nu_\omega \Phi)_{\text{read}} \cos(\mu_{2\omega} \Phi)_{\text{read}} \\ & \times \cos^2(\alpha_\omega \Phi)_{\text{write}} \cos(\beta_{2\omega} \Phi)_{\text{write}} \, \mathrm{d}\Phi \end{aligned} \tag{62}$$

determines the selection rule for proper choice of the writing and reading fiber modes to obtain maximum SH conversion efficiency. More detailed discussion, as well as proposals of the most promising combinations of modes for efficient SHG can be found in [118]. Similar considerations can be also made if writing and reading $\chi^{(2)}$ by the nondegenerate three-wave interaction or when using the poling techniques.

Though the above overlap integral considerations provide certain pragmatic results, they might lead to misleading conclusions in some cases because a strong internal dc electric field can appear between differently polarized regions of the various wings of interacting radiation modes. In the above considerations we have assumed for simplicity that the interacting radiation modes are polarized parallel in the x direction in both the writing and the reading operations, which is the most common situation. In a more general case it would be necessary to assume that the polarization directions of fundamental and SH radiation modes are not identical, and also that the writing and the reading operations are made with differently polarized radiation [42, 45]. The definitions of effective susceptibilities would then be somewhat more complicated, but the form of corresponding differential equations for complex field amplitudes remains the same.

In the next section we shall simply assume that an effective periodical $\chi^{(2)}$ grating has been induced in the fiber. Since the efficiencies of SHG, SFG, or DFG in prepared optical fibers are usually rather small, it is sufficient to use the approximation of nondepleted generating fields. The effects of Kerr nonlinearities (self-phase and cross-phase modulation) are discussed only briefly. The interaction of monochromatic radiation modes is considered in greater detail.

A. Resonant Second-Harmonic Generation

Let us consider the SHG at periodical regular cosine and sine $\chi^{(2)}$ gratings, respectively:

$$\chi^{(2)}_{\mathrm{eff}\,C}(z) = -\chi^{(2)}_{0\,\mathrm{re}} \cos(\Delta\beta_{\mathrm{write}} z - \Delta\varphi_{\mathrm{write}}) \tag{63a}$$

and

$$\chi^{(2)}_{\mathrm{eff}\,S}(z) = \chi^{(2)}_{0\,\mathrm{im}} \sin(\Delta\beta_{\mathrm{write}} z - \Delta\varphi_{\mathrm{write}}) \tag{63b}$$

The approximation of nondepleted fundamental radiation is assumed and only the SH complex field amplitude is considered to be a function of the fiber length z:

$$\mathbf{E}_{\omega}(t, R)_{\mathrm{read}} \approx \mathbf{e}_{\mathrm{x}} F_{\mathrm{T}\omega}(x, y)_{\mathrm{read}} A_{\omega\,\mathrm{read}} \exp[\mathrm{i}(\omega \mathrm{t} - \beta_{\omega\,\mathrm{read}} z)] \tag{64a}$$

$$\mathbf{E}_{2\omega}(t, R)_{\mathrm{read}} \approx \mathbf{e}_{\mathrm{x}} F_{\mathrm{T}2\omega}(x, y)_{\mathrm{read}} A_{2\omega\,\mathrm{read}}(z) \exp[\mathrm{i}(2\omega t - \beta_{2\omega\,\mathrm{read}} z)] \tag{64b}$$

For the resonant SHG the fundamental and SH radiation modes are assumed to be identical with those used for writing $\chi^{(2)}$. Thus, the phase-mismatch of interacting radiation modes equals the grating constant:

$$\Delta\beta_{\mathrm{read}} = \beta_{2\omega\,\mathrm{read}} - 2\beta_{\omega\,\mathrm{read}} = \Delta\beta_{\mathrm{write}} = \Delta\beta \tag{65}$$

In the approximation of nondepleted pump ($A_{\omega\,\mathrm{read}} = \mathrm{const.}$), and if ignoring the Kerr nonlinearities, the evolution of SH complex field amplitude can be described by the first-order differential equation [1, 3, 4]

$$\frac{\mathrm{d}A_{2\omega\,\mathrm{read}}}{\mathrm{d}z} = \mathrm{i}\left(\frac{\mu_0}{\varepsilon_0}\right)^{1/2} \frac{\omega}{n_{2\omega}} A^2_{\omega\,\mathrm{read}} \exp(\mathrm{i}\Delta\beta z) \times \begin{cases} \chi^{(2)}_{0\,\mathrm{Re}} \cos(\Delta\beta z - \Delta\varphi_{\mathrm{write}}) \\ [-\chi^{(2)}_{0\,\mathrm{Im}} \sin(\Delta\beta z - \Delta\varphi_{\mathrm{write}})] \end{cases} \tag{66}$$

For the boundary condition $A_{2\omega\,\text{read}}(0) = 0$ one easily obtains the solution of Eq. (66) for cosine $\chi^{(2)}$ grating:

$$A^{\mathrm{C}}_{2\omega\,\text{read}}(z) = \left(\frac{\mu_0}{\varepsilon_0}\right)^{1/2} \frac{\omega}{\Delta\beta n_{2\omega}} A^2_{\omega\,\text{read}} \chi^{(2)}_{0\,\mathrm{Re}} \times \left\{ 2\Delta\beta z \exp\left[\mathrm{i}\left(\Delta\varphi_{\text{write}} + \frac{\pi}{2}\right)\right] + \exp[\mathrm{i}(2\Delta\beta z - \Delta\varphi_{\text{write}})] - \exp(-\mathrm{i}\Delta\varphi_{\text{write}}) \right\} \tag{67a}$$

and for sine $\chi^{(2)}$ grating:

$$A^{\mathrm{S}}_{2\omega\,\text{read}}(z) = \left(\frac{\mu_0}{\varepsilon_0}\right)^{1/2} \frac{\omega}{\Delta\beta n_{2\omega}} A^2_{\omega\,\text{read}} \chi^{(2)}_{0\,\mathrm{Im}} \times \left\{ 2\Delta\beta z \exp(\mathrm{i}\Delta\varphi_{\text{write}}) - \exp\left[\mathrm{i}\left(2\Delta\beta z + \Delta\varphi_{\text{write}} - \frac{\pi}{2}\right)\right] + \exp\left[-\mathrm{i}\left(2\Delta\varphi_{\text{write}} + \frac{\pi}{2}\right)\right] \right\} \tag{67b}$$

For larger distances, for which $|\Delta\beta| z \gg 1$ ($z \gg 10^{-2}$ mm), the solutions of Eqs. (66) can be simplified as follows:

$$|A^{\mathrm{C}}_{2\omega\,\text{read}}(z)| \exp(\mathrm{i}\Delta\varphi_{\text{read}}) = \left(\frac{\mu_0}{\varepsilon_0}\right)^{1/2} \frac{\omega z}{2 n_{2\omega}} |A_{\omega\,\text{read}}|^2 \chi^{(2)}_{0\,\mathrm{Re}} \exp\left[\mathrm{i}\left(\Delta\varphi_{\text{write}} - \frac{\pi}{2}\right)\right] \tag{68a}$$

and

$$|A^{\mathrm{S}}_{2\omega\,\text{read}}(z)| \exp(\mathrm{i}\Delta\varphi_{\text{read}}) = \left(\frac{\mu_0}{\varepsilon_0}\right)^{1/2} \frac{\omega z}{2 n_{2\omega}} |A_{\omega\,\text{read}}|^2 \chi^{(2)}_{0\,\mathrm{Im}} \exp(\mathrm{i}\Delta\varphi_{\text{write}}) \tag{68b}$$

where $\Delta\varphi_{\text{read}} = \varphi_{2\omega\,\text{read}} - 2\varphi_{\omega\,\text{read}}$ is the relative phase shift of interacting (reading) radiation modes. The total power of generated SH radiation is then

$$W^{\mathrm{C,S}}_{2\omega\,\text{read}}(z) = \left(\frac{\mu_0}{\varepsilon_0}\right)^{3/2} \frac{\omega^2 W^2_{\omega\,\text{read}} O_{2\omega} \chi^{(2)^2}_{0\,{\mathrm{Re} \atop \mathrm{Im}}}}{2 n^2_{\omega} n_{2\omega} O^2_{\omega}} z^2 \tag{69}$$

where the overlap integrals O_ω and $O_{2\omega}$ are given by Eqs. (26).

It is evident from Eqs. (68) that the SH radiation generated at the cosine $\chi^{(2)}$ grating is shifted by $\pi/2$ with respect to the correspondent $\chi^{(2)}$ grating, while the SH radiation generated at the sine $\chi^{(2)}$ grating is in phase with the respective $\chi^{(2)}$ grating. For the simultaneous writing and reading process, this means that the radiation generated at the cosine $\chi^{(2)}$ grating is shifted by $\pi/2$ with respect to the SH seeding radiation, and the SH radiation generated at the sine $\chi^{(2)}$ grating is in phase with the SH seed.

This behavior of SHG in the course of $\chi^{(2)}$ grating formation process was reported for the first time by Margulis et al. [37, 54] and was called "puzzle of the phase." The $\pi/2$ shift between the SH seed and the SH light generated in the fiber was observed in the early stages of the fiber preparation process, but in the later stages of writing $\chi^{(2)}$ no phase shift between the SH seed and the SH radiation generated in the fiber was identified [54]. It follows that initially the cosine $\chi^{(2)}$ grating is formed in the fiber, but later on, after a period of competition, the formation of the sine $\chi^{(2)}$ grating predominates and represents the vital process in preparing optical fibers for the efficient SHG.

It seems very likely that the formation of the cosine $\chi^{(2)}$ grating in the early stage of the preparation process is connected with the alignment of preexisting dipoles, which are present in the core of unexposed fiber [38], owing to the simple optical rectification, while the sine $\chi^{(2)}$ grating arises as a consequence of the excitation of dopant defects due to the higher-order photon absorptions and the creation of an elementary-dipole structure owing to more complicated driving mechanisms of nonlinear interaction, also involving the contributions of longitudinal field components, in the later stages of the fiber preparation process [121].

It is interesting to note in this context that Kamal et al. [135] recently reported slow fluctuations of $\chi^{(2)}_{\text{eff}}$ in the course of fiber preparation with weak pump and SH seed. Slight decreasing $\chi^{(2)}_{\text{eff}}$ in the steady state for relatively high SH seed power was also observed. While the fluctuations of $\chi^{(2)}_{\text{eff}}$ can be attributed to the competition between the cosine and the tertiary $\chi^{(2)}$ gratings (the SH radiation generated at the primary cosine $\chi^{(2)}$ grating acts as SH seed for forming secondary $\chi^{(2)}$ grating, being shifted by $\pi/2$; and the SH radiation generated at the secondary $\chi^{(2)}$ grating creates, in a similar way, the tertiary $\chi^{(2)}$ grating, being π out of phase with respect to the primary grating) [135], the decline of $\chi^{(2)}_{\text{eff}}$ in the steady state seems to be the consequence of competition between the cosine and sine $\chi^{(2)}$ gratings being formed with the external SH seed.

It follows from Eq. (69) that SH radiation is systematically amplified in resonant SHG. The mechanism of efficient SHG at periodical $\chi^{(2)}$ grating is similar to the method of quasi-phase-matching by periodical laminar structures in which the sign of effective quadratic susceptibility is reversed from one lamina to the next [85, 122–124]. The phase shift of interacting waves due to the phase mismatch (difference of refractive indices owing to the frequency dispersion) is balanced by the alternating sign of effective quadratic susceptibility that causes the periodical mutual phase shift of interacting waves by $\pi/2$.

In the next section only one periodical $\chi^{(2)}$ grating will be considered and the subscripts "write" and "read" will be omitted.

B. Nonresonant Second-Harmonic Generation

Let us consider now the SHG at the periodical regular sine $\chi^{(2)}$ grating:

$$\chi_{\text{eff}}^{(2)}(z) = \chi_0^{(2)} \sin(Gz - \psi) \tag{70}$$

where the grating constant slightly differs from the phase mismatch of reading radiation modes $\Delta\beta = \beta_{2\omega\,\text{read}} - 2\beta_{\omega\,\text{read}}$ and it holds that

$$G = \Delta\beta - \delta \tag{71}$$

The first-order differential equation describing the evolution of SH complex field amplitude in the approximation of a nondepleted pump then has the following form:

$$\frac{\mathrm{d}A_{2\omega}}{\mathrm{d}z} = -\mathrm{i}\left(\frac{\mu_0}{\varepsilon_0}\right)^{1/2} \frac{\omega}{n_{2\omega}} A_\omega^2 \chi_0^{(2)} \sin(Gz - \psi)\exp(\mathrm{i}\Delta\beta z) \tag{72}$$

Assuming small detuning, $|G| \gg |\delta|$ and $|\Delta\beta| \gg |\delta|$, as well as larger distances, $|\Delta\beta|z \gg 1$, the solution of Eq. (72) for the boundary condition $A_{2\omega}(0) = 0$ is

$$|A_{2\omega}(z)|\exp(\mathrm{i}\Delta\varphi) = -\mathrm{i}\left(\frac{\mu_0}{\varepsilon_0}\right)^{1/2} \frac{\omega}{2n_{2\omega}\delta} |A_\omega|^2 \chi_0^{(2)}[\exp(\mathrm{i}\delta z) - 1]\exp(\mathrm{i}\psi) \tag{73}$$

The total power of SH radiation generated is then given by

$$W_{2\omega}(z) = \left(\frac{\mu_0}{\varepsilon_0}\right)^{3/2} \frac{2\omega^2 W_\omega^2 O_{2\omega} \chi_0^{(2)^2}}{n_\omega^2 n_{2\omega} O_\omega^2 \delta^2} \sin^2\left(\frac{\delta z}{2}\right) \tag{74}$$

Evidently, the nonresonant SHG at the regular periodic $\chi^{(2)}$ grating is a spatially periodical process with the spatial period of SH light intensity

$$l_{\text{per}} = \frac{2\pi}{\delta} \tag{75}$$

There can be various reasons for the nonresonant SHG here. For example, the SHG can be read by different radiation fiber modes than those used for $\chi^{(2)}$ writing process, or $\chi^{(2)}$ can be formed by the poling technique and its grating constant is not exactly equal to the phase mismatch of reading modes.

An interesting problem is presented by the SHG with nonmonochromatic (pulse-form) radiation. We shall consider two cases separately. First, we assume that the Kerr nonlinearities are negligibly small and the spectral width of fundamental radiation is $\Delta\lambda_{\omega}$. The characteristic detuning constant is then given by

$$\delta^{\text{disp}} = \Delta\beta \frac{\Delta\lambda_{\omega}}{\lambda_{\omega}} \tag{76}$$

Considering the nonmonochromatic fundamental radiation, it is clear that the SH radiation maxima of various wavelengths arise at different fiber lengths because of diffractive-index dispersion. Consequently, the SHG process must be attenuated due to the destructive interference. The effective SHG can occur at the lengths that are shorter than the effective fiber length $l_{\text{eff}}^{\text{disp}}$, corresponding to the first maximum related to the characteristic detuning constant,

$$l_{\text{eff}}^{\text{disp}} \approx \frac{\pi}{\delta^{\text{disp}}} = \frac{\lambda_{\omega}^{2}}{4|\Delta n \Delta\lambda_{\omega}|} \tag{77}$$

where $\Delta n = n_{2\omega} - n_{\omega}$. Taking for the 100 ps pumping pulse $\Delta\lambda_{\omega}/\lambda_{\omega} \approx 5 \times 10^{-6}$, $\lambda_{\omega} \approx 1\ \mu\text{m}$, and $\Delta n \approx 2.5 \times 10^{-2}$ [12, 13, 77], we obtain $l_{\text{eff}}^{\text{disp}} \approx 2$ m, but the experimentally determined effective fiber length is only about from 0.2 to 0.4 m [12, 43]. Even taking into account that the real $\chi^{(2)}$ grating is usually induced by the same nonmonochromatic radiation (which causes a gradual fuzzing of the $\chi^{(2)}$ grating), it is evident that the linear dispersion cannot be considered the main reason for the limitation of the effective fiber length in the interaction of nonmonochromatic short light pulses.

Second, let us consider the SHG with the predominant Kerr nonlinearities. Owing to the nonuniform spectral distribution of fundamental light power $W_\omega(\omega)$, there is a frequency-dependent induced Kerr phase mismatch,

$$\delta^{\text{Kerr}} \approx \frac{2\mu_0 \omega W_\omega(\omega) \chi_{\text{eff}}^{\text{Kerr}}}{\varepsilon_0 n_{\text{average}}^2 O_\omega} \tag{78}$$

where n_{average} is the average refractive index and $\chi_{\text{eff}}^{\text{Kerr}}$ represents the effective Kerr susceptibility given for strong pump and weak SH seed, with respect to Eqs. (18), (20), (23a), (23d), and (25), by

$$\chi_{\text{eff}}^{\text{Kerr}} = \left(2\frac{O_{\omega,2\omega}^{\text{Kerr}}}{O_{2\omega}} - \frac{O_{\omega,\omega}^{\text{Kerr}}}{O_\omega} \right) \chi_{xxxx}^{(3)}(\omega = \omega - \omega + \omega) \tag{79}$$

and O_ω, $O_{2\omega}$, $O_{\omega,\omega}^{\text{Kerr}}$ and $O_{\omega,2\omega}^{\text{Kerr}}$ are given by Eqs. (26a), (26b), (28a), and (28c), respectively.

The Kerr-induced phase mismatch (power-dependent self-phase modulation of fundamental radiation and cross-phase modulation of SH radiation) causes the individual wavelengths to be converted to various levels of effectiveness and phase [42, 64]. The effective fiber length can be approximately defined, with respect to Eqs. (74) and (78), as the distance at which the first maximum of the wavelength corresponding to the average fundamental light power $W_{\omega\ \text{average}}$ occurs:

$$l_{\text{eff}}^{\text{Kerr}} \approx \frac{\varepsilon_0 n_{\text{average}}^2 \lambda_\omega O_\omega}{4\mu_0 c W_{\omega\ \text{average}} \chi_{\text{eff}}^{\text{Kerr}}} \tag{80}$$

Conforming with [12], $\lambda_\omega \approx 1\ \mu\text{m}$, $n_{\text{average}} \approx 1.5$, $W_{\omega\ \text{average}} \approx 10$ kW, and $\chi_{xxxx}^{(3)}(\omega = \omega - \omega + \omega) \approx 1.2 \times 10^{-33}$ Fm/V^2 [34], and computing the overlap integrals for $\text{LP}_{01}^{\omega} \to \text{LP}_{11}^{2\omega}$ interaction with experimental data of [12] to be $O_{\omega,\omega}^{\text{Kerr}}/O_\omega \approx 0.57$, $O_{\omega,2\omega}^{\text{Kerr}}/O_{2\omega} \approx 0.43$, and $O_\omega \approx 1.1 \times 10^{-10}$ m^2, Eq. 80 yields $l_{\text{eff}}^{\text{Kerr}} \approx 0.4$ m. This result is in excellent agreement with the experimentally determined value of effective fiber length by Österberg and Margulis [12], and it also conforms well with measurements by Batdorf et al. [43], who found that $l_{\text{eff}} \approx 0.2$ m in most cases.

It follows that the Kerr nonlinearities represent the dominant factor limiting the fiber length for SHG with a pulse-form pump. However, the above description represents a first approximation only. In the more realistic picture it would be necessary to consider the SHG with nonmonochromatic pulses at the $\chi^{(2)}$ grating that was also written by the nonlinear interaction of nonmonochromatic pulse-form radiation, including the Kerr nonlinearities.

C. Sum-Frequency Generation

As a more general case we shall consider the nonlinear optical mixing of two strong nondepleted pump radiation modes:

$$\mathbf{E}_1(t, R) \approx \mathbf{e}_1 F_{T1}(x, y) A_1 \exp[i(\omega_1 t - \beta_1 z)] \tag{81a}$$

$$\mathbf{E}_2(t, R) \approx \mathbf{e}_2 F_{T2}(x, y) A_2 \exp[i(\omega_2 t - \beta_2 z)] \tag{81b}$$

which generate a relatively weak sum-frequency radiation mode at $\omega_3 = \omega_1 + \omega_2$:

$$\mathbf{E}_3(t, R) \approx \mathbf{e}_3 F_{T3}(x, y) A_3(z) \exp[i(\omega_3 t - \beta_3 z)] \tag{82}$$

at a regular periodical $\chi^{(2)}$ grating with the effective second-order susceptibility

$$\chi_{\text{eff}}^{(2)}(z) = \chi_0^{(2)} \sin(Gz - \psi) \tag{83}$$

The distribution functions $F_{Tj}(x, y)$ $(j = 1, 2, 3)$ are assumed to satisfy the normalization conditions of Eq. 18b.

The first-order differential equation describing the evolution of SF complex field amplitude $A_3(z)$ is rather similar to Eq. (72):

$$\frac{dA_3}{dz} = -i\left(\frac{\mu_0}{\varepsilon_0}\right)^{1/2} \frac{\omega_3}{2n_3} A_1 A_2 \chi_0^{(2)} \sin(Gz - \psi) \exp(i\Delta\beta z) \tag{84}$$

where $\Delta\beta = \beta_3 - \beta_2 - \beta_1$ represents the phase mismatch of interacting radiation modes.

Assuming a small detuning $\delta = \Delta\beta - G$, $|G| \gg |\delta|$, and $|\Delta\beta| \gg |\delta|$, and also larger distances for which $|\Delta\beta| z \gg 1$, it is easy to find the solution of Eq. (84) for the boundary condition $A_3(0) = 0$ to be

$$|A_3(z)| \exp(i\Delta\varphi) = -i\left(\frac{\mu_0}{\varepsilon_0}\right)^{1/2} \frac{\omega_3}{4n_3\delta} |A_1|\,|A_2| \chi_0^{(2)} [\exp(i\delta z) - 1] \exp(i\psi) \tag{85}$$

where $\Delta\varphi = \varphi_3 - \varphi_2 - \varphi_1$ is the relative phase shift of interacting radiation modes. The total power of generated SH radiation is given by

$$W_3(z) = \left(\frac{\mu_0}{\varepsilon_0}\right)^{3/2} \frac{\omega_3^2 W_1 W_2 O_3 \chi_0^{(2)^2}}{2 n_1 n_2 n_3 O_1 O_2 \delta^2} \sin^2\left(\frac{\delta z}{2}\right) \tag{86}$$

where W_1 and W_2 are the total powers of generating subfrequency radiation modes, and O_1, O_2, O_3 are the overlap integrals:

$$O_j = \iint_{(S)} F_{Tj}^2(x, y)\, dx\, dy \qquad j = 1, 2, 3 \tag{87}$$

For the resonant case, if $\delta = 0$, Eqs. (85) and (86) get a more simple form:

$$|A_3(z)| \exp(i\Delta\varphi) = \left(\frac{\mu_0}{\varepsilon_0}\right)^{1/2} \frac{\omega_3 z}{4 n_3} |A_1|\, |A_2| \chi_0^{(2)} \exp(i\psi) \tag{88}$$

and

$$W_3(z) = \left(\frac{\mu_0}{\varepsilon_0}\right)^{3/2} \frac{\omega_3^2 W_1 W_2 O_3 \chi_0^{(2)^2}}{8 n_1 n_2 n_3 O_1 O_2} z^2 \tag{89}$$

The behavior of SFG is rather similar to that of SHG in the previous cases. The situation is somewhat more complicated for the nonmonochromatic radiation pulses, because the intensities of subfrequency radiation modes can be different and cross-phase modulations occur in the subfrequency field.

D. Difference-Frequency Generation

Let us consider again the interaction of three monochromatic radiation modes at $\omega_1, \omega_2, \omega_3$ that satisfy the frequency-resonance condition $\omega_1 + \omega_2 = \omega_3$ (Eqs. (81) and (82)). Unlike in the previous case we shall assume that the sum-frequency mode at ω_3 is very strong and that the other subfrequency modes are relatively weak. Ignoring the depletion of pump at ω_3, we can describe the evolution of subfrequency complex field amplitudes in the nonlinear optical interaction at periodical second-order susceptibility,

$$\chi_{\text{eff}1,2}^{(2)}(z) = \chi_{01,2}^{(2)} \sin(Gz - \psi) \tag{90}$$

by means of two coupled first-order differential equations:

$$\frac{\mathrm{d}A_1}{\mathrm{d}z} = -\mathrm{i}\left(\frac{\mu_0}{\varepsilon_0}\right)^{1/2} \frac{\omega_1}{2n_1} A_3 A_2^* \chi_{01}^{(2)} \sin(Gz - \psi)\exp(-\mathrm{i}\Delta\beta z) \quad (91a)$$

$$\frac{\mathrm{d}A_2}{\mathrm{d}z} = -\mathrm{i}\left(\frac{\mu_0}{\varepsilon_0}\right)^{1/2} \frac{\omega_2}{2n_2} A_3 A_1^* \chi_{02}^{(2)} \sin(Gz - \psi)\exp(-\mathrm{i}\Delta\beta z) \quad (91b)$$

where $\Delta\beta = \beta_3 - \beta_2 - \beta_1$ describes the phase mismatch of interacting waves. As in the previous case we shall assume a small detuning $\delta = \Delta\beta - G$, $|\Delta\beta| \gg |\delta|$, and $|G| \gg |\delta|$, and larger distances will be considered, $|\Delta\beta|z \gg 1$.

The amplitudes of effective quadratic susceptibilities $\chi_{01}^{(2)}$ and $\chi_{02}^{(2)}$ are not equal, but they satisfy, with respect to Manley-Rowe relations [3, 4] (photon number conservation laws [125]), the following relation:

$$\chi_{01}^{(2)} O_1 = \chi_{02}^{(2)} O_2 = K_0^{(2)} \quad (92)$$

where O_1 and O_2 are the overlap integrals given by Eq. 87.

It is well known that the decay of pump sum-frequency photons (parametric down conversion) cannot start with zero amplitudes of subfrequency field, $A_1(0) = 0$ and $A_2(0) = 0$, from the point of view of classical theory [85]. That is why we shall assume that one subfrequency field (ω_2) is not zero at the beginning of nonlinear interaction, while the other (ω_1) equals zero. Thus, the boundary conditions are

$$A_1(0) = 0 \quad \text{and} \quad A_2(0) = A_{2,0} \quad (93)$$

Since the analytical solution of DFG in media with the uniform $\chi^{(2)}$ represents a rather exacting mathematical task in itself [1, 4, 85], we shall introduce an approximate first-step iterative solution of Eqs. (91) only. Considering a weak nonlinear coupling, for which it holds that

$$\left(\frac{\mu_0}{\varepsilon_0}\right)^{3/4} \left(\frac{\omega_j \omega_k W_3}{8 n_1 n_2 n_3 O_1 O_2 O_3}\right)^{1/2} K_0^{(2)} z \ll 1 \qquad j,k = 1,2 \quad (94)$$

we obtain for small detuning, with respect to the boundary conditions (93), the first-step iterative solution of Eqs. (91) in the following form:

$$|A_1(z)|\exp(-i\Delta\varphi) = \left(\frac{\mu_0}{\varepsilon_0}\right)^{1/2} \frac{\omega_1}{4n_1} |A_3|\,|A_{2,0}|\chi_{01}^{(2)} \times \frac{1}{\delta}[\exp(-i\delta z) - 1]\exp\left[-i\left(\psi + \frac{\pi}{2}\right)\right] \tag{95a}$$

$$A_2(z) = A_{2,0}\left(1 - i\frac{\mu_0\omega_1\omega_2}{16\varepsilon_0 n_1 n_2}|A_3|^2\chi_{01}^{(2)}\chi_{02}^{(2)} \times \left\{\frac{z}{\delta} - i\frac{1}{\delta^2}[\exp(-i\delta z) - 1]\right\}\right) \tag{95b}$$

where $\Delta\varphi = \varphi_3 - \varphi_2 - \varphi_1$ is the relative phase shift of interacting radiation modes.

The total powers of generated subfrequency radiations are

$$W_1(z) = \left(\frac{\mu_0}{\varepsilon_0}\right)^{3/2} \frac{\omega_1^2 W_3 W_{2,0} K_0^{(2)^2}}{2n_1 n_2 n_3 O_1 O_2 O_3 \delta^2} \sin^2\left(\frac{\delta z}{2}\right) \tag{96a}$$

and

$$W_2(z) = W_{2,0}\left[1 + \left(\frac{\mu_0}{\varepsilon_0}\right)^{3/2} \frac{\omega_1\omega_2 W_3 K_0^{(2)^2}}{2n_1 n_2 n_3 O_1 O_2 O_3 \delta^2} \sin^2\left(\frac{\delta z}{2}\right)\right] \tag{96b}$$

Similarly as in the previous cases, the DFG is periodic when considering the nonresonant nonlinear optical interaction of monochromatic radiation modes, and its effectiveness decreases with increasing detuning δ.

The maximum efficiency of DFG occurs for the resonant interaction, if $\delta = 0$. For the resonant case, Eqs. (95) and (96) get the following form:

$$|A_1(z)|\exp(-i\Delta\varphi) = -\left(\frac{\mu_0}{\varepsilon_0}\right)^{1/2} |A_3|\,|A_{2,0}|\chi_{01}^{(2)} z \exp(-i\psi) \tag{97a}$$

$$A_2(z) = A_{2,0}\left(1 + \frac{\mu_0\omega_1\omega_2}{32\varepsilon_0 n_1 n_2}|A_3|^2\chi_{01}^{(2)}\chi_{02}^{(2)} z^2\right) \tag{97b}$$

and

$$W_1(z) = \left(\frac{\mu_0}{\varepsilon_0}\right)^{3/2} \frac{\omega_1^2 W_3 W_{2,0} K_0^{(2)^2}}{8n_1 n_2 n_3 O_1 O_2 O_3} z^2 \tag{98a}$$

$$W_2(z) = W_{2,0}\left[1 + \left(\frac{\mu_0}{\varepsilon_0}\right)^{3/2} \frac{\omega_1 \omega_2 W_3 K_0^{(2)^2}}{8n_1 n_2 n_3 O_1 O_2 O_3} z^2\right] \tag{98b}$$

When considering the SFG or DFG with bright-spectrum radiation at periodical $\chi^{(2)}$ grating, the only frequency components that can be considerably enhanced are those that satisfy the resonant condition (for which the detuning δ approaches zero).

VI. GENERATION OF BRIGHT VISIBLE SPECTRA IN OPTICAL FIBERS

The generation of bright visible spectrum exhibiting 15 discrete peaks in 5- to 10-m-long optical fibers, which were prepared for SHG, was reported in an early paper by Österberg and Margulis [12]. Unfortunately, no details were given. Because the experimental data are poor, the next treatment will be rather speculative.

The prepared fiber behaves as any usual quadratic medium, possessing a normal dispersion of $\boldsymbol{\chi}^{(2)}(\omega_j, \omega_k)$ in the visible and close to the visible spectrum. Assuming propagation of fundamental (infrared) and SH (green) radiation in the fiber, the only known nonlinear mechanism that can give rise to a bright-frequency spectrum is the parametric down conversion from quantum noise [126–129].

For the second-order nonlinear optical interaction of a strong pump at ω_p (in either the fundamental or the SH radiation mode) with two weak subfrequency radiation modes at ω_1 and ω_2, satisfying the frequency resonance condition,

$$\omega_p = \omega_1 + \omega_2 \tag{99}$$

the mutual interaction of the radiation modes is described by the following nonlinear quadratic polarizations:

$$\mathbf{P}_j^{(2)} = \boldsymbol{\chi}^{(2)}(\omega_j = \omega_p - \omega_k):\mathbf{E}_p \mathbf{E}_k^* \qquad j, k = 1, 2 \quad j \neq k \tag{100a}$$

and

$$\mathbf{P}_p^{(2)} = \boldsymbol{\chi}^{(2)}(\omega_p = \omega_1 + \omega_2):\mathbf{E}_1\mathbf{E}_2 \tag{100b}$$

It is possible to see from Eqs. (100) that the parametric down conversion cannot start with zero intensities of both subfrequency radiation modes from the point of view of classical theory [1, 4, 85]. However, the spontaneous decay of one pumping photon into two subfrequency photons, $\gamma_p \to \gamma_1 + \gamma_2$, can begin, owing to the vacuum fluctuations with the energy $\frac{1}{2}\hbar\omega_j$ [126].

In the quantum picture the nonlinear optical interaction of one strong nondepleted classical pump and two subfrequency radiation modes can be described by the interaction Hamiltonian [128]

$$\begin{aligned}\hat{H}_{\text{int}}(t) = -\hbar\big[&\kappa\hat{a}_1(t)\hat{a}_2(t)\exp(i\varphi_p)\\ &+\kappa\hat{a}_1^+(t)\hat{a}_2^+(t)\exp(-i\varphi_p)\big]\end{aligned} \tag{101}$$

where $\hat{a}_j$ and $\hat{a}_j^+$ $(j = 1, 2)$ label the time-dependent annihilation and creation operators relative to the jth mode, $\hbar$ is Planck's constant divided by 2π, κ is the nonlinear coupling constant involving the effective quadratic susceptibility and the classical field amplitude of the pumping radiation, and φ_p is the phase constant of pump.

The equations of motion in a Dirac representation can be found, using the Heisenberg equation, to be

$$\frac{d\hat{a}_1}{dt} = i\kappa\hat{a}_2^+ \exp(i\varphi_p) \tag{102a}$$

$$\frac{d\hat{a}_2}{dt} = i\kappa\hat{a}_1^+ \exp(i\varphi_p) \tag{102b}$$

Considering the coherent strong classical pump and assuming that the parametric down conversion starts from vacuum state, $\langle 0|\hat{a}_j(0)|0\rangle = \langle 0|\hat{n}_j|0\rangle = 0$ $(j = 1, 2)$, the mean photon numbers in subfrequency radiation modes are [127, 128]

$$\begin{aligned}\langle n_1(t)\rangle &= \langle \hat{a}_1^+(t)\hat{a}_1(t)\rangle\\ &= \langle n_2(t)\rangle = \langle \hat{a}_2^+(t)\hat{a}_2(t)\rangle = \text{sh}^2(\kappa t)\end{aligned} \tag{103}$$

It follows from Eq. (103) that, apparently, a bright spectrum of subfrequency radiation can be generated in parametric down conversion from quantum noise. Each couple of generated subfrequency radiation modes must satisfy only the frequency-resonance condition of Eq. (99).

Interpreting the above result is not an easy task [130]. The simple quantum model of Eq. (101) involves great simplification, namely only the time evolution of single monochromatic light modes at perfect phase matching is considered. The dispersion effects, as well as spatial modification of $\chi^{(2)}$ are not taken into account. A more adequate quantum description of the process, suitable for our purposes, would be substantionally more complicated [129].

In the following we shall simply assume that a bright subfrequency radiation spectrum can be generated due to the parametric generation from quantum noise. In fact, two bright spectra are initially generated in the SHG-conditioned effective region of the fiber, which arise as a consequence of the decay of fundamental photons $\gamma_{\omega} \rightarrow \gamma_1^{\mathrm{F}} + \gamma_2^{\mathrm{F}}$ and SH photons $\gamma_{2\omega} \rightarrow \gamma_1^{\mathrm{SH}} + \gamma_2^{\mathrm{SH}}$, respectively. The subfrequency field of the fundamental radiation is in the infrared region $(\omega_{1,2}^{\mathrm{F}} < \omega_p^{\mathrm{F}})$, while the subfrequency field of the SH radiation is extended from green to the far-infrared spectral region $(\omega_{1,2}^{\mathrm{SH}} < 2\omega_p^{\mathrm{F}})$.

The generated subfrequency spectra do not possess uniform intensity distribution. Peaks appear at those frequencies that satisfy the resonance with $\chi^{(2)}$ grating for the interaction with either fundamental or SH pumping radiation:

$$\beta_p^{\mathrm{F}} - \beta_2^{\mathrm{F}} - \beta_1^{\mathrm{F}} = G \tag{104a}$$

$$\beta_p^{\mathrm{SH}} - \beta_2^{\mathrm{SH}} - \beta_1^{\mathrm{SH}} = G \tag{104b}$$

G being the $\chi^{(2)}$ grating constant. The bright peak-shaped subfrequency spectra of fundamental and SH radiation, owing to the parametric down conversion at periodical $\chi^{(2)}$ grating, can be generated only in the relatively short region of fiber from the input end side, which was conditioned to the effective SHG. The length of this region (being approximately equal to the effective fiber length) is less than 10% of the total fiber length [12]. The remaining part of the fiber is assumed not to be conditioned to SHG because of self-phase and cross-phase modulation of nonmonochromatic (pulse-form) forming radiation, meaning that the various wavelengths of the fundamental and SH forming field induce dc polarizations with different spatial periodicities due to the power-dependent Kerr-induced dispersion, which results in fuzzing of the written $\chi^{(2)}$. Of course, there is not a sharp boundary between the SHG conditioned and the SHG unconditioned parts of the fiber there.

Let us consider now what probably happens, if the very weak peak-shaped subfrequency radiation generated in the SHG conditioned region of the fiber enters, together with the relatively strong fundamental and SH radiation, the unconditioned part of the fiber. In fact, each subfrequency radiation peak represents a new seeding radiation, which can act in writing the new $\chi^{(2)}$ grating by three-wave interaction with either fundamental or SH pump. Consequently, a lot of new $\chi^{(2)}$ gratings start to be formed in the further (originally unconditioned) region of the fiber. However, there must be competition there and, thus, only those gratings are promoted that correspond to the frequencies with the largest peak powers of seeding radiation. The newly written $\chi^{(2)}$ gratings are expected to be formed in different regions of the fiber. The parametric down conversion with either the fundamental (infrared) or SH (green) pump is expected to take place at the new structure of $\chi^{(2)}$ gratings, which gives rise to the generation of bright infrared and long-wavelength ($\omega < 2\omega_p^{\mathrm{F}}$) visible peak-shaped spectra.

Further frequencies above the SH pump frequency ($\omega > 2\omega_p^{\mathrm{F}}$) can also be generated due to the forming of proper $\chi^{(2)}$ structures for SFG with the subfrequency radiation peaks, being in the visible and near-infrared regions, and the SH pump in the far distant regions of the fiber.

It follows from the above outlined model that, besides the visible spectrum [12], a bright peak-shaped infrared spectrum is expected to be generated in long SHG conditioned optical fibers.

VII. CONCLUDING REMARKS

The aim of this chapter was to give a review of basic self-organized nonlinear optical effects in optical fibers and to outline a plausible phenomenological model. It was not possible to mention all the experiments that have been performed. We cited only the experimental results that are supposed to be significant for finding the basic physics behind these phenomena.

The key point of the proposed phenomenological model is the response tensor function $\boldsymbol{\chi}^{(3)\mathrm{resp}}(I_j, x, y, \tau)$. Indeed, this function involves all microscopic processes that create the new second-order susceptibility structure in the fiber and, thus, it represents the link between the microscopic and macroscopic models. In principle, $\boldsymbol{\chi}^{(3)\mathrm{resp}}(I_j, x, y, \tau)$ could be determined by pure phenomenological procedure, when elaborating the model outlined above in a more precise way and measuring the time evolution of $\boldsymbol{\chi}^{(2)}$ for various parameters (dopant concentration, fiber geometry, light intensities and modal structure of forming radiation, etc.). However, such a procedure comprises a certain amount of treachery. Namely, we can measure the effective quantities only, such as the total power of radiation or the effective susceptibilities, but we are not able to determine the spatial distribution of the induced dc polarization and the appurtenant dc electric field intensity in the fiber. The measurement of mutual phase shift between the seeding radiation and the radiation generated in fiber in the course of fiber preparation also represents a difficult task, especially in the later stages of the preparation process. Thus, it may happen that the results of one measurement could be in contradiction to those of another with different parameters. It follows that certain microscopic conception is unavoidable when evaluating the results of phenomenological measurements. As a possible link the photovoltaic model [57, 60–62, 131–134] seems very promising.

Another serious problem is the wave-packet interaction. In the above treatment the interaction of monochromatic waves was considered and the linear and Kerr-induced dispersion effects were described, in a heuristic approach, in terms of average quantities (spectral width and average light power). In a more precise description, the interaction of wave packets or, alternatively, Fourier spectra of radiation modes must be considered in both the writing and the reading operation. It is not difficult to deduce that the mutual phase shift and, consequently, the phase of created $\chi^{(2)}$ grating must be considerably affected by the Kerr-induced self-phase and cross-phase modulations because of nonuniform time or spectral light-intensity distributions of interacting radiations. Since the SHG with pulse-form pump at the uniform second-order susceptibility represents a very difficult mathematical task in itself [136], the description of self-organized SHG (SFG, DFG) with short light pulses, including the Kerr nonlinearities, would be extremely exacting and numerical solutions of simplified models only are expected to be attainable. It is anticipated, however, that such a treatment might enlighten not only the problem of limitation of the effective fiber length, but also other phenomena, such as SH intensity fluctuation in the early stage of fiber preparation process [46].

The self-organized nonlinear optical phenomena in fibers represent a complex set of problems that are still far from being completely understood with respect to their underlying physics or with respect to practical applications in useful nonlinear devices. This new field of research is in a fairly early stage of development and offers a unique opportunity for the cooperation of scientists from various branches of physics.

References

1. S. A. Akhmanov and R. V. Khokhlov, *Problemy Nelineynoy Optiki*, Itogi Nauki, Moscow, 1964.
2. P. N. Butcher, *Nonlinear Optical Phenomena*, Bull. 200, Ohio State University, Columbus, 1965.

3. F. Zernike and J. E. Midwinter, *Applied Nonlinear Optics*, Wiley, New York, 1973.
4. Y. R. Shen, *The Principles of Nonlinear Optics*, Wiley, New York, 1984.
5. R. H. Stolen: in S. E. Miller and A. G. Chynoweth (Eds.), *Optical Fiber Telecommunications*, Academic Press, New York, 1974, Chap. 5.
6. R. H. Stolen and J. E. Bjorkholm, *IEEE J. Quant. Electron.* **QE-18**, 1062 (1982).
7. R. K. Bullough and P. J. Baudrey (Eds.), *Solitons*, Springer, Berlin, 1980.
8. Y. Fujii, B. S. Kawasaki, K. O. Hill, and D. C. Johnson, *Opt. Lett.* **5**, 48 (1980).
9. Y. Sasaki and Y. Ohmori, *Appl. Phys. Lett.* **39**, 466 (1981).
10. Y. Ohmori and Y. Sasaki, *IEEE J. Quant. Electron.* **QE-18**, 758 (1982).
11. U. Österberg and W. Margulis, *Opt. Lett.* **11**, 516 (1986).
12. U. Österberg and W. Margulis, *Opt. Lett.* **12**, 57 (1987).
13. R. H. Stolen and H. W. K. Tom, *Opt. Lett.* **12** 585 (1987).
14. M. C. Farries, *Laser Focus* **9** 12 (1988).
15. R. H. Stolen, in A. D. Boardman, T. Twardowski, and M. Bertolotti (Eds.), *Nonlinear Waves in Solid State Physics*, Plenum, New York, 1990.
16. P. St. J. Russell, L. J. Poyntz-Wright, and D. P. Hand, in *Fiber Laser Sources and Amplifiers II*, *SPIE Proc.* **1373**, 126 (1990).
17. R. Kashyap, in *Symposium Nonlinear Optical Devices*, University of Twente, Enschede, 1990, p. 60.
18. L. J. Poyntz-Wright and P. St. J. Russell, *Electron. Lett.* **24**, 1054 (1988).
19. L. J. Poyntz-Wright, M. E. Fermann, and P. St. J. Russell, *Opt. Lett.* **13**, 1023 (1988).
20. L. J. Poyntz-Wright and P. St. J. Russell, *Electron. Lett.* **25**, 478 (1989).
21. S. La Rochelle, W. Mizrahi, G. I. Stegeman, and J. E. Sipe, *Appl. Phys. Lett.* **57**, 747 (1990).
22. V. Mizrahi, S. La Rochelle, G. I. Stegeman, and J. E. Sipe, *Phys. Rev. A* **43**, 433 (1991).
23. J. P. Bernandin and N. M. Lawandy: *Opt. Commun.* **79**, 194 (1990).
24. G. Meltz, J. R. Dunphy, W. H. Glenn, J. D. Farina, and F. J. Leonberger, in *Fiber Optic Sensors II*, *SPIE Proc.* **798**, 104 (1987).
25. R. Kashyap, J. R. Armitage, R. Wyatt, S. T. Davey, and D. L. Williams, *Electron. Lett.* **26**, 730 (1990).
26. R. J. Campbell and R. Kashyap, *Opt. Lett.* **16**, 898 (1991).
27. M. C. Farries and M. E. Fermann, *Electron. Lett.* **24**, 294 (1987).
28. W. Margulis and U. Österberg: *J. Opt. Soc. Am. B* **5**, 312 (1988).
29. A. Krotkus and W. Margulis, *Appl. Phys. Lett.* **52**, 1942 (1988).
30. H. W. K. Tom, R. H. Stolen, G. D. Aumiller, and W. Pleibel, *Opt. Lett.* **13**, 512 (1988).
31. F. Ouellette, K. O. Hill, and D. C. Johnson, *Opt. Lett.* **13**, 515 (1988).
32. T. E. Tsai, D. L. Griscom, and D. L. Griscom, *Phys. Rev. Lett.* **61**, 444 (1988).
33. A. M. Saifi and M. J. Andrejco, *Opt. Lett.* **13**, 773 (1988).
34. V. Mizrahi, U. Österberg, J. E. Sipe, and G. I. Stegeman, *Opt. Lett.* **13**, 279 (1988).
35. V. Mizrahi, U. Österberg, C. Krautschik, G. I. Stegeman, J. E. Sipe, and T. F. Morse: *Appl. Phys. Lett.* **53**, 557 (1988).
36. R. Kashyap, in *Nonlinear Guided-Wave Phenomena: Physics and Applications*, Technical Digest Series, Vol. 2, Optical Society of America, Washington, DC, 1989, p. 255.
37. W. Margulis, I. C. S. Carvalho, and J. P. von der Weid, *Opt. Lett.* **14**, 700 (1989).

38. T. E. Tsai, M. A. Saifi, E. J. Frieble, D. L. Griscom, and U. Österberg, *Opt. Lett.* **14**, 1023 (1989).
39. L. J. Poyntz-Wright and P. St. J. Russell, *Electron. Lett.* **25**, 7 (1989).
40. M. E. Fermann, L. Li, M. C. Farries, and D. N. Payne, *Electron. Lett.* **24**, 894 (1988).
41. M. E. Fermann, L. Li, M. C. Farries, J. J. Poyntz-Wright, and L. Dong: *Opt. Lett.* **14**, 748 (1989).
42. F. Ouellette, *Opt. Lett.* **14**, 964 (1989).
43. B. Batdorf, C. Krautschik, U. Österberg, G. Stegeman, J. W. Leitch, J. R. Rotgé, and T. F. Morse, *Opt. Commun.* **73**, 393 (1989).
44. D. M. Selker and N. M. Lawandy, *Electron. Lett.* **25**, 1440 (1989).
45. V. Mizrahi, Y. Hibino, and G. I. Stegeman, *Opt. Commun.* **78**, 283 (1990).
46. J. K. Lucek, R. Kashyp, S. T. Davey, and D. L. Williams, *J. Mod. Opt.* **37**, 533 (1990).
47. N. M. Lawandy and M. D. Selker, *Opt. Commun.* **77**, 339 (1990).
48. N. M. Lawandy, *Phys. Rev. Lett.* **65**, 1745 (1990).
49. A. Kamal, D. A. Weinberger, and W. H. Weber, *Opt. Lett.* **15**, 613 (1990).
50. Y. Hibino, V. Mizrahi, and G. I. Stegeman, *Appl. Phys. Lett.* **57**, 656 (1990).
51. F. Ouellette, *Electron. Lett.* **26**, 740 (1990).
52. E. V. Anoikin, E. M. Dianov, P. G. Kazansky, and D. Yu. Stepanov, *Opt. Lett.* **15**, 834 (1990).
53. U. Österberg, R. I. Lawconnell, L. A. Brambani, C. G. Askins, and E. J. Friebele, *Opt. Lett.* **16**, 132 (1991).
54. W. Margulis, I. C. S. Carvalho, and B. Lesche, in F. Ouellette (Ed.), *International Workshop on Photoinduced Self-Organization Effects in Optical Fiber*, *SPIE Proc.* **1516**, 60 (1992).
55. M. C. Farries, P. St. J. Russell, M. E. Fermann, and D. N. Payne, *Electron. Lett.* **23**, 322 (1987).
56. J. M. Gabriagues and H. Février, *Opt. Lett.* **12**, 720 (1987).
57. N. B. Baranova and B. Ya. Zeldovich, *JETP Lett.* **45**, 717 (1987).
58. M. A. Saifi and M. J. Andrejco, *Opt. Lett.* **13**, 773 (1988).
59. P. Chmela, *Opt. Lett.* **13**, 669 (1988).
60. E. M. Dianov, P. G. Kazansky, and D. Yu. Stepanov, *Sov. J. Quant. Electron.* **19**, 575 (1989).
61. E. M. Dianov, A. M. Prokhorov, V. O. Sokolov, and V. B. Sulimov, *JETP Lett.* **50**, 13 (1989).
62. N. M. Lawandy, *Opt. Commun.* **74**, 180 (1989).
63. B. Lesche, *J. Opt. Soc. Am. B* **7**, 53 (1989).
64. P. Chmela, *J. Mod. Opt.* **37**, 327 (1990).
65. N. M. Lawandy and R. L. MacDonald, *J. Opt. Soc. Am. B* **8**, 1307 (1991).
66. F. Ouellette, D. Gagnon, S. Larochelle, and M. Poirier, in F. Ouellette (Ed.) *International Workshop on Photoinduced Self-Organization Effects in Optical Fiber*, *SPIE Proc.* **1516**, 2 (1992).
67. T. E. Tsai and D. L. Griscom, in F. Ouellette (Ed.), *International Workshop on Photoinduced Self-Organization Effects in Optical Fiber*, *SPIE Proc.* **1516**, 14 (1992).
68. W. Margulis and U. Österberg, *J. Opt. Soc. Am. B* **5**, 312 (1988).

69. B. Poumellec, J. M. Gabriagues, and D. Gardin, *Opt. Commun.* **81**, 80 (1991).
70. M. V. Bergot, M. C. Farries, M. E. Fermann, L. Li, L. J. Poyntz-Wright, P. St. J. Russell, and A. Smithson, *Opt. Lett.* **13**, 592 (1988).
71. M. C. Farries, M. E. Ferrmann, and P. St. J. Russell, in *Nonlinear Guided-Wave Phenomena: Physics and Applications*, Technical Digest Series, Vol. 2. Washington, DC, 1989, p. 246.
72. D. M. Krol, R. M. Atkins, and P. J. Lemaire, in F. Ouellette (Ed.), *International Workshop on Photoinduced Self-Organization Effects in Optical Fiber*, *SPIE Proc.* **1516**, 38 (1992).
73. W. Margulis, private communication.
74. M. E. Fermann, *Characterisation Techniques for Special Optical Fibers*, Thesis, University of Southampton, Southampton, UK, 1988.
75. Y. Hibino, V. Mizrahi, G. I. Stegeman, and S. Sudo, *Appl. Phys. Lett.* **57**, 656 (1990).
76. F. P. Payne, *Electron. Lett.* **23**, 1214 (1987).
77. R. W. Terhune and D. A. Weinberger, *J. Opt. Soc. Am. B* **4**, 661 (1987).
78. N. Bloembergen, R. K. Chang, S. S. Jha, and C. H. Lee, *Phys. Rev.* **174**, 813 (1968).
79. C. C. Wang, *Phys. Rev.* **178**, 1457 (1969).
80. P. S. Pershan, *Phys. Rev.* **130**, 919 (1963).
81. A. W. Snyder and J. D. Love, *Optical Waveguide Theory*, Chapman & Hall, New York, 1983.
82. D. Stoler, *Phys. Rev. Lett.* **33**, 1397 (1974).
83. M. Parent, J. Bures, S. Lacroix, and J. Lapierre, *Appl. Opt.* **24**, 354 (1985).
84. F. Ouellette, D. Gagnon, and M. Poirier, *Appl. Phys. Lett.* **58**, 1813 (1991).
85. J. A. Armstrong, N. Bloembergen, J. Ducuing, and P. S. Pershan, *Phys. Rev.* **127**, 1918 (1962).
86. E. J. Friebele, D. L. Gricsom, and G. H. Sigel, *J. Appl. Phys.* **45**, 3424 (1974).
87. D. L. Griscom, E. J. Friebele, and K. J. Long, *J. Appl. Phys.* **54**, 3743 (1983).
88. R. N. Schwartz, G. L. Tangonan, G. R. Blair, W. Chamulitrat, and L. Kevan, *Mat. Res. Symp. Proc.* **61**, 197 (1986).
89. Y. Chen, *Appl. Phys. Lett.* **54**, 1195 (1989).
90. E. J. Friebele and D. L. Griscom, *Mat. Res. Symp. Proc.* **61**, 319 (1986).
91. M. J. Yeun, *Appl. Opt.* **21**, 136 (1982).
92. J. E. Rowe, *Appl. Phys. Lett.* **10**, 576 (1984).
93. P. St. J. Russell, D. P. Hand, Y. T. Chow, and L. J. Poyntz-Wright, in F. Ouellette (Ed.), *International Workshop on Photoinduced Self-Organization Effects in Optical Fiber*, *SPIE Proc.* **1516** 47 (1992).
94. D. L. Williams, S. T. Davey, R. Kashyap, J. R. Armitage, and B. J. Ainslie, in F. Ouellette (Ed.), *International Workshop on Photoinduced Self-Organization Effects in Optical Fiber*, *SPIE Proc.* **1516**, 29 (1992).
95. Y. Watanable, H. Kawazone, K. Shibuya, and K. Muta, *Jpn. J. Appl. Phys.* **25**, 425 (1986).
96. T. E. Tsai, D. L. Griscom, and E. J. Friebele, *Diffusion and Defect Data* **53–54**, 469 (1987).
97. T. Purcel and R. A. Weeks, *J. Chem. Phys.* **43**, 483 (1965).

98. E. J. Friebele and D. L. Griscom, in F. L. Galeener, D. L. Griscom and M. J. Weber, (Eds.), *Defect in Glasses*, Materials Research Society, Pittsburgh, 1985, Vol. 61, p. 319.
99. J. H. Stathis and M. A. Kastner, *Mat. Res. Soc. Symp. Proc.* **61**, 161 (1986).
100. G. Meltz, W. W. Morey, and W. H. Glenn, *Opt. Lett.* **14**, 823 (1989).
101. P. A. Franken and J. F. Ward, *Rev. Mod. Phys.* **35**, 23 (1963).
102. S. Kielich, *Chem. Phys. Lett.* **2**, 569 (1968).
103. S. Kielich, *IEEE J. Quant. Electron.* **QE-5**, 562 (1969).
104. S. Kielich, *J. Opt. Electron.* **2**, 5 (1970).
105. S. Kielich, *Molekulyarnaya nelineynaya optika*, Nauka, Moscow, 1981.
106. L. Li et al., in *Technical Digest on Integrated Photonic Research*, Optical Society of America, Washington, DC, 1990, paper MJ5.
107. M. E. Fermann, L. Li, M. C. Farries, D. N. Payne, and P. St. J. Russell, in *Nonlinear Guided-Wave Phenomena: Physics and Applications*, Technical Digest Series, Vol. 2, Optical Society of America, Washington, DC, 1989, paper PD6.
108. A. Kamal, M. L. Stock, A. Szpak, C. H. Thomas, D. A. Weinberger, M. Frankel, J. Nees, K. Ozaki, and J. A. Valdmanis, in Technical Digest of OSA Annual meeting, Optical Society of America, Washington, DC, 1990, paper PD25.
109. P. Chmela, *Opt. Lett.* **16**, 443 (1991).
110. P. Chmela, in F. Ouellette (Ed.), *International Workshop on Photoinduced Self-Organization Effects in Optical Fiber*, *SPIE Proc.* **1516**, 116 (1992).
111. M. G. Sceats and S. B. Poole, in *Nonlinear Guided-Wave Phenomena*, Optical Society of America, Cambridge, MA, 1991, paper Pd5-1.
112. R. Kashyap, *Appl. Phys. Lett.* **58**, 1233 (1991).
113. J. D. Jackson, *Classical Electrodynamics*, Wiley, New York, 1975.
114. M. Schubert and B. Wilhelmi, Einführung in die nichtlineare Optik, Teil I-Klassische Beschreibung, BSB B. G. Teubner Verlagsgesellschaft, Leipzig, 1971.
115. P. Chmela and P. Dub, *Czech. J. Phys.* **41**, 258 (1991).
116. R. Kashyap, in *Digest of XVI International Conference on Quantum Electronics*, Japanese Society of Applied Physics, Tokyo, 1988, p. 110.
117. R. Kashyap, *J. Opt. Soc. Am. B* **6**, 313 (1989).
118. R. Kashyap, S. T. Davey, and D. L. Williams, in F. Ouellette (Ed.), *International Workshop on Photoinduced Self-Organization Effects in Optical Fiber*, *SPIE Proc.* **1516**, 164 (1992).
119. D. Z. Anderson, V. Mizrahi, and J. E. Sipe, *Opt. Lett.* **16**, 796 (1991).
120. R. Becker, *Electromagnetic Field and Interactions*, Blackie, London, 1964.
121. P. Chmela, *Opt. Commun.* **89**, 189 (1992).
122. A. Ashkin, G. D. Boyd, and D. A. Kleinman, *Appl. Phys. Lett.* **6**, 179 (1965).
123. S. Somekh and A. Yariv, *Opt. Commun.* **6**, 301 (1972).
124. M. Okada, K. Kakizawa, and S. Ieiri, *Opt. Commun.* **18**, 331 (1976); *Jpn. J. Appl. Phys.* **16**, 55 (1977).
125. P. Chmela, *Czech. J. Phys. B* **23**, 719 (1973).
126. T. G. Giallorenzi and C. L. Tang, *Phys. Rev.* **166**, 225 (1968).
127. P. Chmela, *Acta Phys. Polon. A* **52**, 835 (1977).

128. J. Peřina, *Quantum Statistics of Linear and Nonlinear Optical Phenomena*, Kluwer, Dordrecht-Boston, 1991.

129. C. K. Hong and L. Mandel: *Phys. Rev. A* **31**, 2409 (1985).

130. P. Chmela, *Czech. J. Phys. B* **37**, 1130 (1987).

131. D. Z. Anderson, V. Mizrahi, and J. E. Sipe, in *Technical Digest of OSA Annual Meeting*, Optical Society of America, Washington, DC, 1990, paper PD24.

132. E. M. Dianov, P. G. Kazansky, C. D. Krautschik, and D. Yu. Stepanov, in F. Ouellette (Ed.), *International Workshop on Photoinduced Self-Organization Effects in Optical Fiber*, *SPIE Proc.* **1516**, 75 (1992).

133. E. M. Dianov, P. G. Kazansky, and D. Yu. Stepanov, in F. Ouellette (Ed.), in *International Workshop on Photoinduced Self-Organization Effects in Optical Fiber*, *SPIE Proc.* **1516**, 81 (1992).

134. D. Anderson, V. Mizrahi, and J. E. Sipe, in F. Ouellette (Ed.), *International Workshop on Photoinduced Self-Organization Effects in Optical Fiber*, *SPIE Proc.* **1516**, 154 (1992).

135. A. Kamal, R. W. Terhune, and D. A. Weinberger, in F. Ouellette (Ed.), *International Workshop on Photoinduced Self-Organization Effects in Optical Fiber*, *SPIE Proc.* **1516**, 137 (1992).

136. R. C. Eckard and J. Reintjes, *IEEE J. Quant. Electron.* **QE-20**, 1178 (1984).

NONLINEAR MAGNETO-OPTICS OF MAGNETICALLY ORDERED CRYSTALS

R. ZAWODNY

Laboratory of Theoretical Physics, Joint Institute for Nuclear Research, Dubna, Moscow, Russia

CONTENTS

I. INTRODUCTION

A. Linear Magneto-optical Effects

By his observations of the rotation of the polarization plane of linearly polarized light on the traversal of a path in quartz, Arago in 1811 laid the foundations for the study of the physical properties of matter in the

Permanent address: Nonlinear Optics Division, Institute of Physics, Adam Mickiewicz University, 60-780 Poznań, Poland.

Modern Nonlinear Optics, Part 1, Edited by Myron Evans and Stanisław Kielich. Advances in Chemical Physics Series, Vol. LXXXV.
ISBN 0-471-57546-1

optical region. Thus, he opened up the field of polarization optics (ellipsometry) [1, 2]. Thirty-five years later, Faraday [3] initiated the domain of what is now referred to as magneto-optics with his observations of the rotation plane in lead glass under the action of a static magnetic field of light linearly polarized propagating parallel to the field. Righi [4] and Becquerel [5] applied Fresnel's [6] interpretation of Arago rotation to the effect observed by Faraday. After Fresnel, linearly polarized light is a superposition of two circularly, respectively right- and left-polarized light waves, which, in an optically active medium, propagate with different velocities. On emerging from the active medium, the two waves again superpose to give a linearly polarized wave, albeit with a polarization plane rotated by an angle Ψ equal to one-half of the difference in phase produced by the traversal of the two circularly polarized waves along the same path in the active medium. Becquerel [5], applying classical electron theory [7], showed that the situation described above exists in the case of media transparent to optical frequencies: In a magnetic field, the electrons perform Larmor precession [8] so that their single frequency is replaced by two frequencies, one for the right-circularly and the other for the left-circularly polarized wave, and the band of the light-absorbing electron splits into two components. Consequently, at the output, the two waves have different amplitudes—the light wave is now polarized elliptically, not linearly. The ellipticity observed in Faraday's effect is known as magnetic circular dichroism; like the angle of rotation of the polarization plane, it is a linear function of the static magnetic field strength. The ellipticity accompanying Arago rotation first observed by Cotton [9] is referred to as circular dichroism, and Arago rotation is currently termed natural optical activity.

In 1907, Cotton and Mouton [10] found that liquids in a static magnetic field applied at right angles to the light propagation direction become birefringent for linearly polarized light. The essence of the effect is this: The linearly polarized wave gives rise to a wave of the same frequency polarized linearly at right angles to the incident wave; the two waves propagate with different velocities; on traversal of the same path in the medium they emerge with a difference in phase, thus giving a superposition that is elliptically (not linearly) polarized, with major axis rotated by an angle Ψ with respect to the polarization direction of the wave at the input to the medium. The two parameters of the effect, namely the rotation Ψ and the ellipticity Φ, are now functions of the second power of the external magnetic field strength. Voigt [11] proposed an interpretation for gases in a molecular approach on the assumption that the field modifies the polarizability of the molecule (deformational nonlinear Voigt effect). Langevin [12] considered the magnetic field as causing anisotropic molecules in a gas to exhibit a tendency to reorient with their axis of maximal polarizability into the direction of the magnetic field, complete ordering being unattainable because of the opposite tendency due to thermal (Brownian) motion. Born [13] and Buckingham and Pople [14], taking the two mechanisms into account, proposed a molecular theory of the Cotton–Mouton effect for a diamagnetic gas of identical molecules, for different cases of molecular symmetry. A comprehensive list of the literature with a detailed discussion of the various models used in attempts to explain the effects of Faraday, Cotton–Mouton, as well as magnetic circular and magnetic linear dichroism in one- and many-component liquids and their solutions in nondipolar solvents are to be found in the work of Piekara and Kielich [15–17] and Kielich [18–21]. To Kielich [18–21] is due the most general statistical-molecular theory of these effects.

The quantum-mechanical theory of the Faraday and Cotton–Mouton effects as well as magnetic circular and magnetic linear dichroism was first proposed by Kronig [22] for diamagnetic diatomic molecules and extended by Carrol [23] and Rosenfeld [24] to atoms, by Kroll [25] to one-electron systems in relativistic approximation, and by Serber [26], Groenewege [27], and Hameka [28] to arbitrary molecules. Jørgensen and co-workers [29] determined numerically the Verdet constants for the atoms of helium and beryllium as well as for the molecules CO and FH. Whereas Manakov et al. [30] determined them for inert gases and atomic Hg, Sr, Tl and Ga.

Lubchenko [31], Boswarva et al. [32, 33], and Murao and Ebina [34] proposed a quantum-mechanical theory of the Faraday, Cotton–Mouton effects as well as magnetic circular and magnetic linear dichroism in diamagnetic crystals. Kramers [35] proposed a quantum theory of these effects for paramagnetic ions which Shen and Bloembergen [36, 37] extended to rare earth ions. Druzhinin and Tatsenko [38], in a quantum approach, showed that in transparent diamagnetic crystals the Faraday rotation angle is a linear function of the magnetic field strength up to 10^7 Oe, whereas in paramagnetic crystals only up to 10^4 Oe.

Extensive reviews of the literature concerning magneto-optical effects in nonconducting materials consisting of dia- and paramagnetic molecules are to be found in the papers of Buckingham and Stephens [39], Palik and Henvis [40], and, for crystals, in Starostin and Feofilov [41], Zapaskii and Feofilov [42], Ramasechan and Sivaramakrishnan [43], and Ramachandran and Ramasechan [44]. In the 1970s Koralewski et al. [45–47] measured the diamagnetic Verdet constants in ferroelectric crystals of the KDP, $NaClO_3$, and $NaBrO_3$ types for several light wavelengths. In 1966, Kielich [48] predicted theoretically for liquids and gases the occurrence of birefringence proportional to the product of the magnetic field strength and the wave vector of the light wave. In 1971 Portigal and Burstein [49] predicted this effect for crystals and referred to it as "nonreciprocal linear birefringence." Six years later, Markelov and co-workers [50] measured it in crystal of $LiIO_3$. The effect has been discussed for liquids and gaseous media by Baranova and Zel'dovich [51, 52] and Woźniak and Zawodny [53].

Some years earlier Kielich [21], subsequently Kielich and Zawodny [54], and more recently Ross and co-workers [55] discussed optical birefringence in crossed dc electric and magnetic fields. In 1956 Tavger and Zaitsev [56] showed that by adjoining time inversion $\underline{1}$ (electric current reversal) as an element of symmetry to the well-known 32 point groups $G(P)$ with crystallographically limited foldness of their symmetry axes, one can construct three kinds of groups G (122 groups in all) as follows: $G(NM) = G(P) \otimes G(\underline{1})$, $G(P)$ and $G(M) = G(P') + \{G(P) - G(P')\} \otimes \underline{1}$, where $G(P')$ is a subgroup of the group $G(P)$ (it has to contain one-half of the elements of the group $G(P)$); whereas $G(P) - G(P')$ denotes the set of those elements of $G(P)$ that did not enter the subgroup $G(P')$. The nonmagnetic groups $G(NM)$ describe the symmetry of nonmagnetic molecules and crystals, whereas the remaining 90 magnetic point groups describe that of magnetically ordered crystals or paramagnetic molecules.

In 1958, Porter et al. [57] discovered ferro- and ferrimagnets with "transparency windows" in the region of optical frequencies, and work started on magneto-optical effects in these materials applying transmitted light. Earlier, studies of the polarization state of light reflected from magnetized surfaces of crystals revealed the existence of a linear magneto-optical Kerr effect [58–62]. The effect has since been widely applied in studies on magnetically ordered crystals [8]. The inclusion of magnetic symmetries into the theory has permitted the prediction of a variety of novel optical phenomena. In magnetically ordered crystals (with directional symmetry described by the groups $G(P)$ and $G(M)$) with transparency windows [57] in the region of optical frequencies, even in the absence of perturbating factors, such as an external electric or magnetic field, or mechanical stress, in addition to natural optical birefringence [1, 2] and natural optical activity [48, 63–67], which also occur in nonmagnetic crystals (groups $G(NM)$), two novel optical effects take place: natural gyrotropic birefringence (with a change in sign on reversal of the light propagation direction) [66–70] and natural gyrotropic rotation [71–73].

Krinchik and Chetkin [74] have shown that in certain magnetically ordered crystals the omission of magnetic susceptibility for some frequencies as postulated by Landau and Lifshitz [75] is unjustified. If taken into account, it in fact contributes to the optical effects considered above and can modify the propagation of the light wave [76, 77]. Moreover, if the crystal possesses spontaneous magnetization **M** it can exhibit spontaneous Faraday, Cotton–Mouton, and nonreciprocal linear birefringence (birefringence proportional to the product of the magnetization vector **M** and the wave vector **k** of the light wave) [78, 79] and the spontaneous magnetic circular, magnetic linear, and spontaneous nonreciprocal linear dichroisms related thereto.

Crystals with spontaneous magnetization **M** can exhibit other effects, having no counterparts in nonmagnetic media under the action of a constant magnetic field. The most noteworthy are optical birefringence proportional to the first power of **M** and rotation proportional to the second power of **M** as well as to the product of **M** and **k**. The theoretical and experimental studies of spontaneous magneto-optics have been reviewed repeatedly [78, 80, 81].

The past 10 to 20 years have brought considerable developments in the optical study of magnetically ordered crystals in static magnetic fields [76, 82–89]. Here the magnetic symmetry admits quite new effects, such as birefringence proportional to the first power of the static magnetic field—the magnetic analog of Pockels' effect [76, 84–86, 90, 91]—as well as rotation proportional to the square of the dc magnetic field [76, 88, 89] and proportional to the product of the static magnetic field and the wave vector of the light beam [76]. Dillon and co-workers [85] were the first to show that birefringence linear in a dc magnetic field with induction **B** can be expected in the longitudinal geometry $\mathbf{B}\|\mathbf{e}_z\|\mathbf{k}$ (Faraday configuration) in $Dy_3Al_5O_{12}$ crystals (symmetry $m3\underline{m}$). The effect has already been measured for the antiferromagnets $DyFeO_3$ (magnetic point group $\underline{mmm}$), $Ca_3Mn_2Ge_3O_{12}$ ($\underline{4}/m$), CoF_2 ($\underline{4}/mm\underline{m}$), α-Fe_2O_3, and $CoCO_3$ ($\bar{3}m$). An extensive discussion of the experimental work on the above and other optical effects occurring in antiferromagnetic crystals subjected to external magnetic fields is due to Eremenko and Kharchenko [89].

B. Nonlinear Magneto-optical Effects

Even prior to the coming of lasers, Piekara and Kielich [19, 92] proposed a theory of the modifications to be expected in the magnetic properties of matter under the influence of intense light. These modifications are accessible to observation as optically induced magnetic anisotropy and the inverse Cotton–Mouton effect [21, 93]. However, the detection and investigation of the nonlinear magneto-optical effects required developments in the field of laser technique. In this respect, we essentially have in mind the inverse Faraday effect [94]. The theory of optical magnetization of matter has been worked out by Pershan and co-workers [94] in a phenomenological and quantum-mechanical approach, and by Kielich [21, 93, 95] in a molecular statistical treatment for diamagnetic and paramagnetic liquids. The effect was first observed in ruby [96] and in the antiferromagnet MnF_2 [97]. Quite recently, Woźniak, Evans, Wagnière, and Zawodny discussed the possibility of optically induced magnetization in liquids and gaseous media with molecules having various symmetries, for optical frequencies remote from [98–100] and close to [101] that of an optical transition of the molecule.

Atkins and Miller [102] have developed the theory of magneto-optical effects applying methods of quantum electrodynamics. Courtens [103] has predicted the possibility of a giant Faraday effect, both normal and inverse, in materials experiencing self-induced transparency.

Another interesting nonlinear magneto-optical effect with no counterpart in linear magneto-optics consists of the generation of second and higher harmonics of light in media acted on by a dc magnetic field and in crystals exhibiting magnetic order. To Cohan and Hameka [104] is due a quantum-mechanical theory of second-harmonic generation in magnetized gases and liquids. Kielich and Zawodny [105] have discussed the feasibility of observing the effect for the 32 crystallographic classes. Hafele and co-workers [106] have considered the influence of a dc magnetic field on the process of second-harmonic generation in InSb close to resonance, whereas Van Tran Nguyen and Bridges [107] have studied the mixing of two laser beams in InSb in the presence of a magnetic field. In 1985, Akhmediev [108] investigated the feasibility of second-harmonic generation for ferromagnetic crystals in transmitted light, whereas Reif and co-workers [109] studied the effect in the case of reflected light.

However, there exists a class of nonlinear optical effects, including magneto-optical ones, that, in contradistinction to those discussed above, possess their counterparts in the domain of linear optics. Inasmuch as they are related with a change in refractive index dependent on the light intensity they obviously have no linear counterparts; the correspondence concerns the changes in the light polarization state defined by two well-known parameters: the ellipticity and the rotation of the major axis of the polarization ellipse in the plane perpendicular to the light propagation direction. To these effects belongs rotation of the polarization ellipse of light first observed by Maker and co-workers [110] in isotropic medium. In the pre-laser epoch Buckingham [111], starting from the classical work of Voigt [112] and Langevin [12], succeeded in showing that very intense light can cause optical birefringence in a gas or liquid; his prediction was later confirmed experimentally by Mayer and Gires [113]. Kielich and Zawodny

[114] have discussed the possibility of self-induced optical circular birefringence in crystals, whereas Sala [115] and Roman et al. [116] on the theoretical level and Nguen Phu-Xuan and Rivoire [117] experimentally have investigated the evolution of the state of light polarization in crystals versus the state of polarization at the input.

If, additionally, a dc magnetic field is applied, intensity-dependent magneto-optical rotation appears in Faraday or, respectively, Voigt configuration according to whether the light beam propagates parallel or perpendicular to the magnetic field [118, 119]. A nonlinear Verdet constant was first measured in the experiments of Kubota [120] in semiconductor crystals of CdS and ZnS. Kielich [121–123] proposed a molecular-statistical theory of nonlinear variation in the Verdet constant of gases and liquids. Later, in cooperation with Manakov and Ovsiannikov [124], he determined the nonlinear Verdet constant for atoms and molecules, for frequencies of intense light waves strongly remote from those of optical transitions; for atoms in the case of resonance see Refs. 125–127.

Recently, Kielich et al. [128] have discussed the feasibility of exploiting the nonlinear Faraday effect for the generation of squeezed states of the electromagnetic field in isotropic media as well as crystals; their discussion comprises the role of spatial dispersion in the nonlinear Faraday effect. They determine the Verdet constant for different symmetries not only in the electric-dipole, but also in the magnetic-dipole and electric-quadrupole approximations. In the course of the last five years the nonlinear Faraday effect near to and far from resonance has been the subject of extensive studies for a great variety of materials [129–133]. Detailed discussions of nonlinear magneto-optical phenomena, with extensive reference lists, are found in the review articles of Refs. 134 and 135.

As yet there is a lack of theoretical work on nonlinear magneto-optical effects in magnetically ordered crystals, especially ferro- and antiferromagnets, in the presence and absence of a dc magnetic field. The treatment of these effects is the chief aim of the present paper. We study the joint influence of an intense light wave, spatial dispersion, a dc magnetic field, and ferro-/antiferromagnetic order on the state of polarization of the wave, restricting ourselves to the Faraday configuration, i.e., to the case when the wave propagates along the direction of the static magnetic field. In Section II we give a phenomenological treatment of the elements of linear and nonlinear magneto-optics with a detailed discussion of the permutational symmetry and transformational properties of the linear and nonlinear electro-electric and electro-magnetic multipolar susceptibilities. In Section III we derive nonlinear differential equations for the Stokes parameters and solve them analytically for three cases. In Section IV we discuss the physical meaning of the solutions and the experimental feasibility of observations of the respective effects in two sublattice ferro- and antiferromagnetic crystals with symmetry described by magnetic point groups belonging to the tetragonal, trigonal, and hexagonal systems.

II. CLASSICAL MAGNETO-OPTICS IN A PHENOMENOLOGICAL TREATMENT

The electric and magnetic properties of a medium acted on by a time-variable electric field $\mathbf{E}(\mathbf{r}, t)$ and magnetic field $\mathbf{H}(\mathbf{r}, t)$ are described by the electric induction vector $\mathbf{D}(\mathbf{r}, t)$ and the magnetic induction vector $\mathbf{B}(\mathbf{r}, t)$, respectively. In SI units, they take the well known form [75]

$$\mathbf{D}(\mathbf{r}, t) = \varepsilon_0 \mathbf{E}(\mathbf{r}, t) + \mathbf{P}_e(\mathbf{r}, t) \qquad \mathbf{B}(\mathbf{r}, t) = \mu_0[\mathbf{H}(\mathbf{r}, t) + \mathbf{P}_m(\mathbf{r}, t)] \quad (1)$$

where ε_0 and μ_0 are the electric and magnetic permittivity, and $\mathbf{P}_e(\mathbf{r}, t)$, $\mathbf{P}_m(\mathbf{r}, t)$ are the electric and magnetic polarization vectors of the medium at the moment of time t and the point $\mathbf{r}$.

We consider an arbitrary nondissipative magnetically ordered crystal in a homogeneous dc magnetic field of induction $\mathbf{B}^0$ in which a monochromatic light wave with the electric vector

$$\mathbf{E}(\mathbf{r}, t) = \mathbf{E}(\omega, \mathbf{k}) \exp\{-i\omega[t - (n/c)\mathbf{s} \cdot \mathbf{r}]\} + \text{c.c.} \quad (2)$$

oscillating with the circular frequency ω, propagates in the $\mathbf{s}$ direction with an amplitude $|\mathbf{E}(\omega, \mathbf{k})|$ comparable to the strength of the atomic field. Here, n is the light refractive index of the crystal in the absence of absorption, c the velocity of light in vacuum, $\mathbf{s}k$ the wave vector of length $k = \omega n/c$, and c.c. stands for complex conjugate. The other vectors of Eq. (1) have the same form as $\mathbf{E}(\mathbf{r}, t)$.

The phenomenological approach to nonlinear magneto-optics starts from the amplitudes $\mathbf{P}_e(\omega, \mathbf{k})$ and $\mathbf{P}_m(\omega, \mathbf{k})$ of the electric and magnetic polarization vector, which can be written as the sum of two parts [48, 63, 64, 110, 111, 113, 136–139]:

$$\mathbf{P}_A(\omega, \mathbf{k}) = \mathbf{P}_A^L(\omega, \mathbf{k}) + \mathbf{P}_A^{NL}(\omega, \mathbf{k}) \quad (3)$$

for $A = e$ or m. The linear $\mathbf{P}_A^L(\omega, \mathbf{k})$ and nonlinear $\mathbf{P}_A^{NL}(\omega, \mathbf{k})$ polarization vectors have the form [48, 63, 64, 139]

$$\mathbf{P}_A^L(\omega,\mathbf{k}) = {}_A\boldsymbol{\chi}_e(\omega,\mathbf{k},\mathbf{B}^0)\cdot\mathbf{E}(\omega,\mathbf{k}) + \mu_0\,{}_A\boldsymbol{\chi}_m(\omega,\mathbf{k},\mathbf{B}^0)\cdot\mathbf{H}(\omega,\mathbf{k}) \quad (4)$$

$$\begin{aligned}\mathbf{P}_A^{NL}(\omega,k) = {}&{}_A\boldsymbol{\chi}_{eee}(\omega,\mathbf{k},\mathbf{B}^0)\cdots\mathbf{E}^*(\omega,\mathbf{k})\mathbf{E}(\omega,\mathbf{k})\mathbf{E}(\omega,\mathbf{k})\\ &+\mu_0\{{}_A\boldsymbol{\chi}_{eem}(\omega,\mathbf{k},\mathbf{B}^0)\cdots\mathbf{E}^*(\omega,\mathbf{k})\mathbf{E}(\omega,\mathbf{k})\mathbf{H}(\omega,\mathbf{k})\\ &+{}_A\boldsymbol{\chi}_{eme}(\omega,\mathbf{k},\mathbf{B}^0)\cdots\mathbf{E}^*(\omega,\mathbf{k})\mathbf{H}(\omega,\mathbf{k})\mathbf{E}(\omega,\mathbf{k})\\ &+{}_A\boldsymbol{\chi}_{mee}(\omega,\mathbf{k},\mathbf{B}^0)\cdots\mathbf{H}^*(\omega,\mathbf{k})\mathbf{E}(\omega,\mathbf{k})\mathbf{E}(\omega,\mathbf{k})\}\end{aligned} \quad (5)$$

One, two, and three dots denote respectively the single, double, and triple scalar product and we apply the following notation:

$$\begin{aligned}{}_A\boldsymbol{\chi}_B(\omega,\mathbf{k},\mathbf{B}^0) &= {}_A\boldsymbol{\chi}_B(-\omega,\omega;-k,k;\mathbf{B}^0)\\ {}_A\boldsymbol{\chi}_{BCD}(\omega,\mathbf{k},\mathbf{B}^0) &= {}_A\boldsymbol{\chi}_{BCD}(-\omega,-\omega,\omega,\omega;-\mathbf{k},-\mathbf{k},\mathbf{k},\mathbf{k};\mathbf{B}^0)\end{aligned} \quad (6)$$

The polar tensors of second rank ${}_e\boldsymbol{\chi}_e(\omega,\mathbf{k},\mathbf{B}^0)$ and ${}_m\boldsymbol{\chi}_m(\omega,\mathbf{k},\mathbf{B}^0)$ describe the linear electro-electric and magneto-magnetic susceptibilities of the crystal under the action of the static magnetic field of induction $\mathbf{B}^0$. The axial tensors ${}_e\boldsymbol{\chi}_m(\omega,\mathbf{k},\mathbf{B}^0)$ and ${}_m\boldsymbol{\chi}_e(\omega,\mathbf{k},\mathbf{B}^0)$ describe its linear electromagnetic and magneto-electric susceptibilities. The polar tensor of fourth rank ${}_e\boldsymbol{\chi}_{eee}(\omega,\mathbf{k},\mathbf{B}^0)$ describes the third-order nonlinear electro-electric susceptibility. Similarly, ${}_e\boldsymbol{\chi}_{eee}(\omega,\mathbf{k},\mathbf{B}^0)$ denotes the variation of the electro-electric susceptibility proportional to the light intensity. The remaining axial tensors ${}_e\boldsymbol{\chi}_{eem}(\omega,\mathbf{k},\mathbf{B}^0)$, ${}_e\boldsymbol{\chi}_{eme}(\omega,\mathbf{k},\mathbf{B}^0)$, ${}_e\boldsymbol{\chi}_{mee}(\omega,\mathbf{k},\mathbf{B}^0)$, and ${}_m\boldsymbol{\chi}_{eee}(\omega,\mathbf{k},\mathbf{B}^0)$ describe the third-order nonlinear electromagnetic and magneto-electric susceptibilities, respectively.

A. Linear and Nonlinear Electric and Magnetic Multipolar Susceptibilities

The quantum-mechanical form of the above linear and nonlinear susceptibilities in the multipole approach and in the absence of a static magnetic field have been given by Kielich [48, 63, 64] as follows:

$${}_A\boldsymbol{\chi}_B(\omega,\mathbf{k}) = \sum_{a=1}\sum_{b=1} W(a,b)(\mathbf{s})^{a-1}[a-1]^{(a)}_A\,\boldsymbol{\chi}^{(b)}_B(\omega)[b-1]\mathbf{s}^{b-1} \quad (7)$$

$$\begin{aligned}{}_A\,\boldsymbol{\chi}_{BCD}(\omega,\mathbf{k}) = {}&\sum_{a=1}\sum_{b=1}\sum_{c=1}\sum_{d=1} W(a,b,c,d)(\mathbf{s})^{a-1}[a-1]\\ &\times{}^{(\alpha)}_A\boldsymbol{\chi}^{(b,c,d)}_{BCD}(\omega)[b+c+d-3]\mathbf{s}^{b-1}\mathbf{s}^{c-1}\mathbf{s}^{d-1}\end{aligned} \quad (8)$$

where

$$
{}_{A}^{(a)}\boldsymbol{\chi}_{B}^{(b)}(\omega) = \frac{\rho_{kk}}{\hbar} \sum_{r \neq k} \left\{ \frac{\langle k|\mathbf{M}_{A}^{(a)}|r\rangle\langle r|\mathbf{M}_{B}^{(b)}|k\rangle}{\omega + \omega_{rk}} + \frac{\langle k|\mathbf{M}_{B}^{(b)}|r\rangle\langle r|\mathbf{M}_{A}^{(a)}|k\rangle}{-\omega + \omega_{rk}} \right\} \tag{9}
$$

$$
{}_{A}^{(a)}\boldsymbol{\chi}_{BCD}^{(b,c,d)}(\omega) = \frac{S\left\{\left[\mathbf{M}_{A}^{(a)}, \mathbf{M}_{B}^{(b)}\right], \left[\mathbf{M}_{C}^{(c)}, \mathbf{M}_{D}^{(d)}\right]\right\}\rho_{kk}}{3!\hbar} \sum_{p,q,r \neq k} \langle k|\mathbf{F}_{pqr}|k\rangle
$$

$$
\begin{aligned}
\mathbf{F}_{pqr} = {} & \frac{\mathbf{M}_{A}^{(a)}|p\rangle\langle p|\mathbf{M}_{B}^{(b)}|q\rangle\langle q|\mathbf{M}_{C}^{(c)}|r\rangle\langle r|\mathbf{M}_{D}^{(d)}}{(\omega + \omega_{pk})(2\omega + \omega_{qk})(\omega + \omega_{rk})} \\
& + \frac{\mathbf{M}_{A}^{(a)}|p\rangle\langle p|\mathbf{M}_{C}^{(c)}|q\rangle\langle q|\mathbf{M}_{B}^{(b)}|r\rangle\langle r|\mathbf{M}_{D}^{(d)}}{(\omega + \omega_{pk})\omega_{qk}(\omega + \omega_{rk})} \\
& + \frac{\mathbf{M}_{B}^{(b)}|p\rangle\langle p|\mathbf{M}_{C}^{(c)}|q\rangle\langle q|\mathbf{M}_{D}^{(d)}|r\rangle\langle r|\mathbf{M}_{A}^{(a)}}{(\omega + \omega_{pk})\omega_{qk}(-\omega + \omega_{rk})} \\
& + \frac{\mathbf{M}_{D}^{(d)}|p\rangle\langle p|\mathbf{M}_{A}^{(a)}|q\rangle\langle q|\mathbf{M}_{B}^{(b)}|r\rangle\langle r|\mathbf{M}_{C}^{(c)}}{(-\omega + \omega_{pk})\omega_{qk}(\omega + \omega_{rk})} \\
& + \frac{\mathbf{M}_{C}^{(c)}|p\rangle\langle p|\mathbf{M}_{B}^{(b)}|q\rangle\langle q|\mathbf{M}_{D}^{(d)}|r\rangle\langle r|\mathbf{M}_{A}^{(a)}}{(-\omega + \omega_{pk})\omega_{qk}(-\omega + \omega_{rk})} \\
& + \frac{\mathbf{M}_{C}^{(c)}|p\rangle\langle p|\mathbf{M}_{D}^{(d)}|q\rangle\langle q|\mathbf{M}_{A}^{(a)}|r\rangle\langle r|\mathbf{M}_{B}^{(b)}}{(-\omega + \omega_{pk})(-2\omega + \omega_{qk})(-\omega + \omega_{rk})}
\end{aligned} \tag{10}
$$

whereas

$$
W(a,b) = (-1)^{a-1}\left\{\mathrm{i}\frac{\omega n}{c}\right\}^{a+b-2} \frac{2^{a+b}(a)!(b)!}{(2a)!(2b)!}
$$

$$
W(a,b,c,d) = (-1)^{a-1}\left\{\mathrm{i}\frac{\omega n}{c}\right\}^{a+b+c+d-4} \times \frac{2^{a+b+c+d}(a)!(b)!(c)!(d)!}{(2a)!(2b)!(2c)!(2d)!} \tag{11}
$$

The tensor ${}_{A}^{(a)}\boldsymbol{\chi}_{B}^{b}(\omega)$ of rank $a + b$ describes the linear electric multipole

$A = e$ or magnetic multipole $A = m$ susceptibility of the ath order (for $a = 1, 2$, we have, respectively, the dipole and quadrupole moment) related to an electric $B = e$ or magnetic $B = m$ multipole transition of order b [64] between the stationary ground state $|k\rangle$ and the virtual states $|r\rangle$. For $B = e$ and $b = 1$ or 2 we have a transition $E1$ or $E2$, whereas for $B = m$ and $b = 1$ we have a transition $M1$ [140]. Similarly, the tensor ${}^{(a)}_{A}\boldsymbol{\chi}^{(b,c,d)}_{BCD}(\omega)$ of rank $a + b + c + d$ is the nonlinear multipolar electric $A = e$ or, respectively, magnetic $A = m$ susceptibility of the ath order due to three multipolar electric or magnetic transitions of order b, c and d, respectively. The symbol $[a - 1]$ denotes $(a - 1)$-fold contraction of two tensors, ρ_{kk} is the quantum mean value of the unperturbed density matrix in the ground state $|k\rangle$ and $\omega_{rk} = \omega_r - \omega_k$ the transition frequency between the states $|k\rangle$ and $|r\rangle$.

Formulas (7)–(10) can, moreover, be taken as the quantum-mechanical definitions of the linear ${}_{A}\boldsymbol{\chi}_{B}(\omega, \mathbf{k}, \mathbf{B}^0)$ and nonlinear ${}_{A}\boldsymbol{\chi}_{BCD}(\omega, \mathbf{k}, \mathbf{B}^0)$ susceptibilities provided that, in Eqs. (9) and (10), $|k\rangle$ and $\hbar\omega_k$ are meant to denote the wave function and energy of the state of the crystal in the presence of a static magnetic field [140] signifying that the crystal is now acted on by the perturbation

$$V = -\mathbf{M}^{(1)}_{m} \cdot \mathbf{B}^0 \tag{12}$$

where $\mathbf{M}^{(1)}_{m}$ is an operator representing the magnetic dipole moment.

By stationary perturbation calculus [140], $|k\rangle$ and ω_k are expressed as follows by the functions $|k'\rangle$ and $|l'\rangle$ and frequencies $\omega_{k'}$ and $\omega_{l'}$ of the stationary states of the crystal in the absence of a perturbation:

$$|k\rangle = |k'\rangle + \sum_{l' \neq k'} |l'\rangle \frac{\langle l'|V|k'\rangle}{\hbar(\omega_{k'} - \omega_{l'})} + \cdots$$

$$\omega_k = \omega_{k'} + \frac{\langle k'|V|l'\rangle}{\hbar} + \frac{1}{\hbar^2} \sum_{l' \neq k'} \frac{|\langle k'|V|l'\rangle|^2}{\omega_{k'} - \omega_{l'}} + \cdots \tag{13}$$

Equations (9) and (10), with (11) and (12) taken into account and summation performed over p', r', and q' instead of p, r, and q, define linear ${}^{(a)}_{A}\boldsymbol{\chi}^{(b)}_{B}(\omega, \mathbf{B}^0)$ and nonlinear ${}^{(a)}_{A}\boldsymbol{\chi}^{(b,c,d)}_{BCD}(\omega, \mathbf{B}^0)$ susceptibilities. If the static magnetic field is not too intense, the linear ${}^{(a)}_{A}\boldsymbol{\chi}^{(b)}_{B}(\omega, \mathbf{B}^0)$ and nonlinear

${}^{(\alpha)}_{A}\boldsymbol{\chi}^{(b,c,d)}_{BCD}(\omega, \mathbf{B}^0)$ susceptibilities can be written as follows:

$$
\begin{aligned}
{}^{(a)}_{A}\boldsymbol{\chi}^{(b)}_{B}(\omega, \mathbf{B}^0) = {}^{(a)}_{A}\boldsymbol{\chi}^{(b)}_{B}(\omega) + {}^{(a)}_{A}\boldsymbol{\chi}^{(b)m}_{B}(\omega) \cdot \mathbf{B}^0 \\
+ {}^{(a)}_{A}\boldsymbol{\chi}^{(b)mm}_{B}(\omega) \cdot\cdot \mathbf{B}^0\mathbf{B}^0 + \cdots
\end{aligned} \tag{14}
$$

$$
\begin{aligned}
{}^{(a)}_{A}\boldsymbol{\chi}^{(b,c,d)}_{BCD}(\omega, \mathbf{B}^0) = {}^{(a)}_{A}\boldsymbol{\chi}^{(b,c,d)}_{BCD}(\omega) + {}^{(a)}_{A}\boldsymbol{\chi}^{(b,c,d)m}_{BCD}(\omega) \cdot \mathbf{B}^0 \\
+ {}^{(a)}_{A}\boldsymbol{\chi}^{(b,c,d)mm}_{BCD}(\omega) \cdot\cdot \mathbf{B}^0\mathbf{B}^0 + \cdots
\end{aligned} \tag{15}
$$

The tensors ${}^{(a)}_{A}\boldsymbol{\chi}^{(b)m}_{B}(\omega)$ and ${}^{(a)}_{A}\boldsymbol{\chi}^{(b)mm}_{B}(\omega)$ represent the changes in ${}^{(a)}_{A}\boldsymbol{\chi}^{(b)}_{B}(\omega)$ induced by the static magnetic field in a linear (first order of quantum-mechanical stationary perturbation calculus) and quadratic (second order of stationary perturbation calculus) approximation. Similarly, the linear and quadratic magnetic variation in the ${}^{(a)}_{A}\boldsymbol{\chi}^{(b,c,d)}_{BCD}(\omega)$ is given by ${}^{(a)}_{A}\boldsymbol{\chi}^{(b,c,d)m}_{BCD}(\omega)$ and ${}^{(a)}_{A}\boldsymbol{\chi}^{(b,c,d)mm}_{BCD}(\omega)$, respectively.

In the case of crystals, where spatial dispersion is not excessively great [65], the linear ${}_{A}\boldsymbol{\chi}_{B}(\omega, \mathbf{k}, \mathbf{B}^0)$ and nonlinear ${}_{A}\boldsymbol{\chi}_{BCD}(\omega, \mathbf{k}, \mathbf{B}^0)$ susceptibilities can be written in expansion form [48, 63, 64, 66] as follows (we apply the Einstein summation convention):

$$
\begin{aligned}
{}_{e}\chi_{e\,ij}(\omega, k, \mathbf{B}^0) = {}^{(1)}_{e}\chi^{(1)}_{e\,ij}(\omega, \mathbf{B}^0) + \mathrm{i}\frac{n\omega}{3c}\Big[{}^{(1)}_{e}\chi^{(2)}_{e\,i(jp)}(\omega, \mathbf{B}^0) \\
- {}^{(2)}_{e}\chi^{(1)}_{e\,(ip)j}(\omega, \mathbf{B}^0)\Big] s_p + \cdots
\end{aligned} \tag{16}
$$

$$
{}_{e}\chi_{m\,ij}(\omega, k, \mathbf{B}^0) = {}^{(1)}_{e}\chi^{(1)}_{m\,ij}(\omega, \mathbf{B}^0) + \cdots \tag{17}
$$

$$
{}_{m}\chi_{e\,ij}(\omega, k, \mathbf{B}^0) = {}^{(1)}_{m}\chi^{(1)}_{e\,ij}(\omega, \mathbf{B}^0) + \cdots \tag{18}
$$

$$
\begin{aligned}
{}_{e}\chi_{eee(ij)(kl)}(\omega, k, \mathbf{B}^0) = {}^{(1)}_{e}\chi^{(1,1,1)}_{eee(ij)(kl)}(\omega, \mathbf{B}^0) + \mathrm{i}\frac{n\omega}{3c}\Big[{}^{(1)}_{e}\chi^{(1,1,2)}_{eee(ij)k(lp)}(\omega, \mathbf{B}^0) \\
+ {}^{(1)}_{e}\chi^{(1,2,1)}_{eee\,(ij)(kp)l}(\omega, \mathbf{B}^0) - {}^{(1)}_{e}\chi^{(2,1,1)}_{eee\,i(jp)(kl)}(\omega, \mathbf{B}^0) \\
- {}^{(2)}_{e}\chi^{(1,1,1)}_{eee\,(ip)j(kl)}(\omega, \mathbf{B}^0)\Big] s_p + \cdots
\end{aligned} \tag{19}
$$

$$
{}_{e}\chi_{eem\,(ij)kl}(\omega, k, \mathbf{B}^0) = {}^{(1)}_{e}\chi^{(1,1,1)}_{eem\,(ij)kl}(\omega, \mathbf{B}^0) + \cdots \tag{20}
$$

$$
{}_{e}\chi_{eme\,(ij)kl}(\omega, k, \mathbf{B}^0) = {}^{(1)}_{e}\chi^{(1,1,1)}_{eme\,(ij)kl}(\omega, \mathbf{B}^0) + \cdots \tag{21}
$$

$$
{}_{e}\chi_{mee\,ij(kl)}(\omega, k, \mathbf{B}^0) = {}^{(1)}_{e}\chi^{(1,1,1)}_{mee\,ij(kl)}(\omega, \mathbf{B}^0) + \cdots \tag{22}
$$

$$
{}_{m}\chi_{eee\,ij(kl)}(\omega, k, \mathbf{B}^0) = {}^{(1)}_{m}\chi^{(1,1,1)}_{eee\,ij(kl)}(\omega, \mathbf{B}^0) + \cdots \tag{23}
$$

with ${}_{m}\chi_{m}(\omega, k, \mathbf{B}^0)$ and ${}_{m}\chi_{mee}(\omega, k, \mathbf{B}^0)$ equal to zero in the optical region [75].

The tensor component indices i, j, k, and l refer to laboratory coordinates and take the values x, y, and z. In the above formulas the subscripts in parentheses $(\ldots)$ label the components of the electric quadrupole moment; the parentheses also denote the invariancy (symmetry) of the respective components with respect to transposition of the subscripts.

In accordance with the quantum theory of radiation [140], expressions (16)–(23) have been broken off at terms giving contributions of the same order of magnitude to the linear $\mathbf{P}_A^L(\omega, \mathbf{k})$ and nonlinear $\mathbf{P}_A^{NL}(\omega, \mathbf{k})$ electric $A = e$ and magnetic $A = m$ polarization induced in the crystal. More precisely, we have restricted ourselves in Eq. (7) to terms with $a + b = 3$ for $A = B = e$ and $a + b = 2$ for $A = e$, $B = m$ whereas, in (8), $a + b + c + d = 5$ for $A = B = C = D = e$ and $a + b + c + d = 4$ for $A = B = C = e$, $D = m$. The quantum theory of radiation [140] states that, for optical transitions, the transition matrix element $\langle k'|\mathbf{M}_e^{(a)}|l'\rangle$ of the electric multipole moment of order a is of the same order of magnitude as the matrix element $\langle k'|\mathbf{M}_m^{(a-1)}|l'\rangle$ of the magnetic multipole moment of order $a - 1$ and each of them is six orders of magnitude smaller than the matrix element $\langle k'|M_e^{(a-1)}|l'\rangle$ of the electric multipole moment of order $a - 1$. Hence, we are justified in putting ${}^{(1)}_{m}\chi_m^{(1)}(\omega, \mathbf{B}^0) = 0$ and ${}^{(1)}_{m}\chi_{mee}^{(1,1,1)}(\omega, \mathbf{B}^0) = 0$ on neglecting components ${}^{(a)}_{e}\chi_e^{(b)}(\omega, \mathbf{B}^0)$ with $a + b \geq 4$ in Eq. (7) and ${}^{(a)}_{e}\chi_{eee}^{(b,c,d)}(\omega, \mathbf{B}^0)$ with $a + b + c + d \geq 6$ in Eq. 8. Moreover, Landau and Lifshitz [75] have shown that in the absence of a magnetic field the linear magneto-magnetic susceptibility ${}^{(1)}_{m}\chi_m^{(1)}(\omega)$ tends to zero with increasing frequency much more steeply than the linear electro-electric susceptibility ${}^{(1)}_{e}\chi_e^{(1)}(\omega)$. Their calculations prove that, throughout the entire range of optical frequencies, one may put ${}^{(1)}_{m}\chi_m^{(1)}(\omega) = 0$.

B. Permutational Symmetry for Linear and Nonlinear Multipolar Electric and Magnetic Susceptibilities

With regard to the quantum-mechanical definition of linear and nonlinear multipolar susceptibilities [48, 63, 64], it can be shown that the ${}^{(a)}_{A}\chi_B^{(b)}(\omega)$ and ${}^{(a)}_{A}\chi_{BCD}^{(b,c,d)}(\omega)$ as well as their magnetically induced variations fulfill the following relations [70, 72, 138, 141]:

$$
\begin{aligned}
{}^{(a)}_{A}\chi_B^{(b)}(\omega)^* &= {}^{(a)}_{A}\chi_B^{(b)}(-\omega) & \quad {}^{(a)}_{A}\chi_{BCD}^{(b,c,d)}(\omega)^* &= {}^{(a)}_{A}\chi_{BCD}^{(b,c,d)}(-\omega) \\
{}^{(a)}_{A}\chi_B^{(b)m}(\omega)^* &= {}^{(a)}_{A}\chi_B^{(b)m}(-\omega) & \quad {}^{(a)}_{A}\chi_{BCD}^{(b,c,d)m}(\omega)^* &= {}^{(a)}_{A}\chi_{BCD}^{(b,c,d)m}(-\omega) \\
{}^{(a)}_{A}\chi_B^{(b)mm}(\omega)^* &= {}^{(a)}_{A}\chi_B^{(b)mm}(-\omega) & \quad {}^{(a)}_{A}\chi_{BCD}^{(b,c,d)mm}(\omega)^* &= {}^{(a)}_{A}\chi_{BCD}^{(b,c,d)mm}(-\omega)
\end{aligned}
\tag{24}
$$

and invariance of nonlinear multipole susceptibilities with respect to pairwise permutation of aA and bB, and of cC and dD [136–138, 142]:

$$
\begin{aligned}
{}^{(a)}_{A}\boldsymbol{\chi}^{(b,c,d)}_{BCD}(\omega) &= {}^{(a)}_{A}\boldsymbol{\chi}^{(b,d,c)}_{BDC}(\omega) = {}^{(b)}_{B}\boldsymbol{\chi}^{(a,c,d)}_{ACD}(\omega) = {}^{(b)}_{B}\boldsymbol{\chi}^{(a,d,c)}_{ADC}(\omega) \\
{}^{(a)}_{A}\boldsymbol{\chi}^{(b,c,d)m}_{BCD}(\omega) &= {}^{(a)}_{A}\boldsymbol{\chi}^{(b,d,c)m}_{BDC}(\omega) = {}^{(b)}_{B}\boldsymbol{\chi}^{(a,c,d)m}_{ACD}(\omega) = {}^{(b)}_{B}\boldsymbol{\chi}^{(a,d,c)m}_{ADC}(\omega) \\
{}^{(a)}_{A}\boldsymbol{\chi}^{(b,c,d)mm}_{BCD}(\omega) &= {}^{(a)}_{A}\boldsymbol{\chi}^{(b,d,c)mm}_{BDC}(\omega) = {}^{(b)}_{B}\boldsymbol{\chi}^{(a,c,d)mm}_{ACD}(\omega) = {}^{(b)}_{B}\boldsymbol{\chi}^{(a,d,c)mm}_{ADC}(\omega)
\end{aligned}
\tag{25}
$$

and additionally

$$
\begin{aligned}
{}^{(a)}_{A}\boldsymbol{\chi}^{(b)}_{B}(\omega)^* &= {}^{(b)}_{B}\boldsymbol{\chi}^{(a)}_{A}(\omega) & {}^{(a)}_{A}\boldsymbol{\chi}^{(b,c,d)}_{BCD}(\omega)^* &= {}^{(c)}_{C}\boldsymbol{\chi}^{(d,a,b)}_{DAB}(\omega) \\
{}^{(a)}_{A}\boldsymbol{\chi}^{(b)m}_{B}(\omega)^* &= {}^{(b)}_{B}\boldsymbol{\chi}^{(a)m}_{A}(\omega) & {}^{(a)}_{A}\boldsymbol{\chi}^{(b,c,d)m}_{BCD}(\omega)^* &= {}^{(c)}_{C}\boldsymbol{\chi}^{(d,a,b)m}_{DAB}(\omega) \\
{}^{(a)}_{A}\boldsymbol{\chi}^{(b)mm}_{B}(\omega)^* &= {}^{(b)}_{B}\boldsymbol{\chi}^{(a)mm}_{A}(\omega) & {}^{(a)}_{A}\boldsymbol{\chi}^{(b,c,d)mm}_{BCD}(\omega)^* &= {}^{(c)}_{C}\boldsymbol{\chi}^{(d,a,b)mm}_{DAB}(\omega)
\end{aligned}
\tag{26}
$$

if the widths of the energy levels are neglected. It is easy to show that the last relation involves the vanishing of the time-averaged divergence of the Poynting vector meaning transparency of the medium in the optical region.

The realness of the electric field $\mathbf{E}(\mathbf{r}, t)$ of the optical wave and the electric $A = e$ and magnetic $A = m$ polarization $\mathbf{P}_A(\mathbf{r}, t)$ requires that their spectral amplitudes obey the condition [143] $\mathbf{E}(\omega, \mathbf{k})^* = \mathbf{E}(-\omega, -\mathbf{k})$ and $\mathbf{P}_A(\omega, \mathbf{k})^* = \mathbf{P}_A(-\omega, -\mathbf{k})$. Relation (26) ensures that the linear and nonlinear electric $A = e$ and magnetic $A = m$ polarizations $\mathbf{P}_A(\mathbf{r}, t)$ are real for real electric and magnetic fields.

When deriving the last three relations, all parameters except the wave functions were assumed real. With regard to Eqs. (12) and (13), these assumptions still hold in the presence of a magnetic field: For a nonmagnetic crystal, the wave functions $|k'\rangle$ and $|l'\rangle$ of stationary states in the absence of a field are real, but are complex for a magnetic crystal. For materials with complex wave functions, the linear and nonlinear multipole susceptibilities as well as their magnetically induced variations are complex so that each can be expressed in the form

$$
\begin{aligned}
{}^{(a)}_{A}\boldsymbol{\chi}^{(b)}_{B}(\omega) &= {}^{(a)}_{A}\boldsymbol{\alpha}^{(b)}_{B}(\omega) + \mathrm{i}\,{}^{(a)}_{A}\boldsymbol{\gamma}^{(b)}_{B}(\omega) \\
{}^{(a)}_{A}\boldsymbol{\chi}^{(b,c,d)}_{BCD}(\omega) &= {}^{(a)}_{A}\boldsymbol{\alpha}^{(b,c,d)}_{BCD}(\omega) + \mathrm{i}\,{}^{(a)}_{A}\boldsymbol{\gamma}^{(b,c,d)}_{BCD}(\omega)
\end{aligned}
\tag{27}
$$

and similarly for ${}^{(a)}_{A}\boldsymbol{\chi}^{(b)m}_{B}(\omega)$, ${}^{(a)}_{A}\boldsymbol{\chi}^{(b)mm}_{B}(\omega)$, ${}^{(a)}_{A}\boldsymbol{\chi}^{(b,c,d)m}_{BCD}(\omega)$, and ${}^{(a)}_{A}\boldsymbol{\chi}^{(b,c,d)mm}_{BCD}(\omega)$. Equations (26) and (27) lead to the following transposition relations:

$$
\begin{aligned}
{}^{(\alpha)}_{A}\boldsymbol{\alpha}^{(b)}_{B}(\omega) &= {}^{(b)}_{B}\boldsymbol{\alpha}^{(a)}_{A}(\omega) & \qquad {}^{(a)}_{A}\boldsymbol{\gamma}^{(b)}_{B}(\omega) &= -{}^{(b)}_{B}\boldsymbol{\gamma}^{(\alpha)}_{A}(\omega)\\
{}^{(\alpha)}_{A}\boldsymbol{\alpha}^{(b)m}_{B}(\omega) &= {}^{(b)}_{B}\boldsymbol{\alpha}^{(a)m}_{A}(\omega) & {}^{(a)}_{A}\boldsymbol{\gamma}^{(b)m}_{B}(\omega) &= -{}^{(b)}_{B}\boldsymbol{\gamma}^{(a)m}_{A}(\omega)\\
{}^{(a)}_{A}\boldsymbol{\alpha}^{(b)mm}_{B}(\omega) &= {}^{(b)}_{B}\boldsymbol{\alpha}^{(a)mm}_{A}(\omega) & {}^{(a)}_{A}\boldsymbol{\gamma}^{(b)mm}_{B}(\omega) &= -{}^{(b)}_{B}\boldsymbol{\gamma}^{(a)mm}_{A}(\omega)\\
{}^{(a)}_{A}\boldsymbol{\alpha}^{(b,c,d)}_{BCD}(\omega) &= {}^{(c)}_{C}\boldsymbol{\alpha}^{(d,a,b)}_{DAB}(\omega) & {}^{(a)}_{A}\boldsymbol{\gamma}^{(b,c,d)}_{BCD}(\omega) &= -{}^{(c)}_{C}\boldsymbol{\gamma}^{(d,a,b)}_{DAB}(\omega)\\
{}^{(a)}_{A}\boldsymbol{\alpha}^{(b,c,d)m}_{BCD}(\omega) &= {}^{(c)}_{C}\boldsymbol{\alpha}^{(d,a,b)m}_{DAB}(\omega) & {}^{(a)}_{A}\boldsymbol{\gamma}^{(b,c,d)m}_{BCD}(\omega) &= -{}^{(c)}_{C}\boldsymbol{\gamma}^{(d,a,b)m}_{DAB}(\omega)\\
{}^{(a)}_{A}\boldsymbol{\alpha}^{(b,c,d)mm}_{BCD}(\omega) &= {}^{(c)}_{C}\boldsymbol{\alpha}^{(d,a,b)mm}_{DAB}(\omega) & {}^{(a)}_{A}\boldsymbol{\gamma}^{(b,c,d)mm}_{BCD}(\omega) &= -{}^{(c)}_{C}\boldsymbol{\gamma}^{(d,a,b)mm}_{DAB}(\omega)
\end{aligned}
\tag{28}
$$

C. Time-Reversal Symmetry for Linear and Nonlinear Multipolar Electric and Magnetic Susceptibilities

Making use of the transposition relations we can express the electric polarization vector $\mathbf{P}_e(\mathbf{r}, t)$ and magnetic polarization vector $\mathbf{P}_m(\mathbf{r}, t)$ in a form involving the electric and magnetic field strength $\mathbf{E}(\mathbf{r}, t)$ and $\mathbf{H}(\mathbf{r}, t)$ as well as their time derivatives $\dot{\mathbf{E}}(\mathbf{r}, t)$ and $\dot{\mathbf{H}}(\mathbf{r}, t)$. With the respective expressions, and keeping in mind that $\mathbf{E}(\mathbf{r}, t)$, $\dot{\mathbf{H}}(\mathbf{r}, t)$, and $\mathbf{P}_e(\mathbf{r}, t)$ are invariant with respect to time inversion, whereas $\dot{\mathbf{E}}(\mathbf{r}, t)$, $\mathbf{H}(\mathbf{r}, t)$, $\mathbf{B}(\mathbf{r}, t)$, $\mathbf{P}_m(\mathbf{r}, t)$, $\mathbf{B}^0$, and $\mathbf{k}$ undergo a change in sign if $t \to -t$, we are immediately in a position to determine how the linear and nonlinear multipole susceptibilities transform on time inversion. In this way we find that the components of $\boldsymbol{\alpha}$ with an even number of lower and upper indices m (subscripts and superscripts jointly) and components of γ with an odd number of indices m;

$$
\begin{array}{llll}
{}^{(a)}_{e}\boldsymbol{\alpha}^{(b)}_{e}(\omega) & {}^{(a)}_{e}\boldsymbol{\gamma}^{(b)}_{m}(\omega) & {}^{(a)}_{e}\boldsymbol{\alpha}^{(b,c,d)}_{eee}(\omega) & {}^{(a)}_{e}\boldsymbol{\gamma}^{(b,c,d)}_{eem}(\omega)\\
{}^{(a)}_{e}\boldsymbol{\gamma}^{(b)m}_{e}(\omega) & {}^{(a)}_{e}\boldsymbol{\alpha}^{(b)m}_{m}(\omega) & {}^{(a)}_{e}\boldsymbol{\gamma}^{(b,c,d)m}_{eee}(\omega) & {}^{(\alpha)}_{e}\boldsymbol{\alpha}^{(b,c,d)m}_{eem}(\omega)\\
{}^{(a)}_{e}\boldsymbol{\alpha}^{(b)mm}_{e}(\omega) & {}^{(a)}_{e}\boldsymbol{\gamma}^{(b)mm}_{m}(\omega) & {}^{(a)}_{e}\boldsymbol{\alpha}^{(b,c,d)mm}_{eee}(\omega) & {}^{(a)}_{e}\gamma^{(b,c,d)mm}_{eem}(\omega)
\end{array}
\tag{29}
$$

are invariant with respect to time inversion. After Birss [144], we refer to them as i-tensors. The others ($\boldsymbol{\alpha}$ with an odd number of indices m and γ

with an even number of indices m);

$$
\begin{array}{llll}
{}^{(a)}_{e}\boldsymbol{\gamma}^{(b)}_{e}(\omega) & {}^{(a)}_{e}\boldsymbol{\alpha}^{(b)}_{m}(\omega) & {}^{(a)}_{e}\boldsymbol{\gamma}^{(b,c,d)}_{eee}(\omega) & {}^{(a)}_{e}\boldsymbol{\alpha}^{(b,c,d)}_{eem}(\omega) \\
{}^{(a)}_{e}\boldsymbol{\alpha}^{(b)m}_{e}(\omega) & {}^{(a)}_{e}\boldsymbol{\gamma}^{(b)m}_{m}(\omega) & {}^{(a)}_{e}\boldsymbol{\alpha}^{(b,c,d)m}_{eee}(\omega) & {}^{(a)}_{e}\boldsymbol{\gamma}^{(b,c,d)m}_{eem}(\omega) \\
{}^{(a)}_{e}\boldsymbol{\gamma}^{(b)mm}_{e}(\omega) & {}^{(a)}_{e}\boldsymbol{\alpha}^{(b)mm}_{m}(\omega) & {}^{(a)}_{e}\boldsymbol{\gamma}^{(b,c,d)mm}_{eee}(\omega) & {}^{(a)}_{e}\boldsymbol{\alpha}^{(b,c,d)mm}_{eem}(\omega)
\end{array}
\tag{30}
$$

undergo a change in sign (c-tensors) for arbitrary a, b, c, and d.

D. Neumann's Principle as a Selection Rule for Magneto-Optics

It follows from Neumann's principle [144, 145] that i-tensors can exist in both magnetic and nonmagnetic crystals, whereas c-tensors can exist only in magnetic crystals. Hence, expressions (16)–(23) together with Eqs. (14) and (15) determining the linear and nonlinear electro-electric and, respectively, electro-magnetic susceptibilities of nonmagnetic crystals will involve only i-tensors. For this case these susceptibilities will fulfill the relations

$$
\begin{aligned}
{}_{A}\chi_{Bij}(\omega, \mathbf{k}, \mathbf{B}^0) &= w_A w_{BB} \chi_{Aji}(-\omega, -\mathbf{k}, -\mathbf{B}^0) \\
{}_{A}\chi_{BCDijkl}(\omega, \mathbf{k}, \mathbf{B}^0) &= w_A w_B w_C w_{DC} \chi_{DABklij}(-\omega, -\mathbf{k}, -\mathbf{B}^0)
\end{aligned}
\tag{31}
$$

where $w_A = 1$ for $A = e$ and $w_A = -1$ for $A = m$ (likewise w_B, w_C, and w_D) in complete agreement with the result of Onsager's symmetry principle for kinetic coefficients [75], which is considered in the literature to be the selection rule for the existence of linear and nonlinear magneto-optical effects in crystals.

Kleiner [146] has shown that relations (31) are valid only for nonmagnetic crystals. For magnetic crystals whose directional symmetry is described by 32 magnetic symmetry classes $G(P)$ no Onsager principle can be enounced, whereas for the remaining 58 magnetic symmetry classes $G(M)$ a generalized Onsager principle can be formulated. From what has been said we draw the conclusion that with regard to magnetic crystals one has to drop the Onsager principle as a selection rule for the existence of linear and nonlinear magneto-optical effects. Here, the Neumann principle is a more adequate selection rule. It is applicable to nonmagnetic as well as magnetic crystals provided that the transformation properties of the tensor under time inversion are known.

Expansions (16)–(23) contain the linear ${}^{(a)}_{A}\boldsymbol{\chi}^{(b)}_{B}(\omega, \mathbf{B}^0)$ and nonlinear ${}^{(a)}_{A}\boldsymbol{\chi}^{(b,c,d)}_{BCD}(\omega, \mathbf{B}^0)$ multipolar susceptibilities, which with regard to the Eqs.

(14), (15), and (27) are complex, so that each can be written in the form

$$
\begin{aligned}
{}^{(a)}_{A}\boldsymbol{\chi}^{(b)}_{B}(\omega, \mathbf{B}^0) &= {}^{(a)}_{A}\boldsymbol{\alpha}^{(b)}_{B}(\omega, \mathbf{B}^0) + \mathrm{i}\,{}^{(a)}_{A}\boldsymbol{\gamma}^{(b)}_{B}(\omega, \mathbf{B}^0) \\
{}^{(a)}_{A}\boldsymbol{\chi}^{(b,c,d)}_{BCD}(\omega, \mathbf{B}^0) &= {}^{(a)}_{A}\boldsymbol{\alpha}^{(b,c,d)}_{BCD}(\omega, \mathbf{B}^0) + \mathrm{i}\,{}^{(a)}_{A}\boldsymbol{\gamma}^{(b,c,d)}_{BCD}(\omega, \mathbf{B}^0)
\end{aligned}
\tag{32}
$$

where

$$
\begin{aligned}
{}^{(a)}_{A}\boldsymbol{\alpha}^{(b)}_{B}(\omega, \mathbf{B}^0) = {}^{(a)}_{A}\boldsymbol{\alpha}^{(b)}_{B}(\omega) &+ {}^{(a)}_{A}\boldsymbol{\alpha}^{(b)m}_{B}(\omega) \cdot \mathbf{B}^0 \\
&+ {}^{(a)}_{A}\boldsymbol{\alpha}^{(b)mm}_{B}(\omega) \cdot\cdot\, \mathbf{B}^0\mathbf{B}^0 + \cdots
\end{aligned}
\tag{33}
$$

$$
\begin{aligned}
{}^{(a)}_{A}\boldsymbol{\gamma}^{(b)}_{B}(\omega, \mathbf{B}^0) = {}^{(a)}_{A}\boldsymbol{\gamma}^{(b)}_{B}(\omega) &+ {}^{(a)}_{A}\boldsymbol{\gamma}^{(b)m}_{B}(\omega) \cdot \mathbf{B}^0 \\
&+ {}^{(a)}_{A}\boldsymbol{\gamma}^{(b)mm}_{B}(\omega) \cdot\cdot\, \mathbf{B}^0\mathbf{B}^0 + \cdots
\end{aligned}
\tag{34}
$$

$$
\begin{aligned}
{}^{(a)}_{A}\boldsymbol{\alpha}^{(b,c,d)}_{BCD}(\omega, \mathbf{B}^0) = {}^{(a)}_{A}\boldsymbol{\alpha}^{(b,c,d)}_{BCD}(\omega) &+ {}^{(a)}_{A}\boldsymbol{\alpha}^{(b,c,d)m}_{BCD}(\omega) \\
\cdot\, \mathbf{B}^0 &+ {}^{(a)}_{A}\boldsymbol{\alpha}^{(b,c,d)mm}_{BCD}(\omega) \cdot\cdot\, \mathbf{B}^0\mathbf{B}^0 + \cdots
\end{aligned}
\tag{35}
$$

$$
\begin{aligned}
{}^{(a)}_{A}\gamma^{(b,c,d)}_{BCD}(\omega, \mathbf{B}^0) = {}^{(a)}_{A}\gamma^{(b,c,d)}_{BCD}(\omega) &+ {}^{(a)}_{A}\gamma^{(b,c,d)m}_{BCD}(\omega) \cdot \mathbf{B}^0 \\
&+ {}^{(a)}_{A}\boldsymbol{\gamma}^{(b,c,d)mm}_{BCD}(\omega) \cdot\cdot\, \mathbf{B}^0\mathbf{B}^0 + \cdots
\end{aligned}
\tag{36}
$$

With regard to Neumann's principle, in the case of nonmagnetic crystals the above expressions can contain only i-tensors, whereas for magnetic crystals they can contain c-tensors too. The latter lead to new magneto-optical effects, forbidden by Onsager's principle with regard to nonmagnetic crystals (Eqs. (31)), one of which is the magnetic analog of Pockels' effect [85, 89–91].

III. STOKES PARAMETERS

Let the light wave propagate in the crystal along the z axis taken as parallel to the highest of its axes of symmetry. We accordingly have $s_x = s_y = 0$, $s_z = 1$, and $E_z(\omega, \mathbf{k}) = 0$, meaning that the optical field is transversal. For magnetic insulators, on replacing $\mathbf{D}(\mathbf{r}, t)$ and $\mathbf{B}(\mathbf{r}, t)$ in the Maxwell equations [75],

$$
\operatorname{curl}\mathbf{E}(\mathbf{r}, t) = -\frac{\partial \mathbf{B}(\mathbf{r}, t)}{\partial t} \qquad \operatorname{curl}\mathbf{H}(\mathbf{r}, t) = \frac{\partial \mathbf{D}(\mathbf{r}, t)}{\partial t}
\tag{37}
$$

by expression (1), and taking into account Eq. (2), we have the following equations:

$$
\begin{aligned}
\frac{\partial E_x(\omega, k_z)}{\partial z} &= \mathrm{i}\frac{\omega}{2nc}\Bigg\{(1-n^2)E_x(\omega, k_z) \\
&\qquad + \frac{1}{\varepsilon_0}\left[P_{e\,x}(\omega, k_z) + \frac{n}{c}P_{m\,y}(\omega, k_z)\right]\Bigg\} \\
\frac{\partial E_y(\omega, k_z)}{\partial z} &= \mathrm{i}\frac{\omega}{2nc}\Bigg\{(1-n^2)E_y(\omega, k_z) \\
&\qquad + \frac{1}{\varepsilon_0}\left[P_{e\,y}(\omega, k_z) - \frac{n}{c}P_{m\,y}(\omega, k_z)\right]\Bigg\}
\end{aligned}
\tag{38}
$$

When deriving the last two equations we made the standard assumption [143] that

$$
\frac{\partial E_x(\omega, k_z)}{\partial x} = \frac{\partial E_x(\omega, k_z)}{\partial y} = \frac{\partial E_y(\omega, k_z)}{\partial x} = \frac{\partial E_y(\omega, k_z)}{\partial y} = 0 \tag{39}
$$

meaning that the intensity of the electric field of the light wave is constant perpendicularly to the propagation direction of the wave. Our aim is, essentially, to determine the state of polarization of light on its traversal of a path z in the crystal. The state of polarization is completely determined by the Stokes parameters which, in Cartesian coordinates, take the well-known form [2]

$$
\begin{aligned}
S_0(z) &= |E_x(\omega, k_z)|^2 + |E_y(\omega, k_z)|^2 \\
S_1(z) &= |E_x(\omega, k_z)|^2 - |E_y(\omega, k_z)|^2 \\
S_2(z) &= E_x(\omega, k_z)E_y^*(\omega, k_z) + \text{c.c.} \\
S_3(z) &= \mathrm{i}[E_x(\omega, k_z)E_y^*(\omega, k_z) - \text{c.c.}]
\end{aligned}
\tag{40}
$$

In practice, the state of polarization is determined by two quantities: the ellipticity Φ and the azimuth Ψ [2]. Φ is the ratio of the minor axis and major axis of the polarization ellipse, whereas Ψ is the angle between the major axis of the polarization ellipse and the x axis. The two quantities are related to the Stokes parameters as follows [2]:

$$\Psi = \frac{1}{2}\arctan\frac{S_2}{S_1} \quad \text{and} \quad \Phi = \frac{1}{2}\arcsin\frac{S_3}{S_0} \tag{41}$$

Rather than solve the set of equations (38) and (39) to determine $E_x(\omega, k_z)$ and $E_y(\omega, k_z)$ and then calculate the Stokes parameters, we make use of Eqs. (38)–(40) to determine the equations for S_0, S_1, S_2, and S_3, which, on taking into account expressions (3)–(5) together with (14)–(23) for the components x and y of electric and magnetic polarization, take the form

$$\frac{\partial S_0}{\partial z} = 0 \tag{42}$$

$$\begin{aligned}\frac{\partial S_1}{\partial z} = \frac{\omega}{2n}\sqrt{\frac{\mu_0}{\varepsilon_0}}\,\Big[&(\chi_A^L + \chi_G^{NL}S_0)S_2 + (\kappa_A^L - \varepsilon^{NL}S_0)S_3 \\ &+ (\chi_R^{NL} - \kappa_R^{NL} + \Theta^{NL})S_2S_3 \\ &+ (\Lambda^{NL}S_3 + \Sigma^{NL}S_2)S_1 - \Omega_A^{NL}(S_2^2 - S_3^2)\Big]\end{aligned} \tag{43}$$

$$\begin{aligned}\frac{\partial S_2}{\partial z} = \frac{\omega}{2n}\sqrt{\frac{\mu_0}{\varepsilon_0}}\,\Big[&-(\chi_A^L + \chi_G^{NL}S_0)S_1 \\ &+ (\kappa_R^L + \Omega_R^{NL}S_0)S_3 \\ &- (\chi_R^{NL} - \kappa_R^{NL} - \Theta^{NL})S_1S_3 \\ &- (\Lambda^{NL}S_3 - \Omega_A^{NL}S_2 - \Sigma^{NL}(S_1^2 - S_3^2)\Big]\end{aligned} \tag{44}$$

$$\begin{aligned}\frac{\partial S_3}{\partial z} = \frac{\omega}{2n}\sqrt{\frac{\mu_0}{\varepsilon_0}}\,\Big[&-(\kappa_A^L - \varepsilon^{NL}S_0)S_1 \\ &- (\kappa_R^L + \Omega_R^{NL}S_0)S_2 - 2\Theta^{NL}S_1S_2 \\ &- (\Omega_A^{NL}S_1 + \Sigma^{NL}S_2)S_3 - \Lambda^{NL}(S_1^2 - S_2^2)\Big]\end{aligned} \tag{45}$$

Above, we introduced the following notation:

$$\kappa_R^L = {}^{(1)}_{e}\alpha^{(1)}_{e\,xx}(\omega, \mathbf{B}^0) - {}^{(1)}_{e}\alpha^{(1)}_{e\,yy}(\omega, \mathbf{B}^0) + \frac{2n}{c}\left\{\frac{\omega}{3}\left[{}^{(1)}_{e}\gamma^{(2)}_{e\,x(xz)}(\omega, \mathbf{B}^0) - {}^{(1)}_{e}\gamma^{(2)}_{ey(yz)}(\omega, \mathbf{B}^0)\right] + {}^{(1)}_{e}\alpha^{(1)}_{m\,xy}(\omega, \mathbf{B}^0) + {}^{(1)}_{e}\alpha^{(1)}_{m\,yx}(\omega, \mathbf{B}^0)\right\} s_z + \cdots \tag{46}$$

$$\chi_A^L = 2\,{}^{(1)}_{e}\gamma^{(1)}_{e[xy]}(\omega, \mathbf{B}^0) + \frac{2n}{c}\left\{-\frac{\omega}{3}\left[{}^{(1)}_{e}\alpha^{(2)}_{e\,x(yz)}(\omega, \mathbf{B}^0) - {}^{(1)}_{e}\alpha^{(2)}_{e\,y(xz)}(\omega, \mathbf{B}^0)\right] + {}^{(1)}_{e}\gamma^{(1)}_{m\,xx}(\omega, \mathbf{B}^0) + {}^{(1)}_{e}\gamma^{(1)}_{m\,yy}(\omega, \mathbf{B}^0)\right\} s_z + \cdots \tag{47}$$

$$\kappa_A^L = -2\,{}^{(1)}_{e}\alpha^{(1)}_{e(xy)}(\omega, \mathbf{B}^0) + \frac{2n}{c}\left\{\frac{\omega}{3}\left[{}^{(1)}_{e}\gamma^{(2)}_{e\,x(yz)}(\omega, \mathbf{B}^0) + {}^{(1)}_{e}\gamma^{(2)}_{e\,y(xz)}(\omega, \mathbf{B}^0)\right] + {}^{(1)}_{e}\alpha^{(1)}_{m\,xx}(\omega, \mathbf{B}^0) - {}^{(1)}_{e}\alpha^{(1)}_{m\,yy}(\omega, \mathbf{B}^0)\right\} s_z + \cdots \tag{48}$$

$$\begin{aligned}\chi_R^{NL} = {}& \tfrac{3}{4}\Big[{}^{(1)}_{e}\alpha^{(1,1,1)}_{eee\,xxxx}(\omega, \mathbf{B}^0) + {}^{(1)}_{e}\alpha^{(1,1,1)}_{eee\,yyyy}(\omega, \mathbf{B}^0) - 2\,{}^{(1)}_{e}\alpha^{(1,1,1)}_{eee\,xxyy}(\omega, \mathbf{B}^0) \\ & + 4\,{}^{(1)}_{e}\alpha^{(1,1,1)}_{eee\,xyxy}(\omega, \mathbf{B}^0)\Big] + \frac{ns_z}{c}\Big\{\omega\Big[-{}^{(1)}_{e}\gamma^{(1,1,2)}_{eee\,xxx(xz)}(\omega, \mathbf{B}^0) \\ & + {}^{(1)}_{e}\gamma^{(1,1,2)}_{eee\,xxy(yz)}(\omega, \mathbf{B}^0) - 2\,{}^{(1)}_{e}\gamma^{(1,1,2)}_{eee\,xyx(yz)}(\omega, \mathbf{B}^0) \\ & - {}^{(1)}_{e}\gamma^{(1,1,2)}_{eee\,yyy(yz)}(\omega, \mathbf{B}^0) + {}^{(1)}_{e}\gamma^{(1,1,2)}_{eee\,yyx(xz)}(\omega, \mathbf{B}^0) - 2\,{}^{(1)}_{e}\gamma^{(1,1,2)}_{eee\,yxy(xz)}(\omega, \mathbf{B}^0)\Big] \\ & + 3\Big[{}^{(1)}_{e}\alpha^{(1,1,1)}_{eem\,xxxy}(\omega, \mathbf{B}^0) + {}^{(1)}_{e}\alpha^{(1,1,1)}_{eem\,xxyx}(\omega, \mathbf{B}^0) - 2\,{}^{(1)}_{e}\alpha^{(1,1,1)}_{eem\,yxxx}(\omega, \mathbf{B}^0) \\ & - {}^{(1)}_{e}\alpha^{(1,1,1)}_{eem\,yyyx}(\omega, \mathbf{B}^0) - {}^{(1)}_{e}\alpha^{(1,1,1)}_{eem\,yyxy}(\omega, \mathbf{B}^0) + 2\,{}^{(1)}_{e}\alpha^{(1,1,1)}_{eem\,xyyy}(\omega, \mathbf{B}^0)\Big]\Big\} \cdots\end{aligned} \tag{49}$$

$$\kappa_R^{NL} = \tfrac{3}{2}\Big[{}^{(1)}_{e}\alpha^{(1,1,1)}_{eee\,xxxx}(\omega,\mathbf{B}^0) + {}^{(1)}_{e}\alpha^{(1,1,1)}_{eee\,yyyy}(\omega,\mathbf{B}^0) + 2\,{}^{(1)}_{e}\alpha^{(1,1,1)}_{eee\,xxyy}(\omega,\mathbf{B}^0)\Big]$$

$$+ \frac{2ns_z}{c}\Big\{-\omega\Big[{}^{(1)}_{e}\gamma^{(1,1,2)}_{eee\,xxx(xz)}(\omega,\mathbf{B}^0) + {}^{(1)}_{e}\gamma^{(1,1,2)}_{eee\,xxy(yz)}(\omega,\mathbf{B}^0)$$

$$+ {}^{(1)}_{e}\gamma^{(1,1,2)}_{eee\,yyy(yz)}(\omega,\mathbf{B}^0) + {}^{(1)}_{e}\gamma^{(1,1,2)}_{eee\,yyx(xz)}(\omega,\mathbf{B}^0)\Big] \tag{50}$$

$$+3\Big[{}^{(1)}_{e}\alpha^{(1,1,1)}_{eem\,xxxy}(\omega,\mathbf{B}^0) - {}^{(1)}_{e}\alpha^{(1,1,1)}_{eem\,xxyx}(\omega,\mathbf{B}^0)$$

$$- {}^{(1)}_{e}\alpha^{(1,1,1)}_{eem\,yyyx}(\omega,\mathbf{B}^0) + {}^{(1)}_{e}\alpha^{(1,1,1)}_{eem\,yyxy}(\omega,\mathbf{B}^0)\Big]\Big\} + \cdots$$

$$\Theta^{NL} = \tfrac{3}{4}\Big[{}^{(1)}_{e}\alpha^{(1,1,1)}_{eee\,xxxx}(\omega,\mathbf{B}^0) + {}^{(1)}_{e}\alpha^{(1,1,1)}_{eee\,yyyy}(\omega,\mathbf{B}^0) - 2\,{}^{(1)}_{e}\alpha^{(1,1,1)}_{eee\,xxyy}(\omega,\mathbf{B}^0)$$

$$-4\,{}^{(1)}_{e}\alpha^{(1,1,1)}_{eee\,xyxy}(\omega,\mathbf{B}^0)\Big] + \frac{ns_z}{c}\Big\{\omega\Big[-{}^{(1)}_{e}\gamma^{(1,1,2)}_{eee\,xxx(xz)}(\omega,\mathbf{B}^0)$$

$$+ {}^{(1)}_{e}\gamma^{(1,1,2)}_{eee\,xxy(yz)}(\omega,\mathbf{B}^0) + 2\,{}^{(1)}_{e}\gamma^{(1,1,2)}_{eee\,xyx(yz)}(\omega,\mathbf{B}^0) - {}^{(1)}_{e}\gamma^{(1,1,2)}_{eee\,yyy(yz)}(\omega,\mathbf{B}^0)$$

$$+ {}^{(1)}_{e}\gamma^{(1,1,2)}_{eee\,yyx(xz)}(\omega,\mathbf{B}^0) + 2\,{}^{(1)}_{e}\gamma^{(1,1,2)}_{eee\,yxy(xz)}(\omega,\mathbf{B}^0)\Big]$$

$$+3\Big[{}^{(1)}_{e}\alpha^{(1,1,1)}_{eem\,xxxy}(\omega,\mathbf{B}^0) + {}^{(1)}_{e}\alpha^{(1,1,1)}_{eem\,xxyx}(\omega,\mathbf{B}^0)$$

$$+2\,{}^{(1)}_{e}\alpha^{(1,1,1)}_{eem\,yxxx}(\omega,\mathbf{B}^0) - {}^{(1)}_{e}\alpha^{(1,1,1)}_{eem\,yyyx}(\omega,\mathbf{B}^0) - {}^{(1)}_{e}\alpha^{(1,1,1)}_{eem\,yyxy}(\omega,\mathbf{B}^0)$$

$$-2\,{}^{(1)}_{e}\alpha^{(1,1,1)}_{eem\,xyyy}(\omega,\mathbf{B}^0)\Big]\Big\} + \cdots \tag{51}$$

$$\Omega_R^{NL} = \frac{3}{2}\Big[{}^{(1)}_{e}\alpha^{(1,1,1)}_{eee\,xxxx}(\omega,\mathbf{B}^0) - {}^{(1)}_{e}\alpha^{(1,1,1)}_{eee\,yyyy}(\omega,\mathbf{B}^0)\Big]$$

$$+ \frac{2ns_z}{c}\Big\{-\omega\Big[{}^{(1)}_{e}\gamma^{(1,1,2)}_{eee\,xxx(xz)}(\omega,\mathbf{B}^0) - {}^{(1)}_{e}\gamma^{(1,1,2)}_{eee\,yyy(yz)}(\omega,\mathbf{B}^0)\Big] \tag{52}$$

$$+3\Big[{}^{(1)}_{e}\alpha^{(1,1,1)}_{eem\,xxxy}(\omega,\mathbf{B}^0) + {}^{(1)}_{e}\alpha^{(1,1,1)}_{eem\,yyyx}(\omega,\mathbf{B}^0)\Big]\Big\} + \cdots$$

$$\Omega_A^{NL} = 3\Big[{}^{(1)}_{e}\gamma^{(1,1,1)}_{eee\,xxyy}(\omega,\mathbf{B}^0)\Big]$$

$$+ \frac{2ns_z}{c}\Big\{-\omega\Big[{}^{(1)}_{e}\alpha^{(1,1,2)}_{eee\,xxy(yz)}(\omega,\mathbf{B}^0) - {}^{(1)}_{e}\alpha^{(1,1,2)}_{eee\,yyx(xz)}(\omega,\mathbf{B}^0)\Big] \tag{53}$$

$$-3\Big[{}^{(1)}_{e}\gamma^{(1)}_{eem\,xxyx}(\omega,\mathbf{B}^0) + {}^{(1)}_{e}\alpha^{(1,1,1)}_{eem\,yyxy}(\omega,\mathbf{B}^0)\Big]\Big\} + \cdots$$

$$
\begin{aligned}
\chi_G^{NL} = & -3\left[{}^{(1)}_{e}\gamma^{(1,1,1)}_{eee\,xxxy}(\omega,\mathbf{B}^0) - {}^{(1)}_{e}\gamma^{(1,1,1)}_{eee\,yyyx}(\omega,\mathbf{B}^0)\right] \\
& -\frac{ns_z}{c}\Big\{\omega\Big[{}^{(1)}_{e}\alpha^{(1,1,2)}_{eee\,xxx(yz)}(\omega,\mathbf{B}^0) + {}^{(1)}_{e}\alpha^{(1,1,2)}_{eee\,xxy(xz)}(\omega,\mathbf{B}^0) \\
& \quad -2{}^{(1)}_{e}\alpha^{(1,1,2)}_{eee\,xyx(xz)}(\omega,\mathbf{B}^0) - {}^{(1)}_{e}\alpha^{(1,1,2)}_{eee\,yyy(xz)}(\omega,\mathbf{B}^0) \\
& \quad - {}^{(1)}_{e}\alpha^{(1,1,2)}_{eee\,yyx(yz)}(\omega,\mathbf{B}^0) + 2{}^{(1)}_{e}\alpha^{(1,1,2)}_{eee\,yxy(yz)}(\omega,\mathbf{B}^0)\Big] \\
& \quad +3\Big[-{}^{(1)}_{e}\gamma^{(1,1,1)}_{eem\,xxxx}(\omega,\mathbf{B}^0) + {}^{(1)}_{e}\gamma^{(1,1,1)}_{eem\,xxyy}(\omega,\mathbf{B}^0) \\
& \quad -2{}^{(1)}_{e}\gamma^{(1,1,1)}_{eem\,xyxy}(\omega,\mathbf{B}^0) - {}^{(1)}_{e}\gamma^{(1,1,1)}_{eem\,yyyy}(\omega,\mathbf{B}^0) \\
& \quad + {}^{(1)}_{e}\gamma^{(1,1,1)}_{eem\,yyxx}(\omega,\mathbf{B}^0) - 2{}^{(1)}_{e}\alpha^{(1,1,1)}_{eem\,yxyx}(\omega,\mathbf{B}^0)\Big]\Big\} + \cdots
\end{aligned}
\tag{54}
$$

$$
\begin{aligned}
\Sigma^{NL} = & -3\left[{}^{(1)}_{e}\gamma^{(1,1,1)}_{eee\,xxxy}(\omega,\mathbf{B}^0) + {}^{(1)}_{e}\gamma^{(1,1,1)}_{eee\,yyyx}(\omega,\mathbf{B}^0)\right] \\
& -\frac{ns_z}{c}\Big\{\omega\Big[{}^{(1)}_{e}\alpha^{(1,1,2)}_{eee\,xxx(yz)}(\omega,\mathbf{B}^0) + {}^{(1)}_{e}\alpha^{(1,1,2)}_{eee\,xxy(xz)}(\omega,\mathbf{B}^0) \\
& \quad -2{}^{(1)}_{e}\alpha^{(1,1,2)}_{eee\,xyx(xz)}(\omega,\mathbf{B}^0) + {}^{(1)}_{e}\alpha^{(1,1,2)}_{eee\,yyy(xz)}(\omega,\mathbf{B}^0) \\
& \quad + {}^{(1)}_{e}\alpha^{(1,1,2)}_{eee\,yyx(yz)}(\omega,\mathbf{B}^0) - 2{}^{(1)}_{e}\alpha^{(1,1,2)}_{eee\,yxy(yz)}(\omega,\mathbf{B}^0)\Big] \\
& \quad +3\Big[-{}^{(1)}_{e}\gamma^{(1,1,1)}_{eem\,xxxx}(\omega,\mathbf{B}^0) + {}^{(1)}_{e}\gamma^{(1,1,1)}_{eem\,xxyy}(\omega,\mathbf{B}^0) \\
& \quad -2{}^{(1)}_{e}\gamma^{(1,1,1)}_{eem\,xyxy}(\omega,\mathbf{B}^0) + {}^{(1)}_{e}\gamma^{(1,1,1)}_{eem\,yyyy}(\omega,\mathbf{B}^0) \\
& \quad - {}^{(1)}_{e}\gamma^{(1,1,1)}_{eem\,yyxx}(\omega,\mathbf{B}^0) + 2{}^{(1)}_{e}\gamma^{(1,1,1)}_{eem\,yxyx}(\omega,\mathbf{B}^0)\Big]\Big\} + \cdots
\end{aligned}
\tag{55}
$$

$$
\begin{aligned}
\Lambda^{NL} = & -3\left[{}^{(1)}_{e}\alpha^{(1,1,1)}_{eee\,xxxy}(\omega,\mathbf{B}^0) - {}^{(1)}_{e}\alpha^{(1,1,1)}_{eee\,yyyx}(\omega,\mathbf{B}^0)\right] \\
& +\frac{ns_z}{c}\Big\{\omega\Big[{}^{(1)}_{e}\gamma^{(1,1,2)}_{eee\,xxx(yz)}(\omega,\mathbf{B}^0) + {}^{(1)}_{e}\gamma^{(1,1,2)}_{eee\,xxy(xz)}(\omega,\mathbf{B}^0) \\
& +2{}^{(1)}_{e}\gamma^{(1,1,2)}_{eee\,xyx(xz)}(\omega,\mathbf{B}^0) \\
& - {}^{(1)}_{e}\gamma^{(1,1,2)}_{eee\,yyy(xz)}(\omega,\mathbf{B}^0) - {}^{(1)}_{e}\gamma^{(1,1,2)}_{eee\,yyx(yz)}(\omega,\mathbf{B}^0) \\
& -2{}^{(1)}_{e}\gamma^{(1,1,2)}_{eee\,yxy(yz)}(\omega,\mathbf{B}^0)\Big] \\
& +3\Big[{}^{(1)}_{e}\alpha^{(1,1,1)}_{eem\,xxxx}(\omega,\mathbf{B}^0) - {}^{(1)}_{e}\alpha^{(1,1,1)}_{eem\,xxyy}(\omega,\mathbf{B}^0) \\
& -2{}^{(1)}_{e}\alpha^{(1,1,1)}_{eem\,xyxy}(\omega,\mathbf{B}^0) \\
& + {}^{(1)}_{e}\alpha^{(1,1,1)}_{eem\,yyyy}(\omega,\mathbf{B}^0) - {}^{(1)}_{e}\alpha^{(1,1,1)}_{eem\,yyxx}(\omega,\mathbf{B}^0) \\
& -2{}^{(1)}_{e}\alpha^{(1,1,1)}_{eem\,yxyx}(\omega,\mathbf{B}^0)\Big]\Big\} + \cdots
\end{aligned}
\tag{56}
$$

$$
\begin{aligned}
\varepsilon^{NL} = {} & 3\left[{}^{(1)}_{e}\alpha^{(1,1,1)}_{eee\,xxxy}(\omega, \mathbf{B}^0) + {}^{(1)}_{e}\alpha^{(1,1,1)}_{eee\,yyyx}(\omega, \mathbf{B}^0)\right] \\
& - \frac{ns_z}{c}\Big\{\omega\Big[{}^{(1)}_{e}\gamma^{(1,1,2)}_{eee\,xxx(yz)}(\omega, \mathbf{B}^0) + {}^{(1)}_{e}\gamma^{(1,1,2)}_{eee\,xxy(xz)}(\omega, \mathbf{B}^0) \\
& + 2{}^{(1)}_{e}\gamma^{(1,1,2)}_{eee\,xyx(xz)}(\omega, \mathbf{B}^0) \\
& + {}^{(1)}_{e}\gamma^{(1,1,2)}_{eee\,yyy(xz)}(\omega, \mathbf{B}^0) + {}^{(1)}_{e}\gamma^{(1,1,2)}_{eee\,yyx(yz)}(\omega, \mathbf{B}^0) + 2{}^{(1)}_{e}\gamma^{(1,1,2)}_{eee\,yxy(yz)}(\omega, \mathbf{B}^0)\Big] \\
& + 3\Big[{}^{(1)}_{e}\alpha^{(1,1,1)}_{eem\,xxxx}(\omega, \mathbf{B}^0) - {}^{(1)}_{e}\alpha^{(1,1,1)}_{eem\,xxyy}(\omega, \mathbf{B}^0) \\
& - 2{}^{(1)}_{e}\alpha^{(1,1,1)}_{eem\,xyxy}(\omega, \mathbf{B}^0) \\
& - {}^{(1)}_{e}\alpha^{(1,1,1)}_{eem\,yyyy}(\omega, \mathbf{B}^0) + {}^{(1)}_{e}\alpha^{(1,1,1)}_{eem\,yyxx}(\omega, \mathbf{B}^0) \\
& + 2{}^{(1)}_{e}\alpha^{(1,1,1)}_{eem\,yxyx}(\omega, \mathbf{B}^0)\Big]\Big\} + \cdots
\end{aligned} \tag{57}
$$

The set of equations (42)–(45) has to be solved for well-determined boundary conditions. In our case, these are given by the values of the Stokes parameters and their derivatives with respect to the coordinate z at the input to the crystal, i.e., at $z = 0$. We denote the Stokes parameters at $z = 0$ as follows:

$$
S_i(z = 0) = S_{i0} \tag{58}
$$

with $i = 0$, 1, 2, and 3. The values of S_1, S_2, and S_3 are different for different kinds of polarization of the light wave, and some can vanish. It is well known that linear polarization is described on a Cartesian basis, whereas circular and elliptical polarization is described on a circular basis; the latter two present eigenstates, one right-handed and the other left-handed. We denote the intensity vector component of the electric field of a right-elliptically polarized light wave by $E_R = E_-$ (in conformity with the angular momentum convention), whereas for the left-elliptical polarization we write $E_L = E_+$. The Cartesian and circular components are interrelated as follows:

$$
E_\pm = \frac{1}{\sqrt{2}}(E_x \mp \mathrm{i}E_y) \tag{59}
$$

For circular components, the Stokes parameters have the well-known form [2]

$$
\begin{aligned}
S_0(z) &= |E_+(\omega, k_z)|^2 + |E_-(\omega, k_z)|^2 \\
S_1(z) &= E_+(\omega, k_z)E_-^*(\omega, k_z) + \text{c.c.} \\
S_2(z) &= \mathrm{i}[E_+(\omega, k_z)E_-^*(\omega, k_z) - \text{c.c.}] \\
S_3(z) &= |E_+(\omega, k_z)|^2 - |E_-(\omega, k_z)|^2
\end{aligned} \tag{60}
$$

In the case of circular polarization, we have moreover

$$E_x = E_y \tag{61}$$

The only Stokes parameter insensitive to the state of polarization is S_0. Irrespective of the polarization, it is always nonzero and equal to the light intensity.

By Eq. (42), S_0 is a constant (it is independent of z); thus, at all points in the crystal, its value is equal to its value at $z = 0$:

$$S_0(z) = S_{00} \tag{62}$$

The other three equations for S_1, S_2, and S_3 (Eqs. (43)–(45)) are highly complicated. In fact, we deal here with differential equations of the first order, which, moreover are nonlinear in S_1, S_2, and S_3 (they contain products of two of the parameters or their second powers). These equations are not strictly solvable because the Stokes parameters cannot be expressed analytically as functions of z.

Only the following two sets of equations are accessible to being solved analytically. One set is of the form

$$\frac{\partial S_1}{\partial z} = \frac{\omega}{2n}\sqrt{\frac{\mu_0}{\varepsilon_0}}\left(\chi_A^L S_2 + \kappa_A^L S_3\right) \tag{63}$$

$$\frac{\partial S_2}{\partial z} = \frac{\omega}{2n}\sqrt{\frac{\mu_0}{\varepsilon_0}}\left(-\chi_A^L S_1 + \kappa_R^L S_3\right) \tag{64}$$

$$\frac{\partial S_3}{\partial z} = -\frac{\omega}{2n}\sqrt{\frac{\mu_0}{\varepsilon_0}}\left(\kappa_A^L S_1 + \kappa_R^L S_2\right) \tag{65}$$

It involves coupling constants related to linear polarization only. The other set is

$$\frac{\partial S_1}{\partial z} = \frac{\omega}{2n}\sqrt{\frac{\mu_0}{\varepsilon_0}}\,\chi S_2 \tag{66}$$

$$\frac{\partial S_2}{\partial z} = -\frac{\omega}{2n}\sqrt{\frac{\mu_0}{\varepsilon_0}}\,\chi S_1 \tag{67}$$

$$\frac{\partial S_3}{\partial z} = 0 \tag{68}$$

where

$$\chi = \chi_A^L + \chi_G^{NL} S_0 + \left(\chi_R^{NL} - \kappa_R^{NL}\right) S_{30} \tag{69}$$

Here, nonlinear coupling constants occur also. The complete set of Eqs. (43)–(45) can be solved approximately by applying the iteration method, or numerically.

If the intensity of the light wave incident on the crystal is moderate (the intensity of the electric field of light is small compared with that of the electric field existing in atoms) the coupling constants related to the nonlinear electric and magnetic polarization of the medium (marked with the superscript *NL* to the right) are much smaller than the coupling constants for linear polarization (marked with the superscript L) and thus are negligible, causing Eqs. (43)–(45) to go over into (63)–(65), respectively. Hence, the set (63)–(65) is valid in linear optics; it enables us to study how the medium alone or subjected to external factors (an external constant magnetic or electric field, or externally applied stress) affects the state of polarization of light incident thereon.

The set of Eqs. (63)–(65) is easily solvable even in the general case, when all four Stokes parameters are nonzero at the input to the crystal corresponding to elliptical polarization of the light wave. The solutions are of the form

$$\begin{aligned} S_1^L(z) = {} & \frac{\cos \alpha z}{\alpha^2}\left\{S_{10}\left[\left(\tilde{\chi}_A^L\right)^2 + \left(\tilde{\kappa}_A^L\right)^2\right] + S_{20}\tilde{\kappa}_R^L\tilde{\kappa}_A^L - S_{30}\tilde{\chi}_A^L\tilde{\kappa}_R^L\right\} + \frac{\sin \alpha z}{\alpha} \\ & \times\left\{S_{20}\tilde{\chi}_A^L + S_{30}\tilde{\kappa}_A^L\right\} + \frac{1}{\alpha^2}\left\{S_{10}\left(\tilde{\kappa}_R^L\right)^2 - S_{20}\tilde{\kappa}_R^L\tilde{\kappa}_A^L + S_{30}\tilde{\chi}_A^L\tilde{\kappa}_R^L\right\} \end{aligned} \tag{70}$$

$$\begin{aligned} S_2^L(z) = {} & \frac{\cos \alpha z}{\alpha^2}\left\{S_{10}\tilde{\kappa}_R^L\tilde{\kappa}_A^L + S_{20}\left[\left(\tilde{\chi}_A^L\right)^2 + \left(\tilde{\kappa}_R^L\right)^2\right] + S_{30}\tilde{\chi}_A^L\tilde{\kappa}_A^L\right\} + \frac{\sin \alpha z}{\alpha} \\ & \times\left\{-S_{10}\tilde{\chi}_A^L + S_{30}\tilde{\kappa}_R^L\right\} + \frac{1}{\alpha^2}\left\{-S_{10}\tilde{\kappa}_R^L\tilde{\kappa}_A^L + S_{20}\left(\tilde{\kappa}_A^L\right)^2 - S_{30}\tilde{\chi}_A^L\tilde{\kappa}_A^L\right\} \end{aligned} \tag{71}$$

$$\begin{aligned} S_3^L(z) = {} & \frac{\cos \alpha z}{\alpha^2}\left\{-S_{10}\tilde{\kappa}_R^L\tilde{\chi}_A^L + S_{20}\tilde{\kappa}_A^L\tilde{\chi}_A^L + S_{30}\left[\left(\tilde{\kappa}_R^L\right)^2 + \left(\tilde{\kappa}_A^L\right)^2\right]\right\} \\ & - \frac{\sin \alpha z}{\alpha}\left\{S_{10}\tilde{\kappa}_A^L + S_{20}\tilde{\kappa}_R^L\right\} + \frac{1}{\alpha^2}\tilde{\chi}_A^L\left\{S_{10}\tilde{\kappa}_R^L - S_{20}\tilde{\kappa}_A^L + S_{30}\tilde{\chi}_A^L\right\} \end{aligned} \tag{72}$$

where we have introduced the notation

$$
\begin{aligned}
\tilde{\chi}_A^L &= \frac{\omega}{2n}\sqrt{\frac{\mu_0}{\varepsilon_0}}\,\chi_A^L \\
\tilde{\kappa}_R^L &= \frac{\omega}{2n}\sqrt{\frac{\mu_0}{\varepsilon_0}}\,\kappa_R^L \\
\tilde{\kappa}_A^L &= \frac{\omega}{2n}\sqrt{\frac{\mu_0}{\varepsilon_0}}\,\kappa_A^L \\
\alpha &= \sqrt{\left(\tilde{\chi}_A^L\right)^2 + \left(\tilde{\kappa}_R^L\right)^2 + \left(\tilde{\kappa}_A^L\right)^2}
\end{aligned}
\tag{73}
$$

The other simplified set, Eqs. (66)–(68), refers to the case when the incident wave is of high intensity. Obviously, this set results from Eqs. (43)–(45) on the assumption that all the coupling constants of Eq. (45) vanish:

$$
\begin{array}{llll}
\kappa_R^L = 0 & \Omega_R^{NL} = 0 & \Theta^{NL} = 0 & \Sigma^{NL} = 0 \\
\kappa_A^L = 0 & \Omega_A^{NL} = 0 & \Lambda^{NL} = 0 & \varepsilon^{NL} = 0
\end{array}
\tag{74}
$$

Not all the coupling constants of Eqs. (43)–(45) are necessarily nonzero. Neumann's principle states that, in a medium with well-defined symmetry, only those coupling constants can be nonzero that are invariant with respect to the symmetry of the medium. Having recourse to tables [142, 144, 145, 147–149] giving the exact forms of the polar and axial i- and c-tensors of rank 2, 3, 4, 5, and 6 and performing a considerable amount of cumbersome and lengthy work, one finds that when the dc magnetic field is applied along the z axis ($\mathbf{B}^0 = \mathbf{e}_z\, B^0$), all the coupling constants of Eq. (45) are forbidden (vanish) for crystals with symmetry of the following magnetic classes:

$$
\begin{gathered}
(3,\bar{3}), \underline{\bar{3}}, 32, (3\underline{2}), 3m, \bar{3}m, (3\underline{m}, \bar{3}\underline{m}), \underline{\bar{3}m}, \underline{\bar{3}}m, (6,\bar{6}), \underline{6}, \underline{\bar{6}}, (6/m), \underline{6}/m, \\
6/\underline{m}, \underline{6}/\underline{m}, 622, \underline{6}2\underline{2}, (6\underline{22}), 6mm, \underline{6}m\underline{m}, (6\underline{mm}), \bar{6}m2, \underline{\bar{6}}2\underline{m}, \underline{\bar{6}}m\underline{2}, \\
(\bar{6}\underline{m2}), 6/mmm, \underline{6}/\underline{m}m\underline{m}, (6/m\underline{mm}), 6/\underline{mmm}, 6/\underline{m}mm, \underline{6}/m\underline{m}m
\end{gathered}
\tag{75}
$$

and, consequently, Eqs. (43)–(45) go over into (66)–(68). Above, we have put the classes that admit ferromagnetic ordering in parentheses; the remaining classes are antiferromagnets. In crystals with the symmetries

$$(4\underline{22}, 4/m\underline{mm}), 4/\underline{m}mm \tag{76}$$

we moreover have the constant Θ^{NL}, whereas in crystals with the symmetries

$$(4, 4/m), 4/\underline{m}, 422, 4mm, (4\underline{mm}), 4/mmm \text{ and } 4/\underline{mmm} \tag{77}$$

we deal with the constants Θ^{NL} and Λ^{NL}.

The set (66)–(68) is also strictly solvable even in the case of elliptically polarized incident light, i.e., for all four Stokes parameters nonzero at $z = 0$. By the standard method used in solving sets of differential equations we get

$$S_1^{NL}(z) = S_{10}\cos(\tilde{\chi}z) + S_{20}\sin(\tilde{\chi}z) \tag{78}$$

$$S_2^{NL}(z) = -S_{10}\sin(\tilde{\chi}z) + S_{20}\cos(\tilde{\chi}z) \tag{79}$$

$$S_3^{NL}(z) = S_{30} \tag{80}$$

where

$$\tilde{\chi} = \frac{\omega}{2n}\sqrt{\frac{\mu_0}{\varepsilon_0}}\,\chi \tag{81}$$

On neglecting all the nonlinear coupling constants in Eq. (81) (or, more precisely, in (69)), solutions (78)–(80) will refer to the case when the light wave incident on the crystal is weak; in conformity with condition (74), the crystal is now characterized by $\kappa_R^L = 0$ and $\kappa_A^L = 0$. Obviously, on putting $\kappa_R^L = 0$ and $\kappa_A^L = 0$ and neglecting the nonlinear coupling constants, solutions (70)–(72) go over into (78)–(80).

For ferro- and antiferromagnets with symmetry described by the magnetic point groups enumerated in (76) and (77), the set of equations to be solved is much simpler than the set (43)–(45) where, besides the coupling constants χ_A^L, χ_G^{NL}, χ_R^{NL}, and κ_R^{NL}, the constants Θ^{NL} and Λ^{NL} intervene:

$$\frac{\partial S_1}{\partial z} = \frac{\omega}{2n}\sqrt{\frac{\mu_0}{\varepsilon_0}}\left[\chi S_2 + (\Theta^{NL}S_2 + \Lambda^{NL}S_1)S_3\right] \tag{82}$$

$$\frac{\partial S_2}{\partial z} = -\frac{\omega}{2n}\sqrt{\frac{\mu_0}{\varepsilon_0}}\left[\chi S_1 + (-\Theta^{NL}S_1 + \Lambda^{NL}S_2)S_3\right] \tag{83}$$

$$\frac{\partial S_3}{\partial z} = -\frac{\omega}{2n}\sqrt{\frac{\mu_0}{\varepsilon_0}}\left[2\Theta^{NL}S_1S_2 + \Lambda^{NL}(S_1^2 - S_2^2)\right] \tag{84}$$

The above set of equations can be solved only approximately using the iteration method. The procedure consists of the following: The zeroth-order solution is obtained on solving the simplified set of Eqs. (66)–(68), thus giving expressions (78)–(80). In the next step, we insert (78)–(80) into the right-hand term of the equations for S_1, S_2, and S_3 and perform integration over z' in the limits:

$$0 \le z' \le z \quad \text{and} \quad S_{jo} \le S_j(z') \le S_j(z) \tag{85}$$

We thus arrive at

$$S_1(z) = S_1^{NL}(z) + S_{30}\left\{\frac{\tilde{\Theta}^{NL}\left[S_1^{NL}(z) - S_{10}\right]}{\tilde{\chi}} - \frac{\tilde{\Lambda}^{NL}\left[S_2^{NL}(z) - S_{20}\right]}{\tilde{\chi}}\right\} \tag{86}$$

$$S_2(z) = S_2^{NL}(z) - S_{30}\left\{\frac{\tilde{\Theta}^{NL}\left[S_2^{NL}(z) - S_{20}\right]}{\tilde{\chi}} + \frac{\tilde{\Lambda}^{NL}\left[S_1^{NL}(z) - S_{10}\right]}{\tilde{\chi}}\right\} \tag{87}$$

$$S_3(z) = S_{30} - \frac{\tilde{\Theta}^{NL}\left[(S_{10}^2 - S_{20}^2)(\cos 2\tilde{\chi}z - 1) + 4S_{10}S_{20}\sin 2\tilde{\chi}z\right]}{2\tilde{\chi}} - \frac{\tilde{\Lambda}^{NL}\left[(S_{10}^2 - S_{20}^2)\sin 2\tilde{\chi}z - 2S_{10}S_{20}(\cos 2\tilde{\chi}z - 1)\right]}{2\tilde{\chi}} \tag{88}$$

IV. RESULTS

A. Rotation of the Polarization Ellipse of Light

On insertion of (78)–(80) into (41), we get

$$\Psi(z) = \frac{1}{2}\arctan\frac{-S_{10}\sin(\tilde{\chi}z) + S_{20}\cos(\tilde{\chi}z)}{S_{10}\cos(\tilde{\chi}z) + S_{20}\sin(\tilde{\chi}z)} \quad \text{and}$$

$$\Phi(z) = \frac{1}{2}\arcsin\frac{S_{30}}{S_0} \tag{89}$$

where $\tilde{\chi}$ is given by (81) and (69). Thus, we deal with an effect in which elliptically polarized light, $\Phi(z = 0) \neq 0$, propagating along the z axis in a crystal having the symmetry given by (75), conserves its ellipticity, $\Phi(z) =$ const., whereas the ellipse changes its position with respect to the x axis. More precisely, the polarization ellipse rotates by an angle $\Psi(z)$ about the z axis. Rotation of the ellipse can occur only if the parameter $S_{30} \neq 0$ and at least one of the parameters S_{10} or S_{20} differs from zero. With regard to the definition of Stokes parameters on circular basis (60), S_{10} and S_{20} can be nonzero only if the light wave at the input to the crystal is a superposition of two elliptically polarized modes with opposite handedness. The parameter S_{30} can differ from zero even if the light wave conveys but a single right- or left-handed mode; the polarization ellipse then remains unchanged, $\Phi(z) =$ const., and so does its position, $\Psi(z) = 0$.

Using tables [142, 144, 147–149] of the polar and axial i- and c-tensors of rank 2, 3, 4, 5, and 6, we determine the coupling constants χ_A^L, χ_G^{NL} and $\chi_R^{NL} - \kappa_R^{NL}$ for crystals with the symmetries enumerated in (75). Our results are assembled in Table I, where we have introduced the following notation:

$$\chi_A^L = f + f' + (f_1 + f_1')B_z^0 + (f_2 + f_2')(B_z^0)^2 + \cdots \tag{90}$$

$$\chi_G^{NL} = F + F' + (F_1 + F_1')B_z^0 + (F_2 + \cdots)(B_z^0)^2 + \cdots \tag{91}$$

$$\chi_R^{NL} - \kappa_R^{NL} = R + R' + (R_1 + R_1')B_z^0 + (R_2 + \cdots)(B_z^0)^2 + \cdots \tag{92}$$

TABLE I
The Coupling Constants χ_A^L, χ_G^{NL} and $\chi_R^{NL} - \kappa_R^{NL}$ for Ferro- and Antiferromagnetic Crystals with Symmetries Stated in Eq. 75

$$\tilde{\chi} = \frac{\omega}{2n}\sqrt{\frac{\mu_0}{\varepsilon_0}}\,[\chi_A^L + \chi_G^{NL} S_0 + (\chi_R^{NL} - \kappa_R^{NL}) S_{30}]$$

Magnetic Point Group	χ_A^L			χ_G^{NL}			$\chi_R^{NL} - \kappa_R^{NL}$		
	$f + f'$	$f_1 + f_1'$	$f_2 + f_2'$	$F + F'$	$F_1 + F_1'$	F_2	$R + R'$	$R_1 + R_1'$	R_2
Ferromagnetics									
$3, 3\underline{2}, 6, 6\underline{22}$	$f + f'$	$f_1 + f_1'$	$f_2 + f_2'$	$F + F'$	$F_1 + F_1'$	F_2	$R + R'$	$R_1 + R_1'$	R_2
$\bar{3}, 3\underline{m}, \bar{3}\underline{m}, \bar{6}, 6/m, 6\underline{mm}, \bar{6}\underline{m2}, 6/m\underline{mm}$	f	f_1	f_2	F	F_1	F_2	R	R_1	R_2
Antiferromagnetics									
$\underline{\bar{3}}, 3m, \underline{\bar{3}}m, \underline{\bar{6}}, 6/\underline{m}, 6mm, \underline{\bar{6}}m\underline{2}, 6/\underline{m}mm$	0	$f_1 + f_1'$	0	0	$F_1 + F_1'$	0	$R + R'$	0	R_2
$32, \underline{6}, 622, \underline{6}2\underline{2}$	f'	f_1	f_2'	F'	F_1	0	R	R_1'	R_2
$\bar{3}m, \underline{\bar{3}m}, \underline{6}/m, \underline{6}/\underline{m}, \underline{6}m\underline{m}, \bar{6}\underline{m}2, \underline{\bar{6}}2\underline{m}, \underline{\bar{6}}/\underline{mmm}, \underline{6}/\underline{m}m\underline{m}, 6/\underline{mmm}, \underline{6}/m\underline{m}m$	0	f_1	0	0	F_1	0	R	0	R_2

with

$$f = 2\,{}^{(1)}_{e}\gamma^{(1)}_{e[xy]}(\omega) = 2\left\{{}^{(1)}_{e}S^{(1)M}_{e[xy]z}(\omega)M_z + {}^{(1)}_{e}S^{(1)MMM}_{e[xy]zzz}(\omega)M_z^3; 0\right\} \tag{93}$$

$$\begin{aligned} f' &= \frac{4n}{c}\left[-\frac{\omega}{3}\,{}^{(1)}_{e}\alpha^{(2)}_{e\,x(yz)}(\omega) + {}^{(1)}_{e}\gamma^{(1)}_{m\,xx}(\omega)\right] \\ &= \frac{4n}{c}\left\{-\frac{\omega}{3}\left[\underline{{}^{(1)}_{e}Q^{(2)}_{e\,x(yz)}(\omega)} + {}^{(1)}_{e}Q^{(2)MM}_{e\,x(yz)zz}(\omega)M_z^2\right]\right. \\ &\qquad + \underline{{}^{(1)}_{e}S^{(1)}_{m\,xx}(\omega)} + {}^{(1)}_{e}S^{(1)MM}_{m\,xxzz}(\omega)M_z^2; \\ &\qquad -\frac{\omega}{3}\left[{}^{(1)}_{e}Q^{(2)}_{e\,x(yz)}(\omega) + {}^{(1)}_{e}Q^{(2)LL}_{e\,x(yz)zz}(\omega)L_z^2\right] \\ &\qquad \left. + {}^{(1)}_{e}S^{(1)}_{m\,xx}(\omega) + {}^{(1)}_{e}S^{(1)LL}_{m\,xxzz}(\omega)L_z^2\right\} \end{aligned} \tag{94}$$

$$
\begin{aligned}
f_1 &= 2\,{}^{(1)}_{e}\gamma^{(1)m}_{e[xy]z}(\omega) \\
&= 2\Big\{\Big[\underline{{}^{(1)}_{e}S^{(1)M}_{e[xy]z)}(\omega)} + 3\,{}^{(1)}_{e}S^{(1)MMM}_{e[xy]zzz}(\omega)M_z^2\Big]\alpha^{Mm}_{zz} \\
&\qquad + \Big[{}^{(1)}_{e}S^{(1)L}_{e[xy]z)}(\omega) + {}^{(1)}_{e}S^{(1)LMM}_{e[xy]zzz}(\omega)M_z^2\Big]\alpha^{Lm}_{zz}; \\
&\qquad \Big[{}^{(1)}_{e}S^{(1)M}_{e[xy]z)}(\omega) + {}^{(1)}_{e}S^{(1)MLL}_{e[xy]zzz}(\omega)L_z^2\Big]\alpha^{Mm}_{zz}\Big\}
\end{aligned} \tag{95}
$$

$$
\begin{aligned}
f_1' &= \frac{4n}{c}\Big[-\frac{\omega}{3}\,{}^{(1)}_{e}\alpha^{(2)m}_{e\,x(yz)z}(\omega) + {}^{(1)}_{e}\gamma^{(1)m}_{m\,xxz}(\omega)\Big] \\
&= \frac{4n}{c}\Big(\Big\{-\frac{\omega}{3}\Big[2\,{}^{(1)}_{e}Q^{(2)MM}_{e\,x(yz)zz}(\omega)\alpha^{Mm}_{zz} + {}^{(1)}_{e}Q^{(2)ML}_{e\,x(yz)zz}(\omega)\alpha^{Lm}_{zz}\Big] \\
&\qquad + 2\,{}^{(1)}_{e}S^{(1)MM}_{m\,xxzz}(\omega)\alpha^{Mm}_{zz} + {}^{(1)}_{e}S^{(1)ML}_{m\,xxzz}(\omega)\alpha^{Lm}_{zz}\Big\}M_z; \\
&\qquad \Big[-\frac{\omega}{3}\,{}^{(1)}_{e}Q^{(2)ML}_{e\,x(yz)zz}(\omega) + {}^{(1)}_{e}S^{(1)ML}_{m\,xxzz}(\omega)\Big]\alpha^{Mm}_{zz}L_z\Big)
\end{aligned} \tag{96}
$$

$$
\begin{aligned}
f_2 &= 2\,{}^{(1)}_{e}\gamma^{(1)mm}_{e[xy]zz}(\omega) \\
&= 2\Big\{\Big[{}^{(1)}_{e}S^{(1)M}_{e[xy]z}(\omega) + 3\,{}^{(1)}_{e}S^{(1)MMM}_{e[xy]zzz}(\omega)M_z^2\Big]\alpha^{Mmm}_{zzz} + \Big[{}^{(1)}_{e}S^{(1)L}_{e[xy]z)}(\omega) \\
&\qquad + {}^{(1)}_{e}S^{(1)LMM}_{e[xy]zzz}(\omega)M_z^2\Big]\alpha^{Lmm}_{zzz} + \Big[3\,{}^{(1)}_{e}S^{(1)MMM}_{e[xy]zzz}(\omega)\big(\alpha^{Mm}_{zz}\big)^2 \\
&\qquad + {}^{(1)}_{e}S^{(1)MLL}_{e[xy]zzz}(\omega)\big(\alpha^{Lm}_{zz}\big)^2 + 2\,{}^{(1)}_{e}S^{(1)LMM}_{e[xy]zzz}(\omega)\alpha^{Mm}_{zz}\alpha^{Lm}_{zz}\Big]M_z; 0\Big\}
\end{aligned} \tag{97}
$$

$$
\begin{aligned}
f_2' &= \frac{4n}{c}\Big[-\frac{\omega}{3}\,{}^{(1)}_{e}\alpha^{(2)mm}_{e\,x(yz)zz}(\omega) + {}^{(1)}_{e}\gamma^{(1)mm}_{m\,xxzz}(\omega)\Big] \\
&= \frac{4n}{c}\Big(-\frac{\omega}{3}\Big\{\underline{{}^{(1)}_{e}Q^{(2)MM}_{e\,x(yz)zz}(\omega)}\Big[\big(\alpha^{Mm}_{zz}\big)^2 + 2\alpha^{Mmm}_{zzz}M_z\Big] \\
&\qquad + {}^{(1)}_{e}Q^{(2)LL}_{e\,x(yz)zz}(\omega)\big(\alpha^{Lm}_{zz}\big)^2 + {}^{(1)}_{e}Q^{(2)ML}_{e\,x(yz)zz}(\omega)\Big[\alpha^{Mm}_{zz}\alpha^{Lm}_{zz} + 2\alpha^{Lmm}_{zzz}M_z\Big]\Big\} \\
&\qquad + \underline{{}^{(1)}_{e}S^{(1)MM}_{m\,xxzz}(\omega)}\Big[\big(\alpha^{Mm}_{zz}\big)^2 + 2\alpha^{Mmm}_{zzz}M_z\Big] + {}^{(1)}_{e}S^{(1)LL}_{m\,xxzz}(\omega)\big(\alpha^{Lm}_{zz}\big)^2 \\
&\qquad + {}^{(1)}_{e}S^{(1)ML}_{m\,xxzz}(\omega)\Big[\alpha^{Mm}_{zz}\alpha^{Lm}_{zz} + \alpha^{Lmm}_{zzz}M_z\Big]; \\
&\qquad -\frac{\omega}{3}\Big[{}^{(1)}_{e}Q^{(2)MM}_{e\,x(yz)zz}(\omega)\big(\alpha^{Mm}_{zz}\big)^2 + 2\,{}^{(1)}_{e}Q^{(2)LL}_{e\,x(yz)zz}(\omega)\alpha^{Lmm}_{zzz}L_z\Big] \\
&\qquad + {}^{(1)}_{e}S^{(1)MM}_{m\,xxzz}(\omega)\big(\alpha^{Mm}_{zz}\big)^2 + {}^{(1)}_{e}S^{(1)LL}_{m\,xxzz}(\omega)\alpha^{Lmm}_{zzz}L_z\Big)
\end{aligned} \tag{98}
$$

$$F = -6{}^{(1)}_{e}\gamma^{(1,1,1)}_{eee\,xxxy}(\omega) = -\left[6{}^{(1)}_{e}S^{(1,1,1)M}_{eee\,xxxyz}(\omega)M_z + \cdots ;0\right] \quad (99)$$

$$F' = \frac{24n}{c}\left[\frac{\omega}{3}{}^{(1)}_{e}\alpha^{(1,1,2)}_{eee\,xyx(xz)}(\omega) + {}^{(1)}_{e}\gamma^{(1,1,1)}_{eem\,xyxy}(\omega)\right]$$

$$= \frac{24n}{c}\left\{\frac{\omega}{3}\left[\underline{{}^{(1)}_{e}Q^{(1,1,2)}_{eee\,xyx(xz)}(\omega)} + {}^{(1)}_{e}Q^{(1,1,2)MM}_{eee\,xyx(xz)zz}(\omega)M_z^2\right] + \underline{{}^{(1)}_{e}S^{(1,1,1)}_{eem\,xyxy}(\omega)}\right.$$

$$+ {}^{(1)}_{e}S^{(1,1,1)MM}_{eem\,xyxyzz}(\omega)M_z^2; \quad (100)$$

$$+ \frac{\omega}{3}\left[{}^{(1)}_{e}Q^{(1,1,2)}_{eee\,xyx(xz)}(\omega) + {}^{(1)}_{e}Q^{(1,1,2)LL}_{eee\,xyx(xz)zz}(\omega)L_z^2\right] + {}^{(1)}_{e}S^{(1,1,1)}_{eem\,xyxy}(\omega)$$

$$\left. + {}^{(1)}_{e}S^{(1,1,1)LL}_{eem\,xyxyzz}(\omega)L_z^2\right\}$$

$$F_1 = -6{}^{(1)}_{e}\gamma^{(1,1,1)m}_{eee\,xxxyz}(\omega)$$

$$= -6\left[\underline{{}^{(1)}_{e}S^{(1,1,1)M}_{eee\,xxxyz}(\omega)}\alpha^{Mm}_{zz} + {}^{(1)}_{e}S^{(1,1,1)L}_{eee\,xxxyz}(\omega)\alpha^{Lm}_{zz}; \right. \quad (101)$$

$$\left. {}^{(1)}_{e}S^{(1,1,1)L}_{eee\,xxxyz}(\omega)\alpha^{Lm}_{zz}\right]$$

$$F_1' = \frac{24n}{c}\left[\frac{\omega}{3}{}^{(1)}_{e}\alpha^{(1,1,2)m}_{eee\,xyx(xz)z}(\omega) + {}^{(1)}_{e}\gamma^{(1,1,1)m}_{eem\,xyxyz}(\omega)\right]$$

$$= \frac{24n}{c}\left\{\frac{\omega}{3}\left[2\,{}^{(1)}_{e}Q^{(1,1,2)MM}_{eee\,xyx(xz)zz}(\omega)\alpha^{Mm}_{zz} + {}^{(1)}_{e}Q^{(1,1,2)ML}_{eee\,xyx(xz)zz}(\omega)\alpha^{Lm}_{zz}\right]M_z\right.$$

$$+ \left[2\,{}^{(1)}_{e}S^{(1,1,1)MM}_{eem\,xyxyzz}(\omega)\alpha^{Mm}_{zz} + {}^{(1)}_{e}S^{(1,1,1)ML}_{eem\,xyxyzz}(\omega)\alpha^{Lm}_{zz}\right]M_z;$$

$$\left. \frac{\omega}{3}\left[{}^{(1)}_{e}Q^{(1,1,2)ML}_{eee\,xyx(xz)zz}(\omega)\alpha^{Mm}_{zz} + {}^{(1)}_{e}S^{(1,1,1)ML}_{eem\,xyxyzz}(\omega)\alpha^{Mm}_{zz}\right]L_z\right\}$$

(102)

$$F_2 = -6{}^{(1)}_{e}\gamma^{(1,1,1)mm}_{eee\,xxxyzz}(\omega)$$

$$= -6\left[{}^{(1)}_{e}S^{(1,1,1)M}_{eee\,xxxyz}(\omega)\alpha^{Mmm}_{zzz} + {}^{(1)}_{e}S^{(1,1,1)L}_{eee\,xxxyz}(\omega)\alpha^{Lmm}_{zzz};0\right] \quad (103)$$

$$R = -6\,{}^{(1)}_{e}\alpha^{(1,1,1)}_{eee\,xxyy}(\omega)$$

$$= -6\left[\underline{{}^{(1)}_{e}Q^{(1,1,1)}_{eee\,xxyy}(\omega)} + {}^{(1)}_{e}Q^{(1,1,1)MM}_{eee\,xxyyzz}(\omega)M_z^2; \right. \quad (104)$$

$$\left. {}^{(1)}_{e}Q^{(1,1,1)}_{eee\,xxyy}(\omega) + {}^{(1)}_{e}Q^{(1,1,1)LL}_{eee\,xxyyzz}(\omega)L_z^2\right]$$

$$R' = \frac{24n}{c}\left[\frac{\omega}{3}\,{}^{(1)}_{e}\gamma^{(1,1,2)}_{eee\,xxy(yz)}(\omega) + {}^{(1)}_{e}\alpha^{(1,1,1)}_{eem\,xxyx}(\omega)\right]$$

$$= \frac{24n}{c}\left\{\left[\frac{\omega}{3}\,{}^{(1)}_{e}S^{(1,1,2)M}_{eee\,xxy(yz)z}(\omega) + {}^{(1)}_{e}Q^{(1,1,1)M}_{eem\,xxyxz}(\omega)\right]M_z;\right. \tag{105}$$

$$\left.\left[\frac{\omega}{3}\,{}^{(1)}_{e}S^{(1,1,2)L}_{eee\,xxy(yz)z}(\omega) + {}^{(1)}_{e}Q^{(1,1,1)L}_{eem\,xxyxz}(\omega)\right]L_z\right\}$$

$$R_1 = -6\,{}^{(1)}_{e}\alpha^{(1,1,1)m}_{eee\,xxyyz}(\omega)$$

$$= -6\left\{\left[2\,{}^{(1)}_{e}Q^{(1,1,1)MM}_{eee\,xxyyzz}(\omega)\alpha^{Mm}_{zz} + {}^{(1)}_{e}Q^{(1,1,1)ML}_{eee\,xxyyzz}(\omega)\alpha^{Lm}_{zz}\right]M_z; O\right\} \tag{106}$$

$$R_1' = \frac{24n}{c}\left[\frac{\omega}{3}\,{}^{(1)}_{e}\gamma^{(1,1,2)m}_{eee\,xxy(yz)z}(\omega) + {}^{(1)}_{e}\alpha^{(1,1,1)m}_{eem\,xxyxz}(\omega)\right]$$

$$= \frac{24n}{c}\left\{\left[\underline{\frac{\omega}{3}\,{}^{(1)}_{e}S^{(1,1,2)M}_{eee\,xxy(yz)z}(\omega)} + \underline{{}^{(1)}_{e}Q^{(1,1,1)M}_{eem\,xxyxz}(\omega)}\right]\alpha^{Mm}_{zz}\right.$$

$$+\left[\frac{\omega}{3}\,{}^{(1)}_{e}S^{(1,1,2)L}_{eee\,xxy(yz)z}(\omega) + {}^{(1)}_{e}Q^{(1,1,1)L}_{eem\,xxyxz}(\omega)\right]\alpha^{Lm}_{zz}; \tag{107}$$

$$\left.+\left[\frac{\omega}{3}\,{}^{(1)}_{e}S^{(1,1,2)M}_{eee\,xxy(yz)z}(\omega) + {}^{(1)}_{e}Q^{(1,1,1)M}_{eem\,xxyxz}(\omega)\right]\alpha^{Mm}_{zz}\right\}$$

$$R_2 = -6\,{}^{(1)}_{e}\alpha^{(1,1,1)mm}_{eee\,xxyyzz}(\omega)$$

$$= -6\left\{\underline{{}^{(1)}_{e}Q^{(1,1,1)MM}_{eee\,xxyyzz}(\omega)}\left[\left(\alpha^{Mm}_{zz}\right)^2 + 2\alpha^{Mmm}_{zzz}M_z\right]\right.$$

$$+\,{}^{(1)}_{e}Q^{(1,1,1)LL}_{eee\,xxyyzz}(\omega)\left(\alpha^{Lm}_{zz}\right)^2 \tag{108}$$

$$+\,{}^{(1)}_{e}Q^{(1,1,1)ML}_{eee\,xxyyzz}(\omega)\left[\alpha^{Mm}_{zz}\alpha^{Lm}_{zz} + \alpha^{Lmm}_{zzz}M_z\right];$$

$$\left.{}^{(1)}_{e}Q^{(1,1,1)MM}_{eee\,xxyyzz}(\omega)\left(\alpha^{Mm}_{zz}\right)^2\right\}$$

To gain better insight into the various physical mechanisms, we express the coupling constants of Table I in terms of the parameters characterizing magnetic ordering (see Appendix A). With this in mind, we make use of expressions (A.14)–(A.25), relating the linear

$${}^{(a)}_{A}\boldsymbol{\alpha}^{(b)}_{B}(\omega),\ {}^{(a)}_{A}\boldsymbol{\alpha}^{(b)m}_{B}(\omega),\ {}^{(a)}_{A}\boldsymbol{\alpha}^{(b)mm}_{B}(\omega),\ {}^{(a)}_{A}\boldsymbol{\gamma}^{(b)}_{B}(\omega),$$

$${}^{(a)}_{A}\boldsymbol{\gamma}^{(b)m}_{B}(\omega),\ {}^{(a)}_{A}\boldsymbol{\gamma}^{(b)mm}_{B}(\omega)$$

and nonlinear

$$
{}^{(a)}_{A}\boldsymbol{\alpha}^{(b,c,d)}_{BCD}(\omega),\ {}^{(a)}_{A}\boldsymbol{\alpha}^{(b,c,d)m}_{BCD}(\omega),\ {}^{(a)}_{A}\boldsymbol{\alpha}^{(b,c,d)mm}_{BCD}(\omega),
$$

$$
{}^{(a)}_{A}\boldsymbol{\gamma}^{(b,c,d)}_{BCD}(\omega),\ {}^{(a)}_{A}\boldsymbol{\gamma}^{(b,c,d)m}_{BCD}(\omega),\ {}^{(a)}_{A}\boldsymbol{\gamma}^{(b,c,d)mm}_{BCD}(\omega)
$$

multipolar susceptibilities with the parameters characterizing ferro- and antiferromagnetic ordering

$$
\mathbf{M}, \boldsymbol{\alpha}^{Mm}, \boldsymbol{\alpha}^{Mmm} \quad \text{and} \quad \mathbf{L}, \boldsymbol{\alpha}^{Lm}, \boldsymbol{\alpha}^{Lmm} \tag{109}
$$

and Tables IV–XII (Appendix B) giving the explicit forms of the tensors occurring in expressions (A.12) and (A.13).

Following the procedure outlined above we express the coupling constants in terms of the parameters characterizing ferro- and antiferromagnetic ordering. The results are given in expressions (93)–(108) after the second sign of equality. Expressions preceding the semicolon refer to ferromagnets, whereas expressions following the semicolon concern antiferromagnets.

In the antiferromagnetic case the relationship between the coupling constants and the antiferromagnetic ordering parameter **L** is not so simple as in the ferromagnetic case. This is so because under symmetry operations involving an interchange of magnetic sublattices the vector **L** undergoes a change in sign. In consequence, tables of i- and c-tensors often published in papers [54, 105, 118, 119, 147–149] and monographs [141, 142, 144, 145] are inapplicable to the multipolar susceptibilities **S**, **Q**, and $\boldsymbol{\alpha}$ with odd-fold repeated superscript L.

The relation between the linear and nonlinear multipole susceptibilities and the parameters describing ferro- and antiferromagnetic ordering are discussed in full detail in Appendix A, whereas Appendix B gives step by step the procedure leading to the tensors occurring in Eqs. (109), (A.12), and (A.13) for antiferromagnets belonging to the tetragonal, trigonal, and hexagonal systems composed of two sublattices with antiferromagnetic vector lying along the crystallographical z axis. Our results are assembled in Tables IV–XII.

From Table I, rotation of the polarization ellipse can also take place if the light wave incident on a magnetically ordered crystal is of weak intensity. The coupling constant $\tilde{\chi}$ then reduces to

$$
\tilde{\chi} = \frac{\omega}{2n}\sqrt{\frac{\mu_0}{\varepsilon_0}}\,\chi^{L}_{A} \tag{110}
$$

Strictly speaking, this effect occurs in ferro- and antiferromagnetic crystals already in the absence of a dc magnetic field (coupling constants f and f'). The parameter f differs from zero in ferromagnetic crystals only and is proportional to the first power of the z component of the ferromagnetic vector component M_z. It thus describes the contribution from ferromagnetic ordering in a linear approximation with respect to M_z to the linear electric-dipole susceptibility ${}^{(1)}_{e}\gamma^{(1)}_{e[xy]}(\omega)$ due to an electric-dipole transition.

The constant f' consists of two components: the linear electric-dipole susceptibility ${}^{(1)}_{e}\alpha^{(2)}_{e\,x(yz)}(\omega)$ induced by an electric-quadrupole transition and the linear magnetic-dipole susceptibility ${}^{(1)}_{e}\gamma^{(1)}_{e\,xx}(\omega)$ related to a magnetic-dipole transition. By Eq. (94), either component is the sum of two components that can occur not only in ferro- or antiferromagnetically ordered crystals but also in crystals lacking magnetic ordering. More precisely, the components $\underline{{}^{(1)}_{e}Q^{(2)}_{e\,x(yz)}(\omega)}$ and $\underline{{}^{(1)}_{e}S^{(1)}_{m\,xx}(\omega)}$ determine the linear electric-dipole susceptibility due, respectively, to an electric-quadrupole and magnetic-dipole transition; they can differ from zero in magnetically ordered and magnetically unordered crystals as well as in gaseous and liquid media. The other components, ${}^{(1)}_{e}Q^{(2)MM}_{e\,x(yz)zz}(\omega)$ and ${}^{(1)}_{e}S^{(1)MM}_{m\,xxzz}(\omega)$, describe the contribution to linear susceptibility from, respectively, ${}^{(1)}_{e}\alpha^{(2)}_{e\,x(yz)}(\omega)$ and ${}^{(1)}_{e}\gamma^{(1)}_{m\,xx}(\omega)$ related with ferromagnetic order, whereas ${}^{(1)}_{e}Q^{(2)LL}_{e\,x(yz)zz}(\omega)$ and ${}^{(1)}_{e}S^{(1)LL}_{m\,xxzz}(\omega)$ describe contributions related to antiferromagnetic order in an approximation quadratic in M_z and L_z.

The parameters f_1 and f_1' describe the linear (proportional to the first power of the dc magnetic field), whereas the parameters f_2 and f_2' describe the quadratic (proportional to the square of the magnetic field) magnetic changes in the linear electric-dipole susceptibility due, respectively, to electric-dipole transitions ${}^{(1)}_{e}\gamma^{(1)m}_{e[xy]z}(\omega)$ and ${}^{(1)}_{e}\gamma^{(1)mm}_{e[xy]zz}(\omega)$, electric-quadrupole transitions ${}^{(1)}_{e}\alpha^{(2)m}_{e\,x(yz)z}(\omega)$ and ${}^{(1)}_{e}\alpha^{(2)mm}_{e\,x(yz)zz}(\omega)$, and magnetic-dipole transitions ${}^{(1)}_{e}\gamma^{(1)m}_{m\,xxz}(\omega)$ and ${}^{(1)}_{e}\gamma^{(1)mm}_{m\,xxzz}(\omega)$.

In the case of an incident light wave of high intensity, contributions to the rotation of the polarization ellipse will also come from the parameters F and F' in the absence of a magnetic field and from F_1 and F_1' in the presence of a magnetic field in the linear approximation, and from F_2 in a quadratic approximation; one readily notes that the five F's are nonlinear counterparts of the parameters f, f', f_1, f_1', and f_2.

The coupling constant R consists of two components. The first, ${}^{(1)}_{e}Q^{(1,1,1)}_{eee\,xxyy}(\omega)$, accounts for the contribution to the nonlinear electric-dipole susceptibility ${}^{(1)}_{e}\alpha^{(1,1,1)}_{eee\,xxyy}(\omega)$ coming from the paramagnetic (unordered) phase, and the other, ${}^{(1)}_{e}Q^{(1,1,1)MM}_{eee\,xxyyzz}(\omega)\ M_z^2$ and ${}^{(1)}_{e}Q^{(1,1,1)LL}_{eee\,xxyyzz}(\omega)$ L_z^2, accounts for the contribution to ${}^{(1)}_{e}\alpha^{(1,1,1)}_{eee\,xxyy}(\omega)$ due to ferro- or respectively antiferromagnetic order. The parameter R' is related to the nonlinear electric-dipole susceptibility ${}^{(1)}_{e}\gamma^{(1,1,2)}_{eee\,xxy(yz)}(\omega)$ and ${}^{(1)}_{e}\alpha^{(1,1,1)}_{eem\,xxyx}(\omega)$. The coupling constants R_1 and R'_1 determine the linear magnetically induced change in the nonlinear electric-dipole susceptibilities ${}^{(1)}_{e}\alpha^{(1,1,1)m}_{eee\,xxyyz}(\omega)$, ${}^{(1)}_{e}\gamma^{(1,1,2)m}_{eee\,xxy(yz)z}(\omega)$, and ${}^{(1)}_{e}\alpha^{(1,1,1)m}_{eem\,xxyxz}(\omega)$, respectively, whereas R_2 determines the quadratic magnetically induced change in the nonlinear susceptibility ${}^{(1)}_{e}\alpha^{(1,1,1)mm}_{eee\,xyxyzz}(\omega)$, and has no linear counterparts.

Rotation of the polarization ellipse can also take place in paramagnetic media, where we include magnetically unordered crystals as well as liquids and gases. Such materials can exhibit only the linear and nonlinear multipolar susceptibilities underlined in Eqs. (93)–(108) and thus the linear parameters f', f_1, and f'_2 and the nonlinear parameters F', F_1, R, R'_1, and R_2. Strictly speaking, in gases and liquids whose molecules do not interact, the parameter f' will be dependent on the linear susceptibility ${}^{(1)}_{e}S^{(1)}_{e\,xx}(\omega)$ only, because ${}^{(1)}_{e}Q^{(2)}_{e\,x(yz)}(\omega)$ vanishes as the result of isotropic averaging of the electric-dipole polarizability due to the electric-quadrupole transition, which is a third-rank polar tensor, symmetric in its indices relating to the quadrupole moment.

B. Rotation of the Polarization Plane of Linearly Polarized Light

The plane determined by the light propagation direction and the direction of oscillations of the electric vector of the linearly polarized light wave is referred to as the plane of polarization of linearly polarized light.

Let us assume a light wave of considerable intensity, polarized linearly $\Psi(0) = \pi/4$, as incident on the medium. It can generally be dealt with as the superposition of two modes, polarized circularly with opposite handednesses. We write

$$|E_+| = |E_-| \quad \text{and} \quad E_+^* E_- = -(E_+ E_-^*) \tag{111}$$

leading to $S_{30} = 0$ and $S_{10} = 0$. Accordingly, Eq. (89) reduces to

$$\Psi(z) = \frac{\pi}{4} - \frac{z\omega}{4n}\sqrt{\frac{\mu_0}{\varepsilon_0}}\left\{\chi_A^L + \chi_G^{NL} S_0\right\} \qquad \Phi(z) = 0 \tag{112}$$

The situation we are now considering is this: The high-intensity linearly polarized light wave, on traversing a path z along the z axis of the crystal with symmetry defined by (75), conserves its state of linear polarization, but its plane of polarization undergoes a rotation by an angle $\Psi(z)$ about the z axis. The same occurs if the incident light wave is of moderate intensity ($\chi_A^L \gg \chi_G^{NL} S_0$). Then

$$\Psi(z) = \frac{\pi}{4} - \frac{z\omega}{4n}\sqrt{\frac{\mu_0}{\varepsilon_0}}\,\chi_A^L \tag{113}$$

Thus, the rotation of the plane of polarization is governed by the constants χ_A^L and $\chi_G^{NL} S_0$. The magnetic classes exhibiting ferro- and antiferromagnetic ordering for which rotation of the plane of polarization of light can take place are listed in Table I.

The coupling constant f determines the so-called spontaneous Faraday effect. By Eq. (93) it can take place in ferromagnetics only [78], whereas f' determines natural optical activity [48, 63, 64, 66, 67] in an electric-quadrupole (susceptibility ${}^{(1)}_{e}\alpha^{(2)}_{e\,x(yz)}(\omega)$) and magnetic-dipole (susceptibility ${}^{(1)}_{e}\gamma^{(1)}_{m\,xx}(\omega)$) approximation, respectively. From Eq. (94), we note that natural optical activity can occur in magnetically unordered media (${}^{(1)}_{e}S^{(1)}_{m\,xx}(\omega)$ and ${}^{(1)}_{e}Q^{(2)}_{m\,x(yz)}(\omega)$; the latter susceptibility vanishes for gases and liquids consisting of noninteracting molecules [53]), as well as in magnetically ordered media. Magnetic order gives rise to an additional contribution, proportional to the square of the ferromagnetic moment in ferromagnetics and to the square of the antiferromagnetic moment in antiferromagnets. Moreover, one notes immediately the similarity between these contributions regarding their dependence on the moments characterizing the type of ordering.

The coupling constant f_1 is responsible for the normal Faraday effect present in all materials (susceptibility ${}^{(1)}_{e}S^{(1)M}_{e[xy]z}(\omega)$), whereas, ferro- and antiferromagnets give rise additionally to contributions proportional to the square of the ferromagnetic or antiferromagnetic moment. These contributions are described by functions of different forms. The parameter f_1' determines the linear magnetic variation in natural optical activity [76], an effect that is restricted to magnetically ordered crystals. The parameter, f_2, accounts for rotation of the plane of light polarization proportional to the square of the dc magnetic field strength. The effect is restricted to ferromagnetic crystals [76, 88, 89]. The parameter f_2' describes the quadratic magnetic variation of the natural optical activity. The effect is exhibited by magnetically ordered and unordered crystals as well as by gases and liquids with noninteracting molecules [76]. The parameters F, F', F_1, F_1', and F_2 describe successively the nonlinear (self-induced, proportional to the light intensity) counterparts of the effects described by the constants f, f', f_1, f_1', and f_2 [114, 118, 119, 128].

C. Elliptization of Linearly Polarized Light

Let us assume that the light wave incident on the medium is linearly polarized with the electric vector oscillating along the x axis, $\mathbf{E}(0, t) = \mathbf{e}_x E(0, t)$. From definition (40) of Stokes parameters we now find that $S_{20} = 0$ and $S_{30} = 0$, so that solutions (86)–(88) reduce to

$$\begin{aligned} S_1(z) &= S_1^{NL}(z) \\ S_2(z) &= S_2^{NL}(z) \\ S_3(z) &= \frac{\tilde{\Theta}^{NL}(\cos 2\tilde{\chi} z - 1) + \Lambda^{NL} \sin^2 \tilde{\chi} z}{2\tilde{\chi}} S_{10}^2 \end{aligned} \tag{114}$$

with

$$\begin{aligned} S_1^{NL}(z) &= S_{10} \cos \tilde{\chi} z \\ S_2^{NL}(z) &= -S_{10} \sin \tilde{\chi} z \\ \tilde{\chi} &= \frac{\omega}{2n} \sqrt{\frac{\mu_0}{\varepsilon_0}} \{\chi_A^L + \chi_G^{NL} S_0\} \end{aligned} \tag{115}$$

On inserting of these solutions into (41) we get

$$\Phi(z) = \tfrac{1}{2} \arcsin\left(\Lambda^{NL} S_{10} z\right) + \cdots \tag{116}$$

$$\Psi(z) = -\frac{\tilde{\chi} z}{2} \tag{117}$$

When deriving (116) we took into account only the first term of the series expansion of $\sin 2\tilde{\chi} z$ and $(\cos 2\tilde{\chi} z - 1)$, since the further terms, being products of the form $\Theta^{NL}(\tilde{\chi})^{2n-1}$ and $\Lambda^{NL}(\tilde{\chi})^{2n}$ with $n = 1, 2, \ldots$, are much smaller than the ones retained.

The effect now under consideration is this: The linear state of polarization of the light wave propagating in the crystal along the z axis becomes elliptic. The major axis of the ellipse lies in the xy plane and subtends an angle $\Psi(z)$ with the x axis. The ellipticity $\Phi(z)$ and rotation angle $\Psi(z)$ are dependent on different susceptibilities.

TABLE II
The Coupling Constants χ_A^L, χ_G^{NL} and Λ^{NL} for Ferro- and Antiferromagnetic Crystals with Symmetries Stated in (76) and (77)

$$\tilde{\chi} = \frac{\omega}{2n}\sqrt{\frac{\mu_0}{\varepsilon_0}}\{\chi_A^L + \chi_G^{NL} S_0\}$$

Magnetic Point Groups	χ_A^L			χ_G^{NL}			Λ^{NL}		
	$f+f'$	f_1+f_1'	f_2+f_2'	$F+\tilde{F}'$	$F_1+\tilde{F}_1'$	F_2	$\Lambda+\Lambda'$	$\Lambda_1+\Lambda_1'$	Λ_2
Ferromagnetics									
4	$f+f'$	f_1+f_1'	f_2+f_2'	$F+\tilde{F}'$	$F_1+\tilde{F}_1'$	F_2	$\Lambda+\Lambda'$	$\Lambda_1+\Lambda_1'$	Λ_2
$4\underline{22}$	$f+f'$	f_1+f_1'	f_2+f_2'	$F+\tilde{F}'$	$F_1+\tilde{F}_1'$	F_2	0	0	0
$4/m$	f	f_1	f_2	F	F_1	F_2	Λ	Λ_1	Λ_2
$4\underline{mm}$	f	f_1	f_2	F	F_1	F_2	Λ'	Λ_1'	0
$4/m\underline{mm}$	f	f_1	f_2	F	F_1	F_2	0	0	0
Antiferromagnetics									
$4/\underline{m}$	0	f_1+f_1'	0	0	$F_1+\tilde{F}_1'$	0	$\Lambda+\Lambda'$	0	Λ_2
$4mm$	0	f_1+f_1'	0	0	$F_1+\tilde{F}_1'$	0	0	$\Lambda_1+\Lambda_1'$	0
$4/\underline{m}mm$	0	f_1+f_1'	0	0	$F_1+\tilde{F}_1'$	0	0	0	0
422	f'	f_1	f_2'	$\tilde{F}'$	F_1	0	Λ'	Λ_1	0
$4/mmm$	0	f_1	0	0	F_1	0	0	Λ_1	0
$4/\underline{mmm}$	0	f_1	0	0	F_1	0	Λ'	0	0

By having recourse to the tables [142, 144] of the polar and axial i- and c-tensors of ranks 2 to 6, we determined the coupling constants χ_A^L and χ_G^{NL} responsible for the rotation $\Psi(z)$ of the polarization ellipse and the constant Λ^{NL} responsible for elliptization, for crystals with the symmetries enumerated in (76) and (77). Our results are assembled in Table II, where we have introduced the following notation:

$$\chi_A^L = f + f' + (f_1 + f_1')B_z^0 + (f_2 + \cdots)\left(B_z^0\right)^2 \tag{118}$$

$$\chi_G^{NL} = F + \tilde{F}' + \left(F_1 + \tilde{F}_1'\right)B_z^0 + (F_2 + \cdots)\left(B_z^0\right)^2 \tag{119}$$

$$\Lambda^{NL} = \Lambda + \Lambda' + (\Lambda_1 + \Lambda_1')B_z^0 + (\Lambda_2 + \cdots)\left(B_z^0\right)^2 \tag{120}$$

and where the linear parameter f, f', f_1, f_1', and f_2 and nonlinear parameters F, F_1, and F_2 are defined by expressions (93)–(99), (101), and

(103), whereas the parameters $\tilde{F}'$, $\tilde{F}_1'$, Λ, Λ', Λ_1, Λ_1', and Λ_2 have the form

$$\tilde{F}' = \frac{6ns_z}{c}\Big\{-\frac{\omega}{3}\Big[{}_e^{(1)}\alpha^{(1,1,2)}_{eee\,xxx(yz)}(\omega) + {}_e^{(1)}\alpha^{(1,1,2)}_{eee\,xxy(xz)}(\omega) - 2\,{}_e^{(1)}\alpha^{(1,1,2)}_{eee\,xyx(xz)}(\omega)\Big] + {}_e^{(1)}\gamma^{(1,1,1)}_{eem\,xxxx}(\omega) - {}_e^{(1)}\gamma^{(1,1,1)}_{eem\,xxyy}(\omega) + 2\,{}_e^{(1)}\gamma^{(1,1,1)}_{eem\,xyxy}(\omega)\Big\}$$

$$= \frac{6ns_z}{c}\Big(-\frac{\omega}{3}\Big[\underline{{}_e^{(1)}Q^{(1,1,2)}_{eee\,xxx(yz)}(\omega)} + \underline{{}_e^{(1)}Q^{(1,1,2)}_{eee\,xxy(xz)}(\omega)} - 2\underline{{}_e^{(1)}Q^{(1,1,2)}_{eee\,xyx(xz)}(\omega)}\Big] + \underline{{}_e^{(1)}S^{(1,1,1)}_{eem\,xxxx}(\omega)} - \underline{{}_e^{(1)}S^{(1,1,1)}_{eem\,xxyy}(\omega)} + 2\underline{{}_e^{(1)}S^{(1,1,1)}_{eem\,xyxy}(\omega)}$$
$$+\Big\{-\frac{\omega}{3}\Big[{}_e^{(1)}Q^{(1,1,2)MM}_{eee\,xxx(yz)zz}(\omega) + {}_e^{(1)}Q^{(1,1,2)MM}_{eee\,xxy(xz)zz}(\omega) - 2\,{}_e^{(1)}Q^{(1,1,2)MM}_{eee\,xyx(xz)zz}(\omega)\Big] + {}_e^{(1)}S^{(1,1,1)MM}_{eem\,xxxxzz}(\omega) - {}_e^{(1)}S^{(1,1,1)MM}_{eem\,xxyyzz}(\omega) + 2\,{}_e^{(1)}S^{(1,1,1)MM}_{eem\,xyxyzz}(\omega)\Big\}M_z^2; \tag{121}$$

above, perform the interchange $M \to L\Big)$

$$\tilde{F}_1' = \frac{6ns_z}{c}\Big\{-\frac{\omega}{3}\Big[{}_e^{(1)}\alpha^{(1,1,2)m}_{eee\,xxx(yz)z}(\omega) + {}_e^{(1)}\alpha^{(1,1,2)m}_{eee\,xxy(xz)z}(\omega) - 2\,{}_e^{(1)}\alpha^{(1,1,2)m}_{eee\,xyx(xz)z}(\omega)\Big] + {}_e^{(1)}\gamma^{(1,1,1)m}_{eem\,xxxxz}(\omega) - {}_e^{(1)}\gamma^{(1,1,1)m}_{eem\,xxyyz}(\omega) + 2\,{}_e^{(1)}\gamma^{(1,1,1)m}_{eem\,xyxyz}(\omega)\Big\}$$

$$= \frac{6ns_z}{c}\Big[\Big(\Big\{-\frac{2\omega}{3}\Big[{}_e^{(1)}Q^{(1,1,2)MM}_{eee\,xxx(yz)zz}(\omega) + {}_e^{(1)}Q^{(1,1,2)MM}_{eee\,xxy(xz)zz}(\omega) - 2\,{}_e^{(1)}Q^{(1,1,2)MM}_{eee\,xyx(xz)zz}(\omega)\Big] + 2\Big[{}_e^{(1)}S^{(1,1,1)MM}_{eem\,xxxxzz}(\omega) - {}_e^{(1)}S^{(1,1,1)MM}_{eem\,xxyyzz}(\omega) + 2\,{}_e^{(1)}S^{(1,1,1)MM}_{eem\,xyxyzz}(\omega)\Big]\Big\}\alpha^{Mm}_{zz}$$
$$+\Big\{-\frac{\omega}{3}\Big[{}_e^{(1)}Q^{(1,1,2)ML}_{eee\,xxx(yz)zz}(\omega) + {}_e^{(1)}Q^{(1,1,2)ML}_{eee\,xxy(xz)zz}(\omega) - 2\,{}_e^{(1)}Q^{(1,1,2)ML}_{ee\,xyx(xz)zz}(\omega)\Big] + {}_e^{(1)}S^{(1,1,1)ML}_{eem\,xxxxzz}(\omega) - {}_e^{(1)}S^{(1,1,1)ML}_{eem\,xxyyzz}(\omega) + 2\,{}_e^{(1)}S^{(1,1,1)ML}_{eem\,xyxyzz}(\omega)\Big\}\alpha^{Lm}_{zz}\Big)M_z; \tag{122}$$

$$\left\{-\frac{\omega}{3}\left[{}_e^{(1)}Q_{eee\,xxx(yz)zz}^{(1,1,2)ML}(\omega) + {}_e^{(1)}Q_{eee\,xxy(xz)zz}^{(1,1,2)ML}(\omega) - 2\,{}_e^{(1)}Q_{eee\,xyx(xz)zz}^{(1,1,2)ML}(\omega)\right]\right.$$

$$\left. + {}_e^{(1)}S_{eem\,xxxxzz}^{(1,1,1)ML}(\omega) - {}_e^{(1)}S_{eem\,xxyyzz}^{(1,1,1)ML}(\omega) + 2\,{}_e^{(1)}S_{eem\,xyxyzz}^{(1,1,1)ML}(\omega)\right\}\alpha_{zz}^{Mm}L_z\Bigg]$$

$$\begin{aligned}\Lambda &= -6\,{}_e^{(1)}\alpha_{eee\,xxxy}^{(1,1,1)}(\omega), \\ &= -6\Big[\underline{{}_e^{(1)}Q_{eee\,xxxy}^{(1,1,1)}(\omega)} + {}_e^{(1)}Q_{eee\,xxxyzz}^{(1,1,1)MM}(\omega)M_z^2; \\ &\qquad {}_e^{(1)}Q_{eee\,xxxy}^{(1,1,1)}(\omega) + {}_e^{(1)}Q_{eee\,xxxyzz}^{(1,1,1)LL}(\omega)L_z^2\Big]\end{aligned} \tag{123}$$

$$\begin{aligned}\Lambda' &= \frac{6ns_z}{c}\left\{\frac{\omega}{3}\left[{}_e^{(1)}\gamma_{eee\,xxx(yz)}^{(1,1,2)}(\omega) + {}_e^{(1)}\gamma_{eee\,xxy(xz)}^{(1,1,2)}(\omega) + 2\,{}_e^{(1)}\gamma_{eee\,xyx(xz)}^{(1,1,2)}(\omega)\right]\right. \\ &\qquad \left. + {}_e^{(1)}\alpha_{eem\,xxxx}^{(1,1,1)}(\omega) - {}_e^{(1)}\alpha_{eem\,xxyy}^{(1,1,1)}(\omega) - 2\,{}_e^{(1)}\alpha_{eem\,xyxy}^{(1,1,1)}(\omega)\right\} \\ &= \frac{6ns_z}{c}\left(\left\{\frac{\omega}{3}\left[{}_e^{(1)}S_{eee\,xxx(yz)z}^{(1,1,2)M}(\omega) + {}_e^{(1)}S_{eee\,xxy(xz)z}^{(1,1,2)M}(\omega) + 2\,{}_e^{(1)}S_{eee\,xyx(xz)z}^{(1,1,2)M}(\omega)\right]\right.\right. \\ &\qquad + {}_e^{(1)}Q_{eem\,xxxxz}^{(1,1,1)M}(\omega) - {}_e^{(1)}Q_{eem\,xxyyz}^{(1,1,1)M}(\omega) \\ &\qquad \left. - 2\,{}_e^{(1)}Q_{eem\,xyxyz}^{(1,1,1)M}(\omega)\right\}M_z + \cdots; \\ &\qquad \text{above, perform the interchange } M \to L\Big)\end{aligned} \tag{124}$$

$$\begin{aligned}\Lambda_1 &= -6\,{}_e^{(1)}\alpha_{eee\,xxxyz}^{(1,1,1)m}(\omega) \\ &= -6\Big\{\Big[2\,{}_e^{(1)}Q_{eee\,xxxyzz}^{(1,1,1)MM}(\omega)\alpha_{zz}^{Mm} \\ &\qquad + {}_e^{(1)}Q_{eee\,xxxyzz}^{(1,1,1)ML}(\omega)\alpha_{zz}^{Lm}\Big]M_z + \cdots; \\ &\qquad {}_e^{(1)}Q_{eee\,xxxyzz}^{(1,1,1)ML}(\omega)\alpha_{zz}^{Mm}L_z + \cdots\Big\}\end{aligned} \tag{125}$$

$$\Lambda'_1 = \frac{6ns_z}{c}\left\{\frac{\omega}{3}\left[{}^{(1)}_{e}\gamma^{(1,1,2)m}_{eee\,xxx(yz)z}(\omega) + {}^{(1)}_{e}\gamma^{(1,1,2)m}_{eee\,xxy(xz)z}(\omega) + 2\,{}^{(1)}_{e}\gamma^{(1,1,2)m}_{eee\,xyx(xz)z}(\omega)\right]\right.$$
$$\left. + {}^{(1)}_{e}\alpha^{(1,1,1)m}_{eem\,xxxxz}(\omega) - {}^{(1)}_{e}\alpha^{(1,1,1)m}_{eem\,xxyyz}(\omega) - 2\,{}^{(1)}_{e}\alpha^{(1,1,1)m}_{eem\,xyxyz}(\omega)\right\}$$

$$= \frac{6ns_z}{c}\Bigg(\Bigg\{\frac{\omega}{3}\left[\underline{{}^{(1)}_{e}S^{(1,1,2)M}_{eee\,xxx(yz)z}(\omega)} + \underline{{}^{(1)}_{e}S^{(1,1,2)M}_{eee\,xxy(xz)z}(\omega)}\right.$$
$$\left. + 2\,\underline{{}^{(1)}_{e}S^{(1,1,2)M}_{eee\,xyx(xz)z}(\omega)}\right]$$
$$+ \underline{{}^{(1)}_{e}Q^{(1,1,1)M}_{eem\,xxxxz}(\omega)} - \underline{{}^{(1)}_{e}Q^{(1,1,1)M}_{eem\,xxyyz}(\omega)}$$
$$- 2\,\underline{{}^{(1)}_{e}Q^{(1,1,1)M}_{eem\,xyxyz}(\omega)}\Bigg\}\alpha^{Mm}_{zz}$$
$$+ \Bigg\{\frac{\omega}{3}\left[{}^{(1)}_{e}S^{(1,1,2)L}_{eee\,xxx(yz)z}(\omega) + {}^{(1)}_{e}S^{(1,1,2)L}_{eee\,xxy(xz)z}(\omega)\right. \tag{126}$$
$$\left. + 2\,{}^{(1)}_{e}S^{(1,1,2)L}_{eee\,xyx(xz)z}(\omega)\right]$$
$$+ {}^{(1)}_{e}Q^{(1,1,1)L}_{eem\,xxxxz}(\omega) - {}^{(1)}_{e}Q^{(1,1,1)L}_{eem\,xxyyz}(\omega)$$
$$- 2\,{}^{(1)}_{e}Q^{(1,1,1)L}_{eem\,xyxyz}(\omega)\Bigg\}\alpha^{Lm}_{zz} + \cdots;$$
$$+ \Bigg\{\frac{\omega}{3}\left[{}^{(1)}_{e}S^{(1,1,2)M}_{eee\,xxx(yz)z}(\omega) + {}^{(1)}_{e}S^{(1,1,2)M}_{eee\,xxy(xz)z}(\omega)\right.$$
$$\left. + 2\,{}^{(1)}_{e}S^{(1,1,2)M}_{eee\,xyx(xz)z}(\omega)\right]$$
$$+ {}^{(1)}_{e}Q^{(1,1,1)M}_{eem\,xxxxz}(\omega) - {}^{(1)}_{e}Q^{(1,1,1)M}_{eem\,xxyyz}(\omega)$$
$$- 2\,{}^{(1)}_{e}Q^{(1,1,1)M}_{eem\,xyxyz}(\omega)\Bigg\}\alpha^{Mm}_{zz} + \cdots;\Bigg)$$

$$\Lambda_2 = -6\,{}^{(1)}_{e}\alpha^{(1,1,1)mm}_{eee\,xxxyzz}(\omega)$$
$$= -6\Big\{\underline{{}^{(1)}_{e}Q^{(1,1,1)MM}_{eee\,xxxyzz}(\omega)}\left[\left(\alpha^{Mm}_{zz}\right)^2 + 2M_z\alpha^{Mmm}_{zzz}\right]$$
$$+ {}^{(1)}_{e}Q^{(1,1,1)LL}_{eee\,xxxyzz}(\omega)\left(\alpha^{Lm}_{zz}\right)^2 \tag{127}$$
$$+ {}^{(1)}_{e}Q^{(1,1,1)ML}_{eee\,xxxyzz}(\omega)\left[\alpha^{Mm}_{zz}\alpha^{Lm}_{zz} + M_z\alpha^{Lmm}_{zzz}\right];$$
$$+ {}^{(1)}_{e}Q^{(1,1,1)MM}_{eee\,xxxyzz}(\omega)\,\alpha^{Mm}_{zz}\Big\}$$

The constants Λ and Λ' describe the elliptization self-induced in an intense light wave, linearly polarized in the absence of a dc magnetic field.

The constant Λ can intervene in magnetically ordered as well as unordered crystals, whereas Λ' can intervene only in ordered crystals. Λ_1 and Λ'_1 determine the linear magnetic change in self-induced ellipticity in the electric-dipole and, respectively, magnetic-dipole and electric-quadrupole approximations. Strictly speaking, Λ_1 describes the magnetic analog of the nonlinear longitudinal Pockels effect (the nonlinear counterpart of the magnetic analog of the longitudinal Pockels effect [76, 84–86, 90, 91]). The effect can take place in magnetically ordered crystals only. Λ'_1 describes the nonlinear counterpart of nonreciprocal linear birefringence [53] and can take place in magnetically ordered as well as unordered crystals. The constant Λ_2 describes the quadratic magnetic variation in self-induced ellipticity possible in magnetically ordered and unordered materials.

V. CONCLUSION

Our paper gives an analysis of the self-induced changes in the state polarization of a strong light wave propagating along the z axis of a ferro- or antiferromagnetically ordered crystal with spatial dispersion acted on by a dc magnetic field applied parallel to the light propagation direction (Faraday configuration). We show (see Table I) that if the incident light wave, of high or moderate intensity, is elliptically polarized, it retains its ellipticity in the medium but the major axis of the ellipse is rotated by an angle $\Psi(z)$ in the xy plane. The effect can take place in magnetically ordered as well as unordered crystals in the presence or absence of a static magnetic field. If the incident light wave is linearly polarized and the crystal belongs to the trigonal or hexagonal system, the light wave of high or moderate intensity is still polarized linearly at the output from the crystal but its plane of polarization has undergone a rotation by $\Psi(z)$ in the xy plane. On traversal of a crystal belonging to the tetragonal system, the initially linearly polarized light wave moreover undergoes elliptization (see Table II).

Neither self-induced elliptization or its linear and quadratic magnetically induced variations have as yet been observed in experiment.

APPENDIX A

Dzialoshinskii [150] has shown that the magnetic properties of magnetically ordered crystals are more conveniently described in terms of certain linear combinations of the magnetic moments $\mathbf{S}_i$ for the individual magnetic sublattices (the index i labels the sublattices) rather than in terms of the moments $\mathbf{S}_i$ themselves. The combinations are constructed so as to transform according to the irreducible representation of the magnetic point group (class) describing the directional symmetry of the crystal. In the case of crystals with two magnetic sublattices the lattice vectors admit of two combinations only:

$$\mathbf{M} = \mathbf{S}_1 + \mathbf{S}_2 \quad \text{and} \quad \mathbf{L} = \mathbf{S}_1 - \mathbf{S}_2, \tag{A.1}$$

where $\mathbf{M}$ describes ferromagnetic ordering and thus is referred to as the ferromagnetic vector and, respectively, the antiferromagnetic vector $\mathbf{L}$ describes antiferromagnetic order. For ferromagnetics $\mathbf{M} \neq 0$ and $\mathbf{L} = 0$, whereas for antiferromagnetics $\mathbf{M} = 0$ and $\mathbf{L} \neq 0$. The two vectors, like the lattice vectors $\mathbf{S}_1$ and $\mathbf{S}_2$, are axial vectors antisymmetric with respect to time inversion. In spite of this, the transformation properties of $\mathbf{L}$ can differ quite considerably from those of the other three. The essential difference emerges if the operations of symmetry of the magnetic point group (class) include that of interchange of the sublattices. This operation transforms $\mathbf{M}$ into itself, whereas the vector $\mathbf{L}$ experiences a change in sign [151]. If the operation of symmetry does not cause an interchange of the lattices, the vector $\mathbf{L}$ transforms as $\mathbf{M}$ and the lattice vectors $\mathbf{S}_i$. In the case of crystals with two magnetic sublattices, the operations involving an interchange in positions of the sublattices are the product of space inversion and time inversion, and that of time inversion and reflection with respect to a plane of reflection to which the lattic vectors $\mathbf{S}_1$ and $\mathbf{S}_2$ are perpendicular.

A dc magnetic field acting on a magnetically ordered crystal affects the vectors $\mathbf{M}$ and $\mathbf{L}$ characterizing its magnetic ordering. In a dc magnetic field they become [89]

$$M_i(\mathbf{B}^0) = M_i + \alpha_{ij}^{Mm} B_j^0 + \alpha_{i(jk)}^{Mmm} B_j^0 B_k^0 + \cdots \tag{A.2}$$

$$L_i(\mathbf{B}^0) = L_i + \alpha_{ij}^{Lm} B_j^0 + \alpha_{i(jk)}^{Lmm} B_j^0 B_k^0 + \cdots \tag{A.3}$$

where M_i and L_i are the ith component of the ferromagnetic and antiferromagnetic vector in the absence of the field, whereas α_{ij}^{Mm} and α_{ij}^{Lm} are polar i-tensors of rank 2 describing the change in the ferro- and, respectively, antiferromagnetic vector caused by the dc magnetic field in the linear approximation (the first order of stationary perturbation calculus). The c-tensors of rank 3 $\alpha_{i(jk)}^{Mmm}$ and $\alpha_{i(jk)}^{Lmm}$, symmetric in the indices j, k describe, respectively, the quadratic (second order of stationary perturbation calculus) magnetic variation of the ferromagnetic and antiferromagnetic vector.

When a magnetically ordered crystal is acted on by a moderately strong magnetic field, the linear ${}^{(a)}_{A}\boldsymbol{\alpha}^{(b)}_{B}(\omega, \mathbf{B}^0), {}^{(a)}_{A}\boldsymbol{\gamma}^{(b)}_{B}(\omega, \mathbf{B}^0)$, and nonlinear ${}^{(a)}_{A}\boldsymbol{\alpha}^{(b,c,d)}_{BCD}(\omega, \mathbf{B}^0), {}^{(a)}_{A}\boldsymbol{\gamma}^{(b,c,d)}_{BCD}(\omega, \mathbf{B}^0)$ multipolar susceptibilities are expandable in series in $\mathbf{M}(\mathbf{B}^0)$ and $\mathbf{L}(\mathbf{B}^0)$. Following Pisarev [78], we get for $A = B$

$$
\begin{aligned}
{}^{(a)}_{A}\boldsymbol{\alpha}^{(b)}_{A}(\omega, \mathbf{B}^0) = {}^{(a)}_{A}\mathbf{Q}^{(b)}_{A}(\omega) &+ {}^{(a)}_{A}\mathbf{Q}^{(b)MM}_{A}(\omega) \cdot\cdot\, \mathbf{M}(\mathbf{B}^0)\mathbf{M}(\mathbf{B}^0) \\
&+ {}^{(a)}_{A}\mathbf{Q}^{(b)LL}_{A}(\omega) \cdot\cdot\, \mathbf{L}(\mathbf{B}^0)\mathbf{L}(\mathbf{B}^0) \\
&+ {}^{(a)}_{A}\mathbf{Q}^{(b)ML}_{A}(\omega) \cdot\cdot\, \mathbf{M}(\mathbf{B}^0)\mathbf{L}(\mathbf{B}^0) + \cdots
\end{aligned}
\tag{A.4}
$$

$$
\begin{aligned}
{}^{(a)}_{A}\boldsymbol{\gamma}^{(b)}_{A}(\omega, \mathbf{B}^0) = {}^{(a)}_{A}\mathbf{S}^{(b)M}_{A}(\omega) \cdot \mathbf{M}(\mathbf{B}^0) &+ {}^{(a)}_{A}\mathbf{S}^{(b)MMM}_{A}(\omega) \cdot\cdot\cdot\, \mathbf{M}(\mathbf{B}^0)^3 \\
&+ {}^{(a)}_{A}\mathbf{S}^{(b)MLL}_{A}(\omega) \cdot\cdot\cdot\, \mathbf{M}(\mathbf{B}^0)\mathbf{L}(\mathbf{B}^0)^2 \\
&+ {}^{(a)}_{A}\mathbf{S}^{(b)L}_{A}(\omega) \cdot \mathbf{L}(\mathbf{B}^0) + {}^{(a)}_{A}\mathbf{S}^{(b)LLL}_{A}(\omega) \cdot\cdot\cdot\, \mathbf{L}(\mathbf{B}^0)^3 \\
&+ {}^{(a)}_{A}\mathbf{S}^{(b)LMMM}_{A}(\omega) \cdot\cdot\cdot\, \mathbf{L}(\mathbf{B}^0)\mathbf{M}(\mathbf{B}^0)^2 + \cdots
\end{aligned}
\tag{A.5}
$$

whereas for $A \neq B$

$$
\begin{aligned}
{}^{(a)}_{A}\boldsymbol{\alpha}^{(b)}_{B}(\omega, \mathbf{B}^0) = {}^{(a)}_{A}\mathbf{Q}^{(b)M}_{B}(\omega) \cdot \mathbf{M}(\mathbf{B}^0) &+ {}^{(a)}_{A}\mathbf{Q}^{(b)MMM}_{B}(\omega) \cdot\cdot\cdot\, \mathbf{M}(\mathbf{B}^0)^3 \\
&+ {}^{(a)}_{A}\mathbf{Q}^{(b)MLL}_{B}(\omega) \cdot\cdot\cdot\, \mathbf{M}(\mathbf{B}^0)\mathbf{L}(\mathbf{B}^0)^2 \\
&+ {}^{(a)}_{A}\mathbf{Q}^{(b)L}_{B}(\omega) \cdot \mathbf{L}(\mathbf{B}^0) + {}^{(a)}_{A}\mathbf{Q}^{(b)LLL}_{B}(\omega) \cdot\cdot\cdot\, \mathbf{L}(\mathbf{B}^0)^3 \\
&+ {}^{(a)}_{A}\mathbf{Q}^{(b)LMM}_{B}(\omega) \cdot\cdot\cdot\, \mathbf{L}(\mathbf{B}^0)\mathbf{M}(\mathbf{B}^0)^2 + \cdots
\end{aligned}
\tag{A.6}
$$

$$
\begin{aligned}
{}^{(a)}_{A}\boldsymbol{\gamma}^{(b)}_{B}(\omega, \mathbf{B}^0) = {}^{(a)}_{A}\mathbf{S}^{(b)}_{B}(\omega) &+ {}^{(a)}_{A}\mathbf{S}^{(b)MM}_{B}(\omega) \cdot\cdot\, \mathbf{M}(\mathbf{B}^0)\mathbf{M}(\mathbf{B}^0) \\
&+ {}^{(a)}_{A}\mathbf{S}^{(b)LL}_{B}(\omega) \cdot\cdot\, \mathbf{L}(\mathbf{B}^0)\mathbf{L}(\mathbf{B}^0) \\
&+ {}^{(a)}_{A}\mathbf{S}^{(b)ML}_{B}(\omega) \cdot\cdot\, \mathbf{M}(\mathbf{B}^0)\mathbf{L}(\mathbf{B}^0) + \cdots
\end{aligned}
\tag{A.7}
$$

Similarly, we express the nonlinear multipole susceptibilities ${}^{(a)}_{A}\boldsymbol{\alpha}^{(b,c,d)}_{BCD}(\omega, \mathbf{B}^0)$ and ${}^{(a)}_{A}\boldsymbol{\gamma}^{(b,c,d)}_{BCD}(\omega, \mathbf{B}^0)$. For $A = B = C = D$ or $A = B \neq C = D$, $A = C \neq B = D$, and $A = D \neq B = C$ we have

$$\begin{aligned}{}^{(a)}_{A}\boldsymbol{\alpha}^{(b,c,d)}_{AAA}(\omega, \mathbf{B}^0) = {}^{(a)}_{A}\mathbf{Q}^{(b,c,d)}_{AAA}(\omega) &+ {}^{(a)}_{A}\mathbf{Q}^{(b,c,d)MM}_{AAA}(\omega) \cdot\cdot\, \mathbf{M}(\mathbf{B}^0)\mathbf{M}(\mathbf{B}^0)\\ &+ {}^{(a)}_{A}\mathbf{Q}^{(b,c,d)LL}_{AAA}(\omega) \cdot\cdot\, \mathbf{L}(\mathbf{B}^0)\mathbf{L}(\mathbf{B}^0)\\ &+ {}^{(a)}_{A}\mathbf{Q}^{(b,c,d)ML}_{AAA}(\omega) \cdot\cdot\, \mathbf{M}(\mathbf{B}^0)\mathbf{L}(\mathbf{B}^0) + \cdots\end{aligned} \tag{A.8}$$

$$\begin{aligned}&{}^{(a)}_{A}\boldsymbol{\gamma}^{(b,c,d)}_{AAA}(\omega, \mathbf{B}^0)\\ &\quad= {}^{(a)}_{A}\mathbf{S}^{(b,c,d)M}_{AAA}(\omega) \cdot \mathbf{M}(\mathbf{B}^0) + {}^{(a)}_{A}\mathbf{S}^{(b,c,d)MMM}_{AAA}(\omega) \cdot\cdot\cdot\, \mathbf{M}(\mathbf{B}^0)^3\\ &\quad+ {}^{(a)}_{A}\mathbf{S}^{(b,c,d)MLL}_{AAA}(\omega) \cdot\cdot\cdot\, \mathbf{M}(\mathbf{B}^0)\mathbf{L}(\mathbf{B}^0)^2\\ &\quad+ {}^{(a)}_{A}\mathbf{S}^{(b,c,d)L}_{AAA}(\omega) \cdot \mathbf{L}(\mathbf{B}^0) + {}^{(a)}_{A}\mathbf{S}^{(b,c,d)LLL}_{AAA}(\omega) \cdot\cdot\cdot\, \mathbf{L}(\mathbf{B}^0)^3\\ &\quad+ {}^{(a)}_{A}\mathbf{S}^{(b,c,d)LMM}_{AAA}(\omega) \cdot\cdot\cdot\, \mathbf{L}(\mathbf{B}^0)\mathbf{M}(\mathbf{B}^0)^2 + \cdots\end{aligned} \tag{A.9}$$

whereas for $A = B = C \neq D$ (odd recurrences of e or, respectively, m) we obtain

$$\begin{aligned}&{}^{(a)}_{A}\boldsymbol{\alpha}^{(b,c,d)}_{AAD}(\omega, \mathbf{B}^0)\\ &\quad= {}^{(a)}_{A}\mathbf{Q}^{(b,c,d)M}_{AAD}(\omega) \cdot \mathbf{M}(\mathbf{B}^0) + {}^{(a)}_{A}\mathbf{Q}^{(b,c,d)MMM}_{AAD}(\omega) \cdot\cdot\cdot\, \mathbf{M}(\mathbf{B}^0)^3\\ &\quad+ {}^{(a)}_{A}\mathbf{Q}^{(b,c,d)MLL}_{AAD}(\omega) \cdot\cdot\cdot\, \mathbf{M}(\mathbf{B}^0)\mathbf{L}(\mathbf{B}^0)^2\\ &\quad+ {}^{(a)}_{A}\mathbf{Q}^{(b,c,d)L}_{AAD}(\omega) \cdot \mathbf{L}(\mathbf{B}^0) + {}^{(a)}_{A}\mathbf{Q}^{(b,c,d)LLL}_{AAD}(\omega) \cdot\cdot\cdot\, \mathbf{L}(\mathbf{B}^0)^3\\ &\quad+ {}^{(a)}_{A}\mathbf{Q}^{(b,c,d)LMM}_{AAD}(\omega) \cdot\cdot\cdot\, \mathbf{L}(\mathbf{B}^0)\mathbf{M}(\mathbf{B}^0)^2 + \cdots\end{aligned} \tag{A.10}$$

$$\begin{aligned}&{}^{(a)}_{A}\boldsymbol{\gamma}^{(b,c,d)}_{AAD}(\omega, \mathbf{B}^0)\\ &\quad= {}^{(a)}_{A}\mathbf{S}^{(b,c,d)}_{AAD}(\omega) + {}^{(a)}_{A}\mathbf{S}^{(b,c,d)MM}_{AAD}(\omega) \cdot\cdot\, \mathbf{M}(\mathbf{B}^0)\mathbf{M}(\mathbf{B}^0)\\ &\quad+ {}^{(a)}_{A}\mathbf{S}^{(b,c,d)LL}_{AAD}(\omega) \cdot\cdot\, \mathbf{L}(\mathbf{B}^0)\mathbf{L}(\mathbf{B}^0)\\ &\quad+ {}^{(a)}_{A}\mathbf{S}^{(b,c,d)ML}_{AAD}(\omega) \cdot\cdot\, \mathbf{M}(\mathbf{B}^0)\mathbf{L}(\mathbf{B}^0) + \cdots\end{aligned} \tag{A.11}$$

Above, all the linear

$$
\begin{array}{llll}
{}^{(a)}_{A}\mathbf{Q}^{(b)M}_{B}(\omega) & {}^{(a)}_{A}\mathbf{Q}^{(b)MM}_{B}(\omega) & {}^{(a)}_{A}\mathbf{Q}^{(b)MMM}_{B}(\omega) & {}^{(a)}_{A}\mathbf{Q}^{(b)MLL}_{B}(\omega) \\
{}^{(a)}_{A}\mathbf{Q}^{(b)L}_{B}(\omega) & {}^{(a)}_{A}\mathbf{Q}^{(b)LL}_{B}(\omega) & {}^{(a)}_{A}\mathbf{Q}^{(b)LLL}_{B}(\omega) & {}^{(a)}_{A}\mathbf{Q}^{(b)LMM}_{B}(\omega) \\
{}^{(a)}_{A}\mathbf{S}^{(b)M}_{B}(\omega) & {}^{(a)}_{A}\mathbf{S}^{(b)MM}_{B}(\omega) & {}^{(a)}_{A}\mathbf{S}^{(b)MMM}_{B}(\omega) & {}^{(a)}_{A}\mathbf{S}^{(b)MLL}_{B}(\omega) \\
{}^{(a)}_{A}\mathbf{S}^{(b)L}_{B}(\omega) & {}^{(a)}_{A}\mathbf{S}^{(b)LL}_{B}(\omega) & {}^{(a)}_{A}\mathbf{S}^{(b)LLL}_{B}(\omega) & {}^{(a)}_{A}\mathbf{S}^{(b)LMM}_{B}(\omega)
\end{array}
\tag{A.12}
$$

and nonlinear multipole susceptibilities

$$
\begin{array}{lll}
{}^{(a)}_{A}\mathbf{Q}^{(b,c,d)M}_{BCD}(\omega) & {}^{(a)}_{A}\mathbf{Q}^{(b,c,d)MM}_{BCD}(\omega) & {}^{(a)}_{A}\mathbf{Q}^{(b,c,d)MMM}_{BCD}(\omega) \\
{}^{(a)}_{A}\mathbf{Q}^{(b,c,d)MLL}_{BCD}(\omega) & {}^{(a)}_{A}\mathbf{Q}^{(b,c,d)L}_{BCD}(\omega) & {}^{(a)}_{A}\mathbf{Q}^{(b,c,d)LL}_{BCD}(\omega) \\
{}^{(a)}_{A}\mathbf{Q}^{(b,c,d)LLL}_{BCD}(\omega) & {}^{(a)}_{A}\mathbf{Q}^{(b,c,d)LMM}_{BCD}(\omega) & {}^{(a)}_{A}\mathbf{S}^{(b,c,d)M}_{BCD}(\omega) \\
{}^{(a)}_{A}\mathbf{S}^{(b,c,d)MM}_{BCD}(\omega) & {}^{(a)}_{A}\mathbf{S}^{(b,c,d)MMM}_{BCD}(\omega) & {}^{(a)}_{A}\mathbf{S}^{(b,c,d)MLL}_{BCD}(\omega) \\
{}^{(a)}_{A}\mathbf{S}^{(b,c,d)L}_{BCD}(\omega) & {}^{(a)}_{A}\mathbf{S}^{(b,c,d)LL}_{BCD}(\omega) & {}^{(a)}_{A}\mathbf{S}^{(b,c,d)LLL}_{BCD}(\omega) \\
{}^{(a)}_{A}\mathbf{S}^{(b,c,d)LMM}_{BCD}(\omega) & &
\end{array}
\tag{A.13}
$$

dependent on **M** and **L** are real; those denoted as **Q** are polar i-tensors, whereas those denoted as **S** are axial i-tensors.

With expansions A.2 and A.3 taken into account, a comparison of Eqs. 33–36 and, respectively, A.4–A.11 enables us to express the linear

$$
{}^{(a)}_{A}\boldsymbol{\alpha}^{(b)}_{B}(\omega), \quad {}^{(a)}_{A}\boldsymbol{\alpha}^{(b)m}_{B}(\omega), \quad {}^{(a)}_{A}\boldsymbol{\alpha}^{(b)mm}_{B}(\omega),
$$
$$
{}^{(a)}_{A}\boldsymbol{\gamma}^{(b)}_{B}(\omega), \quad {}^{(a)}_{A}\boldsymbol{\gamma}^{(b)m}_{B}(\omega), \quad {}^{(a)}_{A}\boldsymbol{\gamma}^{(b)mm}_{B}(\omega)
$$

and nonlinear multipole susceptibilities

$$
{}^{(a)}_{A}\boldsymbol{\alpha}^{(b,c,d)}_{BCD}(\omega), \quad {}^{(a)}_{A}\boldsymbol{\alpha}^{(b,c,d)m}_{BCD}(\omega), \quad {}^{(a)}_{A}\boldsymbol{\alpha}^{(b,c,d)mm}_{BCD}(\omega),
$$
$$
{}^{(a)}_{A}\boldsymbol{\gamma}^{(b,c,d)}_{BCD}(\omega), \quad {}^{(a)}_{A}\boldsymbol{\gamma}^{(b,c,d)m}_{BCD}(\omega), \quad {}^{(a)}_{A}\boldsymbol{\gamma}^{(b,c,d)mm}_{BCD}(\omega)
$$

in terms of the quantities:

$$
\mathbf{M}, \mathbf{L}, \boldsymbol{\alpha}^{Mm}, \boldsymbol{\alpha}^{Mmm}, \boldsymbol{\alpha}^{Lm} \quad \text{and} \quad \boldsymbol{\alpha}^{Lmm}
$$

characterizing the type of magnetic ordering. We thus obtain the following:

For the linear multipole susceptibilities with $A = B$

$$
\begin{aligned}
{}^{(a)}_{A}\boldsymbol{\alpha}^{(b)}_{A}(\omega) &= \underline{{}^{(a)}_{A}\mathbf{Q}^{(b)}_{A}(\omega)} + {}^{(a)}_{A}\mathbf{Q}^{(b)MM}_{A}(\omega) \cdot\cdot \mathbf{MM} \\
&\quad + {}^{(a)}_{A}\mathbf{Q}^{(b)LL}_{A}(\omega) \cdot\cdot \mathbf{LL} + {}^{(a)}_{A}\mathbf{Q}^{(b)ML}_{A}(\omega) \cdot\cdot \mathbf{ML} + \cdots
\end{aligned} \tag{A.14}
$$

$$
\begin{aligned}
{}^{(a)}_{A}\boldsymbol{\alpha}^{(b)m}_{A}(\omega) &= 2\,{}^{(a)}_{A}\mathbf{Q}^{(b)MM}_{A}(\omega) \cdot\cdot \mathbf{M}\boldsymbol{\alpha}^{Mm} \\
&\quad + 2\,{}^{(a)}_{A}\mathbf{Q}^{(b)LL}_{A}(\omega) \cdot\cdot \mathbf{L}\boldsymbol{\alpha}^{Lm} \\
&\quad + {}^{(a)}_{A}\mathbf{Q}^{(b)ML}_{A}(\omega) \cdot\cdot [\mathbf{M}\boldsymbol{\alpha}^{Lm} + \mathbf{L}\boldsymbol{\alpha}^{Mm}] + \cdots
\end{aligned} \tag{A.15}
$$

$$
\begin{aligned}
{}^{(a)}_{A}\boldsymbol{\alpha}^{(b)mm}_{A}(\omega) &= \underline{{}^{(a)}_{A}\mathbf{Q}^{(b)MM}_{A}(\omega)} \cdot\cdot \left[(\boldsymbol{\alpha}^{Mm})^2 + 2\mathbf{M}\boldsymbol{\alpha}^{Mm}\right] \\
&\quad + {}^{(a)}_{A}\mathbf{Q}^{(b)LL}_{A}(\omega) \cdot\cdot \left[(\boldsymbol{\alpha}^{Lm})^2 + 2\mathbf{L}\boldsymbol{\alpha}^{Lmm}\right] \\
&\quad + {}^{(a)}_{A}\mathbf{Q}^{(b)ML}_{A}(\omega) \cdot\cdot [\boldsymbol{\alpha}^{Mm}\boldsymbol{\alpha}^{Lm} + \mathbf{M}\boldsymbol{\alpha}^{Lmm} + \mathbf{L}\boldsymbol{\alpha}^{Mmm}] + \cdots
\end{aligned} \tag{A.16}
$$

$$
\begin{aligned}
{}^{(a)}_{A}\boldsymbol{\gamma}^{(b)}_{A}(\omega) &= {}^{(a)}_{A}\mathbf{S}^{(b)M}_{A}(\omega) \cdot \mathbf{M} + {}^{(a)}_{A}\mathbf{S}^{(b)MMM}_{A}(\omega) \cdots \mathbf{M}^3 \\
&\quad + {}^{(a)}_{A}\mathbf{S}^{(b)MLL}_{A}(\omega) \cdots \mathbf{ML}^2 \\
&\quad + {}^{(a)}_{A}\mathbf{S}^{(b)L}_{A}(\omega) \cdot \mathbf{L} + {}^{(a)}_{A}\mathbf{S}^{(b)LLL}_{A}(\omega) \cdots \mathbf{L}^3 \\
&\quad + {}^{(a)}_{A}\mathbf{S}^{(b)LMM}_{A}(\omega) \cdots \mathbf{LM}^2 + \cdots
\end{aligned} \tag{A.17}
$$

$$
\begin{aligned}
{}^{(a)}_{A}\boldsymbol{\gamma}^{(b)m}_{A}(\omega) &= \underline{{}^{(a)}_{A}\mathbf{S}^{(b)M}_{A}(\omega)} \cdot \boldsymbol{\alpha}^{Mm} + 3\,{}^{(a)}_{A}\mathbf{S}^{(b)MMM}_{A}(\omega) \cdots \mathbf{M}^2\boldsymbol{\alpha}^{Mm} \\
&\quad + {}^{(a)}_{A}\mathbf{S}^{(b)MLL}_{A}(\omega) \cdots [\mathbf{L}^2\boldsymbol{\alpha}^{Mm} + 2\mathbf{ML}\boldsymbol{\alpha}^{Lm}] \\
&\quad + {}^{(a)}_{A}\mathbf{S}^{(b)L}_{A}(\omega) \cdot \boldsymbol{\alpha}^{lm} + 3\,{}^{(a)}_{A}\mathbf{S}^{(b)LLL}_{A}(\omega) \cdots \mathbf{L}^2\boldsymbol{\alpha}^{Lm} \\
&\quad + {}^{(a)}_{A}\mathbf{S}^{(b)LMM}_{A}(\omega) \cdots [\mathbf{M}^2\boldsymbol{\alpha}^{lm} + 2\mathbf{LM}\boldsymbol{\alpha}^{Mm}] + \cdots
\end{aligned} \tag{A.18}
$$

$$
\begin{aligned}
{}^{(a)}_{A}\boldsymbol{\gamma}^{(b)mm}_{A}(\omega) &= {}^{(a)}_{A}\mathbf{S}^{(b)M}_{A}(\omega) \cdot \boldsymbol{\alpha}^{Mmm} \\
&\quad + 3\,{}^{(a)}_{A}\mathbf{S}^{(b)MMM}_{A}(\omega) \cdots \left[\mathbf{M}(\boldsymbol{\alpha}^{Mm})^2 + \mathbf{M}^2\boldsymbol{\alpha}^{Mmm}\right] \\
&\quad + {}^{(a)}_{A}\mathbf{S}^{(b)MLL}_{A}(\omega) \cdots \Big\{2[\mathbf{L}\boldsymbol{\alpha}^{Mm}\boldsymbol{\alpha}^{Lm} + \mathbf{ML}\boldsymbol{\alpha}^{Lmm}] \\
&\qquad\qquad + M(\boldsymbol{\alpha}^{Lm})^2 + \mathbf{L}^2\boldsymbol{\alpha}^{Mmm}\Big\} \\
&\quad + {}^{(a)}_{A}\mathbf{S}^{(b)L}_{A}(\omega) \cdot \boldsymbol{\alpha}^{Lmm} \\
&\quad + 3\,{}^{(a)}_{A}\mathbf{S}^{(b)LLL}_{A}(\omega) \cdots \left[\mathbf{L}(\boldsymbol{\alpha}^{Lm})^2 + \mathbf{L}^2\boldsymbol{\alpha}^{Lmm}\right] \\
&\quad + {}^{(a)}_{A}\mathbf{S}^{(b)LMM}_{A}(\omega) \cdots \Big\{2[\mathbf{M}\boldsymbol{\alpha}^{Lm}\boldsymbol{\alpha}^{Mm} + \mathbf{LM}\boldsymbol{\alpha}^{Mmm}] \\
&\qquad\qquad + \mathbf{L}(\boldsymbol{\alpha}^{Mm})^2 + \mathbf{M}^2\boldsymbol{\alpha}^{Lmm}\Big\} + \cdots
\end{aligned} \tag{A.19}
$$

For the linear multipole susceptibilities with $A \neq B$

$$
\begin{aligned}
{}^{(a)}_{A}\boldsymbol{\alpha}^{(b)}_{B}(\omega) = {}& {}^{(a)}_{A}\mathbf{Q}^{(b)M}_{B}(\omega) \cdot \mathbf{M} + {}^{(a)}_{A}\mathbf{Q}^{(b)MMM}_{B}(\omega) \cdot\cdot\cdot \mathbf{M}^{3} \\
& + {}^{(a)}_{A}\mathbf{Q}^{(b)MLL}_{B}(\omega) \cdot\cdot\cdot \mathbf{M}\mathbf{L}^{2} \\
& + {}^{(a)}_{A}\mathbf{Q}^{(b)L}_{B}(\omega) \cdot \mathbf{L} + {}^{(a)}_{A}\mathbf{Q}^{(b)LLL}_{B}(\omega) \cdot\cdot\cdot \mathbf{L}^{3} \\
& + {}^{(a)}_{A}\mathbf{Q}^{(b)LMM}_{B}(\omega) \cdot\cdot\cdot \mathbf{L}\mathbf{M}^{2} + \cdots
\end{aligned}
\tag{A.20}
$$

$$
\begin{aligned}
{}^{(a)}_{A}\boldsymbol{\alpha}^{(b)m}_{B}(\omega) = {}& \underline{{}^{(a)}_{A}\mathbf{Q}^{(b)M}_{B}(\omega)} \cdot \boldsymbol{\alpha}^{Mm} + 3\,{}^{(a)}_{A}\mathbf{Q}^{(b)MMM}_{B}(\omega) \cdot\cdot\cdot \mathbf{M}^{2}\boldsymbol{\alpha}^{Mm} \\
& + {}^{(a)}_{A}\mathbf{Q}^{(b)MLL}_{B}(\omega) \cdot\cdot\cdot [\mathbf{L}^{2}\boldsymbol{\alpha}^{Mm} + 2\mathbf{M}\mathbf{L}\boldsymbol{\alpha}^{Lm}] \\
& + {}^{(a)}_{A}\mathbf{Q}^{(b)L}_{B}(\omega) \cdot \boldsymbol{\alpha}^{Lm} + 3\,{}^{(a)}_{A}\mathbf{Q}^{(b)LLL}_{B}(\omega) \cdot\cdot\cdot \mathbf{L}^{2}\boldsymbol{\alpha}^{Lm} \\
& + {}^{(a)}_{A}\mathbf{Q}^{(b)LMM}_{B}(\omega) \cdot\cdot\cdot [\mathbf{M}^{2}\boldsymbol{\alpha}^{Lm} + 2\mathbf{L}\mathbf{M}\boldsymbol{\alpha}^{Mm}] + \cdots
\end{aligned}
\tag{A.21}
$$

$$
\begin{aligned}
{}^{(a)}_{A}\boldsymbol{\alpha}^{(b)mm}_{B}(\omega) = {}& {}^{(a)}_{A}\mathbf{Q}^{(b)M}_{B}(\omega) \cdot \boldsymbol{\alpha}^{Mmm} \\
& + 3\,{}^{(a)}_{A}\mathbf{Q}^{(b)MMM}_{B}(\omega) \cdot\cdot\cdot \left[\mathbf{M}(\boldsymbol{\alpha}^{Mm})^{2} + \mathbf{M}^{2}\boldsymbol{\alpha}^{Mmm}\right] \\
& + {}^{(a)}_{A}\mathbf{Q}^{(b)MLL}_{B}(\omega) \cdot\cdot\cdot \Big\{2[\mathbf{L}\boldsymbol{\alpha}^{Mm}\boldsymbol{\alpha}^{Lm} + \mathbf{M}\mathbf{L}\boldsymbol{\alpha}^{Lmm}] \\
& \qquad + \mathbf{M}(\boldsymbol{\alpha}^{Lm})^{2} + \mathbf{L}^{2}\boldsymbol{\alpha}^{Mmm}\Big\} \\
& + {}^{(a)}_{A}\mathbf{Q}^{(b)L}_{B}(\omega) \cdot \boldsymbol{\alpha}^{Lmm} \\
& + 3\,{}^{(a)}_{A}\mathbf{Q}^{(b)LLL}_{B}(\omega) \cdot\cdot\cdot \left[\mathbf{L}(\boldsymbol{\alpha}^{Lm})^{2} + \mathbf{L}^{2}\boldsymbol{\alpha}^{Lmm}\right] \\
& + {}^{(a)}_{A}\mathbf{Q}^{(b)LMM}_{B}(\omega) \cdot\cdot\cdot \Big\{2[\mathbf{M}\boldsymbol{\alpha}^{Lm}\boldsymbol{\alpha}^{Mm} + \mathbf{L}\mathbf{M}\boldsymbol{\alpha}^{Mmm}] \\
& \qquad + \mathbf{L}(\boldsymbol{\alpha}^{Mm})^{2} + \mathbf{M}^{2}\boldsymbol{\alpha}^{Lmm}\Big\} + \cdots
\end{aligned}
\tag{A.22}
$$

$$
\begin{aligned}
{}^{(a)}_{A}\boldsymbol{\gamma}^{(b)}_{B}(\omega) = {}& \underline{{}^{(a)}_{A}\mathbf{S}^{(b)}_{B}(\omega)} + {}^{(a)}_{A}\mathbf{S}^{(b)MM}_{B}(\omega) \cdot\cdot \mathbf{M}\mathbf{M} \\
& + {}^{(a)}_{A}\mathbf{S}^{(b)LL}_{B}(\omega) \cdot\cdot \mathbf{L}\mathbf{L} \\
& + {}^{(a)}_{A}\mathbf{S}^{(b)ML}_{B}(\omega) \cdot\cdot \mathbf{M}\mathbf{L} + \cdots
\end{aligned}
\tag{A.23}
$$

$$
\begin{aligned}
{}^{(a)}_{A}\boldsymbol{\gamma}^{(b)m}_{B}(\omega) = {}& 2\,{}^{(a)}_{A}\mathbf{S}^{(b)MM}_{B}(\omega) \cdot\cdot \mathbf{M}\boldsymbol{\alpha}^{Mm} \\
& + 2\,{}^{(a)}_{A}\mathbf{S}^{(b)LL}_{B}(\omega) \cdot\cdot \mathbf{L}\boldsymbol{\alpha}^{Lm} \\
& + {}^{(a)}_{A}\mathbf{S}^{(b)ML}_{B}(\omega) \cdot\cdot [\mathbf{M}\boldsymbol{\alpha}^{Lm} + \mathbf{L}\boldsymbol{\alpha}^{Mm}] + \cdots
\end{aligned}
\tag{A.24}
$$

$$
\begin{aligned}
{}^{(a)}_{A}\boldsymbol{\gamma}^{(b)mm}_{B}(\omega) = \underline{{}^{(a)}_{A}\mathbf{S}^{(b)MM}_{B}(\omega)} \cdot \cdot \left[\left(\boldsymbol{\alpha}^{Mm} \right)^{2} + 2\mathbf{M}\boldsymbol{\alpha}^{Mmm} \right] \\
+ {}^{(a)}_{A}\mathbf{S}^{(b)LL}_{B}(\omega) \cdot \cdot \left[\left(\boldsymbol{\alpha}^{Lm} \right)^{2} + 2\mathbf{L}\boldsymbol{\alpha}^{Lmm} \right] \\
+ {}^{(a)}_{A}\mathbf{S}^{(b)ML}_{B}(\omega) \cdot \cdot \left[\boldsymbol{\alpha}^{Mm}\boldsymbol{\alpha}^{Lm} + \mathbf{M}\boldsymbol{\alpha}^{Lmm} + \mathbf{L}\boldsymbol{\alpha}^{Mmm} \right] \\
+ \cdots
\end{aligned}
\tag{A.25}
$$

Expressions for the nonlinear multipole susceptibilities ${}^{(a)}_{A}\boldsymbol{\alpha}^{(b,c,d)}_{AAA}(\omega)$, ${}^{(a)}_{A}\boldsymbol{\alpha}^{(b,c,d)m}_{AAA}(\omega)$, ${}^{(a)}_{A}\boldsymbol{\alpha}^{(b,c,d)mm}_{AAA}(\omega)$, ${}^{(a)}_{A}\boldsymbol{\gamma}^{(b,c,d)}_{AAA}(\omega)$, ${}^{(a)}_{A}\boldsymbol{\gamma}^{(b,c,d)m}_{AAA}(\omega)$, and ${}^{(a)}_{A}\boldsymbol{\gamma}^{(b,c,d)mm}_{AAA}(\omega)$ are easily obtained from (A.14)–(A.19) on replacing (b) by (b,c,d) and A by AAA, respectively. Likewise, on replacing (b) by (b,c,d) and B by AAB in (A.20)–(A.25) we get the nonlinear multipole susceptibilities: ${}^{(a)}_{A}\boldsymbol{\alpha}^{(b,c,d)}_{AAB}(\omega)$, ${}^{(a)}_{A}\boldsymbol{\alpha}^{(b,c,d)m}_{AAB}(\omega)$, ${}^{(a)}_{A}\boldsymbol{\alpha}^{(b,c,d)mm}_{AAB}(\omega)$, ${}^{(a)}_{A}\boldsymbol{\gamma}^{(b,c,d)}_{AAB}(\omega)$, ${}^{(a)}_{A}\boldsymbol{\gamma}^{(b,c,d)m}_{AAB}(\omega)$, and ${}^{(a)}_{A}\boldsymbol{\gamma}^{(b,c,d)mm}_{AAB}(\omega)$ as functions of the parameters characterizing the magnetic ordering.

Expression (A.2) without its first ($M_i = 0$) and third ($\alpha^{(Mmm}_{i(jk)} = 0$) components is applicable to magnetically unordered media. It then describes the magnetization induced by a dc magnetic field. With regard to the above, expressions (A.14)–(A.25) are applicable as well to magnetically unordered media; they then contain only independently occurring tensors ${}^{(a)}_{A}\mathbf{Q}^{(b)}_{B}(\omega)$ and ${}^{(a)}_{A}\mathbf{S}^{(b)}_{B}(\omega)$ and terms proportional to $\boldsymbol{\alpha}^{Mm}$. The susceptibilities ${}^{(a)}_{A}\mathbf{Q}^{(b)}_{B}(\omega)$ and ${}^{(a)}_{A}\mathbf{S}^{(b)}_{B}(\omega)$ which generate the above terms are underlined.

APPENDIX B

The structure of an antiferromagnet involves the presence of at least two magnetized sublattices. In the general case, the order existing in a crystal with two sublattices with the magnetic moments $\mathbf{S}_1$ and $\mathbf{S}_2$ can be characterized in terms of two vectors, the one ferromagnetic, $\mathbf{M}$, and the other antiferromagnetic, $\mathbf{L}$ [150]. In an ideal antiferromagnet $\mathbf{M} = 0$ and $\mathbf{L} \neq 0$, whereas in a ferromagnet the inverse holds.

The elements of symmetry determining the magnetic point groups fall in two categories: operations R_Z interchanging the positions of magnetic sublattices [151], and operations R_{NZ}, which do not interchange them. Obviously, the ferromagnetic moment, being the sum (A.1) of two axial vectors, is insensitive to the operation R_Z of interchange in labeling of the sublattices, whereas the antiferromagnetic vector undergoes a change in sign under R_Z. One notes immediately that the diad $\mathbf{LL}$ exhibits no change in sign under R_Z, whereas the triad $\mathbf{LLL}$ does. Hence, in the case of point groups involving at least one operation R_Z the vectors **M** and **L**, though axial, belong to (transform according to) different irreducible representations of the respective point group meaning that, from the viewpoint of their transformation properties, they are not identical vectors.

The well-known tables [142, 147–149] listing the shapes of the Cartesian polar and axial tensors of the first, second, third, fourth, fifth, and sixth ranks for the 32 point groups $G(P)$ (not containing time inversion) have been established on the basis of their transformational properties (transformation invariants—in fact, linear combinations of the tensor components) as proposed by Fumi and Fieschi [147–149]. They comprise one-lattice crystals only. In practice, to find the form of a Cartesian tensor of rank n (its nonzero components and the relations between them) for a given point group, one has to solve a set of 3^n equations. Those equations are linear combinations of the tensor components, transforming according to the irreducible representations of the point group considered. In the case of a polar tensor all linear combinations belonging to the fully symmetric representation are nonzero, whereas the others vanish. The set of equations obtained in this way permits the determination of the form of the polar tensor. However, in the axial case, the only nonzero combinations of the components are those belonging to the antisymmetric repre-

TABLE III
List of Representations P for All the Magnetic Point Groups $G(AFM)$ Admitting Antiferromagnetic Order and Simultaneously Belonging to the Tetragonal, Trigonal, or Hexagonal System

$G(AFM)$	$G(P)$	P	$G(AFM)$	$G(P)$	P	$G(AFM)$	$G(P)$	P
$\underline{4}$	4	B	$4/\underline{mmm}$	$4/mmm$	A_{2u}	$\underline{6}/\underline{m}$	$6/m$	B_g
$\underline{\bar{4}}$	$\bar{4}$	B	$4/\underline{m}mm$	$4/mmm$	A_{1u}	622	622	A_2
$\underline{4}/m$	$4/m$	B_g	$\underline{4}/\underline{mm}m$	$4/mmm$	B_{2u}	$\underline{6}2\underline{2}$	622	B_2
$4/\underline{m}$	$4/m$	A_u	$\underline{\bar{3}}$	$\bar{3}$	A_u	$6mm$	$6mm$	A_2
$\underline{4}/\underline{m}$	$4/m$	B_u	$\underline{\bar{3}m}$	$\bar{3}m$	A_{2u}	$\underline{6}m\underline{m}$	$6mm$	B_2
422	422	A_2	$\underline{\bar{3}}m$	$\bar{3}$	A_{1u}	$\bar{6}m2$	$\bar{6}m2$	A'_2
$4mm$	$4mm$	A_2	$\bar{3}m$	$\bar{3}m$	A_{2g}	$\underline{\bar{6}}2\underline{m}$	$\bar{6}m2$	A''_2
$\bar{4}2m$	$\bar{4}2m$	A_2	32	32	A_2	$\underline{\bar{6}}m\underline{2}$	$\bar{6}m2$	A''_1
$4/mmm$	$4/mmm$	A_{2g}	$3m$	$3m$	A_2	$6/mmm$	$6/mmm$	A_{2g}
$\underline{4}2\underline{2}$	422	B_2	6	6	B	$\underline{6}/\underline{m}m\underline{m}$	$6/mmm$	B_{2g}
$\underline{4}m\underline{m}$	$4mm$	B_2	$\underline{\bar{6}}$	$\bar{6}$	A''	$6/\underline{mmm}$	$6/mmm$	A_{2u}
$\underline{\bar{4}}2\underline{m}$	$\bar{4}2m$	B_2	$6/\underline{m}$	$6/m$	A_u	$6/\underline{m}mm$	$6/mmm$	A_{1u}
$\underline{\bar{4}}m\underline{2}$	$\bar{4}2m$	B_1	$\underline{6}/m$	$6/m$	B_u	$\underline{6}/m\underline{m}m$	$6/mmm$	B_{2u}
$\underline{4}/mm\underline{m}$	$4/mmm$	B_{2g}						

sentation. Tavger and Zaitsev [56] have shown that the correct description of the directional symmetry of magnetically ordered crystals requires the adjoining of time inversion to the elements of symmetry; including time inversion among the elements of symmetry forming the 32 well-known point groups $G(P)$, they constructed the 90 magnetic point groups G.

Subsequently, Birss [144] proposed to distinguish tensors according to their behavior, symmetrical or antisymmetrical, under time inversion, thus introducing the concepts of i- and c-tensors. To Birss [144] also is due a procedure on the basis of which and applying the results of Fumi and Fieschi [147–149] he determined the forms of the polar and axial i- and c-tensors of rank 1 up to 4; and Zawodny [142], of ranks 5 and 6 for the 90 magnetic point groups. Regrettably, the work of Fumi and Fieschi and, consequently, Birss concerned one-lattice crystals only and is not applicable to tensors that are functions of odd powers of the antiferromagnetic vector **L** in crystals, which include the element R_Z among their elements of symmetry, as is the case for all magnetic point groups admitting of antiferromagnetic ordering, i.e., with $\mathbf{M} = 0$ and $\mathbf{L} \neq 0$. More exactly, owing to the circumstance that the operation R_Z does not change sign at the ferromagnetic vector **M** or at any even power of the antiferromagnetic vector **L**, the results of the preceding authors are also valid for tensors that are functions of arbitrary powers of **M**, and even powers of **L**, in any of the magnetic point groups.

TABLE IV

The Form of the Polar Second-Rank i-Tensor α_{ij}^{Lm} for Antiferromagnetic Crystals Belonging to the Tetra-, Tri-, and Hexagonal Systems and Consisting of Two Magnetic Sublattices, the Antiferromagnetic Vector **L** Lying Parallel to the Crystallographical z Axis

Antiferromagnetic Point Group	Components α_{ij}^{Lm}
$\underline{4}, \underline{\bar{4}}, \underline{4}/m$	$xx = -yy,\ xy = yx$
$422, 4mm, \bar{4}2m, 4/mmm, 32, 3m, \bar{3}m, 622, 6mm, \bar{6}m2$ $6/mmm$	$xy = -yx$
$\underline{4}\underline{2}\underline{2}, 4m\underline{m}, \underline{\bar{4}}2\underline{m}, \underline{4}/mm\underline{m}$	$xy = yx$
$\underline{\bar{4}}m\underline{2}$	$xx = -yy$
$4/\underline{m}, \underline{4}/\underline{m}, 4/\underline{mmm}, 4/\underline{m}mm, \underline{4}/\underline{mm}m, \underline{\bar{3}}, \underline{\bar{3}m}, \underline{6}, \underline{\bar{6}}$, $6/\underline{m}, \underline{6}/m, \underline{6}/\underline{m}, \underline{6}2\underline{2}, \underline{6}m\underline{m}, \underline{\bar{6}}2\underline{m}, \underline{\bar{6}}m\underline{2}, \underline{6}/\underline{m}mm, 6/mm\underline{m}, 6/\underline{mmm}$, $6/\underline{m}mm, \underline{6}/m\underline{m}m$	All components vanish

Note. The components α_{ij}^{Lm} are denoted by the subscripts i, j, taking values x, y, and z in the laboratory frame of reference.

TABLE V

The Form of the Axial Third-Rank c-Tensor $\alpha_{i(jk)}^{Lmm}$ for Antiferromagnetic Crystals Belonging to the Tetra-, Tri-, and Hexagonal Systems and Consisting of Two Magnetic Sublattices, the Antiferromagnetic Vector **L** Lying Parallel to the Crystallographical z Axis

Antiferromagnetic Point Group	Components $\alpha_{i(jk)}^{Lmm}$
$\underline{4}, \underline{\bar{4}}, \underline{4}/m, 4/\underline{m}, \underline{4}/\underline{m}, \underline{6}, \underline{\bar{6}}, \underline{6}/m, 6/\underline{m}, \underline{6}/\underline{m}$	$d \equiv zzz,\ zxx = zyy$ $xxz = xzx = yyz = yzy$ $e \equiv xyz = -yxz = xzy = -yzx$
$422, 4\underline{2}\underline{2}, 4mm, \underline{4}m\underline{m}, \bar{4}2m, \underline{\bar{4}}2\underline{m}, \underline{\bar{4}}m\underline{2}, 4/mmm,$ $\underline{4}/mm\underline{m}, 4/\underline{mmm}, 4/\underline{m}mm, \underline{4}/\underline{mm}m, 622, 6\underline{2}\underline{2},$ $6mm, \underline{6}m\underline{m}, \bar{6}m2, \underline{\bar{6}}2\underline{m}, \underline{\bar{6}}m\underline{2}, 6/mmm, \underline{6}/\underline{m}m\underline{m},$ $6/\underline{mmm}, 6/\underline{m}mm, \underline{6}/m\underline{m}m,$	d
$\underline{\bar{3}}$	$d,\ h \equiv xxx = -xyy = -yxy = -yyx$ $e,\ j \equiv yyy = -yxx = -xyx = -xxy$
$32, 3m, \bar{3}m, \underline{\bar{3}m}, \bar{3}\underline{m}$	d, h

Note. The components $\alpha_{i(jk)}^{Lmm}$ are denoted by the subscripts i, j, k, taking values x, y, and z in the laboratory frame of reference. Sets of components recurring in various point groups are denoted by lowercase letters.

TABLE VI

The Form of the Axial Third-Rank i-Tensor ${}_{e}^{(1)}S_{eijk}^{(1)L}(\omega)$ for Antiferromagnetic Crystals Belonging to the Tetra-, Tri-, and Hexagonal Systems and Consisting of Two Magnetic Sublattices, the Antiferromagnetic Vector **L** Lying Parallel to the Crystallographical z Axis

Antiferromagnetic Point Group	Components ${}_{e}^{(1)}S_{eijk}^{(1)L}(\omega)$
$\underline{4}, \underline{\bar{4}}, \underline{4}/m$	$\tilde{g} \equiv xyz = yxz,\ xzy = yzx,$ $zxy = zyx$ $\tilde{f} \equiv xxz = -yyz,\ xzx = -yzy,$ $zxx = -zyy$
$422, 4mm, \bar{4}2m, 4/mmm, 622, 6mm, \bar{6}m2, 6/mmm,$ $32, 3m, \bar{3}m$	$\tilde{d} \equiv xxz = yyz,\ xzx = yzy,$ $zxx = zyy$
$\underline{4}2\underline{2}, \underline{4}m\underline{m}, \underline{\bar{4}}2\underline{m}, \underline{4}/mm\underline{m}$	$\tilde{f}$
$\underline{\bar{4}}m\underline{2}$	$\tilde{g}$
$4/\underline{m}, \underline{4}/\underline{m}, 4/\underline{mmm}, 4/\underline{m}mm, \underline{4}/\underline{mm}m, \underline{\bar{3}}, \underline{\bar{3}m}, \underline{\bar{3}}m, \underline{6},$ $\underline{\bar{6}}, \underline{6}/m, 6/\underline{m}, \underline{6}/\underline{m}, \underline{6}2\underline{2}, \underline{6}m\underline{m}, \underline{\bar{6}}2\underline{m}, \underline{\bar{6}}m\underline{2}, \underline{6}/\underline{m}m\underline{m},$ $6/\underline{mmm}, 6/\underline{m}mm, \underline{6}/\underline{m}mm$	All components vanish

Note. The components ${}_{e}^{(1)}S_{eijk}^{(1)L}(\omega)$ are denoted only by the subscripts i, j, k, taking values x, y, and z in the laboratory frame of reference. Sets of components recurring in various point groups are denoted by lowercase letters with tildes.

Provided that, for a given magnetic point group G, we are able to determine a representation P which transforms the representation P_M (according to which the ferromagnetic vector transforms) into the representation P_L comprising the antiferromagnetic vector, which amounts to the fulfillment of the condition

$$P_L = P \otimes P_M \tag{B.1}$$

we are in a position to make direct use of the results of Fumi and Fieschi [147–149] for the determination of the form of tensors dependent on odd powers of **L**. The representation P has to be unidimensional and contain values of characters equal to minus unity for elements R_Z. In Table III we give a list of representations P for magnetic point groups admitting

TABLE VII
The Form of the Polar Third-Rank i-Tensor ${}^{(1)}_{e}Q^{(1)L}_{eijk}(\omega)$ for Antiferromagnetic Crystals Belonging to the Tetra-, Tri-, and Hexagonal Systems and Consisting of Two Magnetic Sublattices, the Antiferromagnetic Vector **L** Lying Parallel to the Crystallographical z Axis

Antiferromagnetic Point Group	Components ${}^{(1)}_{e}Q^{(1)L}_{mijk}(\omega)$
$\underline{\bar{4}}, 4/\underline{m}$	$\underline{d} \equiv zzz,\ xxz = yyz,\ xzx = yzy,\ zxx = zyy$
	$\underline{e} \equiv xyz = -yxz,\ xzy = -yzx,\ zxy = -zyx$
$\underline{4}, \underline{4}/\underline{m}$	$\underline{f} \equiv xxz = -yyz,\ xzx = -yzy,\ zxx = -zyy$
	$\underline{g} \equiv xyz = yxz,\ xzy = yzx,\ zxy = zyx$
$422, \underline{\bar{4}}2\underline{m}, 4/\underline{mmm}, 622, \underline{\bar{6}}2\underline{m}, 6/\underline{mmm}$	$\underline{d}$
$4mm, \underline{\bar{4}}m\underline{2}, 4/\underline{m}mm, 6mm, \underline{\bar{6}}m\underline{2}, 6/\underline{m}mm$	$\underline{e}$
$\bar{4}2m, \underline{4}2\underline{2}, \underline{4}/\underline{mmm}$	$\underline{f}$
$\underline{4}m\underline{m}$	$\underline{g}$
$\underline{\bar{3}}$	$\underline{d}, \underline{h} \equiv xxx = -xyy = -yxy = -yyx$
	$\underline{e}, \underline{j} \equiv yyy = -yxx = -xyx = -xxy$
$32, \underline{\bar{3}m}$	$\underline{d}, \underline{h}$
$3m, \underline{\bar{3}}m$	$\underline{e}, \underline{j}$
$\underline{6}, \underline{6}/m$	$\underline{h}, \underline{j}$
$\underline{\bar{6}}, 6/\underline{m}$	$\underline{e}, \underline{d}$
$\underline{6}2\underline{2}, \underline{6}m\underline{m}, \bar{6}m2, \underline{6}/m\underline{m}m$	$\underline{j}$
$\underline{4}/m, 4/mmm, \underline{4}/mm\underline{m}, \bar{3}m$	All components vanish
$\underline{6}/\underline{m}, 6/mmm, \underline{6}/\underline{m}m\underline{m}$	

Note. The components ${}^{(1)}_{e}Q^{(1)L}_{eijk}(\omega)$ are denoted by the subscripts i, j, k, taking values x, y, and z in the laboratory frame of reference. Sets of components recurring in various point groups are denoted by underlined lowercase letters.

antiferromagnetic ordering belonging to the tetragonal, trigonal, and hexagonal systems. There, too, we list the magnetic point groups $G(P)$ that generate magnetic point groups $G(M)$ (the respective symbols denoting the names of the groups $G(M)$ in the international convention are underlined). Moreover, the representations P are irreducible representations of the groups $G(P)$ because, in agreement with Birss' criterion [144], the form of an i-tensor for the group $G(M)$ is the same as for a group $G(P)$ generating a group $G(M)$; in practice, this means that when it comes to determining the form of an i-tensor for $G(M)$ we may have recourse to the linear combinations of components for the group $G(P)$ already determined by Fumi and Fieschi [147–149].

TABLE VIII
The Form of the Axial Fourth-Rank, the Polar Fifth-Rank, and the Axial Sixth-Rank i-Tensors for Antiferromagnetic Crystals Belonging to the Tetra-, Tri-, and Hexagonal Systems and Consisting of Two Magnetic Sublattices, the Antiferromagnetic Vector **L** Lying Parallel to the Crystallographical z Axis

Antiferromagnetic Point Group	Tensor Components		
	${}^{(1)}_{e}S^{(1)}_{eijkl}(\omega)$	${}^{(1)}_{e}Q^{(1)}_{eijklm}(\omega)$	${}^{(1)}_{e}S^{(1)}_{eijklmn}(\omega)$
$\underline{\bar{4}}, 4/\underline{m}$	d_{40}, d_{4+}, e_{4-}	d_{5-}, e_{50}, e_{5+}	d_6, e_6
$\underline{4}, \underline{4}/\underline{m}$	d_{4-}, e_{4+}	d_{5+}, e_{5-}	f_6, g_6
$422, \underline{\bar{4}}2\underline{m}, 4/\underline{mmm}$	e_{4-}	e_{50}, e_{5+}	e_6
$4mm, \underline{\bar{4}}m\underline{2}, 4/\underline{m}mm$	d_{40}, d_{4+}	d_{5-}	d_6
$\bar{4}2m, \underline{4}2\underline{2}, \underline{4}/\underline{mm}m$	e_{4+}	e_{5-}	g_6
$\underline{4}m\underline{m}$	d_{4-}	d_{5+}	f_6
$\underline{\bar{3}}$	h_4, i_4, j_4, k_4	h_5, i_5, j_5, k_5	h_6, i_6, j_6, k_6
$32, \underline{\bar{3}m}$	i_4, k_4	i_5, j_5	i_6, j_6
$3m, \underline{\bar{3}}m$	h_4, j_4	h_5, k_5	h_6, k_6
$\underline{6}, \underline{6}/m$	j_4, k_4	i_5, k_5	j_6, k_6
$\underline{\bar{6}}, 6/\underline{m}$	h_4, i_4	h_5, j_5	h_6, i_6
$622, \underline{\bar{6}}2\underline{m}, 6/\underline{mmm}$	i_4	j_5	i_6
$\underline{6}2\underline{2}, \underline{6}m\underline{m}, \bar{6}m2, \underline{6}/m\underline{m}m$	j_4	k_5	k_6
$6mm, \underline{\bar{6}}m\underline{2}, 6/\underline{m}mm$	h_4	h_5	h_6
$\underline{4}/m, 4/mmm, \underline{4}/m\underline{m}, \bar{3}m,$ $\underline{6}/\underline{m}, 6/mmm, \underline{6}/\underline{m}m\underline{m}$		All components vanish	

Note. The tensor components are denoted by the subscripts i, j, k, l, m, n, taking values x, y, and z in the laboratory frame of reference. Sets of components recurring in various point groups are denoted by lowercase letters with index p, which takes the values 4, 5, and 6 for the tensor of the fourth, fifth, and sixth rank, respectively.

TABLE IX

The Form of the Polar Fourth-Rank, the Axial Fifth-Rank, and the Polar Sixth-Rank i-Tensors for Antiferromagnetic Crystals Belonging to the Tetra-, Tri-, and Hexagonal Systems and Consisting of Two Magnetic Sublattices, the Antiferromagnetic Vector **L** Lying Parallel to the Crystallographical z Axis

Antiferromagnetic Point Group	Tensor Components		
	${}^{(1)}_{e}S^{(1)}_{eijkl}(\omega)$	${}^{(1)}_{e}Q^{(1)}_{eijklm}(\omega)$	${}^{(1)}_{e}S^{(1)}_{eijklmn}(\omega)$
$\underline{4}, \underline{\bar{4}}, \underline{4}/m$	d_{4-}, e_{r+}	d_{5+}, e_{5-}	d_{6-}, e_{6+}
$422, 4mm, \bar{4}2m, 4/mmm$	e_{4-}	e_{50}, e_{5+}	e_{6-}
$\underline{4}2\underline{2}, \underline{4}m\underline{m}, \underline{\bar{4}}2\underline{m}, \underline{4}/mm\underline{m}$	e_{4+}	e_{5-}	e_{6+}
$\underline{\bar{4}}m\underline{2}$	d_{4-}	d_{5+}	d_{6-}
$32, 3m, \bar{3}m$	i_4, k_4	i_5, j_5	i_6, k_6
$\underline{6}, \underline{\bar{6}}, \underline{6}/\underline{m}$	j_4, k_4	i_5, k_5	j_6, k_6
$622, 6mm, \bar{6}m2, 6/mmm$	i_4	j_5	i_6
$\underline{6}2\underline{2}, \underline{\bar{6}}2\underline{m}, \underline{6}/\underline{m}m\underline{m}$	j_4	k_5	k_6
$\underline{6}m\underline{m}, \underline{\bar{6}}m\underline{2}$	k_4	i_5	j_6
$4/\underline{m}, \underline{4}/\underline{m}, 4/\underline{mmm}, 4/\underline{m}mm$			
$\underline{4}/\underline{mmm}, \underline{\bar{3}}, \underline{\bar{3}m}, \underline{\bar{3}}m, 6/\underline{m}, \underline{6}/m$		All components vanish	
$6/\underline{mmm}, 6/\underline{m}mm, \underline{6}/m\underline{m}m$			

Note. The tensor components are denoted by the subscripts i, j, k, l, m, n, taking values x, y, and z in the laboratory frame of reference. Sets of components recurring in various point groups are denoted by lowercase letters with index p, which takes the values 4, 5, and 6 for the tensor of the fourth, fifth, and sixth rank, respectively.

TABLE X

The Sets of the Polar and Axial Fourth-Rank i-Tensor Components Denoted by Lowercase Letters in Tables VIII and IX

$d_4 \ldots$ $\ldots k_4$	N	I	Components $ijkl$
d_{40}	1	1	$zzzz$
$d_{4\pm}$	20	10	$xxxx = \pm yyyy$,
			$xxyy = \pm yyxx$, $xyxy = \pm yxyx$, $xyyx = \pm yxxy$,
			$xxzz = \pm yyzz$, $xzxz = \pm yzyz$, $xzzx = \pm yzzy$,
			$zzxx = \pm zzyy$, $zxzx = \pm zyzy$, $zxxz = \pm zyyz$
$e_{4\pm}$	20	10	$xxxy = \pm yyyx$, $xxyx = \pm yyxy$, $xyxx = \pm yxyy$, $yxxx = \pm xyyy$,
			$xyzz = \pm yxzz$, $xzyz = \pm yzxz$, $xzzy = \pm yzzx$, $zxyz = \pm zyxz$,
			$zxzy = \pm zyzx$, $zzxy = \pm zzyx$
h_4	21	10	d_{40}, d_{4+} and $xxxx = yyyy = xxyy + xyxy + xyyx$
i_4	20	9	e_{4-} and $xxxy = -yyyx = -[xxyx + xyxx + yxxx]$
j_4	16	4	$xxxz = -xyyz = -yxyz = -yyxz$, $xzxx = -xzyy = -yzxy = -yzyx$,
			$xxzx = -xyzy = -yxzy = -yyzx$, $zxxx = -zxyy = -zyxy = -zyyx$
k_4	16	4	$yyyz = -yxxz = -xyxz = -xxyz$, $yzyy = -yzxx = -xzyx = -xzxy$,
			$yyzy = -yxzx = -xyzx = -xxzy$, $zyyy = -zyxx = -zxyx = -zxxy$

Note. The components are denoted by indices $ijkl$, whereas N and I denote the numbers of nonzero and mutually independent components, respectively.

TABLE XI
The Sets of the Polar and Axial Fifth-Rank i-Tensor Components Denoted by Lowercase Letters in Tables VIII and IX

$d_{5\pm}\ldots$ $\ldots k_4$	N	I	Components $ijklm$
$d_{5\pm}$	60	30	$xxxyz = \pm yyyxz,\ xxxzy = \pm yyyzx,\ xxzxy = \pm yyzyx,\ xzxxy = \pm yzyyx,$ $zxxxy = \pm zyyyx,\ xxyxz = \pm yyxyz,\ xxyzx = \pm yyxzy,\ xxzyx = \pm yyzxy,$ $xzxyx = \pm yzyxy,\ zxxyx = \pm zyyxy,\ xyxxz = \pm yxyyz,\ xyxzx = \pm yxyzy,$ $xyzxx = \pm yxzyy,\ xzyxx = \pm yzxyy,\ zxyxx = \pm zyxyy,\ yxxxz = \pm xyyyz,$ $yxxzx = \pm xyyzy,\ yxzxx = \pm xyzyy,\ yzxxx = \pm xzyyy,\ zyxxx = \pm zxyyy,$ $xyzzz = \pm yxzzz,\ xzyzz = \pm yzxzz,\ xzzyz = \pm yzzxz,\ xzzzy = \pm yzzzx,$ $zxzyz = \pm zyzxz,\ zxyzz = \pm zyxzz,\ zxzzy = \pm zyzzx,\ zzxyz = \pm zzyxz,$ $zzxzy = \pm zzyzx,\ zzzxy = \pm zzzyx$
e_{50}	1	1	$zzzzz;$
$e_{5\pm}$	60	30	$xxxxz = \pm yyyyz,\ xxxzx = \pm yyyzy,\ xxzxx = \pm yyzyy,\ xzxxx = \pm yzyyy,$ $zxxxx = \pm zyyyy,$ $xxyyz = \pm yyxxz,\ xxyzy = \pm yyxzx,\ xxzyy = \pm yyzxx,\ xzxyy = \pm yzyxx,$ $zxxyy = \pm zyyxx,\ xyxyz = \pm yxyxz,\ xyxzy = \pm yxyzx,\ xyzxy = \pm yxzyx,$ $xzyxy = \pm yzxyx,\ zxyxy = \pm zyxyx,\ xyyxz = \pm yxxyz,\ xyyzx = \pm yxxzy,$ $xyzyx = \pm yxzxy,\ xzyyx = \pm yzxxy,\ zxyyx = \pm zyxxy,$ $xxzzz = \pm yyzzz,\ xzxzz = \pm yzyzz,\ xzzxz = \pm yzzyz,\ xzzzx = \pm yzzzy,$ $zxzxz = \pm zyzyz,\ zxxzz = \pm zyyzz,\ zxzzx = \pm zyzzy,\ zzxxz = \pm zzyyz,$ $zzxzx = \pm zzyzy,\ zzzxx = \pm zzzyy$
h_5	60	25	d_{5-} and additionally the relations: $xxxyz = -(xxyxz + xyxxz + yxxxz),$ $xxxzy = -(xxyzx + xyxzx + yxxzx),\ xxzxy = -(xxzyx + xyzxx + yxzxx),$ $xzxxy = -(xzxyx + xzyxx + yzxxx),\ zxxxy = -(zxxyx + zxyxx + zyxxx)$
i_5	56	15	$xxxzz = -xyyzz = -yxyzz = -yyxzz,\ xzzxx = -xzzyy = -yzzxy = -yzzyx,$ $xxzxz = -xyzyz = -yxzyz = -yyzxz,\ zxxzx = -zxyzy = -zyxzy = -zyyzx,$ $xxzzx = -xyzzy = -yxzzy = -yyzzx,\ zxxxz = -zxyyz = -zyxyz = -zyyxz,$ $xzxxz = -xzyyz = -yzxyz = -yzyxz,\ zxzxx = -zxzyy = -zyzxy = -zyzyx,$ $xzxzx = -xzyzy = -yzxzy = -yzyzx,\ zzxxx = -zzxyy = -zzyxy = -zzyyx,$ $xxxxx = -\frac{1}{3}(yyyyx + yyyxy + yyxyy + yxyyy + xyyyy),$ $xxxyy = \frac{1}{3}(2yyyyx + 2yyyxy - yyxyy - yxyyy - xyyyy),$ $xxyxy = \frac{1}{3}(2yyyyx - yyyxy + 2yyxyy - yxyyy - xyyyy),$ $xxyyx = \frac{1}{3}(-yyyyx + 2yyyxy + 2yyxyy - yxyyy - xyyyy),$ $xyxxy = \frac{1}{3}(2yyyyx - yyyxy - yyxyy + 2yxyyy - xyyyy),$ $xyyxx = \frac{1}{3}(-yyyyx - yyyxy + 2yyxyy + 2yxyyy - xyyyy),$ $xyxyx = \frac{1}{3}(-yyyyx + 2yyyxy - yyxyy + 2yxyyy - xyyyy),$ $yxxxy = \frac{1}{3}(2yyyyx - yyyxy - yyxyy - yxyyy + 2xyyyy),$ $yxxyx = \frac{1}{3}(-yyyyx + 2yyyxy - yyxyy - yxyyy + 2xyyyy),$ $yxyxx = \frac{1}{3}(-yyyyx - yyyxy + 2yyxyy - yxyyy + 2xyyyy),$ $yyxxx = \frac{1}{3}(-yyyyx - yyyxy - yyxyy + 2yxyyy + 2xyyyy),$ $yyyyx,\ yyyxy,\ yyxyy,\ yxyyy,\ xyyyy$
j_5	61	26	e_{50}, e_{5+} and additionally the relations: $xxxxz = xxyyz + xyxyz + xyyxz,$ $xxxzx = xxyzy + xyxzy + xyyzx,\ xxzxx = xxzyy + xyzxy + xyzyx,$ $xzxxx = xzxyy + xzyxy + xzyyx,\ zxxxx = zxxyy + zxyxy + zxyyx$
k_5	56	15	Components i_5 on performing the transposition $x \rightarrow y$ and $y \rightarrow x$

Note. The components are denoted by indices $ijklm$, whereas N and I denote the numbers of nonzero and mutually independent components, respectively.

In the case of tensors dependent on odd powers of **L** we may continue to make use of their linear combinations of tensor components. These combinations will belong to new irreducible representations of the point group under consideration, representations being a direct product of P and the Fumi representation. This is equivalent to the statement that, now, linear combinations other than those for the case of a one-lattice crystal will belong to the fully symmetric as well as the antisymmetric representation. Here, as previously for a polar tensor, all linear combinations not belonging to the fully symmetric representation have to vanish, whereas for an axial tensor, combinations belonging to the antisymmetric representation will differ from zero. Along these lines, we have determined the form of the polar and axial i-tensors of the second, third, fourth, fifth, and sixth rank for antiferromagnets (consisting of two sublattices) belonging to the tetragonal, trigonal, and hexagonal systems. Our results for tensors with no permutational symmetry are given in Tables IV and VI–XII. We also have determined the form of the axial c-tensor $\boldsymbol{\alpha}^{Lmm}$ of the third rank (Table V).

TABLE XII
The Sets of the Polar and Axial Sixth-Rank i-Tensor Components Denoted by Lowercase Letters in Tables VIII and IX

$d_{6\pm}\ldots$ $\ldots k_6$	N	I	Components *ijklmn*
d_{60}	1	1	*zzzzzz*,
$d_{6\pm}$	182	92	$d_{6\pm1} \equiv$ *xxxxxx* = ±*yyyyyy*,
			xxxxzz = ±*yyyyzz*, *xxxzxz* = ±*yyyzyz*, *xxzxxz* = ±*yyzyyz*,
			xzxxxz = ±*yzyyyz*, *zxxxxz* = ±*zyyyyz*, *xxxzzx* = ±*yyyzzy*,
			xxzxzx = ±*yyzyzy*, *xzxxzx* = ±*yzyyzy*, *zxxxzx* = ±*zyyyzy*,
			xxzzxx = ±*yyzzyy*, *xzxzxx* = ±*yzyzyy*, *zxxzxx* = ±*zyyzyy*,
			xzzxxx = ±*yzzyyy*, *zxzxxx* = ±*zyzyyy*, *zzxxxx* = ±*zzyyyy*,
			xxxxyy = ±*yyyyxx*, *xxxyxy* = ±*yyyxyx*, *xxyxxy* = ±*yyxyyx*,
			xyxxxy = ±*yxyyyx*, *yxxxxy* = ±*xyyyyx*, *xxxyyx* = ±*yyyxxy*,
			xxyxyx = ±*yyxyxy*, *xyxxyx* = ±*yxyyxy*, *yxxxyx* = ±*xyyyxy*,
			xxyyxx = ±*yyxxyy*, *xyxyxx* = ±*yxyxyy*, *yxxyxx* = ±*xyyxyy*,
			xyyxxx = ±*yxxyyy*, *yxyxxx* = ±*xyxyyy*, *yyxxxx* = ±*xxyyyy*
			$d_{6\pm2} \equiv$ *xxzzzz* = ±*yyzzzz*, *xzxzzz* = ±*yzyzzz*, *xzzxzz* = ±*yzzyzz*,
			xzzzxz = ±*yzzzyz*, *xzzzzx* = ±*yzzzzy*, *zxxzzz* = ±*zyyzzz*,
			zxzxzz = ±*zyzyzz*, *zxzzxz* = ±*zyzzyz*, *zxzzzx* = ±*zyzzzy*,
			zzxxzz = ±*zzyyzz*, *zzxzxz* = ±*zzyzyz*, *zzxzzx* = ±*zzyzzy*,
			zzzxxz = ±*zzzyyz*, *zzzxzx* = ±*zzzyzy*, *zzzzxx* = ±*zzzzyy*,
			xxyyzz = ±*yyxxzz*, *xxyzyz* = ±*yyxzxz*, *xxzyyz* = ±*yyzxxz*,
			xzxyyz = ±*yzyxxz*, *zxxyyz* = ±*zyyxxz*, *xxyzzy* = ±*yyxzzx*,
			xxzyzy = ±*yyzxzx*, *xzxyzy* = ±*yzyxzx*, *zxxyzy* = ±*zyyxzx*,
			xxzzyy = ±*yyzzxx*, *xzxzyy* = ±*yzyzxx*, *zxxzyy* = ±*zyyzxx*,
			xzzxyy = ±*yzzyxx*, *zxzxyy* = ±*zyzyxx*, *zzxxyy* = ±*zzyyxx*,

TABLE XII *(Continued)*

$d_{6\pm}\ldots$ $\ldots k_6$	N	I	Components *ijklmn*
			xyxyzz = ±yxyxzz, xyxzyz = ±yxyzxz, xyzxyz = ±yxzyxz,
			xzyxyz = ±yzxyxz, xyxzzy = ±yxyzzx, xyzxzy = ±yxzyzx,
			xzyxzy = ±yzxyzx, zxyxzy = ±zyxyzx, zxyxyz = ±zyxyxz,
			xyzzxy = ±yxzzyx, xzyzxy = ±yzxzyx, zxyzxy = ±zyxzyx,
			xzzyxy = ±yzzxyx, zxzyxy = ±zyzxyx, zzxyxy = ±zzyxyx,
			xyyxzz = ±yxxyzz, xyyzxz = ±yxxzyz, xyzyxz = ±yxzxyz,
			xzyyxz = ±yzxxyz, zxyyxz = ±zyxxyz, xyyzzx = ±yxxzzy,
			xyzyzx = ±yxzxzy, xzyyzx = ±yzxxzy, zxyyzx = ±zyxxzy,
			xyzzyx = ±yxzzxy, xzyzyx = ±yzxzxy, zxyzyx = ±zyxzxy,
			xzzyyx = ±yzzxxy, zxzyyx = ±zyzxxy, zzxyyx = ±zzyxxy
$e_{6\pm}$	182	91	$e_{6\pm1}$ ≡ xxxyzz = ±yyyxzz, xxxzyz = ±yyyzxz, xxzxyz = ±yyzyxz,
			xzxxyz = ±yzyyxz, zxxxyz = ±zyyyxz, xxxzzy = ±yyyzzx,
			xxzxzy = ±yyzyzx, xzxxzy = ±yzyyzx, zxxxzy = ±zyyyzx,
			xxzzxy = ±yyzzyx, xzxzxy = ±yzyzyx, zxxzxy = ±zyyzyx,
			xzzxxy = ±yzzyyx, zxzxxy = ±zyzyyx, zzxxxy = ±zzyyyx,
			xxyxzz = ±yyxyzz, xxyzxz = ±yyxzyz, xxzyxz = ±yyzxyz,
			xzxyxz = ±yzyxyz, zxxyxz = ±zyyxyz, xxyzzx = ±yyxzzy,
			xxzyzx = ±yyzxzy, xzxyzx = ±yzyxzy, zxxyzx = ±zyyxzy,
			xxzzyx = ±yyzzxy, xzxzyx = ±yzyzxy, zxxzyx = ±zyyzxy,
			xzzxyx = ±yzzyxy, zxzxyx = ±zyzyxy, zzxxyx = ±zzyyxy,
			xyxxzz = ±yxyyzz, xyxzxz = ±yxyzyz, xyzxxz = ±yxzyyz,
			xzyxxz = ±yzxyyz, zxyxxz = ±zyxyyz, xyxzzx = ±yxyzzy,
			xyzxzx = ±yxzyzy, xzyxzx = ≡ yzxyzy, zxyxzx = ±zyxyzy,
			xyzzxx = ±yxzzyy, xzyzxx = ±yzxzyy, zxyzxx = ±zyxzyy,
			xzzyxx = ±yzzxyy, zxzyxx = ±zyzxyy, zzxyxx = ±zzyxyy,
			yxxxzz = ±xyyyzz, yxxzxz = ±xyyzyz, yxzxxz = ±xyzyyz,
			yzxxxz = ±xzyyyz, zyxxxz = ±zxyyyz, yxxzzx = ±xyyzzy,
			yxzxzx = ±xyzyzy, yzxxzx = ±xzyyzy, zyxxzx = ±zxyyzy,
			yxzzxx = ±xyzzyy, yzxzxx = ±xzyzyy, zyxzxx = ±zxyzyy,
			yzzxxx = ±xzzyyy, zyzxxx = ±zxzyyy, zzyxxx = ±zzxyyy,
			xyzzzz = ±yxzzzz, xzyzzz = ±yzxzzz, xzzyzz = ±yzzxzz,
			xzzzyz = ±yzzzxz, xzzzzy = ±yzzzzx, zxyzzz = ±zyxzzz,
			zxzyzz = ±zyzxzz, zxzzyz = ±zyzzxz, zxzzzy = ±zyzzzx,
			zzxyzz = ±zzyxzz, zzxzyz = ±zzyzxz, zzxzzy = ±zzyzzx,
			zzzxyz = ±zzzyxz, zzzxzy = ±zzzyzx, zzzzxy = ±zzzzyx
			$e_{6\pm2}$ ≡ xxxxxy = ±yyyyyx, xxxxyx = ±yyyyxy, xxxyxx = ±yyyxyy,
			xxyxxx = ±yyxyyy, xyxxxx = ±yxyyyy, yxxxxx = ±xyyyyy,
			xxxyyy = ±yyyxxx, xxyxyy = ±yyxyxx, xxyyxy = ±yyxxyx,
			xxyyyx = ±yyxxxy, xyxyyx = ±yxyxxy, xyyyxx = ±yxxxyy,
			xyyyxx = ±yxxxyy, xyxyxy = ±yxyxyx, xyxxyy = ±yxyyxx,
			xyyxxy = ±yxxyyx
h_6	183	72	d_{60}, d_{6+2} *and*
			xxxxzz = yyyyzz = xxyyzz + xyxyzz + xyyxzz,
			xxxzxz = yyyzyz = xxyzyz + xyxzyz + xyyzxz,
			xxzxxz = yyzyyz = xxzyyz + xyzxyz + xyzyxz,
			xzxxxz = yzyyyz = xzxyyz + xzyxyz + xzyyxz,
			zxxxxz = zyyyyz = zxxyyz + zxyxyz + zxyyxz,

TABLE XII *(Continued)*

$d_{6\pm}\ldots$ $\ldots k_6$	N	I	Components *ijklmn*
			xxxzzx = yyyzzy = xxyzzy + xyxzzy + xyyzzx,
			xxzxzx = yyzyzy = xxzyzy + xyzxzy + xyzyzx,
			xzxxzx = yzyyzy = xzxyzy + xzyxzy + xzyyzx,
			zxxxzx = zyyyzy = zxxyzy + zxyxzy + zxyyzx,
			xxzzxx = yyzzyy = xxzzyy + xyzzxy + xyzzyx,
			xzxzxx = yzyzyy = xzxzyy + xzyzxy + xzyzyx,
			zxxzxx = zyyzyy = zxxzyy + zxyzxy + zxyzyx,
			xzzxxx = yzzyyy = xzzxyy + xzzyxy − xzzyyx,
			zxzxxx = zyzyyy = zxzxyy + zxzyxy + zxzyyx,
			zzxxxx = zzyyyy = zzxxyy + zzxyxy + zzxyyx,
			xxyyxx = − 2xxxxxx + 3yyyyyy − xxyxxy − xxyxyx
			−xyyxxx − yxyxxx,
			xyxxxy = −2xxxxxx + 3yyyyyy − xxxxyy − xxyxxy − xxyxyx
			−xyxxyx − xyyxxx + yxxyxx,
			yxxxxy = −xxxyxy + xxyxyx + xyxxyx + xyyxxx − yxxyxx,
			yxxxyx = −2xxxxxx + 3yyyyyy − xxxxyy − xxxyyx − xxyxyx
			−xyxxyx,
			xyxyxx = −xxxyxy − xxxyyx + xxyxxy + xxyxyx + xyyxxx − yxxyxx
			+yxyxxx,
			yyxxxx = xxxxyy + xxxyxy + xxxyyx − xyyxxx − yxyxxx,
			xxyyyy = xxxxxx − yyyyyy + xxxyxy + xxxyyx − xyyxxx − yxyxxx
			+xxxxyy,
			yyxxyy = −xxxxxx + 2yyyyyy − xxyxxy − xxyxyx − xyyxxx
			−yxyxxx,
			yyyyxx = xxxxxx − yyyyyy + xxxxyy,
			yyxyxy = xxxxxx − yyyyyy + xxyxyx,
			yyxyyx = xxxxxx − yyyyyy + xxyxxy,
			yyyxxy = xxxxxx − yyyyyy + xxxyyx,
			yyyxyx = xxxxxx − yyyyyy + xxxyxy,
			yxyyxy = xxxxxx − yyyyyy + xyxxyx,
			xyxyyy = xxxxxx − yyyyyy + yxyxxx,
			xyyxyy = xxxxxx − yyyyyy + yxxyxx,
			yxxyyy = xxxxxx − yyyyyy + xyyxxx,
			xyyyxy = −xxxxxx + 2yyyyyy − xxxxyy − xxxyyx − xxyxyx
			−xyxxyx,
			xyyyyx = xxxxxx − yyyyyy − xxxyxy + xxyxyx + xyxxyx + xyyxxx
			−yxxyxx,
			yxyyyx = −xxxxxx + 2yyyyyy − xxxxyy − xxyxxy − xxyxyx
			−xyxxyx − xyyxxx + yxxyxx,
			yxyxyy = xxxxxx − yyyyyy − xxxyxy − xxxyyx + xxyxxy + xxyxyx
			+xyyxxx − yxxyxx + yxyxxx,
			xxxxxx, yyyyyy,
			xxxxyy, xxxyxy, xxyxxy, xxxyyx, xxyxyx, xyxxyx,
			yxxyxx, xyyxxx, yxyxxx
i_6	182	71	e_{6-1} *plus the relations*:
			xxxyzz = −xxyxzz − xyxxzz − yxxxzz,
			xxxzyz = −xxyzxz − xyxzxz − yxxzxz,

TABLE XII (*Continued*)

$d_{6\pm}$ ... ... k_6	N	I	Components *ijklmn*
			xxzxyz = − xxzyxz − xyzxxz − yxzxxz,
			xzxxyz = − xzxyxz − xzyxxz − yzxxxz,
			zxxxyz = − zxxyxz − zxyxxz − zyxxxz,
			xxxzzy = − xxyzzx − xyxzzx − yxxzzx,
			xxzxzy = − xxzyzx − xyzxzx − yxzxzx,
			xzxxzy = − xzxyzx − xzyxzx − yzxxzx,
			zxxxzy = − zxxyzx − zxyxzx − zyxxzx,
			xxzzxy = − xxzzyx − xyzzxx − yxzzxx,
			xzxzxy = − xzxzyx − xzyzxx − yzxzxx,
			zxxzxy = − zxxzyx − zxyzxx − zyxzxx,
			xzzxxy = − xzzxyx − xzzyxx − yzzxxx,
			zxzxxy = − zxzxyx − zxzyxx − zyzxxx,
			zzxxxy = − zzxxyx − zzxyxx − zzyxxx,
			xxxxxy = $\frac{1}{3}$(− 2yyyyyx + yyyyxy + yyyxyy +yyxyyy + yxyyyy + xyyyyy),
			xxxxyx = $\frac{1}{3}$(yyyyyx − 2yyyyxy + yyyxyy + yyxyyy + yxyyyy + xyyyyy),
			xxxyxx = $\frac{1}{3}$(yyyyyx + yyyyxy − 2yyyxyy + yyxyyy + yxyyyy + xyyyyy),
			xxyxxx = $\frac{1}{3}$(yyyyyx + yyyyxy + yyyxyy − 2yyxyyy + yxyyyy + xyyyyy),
			xyxxxx = $\frac{1}{3}$(yyyyyx + yyyyxy + yyyxyy + yyxyyy − 2yxyyyy + xyyyyy),
			yxxxxx = $\frac{1}{3}$(yyyyyx + yyyyxy + yyyxyy + yyxyyy + yxyyyy − 2xyyyyy),
			yyyyyx, yyyyxy, yyyxyy, yyxyyy, yxyyyy, xyyyyy,
			yyyxxx = − xxxyyy − $\frac{1}{3}$(yyyyyx + yyyyxy + yyyxyy + yyxyyy + yxyyyy + xyyyyy),
			yyxyxx = − xxyxyy − $\frac{1}{3}$(yyyyyx + yyyyxy + yyyxyy + yyxyyy + yxyyyy + xyyyyy),
			yyxxyx = − xxyyxy − $\frac{1}{3}$(yyyyyx + yyyyxy + yyyxyy + yyxyyy + yxyyyy + xyyyyy),
			yxxxyy = − xyyyxx − $\frac{1}{3}$(yyyyyx + yyyyxy + yyyxyy + yyxyyy + yxyyyy + xyyyyy),
			yxxyxy = − xyyxyx − $\frac{1}{3}$(yyyyyx + yyyyxy + yyyxyy + yyxyyy + yxyyyy + xyyyyy),
			xyxxyy = − xxxyyy − xxyxyy + xyyyxx − yyyyyx − yyyyxy,
			xyxyxy = − xxxyyy − xxyyxy + xyyxyx − yyyyyx − yyyxyy,
			xyyxxy = xxxyyy − xyyyxx − xyyxyx − yyxyyy − yxyyyy,
			yyxxxy = xxxyyy + xxyyxy + xxyxyy + $\frac{1}{3}$(2yyyyyx + 2yyyyxy + 2yyyxyy + 2yyxyyy − yxyyyy − xyyyyy),
			yxyyxx = xxxyyy + xxyxyy − xyyyxx + $\frac{1}{3}$(2yyyyyx + 2yyyyxy − yyyxyy − yyxyyy − yxyyyy − xyyyyy),
			yxxyyx = − xxxyyy + xyyyxx + xyyxyx + $\frac{1}{3}$(− yyyyyx − yyyyxy − yyyxyy + 2yyxyyy + 2yxyyyy − xyyyyy),
			xxyyyx = − xxxyyy − xxyyxy − xxyxyy − yyyyyx − yyyyxy − yyyxyy − yyxyyy,

TABLE XII *(Continued)*

$d_{6\pm}\ldots$ $\ldots k_6$	N	I	Components *ijklmn*
			xyxyyx = xxxyyy + xxyyxy + xxyxyy − xyyyxx − xyyxyx + yyyyyx
			− yxyyyy,
			yxyxyx = xxxyyy + xxyyxy − xyyxyx
			+ $\frac{1}{3}$(2yyyyyx − yyyyxy + 2yyyxyy − yyxyyy − yxyyyy
			− xyyyyy),
			yxyxxy = − xxxyyy − xxyyxy + xyyxyx − xxyxyy + xyyxyx
			+ $\frac{1}{3}$(− 4yyyyyx − yyyyxy − yyyxyy − yyxyyy
			+ 2yxyyyy − xyyyyy),
			xxxyyy, xxyxyy, xxyyxy, xyyxyx, xyyyxx
j_6	176	50	yyyzzz = − yxxzzz = − xyxzzz = − xxyzzz,
			yyzyzz = − yxzxzz = − xyzxzz = − xxzyzz,
			yyzzyz = − yxzzxz = − xyzzxz = − xxzzyz,
			yyzzzy = − yxzzzx = − xyzzzx = − xxzzzy,
			yzyzzy = − yzxzzx = − xzyzzx = − xzxzzy,
			yzzyzy = − yzzxzx = − xzzyzx = − xzzxzy,
			yzzzyy = − yzzzxx = − xzzzyx = − xzzzxy,
			yzyzyz = − yzxzxz = − xzyzxz = − xzxzyz,
			yzyyzz = − yzxxzz = − xzyxzz = − xzxyzz,
			yzzyyz = − yzzxxz = − xzzyxz = − xzzxyz,
			zyzyyz = − zyzxxz = − zxzyxz = − zxzxyz,
			zyyzzy = − zyxzzx = − zxyzzx = − zxxzzy,
			zyzzyy = − zyzzxx = − zxzzyx = − zxzzxy,
			zyzyzy = − zyzxzx = − zxzyzx = − zxzxzy,
			zyyyzz = − zyxxzz = − zxyxzz = − zxxyzz,
			zyyzyz = − zyxzxz = − zxyzxz = − zxxzyz,
			zzyyyz = − zzyxxz = − zzxyxz = − zzxxyz,
			zzyyzy = − zzyxzx = − zzxyzx = − zzxxzy,
			zzyzyy = − zzyzxx = − zzxzyx = − zzxzxy,
			zzzyyy = − zzzyxx = − zzzxyx = − zzzxxy,
			yyyyyz = − $\frac{1}{3}$(xxxxyz + xxxyxz + xxyxxz + xyxxxz + yxxxxz),
			yyyyzy = − $\frac{1}{3}$(xxxxzy + xxxyzx + xxyxzx + xyxxzx + yxxxzx),
			yyyzyy = − $\frac{1}{3}$(xxxzxy + xxxzyx + xxyzxx + xyxzxx + yxxzxx),
			yyzyyy = − $\frac{1}{3}$(xxzxxy + xxzxyx + xxzyxx + xyzxxx + yxzxxx),
			yzyyyy = − $\frac{1}{3}$(xzxxxy + xzxxyx + xzxyxx + xzyxxx + yzxxxx),
			zyyyyy = − $\frac{1}{3}$(zxxxxy + zxxxyx + zxxyxx + zxyxxx + zyxxxx),
			yyyxxz = $\frac{1}{3}$(2xxxxyz + 2xxxyxz − xxyxxz − xyxxxz − yxxxxz),
			yyyxzx = $\frac{1}{3}$(2xxxxzy + 2xxxyzx − xxyxzx − xyxxzx − yxxxzx),
			yyyzxx = $\frac{1}{3}$(2xxxzxy + 2xxxzyx − xxyzxx − xyxzxx − yxxzxx),
			yyzyxx = $\frac{1}{3}$(2xxzxxy + 2xxzxyx − xxzyxx − xyzxxx − yxzxxx),
			yzyyxx = $\frac{1}{3}$(2xzxxxy + 2xzxxyx − xzxyxx − xzyxxx − yzxxxx),
			zyyyxx = $\frac{1}{3}$(2zxxxxy + 2zxxxyx − zxxyxx − zxyxxx − zyxxxx),
			yyxyxz = $\frac{1}{3}$(2xxxxyz − xxxyxz + 2xxyxxz − xyxxxz − yxxxxz),
			yyxyzx = $\frac{1}{3}$(2xxxxzy − xxxyzx + 2xxyxzx − xyxxzx − yxxxzx),
			yyxzyx = $\frac{1}{3}$(2xxxzxy − xxxzyx + 2xxyzxx − xyxzxx − yxxzxx),

TABLE XII *(Continued)*

$d_{6\pm}\ldots$ $\ldots k_6$	N	I	Components *ijklmn*
			yyzxyx = $\frac{1}{3}$(2xxzxxy − xxzxyx + 2xxzyxx − xyzxxx − yxzxxx),
			yzyxyx = $\frac{1}{3}$(2xzxxxy − xzxxyx + 2xzxyxx − xzyxxx − yzxxxx),
			zyyxyx = $\frac{1}{3}$(2zxxxxy − zxxxyx + 2zxxyxx − zxyxxx − zyxxxx),
			yyxxyz = $\frac{1}{3}$(−xxxxyz + 2xxxyxz + 2xxyxxz − xyxxxz − yxxxxz),
			yyxxzy = $\frac{1}{3}$(−xxxxzy + 2xxxyzx + 2xxyxzx − xyxxzx − yxxxzx),
			yyxzxy = $\frac{1}{3}$(−xxxzxy + 2xxxzyx + 2xxyzxx − xyxzxx − yxxzxx),
			yyzxxy = $\frac{1}{3}$(−xxzxxy + 2xxzxyx + 2xxzyxx − xyzxxx − yxzxxx),
			yzyxxy = $\frac{1}{3}$(−xzxxxy + 2xzxxyx + 2xzxyxx − xzyxxx − yzxxxx),
			zyyxxy = $\frac{1}{3}$(−zxxxxy + 2zxxxyx + 2zxxyxx − zxyxxx − zyxxxx),
			yxyxyz = $\frac{1}{3}$(−xxxxyz + 2xxxyxz − xxyxxz + 2xyxxxz − yxxxxz),
			yxyxzy = $\frac{1}{3}$(−xxxxzy + 2xxxyzx − xxyxzx + 2xyxxzx − yxxxzx),
			yxyzxy = $\frac{1}{3}$(−xxxzxy + 2xxxzyx − xxyzxx + 2xyxzxx − yxxzxx),
			yxzyxy = $\frac{1}{3}$(−xxzxxy + 2xxzxyx − xxzyxx + 2xyzxxx − yxzxxx),
			yzxyxy = $\frac{1}{3}$(−xzxxxy + 2xzxxyx − xzxyxx + 2xzyxxx − yzxxxx),
			zyxyxy = $\frac{1}{3}$(−zxxxxy + 2zxxxyx − zxxyxx + 2zxyxxx − zyxxxx),
			yxxyyz = $\frac{1}{3}$(−xxxxyz − xxxyxz + 2xxyxxz + 2xyxxxz − yxxxxz),
			yxxyzy = $\frac{1}{3}$(−xxxxzy − xxxyzx + 2xxyxzx + 2xyxxzx − yxxxzx),
			yxxzyy = $\frac{1}{3}$(−xxxzxy − xxxzyx + 2xxyzxx + 2xyxzxx − yxxzxx),
			yxzxyy = $\frac{1}{3}$(−xxzxxy − xxzxyx + 2xxzyxx + 2xyzxxx − yxzxxx),
			yzxxyy = $\frac{1}{3}$(−xzxxxy − xzxxyx + 2xzxyxx + 2xzyxxx − yzxxxx),
			zyxxyy = $\frac{1}{3}$(−zxxxxy − zxxxyx + 2zxxyxx + 2zxyxxx − zyxxxx),
			yxyyxz = $\frac{1}{3}$(2xxxxyz − xxxyxz − xxyxxz + 2xyxxxz − yxxxxz),
			yxyyzx = $\frac{1}{3}$(2xxxxzy − xxxyzx − xxyxzx + 2xyxxzx − yxxxzx),
			yxyzyx = $\frac{1}{3}$(2xxxzxy − xxxzyx − xxyzxx + 2xyxzxx − yxxzxx),
			yxzyyx = $\frac{1}{3}$(2xxzxxy − xxzxyx − xxzyxx + 2xyzxxx − yxzxxx),
			yzxyyx = $\frac{1}{3}$(2xzxxxy − xzxxyx − xzxyxx + 2xzyxxx − yzxxxx),
			zyxyyx = $\frac{1}{3}$(2zxxxxy − zxxxyx − zxxyxx + 2zxyxxx − zyxxxx),
			xyyyxz = $\frac{1}{3}$(2xxxxyz − xxxyxz − xxyxxz − xyxxxz + 2yxxxxz),
			xyyyzx = $\frac{1}{3}$(2xxxxzy − xxxyzx − xxyxzx − xyxxzx + 2yxxxzx),
			xyyzyx = $\frac{1}{3}$(2xxxzxy − xxxzyx − xxyzxx − xyxzxx + 2yxxzxx),
			xyzyyx = $\frac{1}{3}$(2xxzxxy − xxzxyx − xxzyxx − xyzxxx + 2yxzxxx),
			xzyyyx = $\frac{1}{3}$(2xzxxxy − xzxxyx − xzxyxx − xzyxxx + 2yzxxxx),
			zxyyyx = $\frac{1}{3}$(2zxxxxy − zxxxyx − zxxyxx − zxyxxx + 2zyxxxx),
			xyyxyz = $\frac{1}{3}$(−xxxxyz + 2xxxyxz − xxyxxz − xyxxxz + 2yxxxxz),
			xyyxzy = $\frac{1}{3}$(−xxxxzy + 2xxxyzx − xxyxzx − xyxxzx + 2yxxxzx),

TABLE XII *(Continued)*

$d_{6\pm}\ldots$ $\ldots k_6$	N	I	Components *ijklmn*
			$xyyzxy = \frac{1}{3}(-xxxzxy + 2xxxzyx - xxyzxx - xyxzxx + 2yxxzxx)$,
			$xyzyxy = \frac{1}{3}(-xxzxxy + 2xxzxyx - xxzyxx - xyzxxx + 2yxzxxx)$,
			$xzyyxy = \frac{1}{3}(-xzxxxy + 2xzxxyx - xzxyxx - xzyxxx + 2yzxxxx)$,
			$zxyyxy = \frac{1}{3}(-zxxxxy + 2zxxxyx - zxxyxx - zxyxxx + 2zyxxxx)$,
			$xyxyyz = \frac{1}{3}(-xxxxyz - xxxyxz + 2xxyxxz - xyxxxz + 2yxxxxz)$,
			$xyxyzy = \frac{1}{3}(-xxxxzy - xxxyzx + 2xxyxzx - xyxxzx + 2yxxxzx)$,
			$xyxzyy = \frac{1}{3}(-xxxzxy - xxxzyx + 2xxyzxx - xyxzxx + 2yxxzxx)$,
			$xyzxyy = \frac{1}{3}(-xxzxxy - xxzxyx + 2xxzyxx - xyzxxx + 2yxzxxx)$,
			$xzyxyy = \frac{1}{3}(-xzxxxy - xzxxyx + 2xzxyxx - xzyxxx + 2yzxxxx)$,
			$zxyxyy = \frac{1}{3}(-zxxxxy - zxxxyx + 2zxxyxx - zxyxxx + 2zyxxxx)$,
			$xxyyyz = \frac{1}{3}(-xxxxyz - xxxyxz - xxyxxz + 2xyxxxz + 2yxxxxz)$,
			$xxyyzy = \frac{1}{3}(-xxxxzy - xxxyzx - xxyxzx + 2xyxxzx + 2yxxxzx)$,
			$xxyzyy = \frac{1}{3}(-xxxzxy - xxxzyx - xxyzxx + 2xyxzxx + 2yxxzxx)$,
			$xxzyyy = \frac{1}{3}(-xxzxxy - xxzxyx - xxzyxx + 2xyzxxx + 2yxzxxx)$,
			$xzxyyy = \frac{1}{3}(-xzxxxy - xzxxyx - xzxyxx + 2xzyxxx + 2yzxxxx)$,
			$zxxyyy = \frac{1}{3}(-zxxxxy - zxxxyx - zxxyxx + 2zxyxxx + 2zyxxxx)$,
			xxxxyz, xxxyxz, xxyxxz, xyxxxz, yxxxxz,
			xxxxzy, xxxyzx, xxyxzx, xyxxzx, yxxxzx,
			xxxzxy, xxxzyx, xxyzxx, xyxzxx, yxxzxx,)
			xxzxxy, xxzxyx, xxzyxx, xyzxxx, yxzxxx,)
			xzxxxy, xzxxyx, xzxyxx, xzyxxx, yzxxxx,)
			zxxxxy, zxxxyx, zxxyxx, zxyxxx, zyxxxx)
k_6	176	50	*Components* j_6 *on performing the transposition* $x \to y$ *and* $y \to x$

Note. *The components are denoted by indices* ijklmn, *whereas N and I denote the numbers of nonzero and mutually independent components, respectively.*

Acknowledgments

The author expresses his lasting indebtedness to Stanisław Kielich for the guidance and help that rendered his work possible.

References

1. G. Szivessy, *Handb. Phys.* **20**, 715 (1929).
2. M. Born and E. Wolf, *Principles of Optics*, Pergamon, Oxford, UK, 1968.
3. M. Faraday, *Philos. Mag.* **29**, 153 (1846).
4. A. Righi, *Nuovo Cimento* **3**, 312 (1878).
5. H. Becquerel, *Comptes Rendus* **88**, 334 (1879).
6. A. Fresnel, *Ann. Chim. Phys.* **28**, 147 (1825).
7. H. A. Lorentz, *The Theory of Electrons*, Teubner, Leipzig, 1909.
8. S. V. Vonsovskii, *Magnetizm*, *Izd.* Nauka, Moskva, 1971.

9. A. Cotton, *Ann. Chim. Phys.* **8**, 347 (1896).
10. A. Cotton and H. Mouton, *Comptes Rendus* **145**, 229 (1907).
11. W. Voigt, *Magneto und Electro-Optik*, Teubner, Leipzig, 1908.
12. P. Langevin, *Le Radium* **7**, 249 (1910).
13. M. Born, *Optik*, Springer, Berlin, 1933.
14. A. D. Buckingham and J. A. Pople, *Proc. Phys. Soc.* **B69**, 1133 (1956).
15. A. Piekara and S. Kielich, *J. Phys. Radium* **18**, 490 (1957).
16. A. Piekara and S. Kielich, *Acta Phys. Pol.* **17**, 209 (1958).
17. A. Piekara and S. Kielich, *J. Chem. Phys.* **29**, 1292 (1958).
18. S. Kielich, *Acta Phys. Pol.* **17**, 239 (1958).
19. S. Kielich and A. Piekara, *Acta Phys. Pol.* **18**, 439 (1959).
20. S. Kielich, *Acta Phys. Pol.* **22**; 65, 299 (1962).
21. S. Kielich, *Acta Phys. Pol.* **31**, 929 (1967).
22. R. L. Kronig, *Z. Phys.* **45**, 458, 508 (1927).
23. T. Carrol, *Phys. Rev.* **52**, 822 (1937).
24. L. Rosenfeld, *Z. Phys.* **57**, 835 (1930).
25. W. Kroll, *Z. Phys.* **66**, 69 (1930).
26. R. Serber, *Phys. Rev.* **41**, 489 (1932).
27. M. P. Groenewege, *Mol. Phys.* **5**, 541 (1962).
28. H. F. Hameka, *Advanced Quantum Chemistry*, Addison-Wesley, Reading, MA, 1965.
29. P. Jørgensen, J. Oddershede, P. Albertsen, and N. H. Beebe, *J. Chem. Phys.* **68**, 2533 (1978).
30. N. L. Manakov, V. D. Ovsiannikov, and S. Kielich, *Acta Phys. Pol.* **A53**; 581, 595 (1978); Phys. Rev. **A21**, 1589 (1980).
31. A. F. Lubchenko, *Opt. Spektrosk.* **10**, 379, 477 (1961).
32. I. M. Boswarva, R. E. Howard, and A. B. Lidiard, *Proc. R. Soc. London Ser. A* **269**, 125 (1962).
33. I. M. Boswarva, R. E. Howard, and A. B. Lidiard, *Proc. R. Soc. London Ser. A.* **278**, 588 (1964).
34. T. Murao and A. Ebina, *J. Phys. Soc. J.* **20**, 997 (1965).
35. H. A. Kramers, *Proc. Acad. Sci. Amsterdam* **33**, 959 (1930).
36. Y. R. Shen, *Phys. Rev.* **133**, 511 (1964).
37. Y. R. Shen and N. Bloembergen, *Phys.* v. **A133**, 515 (1964).
38. V. V. Druzhynin and O. M. Tatsenko, *Opt. Spektrosk.* **36**, 733 (1074).
39. A. D. Buckingham and P. J. Stephens, *Annu. Rev. Phys. Chem.* **17**, 389 (1966).
40. E. D. Palik and B. W. Henvis, *Appl. Opt.* **6**, 603 (1967).
41. N. W. Starostin and P. P. Feofilov, *Usp. Fiz. Nauk* **97**, 621 (1969).
42. W. S. Zapaskii and P. P. Feofilov, *Usp. Fiz. Nauk* **116**, 41 (1975).
43. S. Ramasechan and V. Sivaramakrishnan, *Progress in Crystal Physics*, Madras, 1958.
44. G. N. Ramachandran and S. Ramasechan, Crystal Optics in *Handbuch der Physik* **25 / 1**, 1 (1961).
45. M. Koralewski and M. Surma, *Acta Phys. Pol.* **A49**, 803 (1976).
46. M. Koralewski, M. Surma and M. Mróz, *Opt. Appl.* **8**, 171 (1978).

47. M. Koralewski, S. Habrylo, and M. Mróz, *Acta Phys. Pol.* **A56**, 419 (1979).
48. S. Kielich, *Acta Phys. Pol.* **29**, 875 (1966).
49. D. L. Portigal and E. J. Burstein, *J. Phys. Chem. Solids* **32**, 603 (1971).
50. V. A. Markelov, M. A. Novikov, and A. A. Turkin, *Pis'ma Zh. Eksp. Teor. Fiz.* **25**, 404 (1977).
51. N. B. Baranova, Yu. V. Bogdanov, and B. Ya. Zel'dovich, *Opt. Commun.* **22**, 243 (1977).
52. N. B. Baranova and B. Ya. Zel'dovich, *Mol. Phys.* **38**, 1085 (1979).
53. S. Woźniak and R. Zawodny, *Acta Phys. Pol.* **A61**, 175 (1982).
54. S. Kielich and R. Zawodny, *Acta Phys. Pol.* **A42**, 337 (1972).
55. H. J. Ross, B. S. Sherborne, and G. E. Stedman, *J. Phys. B* **22**, 459 (1989).
56. A. C. Tavger and V. I. Zaitsev, *Zh. Eksp. Teor. Fiz.* **30**, 564 (1956).
57. C. S. Porter, E. G. Spencer, and R. S. Le Crow, *J. Appl. Phys.* **29**, 495 (1958).
58. C. Kittel, *Phys. Rev.* **83**, 208 (1951).
59. P. N. Argyres, *Phys. Rev.* **97**, 334 (1955).
60. P. E. Ferguson and R. J. Romagnoli, *J. Appl. Phys.* **40**, 1236 (1969).
61. S. Wittekoek and T. J. A. Popma, *J. Appl. Phys.* **44**, 5560 (1973).
62. J. L. Erskine and E. A. Stern, *Phys. Rev.* **30**, 1329 (1973).
63. S. Kielich, *Proc. Phys. Soc.* **86**, 709 (1965).
64. S. Kielich, *Physica* **32**, 385 (1966).
65. V. M. Agranovich and V. L. Ginzburg, *Spatial Dispersion in Crystal Optics and the Theory of Excitons*, Wiley, New York, 1965.
66. U. Schlagheck, *Z. Phys.* **266**, 312 (1974).
67. G. E. Stedman, *Adv. Phys.* **34**, 513 (1985).
68. W. Brown, S. Shtrikman, and D. Treves, *J. Appl. Phys.* **34**, 1233 (1963).
69. R. Fuchs, *Philos. Mag.* **11**, 647 (1965).
70. R. M. Hornreich and S. Shtrikman, *Phys. Rev.* **171**, 1065 (1968).
71. R. R. Birss and R. G. Shrubsall, *Philos. Mag.* **15**, 687 (1967).
72. P. Chandrasekhar and T. P. Srinivasan, *J. Phys. C* **6**, 1085 (1973).
73. R. Zawodny, *Acta Crystallogr.* **A34**, 357 (1978).
74. G. S. Krinchik and M. W. Chetkin, *Zh. Eksp. Teor. Fiz.* **40**, 729 (1961).
75. L. D. Landau and E. M. Lifshits, *Electrodynamics of Continuous Media*, Pergamon, 1959.
76. S. Kielich and R. Zawodny, *Physica B* **89**, 112 (1977).
77. J. Brandmüller, A. Lehmeyer, K. M. Häussler, and L. Merten, *Phys. Status Solidi B* **117**, 323 (1983).
78. R. V. Pisarev, *Zh. Eksp. Teor. Fiz.* **58**, 1421, 1481 (1970).
79. S. Bhagavantam, *Proc. Ind. Ac. Sci.* **A73**, 269 (1971).
80. G. A. Smolensky, R. V. Pisarev, and I. G. Sinii, *Sov. Phys. Usp.* **18**, 323 (1975).
81. J. Ferre and G. A. Gehring, *Rep. Prog. Phys.* **47**, 513 (1984).
82. S. Shtrikman and D. Treves, *Proc. Int. Conf. Magnetism*, Nottingham, 1965, p. 484.
83. J. V. Dillon, Jr., in *Magnetic Properties of Materials*, McGraw-Hill, New York, 1971, Part 5, p. 149.

84. N. F. Kharchenko, O. P. Tutakina, and L. Bely, *Abstracts of Papers 16 All-Union Conf. on Low Temp. Phys.*, Minsk, 1976, p. 650.
85. J. V. Dillon, Jr., L. D. Tallay, and E. I. Chen, *A. I. P. Conf. Proc.* **34**, 388 (1976).
86. N. F. Kharchenko, V. V. Eremenko, and O. P. Tutakina, *Pis'ma Zh. Eksp. Teor. Fiz.* **27**, 466 (1978).
87. N. N. Akhmedov and A. K. Zvezdin, *Pis'ma Zh. Eksp. Teor. Fiz.* **38**, 167 (1983).
88. N. F. Kharchenko, A. V. Bibik, and V. V. Eremenko, *Pis'ma Zh. Eksp. Teor. Fiz.* **42**, 447 (1985).
89. V. V. Eremenko and N. F. Kharchenko, *Phys. Rep.* **155(6)**, 379 (1987).
90. R. Zawodny and S. Kielich, *Phys. Rev. A* **38**, 3504 (1988).
91. R. Zawodny and S. Kielich, *SPIE Proc.* **1018**, 130 (1988).
92. A. Piekara and S. Kielich, *Arch. Sci.* **11**, 304 (1958).
93. S. Kielich, *Acta Phys. Pol.* **32**, 405 (1967).
94. P. S. Pershan, J. P. Van der Ziel, and L. D. Malmstrom, *Phys. Rev.* **143**, 574 (1966).
95. S. Kielich, *J. Colloid Interface Sci.* **30**, 159 (1969).
96. J. P. Van der Ziel and N. Bloembergen, *Phys. Rev. A* **138**, 1287 (1965).
97. J. F. Holzrichter, R. M. Macfarlane, and A. L. Schaulow, *Phys. Rev. Lett.* **26**, 652 (1971).
98. G. Wagnière, *Phys. Rev. A* **40**, 2437 (1989).
99. S. Woźniak, G. Wagnière, and R. Zawodny, *Phys. Lett. A* **154**, 259 (1991).
100. S. Woźniak, M. W. Evans, and G. Wagnière, *Mol. Phys.* **75**, 81 (1992).
101. S. Woźniak, M. W. Evans, and G. Wagnière, *Mol. Phys.* **75**, 99 (1992).
102. P. W. Atkins and M. H. Miller, *Mol. Phys.* **15**, 491, 503 (1968).
103. E. Courtens, *Phys. Rev. Lett.* **21**, 3 (1968).
104. N. V. Cohan and H. F. Hameka, *Physica* **38**, 320 (1967).
105. S. Kielich and R. Zawodny, *Opt. Commun.* **4**, 132 (1971).
106. H. G. Hafele, R. Grisar, C. Irslinger, H. Wachering, S. D. Smith, R. B. Dennis, and B. S. Wherrett, *J. Phys. C* **4**, 2637 (1971).
107. Van Tran Nguyen and T. J. Bridges, *Phys. Rev. Lett.* **29**, 359 (1972).
108. N. N. Akhmedev, S. B. Borisov, A. K. Zvezdin, I. L. Lyubchansky, and Yu. V. Melikhov, *Sov. Solid State Phys.* **27**, 1075 (1985).
109. J. Reif, J. C. Zink, C. M. Schneider, and J. Kirschner, *Phys. Rev. Lett.* **67**, 2878 (1991).
110. P. D. Maker, R. W. Terhune, and C. M. Savage, *Phys. Rev. Lett.* **12**, 507 (1964).
111. A. D. Buckingham, *Proc. Phys. Soc. B* **69**, 344 (1956).
112. W. Voigt, *Ann. Phy.* **4**, 197 (1901).
113. G. Mayer and F. Gires, *C. R. Acad. Sci. Paris B* **258**, 2039 (1964).
114. S. Kielich and R. Zawodny, *Opt. Commun.* **15**, 267 (1975).
115. K. L. Sala, *Phys. Rev.* **29**, 1944 (1984).
116. M. Roman, S. Kielich, and W. Gadomski, *Phys. Rev. A* **34**, 351 (1986).
117. Nguen Phu-Xuan and G. Rivoire, *Opt. Acta* **25**, 233 (1978).
118. S. Kielich and R. Zawodny, *Opt. Acta* **20**, 867 (1973).
119. S. Kielich and R. Zawodny, *Acta Phys. Pol.* **A43**, 579 (1973).
120. K. Kubota, *J. Phys. Soc. J.* **29**, 986, 998 (1970).

121. S. Kielich, *Phys. Lett. A* **25**, 517 (1967).
122. S. Kielich, *Appl. Phys. Lett.* **13**, 152 (1968).
123. S. Kielich, *Bull. Soc. Amis Sci. Lett. Poznań B* **21**, 47 (1968/69).
124. S. Kielich, N. L. Manakov, and V. D. Ovsiannikov, *Acta Phys. Pol.* **A53**, 737 (1978).
125. B. A. Zon and T. T. Urazbaev, *Sov. Phys. JETP* **41**, 1006 (1976).
126. B. A. Glushko and V. O. Chaltykyan, *Sov. J. Quant. Electron.* **8**, 631 (1978).
127. G. G. Adonts, E. G. Kanetsyan, and M. V. Slobodskoi, *Opt. Spektrosk.* **64**, 258 (1988).
128. S. Kielich, R. Tanaś, and R. Zawodny, *Phys. Rev. A* **36**, 5670 (1987).
129. S. B. Borisov, I. L. Lubchansky, and A. D. Petrenko, *Opt. Spektrosk.* **64**, 1379 (1988).
130. A. A. Golubkov and W. A. Makarov, *Opt. Spektrosk.* **67**, 1134 (1989).
131. S. A. Bakhramov, A. T. Berdikulov, A. M. Kokhkharov, V. V. Tihonenko, and P. K. Habibullaev, Dokl. Akad. Nauk USSR **309**, 607 (1989).
132. J. Frey, R. Frey, C. Flytzanis, and R. Triboulet, *Opt. Commun*, **84**, 76 (1991).
133. J. Frey, R. Frey, C. Flytzanis, and R. Triboulet, *J. Opt. Soc. Am. B* **9**, 132 (1992).
134. S. Kielich, in C. T. O'Konski (Ed.), *Molecular Electro-Optics*, Dekker, New York, 1976, Vol. 1, p. 445.
135. A. G. Fainshtein, N. L. Manakov, V. D. Ovsiannikov, and L. P. Rapaport, *Phys. Rep.* **210**, 111 (1992).
136. J. A. Armstrong, N. Bloembergen, J. Ducuing, and P. S. Pershan, *Phys. Rev.* **127**, 1918 (1962).
137. P. S. Pershan, *Phys. Rev.* **130**, 919 (1963).
138. N. Bloembergen, *Nonlinear Optics*, Benjamin, Reading, MA, 1965.
139. T. H. O'Dell, *The Electrodynamics of Magneto-Optical Media*, North-Holland, Amsterdam, 1970.
140. A. Messiah, *Quantum Mechanics*, North-Holland, Amsterdam, 1964.
141. P. N. Butcher, *Nonlinear Optical Phenomena*, Bulletin 200 Engineering Experiment Station, Ohio State University Press, Columbus, 1965.
142. R. Zawodny, Thesis, UAM Poznań, 1977.
143. M. Schubert and B. Wilhelmi, *Einführung in die Nichtlineare Optik*, Leipzig, 1971.
144. R. R. Birss, *Symmetry and Magnetism*, North-Holland, Amsterdam, 1964.
145. S. Bhagavantam, *Crystal Symmetry and Physical Properties*, Academic, London, 1966.
146. H. Kleiner, *Phys. Rev.* **142**, 318 (1966).
147. F. G. Fumi, *Nuovo Cimento* **9**, 739 (1952).
148. R. Fieschi and F. G. Fumi, *Nuovo Cimento* **10**, 865 (1953).
149. R. Fieschi, *Physica* **24**, 972 (1957).
150. I. E. Dzyaloshinskii, *Zh. Eksp. Teor. Fiz.* **32**, 1547 (1957).
151. E. A. Turov, *Physical Properties of Magnetically Ordered Crystals*, Izd. AN SSSR, Moscow, 1963.

DYNAMICAL QUESTIONS IN QUANTUM OPTICS

ALEXANDER STANISLAW SHUMOVSKY

Joint Institute for Nuclear Research, N. N. Bogolubov Laboratory of Theoretical Physics, Quantum Optics Division, Moscow, Russia

CONTENTS

I. INTRODUCTION

A dedicatory volume always provides an opportunity for an informal and qualitative discussion of the interesting problems in one's field of scientific interest. In this spirit, I should like to report an approach concerning a major question in quantum optics, namely the dynamical behavior of the collective processes of light–matter interaction. The problem can, of course, be formulated in many different ways. One way is connected with the hierarchical equations for the time-dependent correlation functions or

I extend my best wishes to Professor Stanislaw Kielich on the occasion of 40 years of his scientific activity.

Modern Nonlinear Optics, Part 1, Edited by Myron Evans and Stanisław Kielich. Advances in Chemical Physics Series, Vol. LXXXV.
ISBN 0-471-57546-1

Green functions and has close analogy with the general method of kinetic theory introduced by Bogolubov [1].

An important aspect involved in the derivation of kinetic equations is the separation of time scales. A kinetic equation can be expected to hold only if the relaxation time is much larger than the interaction time, which means that only large-scale dynamical behavior can be considered in this way. For example, the superradiance is determined by the following condition (see, e.g., [2, 3]):

$$\tau \ll \tau_c \ll T_2$$

where τ is the time of flight of a photon through the atomic system, τ_c is the collective decay time, and T_2 is the dipole relaxation time. Thus, the kinetic equation can be used to describe that process.In quantum mechanics, complete information about the dynamical and statistical properties of a system is contained in the density matrix ρ given by the Liouville equation

$$i\hbar \frac{\partial \rho}{\partial t} = [H, \rho] \tag{1}$$

where H is the Hamiltonian of the system. Let C be a dynamical variable in the Schrödinger representation. Then its time-dependent mean value is determined in the following manner:

$$\langle C \rangle_t = \mathrm{Tr}[C\rho(t)] \tag{2}$$

The Heisenberg representation is constructed from the Schrödinger representation by the relation

$$C(t) = U^{-1}(t) C U(t)$$

where $U(t)$ is the unitary operator obeying the equation

$$i\hbar \frac{\partial U}{\partial t} = HU(t), U(t_0) = 1$$

Hence, the mean value (2) can be represented in the form

$$\langle C \rangle_t = \mathrm{Tr}[C(t)\rho(t_0)] \tag{3}$$

where $\rho(t_0)$ is the density matrix of the initial state. Operator $C(t)$ obeys the Heisenberg equation of motion

$$i\hbar\frac{\partial C}{\partial t} = [C(t), H] \tag{4}$$

Thus, to obtain the time-dependent mean value (3), one ought to solve the operator equation (4) and determine the density matrix $\rho(t_0)$ of an initial state of the system.

For problems of quantum optics the density matrix of the initial state is usually written in the form

$$\rho(t_0) = \rho_F \otimes \rho_M \tag{5}$$

where ρ_F is the initial density matrix of a radiation field and ρ_M is the same for the matter. Several models of quantum optics are described by the linear interaction Hamiltonians with respect to the photon operators. The multipolar coupling in the dipole approximation and the minimal coupling for the atom–radiation interaction can be shown as an example [4]. So, let us consider the following general Hamiltonian:

$$\begin{aligned} H &= \sum_k \hbar\omega_k a_k^\dagger a_k + H_j + H_{\text{int}} \\ H_{\text{int}} &= \sum_k (j_k^+ a_k + a_k^+ j_k); \qquad [a_k, j_m] = 0 \end{aligned} \tag{6}$$

Here a_k^+ and a_k are the Bose operators corresponding to the creation and annihilation of photons, j_k^+ and j_k are some operators describing a matter, and H_j is the Hamiltonian of free matter. The Heisenberg equations for the field operators

$$i\hbar\frac{da_k}{dt} = \hbar\omega_k a_k + j_k \qquad i\hbar\frac{da_k^+}{dt} = -\hbar\omega_k a_k^+ - j_k^+$$

have the following formal solutions

$$a_k(t) = a_{kf}(t) - i\hbar^{-1}A_k \tag{7}$$

where

$$a_{kf}(t) \equiv a_k(t_0)e^{-i\omega_k(t-t_0)} \qquad A_k = \int_{t_0}^t d\tau\, e^{-i\omega_k(t-\tau)} j_k(\tau)$$

Let J be an arbitrary operator of the j subsystem. Then

$$i\hbar \frac{dJ}{dt} = [J, H_j] + \sum_k ([J, j_k^+] a_k + a_k^+ [J, j_k])$$

Substituting Eq. (7) and averaging with respect to the density matrix of the initial state (5) we get

$$\frac{d\langle J \rangle_t}{dt} = -i\hbar^{-1} \langle [J, H_j] \rangle_t - \hbar^{-2} \sum_k (\langle [J, j_k] A_k \rangle_t - \langle A_k^+ [J, j_k] \rangle_t) \\ - i\hbar^{-1} \sum_k (\langle [J, j_k^+] a_{kf} \rangle_t + \langle a_{kf}^+ [J, j_k] \rangle_t) \tag{8}$$

The first three terms on the right in Eq. (8) contain j-type operators only, while the last two terms are the "mixed" two-time correlation functions. They depend on the initial field amplitudes and on j-type operators at time t. Similar correlation functions arise in different problems of quantum optics. Behavior of these functions depends on the initial state, and the corresponding calculations are usually based on some complicated approximations [5]. Moreover, the exact relations have been established by Knoll and Shumovsky [6] for the initial state, which is some superposition of the chaotic, coherent, and squeezed states of the field. By virtue of those relations it is possible to eliminate entirely the field operators from equations of the type of Eq. (8). Employing this approach then gives the hierarchical integro-differential equation with respect to the time-dependent mean values of j-type operators. The next step in the derivation of the kinetic equation is connected with decoupling the high-order correlations having those or other physical reasons.

It is possible, of course, to use different approaches. Examples are obtained by the Zwanzig method of the projection operator [7, 8], by the adiabatic elimination of the field variables from the master equation [9, 10], and by the use of atomic coherent states to obtain the Fokker–Planck equation for the Markoffian stochastic process [11, 12].

Interest in the present approach lies not only in its application to the relatively simple models of quantum optics and condensed matter physics but also in the fact that it leads to a much clearer understanding of the relations between quantum mechanics and statistical physics. In the present chapter we discuss some aspects of the dynamics of matter–light interaction and some analogies with statistical physics.

II. PHYSICAL MODELS

Let us consider some important realizations of the model problem with the Hamiltonian (6).

A. Dicke Model [13]

The simplest atom–photon interaction model is based on the concept of a two-level atom introduced by Einstein [14]. It is often the case that the experimental conditions permits us to neglect the influence of all atomic states with the exception of some pairs of states, which we call $|g\rangle$ and $|e\rangle$ (see, e.g., [15]). The atomic operators $|i\rangle\langle j|$ $(i, j = e, g)$, characterizing the level populations at $i = j$ and transitions from j to i at $i \neq j$, can be represented in terms of the angular-momentum operators:

$$S^{+} = |e\rangle\langle g| \qquad S^{-} = |g\rangle\langle e| \qquad S^{z} = \frac{|e\rangle\langle e| - |g\rangle\langle g|}{2}$$

whose commutation relations are

$$[S^{+}, S^{-}] = 2S^{z} \qquad [S^{z}, S^{\pm}] = \pm S^{\pm} \tag{9}$$

Then the Hamiltonian of a point-like atomic system interacting with electromagnetic field has dipole approximation in the following form:

$$H = \sum_{k} \hbar\omega_k a_k^{+} a_k + \hbar\omega_0 \sum_{f=1}^{N} S_f^{z} + \sum_{k} \sum_{f=1}^{N} (g_k S_f^{+} - g_k^{*} S_f^{-})(a_k - a_k^{+}) \tag{10}$$

where

$$g_k = \mathrm{i}\left(\frac{2\pi\hbar\omega_k}{V}\right)^{1/2} e_k d_{ge}$$

Here N is the number of atoms, ω_0 is the transition frequency, V is the volume of field quantization, e_k is the unit polarization vector, and d_{ge} is the electric dipole moment of transition $g \leftrightarrow e$.

Comparison of Eqs. (6) and (10) gives us

$$H_j = \hbar\omega_0 \sum_{f=1}^{N} S_f^{z}$$

and

$$j_k = \sum_{f=1}^{N} (g_k S_k^+ - g_k^* S_k^-)$$

A special case of the so-called rotating-wave approximation (RWA), which is obtained by neglecting the energy-nonconserving terms S^+a^+ and S^-a, is usually considered.

B. Raman Scattering

Conception of scattering in optics is standard for the process of interaction between light and matter leading to an appreciable change of the radiation flow characteristics. It can be, for example, spatial intensity distribution, polarization, frequency spectrum, and quantum statistical properties. Raman scattering mainly consists of a change of a scattered light frequency in comparison with an incident light [16], although a violation of the initial quantum statistical properties can be also observed [17].

The simple energy conservation laws for Raman scattering allow us to use here the two-level formalism leading to the Dicke-type Hamiltonian [18]. The model of four one-mode Bose field interactions with the Hamiltonian [19, 20] can also be considered:

$$\begin{aligned} H = \hbar(\omega_I a_I^+ a_I + \omega_S a_S^+ a_S + \omega_{AS} a_{as}^+ a_{as} + \omega_V b^+ b) \\ - \hbar(\kappa_S a_S^+ b^+ a_I + \kappa_{AS} a_{AS}^+ b a_I + \text{h.c.}) \end{aligned} \tag{11}$$

where a_j corresponds to the incident light ($j = I$), Stokes ($j = S$), or anti-Stokes ($j = AS$) component and b describes a vibrational mode. This Hamiltonian also describes the sum and difference generation and frequency conversion.

We note now that Eq. (11) is a particular form of Eq. (6) and that b can be chosen as a-field while $j^+ = a_S^+ a_I, a_{AS}^+ a_I$.

C. Polariton-like System

In ionic crystals the retarded interaction between the ions is propagated by photons created due to the transversal optical vibrations of the lattice [21, 22]. The corresponding Hamiltonian is

$$H = \hbar \sum_k (\omega_k a_k^+ a_k + \Omega_k b_k^+ b_k - D_k[(a_k^+ - a_k)(b_k + b_{-k}^+) + \text{h.c.})] \tag{12}$$

where a^+, a are the photon operators and b^+, b are the phonon operators. Employing the Bogolubov canonical transformation [23] then gives the exact solution describing the bound state of photons and phonons (polariton). This Hamiltonian is also related to Eq. (6) and any Bose field can be chosen as a-field here.

D. High-T_c Superconductivity

Several recent experiments have shown the existence of structural instabilities in $YBa_2Cu_3O_7$ [24] and $Tl_2Ba_2CaCu_2O_8$ [25]. The Cu(1)–O(4) bonds have a double-well structure, which changes in the vicinity of T_c and as a function of oxygen content and doping. Structural studies indicate dynamical oscillations between two possible positions. The two characteristic phonon modes, namely the Raman mode and the infrared mode, demonstrate the hardening and softening, respectively, with the decrease in temperature. The situation can be considered in an isolated O(4)–Cu(1)–O(4) cluster in $YBa_2Cu_3O_7$ with the aid of following model Hamiltonian [26]:

$$\begin{aligned} H = & \sum_{\sigma, i=1}^{3} (\varepsilon_i n_{i\sigma} + U n_{i\uparrow} n_{i\downarrow}) + t \sum_{\sigma} (c_{1\sigma}^+ c_{2\sigma} + c_{2\sigma}^+ c3\sigma + \text{h.c.}) \\ & + \omega_{IR} b_{IR}^+ b_{IR} + \omega_R b_R^+ b_R \\ & - \kappa_{IR}(b_{IR}^+ + b_{IR}) \sum_{\sigma} (n_{3\sigma} - n_{1\sigma}) \\ & - \kappa_R(b_R^+ + b_R) \sum_{\sigma} (n_{1\sigma} - 2n_{2\sigma} + n_{3\sigma}) \end{aligned} \tag{13}$$

where $n_{i\sigma} = c_{i\sigma}^+ c_{i\sigma}$ and c is the electron operator, U is the repulsion, t is the hoping amplitude, and b_j describes the jth phonon mode, which may be considered as a-field in Eq. (6).

As additional examples of the realization of Eq. (6) in condensed matter physics, the Frölich model and spin–phonon interaction can be mentioned [27].

III. EXACT RELATIONS FOR MIXED CORRELATION FUNCTIONS

Following the paper by Knöll and Shumovsky [6], let us suppose that the initial state of the Bose field in (6) can be chaotic, coherent, or squeezed.

For this purpose ρ_F should be chosen in the following form:

$$\rho_F = \prod_k \rho_k; \qquad \rho_k = D_k(\alpha_k) S_k(\xi_k) T_k(\beta_k) S_k^+(\xi_k) D_k^+(\alpha_k) \quad (14)$$

Here

$$D_k(\alpha_k) = \exp[\alpha_k a_k^+(t_0) - \alpha_k^* a_k(t_0)]$$

is the displacement operator,

$$S_k(\xi_k) = \exp\left[\frac{\xi_k^* a_k^2(t_0)}{2} - \frac{\xi_k a_k^{+2}(t_0)}{2}\right] \quad (15)$$

is the squeezing operator, and

$$T_k(\beta_k) = \sum_{n=0}^{\infty} \frac{e^{+\beta_k} - 1}{e^{(n+1)\beta_k}} |n\rangle_{kk}\langle n|$$

$$\beta_k \equiv \omega_k/\theta, \, a_k^+(t_0) a_k(t_0) |n\rangle_k = n|n\rangle_k$$

describes the chaotic state with temperature θ.

The following well-known commutation relations take place (see, e.g., [28]):

$$\begin{aligned} D_k^+ a_k D_k &= a_k + \alpha_k & D_k^+ a_k^+ D_k &= a_k^+ + \alpha_k^* \\ S_k^+ a_k S_k &= \mu_k a_k - \nu_k a_k^+ & S_k^+ a_k^+ S_k &= -\nu_k^* a_k + \mu_k a_k^+ \\ T_k^{-1} a_k T_k &= e^{-\beta_k} a_k & T_k^{-1} a_k^+ T_k &= e^{\beta_k} a_k^+ \end{aligned} \quad (16)$$

where $\mu_k = \cosh r_k$, $\nu_k = e^{i\varphi_k} \sinh r_k$, $\xi_k = r_k e^{i\varphi_k}$.

Let us introduce the following notation: $\delta a_k \equiv a_{kf} - \alpha_k$, where a_{kf} is defined by Eq. (7). Then from Eqs. (14) and (15) one can obtain

$$\delta a_k \rho_k = \rho_k \left\{ \left(\mu_k^2 e^{-\beta_k} - |\nu_k|^2 e^{\beta_k} \right) \delta a_k - 2\mu_k \nu_k \sinh \beta_k \delta a_k^+ \right\}$$

For the mixed correlation function

$$\langle J \delta a_k \rangle = \langle J a_{kf} \rangle - \alpha_k \langle J \rangle$$

it follows that

$$\langle J\delta a_k \rangle = \left(\mu_k^2 \, \mathrm{e}^{-\beta_k} - |\nu_k|^2 \, \mathrm{e}^{\beta_k}\right)\langle \delta a_k J \rangle - 2\mu_k \nu_k \sinh \beta_k \langle \delta a_k^+ J \rangle$$

By analogy we get

$$\langle J\delta a_k^+ \rangle = 2\mu_k \nu_k^* \langle \delta a_k J \rangle + \left(\mu_k^2 \, \mathrm{e}^{\beta_k} - |\nu_k|^2 \, \mathrm{e}^{-\beta_k}\right)\langle \delta a_k J \rangle$$

Solving these two expressions with respect to $\langle \delta a_k J \rangle$ and $\langle \delta a_k^+ J \rangle$, and taking into account the equalities

$$\langle a_{kf}^+, a_{kf} \rangle = |\nu_k|^2 + \frac{\mu_k^2 + |\nu_k|^2}{\mathrm{e}^{\beta_k} - 1}; \qquad \langle a_k(t_0), a_k(t_0) \rangle = -\mu_k \nu_k \frac{\mathrm{e}^{\beta_k} + 1}{\mathrm{e}^{\beta_k} - 1}$$

where $\langle A, B \rangle = \langle AB \rangle - \langle A \rangle \langle B \rangle$, we obtain

$$\begin{aligned} \langle Ja_{kf} \rangle &= \langle J \rangle \langle a_{kf} \rangle + \langle a_{kf}^+, a_{kf} \rangle \langle [a_{kf}, J] \rangle - \langle a_{kf}, a_{kf} \rangle \langle [a_{kf}^+, J] \rangle \\ \langle Ja_{kf}^+ \rangle &= \langle J \rangle \langle a_{kf}^+ \rangle + \langle a_{kf}^+, a_{kf}^+ \rangle \langle [a_{kf}, J] \rangle - \langle a_{kf}, a_{kf}^+ \rangle \langle [a_{kf}^+, J] \rangle \end{aligned} \tag{17}$$

Let us now turn to expression 7. Since

$$[J(t), a_k(t)] = [J(t), a_k^+(t)] = 0$$

then by virtue of Eq. (7) for the right-handed commutators in Eq. (17) we get

$$[a_{kf}, J] = \mathrm{i}[A_k, J], [a_{kf}^+, J] = -\mathrm{i}[A_k^+, J]$$

Thus, instead of Eq. (17) we finally obtain

$$\begin{aligned} \langle Ja_{kf} \rangle &= \langle J \rangle \langle a_{kf} \rangle + \mathrm{i}\langle a_{kf}^+, a_{kf} \rangle \langle [A_k, J] \rangle \\ &\quad + \mathrm{i}\langle a_{kf}, a_{kf} \rangle \langle [A_k^+, J] \rangle \\ \langle Ja_{kf}^+ \rangle &= \langle J \rangle \langle a_{kf}^+ \rangle + \mathrm{i}\langle a_{kf}^+, a_{kf}^+ \rangle \langle [A_k, J] \rangle \\ &\quad + \mathrm{i}\langle a_{kf}, a_{kf}^+ \rangle \langle [A_k^+, J] \rangle \end{aligned} \tag{18}$$

These relations express the mixed correlation functions through the mean values of the field operators at the initial time moment, which are the known functions of the initial state parameters α, ξ, and β, and through the integrals of the two-time correlation functions of j subsystem. Relations (18) coincide with the result obtained by Bogolubov [29] at $\alpha_k = \xi_k = 0$.

IV. HIERARCHICAL EQUATION WITH ELIMINATED BOSE OPERATORS

Let us now substitute exact expressions (18) into the equation of motion for j-type operator (Eq. (8)). We have

$$
\begin{aligned}
\frac{d\langle J\rangle}{dt} &= -i\langle [J, H_j]\rangle \\
&\quad + \sum_k \Big\{\langle [j_k^+, J] A_k\rangle + \langle A_k^+[J, j_k]\rangle - i\langle J\rangle(\langle a_{kf}\rangle + \langle a_{kf}^+\rangle) \\
&\qquad + \langle a_{kf}^+, a_{kf}\rangle(\langle [A_k, [J, j_k^+]]\rangle + \langle [A_k^+, [J, j_k]]\rangle) \\
&\qquad - \langle a_{kf}, a_{kf}\rangle\langle A_k^+, [J, j_k^+]]\rangle + \langle a_{kf}^+, a_{kf}^+\rangle\langle [A_k, [J, j_k]]\rangle\Big\}
\end{aligned}
\tag{19}
$$

This is the exact hierarchical integro-differential equation describing the dynamics of the system with the Hamiltonian (6) at the initial state of the field described by the density matrix (14), and at any initial state of the j subsystem.

The right side of Eq. (19) contains two-time correlation functions. It was established by Bocchieri and Loinger [30] and Percivall [31] that such objects are the almost periodical functions of $t - \tau$ in the case of a finite system (before the thermodynamical limit). This is the quantum analogy of the Poincaré recurrence theorem in classical dynamics. The damping of correlation functions can be obtained either in the thermodynamical limit ($V \to \infty$, $V/N \to$ const.) or in some approximations that demolish the long Poincaré cycles.

We now note that this hierarchical equation can be used not only for the determination of any time-dependent mean value in a j subsystem, but also for the investigation of the dynamics of the Bose field. Let us calculate, for example, the photon number mean value for the kth mode. By virtue of Eq. (7) we have

$$\langle a_k^+ a_k\rangle = \langle a_{kf}^+ a_{kf}\rangle + \langle A_k^+ A_k\rangle + i(\langle A_k^+ a_{kf}\rangle - \langle a_{kf}^+ A_k\rangle)$$

Using here expressions (18) we obtain

$$
\begin{aligned}
\langle a_k^+ a_k \rangle = \langle A_{kf}^+ a_{kf} \rangle &+ \langle a_{kf}^+, a_{kf} \rangle \langle [A_k^+, A_k] \rangle \\
&- \mathrm{i}\big(\langle a_{kf}^+ \rangle \langle A_k \rangle - \langle a_{kf} \rangle A_k^+ \rangle\big) \\
&+ \langle a_{kf}^+, a_{kf}^+ \rangle \langle A_k^+ \rangle\big) + \langle a_{kf}^+, a_{kf}^+ \rangle \langle [A_k^>, A_k^<] \rangle \\
&+ \langle a_{kf}, a_{kf} \rangle \langle [A_k^{+<}, A_k^{+>}] \rangle
\end{aligned}
\tag{20}
$$

where

$$
\begin{aligned}
\langle [A_k^>, A_k^<] \rangle &= \langle [A_k^{+<}, A_k^{+>}] \rangle^* \\
&= -\int_0^t \mathrm{d}\tau \int_0^\tau \mathrm{d}\sigma \, \mathrm{e}^{-\mathrm{i}\omega_k(2t-\tau-\sigma)} \langle [j_k(\tau), j_k(\sigma)] \rangle
\end{aligned}
$$

Thus, the mean number of photons at any time $t > 0$ depends on the initial mean number of photons and on the time-dependent mean values of various combinations of j-subsystem operators. The latter can be calculated with the aid of Eq. (19).

It may be necessary to represent operators j_k and j_k^+ in the following manner:

$$
j_k(t) = \sum_\gamma \psi_{k\gamma} q_\gamma(t), \; j_k^+(t) = \sum_\gamma \varphi_{k\gamma} q_\gamma(t)
$$

Here ψ and φ are the known c-number functions and $q_\gamma(t)$ are the operators. In the absence of the interaction they have the following time evolution:

$$
\mathrm{e}^{\mathrm{i}H_j t} q_\gamma(0) \mathrm{e}^{-\mathrm{i}H_j t} = \mathrm{e}^{-\mathrm{i}\omega_\gamma t} q_\gamma(0)
$$

where ω_γ are the characteristic frequencies of the j subsystem. Let

$$
\Phi_\gamma(\tau) = \sum_k \phi_{k\gamma} a_{kf}(\tau) \qquad \Psi_\gamma(\tau) = \sum_k \psi_{k\gamma} a_{kf}^+(\tau)
$$

$$
\Omega_\gamma(\tau) = \Phi_\gamma(\tau) + \Psi_\gamma(\tau)
$$

Then instead of Eq. (19) we get

$$\frac{d\langle J\rangle}{dt} = -i\langle [J, H_j]\rangle - i\sum_{\gamma}\langle [J, q_{\gamma}]\rangle\langle \Omega_{\gamma}\rangle$$
$$-\int_0^t d\tau \sum_{\gamma}\sum_{\lambda}\Big\{\langle [\Phi_{\gamma}(t), \Phi_{\lambda}(\tau)]\rangle\langle [J(t), q_{\gamma}(t)]q_{\lambda}(\tau)\rangle$$
$$-\langle [\Phi_{\gamma}(\tau), \Psi_{\lambda}(t)]\rangle\langle q_{\gamma}(\tau)[J(t), q_{\lambda}(t)]\rangle$$
$$+\langle [\Phi_{\gamma}(\tau), \Psi_{\gamma}(t)]\rangle\langle [q_{\gamma}(\tau), [J(t), q_{\lambda}(t)]]\rangle$$
$$-\langle \Omega_{\gamma}(\tau), \Omega_{\lambda}(t)\rangle\langle [q_{\gamma}(\tau), q_{\lambda}(t)]]\rangle\Big\} \qquad (21)$$

It should be emphasized that the Bose operators can be eliminated from the dynamical equations in different ways, for example, the Zwanzig method [7, 8]. But the approach considered above permits us to take into account the initial conditions by the most clear way. The question now is how to calculate something with the aid of the exact hierarchical equation (19).

V. MARKOFFIAN APPROXIMATION

Let us turn our attention to the hierarchical equation in the form of Eq. (21). If the relaxation time of the j subsystem is expected to be much longer than ω_{γ}^{-1}, one can replace the two-time correlation function

$$\langle [J(t), q_{\gamma}(t)]q_{\lambda}(\tau)\rangle$$

by a single-time correlation function [32]:

$$e^{-i\omega_{\lambda}(t-\tau)}\langle [J(t), q_{\gamma}(t)]q_{\lambda}(t)\rangle$$

That substitution corresponds to the Markoffian approximation. Then the Markoffian hierarchical equation is

$$\frac{d\langle J\rangle}{dt} = -i\langle [J, H_j]\rangle - i\sum_{\gamma}\langle [J, q_{\gamma}]\rangle\langle \Omega_{\gamma}\rangle$$
$$+\sum_{\gamma,\lambda}\Big\{M_{\gamma\lambda}(t)\langle [J, q_{\gamma}]q_{\lambda}\rangle + N_{\gamma\lambda}(t)\langle q_{\gamma}[J, q_{\lambda}]\rangle \qquad (22)$$
$$+P_{\gamma\lambda}(t)\langle [q_{\gamma}, [J, q_{\lambda}]]\rangle\Big\}$$

Here

$$M_{\gamma\lambda}(t) = -\int_{t_0}^{t} \mathrm{d}\tau \langle [\Phi_\gamma(t), \Psi_\lambda(\tau)] \rangle \mathrm{e}^{-\mathrm{i}\omega_\lambda(\tau - t)}$$

$$N_{\gamma\lambda}(t) = -\int_{t_0}^{t} \mathrm{d}\tau \langle [\Phi_\gamma(\tau), \Psi_\lambda(t)] \rangle \mathrm{e}^{-\mathrm{i}\omega_\gamma(\tau - t)}$$

$$P_{\gamma\lambda}(t) = -\int_{t_0}^{t} \mathrm{d}\tau \langle :\Omega_\gamma(\tau), \Omega_\lambda(t): \rangle \mathrm{e}^{-\mathrm{i}\omega_\gamma(\tau - t)}$$

where $\langle : \ldots : \rangle$ is the mean value of the normally ordered product of operators a_{kf}^{+} and a_{kf}. Expression (22) represents the single-time hierarchical equation with time-dependent coefficients. Usually for $t \gg \omega_\gamma^{-1}$ one can let the upper limit of integration go to infinity in the Markoffian approximation [32].

Let us now introduce the effective Hamiltonian [33]:

$$H_{\mathrm{eff}} = H_j + \sum_\gamma q_\gamma \langle \Omega_\gamma \rangle + \sum_{\gamma, \lambda} Y_{\gamma\lambda} q_\gamma q_\lambda \tag{23}$$

where

$$\begin{aligned} Y_{\gamma\lambda} = &\frac{1}{2} \int_{t_0}^{t} \mathrm{d}\tau\, \mathrm{e}^{\mathrm{i}\omega_\gamma(t-\tau)} \left\{ \langle :\Omega_\gamma(\tau), \Omega_\lambda(t): \rangle + \langle [\Phi_\gamma(\tau), \Psi_\lambda(t)] \rangle \right\} \\ &- \frac{1}{2} \int_{t_0}^{t} \mathrm{d}\tau\, \mathrm{e}^{\mathrm{i}\omega_\lambda(t-\tau)} \left\{ \langle :\Omega_\lambda(\tau), \Omega_\gamma(t): \rangle + \langle [\Phi_\gamma(t), \Psi_\lambda(\tau)] \rangle \right\} \end{aligned}$$

Then instead of Eq. (22) we get

$$\frac{\mathrm{d}}{\mathrm{d}t} \langle J \rangle = -\mathrm{i} \langle [J, H_{\mathrm{eff}}] \rangle + \sum_{\gamma, \lambda} X_{\gamma\lambda} \langle q_\gamma J q_\lambda - J q_\gamma q_\lambda / 2 - q_\gamma q_\gamma J / 2 \rangle \tag{24}$$

where

$$\begin{aligned} X_{\gamma\lambda} = &\int_{t_0}^{t} \mathrm{d}\tau\, \mathrm{e}^{\mathrm{i}\omega_\gamma(t-\tau)} \left\{ \langle :\Omega_\gamma(\tau), \Omega_\lambda(t): \rangle + \langle [\Phi_\gamma(\tau), \Psi_\lambda(t)] \rangle \right\} \\ &+ \int_{t_0}^{t} \mathrm{d}\tau\, \mathrm{e}^{\mathrm{i}\omega_\lambda(t-\tau)} \left\{ \langle :\Omega_\lambda(\tau), \Omega_\gamma(t): \rangle + \langle [\Phi_\gamma(t), \Psi_\lambda(\tau)] \rangle \right\} \end{aligned}$$

This effective Hamiltonian (23) describes the correlations in the j subsystem that are conditioned by the exchange of bosons. Unlike the effective (trial) Hamiltonian in the theory of phase transition [27], it is bilinear with respect to the operators q. The last sum in Eq. (24) describes the damping process conditioned by the Markoffian approximation. In other words, this approximation eliminates the long Poincare cycles.

Let us now consider the Markoffian hierarchical equation for the Dicke model (10) in the rotating wave approximation. Putting $J = S^z$ in the single-atom case we get

$$\mathrm{i}\frac{\mathrm{d}}{\mathrm{d}t}\langle S^z\rangle = -\mathrm{i}\gamma\left(\tfrac{1}{2} + \langle S^z\rangle\right) - 2\mathrm{i}\gamma\langle S^z\rangle\langle a^+_{k_0 f}, a_{k_0 f}\rangle + \langle S^+\rangle\xi - \langle S^-\rangle\xi^* \tag{25}$$

where

$$\gamma = 4\pi \sum_k g_k^2 \delta(\omega_k - \omega_0)$$

$$\xi = \sum_k g_k \langle a_{kf}\rangle$$

and k_0 corresponds to the resonance mode with frequency ω_0. This equation should be added by two Markoffian equations for $\langle S^z\rangle$ and $\langle S^+\rangle$, the right sides of which will also contain the first-order dynamical variables $\langle S^z\rangle$, $\langle S^+\rangle$, and $\langle S^-\rangle$. Hence, in the single-atom case the closed set of the Markoffian equations is very simple. It leads to the famous result obtained by Weisskopf and Wigner [34] in the case of an initial vacuum state of the field. The other possibilities connected with the density matrix (14) describe the induced radiation processes.

In the two-atom case the closed system of the Markoffian equations should contain the equation for the second-order correlation functions $\langle S_i^+ S_j^-\rangle, \langle S_i^\pm S_j^z\rangle$, $i \neq j$. In the general case of N atoms it is necessary to write the equations for all correlation functions up to the Nth order, which is, of course, impossible at large enough N. Therefore, some additional approximations are inevitable.

VI. WEAKENING OF CORRELATIONS

One of the main principles of the kinetic theory in statistical physics, established by Bogolubov [1], declares that after some time, characterizing the interaction process the time dependence of high-order correlations is defined by the evolution of lower correlation functions. Employing this principle gives the closed kinetic equation instead of the hierarchical equation (see e.g., [35]).

Let us illustrate this principle with the aid of the theory of radio-band superradiance generated by a system of nuclear magnetic moments [36]. The active substance with a high proton concentration was propanediol ($C_3H_6(OH)_2$), polarized by the dynamical polarization method, cooled to low temperature, and placed in an external magnetic field B_0 that is antiparallel to the magnetization of the system. It should be emphasized that the proton spines in the magnetic field form just the inverted two-level system with the Zeeman frequency of transition. The state of the system is thermodynamically unstable, but due to the low temperature it has very high relaxation time ($\tau \geq 2^{10}$ s). Since the possibility of the spontaneous emission is small for the radio band, the superradiation was initiated by the noise photons in a passive resonator (vibrational contour) with low quality ($Q < 600$).

Let the initial magnetization be in the direction of the z axis. Then the system can be described by the Hamiltonian [36, 37]:

$$H = \hbar\gamma B_0 S_z + \hbar\gamma B_x S_x + H_F \tag{26}$$

where γ is the gyromagnetic ratio, S is the total spin of the protons, B_x is the radio-band field of radiation, which is in equilibrium with temperature T at the initial time, and H_F describes the free field.

Let us note that we have here a point-like system (linear size is about 1 cm while the wavelength is of the order of 10^3 cm). Taking into consideration only one circularly polarized component of the field, which has a rotation direction coinciding with the Larmor precession direction, one can reduce Eq. (26) to the Dicke Hamiltonian (10) in the rotating wave approximation with $\omega_0 = \gamma B_0$ and

$$g_k = \gamma \left(\frac{2\pi\hbar c^2}{\omega_k V} \right)^{1/2} (k_x e_y - k_y e_x)$$

Thus, the process is described by the hierarchical equation of the type of Eq. (19).

For a process in a resonator with low quality the losses should be taken into account. In this case the Heisenberg equations for the field operators should be replaced by the Heisenberg–Langevin equations [38]:

$$\mathrm{i}\frac{\mathrm{d}}{\mathrm{d}t}a_k = (\omega_k - \mathrm{i}\kappa/2)a_k + g_k S^- \tag{27}$$

where $\kappa = \omega_0/Q$ represents the losses due to the finite quality.

The mean number of the resonance Planck photons (thermal noise) is high enough ($n_0 \sim 300$) under the conditions of the experiment. Therefore, it is necessary to take into account the chaotic initial state of the field. Thus, for S^z we obtain the following Markoffian equation:

$$\frac{\mathrm{d}}{\mathrm{d}t}\langle S^z \rangle = -\sum_k \Gamma_k(\langle S^+S^- \rangle + 2n_k\langle S^z \rangle) \tag{28}$$

where

$$n_k = (\mathrm{e}^{\hbar\omega_k/\Theta} - 1)^{-1}$$

and

$$\Gamma_k = \frac{\kappa g_k^2}{(\kappa/2)^2 + (\omega_k - \omega_0)^2}$$

The proton subsystem at the initial moment contains N inverted and uncorrelated two-level atoms (Zeeman sublevels), which means that $\langle S^z \rangle \approx N/2$ and

$$\langle S^+S^- \rangle = \frac{N}{2} + \langle S^z \rangle + \sum_{\substack{f,f' \\ f \neq f'}} \langle S_f^+ S_{f'}^- \rangle \approx N \tag{29}$$

Due to the factor $2n_k$ the second term on the right side of Eq. (28) plays the main role during the initial stage of the radiation process. This is the stage of thermal noise amplification by the paramagnetic system. In the course of evolution zero inversion $\langle S^z \rangle \approx 0$ can be reached. Then the first term on the right side of Eq. (28) should play the leading role (the stage of collective radiation). According to Eq. (29) this term describes the correlations between the atoms in the radiation process. The corresponding

Markoffian equation is

$$\frac{d}{dt}\langle S^+S^-\rangle = -\sum_k \Gamma_k \Big\{(1+n_k)\langle S^+S_-\rangle + \langle S^+S^-S^z\rangle + n_k\Big\langle (S^z)^2\Big\rangle\Big\} \tag{30}$$

To calculate the mean value $\langle (S^z)^2\rangle$ one can use the following exact integral of motion for the Dicke Hamiltonian:

$$(S^z)^2 + S^+S^- - S^z = \text{const}. \tag{31}$$

which is the Casimire operator of the SU(2) group. The right side of Eq. (30) also contains the third-order correlation function $\langle S^+S^-S^z\rangle$, which can be decoupled in the following way:

$$\langle S^+S^-S^z\rangle \approx \langle S^+S^-\rangle\langle S^z\rangle \tag{32}$$

This is exactly the same as the well-known random phase approximation (RPA) in statistical physics. This decoupling scheme permits us to take into consideration the pair correlations while the triple correlations are considered approximately in the spirit of the weakening of correlations. It should be noted that substitution of $n_k = 0$ together with decoupling into (28) and (30) leads to the result obtained by Rehler and Eberly ([39]; see also Eq. (32) [40, 41]).

Employing expressions (31) and (32) together with (28) and (30) then gives the closed system of nonlinear differential equations. The solution of this system determines the radiation intensity

$$\begin{aligned} I(t) &= \frac{d}{dt}\sum_k \hbar\omega_k \langle a_k^+ a_k\rangle \\ &= \sum_k \hbar\omega_k \{-\kappa\langle a_k^+ a_k\rangle + \kappa n_k + 2\Gamma\langle S^+S^-\rangle + 4n_k\Gamma_k\langle S^z\rangle\langle S^z\rangle\} \end{aligned} \tag{33}$$

The numerical investigation of the radiation intensity gives good agreement with the experimental data [36].

By performing a similar decoupling scheme of RPA the superradiance with the initial squeezed vacuum state of the resonance field have been described [42]. It was shown that the presence of a squeezed resonance field leads to the cleerease of the superradiation pulse delay time and to the appearance of the residual pair correlation of the atoms.

It should be emphasized that analysis of the RPA-type decoupling

$$\langle (S^+S^-)^m (S^z)^l \rangle \approx \langle (S^+S^-)^m \rangle \langle (S^z) \rangle^l$$

within the framework of the Fokker–Planck equation in the theory of superradiance shows the usefulness of that approximation for an adequate description of the process [11].

VII. RAMAN CORRELATION SPECTROSCOPY

The ability to measure the correlation of photons, demonstrated first by Hanbery Brown and Twiss [43], opened a new chapter in the history of quantum optics. In particular, it led to the creation of spectroscopic correlation measurements (see [44]). Correlation spectroscopy can be employed in condensed matter physics to obtain information about the quantum statistical properties of collective excitations (quasi-particles).

Let us show first that the quasi-particles in solids at thermodynamical equilibrium can have nontrivial quantum statistical properties connected with the squeezed states. For this aim we shall examine the squeezed chaotic states, which can be defined by the following way [45, 46].

For simplicity we shall consider the single-mode case. The Bogolubov canonical transformation [23, 47, 48] is

$$B = UA \tag{34}$$

where A and B are the formal columns

$$A = \begin{pmatrix} a \\ a^+ \end{pmatrix} \qquad B = \begin{pmatrix} b \\ b^+ \end{pmatrix}$$

constructed from the Bose operators, and

$$U = \begin{pmatrix} u & v \\ u^* & v^* \end{pmatrix}$$

and

$$\det U = 1$$

The Fock number state of operators b^+, b

$$b^+b|n\rangle_b = n|n\rangle_b$$

is the squeezed number state of the field described by operators a^+ and a:

$$|n\rangle_{sn}(\equiv |n\rangle_b) = S(\xi)|n\rangle \tag{35}$$

where

$$a^+a|n\rangle = n|n\rangle$$

and $S(\xi)$ is the squeezing operator defined by expression (15) with

$$\xi = r\,\mathrm{e}^{\mathrm{i}\phi} \qquad r = \tanh^{-1}\{|v|/\,|u|\} \qquad \phi = \arg u - \arg v$$

Then the squeezed chaotic state of the field described by the operators a^+ and a is defined as the equilibrium state of the quasi-particles. The corresponding density matrix is

$$\rho = \sum_{n=0}^{\infty} \rho(n)S(\xi)|n\rangle\langle n|S^+(\xi) \tag{36}$$

with

$$\rho(n) = \frac{\langle b^+b\rangle^n}{(1+\langle b^+b\rangle)^{n+1}}$$

This state demonstrates the equilibrium squeezing for the "displacement" operator $Q = (a^+ + a)/2$ below some threshold temperature and super-Poissonian statistics at any temperature [46].

By analogy, the multimode squeezed chaotic state with density matrix

$$\rho = \sum_{\{n_m\}} \prod_{m'} \frac{\langle b_{m'}^+ b_{m'}\rangle^{n_{m'}}}{(1+\langle b_{m'}^+ b_{m'}\rangle)^{n_{m'}+1}} |\{n_m\}\rangle_{bb}\langle \{n_m\}|$$

can be defined. As an example, the standard model of the polariton theory with the Hamiltonian (12) can be considered. For simplicity the quasi-resonance case with Hamiltonian

$$H = \hbar\omega_1 a_1^+ a_1 + \hbar\omega_2 a_2^+ a_2 + \mathrm{i}\hbar g(a_1 + a_1^+)(a_2 - a_2^+) \tag{37}$$

will be considered. Here index 1 corresponds to the photons, index 2 corresponds to the phonons, and g is the photon–phonon coupling parameter.

This Hamiltonian is in quadratic form with respect to the Bose operators or two types and it can be exactly diagonalized be means of the Bogolubov canonical transformation of the form [49]

$$b_k = \sum_{m=1}^{2} \left(A_{km} a_m + B_{km} a_m^+ \right) \qquad k = 1, 2$$

where

$$A_{k1} = \frac{\Omega_k + \omega_1}{2\omega_1} C_{k1} \qquad A_{k2} = \frac{(\Omega_k + \omega_2)\left(\Omega_k^2 - \omega_1^2\right) + 4\omega_1 g^2}{2\Omega_k\left(\Omega_k^2 - \omega_1^2\right)} C_{k2}$$

$$B_{k1} = \frac{\Omega_k - \omega_1}{2\omega_1} C_{k1} \qquad B_{k2} = \frac{(\Omega_k - \omega_2)\left(\Omega_k^2 - \omega_1^2\right) + 4\omega_1 g^2}{2\Omega_k\left(\Omega_k^2 - \omega_1^2\right)} C_{k2}$$

$$|C_{k1}|^2 = \pm \frac{\omega_1\left(\Omega_k^2 - \omega_2^2\right)}{\Omega_k\left(\Omega_1^2 - \Omega_2^2\right)} \qquad |C_{k2}|^2 = \pm \frac{\omega_2\left(\Omega_k^2 - \omega_1^2\right)}{\Omega_k\left(\Omega_1^2 - \Omega_2^2\right)}$$

$$\Omega_k^2 = \frac{\omega_1^2 + \omega_2^2}{2} \pm \frac{1}{2}\sqrt{\left(\omega_1^2 - \omega_2^2\right) + 16\omega_1\omega_2 g^2}$$

The diagonalized Hamiltonian (Eq. 37) takes the form

$$H = \sum_{k=1}^{2} \hbar\Omega_k\left(b_k^+ b_k + \tfrac{1}{2}\right) \tag{38}$$

It is easy now to calculate any temperature mean value. Let us turn our attention to the degree of coherence of second order for the phonon field:

$$G_b = \frac{\langle b^{+2} b^2 \rangle}{\langle b^+ b \rangle^2} = 1 + \frac{V(b^+ b)}{\langle b^+ b \rangle^2} - \frac{1}{\langle b^+ b \rangle}$$

where

$$V(b^+ b) = \left\langle (b^+ b)^2 \right\rangle - \langle b^+ b \rangle^2$$

for the standard chaotic (Gaussian) state, which is the equilibrium state of free Bose gas $G = 2$. The degree of coherence in the model system under consideration is shown in Fig. 1 as a function of temperature and detuning

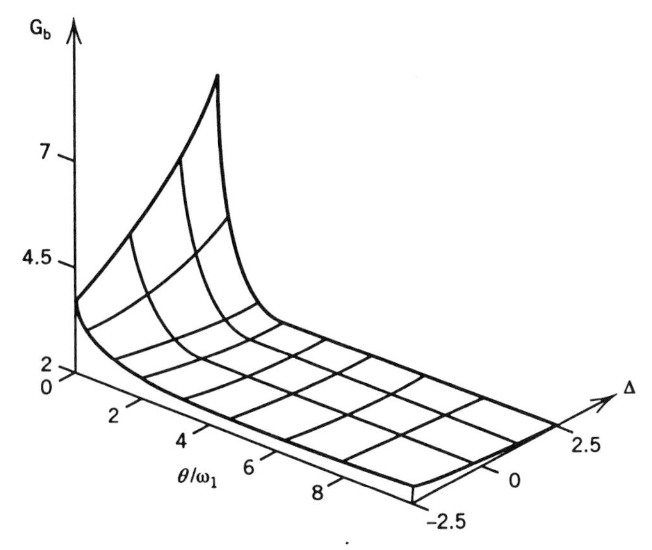

Figure 1. Degree of coherence in the model system.

parameter $\Delta = \omega_1 - \omega_2$. One can conclude that the super-Gaussian probability distribution for the number of phonons takes place. Such a broadening is due to the strong quantum fluctuations created by the photon–phonon interaction. An increase in temperature leads to a decrease of deflexion from the Gaussian distribution. This is quite natural from the physical point of view because an increase in temperature intensifies the temperature fluctuations in comparison with the quantum fluctuations.

The same situation takes place in the case of phonons in the high-T_c superconductors described by the Hamiltonian (13). Here the mean number of phonons is connected with the order parameter, i.e., with the transition temperature [50]. Hence, the strong quantum fluctuations of the number of phonons should lead to a drop in the critical temperature.

Note that the order parameter in ferroelectrics [51] and polaron-like systems [52] is also connected with the mean number of phonons, and investigation of quantum fluctuations is important for the comprehension of the mechanisms and conditions of phase transition.

Information about the quantum fluctuations of the number of Bose-type quasi-particles in solids can be obtained experimentally with the aid of correlation spectroscopy. An example is provided by the Raman scattering process. For simplicity let us limit ourselves to the resonance case with Stokes line generation. We can use the Hamiltonian (11) at $\kappa_{AS} = 0$. This

Hamiltonian has the following conservation laws:

$$\begin{aligned} N_R(t) + N_b(t) &= N_R(0) + N_b(0) \\ N_R(t) + N_S(t) &= N_R(0) \end{aligned} \tag{39}$$

where N_x is the number operator and index x corresponds to the Rayleigh or Stokes line, or to the phonons ($x = b$). We suppose here that initially the Stokes line is in a vacuum state. It follows from the conservation laws (39) that

$$\langle N_b(t)\rangle = \langle N_b(0)\rangle + \langle N_S(t)\rangle \tag{40}$$

and

$$\langle N_b(t)\rangle = \langle N_b(0)\rangle + \langle N_a(0)\rangle - \langle N_a(t)\rangle$$

The functions $\langle N_S(t)\rangle$, $\langle N_a(t)\rangle$, and $\langle N_a(0)\rangle$, which are the intensities of Stokes, Rayleigh, and pumping fields, respectively, can be measured directly. To determine the mean value of phonons at $t = 0$ one can use the approximate solution in powers of t [53, 54]:

$$\begin{aligned} a(t) &= a_f(t)\left\{1 - \tfrac{1}{2}(\kappa t)^2[N_b(0) + 1]\right\} + \mathrm{i}(\kappa t)\mathrm{e}^{-\mathrm{i}\omega t} b(0) a_S(0) \\ a_S(t) &= a_{sf}(t)\left\{1 + \tfrac{1}{2}(\kappa t)^2[N_a(0) - N_b(0)]\right\} \\ &\quad + \mathrm{i}(\kappa t)\mathrm{e}^{-\mathrm{i}\omega_S t} b^+(0) a(0) \end{aligned} \tag{41}$$

where $\kappa t < 1$. It follows that

$$\langle N_S(t)\rangle = (\kappa t)^2 \langle N_a(0)\rangle [\langle N_b(0)\rangle + 1]$$

Thus,

$$\langle N_b(0)\rangle = \frac{\langle N_S(t)\rangle}{(\kappa t)^2 \langle N_a(0)\rangle} - 1 \tag{42}$$

Directly from the conservation laws one can obtain

$$\begin{aligned} V(N_b(t)) &= V(N_b(0)) + V(N_a(0)) + V(N_a(t)) - \langle N_a(0), N_a(t)\rangle \\ &\quad - \langle N_a(t), N_a(0)\rangle - \langle N_b(0), N_a(t)\rangle - \langle N_a(t), N_b(0)\rangle \end{aligned} \tag{43}$$

$$\begin{aligned} V(N_b(t)) &= V(N_b(0)) + V(N_S(t)) \\ &\quad + \langle N_b(0), N_S(t)\rangle + \langle N_S(t), N_b(0)\rangle \end{aligned} \tag{44}$$

By performing a similar analysis to that described in Section III we can calculate here the mixed correlation functions $\langle N_b(0), J\rangle$ and $\langle J, N_b(0)\rangle$.

Let us consider first the chaotic initial state of free phonons with density matrix

$$\rho_b(0) = T(\beta) = \frac{e^{-\beta N_b(0)}}{\mathrm{Tr}\, e^{-\beta N_b(0)}} \qquad \beta = \frac{\omega_b}{kT} \tag{45}$$

Since

$$[n_b(0), \rho_b(0)] = 0$$

it follows that $\langle N_b(0)J\rangle = \langle JN_b(0)\rangle = \langle J\rangle\langle N_b(0)\rangle$ and $\langle N_b(0), J\rangle = \langle J, N_b(0)\rangle = 0$. In this case

$$\begin{aligned} V(N_S(t)) = V(N_a(t)) + V(N_a(0)) \\ -\langle N_a(t), N_a(0)\rangle - \langle N_a(0), N_a(t)\rangle \end{aligned} \tag{46}$$

So far as all quantities in $(x + 3)$ can be measured it is possible to verify directly whether the initial state is different from $(x + 4)$.

We now turn to the case when the initial state of the phonons is a squeezed chaotic state with

$$\rho_b(0) = S(\xi)T(\beta)S^+(\xi) \tag{47}$$

By virtue of relations (17) we get

$$\begin{aligned} N_b(0)\rho &= \rho\left[|v|^2 + (|v|^2 + u^2)N_b(0) - uv^*b^2(0) - uvb^{+2}(0)\right] \\ b^2(0)\rho &= \rho\left[-uv - 2uvN_b(0) + u^2b^2(0) + v^2b^{+2}(0)\right] \\ b^{+2}(0)\rho &= \rho\left[-uv^* - 2uv^*N_b(0) + v^{*2}b^2(0) + u^2b^{+2}(0)\right] \end{aligned}$$

where u, v are the parameters of squeezing. Employing the above relations then gives

$$\begin{aligned} \langle J(t)N_b(0)\rangle = -\frac{1}{2}\langle J\rangle - \frac{1}{2}\langle [N_b(0), J]\rangle \\ + \frac{u}{4|v|\cos\theta}\left(\langle [b^2(0), J]\rangle + \langle [b^{+2}, J]\rangle\right) \end{aligned} \tag{48}$$

where $\theta = \arg v$.

To calculate the mean values of commutators in Eq. (48) we now consider the Heisenberg equations for phonons in the integral representation:

$$b(t) = b_f(t) + \mathrm{i}B(t) \qquad b_f(t) \equiv b(0)\mathrm{e}^{-\mathrm{i}\omega_b t}$$

$$B(t) \equiv \kappa \int_0^t \mathrm{d}\tau\, \mathrm{e}^{-\mathrm{i}\omega_b(t-\tau)} a_S^+(\tau) a(\tau)$$

Then we obtain

$$\langle [N_b(0), J] \rangle = \mathrm{i}\langle [B^+, J] b_f(t) \rangle - \mathrm{i}\langle b_f^+ [B(t), J] \rangle \tag{49}$$

The mixed correlation functions on the right side of Eq. (49) can be calculated with the aid of Eq. (18). Hence,

$$\langle [N_b(0), J] \rangle = |v|^2 \langle [J, [B^+, B]] \rangle \tag{50}$$

In addition,

$$\langle [b^2(0), J] \rangle = \mathrm{e}^{2\mathrm{i}\omega_b t}(1 + 2|v|^2)\langle [[B, J], B] \rangle \tag{51}$$

$$\langle [b^{+2}(0), J] \rangle = \mathrm{e}^{-2\mathrm{i}\omega_b t}(1 + 2|v|^2)\langle [B^+, [B^+, J]] \rangle \tag{52}$$

By virtue of expressions (18) and (49)–(52) one can now express the right side of Eq. (48) in terms of the light correlation functions, which gives us the ability to calculate all mixed correlation functions in Eqs. (43) and (44).

Of course, the initial state does not always coincide with $(x + 4)$. Different model assumptions should be used in that case. The problem of measuring the phonon number variance also needs more detailed investigation.

VIII. CONCLUSION

In this paper we have emphasized a similarity in the description of dynamical processes in quantum optics and statistical physics. In both cases the hierarchical equations for the correlation functions have the same structure. The separation of time scales is the general way to construct the kinetic equation. Although the Markoff approximation is not valid, at least for the short-time behavior [55, 56], the deflections in many cases are negligible [57]. The elimination of Bose fields from the mixed correlation functions with the aid of exact relations (18) leads to simplification of the hierarchical equations and further decoupling.

On the other hand, some analogies with the quantum noise redistribution for light can take place for the Bose field of a different nature with a finite temperature. Examples are provided by the phonons in ionic crystals and high-T_c superconductors. The parameters of quantum probability distributions of the Bose-type excitations in solids can be measured by the quantum correlation spectroscopy methods.

It should be emphasized that the model of interaction of three one-mode Bose fields (11) should be used with caution because it has an unstable vacuum state (see the appendix).

APPENDIX

Let us consider a model of $p + 1$ interacting fields with the Hamiltonian

$$H_p = \sum_{k=0}^{p} \omega_k a_k^+ a_k + \gamma a_0^+ \prod_{k=1}^{p} a_k + \text{h.c.} \tag{A.1}$$

This model describes decay of the pumping mode (with index 0) into p components. When $p = 2$ this Hamiltonian coincides with Eq. (11) at $\kappa_{as} = 0$. We have here p conservation laws:

$$\forall k \geq 1 \left[a_0^+ a_0 + a_k^+ a_k, H_p \right] = 0 \tag{A.2}$$

The vacuum state of the system is defined as

$$\begin{aligned} \psi_0 &= \prod_{k=0}^{p} |0\rangle_k, \forall k \\ a_k |0\rangle_k &= 0 \end{aligned} \tag{A.3}$$

The operator

$$N = a_0^+ a_0 + \frac{1}{p} \sum_{k=1}^{p} a_k^+ a_k \tag{A.4}$$

can be considered as a number of excitations. Any mode with $k \geq 1$ has the same number of excitations. We have $[N, H_p] = 0$. Therefore, the eigenstates of H_p can be considered as a linear combination of the eigenstates of N

$$\psi_n = \sum_{j=0}^{n} \lambda_j^n |n - j\rangle_k \qquad \sum_{j=0}^{n} |\lambda_j^n|^2 = 1 \tag{A.5}$$

where

$$N\psi_n = n\psi_n \tag{A.6}$$

and $|i\rangle_k$ is the Fock number state of kth mode.

The Schrödinger equation

$$H_p\psi_n = E_n\psi_n \tag{A.7}$$

leads to the system of algebraic equations

$$\begin{aligned} \lambda_j^n \left[\omega_0(n - j) + j \sum_{k=1}^{p} \omega_k - E_k \right] \\ = \gamma(n - j)(j + 1)^{p/2} \lambda_{j+1}^n + \gamma\sqrt{n - j + 1}\, j^{p/2} \lambda_{j-1}^n \end{aligned} \tag{A.8}$$

Taking into account the energy conservation law $\omega_0 = \omega_1 + \cdots + \omega_k$ and introducing a new variable

$$x_n = \frac{E_n - \omega_0 n}{\gamma} \tag{A.9}$$

we can represent this equation in the form

$$\lambda_{j+1}^n \sqrt{n - j}\,(j + 1)^{p/2} = -\sqrt{n - j + 1}\, j^{p/2} \lambda_{j-1}^n + x_n \lambda_j^n \tag{A.10}$$

Let $P_j^n(x)$ be the polynomial of the form $P_j^n(x) = \alpha_j^n \lambda_j^n(x)$ where

$$\alpha_j^n = \alpha_{j+1}^n \sqrt{n - j}\,(j + 1)^{p/2} \tag{A.11}$$

Then the equation can be rewritten in the form

$$P_{k+1}^n(x) = x P_k^n(x) - q_k P_{k-1}^n(x) \tag{A.12}$$

where $q_k = k^p(n - k + 1)$.

We should obtain the solutions of equation

$$P_n^n(x) = 0 \tag{A.13}$$

which determine the true solution of the Schrödinger equation. The exact analytical solution is possible for small n only. But one can show that

$$x_{\max}^2 \geq \frac{n^{p+1}}{(p+1)(p+2)} \tag{A.14}$$

while

$$|x_{\max}| \leq \frac{2n^{(p+1)/2}}{\sqrt{p+1}} \left(\frac{p}{p+1} \right)^{p/2} \tag{A.15}$$

It follows from these estimations that

For $p = 0$ or 1 all eigenvalues E_n can have the nonnegative values.

For $p \leq 2$ there is a negative part of the eigenvalue spectrum with $E_n \leq 0$ at high enough n; when $n \to \infty$ then $E_n \to -\infty$.

In other words, we have here some instability with respect to the strong excitations.

References

1. N. N. Bogolubov, *Selected Works*, Par 1, Gordon & Breach, New York, 1991.
2. G. S. Agarwal, *Quantum Statistical Theories of Spontaneous Emission and Their Relation to the Approaches*, Springer, New York, 1977.
3. A. V. Andreev, V. I. Emel'yanov, and Y. A. Il'inskii, *Cooperative Effects in Optics*, IOP, Bristol, UK, 1992.
4. C. Leonardi, F. Persico, and G. Vetri, *Rev. Nuovo Cimento* **9**(4) (1986).
5. C. W. Gardiner and M. J. Collett, *Phys. Rev. A* **31**, 3761 (1985).
6. L. Knöll and A. S. Shumovsky, *Int. J. Mod. Phys. B* **4**, 151 (1990).
7. F. Haake, *Z. Phys.* **227**, 179 (1969).
8. P. Mandel, *Phys. Lett. A* **47**, 307 (1974).
9. R. Bonifacio and P. Schwendimann, *Nuovo Cimento Lett.* **3**, 512 (1970).
10. R. Bonifacio and L. A. Lugiato, *Phys. Rev. A* **12**, 2068 (1975).
11. L. M. Narducci, C. A. Coulter, and C. M. Bowden, *Phys. Rev. A* **9**, 829 (1974).
12. H. Gronchi and L. A. Lugiato, *Phys. Rev. A.* **14**, 502 (1976).
13. R. M. Dicke, *Phys. Rev.* **93**, 99 (1954).
14. A. Einstein, *Phys. Z.* **10**, 185, 817 (1909).
15. L. Allen and J. H. Eberly, *Optical Resonance and Two-Level Atoms*, Wiley, New York, 1975.

16. Y. R. Shen, *The Principles of Nonlinear Optics*, Wiley, New York, 1984.
17. A. S. Shumovsky and Tran Quang, in N. Bogolubov, A. Shumovsky, and V. Yukalov (Eds.), *Interaction of Electromagnetic Field with Condensed Matter*, World Scientific, Teaneck, NJ, 1990, p. 103.
18. G. S. Agarwal and S. S. Iha, *Z. Phys. B* **25**, 391 (1979).
19. R. Graham, *Z. Phys.* **210**, 319 (1968).
20. D. F. Walls, *Z. Phys.* **237**, 224 (1970).
21. V. M. Agranovitch and V. L. Ginzburg, *Spatial Dispersion in Crystal Optics and the Theory of Excitons*, Interscience, New York, 1965.
22. O. Madelung, *Introduction to Solid-State Theory*, Springer, New York, 1987.
23. N. N. Bogolubov, *J. Phys. USSR* **11**, 23 (1947).
24. J. Mustre de Leon, *Phys. Rev. B* **44**, 2422 (1991).
25. B. H. Toby, *Phys. Rev. Lett.* **64**, 2414 (1990).
26. I. Batistic, *Phys. Rev. B* **40**, 6896 (1989).
27. N. N. Bogolubov, Jr., B. I. Sadovnikov, and A. S. Shumovsky, *Mathematical Methods in Statistical Mechanics of Model Systems*, CRC Press, Boca Raton, FL, 1992.
28. R. Loudon and P. L. Knight, *J. Mod. Opt.* **34**, 709 (1987).
29. N. N. Bogolubov, *Commun. JINR*, E17-11822, Dubna, 1978.
30. P. Bocchiery and A. Loinger, *Phys. Rev.* **107**, 337 (1957).
31. I. Percivall, *J. Math. Phys.* **2**, 235 (1961).
32. J. R. Akerhalt and J. H. Eberly, *Phys. Rev. D* **10**, 3350 (1974).
33. L. Knöll and A. S. Shumovsky, *Theor. Math. Phys.* (1992) (in Russian).
34. V. F. Weisskopf and E. Wigner, *Z. Phys.* **63**, 54 (1930).
35. R. Balescu, *Equilibrium and Nonequilibrium Statistical Mechanics*, Wiley-Interscience, New York, 1975.
36. Yu. F. Kiselev, A. S. Shumovsky, and V. I. Yukalov, *Mod. Phys. Lett. B* **3**, 1149 (1989).
37. M. T. Turaev and A. S. Shumovsky, Prep. ICTP, IC/88/227, Trieste, 1988.
38. W H. Louisell, *Radiation and Noise in Quantum Electronics*, McGraw-Hill, New York, 1964.
39. N. E. Rehler and J. Eberly, *Phys. Rev. A* **12**, 2068 (1971).
40. N. N. Bogolubov, Jr., Fam Le Kien, and A. S. Shumovsky, *Physica A* **128**, 82 (1984).
41. N. N. Bogolubov, Jr., Fam Le Kien, and A. S. Shumovsky, *Physica A* **130**, 273 (1985).
42. I. K. Kudryavtsev, A. S. Shumovsky, and N. P. Bogolubov, in E. R. Caianiello (Ed.), *Advances in Theoretical Physics*, World Scientific, Teaneck, NJ, 1991, p. 19.
43. R. Hanbury Brown and R. Q. Twiss, *Nature* **177**, 27 (1956).
44. B. Crosignani, P. DiPorto, and M. Bertolotti, *Statistical Properties of Scattered Light*, Academic, New York, 1975.
45. A. S. Shumovsky, in J. P. Laheurte, M. Le Bellac, and F. Raymond (Eds.), 9th Gen Conf. of the Condensed Matter, Div. of EPS, 1989, p. 223.
46. A. S. Shumovsky, *Theor. Math. Phys.* **89**, 438 (1991) (in Russian).
47. D. Stoler, *Phys. Rev. D* **1**, 3217 (1970).
48. D. Stoler, *Phys. Rev. D* **4**, 1925 (1971).
49. A. V. Chizhov, R. G. Nazmitdinov, and A. S. Shumovsky, *Quant. Opt.* **3**, 1 (1991).

50. J. Mustre de Leon, et al., *A Polaron Origin for Anharmonicity of the Axial Oxygen in Y* $Ba_2Cu_3O_7$ (to be published).
51. R. Blinc and B. Zeks, *Soft Modes in Ferroelectrics and Antiferroelectrics*, North-Holland, Amsterdam, 1974.
52. J. I. Devreese and F. Peeters, *Polarons and Excitons in Polar Semiconductors and Ionic Crystals*, Plenum, New York, 1984.
53. P. Szlachetka, S. Kielich, J. Perina, and V. Perinova, *J. Phys. A* **12**, 1921 (1979).
54. P. Szlachetka, S. Kielich, J. Perina, and V. Perinova, *J. Mol. Spectrosc.* **61**, 281 (1980).
55. L. Fonda, G. C. Ghirardi, and A. Rimini, *Rep. Prog. Phys.* **41**, 587 (1978).
56. P. L. Knight, *Phys. Lett. A* **56**, 11 (1976).
57. P. L. Levy and W. L. Williams, *Phys. Rev. Lett.* **48**, 607 (1982).

PHOTON STATISTICS OF NONCLASSICAL FIELDS

JAN PEŘINA, JIŘÍ BAJER, VLASTA PEŘINOVÁ, AND ZDENĚK HRADIL

Laboratory of Quantum Optics and Department of Optics, Palacký University, Olomouc, Czechoslovakia

CONTENTS

I. INTRODUCTION

In this chapter we discuss some aspects of photon statistics and nonclassical properties of optical radiation in nonlinear optical processes. An introductory part (Section II) provides a basis for the quantum statistical

This work was partially supported by a special grant from Palacký University.

Modern Nonlinear Optics, Part 1, Edited by Myron Evans and Stanisław Kielich. Advances in Chemical Physics Series, Vol. LXXXV.
ISBN 0-471-57546-1

description of nonlinear optical properties in terms of the so-called generalized superposition of coherent fields and quantum noise, which represents a quantum generalization of the classical superposition of signal and noise. Even if such a description is rigorous only for the lowest nonlinearity, it may serve as an approximate description of various aspects of higher order nonlinear optical processes. Moreover, it can provide the possibility of introducing some basic notions, such as principal squeezing, and simple and natural explanations of photocount oscillations.

There are many quantum aspects of light beams, such as photon antibunching, sub-Poissonian behavior (photon number squeezing), quadrature squeezing, oscillations exhibited in the photocount distribution, and violation of various classical inequalities. Relationships of these effects are mostly complicated. Some results of these aspects are included in Section III.

The central part of this paper is Section IV, which is split into two groups of results. In the first group the quantum treatment is given for four-wave mixing processes, phase conjugation, Raman and hyper-Raman scattering, and Nth subharmonic generation using either the assumption of strong (classical) pump beams or a linear operator correction to classical stationary solutions, so that the application of generalized superposition of the coherent fields and quantum noise is appropriate. When treating those nonlinear optical processes, we include effects of nonlinear dynamics, initially nonclassical fields and losses and quantum phase considerations. The second group of results in Section IV is related to non-Gaussian solutions. The anharmonic oscillator is discussed in depth, including arbitrary initial quantum statistical properties of fields and losses. It is possible to solve this physically important and mathematically nontrivial case in a closed form. Finally, we outline a symbolic method for a computer, which is able to provide approximate solutions in powers of the interaction time to any order, in principle.

II. GENERALIZED SUPERPOSITION OF COHERENT FIELDS AND QUANTUM NOISE

We introduce only basic definitions and main results necessary for the following applications of this description, which is equivalent to the two-photon coherent-state technique. More details can be found in Ref. 1, Secs. 8.5, 9.3, and 9.4.

A. Definitions

We include initial nonclassical light into our considerations and consequently adopt the antinormal ordering of field operators in order to describe simply the nonclassical states and to avoid manipulations with generalized functions appropriate for the Glauber–Sudarshan quasidistribution in this case. If solutions of Heisenberg equations are linear in the initial annihilation and creation photon operators $\hat{a}_j$ and $\hat{a}_j^\dagger$, respectively (j being a mode index), i.e., when the interaction Hamiltonian is quadratic in the annihilation and creation operators, we can obtain the antinormal characteristic function for the field in the form

$$\begin{aligned} C_{\mathscr{A}}(\{\beta_j\}, t) &= \left\langle \prod_{j=1}^{M} \exp\left[-\beta_j^* \hat{a}_j(t)\right] \prod_{k=1}^{M} \exp\left[\beta_k \hat{a}_k^\dagger(t)\right] \right\rangle \\ &= \exp\left\{ \tfrac{1}{2} \hat{\beta}_j^\dagger \hat{A}(t) \hat{\beta} + \hat{\beta}\hat{\xi}(t) \right\} \\ &= \exp\left\{ \sum_{j=1}^{M} \left[-B_j(t)|\beta_j|^2 + \left(C_j^*(t)\beta_j^2/2 + \text{c.c.}\right)\right] \right. \\ &\qquad + \sum_{j<k}^{M} \left[D_{jk}(t)\beta_j^*\beta_k^* + \overline{D}_{jk}(t)\beta_j\beta_k^* + \text{c.c.}\right] \\ &\qquad \left. + \sum_{j=1}^{M} \left[\beta_j \xi_j^*(t) - \text{c.c.}\right] \right\} \end{aligned} \tag{1}$$

when c.c. denotes the complex conjugate expressions, β_j denote complex parameters, and the angle bracket $\langle \cdots \rangle$ mean the average with the help of the density matrix $\hat{\rho}(0)$. Here

$$\hat{\beta} = \begin{bmatrix} \beta_1 \\ \beta_1^* \\ \vdots \\ \beta_M \\ \beta_M^* \end{bmatrix} \qquad \hat{\xi} = \begin{bmatrix} -\xi_1 \\ \xi_1^* \\ \vdots \\ -\xi_M \\ \xi_M^* \end{bmatrix}$$

$$\hat{A} = \begin{bmatrix} -B_1 & C_1 & \overline{D}_{12}^* & D_{12} & \cdots & \overline{D}_{1M}^* & D_{1M} \\ C_1^* & -B_1 & D_{12}^* & \overline{D}_{12} & \cdots & D_{1M}^* & \overline{D}_{1M} \\ \overline{D}_{12} & D_{12} & -B_2 & C_2 & \cdots & \overline{D}_{2M}^* & D_{2M} \\ D_{12}^* & \overline{D}_{12}^* & C_2^* & -B_2 & \cdots & D_{2M}^* & \overline{D}_{2M} \\ \cdots & \cdots & \cdots & \cdots & \cdots & \cdots & \cdots \\ \overline{D}_{1M} & D_{1M} & \overline{D}_{2M} & D_{2M} & \cdots & -B_M & C_M \\ D_{1M}^* & \overline{D}_{1M}^* & D_{2M}^* & \overline{D}_{2M}^* & \cdots & C_M^* & -B_M \end{bmatrix}$$

where M is the number of modes, the time-dependent coefficients $B_j, C_j, D_{jk}, \overline{D}_{jk}$ are defined in terms of fluctuations $\Delta\hat{a}_j = \hat{a}_j - \langle\hat{a}_j\rangle$ as follows:

$$B_j = \langle \Delta\hat{a}_j \Delta\hat{a}_j^\dagger \rangle \qquad C_j = \langle (\Delta\hat{a}_j)^2 \rangle$$
$$D_{jk} = \langle \Delta\hat{a}_j \Delta\hat{a}_k \rangle \qquad \overline{D}_{jk} = -\langle \Delta\hat{a}_j^\dagger \Delta\hat{a}_k \rangle \tag{2}$$

$j, k = 1, 2, \ldots, M$, $j \neq k$, and $\xi_j(t)$ are the time-dependent coherent complex amplitudes determined by the interaction. If the normal characteristic function $C_{\mathcal{N}}(\{\beta_j\}, t)$ is needed, one must substitute B_j in (1) by $B_j = \langle \Delta\hat{a}_j^\dagger \Delta\hat{a}_j \rangle = \langle \Delta\hat{a}_j \Delta\hat{a}_j^\dagger \rangle - 1$. Derivatives of (1) with respect to β_j^*, β_k at $\beta_1 = \cdots = \beta_M = \beta_1^* = \cdots = \beta_M^* = 0$ provide the antinormal moments $\langle \hat{a}_1^{k_1} \cdots \hat{a}_M^{k_M} \hat{a}_M^{\dagger l_M} \cdots \hat{a}_1^{\dagger l_1} \rangle$. The corresponding quasidistribution related to antinormal ordering, provided that the coupling of modes is neglected, is

$$\Phi_{\mathscr{A}}(\{\alpha_j\}, t) = \left\langle \prod_{j=1}^{M} \left(\pi K_j^{1/2}\right)^{-1} \exp\left\{ -\frac{B_j}{K_j} |\alpha_j - \xi_j|^2 + \left[\frac{C_j^*}{2K_j} (\alpha_j - \xi_j)^2 + \text{c.c.} \right] \right\} \right\rangle \tag{3}$$

where the $\{\alpha_j\}$ are the complex field amplitudes and $K_j = B_j^2 - |C_j|^2$ is the determinant of the Fourier transformation. The moment generating function $\langle \exp(-\lambda W) \rangle_{\mathcal{N}}$, which is the generating function for the photon-number distribution $p(n, t) = d^n \langle \exp(-\lambda W) \rangle_{\mathcal{N}} / d(-\lambda)^n n!$ at $\lambda = 1$ and its factorial moments $\langle W^k \rangle = d^n \langle \exp(-\lambda W) \rangle_{\mathcal{N}} / d(-\lambda)^n$ at $\lambda = 0$ (W and λ being the integrated intensity and a parameter, respectively), can be obtained directly from the normal characteristic function [1, Eq. (8.58)]:

$$\begin{aligned} \langle \exp[-\lambda W(t)] \rangle_{\mathcal{N}} &= \frac{1}{(\pi\lambda)^M} \int d^2\{\beta_j\} \exp\left(-\frac{1}{\lambda} \sum_j |\beta_j|^2 \right) C_{\mathcal{N}}(\{\beta_j\}, t) \\ &= \left\langle \prod_j \left[1 + \lambda(E_j - 1)\right]^{-1/2} \left[1 + \lambda(F_j - 1)\right]^{-1/2} \right. \\ &\quad \left. \times \exp\left[-\frac{\lambda A_{1j}}{[1 + \lambda(E_j - 1)]} - \frac{\lambda A_{2j}}{[1 + \lambda(F_j - 1)]} \right] \right\rangle \end{aligned} \tag{4}$$

where

$$E_j = B_j - |C_j| \qquad F_j = B_j + |C_j|$$
$$A_{1,2j} = \frac{1}{2} \left[|\xi_j(t)|^2 \mp \frac{1}{2|C_j|} \left(\xi_j^2(t) C_j^* + \text{c.c.} \right) \right] \tag{5}$$

The angle brackets in (3) and (4) mean the average over the initial realizations of the field. We can mention that the expression in brackets in (5) has a form of the product of generating functions for the superposition of signal components $A_{1,2j} \geq 0$ and noise components $E_j - 1 \gtrless 0$, $F_j - 1 \geq 0$. The deviation of the expression within the angle brackets in (4) from the Poisson generating function $\exp[-\lambda \Sigma_j (A_{1j} + A_{2j})]$ reflects the change of the photon statistics caused by the nonlinear dynamics of the optical process under discussion, whereas the average over the initial complex amplitudes of the field, represented by the angle brackets, leads to an additional change of the photon statistics with respect to the initial state of the field (including nonclassical states). The above-mentioned quantum features of optical fields are reflected by negative values of noise $E_j - 1 = B_j - |C_j| - 1$.

Usually we do not need such general multimode statistical characteristics as described by (1). In this paper we consider single modes and pairs of modes (compound modes) involving the effects of coupling of single modes.

1. Single-Mode Case

The corresponding factors in brackets in (4) also represent the generating function for the Laguerre polynomials $L_j^{-1/2}$; consequently, the photon-number distribution and its factorial moments can be expressed in the single-mode case in the form

$$p(n,t) = \Bigg\langle (EF)^{-1/2}(1 - 1/F)^n \exp\left[-\frac{A_1}{E} - \frac{A_2}{F}\right] \times \sum_{k=0}^{n} \frac{1}{\Gamma(k + 1/2)\Gamma(n - k + 1/2)} \left(\frac{1 - 1/E}{1 - 1/F}\right)^k \times L_k^{-1/2}\left[\frac{-A_1}{E(E-1)}\right] L_{n-k}^{-1/2}\left[\frac{-A_2}{F(F-1)}\right] \Bigg\rangle \tag{6}$$

$$\langle W^k \rangle = k! \Bigg\langle (F-1)^k \sum_{l=0}^{k} \frac{1}{\Gamma(l + 1/2)\Gamma(k - l + 1/2)} \left(\frac{E-1}{F-1}\right)^l \times L_l^{-1/2}\left(\frac{-A_1}{E-1}\right) L_{k-l}^{-1/2}\left(\frac{-A_2}{F-1}\right) \Bigg\rangle$$

The mean number of photons is

$$\langle \hat{n} \rangle = \langle W \rangle = \langle A_1 + A_2 + (E + F)/2 - 1 \rangle \tag{7}$$

and the photon variance $\langle (\Delta \hat{n})^2 \rangle = \langle \hat{n} \rangle + \langle (\Delta W)^2 \rangle$, where the normal variance of integrated intensity equals

$$\left\langle (\Delta W)^2 \right\rangle = \left\langle 2A_1(E-1) + 2A_2(F-1) + \left[(E-1)^2 + (F-1)^2\right]/2 \right\rangle \tag{8}$$

The last terms in (7) and (8) represent quantum noise contributions of the physical vacuum and they are nonzero even if the incident light intensity is zero ($A_1 = A_2 = 0$). For nonclassical fields the first term in (8) may be dominant at least for some time intervals; and since $E - 1 < 0$ in this case, the variance $\langle (\Delta W)^2 \rangle < 0$ and the light is called sub-Poissonian because $\langle (\Delta \hat{n})^2 \rangle < \langle \hat{n} \rangle$.

In the particular case of degenerate subharmonic generation with strong classical pumping, the solution is given by the Bogolyubov transformation, $\hat{b} = u\hat{a} + v\hat{a}^\dagger$, $|u|^2 - |v|^2 = 1$, $u = \cosh r$, $v = \mathrm{i} \sinh r \exp(\mathrm{i}\varphi)$, where $r \geq 0$ is the so-called squeeze parameter equal to gt, g is the interaction constant, and φ is a pump phase. Then

$$\begin{aligned} E - 1 &= B - |C| - 1 = [\mathrm{e}^{-2r} - 1]\mathrm{e}^{-\gamma t}/2 + \langle n_d \rangle (1 - \mathrm{e}^{-\gamma t}) \gtreqless 0 \\ F - 1 &= B + |C| - 1 = [\mathrm{e}^{2r} - 1]\mathrm{e}^{-\gamma t}/2 + \langle n_d \rangle (1 - \mathrm{e}^{-\gamma t}) \geq 0 \\ A_{1,2} &= |\xi(0)|^2 \mathrm{e}^{-\gamma t} \mathrm{e}^{\mp 2r}[1 \mp \sin(2\vartheta - \varphi)]/2 \geq 0 \end{aligned} \tag{9}$$

where the initial coherent state $|\xi(0)\rangle$ is assumed (the angle brackets in the above equations can be omitted), ϑ is the phase of the initial complex amplitude $\xi(0)$, γ is the damping constant, and $\langle n_d \rangle$ is the mean number of the reservoir oscillators (cf. [1], Chap. 7).

Finally, it is interesting to note that the entropy $H = -\mathrm{Tr}(\hat{\rho} \ln \hat{\rho})$ can also be expressed for the generalized superposition of coherent fields and quantum noise in the simple form

$$H = -\ln \frac{x^x}{(1+x)^{1+x}} \tag{10}$$

where $x = [(B - 1/2)^2 - |C|^2]^{1/2} - \frac{1}{2}$. For a chaotic state $x = \langle \hat{a}^\dagger \hat{a} \rangle = \langle \hat{n} \rangle$, which is the mean number of chaotic photons. This value is unchanged if an additional coherent component is superimposed. The entropy can also be used to characterize nonclassical states [2].

2. *Compound-Mode Case*

This case is more complicated from the point of view of the photon statistics. The generating function can be written in the form [3]

$$\langle \exp(-\lambda W) \rangle_{\mathcal{N}} = \prod_{k=1}^{4} (1 + \lambda \lambda_k)^{-1/2} \exp\left(-\sum_{j=1}^{4} \frac{\lambda A_k}{1 + \lambda \lambda_k} \right) \tag{11}$$

where λ_k are roots of certain polynomials, and the coherent components A_k are constructed from λ_k and incident field quantities. Because this procedure is rather complicated, we refer the reader to Refs. 3 and 4 for more details. The corresponding photon-number distribution and its factorial moments are expressed as

$$p(n,t) = \exp\left(-\sum_{l=1}^{4} \frac{A_l}{1+\lambda_l}\right) \Sigma' \prod_{l=1}^{4} \frac{\lambda_l^{k_l}}{(1+\lambda_l)^{k_l+1/2}\Gamma(k_l+1/2)} \times L_{k_l}^{-1/2}\left[\frac{-A_l}{\lambda_l(1+\lambda_l)}\right]$$

$$\langle W^k(t)\rangle = k!\, \Sigma' \prod_{l=1}^{4} \frac{\lambda_l^{k_l}}{\Gamma(k_l+1/2)} L_{k_l}^{-1/2}\left(\frac{-A_l}{\lambda_l}\right) \tag{12}$$

where Σ' is taken under the conditions $\Sigma_{l=1}^{4} = n$ or k.

B. Principal Squeezing

We have defined sub-Poissonian behavior of light fields. One of the most important nonclassical effects observed in optical fields is squeezing of fluctuations in the beam below the level corresponding to vacuum fluctuations, or fluctuations in the coherent state. If we define the Hermitian quadrature operators $\hat{q}$, $\hat{p}$, which are related to the generalized coordinate and momentum, or to real and imaginary parts of the complex field amplitude, respectively, $\hat{q} = \hat{a} + \hat{a}^\dagger$, $\hat{p} = -\mathrm{i}(\hat{a} - \hat{a}^\dagger)$, then their fluctuations must fulfill the Heisenberg uncertainty relation

$$\langle(\Delta\hat{q})^2\rangle\langle(\Delta\hat{p})^2\rangle \geq 1 \tag{13}$$

The squeezed state is defined by the condition that $\langle(\Delta\hat{q})^2\rangle$ or $\langle(\Delta\hat{p})^2\rangle < 1$, i.e., the variance of one of the quadrature components is reduced compared to vacuum fluctuations $\langle(\Delta\hat{q})^2\rangle = \langle(\Delta\hat{p})^2\rangle = 1$. Of course, the reduction of fluctuations in one variable must lead to the increase of fluctuations in the other variable to fulfill the uncertainty relation (13). A squeezed state can be the minimum uncertainty state (squeezed coherent state), i.e., the equality sign holds in (13), or it can be a mixed state, i.e. the inequality sign holds in (13). Such reduction of fluctuations is usually detected with the help of homodyne detection, where a coherent component of a local oscillator of the coherent complex amplitude η is superimposed on the signal beam using a beamsplitter. Therefore, detectors placed beyond the splitter will provide the difference of photocurrents in the form

$$\hat{Q} = (\hat{a}^\dagger + \eta^*)(\hat{a} + \eta)/2 - (\hat{a}^\dagger - \eta^*)(\hat{a} - \eta)/2 = \eta\hat{a}^\dagger + \eta^*\hat{a}$$

which can be made extremum with respect to the phase of η. This procedure provides the following rotational invariants for the large and small half-axes of the noise ellipse

$$\begin{aligned}\lambda_{1,2} &= \frac{1}{2}\operatorname{Tr}\hat{M}\left\{1 \pm \left[1 - \frac{4\operatorname{Det}\hat{M}}{(\operatorname{Tr}\hat{M})^2}\right]^{1/2}\right\} \\ &= 1 + 2[B - 1 \pm |C|]\end{aligned} \tag{14}$$

where the matrix

$$\hat{M} = \begin{pmatrix} \langle(\Delta\hat{q})^2\rangle & \langle\{\Delta\hat{q}, \Delta\hat{p}\}\rangle/2 \\ \langle\{\Delta\hat{q}, \Delta\hat{p}\}\rangle/2 & \langle(\Delta\hat{p})^2\rangle \end{pmatrix} \tag{15}$$

$\{\Delta\hat{q}, \Delta\hat{p}\} = \Delta\hat{q}\Delta\hat{p} + \Delta\hat{p}\Delta\hat{q}$ being the anticommutator, which is nonzero and plays an important role particularly in nonlinear processes. One can see that this formulation resembles the formulation of partial polarization [5]. Instead of the Heisenberg inequality (13), now we have the Schrödinger–Robertson inequality

$$\operatorname{Det}\hat{M} = \lambda_1\lambda_2 = \langle(\Delta\hat{q})^2\rangle\langle(\Delta\hat{p})^2\rangle - \langle\{\Delta\hat{q}, \Delta\hat{p}\}\rangle^2/4 \geq 1 \tag{16}$$

and the principal squeezing is defined by the following condition [6, 7]:

$$\lambda_2 = 1 + 2[B - 1 - |C|] < 1 \tag{17}$$

If the anticommutator equals zero, the principal squeezing reduces to the standard squeezing. A graphical representation of principal squeezing with the help of the noise ellipse and its lemniscate has been suggested in Ref. 8 and applied in Refs. 9 and 10. Writing the anticommutator term in (16) on the right side, one can introduce minimum-uncertainty coherent states with correlations [11]. In the above process of subharmonic generation $\lambda_2 = \exp(-2gt) < 1$ and squeezing of vacuum fluctuations always occurs.

If the compound-mode case is assumed, we obtain for the principal squeezing involving the coupling of modes

$$\lambda_2 = 1 + (B_j - 1) + (B_k - 1) - 2\,\mathrm{Re}\,\overline{D}_{jk} - |C_j + C_k + 2D_{jk}| < 1 \qquad j < k \quad (18)$$

C. Oscillations in Photon-Number Distribution

Returning to the degenerate process of subharmonic generation and examining (9) we can see that the existence of principal squeezing in this process for all interaction times is a necessary, but not a sufficient condition for sub-Poissonian (antibunched) light. Assuming initially coherent light and restricting (4) to this single-mode case, we have this generating function in a form of the product of generating functions for the superpositions of coherent fields and quantum noise, so that the resulting photon-number distribution $p(n,t)$ is the discrete convolution of partial distributions p_1, p_2 related to signals A_1, A_2 and noise $E - 1$ and $F - 1$, respectively,

$$p(n,t) = \sum_{k=0}^{n} p_1(n-k,t)\,p_2(k,t) \qquad (19)$$

Neglect losses for a moment ($\gamma = \langle n_d \rangle = 0$). Assume now that the special initial phase condition $2\vartheta - \varphi = -\pi/2$ is fulfilled. Then the signal is $A_1 = |\xi(0)|^2 \exp(-2gt)$, is attenuated with the time, and is dominant since $A_2 = 0$; quantum noise $F - 1$ plays an unimportant role at the beginning, and we have the superposition of the signal A_1 and negative noise $E - 1$. The result is the sub-Poissonian behavior of photon statistics. On the other hand, if $2\vartheta - \varphi = \pi/2$, the dominant behavior at the beginning is determined by the signal $A_2 = |\xi(0)|^2 \exp(2gt)$, which is amplified ($A_1 = 0$), superimposed on the positive quantum noise $F - 1$, and, consequently, the photon statistics are super-Poissonian from the beginning. (In the former case and for later interaction times, the photon-number distributions $p_1(n,t) = p_1(n)$ (curve a), $p_2(n,t) = p_2(n)$ (curve b), and $p(n,t) = p(n)$ (curve c obtained from (19)) are shown in Fig. 1 ($gt = 1$, $|\xi(0)| = 0.8$), demonstrating that the nonclassical behavior of light reflects itself also in negative values of pseudoprobability p_2 ($E - 1 < 0$) and consequently, in oscillating behavior of the resulting photon-number distribution $p(n,t)$ [12].

Another sophisticated explanation of the quantum origin of these oscillations was suggested using interference in phase space [13]. If

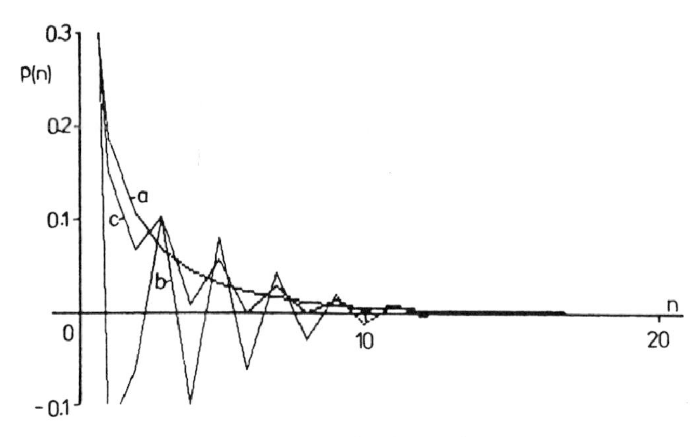

Figure 1. Photon-number distributions $p_1(n)$ (curve a), $p_2(n)$ (curve b) and $p(n)$ (curve c).

dissipation (losses) is included, then we see from (9) that the negative values of the quantum noise $E - 1$ are diminished by the damping factor $e^{-\gamma t}$ and positive additional noise values of $\langle n_d \rangle(1 - e^{-\gamma t})$ degrade oscillations in $p(n, t)$ and nonclassical behavior of light. More complicated relations between sub-Poisson photon behavior and squeezing of vacuum fluctuations are discussed in the following section.

III. RELATION BETWEEN ANTIBUNCHING AND SQUEEZING

To obtain usable information about statistical properties of light, quantum optics deals with moments of the operator of the electromagnetic field $\hat{E}$, instead of using the complete information contained in the density matrix $\hat{\rho}$. Since the density matrix must obey the conditions

$\hat{\rho}$ is normalizable, $\text{Tr}(\hat{\rho}) = 1$

$\hat{\rho}$ is positive-definite Hermitian operator

the necessary condition $\text{Tr}(\hat{\rho}^2) \leq 1$ holds and the equality is valid for pure states. Consequently, some intrinsic restrictions appear, which apply to all the possible moments of the field operators. These are known as generalized Cauchy-Schwartz inequalities [14]. Their origin is the same as that of the violation of Bell inequalities: They are predictions of quantum mechanics. Let us restrict our attention to the relations between moments of the operators $\hat{a}, \hat{a}^\dagger$ up to the fourth order. Then we can obtain some general consequences between squeezing and antibunching of light, because the former effect involves the operators up to the second order, whereas the latter includes operators to the fourth order. This connection was previously mentioned for special cases. Walls [15] showed that ideal squeezed vacuum states are always bunched (for the correlations of light intensities $\langle I(t)I(t+\tau)\rangle$ it holds that $g^{(2)} = \langle I(t)I(t+\tau)\rangle / \langle I\rangle^2 > 1$). Bondurant and Shapiro [16] found maximal antibunching for ideal squeezed states as $\langle(\Delta\hat{n})^2\rangle = (\bar{n})^{2/3}$ for mean photon number $\bar{n}$ large. Stronger antibunching $\langle(\Delta\hat{n})^2\rangle = (\bar{n})^{1/3}$ was reported by Kitagawa and Yamamoto for states generated with the help of Kerr nonlinearity [17]. The extremal behavior of squeezing and antibunching was appreciated from the general point of view in Refs. 18 and 19, where the broad class of extremal states —the so-called noise minimum states (NMS)—was introduced. They are defined as states minimizing the root-mean-square of operator $\hat{n}$ under the constraints of lower-order moments (up to the second order). More mathematically, they can be specified as extremal states of generalized Cauchy-Schwartz inequalities or as solutions of variational equations. Let us briefly describe both these methods.

The generalized Cauchy-Schwartz inequality for two operators $\hat{A}, \hat{B}$ with bounded norms may be specified as the relation

$$|\langle\hat{A}, \hat{B}\rangle|^2 \leq \|\hat{A}\|^2 \|\hat{B}\|^2 \tag{20}$$

where the scalar product is defined as $\langle\hat{A}, \hat{B}\rangle = \mathrm{Tr}(\hat{\rho}\hat{A}^\dagger\hat{B})$ and the norm is given as $\|\hat{A}\|^2 = \mathrm{Tr}(\hat{\rho}\hat{A}^\dagger\hat{A})$. The scalar product admits the possibility that $\|\hat{A}\| = 0$ for $\hat{A} \neq 0$. The equality in (20) occurs only for the states $\hat{\rho}$, where

$\|\hat{A}\| = 0$, $\|\hat{B}\| = 0$

$\|\hat{A}\|\,\|\hat{B}\| \neq 0$ and $(\hat{A} - \lambda\hat{B})\hat{\rho} = 0$ for some parameter λ

We emphasize that inequality (20) cannot be violated in quantum mechanics, in contrast to the classical Cauchy-Schwartz inequality, which can.

Consequences for lower-order moments are quite simple and the following inequalities can be written:

$$|\bar{a}|^2 \leq \bar{n} \tag{21}$$

and

$$\left|\overline{a^2} - (\bar{a})^2\right|^2 \leq \left(\bar{n} - |\bar{a}|^2\right)^2 + \bar{n} - |\bar{a}|^2 \tag{22}$$

where the upper bar denotes the mean value. The equality occurs for coherent states in (21) and for ideal squeezed state in (22), since this is just the Schrödinger-Robertson inequality (16). A much more complicated task is to study the connections between second-order moments and the fourth-order correlation function $g^{(2)} = \langle\hat{a}^{\dagger 2}\hat{a}^2\rangle/(\bar{n})^2$ and $\langle(\Delta\hat{n})^2\rangle =$

$\langle \hat{n}^2 \rangle - (\bar{n})^2 = \langle \hat{a}^{\dagger 2} \hat{a}^2 \rangle - (\bar{n})^2 + \bar{n}$, respectively. Inequality (20) provides for choice of operators $\hat{A} = \hat{a}^2$, $\hat{B} = \hat{1}$ the relation

$$g^{(2)} \geq \frac{|\overline{a^2}|}{\bar{n}^2} \tag{23}$$

and the extremal states are known as SU(1, 1) coherent states or even (odd) coherent states [1]. This condition forbids some fields to be antibunched. All squeezed vacuum states ($\bar{a} = 0$) are an evident example as a consequence of the squeezing condition (17). States described by the generalized superposition of signal and noise also exhibit a simple relation between both nonclassical effects, as was pointed out in the previous section, since all antibunched states are squeezed. These and other consequences together with geometrical interpretation are given in Ref. 18.

Another estimation of photon-number variance may be derived as a consequence of the Heisenberg relation for noncommuting quadrature operators

$$\hat{X}_1(\theta) = \hat{a}\, e^{-i\theta} + \hat{a}^\dagger e^{i\theta} \qquad \hat{X}_2(\theta) = -i(\hat{a}\, e^{-i\theta} - \hat{a}^\dagger e^{i\theta})$$

and photon-number operator

$$\left[\hat{n}, \hat{X}_1(\theta)\right] = -i\hat{X}_2(\theta) \tag{24}$$

The following inequality may be specified [19]:

$$\left\langle (\Delta\hat{n})^2 \right\rangle^{1/2} \geq |\bar{a}| \frac{\lambda}{\lambda_1 \lambda_2} \tag{25}$$

where $\lambda_{1,2}$ are as in (14) half-axes of the noise ellipse,

$$\lambda^2 = \lambda_1^2 \sin^2 \Phi + \lambda_2^2 \cos^2 \Phi$$

and

$$\Phi = \tfrac{1}{2} \arg\left[\overline{a^2} - (\bar{a})^2\right] - \arg \bar{a}$$

The extremal states associated with the inequality may be found among the eigenvectors of the non-Hermitian operator:

$$\left[\hat{n} - i|\xi|\hat{X}_1(\theta)\right] |\psi\rangle = \Omega |\psi\rangle \tag{26}$$

$\xi = \mathrm{i}|\xi|\mathrm{e}^{i\theta}$ being a complex parameter. The solution may be evolved as the state

$$|\psi\rangle = N\exp\left[|\xi|\hat{X}_2(\theta)\right]|M\rangle \tag{27}$$

where the normalization factor N is given as the function of continuous and discrete parameters ξ and M, $N(\xi, M) = \exp(-|\xi|^2)[M!/L_M(-4|\xi|^2)]^{1/2}$. We will call these states crescent states, since their quasidistribution $\Phi_{\mathscr{A}}$ exhibits typical crescent shape. Similar states were discussed as near-photon-number eigenstates generated by the state reduction during down-conversion [19], displaced Fock states, or stationary states in a medium with the Kerr nonlinearity. The above states may also be associated with the number-phase minimum uncertainty states of the Shapiro-Wagner feasible phase concept [20]. The crescent state evidently belongs to the non-Gaussian states, which is discussed in Section IV.B.1.

The generalized Cauchy-Schwartz inequalities enable us to find the minimum value of the fluctuation and correlation function $\langle(\Delta\hat{n})^2\rangle(\bar{a}, \overline{a^2}, \bar{n})$ and $g^{(2)}(\bar{a}, \overline{a^2}, \bar{n})$, respectively, only for the values of moments $\bar{a}, \overline{a^2}, \bar{n}$ corresponding to the extremal states. To address the problem more generally, the variational approach has to be used. All states minimizing the functional $\mathrm{Tr}(\hat{\rho}\hat{n}^2)$ under the restrictions

$$\begin{aligned}
\mathrm{Tr}(\hat{\rho}) &= 1\\
\mathrm{Tr}(\hat{\rho}\hat{a}) &= \bar{a}\\
\mathrm{Tr}(\hat{\rho}\hat{a}^2) &= \overline{a^2}\\
\mathrm{Tr}(\hat{\rho}\hat{n}) &= \bar{n}
\end{aligned}$$

are noise minimum states and obey the operator (Lagrange-Euler) equation

$$\left(\hat{n}^2 + \gamma\hat{n} - \xi^{*2}\hat{a}^2 - \xi^2\hat{a}^{\dagger 2} + \alpha\hat{a}^\dagger + \alpha^*\hat{a}\right)\hat{\rho}_{\min} = \mu\hat{\rho}_{\min} \tag{28}$$

μ, γ being the real Lagrange multipliers and ξ and α complex ones. Equation 28 is nothing else than the steady-state solution for the nonlinear Hamiltonian, including Kerr nonlinearity together with the generation of squeezed states in the rotational-wave approximation. If we know the solution for ground state $\hat{\rho}_{\min}$ of the physical system and $\mu_{\min}$ as functions of parameters ξ, α, γ and if we substitute them with the parameters $\bar{a}, \overline{a^2}, \bar{n}$, we could obtain the desired result for the minimum of photon number variance. Nevertheless, such treatment represents a rather complicated problem, because the above-mentioned Hamiltonian includes a variety of quantum optical effects such as antibunching, squeezing, and bistability. Some particular solutions are given in Ref. 18 together with the relations to other recently solved problems.

IV. NONLINEAR OPTICAL PROCESSES

A. Gaussian Solutions (Classical Pumping and Linear Operator Corrections)

1. *Three- and Four-Wave Mixing*

In this section we consider four- or three-wave mixing with squeezed input signals. Pump laser light is considered to be strong so that it can be described as a classical wave. The phonon reservoirs are assumed to be chaotic, mutually independent, and having a flat spectrum, as is usual. The development of squeezing effects, factorial moments, and photocount distributions are provided for both single-signal modes and for compound modes involving the nonlinear coupling of modes. Nonclassical effects including squeezing of vacuum fluctuations and sub-Poissonian behavior of generated light are obtained, particularly as a result of dynamic coupling of all interacting modes and of initial squeezing of light.

One can easily derive that under the above assumptions the solution is described as a linear combination of annihilation and creation operators of two input fields $\hat{a}_1, \hat{a}_2$:

$$\begin{aligned} \hat{A}_1(z) &= \mu(z)\hat{a}_1 + \nu^*(z)\hat{a}_2^\dagger \\ \hat{A}_2(z) &= \mu(z)\hat{a}_2 + \nu^*(z)\hat{a}_1^\dagger \end{aligned} \tag{29}$$

the influence of the pump modes being contained in the approximation of classical pumping in the space-dependent coefficients $\mu(z)$ and $\nu(z)$. To satisfy the Heisenberg commutation rules $[\hat{A}_i, \hat{A}_k^\dagger] = \delta_{ik}$ it must be fulfilled that $|\mu(z)|^2 - |\nu(z)|^2 = 1$.

Equations of the form (29) can really be obtained [21–26] in four- or three-wave optical mixing with the assumption of strong pumping, which may then be described classically. Equations (29) are symmetric with respect to indices because of the physical symmetry of mixing processes. Then the operators $\hat{A}_1, \hat{A}_2$ and $\hat{a}_1, \hat{a}_2$ are output and input annihilation operators, respectively, and the parameter z represents the interaction length.

The concrete form of coefficients $\mu(z)$ and $\nu(z)$ is dependent on the type of interaction. We can obtain

$$\begin{aligned} \mu(z) &= \cos(\Omega z) - ik\left(|\beta_1|^2 + |\beta_2|^2\right)\sin(\Omega z)/\Omega \\ \nu(z) &= -ig\beta_1\beta_2 \sin(\Omega z)/\Omega \end{aligned} \tag{30}$$

where $\Omega^2 = k^2(|\beta_1|^2 + |\beta_2|^2)^2 - g^2|\beta_1\beta_2|^2$ is assumed to be positive in accordance with the fact that usually $k = g = \hbar\omega^2\chi^{(3)}/\varepsilon^2 V$ [22] provided that we consider the forward four-wave mixing with the interaction Hamiltonian:

$$\hat{H} = \hbar gc\left(\beta_1\beta_2\hat{A}_1^\dagger\hat{A}_2^\dagger + \beta_1^*\beta_2^*\hat{A}_1\hat{A}_2\right) + \hbar kc\left(|\beta_1|^2 + |\beta_2|^2\right)\left(\hat{A}_1^\dagger\hat{A}_1 + \hat{A}_2^\dagger\hat{A}_2\right) \tag{31}$$

The β_1, β_2 are the c-number coherent amplitudes of pumping modes, g, k are the third-order material interaction constants, c is the light velocity in the material, and $\hbar$ is the Planck constant divided by 2π.

Analogously we can get

$$\begin{aligned} \mu(z) &= \cosh(|\kappa|z) \\ \nu(z) &= -\kappa \sinh(|\kappa|z)/|\kappa| \end{aligned} \tag{32}$$

where $\kappa = ig\beta$, for the forward three-wave mixing or

$$\begin{aligned} \mu(z) &= 1/\cos(|\kappa|z) \\ \nu(z) &= -\kappa \tan(|\kappa|z)/|\kappa| \end{aligned} \tag{33}$$

for the backward three-wave mixing using the interaction Hamiltonian

$$\hat{H} = \hbar gc\left(\beta\hat{A}_1^\dagger\hat{A}_2^\dagger + \beta^*\hat{A}_1\hat{A}_2\right)$$

In this brief summary we have assumed the degenerate cases only characterized by the same signal and pump frequencies. Nevertheless, we have shown representative examples of various types of dynamics, e.g., the oscillating solution (30), monotonous solution (32), and diverging solution (33), and the number of sorts of space–time behavior may yet be extended to account for the influence of damping processes.

The influence of losses can be included in a general way using the following procedure. We can adopt the model of the independent phonon reservoirs R_1 and R_2, considering an additional interaction Hamiltonian for the reservoir interaction with radiation modes,

$$\hat{H} = \sum_{k=1}^{\infty} \sum_{j=1}^{2} \hbar c \gamma_k^{(j)} \hat{b}_k^{(j)} \hat{A}_j^{\dagger} \exp\left[\mathrm{i}\left(\omega - \omega_k^{(j)}\right)z/c\right] + \text{h.c.}$$

The exponentials here arise from the interaction representation used, $\gamma_k^{(j)}$ are interaction constants, and $\hat{b}_k^{(j)}$ are phonon reservoir annihilation operators. The solution in the Wigner-Weisskopf approximation is (compare with [27])

$$\begin{aligned} \hat{A}_1(z) &= \tilde{\mu}(z)\hat{a}_1 + \tilde{\nu}^*(z)\hat{a}_2^{\dagger} + \hat{\Lambda}_1 \\ \hat{A}_2(z) &= \tilde{\mu}(z)\hat{a}_2 + \tilde{\nu}^*(z)\hat{a}_1^{\dagger} + \hat{\Lambda}_2 \end{aligned} \tag{34}$$

where $\tilde{\mu}(z) = \mu(z)\exp(-\gamma z)$, $\tilde{\nu}(z) = \nu(z)\exp(-\gamma z)$ and

$$\hat{\Lambda}_1 = \tilde{\mu} * \hat{L}_1 + \tilde{\nu}^* * \hat{L}_2^{\dagger}$$

$$\hat{\Lambda}_2 = \tilde{\mu} * \hat{L}_2 + \tilde{\nu}^* * \hat{L}_1^{\dagger}$$

Here the asterisk $(*)$ denotes convolution, for instance, $\tilde{\mu} * \hat{L}_1 = \int_0^z \tilde{\mu}(z - z')\hat{L}_1(z')\,dz'$, where $\hat{L}_j$ are the usual Langevin forces mutually independent and correlated to the delta function $\langle \hat{L}_j(z)\rangle = 0$, $\langle \hat{L}_j^{\dagger}(z)\hat{L}_k(z')\rangle = \gamma\langle \hat{n}_d\rangle\delta_{jk}$, and $\langle \hat{L}_j(z)\hat{L}_k(z')\rangle = 0$ for $j, k = 1, 2$. The constant γ is the resulting damping constant of both signal modes, $\langle \hat{n}_d\rangle$ is mean number of reservoir phonons.

Equation (34) is the direct generalization of Eq. (29) and therefore we shall examine states described by this equation. One may prove that the commutation rules are valid for all z. Now choosing the particular initial states of subfrequency modes, we can calculate arbitrary statistical parameters of modes at the output of the mixer. We assume the squeezed input states, which are described by the quasidistribution function $\Phi_{\mathscr{A}}(\alpha_1, \alpha_2, \{\eta_k^1\}, \{\eta_k^2\}, 0)$ related to the antinormal ordering in the form

$$\begin{aligned} \Phi_{\mathscr{A}}&\left(\alpha_1, \alpha_2, \{\eta_k^1\}, \{\eta_k^2\}, 0\right) \\ &= \prod_{j=1}^{2} \frac{1}{\pi\sqrt{K_j}} \\ &\quad \times \exp\left[-\frac{(1 + B_j)|\alpha_j - \xi_j|^2}{K_j} + \frac{C_j^*(\alpha_j - \xi_j)^2 + \text{c.c.}}{2K_j}\right] \\ &\quad \times \prod_{s=1}^{2} \prod_{k=1}^{\infty} \frac{1}{\pi\left(1 + B_k^{(s)}\right)} \exp\left(-\frac{\left|\eta_k^{(s)}\right|^2}{1 + B_k^{(s)}}\right) \end{aligned} \tag{35}$$

where

$$B_j = \sinh^2 r_j + \langle \hat{n}_j \rangle, \quad C_j = \sinh r_j \cosh r_j \exp(\mathrm{i}\phi_j),$$

$$K_j = \left(1 + B_j\right)^2 - |C_j|^2, \quad B_j^{(s)} = \langle \hat{n}_d \rangle, j = 1,2, \quad \text{and} \quad s = 1,2; r_j$$

are the corresponding squeeze parameters, ϕ_j are the phases of rotation coefficients C_j, ξ_j are the initial complex amplitudes, $\langle \hat{n}_j \rangle$ are the mean photon numbers of chaotic part of fields, and $\langle \hat{n}_d \rangle$ is the mean number of phonons in every mode of the reservoirs R_1, R_2. As special cases we can describe the superposition of coherent fields ξ_j and noise $\langle \hat{n}_j \rangle (r_j = 0)$, and coherent $(\langle \hat{n}_j \rangle = 0,\ r_j = 0)$ or chaotic $(\xi_j = 0,\ r_j = 0)$ fields. Further details concerning the use of such distributions to describe the photon statistics of nonclassical optical fields can be found in Ref. 1, Chapters 8 and 9.

The initial states defined by (35) have a Gaussian form of characteristic function and one may verify by a direct calculation of the characteristic function that dynamics defined by Eqs. (34) conserve this form. We obtain the normal characteristic function

$$\begin{aligned} C_{\mathcal{N}}(\beta_1, \beta_2, z) = \exp\Big[& -B_1(z)|\beta_1|^2 - B_2(z)|\beta_2|^2 \\ & + \big(C_1^*(z)\beta_1^2/2 + C_2^*(z)\beta_2^2/2 \\ & \quad + D(z)\beta_1\beta_2 + \overline{D}(z)\beta_1\beta_2^* + \text{c.c.}\big) \\ & + \left(\xi_1^*(z)\beta_1 + \xi_2^*(z)\beta_2 - \text{c.c.}\right)\Big] \end{aligned}$$

where all the space-dependent coefficients are

$$\xi_k(z) = \xi_k \tilde{\mu}(z) + \xi_{3-k}^* \tilde{\nu}^*(z)$$

$$\begin{aligned} B_k(z) = {} & B_k |\tilde{\mu}(z)|^2 + (1 + B_{3-k}) |\tilde{\nu}(z)|^2 \\ & + 2\gamma \int_0^z \Big[|\tilde{\mu}(z')|^2 \langle \hat{n}_d \rangle + |\tilde{\nu}(z')|^2 \big(1 + \langle \hat{n}_d \rangle\big) \Big] \, \mathrm{d}z' \end{aligned}$$

$$C_k(z) = C_k \tilde{\mu}^2(z) + C_{3-k}^* \tilde{\nu}^{*2}(z)$$

$$D(z) = (1 + B_1 + B_2)\tilde{\mu}^*(z)\tilde{\nu}(z) + 2\gamma\big(1 + 2\langle \hat{n}_d \rangle\big) \int_0^z \tilde{\mu}^*(z')\tilde{\nu}(z') \, \mathrm{d}z'$$

$$\overline{D}(z) = -C_1^* \tilde{\mu}^*(z)\tilde{\nu}^*(z) - C_2 \tilde{\mu}(z)\tilde{\nu}(z)$$

To illustrate the main properties of squeezing in the processes under discussion we will first restrict ourselves to the lossless case ($\gamma = 0$). In a general case including losses the results must be a little modified.

We can easily prove that the single-mode principal squeezing variance $\lambda_2^{(k)}(z) = 1 + 2B_k(z) - 2|C_k(z)|$ ($k = 1$ or 2) is

$$\lambda_2^{(k)} = 1 + 2\left[|\mu|^2 B_k + |\nu|^2(1 + B_{3-k})\right] - 2|\mu^2 C_k + \nu^{*2} C_{3-k}|$$

which admits the squeezing if at least one of the modes is squeezed, i.e., $|C_k| \geq B_k$. No squeezing may be generated from classical initial states. The situation changes if we superpose both the signal modes A_1, A_2 on a 50%–50% beamsplitter. Then we can measure squeezing in the signal $S = [\hat{A}_1 \exp(-i\delta) + \hat{A}_2 \exp(i\delta)]/\sqrt{2}$. The normalizing coefficient $1/\sqrt{2}$ ensures the validity of the commutation rule and 2δ is the phase mismatch between both the signals A_1, A_2 introduced by the beamsplitter.

If no coupling is present in the input signals a_1, a_2 we can calculate that

$$\lambda_2 = (|\mu| - |\nu|)^2\left(1 + B_1 + B_2 - |C_2 \exp(-2i\delta) + C_2 \exp(2i\delta)|\right). \quad (36)$$

The main meaning of Eq. (36) is that it shows the possibility of generating the squeezing from coherent input states. For instance, let us assume the coherent input states (i.e., all B_k and C_k coefficients are equal to zero); then

$$\lambda_2 = (|\mu| - |\nu|)^2 = \frac{1}{\left(|\nu| + \sqrt{1 + |\nu|^2}\right)^2} \leq 1$$

i.e., squeezing is generated along the interaction length always where $|\nu(z)| > 0$.

From the last formula it is easy to see that for bounded solution $|\nu(z)| \leq |\nu|_{\max}$ the limit exists of maximal possible squeezing [25]

$$\lambda_2(z) \leq \lambda_{2\min}(z) = \frac{1}{\left(|\nu|_{\max} + \sqrt{1 + |\nu|^2_{\max}}\right)^2} \leq 1$$

This corresponds to the oscillating-type solution (30), with $k = g$; minimization with respect to all $m_1 = |\beta_1|^2$, $m_2 = |\beta_2|^2$ leads to $|\nu|_{\max} = 1/\sqrt{3}$ and $\lambda_{2\min} = 1/3$ [25].

The degrading influence of noise components B_1, B_2 on the production of squeezed states was already discussed in Refs. 24 and 26, where the initial signal-plus-noise states, i.e., the superposition of coherent and chaotic states, were assumed.

One may conclude that the squeezing according to formula 36 can arise either as a consequence of input squeezing (i.e., $B_1 + B_2 < |C_1| + |C_2|$) or as a consequence of the suitable evolution of interaction dynamics (i.e., $(|\mu| - |\nu|)$ must tend to zero).

The results introduced above may easily be generalized to include losses,

$$\begin{aligned} \lambda_2(z) = {} & (|\mu| - |\nu|)^2 \exp(-2\gamma z) \\ & \times \left[1 + B_1 + B_2 - |C_1 \exp(-2\mathrm{i}\delta) + C_2 \exp(2\mathrm{i}\delta)|\right] \\ & + 2\gamma(1 + 2\langle \hat{n}_d \rangle) \int_0^z \exp(-2\gamma z') |\mu(z') \exp(\mathrm{i}\theta) \\ & \qquad + \nu(z') \exp(-\mathrm{i}\theta)|^2 \, \mathrm{d}z' \end{aligned} \tag{37}$$

where θ is not arbitrary but defined by formula $|\mu(z)\exp(\mathrm{i}\theta) + \nu(z)\exp(-\mathrm{i}\theta)| = |\mu(z)| - |\nu(z)|$ and it depends on z.

It should be noted that the variance (37) is composed of two different terms. The first term is analogous to the term in (36) with an additional exponentially damped factor $\exp(-2\gamma z)$ and so this term tends to zero for large interaction lengths. The second term containing integration is monotonously increasing with interaction length and is present only if losses are considered.

We note that the asymptotic solution can give some squeezing even if damping is present and mean number of reservoir phonons is not zero, $\langle \hat{n}_d \rangle > 0$. Particularly if we consider the three-wave mixing process (31), we obtain the asymptotic solution

$$\lambda_2 = (1 + 2\langle \hat{n}_d \rangle) \frac{\gamma}{\gamma + |\kappa|}$$

which admits squeezing if $\langle \hat{n}_d \rangle < |\kappa|/2\gamma$, i.e., if the loss parameters are sufficiently small.

More complicated calculations provide the following asymptotic expression for four-wave mixing processes (31):

$$\lambda_2 = (1 + 2\langle \hat{n}_d \rangle) \left[1 - \frac{1}{1 + \sqrt{\dfrac{\gamma^2 + k^2(m_1 + m_2)^2}{g^2 m_1 m_2}}} \right]$$

The maximum effect is obtained if $m_1 = m_2 \gg 1$ and then

$$\lambda_2 = \left(1 + 2\langle \hat{n}_d \rangle\right) \frac{2k}{g + 2k}$$

does not depend on the damping constant γ. The last result admits absolute squeezing (i.e., $\lambda_2 = 0$) if $k \ll g$. So we may conclude that the second term in (31) involving the coefficient k reduces the possibility of generation of squeezed states.

We note that if one assumes that both input signals $\hat{A}_1, \hat{A}_2$ are mutually dependent (e.g., they are derived from one laser mode using a beamsplitter), then one obtains the following interesting result for the principal squeezing variance (neglecting losses):

$$\lambda_2 = \left[1 + B_1 + B_2 + 2\sqrt{B_1 B_2} \cos(2\delta + \phi)\right] \left(|\mu| - |\nu|\right)^2 \tag{38}$$

A new interference term depending on the mutual phase mismatch ϕ appears here. With a suitable choice of the phase ϕ we can significantly reduce (if $B_1 \neq B_2$) or completely eliminate (if $B_1 = B_2$) the negative influence of possible input chaotic components [28]. The obtained result (38) can also be used to calculate alternative kinds of squeezing, for example, amplitude-square squeezing [29] or the higher-order squeezing defined in [30, 31].

The derived results allow us to calculate any statistical properties by using a computer. Many interesting examples were calculated and discussed in Ref. 26. In this short review we restrict ourselves to general results only.

In analogy to the squeezing analysis the single mode output can be sub-Poissonian (i.e., nonclassical) only if inputs are sub-Poissonian. For a compound output mode one can show that sub-Poissonian light may be generated from coherent inputs due to nonlinear coupling of quantum noises. In general, the photon distribution functions are given by discrete convolution containing Laguerre polynomials [4, 31].

Interesting evolutions are obtained in spontaneous processes without any losses. The compound mode state generated from input vacuum states has nonzero probability of even numbers n of detected photons only. This feature can be understood by remembering that the pump photon is split into two signal photons at the same time. The exact photon distribution can be expressed in the explicit form

$$p(2n) = \frac{|\nu|^{2n}}{\left(1 + |\nu|^2\right)^{n+1}} \qquad p(2n + 1) = 0$$

and can be interpreted as if the photon pairs with the Bose-Einstein statistics and the mean number $|\nu|^2$ of photon pairs really existed in examined output light. This compound mode light is nonclassical because it is squeezed $\lambda_2 = (|\mu| - |\nu|)^2 \leq 1$, but super-Poissonian because $\langle(\Delta\hat{n})^2\rangle = \langle\hat{n}\rangle^2 + 2\langle\hat{n}\rangle \geq \langle\hat{n}\rangle^2 + \langle\hat{n}\rangle$, whereas the single modes remain chaotic forever.

2. *Optical Phase Conjugation*

In this section we review some quantum statistical properties of optical phase conjugation in relation to Section IV.A.1 on four-wave mixing. The dynamics of this process is described by the operator relation [32–34]

$$\hat{b} = r\hat{a}^\dagger + \sqrt{1 + |r|^2}\,\hat{L} \tag{39}$$

where $\hat{b}$ is an annihilation operator of a phase-conjugated beam in which wave-front aberrations are removed, r is a reflectivity coefficient ($|r|$ can become larger than unity), $\hat{a}^\dagger$ is a creation operator of an incident beam, and a noise operator $\hat{L}$ fulfils the boson commutation rule $[\hat{L}, \hat{L}^\dagger] = 1$; of course, the commutation rules $[\hat{a}, \hat{a}^\dagger] = [\hat{b}, \hat{b}^\dagger] = 1$ hold, too. Because the first and the second terms in (39) commute, one cannot expect the occurrence of nonclassical behavior of conjugated light arising from the dynamics of phase conjugation.

Calculating the single-mode normal characteristic function, we obtain for the mode b

$$\begin{aligned} C_{\mathcal{N}}^{(b)}(\eta) &= \mathrm{Tr}\left\{\hat{\rho}\exp\left(\eta\hat{b}^\dagger\right)\exp\left(-\eta^*\hat{b}\right)\right\} \\ &= \exp\left(-|r|^2|\eta|^2\right)\int\int\Phi_{\mathcal{N}}^{(a)}(\alpha)\Phi_{\mathcal{N}}^{(L)}(\lambda)\exp\Big[-\eta^*\Big(r\alpha^* \\ &\quad + \sqrt{1 + |r|^2}\,\lambda\Big) + \eta\Big(r^*\alpha + \sqrt{1 + |r|^2}\,\lambda^*\Big)\Big]\, d^2\alpha\, d^2\lambda \end{aligned} \tag{40}$$

where η is the parameter of the characteristic function, α and λ are the complex amplitudes corresponding to the operators $\hat{a}$ and $\hat{L}$ in the coherent states, and $\Phi_{\mathcal{N}}^{(a)}(\alpha)$ and $\Phi_{\mathcal{N}}^{(L)}(\lambda)$ are corresponding Glauber–Sudarshan quasidistributions, respectively. In deriving (40) the Baker-

Hausdorff identity has been used. Performing the Fourier transform, we arrive at the resulting Glauber-Sudarshan quasidistribution for this process:

$$\Phi_{\mathscr{N}}^{(b)}(\beta) = \frac{1}{\pi|r|^2} \int\int \Phi_{\mathscr{N}}^{(a)}(\alpha) \Phi_{\mathscr{N}}^{(L)}(\lambda) \times \exp\left[- \frac{\left| \beta - \left(r\alpha^* + \sqrt{1 + |r|^2}\,\lambda \right) \right|^2}{|r|^2} \right] d^2\alpha \, d^2\lambda \tag{41}$$

where β is the resulting complex amplitude of the mode b; this shows that for initially coherent modes with the complex amplitudes α_0 and λ_0 ($\Phi_{\mathscr{N}}^{(a)}(\alpha) = \delta(\alpha - \alpha_0)$, $\Phi_{\mathscr{N}}^{(L)}(\lambda) = \delta(\lambda - \lambda_0)$), the phase conjugated field is the superposition of the signal $r\alpha_0^* + \sqrt{1 + |r|^2}\,\lambda_0$ and the noise $|r|^2$ determined by the reflection coefficient (if the additional noise is in the vacuum state, $\lambda_0 = 0$). Thus, no quantum effects can arise in this case. The same conclusion is valid for the arbitrary quantum statistical state of the incident radiation, provided that L is coherent or more noisy including the vacuum state. In this last case we have

$$\Phi_{\mathscr{N}}^{(b)}(\beta) = \frac{1}{\pi|r|^2} \int \Phi_{\mathscr{N}}^{(a)}(\alpha) \exp\left[- \frac{|\beta - r\alpha^*|^2}{|r|^2} \right] d^2\alpha = \Phi_{\mathscr{A}}^{(b)}\left(\frac{\beta^*}{r^*} \right) \Big/ |r|^2 \tag{42}$$

where $\Phi_{\mathscr{A}}$ is the quasidistribution related to the antinormal ordering, which is always positively semidefinite ($0 \le \Phi_{\mathscr{A}} \le 1/\pi$; see, e.g., Ref. 1, p. 93). Only when noise mode L is nonclassical, can the conjugated radiation occur in some nonclassical states, as is discussed in the following.

For the principal squeezing variance $\lambda_2^{(b)}$ we derive from its definition

$$\lambda_2^{(b)} = 1 + 2|r|^2(1 + B_a) + 2(1 + |r|^2)B_L - 2\left| r^2 C_a^* + (1 + |r|^2)C_L \right| \tag{43}$$

where B_a, B_L, C_a, and C_L are corresponding noise quantities from Section II (defined for the normal ordering) for the modes a and L, respectively. Analyzing this expression we can conclude that in the case of a coherent incident light and squeezed noise L

$$\lambda_2^{(b)} = |r|^2 + (1 + |r|^2)(1 + 2B_L - 2|C_L|) \ge |r|^2 \tag{44}$$

and squeezing of vacuum fluctuations ($\lambda_2^{(b)} < 1$) is possible in the conjugated mode provided that the radiation is attenuated ($|r|^2 < 1$). If we denote the value of the squeeze parameter as s, then squeezing in the conjugated beam occurs if

$$\exp(-2s) < \frac{1 - |r|^2}{1 + |r|^2} \tag{45}$$

If modes a and L behave generally and assuming an optimal phase condition, then for squeezing it must hold that

$$|r|^2 < \frac{1 - \left\langle \left(\Delta \hat{Q}_L\right)^2 \right\rangle}{\left\langle \left(\Delta \hat{Q}_a\right)^2 \right\rangle + \left\langle \left(\Delta \hat{Q}_L\right)^2 \right\rangle} \tag{46}$$

where $\hat{Q}_j$ is defined in Section II.B.

Similarly, we can investigate sub-Poissonian behavior of conjugated light. In general, we can derive that

$$\left\langle (\Delta W_b)^2 \right\rangle = B_b^2 + |C_b|^2 + 2B_b|\beta|^2 + 2\,\mathrm{Re}\left(C_b^* \beta^2\right) \tag{47}$$

where B_b and C_b are again the noise function (for the mode b) and

$$\begin{aligned} \beta &= r\alpha^* + \sqrt{1 + |r|^2}\,\lambda \\ B_b &= |r|^2(1 + B_a) + \left(1 + |r|^2\right)B_L \\ C_b &= r^2 C_a^* + \left(1 + |r|^2\right)C_L \end{aligned} \tag{48}$$

The photon-number fluctuations are then given by $\langle (\Delta \hat{n}_b)^2 \rangle = \langle \hat{n}_b \rangle + \langle (\Delta W_b)^2 \rangle$. We find again that the sub-Poisson behavior can occur only if the L-mode is squeezed and $|r| < 1$, in agreement with the condition (45) for squeezing. In Fig. 2 we see the influence of the reflectivity parameter r on the photon-number distribution $p(n)$ using equations of Section II. For small $|r|$ the statistics are sub-Poissonian or they oscillate. Both these nonclassical effects rapidly disappear when the reflectivity grows.

To control the statistics of L we can use the standard four-wave mixing with counterpropagating beams in a medium of the length l. In this case, denoting the annihilation operator for the incident signal beam as $\hat{a}_1(0)$

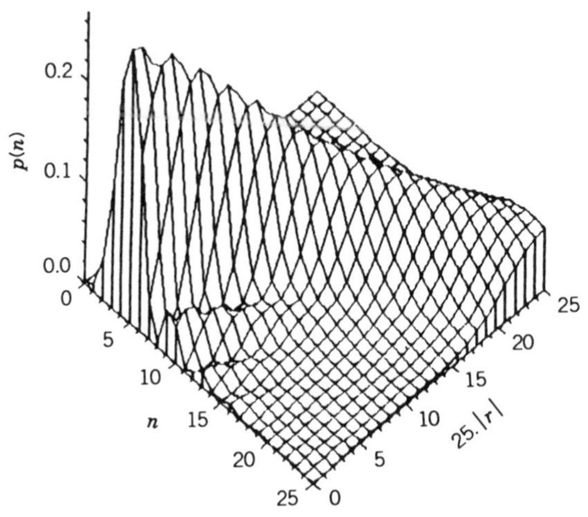

Figure 2. Dependence of the photon-number distribution on the reflectivity $|r|$ for $s = 1$, $\alpha = \lambda_2 = 2$.

and that of the conjugated beam by $\hat{a}_2(0)$, it holds that [35]

$$\hat{a}_2(0) = \frac{\hat{a}_2(l)}{\cos(|\kappa|l)} - \mathrm{i}\frac{\kappa^*}{|\kappa|}\hat{a}_1^\dagger(0)\tan(|\kappa|l) \tag{49}$$

where $\kappa = g\alpha_3^*\alpha_4^*$, g being the coupling constant and α_3 and α_4 are strong (classical) pumping complex amplitudes. Thus, $\hat{a} = \hat{a}_1(0)$, $\hat{b} = \hat{a}_2(0)$, and $\hat{L} = \hat{a}_2(l)$, $r = -\mathrm{i}(\kappa^*/|\kappa|)\tan(|\kappa|l)$. Therefore, we can control the quantum statistics of the noise function $L = a_2(l)$ by injecting a signal $a_2(l)$ of given statistical properties, including nonclassical ones, at the free entrance of the phase conjugating mirror.

Finally, we mention that interesting results can be obtained for radiation of single atoms placed near the phase conjugating mirror [36].

3. *Stimulated Raman and Hyper-Raman Scattering*

Photon statistics of Raman and hyper-Raman scattering were intensively investigated recently since the nonlinear dynamics of these processes permit us to obtain a variety of regimes for generating nonclassical light (for a review, see Ref. 1, Sec. 10.4, and references therein, and the Chapter by A. Miranowicz and S. Kielich in Volume 85, Part 2.) We mention here only some additional results derived quite recently to include effects of initially squeezed light, external noise, and losses [37], the effect of squeezed photons in scattering of nonclassical light [38], and the effect of pump depletion [39, 40].

Starting with the standard effective interaction Hamiltonian

$$\hat{H}_{\text{int}} = \left(-\hbar g \hat{a}_L^k \hat{a}_S^\dagger \hat{a}_V^\dagger - \hbar \kappa^* \hat{a}_L^k \hat{a}_A^\dagger \hat{a}_V \right) + \text{h.c.} \tag{50}$$

g and κ being the Stokes and anti-Stokes coupling constants, L, S, A, V denote the laser, Stokes, anti-Stokes, and vibration phonon modes, respectively, and $k = 1$ for Raman scattering and $k = 2$ for hyper-Raman scattering, we can use the standard quantum theory of losses (e.g., [1], Chap. 7) and in the interaction picture we can write down the operator equations of motion for the operators $\hat{A}_j = \hat{a}_j \exp(\mathrm{i}\omega_j t)$; $j = L, S, A, V$:

$$\begin{aligned}
\frac{\mathrm{d}}{\mathrm{d}t}\hat{A}_L &= -\frac{\gamma_L}{2}\hat{A}_L + \mathrm{i}kg^*\hat{A}_L^{\dagger k-1}\hat{A}_S\hat{A}_V + \mathrm{i}k\kappa\hat{A}_L^{\dagger k-1}\hat{A}_A\hat{A}_V^\dagger + \hat{L}_L + p_L \\
\frac{\mathrm{d}}{\mathrm{d}t}\hat{A}_S &= -\frac{\gamma_S}{2}\hat{A}_S + \mathrm{i}g\hat{A}_L^k\hat{A}_V^\dagger + \hat{L}_S + p_S \\
\frac{\mathrm{d}}{\mathrm{d}t}\hat{A}_A &= -\frac{\gamma_A}{2}\hat{A}_A + \mathrm{i}\kappa^*\hat{A}_L^k\hat{A}_V + \hat{L}_A + p_A \\
\frac{\mathrm{d}}{\mathrm{d}t}\hat{A}_V &= -\frac{\gamma_V}{2}\hat{A}_V + \mathrm{i}g\hat{A}_L^k\hat{A}_S^\dagger + \mathrm{i}\kappa\hat{A}_L^{\dagger k}\hat{A}_A + \hat{L}_V
\end{aligned} \tag{51}$$

where γ_j and $\hat{L}_j$ are the corresponding damping constants and Langevin forces, respectively. Here all modes are described by dynamical variables. In an alternative description, phonon modes form large reservoir system [37].

If the above equations are considered as c-number equations, then by putting $\mathrm{d}A_j/\mathrm{d}t = 0$ and neglecting the Langevin forces ($\langle \hat{L}_j \rangle = 0$) we can find the stationary solutions A_{j0}, $j = L, S, A, V$. Then we can perform the first operator correction to the classical solution A_{j0}, writing $\hat{A}_j = A_{j0} + \delta\hat{A}_j$, $j = L, S, A, V$, which leads to the following linear system of operator equations:

$$\frac{\mathrm{d}}{\mathrm{d}t}\delta\hat{A} = \hat{M}\delta\hat{A} + \hat{L} \tag{52}$$

where $\delta\hat{A}$ and $\hat{L}$ are columns composed of $\delta\hat{A}_L, \delta\hat{A}_L^\dagger, \ldots, \delta\hat{A}_V, \delta\hat{A}_V^\dagger$, and $\hat{L}_L, \hat{L}_L^\dagger, \ldots, \hat{L}_V, \hat{L}_V^\dagger$, respectively, and $\hat{M}$ is a more complicated matrix [39, 40] involving the stationary solutions A_{j0}, the coupling constants g and κ, and the damping constants γ_j. The Langevin forces $\hat{L}_j$ fulfill the standard Markovian properties, $\langle \hat{L}_j^\dagger(t)\hat{L}_k(t')\rangle = \gamma_j \langle \hat{n}_{d_j}\rangle \delta_{jk}\delta(t-t')$, $\langle \hat{n}_{d_j}\rangle$ being the mean number of reservoir oscillators coupled to the jth mode ($j = L, S, A, V$), δ is the Dirac function, and δ_{jk} is the Kronecker symbol.

The system of operator equations (52) can be solved in the matrix form

$$\delta\hat{A}(t) = \exp(\hat{M}t)\delta\hat{A}(0) + \int_0^t \exp\left[\hat{M}(t-t')\right]\hat{L}(t')\,\mathrm{d}t' \tag{53}$$

which can be written in components

$$\delta\hat{A}_i(t) = \sum_{j=L,S,A,V}\left[U_{ij}(t)\delta\hat{A}_j(0) + V_{ij}(t)\delta\hat{A}_j^\dagger(0)\right] + \hat{F}_i(t)$$
$$i = L, S, A, V \tag{54}$$

where the time-dependent functions U_{ij} and V_{ij} and the needed reservoir correlations such as $\langle \hat{F}_i^\dagger(t)\hat{F}_i(t)\rangle$, etc. can be expressed explicitly [39].

The operator solution (54) generates the photon statistics in the form of the generalized superposition of coherent fields and quantum noise, as discussed in Section II. The time-dependent complex amplitudes are expressed as

$$\xi_i(t) = \sum_{j=L,S,A,V}\left[U_{ij}(t)\xi_j + V_{ij}(t)\xi_j^*\right] + A_{i0} \tag{55}$$

where $\xi_i \equiv \xi_i(0)$ are the initial complex amplitudes, and quantum noise functions $B_j(t)$, $C_j(t)$, $D_{ij}(t)$, and $\overline{D}_{ij}(t)$ can be expressed in terms of $U_{jk}(t)$, $V_{jk}(t)$, and the initial values $B_j(0) \equiv B_j$ and $C_j(0) \equiv C_j$ (provided that the initial coupling is neglected, so that $D_{jk}(0) = \overline{D}_{jk}(0) = 0$) [39]. To be able to describe initially squeezed light and external noise, we can choose

$$B_j = \cosh^2 r_j + \bar{n}_{j0} \qquad C_j = \tfrac{1}{2}\exp(\mathrm{i}\phi_j)\sinh(2r_j) \qquad j = L, S, A, V \tag{56}$$

where r_j are squeeze parameters of the initial fields, ϕ_j are the squeeze phases, and $\bar{n}_{j0}$ are the external noise components. If $r_j = 0$, we have again the initial state in form of the superposition of the coherent state $|\{\xi_j\}\rangle$ and noise $\bar{n}_{j0}$ (see [1], Sec. 5.3). If $r_j = \bar{n}_{j0} = 0$, the initial state is coherent ($B_j = 1$, $C_j = 0$). The expression for the other quantum characteristics, such as photon-number distribution, its factorial moments, and principal squeezing for single as well as compound modes are given in Section II.

Now we can review the main conclusions for these processes on the basis of quasidistribution related to the antinormal ordering, photocount distribution, its factorial moments and the principal squeezing variance, as discussed in greater detail in Refs. 37–40:

1. Initial nonclassical behavior of scattered radiation is in general more pronounced if the Stokes mode is less amplified than the anti-Stokes mode is damped. In the opposite case, strong amplification of the Stokes mode leads to an increase of quantum noise in radiation and to fast degradation of initial nonclassical behaviour. If the frequency mismatch is also taken into account, then there exists an optimum value of frequency detuning for a given time for which the nonclassical effects caused by nonlinear dynamics are maximum. Up to a certain bound of the frequency mismatch, nonclassical effects can periodically return; for higher values of frequency mismatch, the nonclassical effects are missing. The phonon system always increases the value of quantum noise in radiation modes, even if it is squeezed. A small initial squeezing can be increased by nonlinear dynamics, whereas strong initial squeezing can only be diminished. Some of the nonclassical effects obtained are also considerable in asymptotic states.
2. If radiation modes are damped, relatively long-time conservation of an initial sub-Poisson behavior can be reached for the corresponding initial phase conditions; however, asymptotically a super-chaotic state is obtained. The initial sub-Poisson behavior can be pronounced by nonlinear dynamics more in the compound photon-phonon modes than in photon-photon modes. The initial sub-Poisson statistics of the compound Stokes–anti-Stokes mode perform asymptotically evolution to super-chaotic statistics. External and reservoir noises generally lead to the degradation of nonclassical behavior of light.
3. Taking the pump depletion into account, we can demonstrate (see Fig. 3) that initially coherent laser mode conserves its coherence for a relatively long time, regardless of its depletion. Only in the region of very strong depletion and the inversion of the process, where the mean photon number again begins to increase, is there a strong increase in quantum noise, which is again reduced on further increase in the laser intensity. The anti-Stokes, Stokes, and compound Stokes–anti-Stokes modes behave similarly. However, the initial coherence is more rapidly destroyed and stationary quantum noise

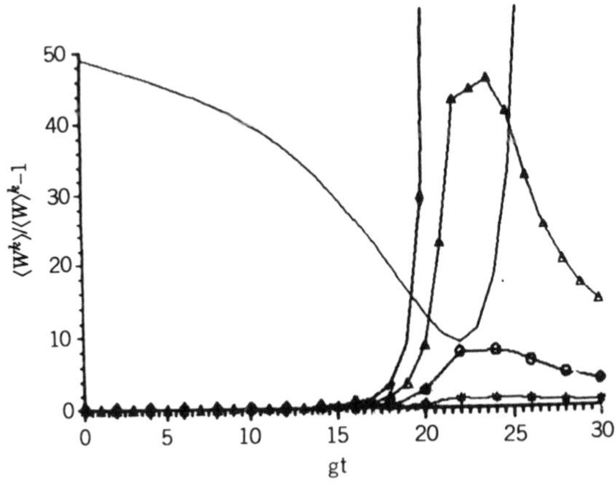

Figure 3. Time evolution of the first moment $\langle W \rangle$ (full curve without notation) and of the reduced factorial moments $\langle W^k \rangle / \langle W \rangle^k - 1$ for $k = 2(*)$, $k = 3(\circ)$ $k = 4(\triangle)$ and $k = 5(\diamond)$ for the depleted laser mode in Raman scattering; $g/\kappa = 2$, $\gamma_L = \gamma_S = \gamma_A = \gamma_V = 1$, $2p_L/\gamma_L = 5$, $2p_S/\gamma_S = 2p_A/\gamma_A = 1$, $\xi_L = 2$, $\xi_S = \xi_A = 1$, $\xi_V = 0$, $\phi_j = r_j = \bar{n}_{j0} = \langle \hat{n}_{dj} \rangle = 0$, $j = L, S, A, V$.

has a substantially higher level in the Stokes and compound modes than in the anti-Stokes mode. In compound photon–phonon modes interesting regimes for squeezing of vacuum fluctuations can occur. If the initial modes are squeezed, the photon-number distribution will oscillate as a result of quantum properties, as discussed in Section II. These oscillations will be smoothed out successively by the nonlinear dynamics, as illustrated in Fig. 4 for the compound Stokes–anti-Stokes mode. Additional external noise degrades such oscillations, which can also be observed in the squeezed phonon mode.

4 Self-interaction of the pump laser light in the hyper-Raman process can substantially reduce the initial Poissonian fluctuations for some time interval after the interaction is switched on. The tendency of the Stokes mode to be amplified and the anti-Stokes mode to be attenuated leads to faster increase of quantum noise in the Stokes mode compared to the anti-Stokes mode in both the stimulated Raman and hyper-Raman scattering.

5. Hyper-Raman scattering provides a broader variety of regimes for generation of nonclassical light compared to Raman scattering as a result of self-interaction of the pumping laser mode.

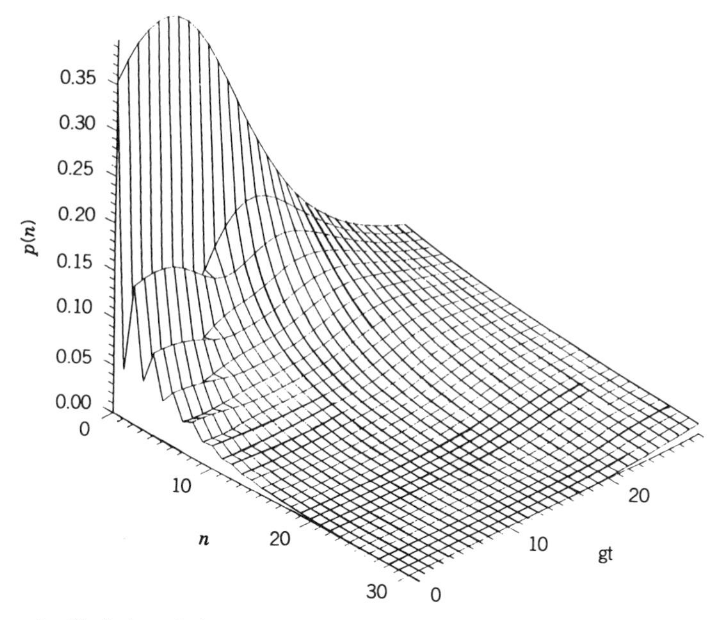

Figure 4. Evolution of photon-number distribution for hyper-Raman scattering and the anti-Stokes mode; $g/\kappa = 2$, $\gamma_j = 1$, $\bar{n}_{j0} = \langle \hat{n}_{dj} \rangle = 0$, $j = L, S, A, V$, $\xi_L = 2$, $\xi_S = \xi_A = \xi_V = 1$, $2p_L/\gamma_L = 5$, $2p_S/\gamma_S = 2p_A/\gamma_A = 1$, $r_S = r_A = 1$, $r_L = r_V = 0$.

A comparison of two-photon absorption, second-harmonic generation and hyper-Raman scattering can be found in Refs. 41 and 42. The quantum theory of hyper-Raman scattering involving three-level systems was discussed in Ref. 43.

4. *Nth Subharmonic Generation*

Using the Heisenberg picture we can describe the interaction of two modes in the process of the Nth subharmonic generation in a resonator by assuming that the modes are pumped by classical light. The simpler case of the degenerate second subharmonic generation has already been discussed from various points of view in greater detail. For example, photocount statistics well above threshold were derived in Ref. 44 using the Schrödinger picture, and the authors of papers 45–49 discussed global physical problems such as stability, phase transition, switching, and spectral behavior. Non-classical statistics and oscillations in photocount distribution below threshold were discussed for this model in Refs. 50 and 51.

Squeezing spectra of nondegenerate parametric processes were obtained in Refs. 52 and 53.

Let $\hat{a}, \hat{c}$ be annihilation operators of subharmonic and pump modes so that the effective Hamiltonian of the interaction is of the form

$$\hat{H} = \frac{\mathrm{i}\hbar}{N}\left(G\hat{a}^{N}\hat{c}^{\dagger} - G^{*}\hat{a}^{\dagger N}\hat{c}\right) + \mathrm{i}\hbar\left(\varepsilon\hat{c}^{\dagger} - \varepsilon^{*}\hat{c}\right) \tag{57}$$

where N is the order of subharmonic generation, G is a complex interaction constant, and ε denotes the complex amplitude of the external pumping field. By including losses we obtain two Heisenberg-Langevin equations:

$$\begin{aligned} \frac{\mathrm{d}}{\mathrm{d}t}\hat{a} &= G\hat{a}^{\dagger N-1}\hat{c} - \gamma_a\hat{a} + \hat{L}_a \\ \frac{\mathrm{d}}{\mathrm{d}t}\hat{c} &= \varepsilon - \frac{G^{*}}{N}\hat{a}^{N} - \gamma_c\hat{c} + \hat{L}_c \end{aligned} \tag{58}$$

where γ_a, γ_c and $\hat{L}_a, \hat{L}_c$ are damping constants and Langevin forces of the two modes, respectively. Their diffusion constants D_a, D_c are given by

$$D_a = \gamma_a\left(1 + \langle\hat{n}_a\rangle\right) \qquad D_c = \gamma_c\left(1 + \langle\hat{n}_c\rangle\right)$$

$\langle\hat{n}_a\rangle, \langle\hat{n}_c\rangle$ being the mean numbers of phonons in each reservoir mode.

An exact solution of (58) cannot be obtained for $N > 1$ due to the nonlinear character of (58), but stationary solutions for a general integer N can be derived using the two following assumptions: We suppose that the pump mode decays rapidly compared to the subharmonic signal, i.e., $\gamma_a \ll \gamma_c$, and therefore we can use an adiabatic elimination of the pump operator; further we suppose that the pump is so strong that we are well above threshold and may linearize Eq. (58) around the stationary solution. We note that a nondegenerate case cannot be solved in this way because it does not have an ordinary stationary solution. Phases of each subharmonic amplitude are then randomly diffusing in analogy to the diffusing phase of free laser amplitude. In such cases one can adopt other methods, for example, the spectral squeezing approach [52, 53].

On assuming $\gamma_a \ll \gamma_c$, the pump mode can be eliminated adiabatically to arrive at

$$\frac{\mathrm{d}}{\mathrm{d}t}\hat{a} = \frac{G\hat{a}^{\dagger N-1}\left[\varepsilon - (G^{*}/N)\hat{a}^{N}\right]}{\gamma_c} - \gamma_a\hat{a} + \hat{L} \tag{59}$$

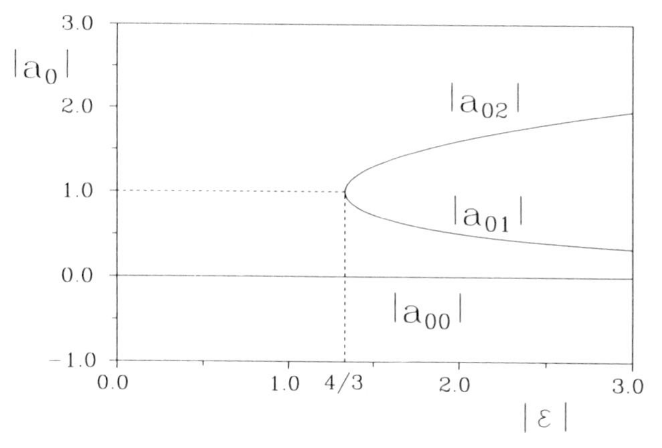

Figure 5. Dependence of the modulus of the mean coherent value $|a_0|$ on the pumping parameter $|\varepsilon|$ for the third subharmonics $N = 3$; $|G| = 1$, $\gamma_a = 0.1$, $\gamma_c = 10$. Above threshold $\varepsilon_{\text{th}} = 4/3$ two stable states $|a_{00}|$ and a triplet $|a_{02}|$ are possible; the state $|a_{01}|$ is not stable.

where $\hat{L} = \hat{L}_a + \hat{L}_c G\hat{a}^{\dagger N-1}/\gamma_c$ and it holds that

$$\left\langle \hat{L}(t_1)\hat{L}^{\dagger}(t_2)\right\rangle = 2D\delta(t_1 - t_2) \qquad \left\langle \hat{L}(t_1)\hat{L}(t_2)\right\rangle = 0$$

Here $D = D_a + D_c|G|^2|a|^{2N-2}/\gamma_c^2$.

Assuming that a is a complex number and neglecting the Langevin force, we can calculate the critical points a_0 as possible stationary solutions of Eq. (59). The trivial solution is $a_0 = 0$ and $c_0 = \varepsilon\gamma_c$. The nontrivial solution $a_0 = |a_0|\exp(i\alpha) \neq 0$ can be solved numerically only [54]; from simple analysis one can deduce that the equation has no real root below threshold $|\varepsilon| < \varepsilon_{\text{th}}$, a single root at the threshold $|\varepsilon| = \varepsilon_{\text{th}}$ and two different roots $|a_{01}| < |a_{02}|$ above threshold $|\varepsilon| > \varepsilon_{\text{th}}$ (see Fig. 5), where

$$\varepsilon_{\text{th}} = \frac{\gamma_a\gamma_c(2N-2)}{|G|N}\left[\frac{|G|^2}{\gamma_a\gamma_c(N-2)}\right]^{(N-2)/(2N-2)}$$

Here we must note that phase α is not unique, because there exist N different phases with the step $2\pi/N$, which are also physically admissible. So we have $2N + 1$ critical points, from which $N + 1$ points are stable, as we will see later.

Now we can perform the stability analysis. We will linearize the c-number equation (59) and then we use the fact that the c-number equation

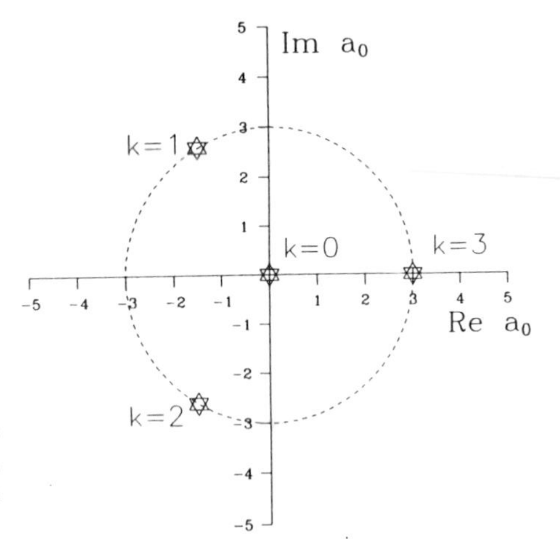

Figure 6. A diagram of stable critical points of Eq. (59) in a complex plane for $N = 3$, $|G| = 1$, $\gamma_a = 0.1$, $\gamma_c = 10$, $\varepsilon = |\varepsilon| = 28/3 = 7\varepsilon_{\text{th}}$. We see single trivial point ($k = 0$) and triplet nontrivial points ($k = 1 - 3$) creating vertices of a regular triangle.

$(\mathrm{d}/\mathrm{d}t)A = -\gamma A - \delta A^*$ has the stable solution $A = 0$ if $\gamma > |\delta|$. This condition is fulfilled only for the largest root $|a_{02}|$ because $\gamma(|a_{02}|) > |\delta|$ and $\gamma(|a_{01}|) < |\delta|$. Note that for threshold pumping $|\varepsilon| = \varepsilon_{\text{th}}$ it holds that $\gamma(|a_{02}|) = |\delta|$.

For the zero amplitude solution $a_0 = 0$ we obtain $\gamma = \gamma_a$ and $\delta = -G\varepsilon/\gamma_c$ for $N = 2$, and $\delta = 0$ for $N > 2$. For $N = 2$ this solution is stable provided that $|\varepsilon| < \varepsilon_{\text{th}} = \gamma_a\gamma_c/|G|$, that is, below threshold only. For $N > 2$ the trivial solution is stable for all pump amplitudes ε, that is, both below and above threshold.

We can conclude that Eq. (59) has $N + 1$ stable solutions. The trivial solution $a_0 = 0$ and N nontrivial solutions $a_0 = |a_{02}|\exp(\mathrm{i}\alpha_k)$, $k = 1, 2 \ldots N$, form the center and the vertices of an N-sided regular polygon in the complex plane (Fig. 6), and so we are studying the example of multistability and hysteresis. In what follows we will see that a_0 is one of the nontrivial stable solutions a_{02}.

We now return to the operator equation (59). Because it cannot be solved exactly, we will linearize it around the stable stationary solution and for strong pump. Writing $\hat{a} = a_0 + \hat{A}$ with $|a_0| \gg |A|$, where $\hat{A}$ is a correction operator to the annihilation operator, we obtain

$$\frac{\mathrm{d}}{\mathrm{d}t}\hat{A} = -\gamma\hat{A} - \delta\hat{A}^\dagger + \hat{L} \tag{60}$$

where $\gamma = \gamma_a + |G|^2|a_0|^{2N-2}/\gamma_c$ is a positive number and $\delta = (N - 1)\gamma_a \exp(\mathrm{i}(2\alpha + \pi))]$. The corresponding Fokker-Planck equation for the quasidistribution function $\Phi_{\mathscr{A}}(\alpha, t) = \langle\alpha|\hat{\rho}|\alpha\rangle/\pi$ related to the antinormal ordering [55] is

$$\frac{d}{dt}\Phi_{\mathscr{A}} = \left[\frac{\partial}{\partial\alpha}(\gamma\alpha + \delta\alpha^*) + \text{c.c.} + \frac{\delta}{2}\frac{\partial^2}{\partial\alpha^2} + \text{c.c.} + 2D\frac{\partial^2}{\partial\alpha\,\partial\alpha^*}\right]\Phi_{\mathscr{A}}$$

and has a stable stationary solution [56] of a Gaussian form if $\gamma > |\delta|$:

$$\Phi_{\mathscr{A}}(\alpha) = \frac{1}{\pi\sqrt{K}}\exp\left[-\frac{B}{K}|\alpha - a_0|^2 + \frac{C^*}{2K}(\alpha - a_0)^2 + \text{c.c.}\right]$$

with

$$B = \frac{2D\gamma - |\delta|^2}{2(\gamma^2 - |\delta|^2)}$$

$$C = -\frac{\delta(2D - \gamma)}{2(\gamma^2 - |\delta|^2)}$$

$$K = B^2 - |C|^2 = \frac{4D^2 - |\delta|^2}{4(\gamma^2 - |\delta|^2)}$$

Owing to the validity of the stability condition $\gamma > |\delta|$ the coefficients B, C, and K, exist and $K > 1$.

Substituting for coefficients B and C in the principal squeezing variance definition $\lambda_2 = -1 + 2B - 2|C|$, we get

$$\lambda_2 = \frac{2D - g}{\gamma + |\delta|} > \frac{\gamma}{\gamma + |\delta|}$$

For the cold reservoirs $\langle \hat{n}_a \rangle = \langle \hat{n}_c \rangle = 0$, and therefore $D = g$. In this case

$$\lambda_2 = \frac{\gamma}{\gamma + |\delta|} < 1$$

and the generated subharmonic mode is squeezed. With respect to the stability condition $\gamma > |\delta|$ it holds that $\lambda_2 > \frac{1}{2}$; that is, the lower bound of squeezing exists. The largest squeezing, $\lambda_2 = \frac{1}{2}$, is obtained near the threshold $\gamma \to |\delta|$; if the pumping power increases, then the level of squeezing decreases (see Fig. 7).

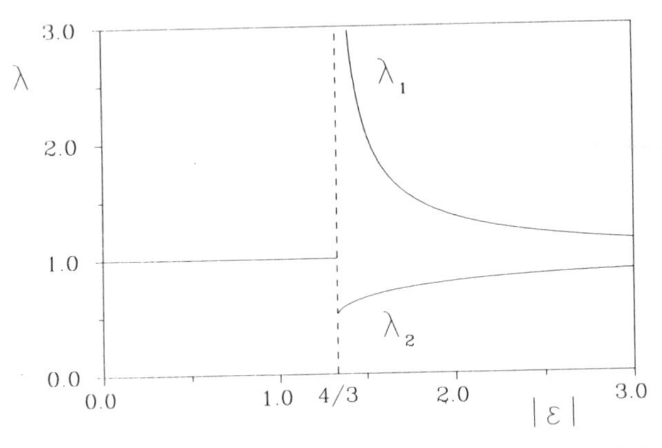

Figure 7. Dependence of the principal squeezing variance λ_2 for the third subharmonics mode $N = 3$ on pumping parameter $|\varepsilon|$; $|G| = 1$, $\gamma_a = 0.1$, $\gamma_c = 10$, $\langle \hat{n}_a \rangle = \langle \hat{n}_c \rangle = 0$. The lower bound of squeezing 1/2 is reached near the threshold, and above threshold squeezing is monotonously decreased.

Assume now that the reservoirs are warm and that $\langle \hat{n}_a \rangle = \langle \hat{n} \rangle$; one can easily get

$$\lambda_2 = \frac{\gamma(1 + 2\langle \hat{n} \rangle)}{\gamma + |\delta|}$$

which gives squeezing only if$\langle \hat{n} \rangle < |\delta|/2\gamma$. Further, we can derive the photocount distribution function by using the standard method of the characteristic and generating functions [1, 54] by using the Laguerre polynomials. Unfortunately, no sub-Poissonian statistics can be obtained,

$$\left\langle (\Delta W)^2 \right\rangle = B^2 + |C|^2 + 2B|a_0|^2 + 2|C^* a_0^2|\cos\phi > 0$$

as follows from the zero phase $\phi = \arg(C^* a_0^2) = 0$.

Any observable moment depends only on the difference of phases and not on a single phase. Therefore, squeezing depends on the phase difference of local oscillator and pump, but the photocount statistics are functions of the amplitude ε only and so they are insensitive to the phase.

B. Non-Gaussian Solutions

1. *Anharmonic Oscillator*

The properties of the Kerr-law media have been studied intensively. In addition to classical nonlinear properties (self-focusing, self-phase modulation), quantum effects have been considered. In quantum optics the

Kerr-law medium is modeled as a one-dimensional quantum anharmonic oscillator [57–59]. This nonlinear oscillator aroused interest from the viewpoint of the quantum statistical physics [60]. It also developed in the context of nonlinear quantum optics with all its requirements and simplifying assumptions. In Refs. 61–63 dissipation was included by coupling the oscillator to a zero-temperature heat bath, and in Ref. 62 the most general input state was considered. The assumption of zero temperature was removed in Refs. 64–66, and the dynamics of the nonlinear dissipative oscillator was determined exactly for the initial coherent state in Ref. 64 and a general input state in papers 65 and 66; the connection to the group theoretical methods was found in Ref. 66.

Modeling dissipation by coupling the third-order nonlinear oscillator to a reservoir of (linear) oscillators, we can formulate the Hamiltonian of the considered system:

$$\hat{H} = \hbar\left[\omega\left(\hat{a}^{\dagger}\hat{a} + \tfrac{1}{2}\right) + \kappa\hat{a}^{\dagger 2}\hat{a}^{2} + \sum_{j}\psi_j\left(\hat{c}_j^{\dagger}\hat{c}_j + \tfrac{1}{2}\right) + \sum_{j}\left(\kappa_j\hat{c}_j\hat{a}^{\dagger} + \text{h.c.}\right)\right] \tag{61}$$

Here $\hat{a}$ $(\hat{a}^{\dagger})$ is the photon annihilation (creation) operator, ω is the frequency of light, κ is a real constant for the intensity dependence of the refractive index, $\hat{c}_j$ $(\hat{c}_j^{\dagger})$ are the boson annihilation (creation) operators of the reservoir oscillators with the frequencies ψ_j, and κ_j are the coupling constants of the interaction with the reservoir. The reservoir noise spectrum is assumed to be flat and the reservoir oscillators are assumed to form a chaotic system.

In the standard treatments of the quantum theory of dissipation [1] we arrive at the master equation for the reduced density operator (in the interaction picture) and using a general procedure proposed in Ref. 67, we obtain the generalized Fokker-Planck equation for the quasidistribution $\Phi_{\mathscr{A}}(\alpha, t)$ related to the antinormal ordering of field operators:

$$\begin{aligned}\frac{\partial}{\partial t}\Phi_{\mathscr{A}}(\alpha, t) = \Bigg\{ & \mathrm{i}\kappa\left[2\alpha|\alpha|^2\frac{\partial}{\partial\alpha} + \alpha^2\frac{\partial^2}{\partial\alpha^2} - \text{c.c.}\right] \\ & + \gamma + \frac{\gamma}{2}\left[\alpha\frac{\partial}{\partial\alpha} + \alpha^*\frac{\partial}{\partial\alpha^*}\right] \\ & + \gamma(\bar{n} + 1)\frac{\partial^2}{\partial\alpha\,\partial\alpha^*}\Bigg\}\Phi_{\mathscr{A}}(\alpha, t)\end{aligned} \tag{62}$$

where γ is the damping constant and $\bar{n}$ reflects the thermal properties of the reservoir. Using the fact that for an arbitrary initial state there exist numbers $f_{mn}(0)$ such that

$$\Phi_{\mathscr{A}}(\alpha, 0) = \exp(-|\alpha|^2) \sum_{m=0}^{\infty} \sum_{n=0}^{\infty} \alpha^m \alpha^{*n} f_{mn}(0) \tag{63}$$

we find the solution of Eq. 62 in the form [65]

$$\Phi_{\mathscr{A}}(\alpha, t) = \exp(-|\alpha|^2) \sum_{m=0}^{\infty} \sum_{n=0}^{\infty} \alpha^m \alpha^{*n} f_{mn}(t) \tag{64}$$

where

$$\begin{aligned} f_{mn}(t) = {} & \exp\left\{\left[-2i\kappa(m-n) + \frac{\gamma}{2}\right]t\right\} E_{m-n}^{m+n+1}(t) \\ & \times \sum_{j=0}^{min(m,n)} \frac{1}{j!(m-j)!(n-j)!} \left[\frac{g_{m-n}(t)}{E_{m-n}^2(t)}\right]^j \\ & \times \sum_{l=0}^{\infty} \frac{1}{l!} \left[\frac{\bar{n}+1}{\bar{n}} g_{m-n}(t)\right]^l (m-j+l)!(n-j+l)! \\ & \times f_{m-j+l, n-j+l}(0) \qquad m \geq n \end{aligned} \tag{65}$$

with

$$E_l(t) = \frac{\Delta_l}{\Omega_l \sinh[(\gamma/2)\Delta_l t] + \Delta_l \cosh[(\gamma/2)\Delta_l t]} \tag{66}$$

$$g_l(t) = \frac{2\bar{n}}{\Omega_l + \Delta_l \coth[(\gamma/2)\Delta_l t]} \qquad \Omega_l = 1 + 2\bar{n} + i\frac{2}{\gamma}\kappa l$$

$$\Delta_l = \sqrt{\Omega_l^2 - 4\bar{n}(\bar{n}+1)}$$

In the case $m < n$ we use the fact that $f_{mn}(t) = [f_{nm}(t)]^*$. There is a simple relationship between the f coefficients and the density matrix elements

$$\rho_{nm}(t) = \pi\sqrt{n!m!}\, f_{mn}(t) \tag{67}$$

The quasidistribution $\Phi_{\mathscr{A}}(\alpha, t)$ plays a substantial role when we determine the photon statistics and squeezing properties of the radiation under consideration. Starting from its form (64), we can determine the photon-number distribution and its factorial moments as

$$p(n,t) = \pi n! f_{nn}(t) \tag{68}$$

and

$$\langle W^k(t)\rangle_{\mathcal{N}} = \sum_{n=k}^{\infty} \frac{n!}{(n-k)!} p(n,t) \tag{69}$$

respectively. They are likewise independent of κ and, consequently, the photon statistics of the dissipative nonlinear oscillator are identical with those of the corresponding dissipative linear oscillator. The explicit expressions for $p(n,t)$ and $\langle W^k(t)\rangle_{\mathcal{N}}$ are of the form

$$\begin{aligned} p(n,t) = {} & \frac{1}{\bar{n}_t + 1} \sum_{m=0}^{\infty} m! f_{mm}(0)(\bar{n}_t + 1)^{-m} \sum_{j=0}^{\min(m,n)} \frac{(m+n-j)!}{j!(n-j)!(m-j)!} \\ & \times (-1)^j [\bar{n}_t - \exp(-\gamma t)]^j \left(\frac{\bar{n}_t}{\bar{n}_t + 1} \right)^{n-j} \\ & \times [\bar{n}_t + 1 - \exp(-\gamma t)]^{m-j} \end{aligned} \tag{70}$$

and

$$\begin{aligned} \langle W^k(t)\rangle_{\mathcal{N}} = {} & \pi \sum_{m=0}^{\infty} m! f_{mm}(0) \sum_{j=0}^{\min(k,m)} \binom{k}{j} \frac{(m+k-j)!}{(m-j)!} (-1)^j \\ & \times [\bar{n}_t - \exp(-\gamma t)]^j \bar{n}_t^{k-j} \end{aligned} \tag{71}$$

where

$$\bar{n}_t = \bar{n}[1 - \exp(-\gamma t)] \tag{72}$$

When investigating squeezed states of the considered radiation, we can apply the expectation values of the antinormally ordered field operators [62]

$$\begin{aligned} \langle \hat{a}^k \hat{a}^{\dagger l}\rangle & = \langle \alpha^k \alpha^{*l}\rangle_{\mathcal{A}} = \int \alpha^k \alpha^{*l} \Phi_{\mathcal{A}}(\alpha,t)\, d^2\alpha \\ & = \pi \sum_{n=0}^{\infty} (n+l)! f_{n+l-k,n}(t) \qquad k \le l \end{aligned} \tag{73}$$

while for $k > l$ we consider the complex conjugate quantity. Formula (73) simplifies in cases $k = 0$, $l \neq 0$ and $k = 1$, $l = 1$ to

$$\langle \hat{a}^{\dagger l} \rangle = \pi \exp\left[\left(-2i\kappa l + \frac{\gamma}{2}\right)t\right]\left[\frac{E_l(t)}{1 - g_l(t)}\right]^{l+1} \sum_{n=0}^{\infty} (n + l)! f_{n+l,n}(0) G_l^n(t) \tag{74}$$

and

$$\langle \hat{a}\hat{a}^{\dagger} \rangle = \exp(-\gamma t)\left[\langle \hat{a}(0)\hat{a}^{\dagger}(0) \rangle - 1\right] + \bar{n}_t + 1 \tag{75}$$

where

$$G_l(t) = \frac{\bar{n} + 1}{\bar{n}} g_l(t) + \frac{E_l^2(t)}{1 - g_l(t)} \qquad l = 0, 1, \ldots, \infty \tag{76}$$

respectively. As follows from the explicit expressions for the standard and principal squeezing variances given in Section II.B, properties of these amplitude squeezed states of the third-order nonlinear oscillator may be assessed and ascertained using the moments (73). The same is valid for other representatives of these states, i.e., those of higher-order squeezing [30] and of the square field amplitude squeezing [29]. The principal squeezing definition may be advantageous in the case of the nonlinear oscillator because the free-field frequency is here modified by self-interaction and depends on the intensity of the field. Thus, the principal squeezing variance λ_2, which is phase independent, is also independent of this effect.

The quasidistribution $\Phi_{\mathscr{A}}(\alpha, t)$ illustrates a number of important features of the third-order nonlinear oscillator. It is well known that without dissipation the initial state repeats after a certain time interval, viz., the period. In Ref. 61 the role of higher-order derivatives in the equation of motion for the Q function (equal to $\pi\Phi_{\mathscr{A}}$) with respect to the revivals of the initial state was explained, and in Ref. 64 the interference in the phase space was pointed out as a possible source of this quantum effect. In Ref. 65 recurrences of the initial state are related to the properties of the quasidistributions $\Phi_l(\phi, t)$, $l = 0, \frac{1}{2}, 1, \frac{3}{2}, 2, \ldots,$ connected with the phase of complex field amplitude $\alpha = r \exp(i\phi)$,

$$\Phi_l(\phi, t) = \frac{\Gamma(l + 1)}{2} \sum_{k=-l}^{l} f_{l+k,l-k}(t) \exp(i2k\phi) \tag{77}$$

and involved in the quasidistribution (64) as follows:

$$\Phi_{\mathscr{A}}(r,\phi,t) = r\Phi_{\mathscr{A}}[r\exp(i\phi),t] = \sum_{l} \frac{2}{\Gamma(l+1)} r^{2l+1} \exp(-r^2)\Phi_l(\phi,t) \tag{78}$$

The equations of motion for $\Phi_l(\phi,t)$ in case without dissipation evoke the picture of the quasidistributions rotating in a clockwise direction with angular velocity $\kappa(2l-1)$, circular frequency 2κ, and period π/κ. All relevant quantum statistics repeat after this time interval. The quantum coherence is sensitive to dissipation since it consists in the harmony of the orbits $|\alpha|^2 = 0, \frac{1}{2}, 1, \frac{3}{2}, 2, \ldots$. The dissipation not only means that the system as described by $\Phi_l(\phi,t)$ descends to the lower values of half-integer intensities l, but also that the quasidistributions $\Phi_l(\phi,t)$ may rotate with velocities that do not preserve the harmony of motion.

Phase properties of the considered radiation can be studied in the framework of the formalisms developed for a single-mode electromagnetic field. The concept of the Hermitian phase operator introduced by D. T. Pegg and S. M. Barnett [68–70] is based on the phase operators constructed with the use of Loudon's phase states [71] forming orthogonal vector systems $|\theta_m, s\rangle$,

$$|\theta_m, s\rangle = (s+1)^{-1/2} \sum_{n=0}^{s} \exp(in\theta_m)|n\rangle \tag{79}$$

where

$$\theta_m = \theta + 2\pi\frac{m}{s+1} \qquad m = 0, 1, \ldots, s \tag{80}$$

and θ is a chosen value. According to Ref. 69, it is recommendable to use a unitary phase operator, which can be defined by means of the orthogonal system (79) and mapping of the following type:

$$\hat{M}_{\theta,s} = \sum_{m=0}^{s} M(\theta_m)|\theta_m, s\rangle\langle s, \theta_m| \tag{81}$$

where θ_m is given in (80). After Ref. 69, the quantum phase-measurement statistics are the limit values obtained by letting $s \to \infty$. If the field is described by the density operator $\hat{\rho}$, the phase distribution is expressed as follows:

$$P(\phi) \equiv P(\phi, t) = \frac{1}{2\pi} \sum_{n=0}^{\infty} \sum_{m=0}^{\infty} \exp[-\mathrm{i}(m-n)\phi] \rho_{nm}(t) \tag{82}$$

With the aid of $P(\phi)$ we establish the expectation values of the operators

$$\hat{\Phi}_{\theta} = \lim_{s \to \infty} \hat{\Phi}_{\theta, s}$$

and $\hat{\Phi}_{\theta}^2$, defined by the similar expression,

$$\langle \hat{\Phi}_{\theta} \rangle = \int_{\theta}^{\theta + 2\pi} \phi P(\phi)\, \mathrm{d}\phi \qquad \langle \hat{\Phi}_{\theta}^2 \rangle = \int_{\theta}^{\theta + 2\pi} \phi^2 P(\phi)\, \mathrm{d}\phi \tag{83}$$

The number operator $\hat{n}$ and the phase operator $\hat{\Phi}_{\theta}$ obey the uncertainty relation

$$\left\langle (\Delta \hat{n})^2 \right\rangle \left\langle \left(\Delta \hat{\Phi}_{\theta}\right)^2 \right\rangle \geq \tfrac{1}{4} \tag{84}$$

where the phase dispersion is measured by the phase variance

$$\left\langle \left(\Delta \hat{\Phi}_{\theta}\right)^2 \right\rangle = \langle \hat{\Phi}_{\theta}^2 \rangle - \langle \hat{\Phi}_{\theta} \rangle^2 \tag{85}$$

Another solution of the quantum phase problem uses the operators [72, 73]

$$\hat{u} = \widehat{\exp}(\mathrm{i}\phi) \qquad \hat{u}^{\dagger} = \widehat{\exp}(-\mathrm{i}\phi) \tag{86}$$

or

$$\hat{u} = (\hat{n} + \hat{1})^{-1/2} \hat{a} \qquad \hat{u}^{\dagger} = \hat{a}^{\dagger} (\hat{n} + \hat{1})^{-1/2} \tag{87}$$

for which it holds that

$$\hat{u}\hat{u}^{\dagger} = \hat{1} \qquad \hat{u}^{\dagger}\hat{u} = \hat{1} - |0\rangle\langle 0| \tag{88}$$

It can be seen that the requirement of working with a unitary phase operator may be fulfilled algebraically by adopting a sort of the antinormal ordering of the operators $\hat{u}, \hat{u}^\dagger$. This leads to the classical-quantum correspondence

$$\hat{M} = \frac{1}{2\pi}\int_\theta^{\theta+2\pi} M(\phi)\,|\phi\rangle\langle\phi|\,\mathrm{d}\phi \tag{89}$$

based on the approximately orthogonal vector system

$$|\phi\rangle = \sum_{n=0}^{\infty} \exp(\mathrm{i}n\phi)\,|n\rangle \tag{90}$$

The Susskind–Glogower cosine and sine operators can be obtained as a special case. All canonical quantum statistics can be determined with the aid of the phase distribution $P(\phi) = \langle\phi|\hat{\rho}|\phi\rangle/(2\pi)$ resulting in (82). The importance of the phase distribution is recognized from the relation for the expectation value of the operator $\hat{M}$:

$$\langle\hat{M}\rangle = \int_\theta^{\theta+2\pi} P(\phi)M(\phi)\,\mathrm{d}\phi = \langle M(\phi)\rangle_a \tag{91}$$

where the subscript a indicates the antinormal ordering of the operators $\hat{u}, \hat{u}^\dagger$. The phase dispersion is measured by the quantity V [74]:

$$V = 1 - \left|\left\langle \widehat{\exp}(\mathrm{i}\phi)\right\rangle\right|^2 \tag{92}$$

and the uncertainty relation holds

$$\left(\left\langle(\Delta\hat{n})^2\right\rangle + \tfrac{1}{4}\right)\left(1 - \left|\left\langle\widehat{\exp}(\mathrm{i}\phi)\right\rangle\right|^2\right) \geq \tfrac{1}{4} \tag{93}$$

The formalism of measured phase operators [75] supposes the operators $\hat{u}, \hat{u}^\dagger$ in the form

$$\hat{u}_M = \hat{a}\left(\langle\hat{n}\rangle + \tfrac{1}{2}\right)^{-1/2} \qquad \hat{u}_M^\dagger = \hat{a}^\dagger\left(\langle\hat{n}\rangle + \tfrac{1}{2}\right)^{-1/2} \tag{94}$$

where $\langle\hat{n}\rangle$ is taken in the considered state. The measured cosine and sine operators $\hat{C}_M$ and $\hat{S}_M$ are expressed as

$$\hat{C}_M = \frac{\hat{u}_M + \hat{u}_M^\dagger}{2} \qquad \hat{S}_M = \frac{\hat{u}_M - \hat{u}_M^\dagger}{2\mathrm{i}} \tag{95}$$

and appropriate measured phase characteristics can be applied to them.

The determination of phase properties of the third-order nonlinear oscillator is now formulated in terms of the density matrix elements $\rho_{mn}(t)$, i.e., in terms of the functions $f_{mn}(t)$ given in (65). The dynamics of the statistical properties of the considered anharmonic oscillator was analyzed in greater detail for the coherent state, the Gaussian pure and mixed states, and the squeezed and displaced Fock states as the input states.

For the initial coherent state $|\xi(0)\rangle$ the quasidistribution $\Phi_{\mathscr{A}}(\alpha, 0)$ is of the form

$$\Phi_{\mathscr{A}}(\alpha, 0) = \frac{1}{\pi} \exp\left(-|\alpha - \xi(0)|^2\right) \tag{96}$$

The quantum dynamics of this optical system was studied by means of the quasidistribution $\Phi_{\mathscr{A}}(\alpha, t)$ in the conservative case in Refs. 9 and 76, in the dissipative case for quiet reservoir in Refs. 61 and 62, and for noisy reservoir in Refs. 62, 64, and 65. The photon statistics are determined by the superposition of coherent and chaotic fields [1, 2, 62]. The standard squeezing was investigated in the lossless system in Refs. 9, 58, 59, and 77, and in the dissipative system in Ref. 62. The principal squeezing variance λ_2 for the conservative case being a rotational invariant was introduced in Ref. 7:

$$\begin{aligned}
\lambda_2 = 1 + 2 + |\xi(0)|^2\Big(1 - \exp\big\{2|\xi(0)|^2[\cos(2\kappa t) - 1]\big\}\Big) \\
- 2|\xi(0)|^2\Big(\exp\big\{2|\xi(0)|^2[\cos(4\kappa t) - 1]\big\} \\
+ \exp\big\{4|\xi(0)|^2[\cos(2\kappa t) - 1]\big\} \\
- 2\exp\big\{|\xi(0)|^2[\cos(4\kappa t) + 2\cos(2\kappa t) - 3]\big\} \\
\times \cos\big\{2\kappa t + |\xi(0)|^2[\sin(4\kappa t) - 2\sin(2\kappa t)]\big\}\Big)^{1/2}
\end{aligned} \tag{97}$$

and thoroughly analyzed in Ref. 9; the dissipative case was treated in Ref. 62. In papers 78 and 79 conditions for the self-squeezing of higher-order were found. In Ref. 80 squeezing of the squared field amplitude was predicted under the same conditions as the standard squeezing and the higher-order squeezing. Because the anharmonic oscillator produces squeezing from the input coherent state in the quadrature $\hat{q}$ (e.g., [58]), it could be expected with respect to the relationship between squeezing and the sub-Poissonian photon statistics (Section III) that it could generate number squeezed states. This is not true because the photon statistics, being insensitive to the nonlinearity, are represented by the superposition of coherent and chaotic fields. However, the photon-number variance $\langle(\Delta\hat{n})^2\rangle$ can be reduced by the interference of the Kerr medium output field, with a reference signal, conserving the variance $\langle(\Delta\hat{q})^2\rangle$. A more general formulation of the problem and an appropriate solution can be found in Refs. 17, 81, and 82.

Phase properties of the coherent light interacting with the lossless Kerr medium were studied using the Pegg-Barnett concept in Refs. 83–85, the Susskind–Glogower approach in Refs. 84 and 86, and the formalism of the measured phase operators in Refs. 84 and 87. The dissipative Kerr medium was analyzed by means of the Pegg-Barnett and Susskind-Glogower phase concepts in Ref. 88. It is shown that dissipation speeds up the phase randomization at the beginning of the evolution and deteriorates its quantum periodicity.

The coherent state $|\xi(0)\rangle$ develops in the lossless Kerr medium into the generalized coherent state:

$$|\psi(t)\rangle = \exp\left(-\frac{1}{2}|\xi(0)|^2\right)\sum_{n=0}^{\infty}\frac{\xi^n(0)}{(n!)^{1/2}}\exp(\mathrm{i}\phi_n)|n\rangle \tag{98}$$

where

$$\phi_n = -\kappa t n(n-1) \tag{99}$$

When the phases ϕ_n obey the periodicity condition

$$\exp(\mathrm{i}\phi_{n+k}) = \exp(\mathrm{i}\phi_n) \tag{100}$$

for each n and for a positive integer k, the state (98) can be expressed as the finite superposition of coherent states [89, 90]:

$$|\psi(t)\rangle = \sum_{j=1}^{k} a_j\left|\exp(\mathrm{i}\theta_j)\xi(0)\right\rangle \tag{101}$$

The determination of phases θ_j and coefficients a_j in the dependence on k is performed in Refs. 76 and 91. The quasidistribution $\Sigma_l\Phi_l(\phi, t)$ (see (77)) of the phase of the field amplitude and the phase distribution $P(\phi, t)$ (see (82)) visualize the formation of superposition states because they exhibit the same k-rotational symmetry for the superposition of k states [85, 92].

The Gaussian mixed state (the synonym for the generalized superposition of coherent and chaotic states) is at $t = 0$ described by the quasidistribution (cf. (3))

$$\Phi_{\mathscr{A}}(\alpha, 0) = \frac{1}{\pi\sqrt{K(0)}} \exp\left(-\frac{B_{\mathscr{A}}(0)}{K(0)} |\alpha - \xi(0)|^2 + \left\{ \frac{C^*(0)}{2K(0)} [\alpha - \xi(0)]^2 + \text{c.c.} \right\} \right) \tag{102}$$

where

$$K(0) = B_{\mathscr{A}}^2(0) - |C(0)|^2 \tag{103}$$

and $\xi(0)$ is the initial field amplitude. The Gaussian pure state (the synonym for the two-photon coherent state $|\beta\rangle_g = |\mu\alpha + \nu\alpha^*\rangle_g$) as a special case of the mixed state is characterized by the parameters $B_{\mathscr{A}}(0) = |\mu|^2$ and $C(0) = -\mu^*\nu$. The quantum dynamics of the dissipative anharmonic oscillator with these initial states was examined in Ref. 93. The statistical properties of the considered radiation are established with the aid of the quasidistribution $\Phi_{\mathscr{A}}(\alpha, t)$ given in (64), with $f_{mn}(t)$ determined in (65). The behavior of the system is fully described by the functions $f_{mn}(t)$. Their explicit expressions confirm that the system exhibits periodic behavior with the period π/κ for $\bar{n} = 0$, $\gamma = 0$. For $\bar{n} = 0$ and $\gamma \neq 0$, this property is not fully conserved. For $\bar{n} \gg 0$ or $\gamma \gg 0$, the periodic behavior of the system is destroyed. A detailed analysis of the case $\bar{n} = 0$ and $\gamma = 0$ would show not only that the initial states reproduce themselves at times $k\pi/\kappa$ during the evolution, but also that they form a superposition of two generalized coherent states [90] at times $(2k + 1)\pi/(2\kappa)$. The effect of damping on the formation of squeezed state superpositions is studied for the initial Gaussian pure state with the parameters $\mu = \cosh r$, $\nu = \sinh r$, and a quiet reservoir in Ref. 63.

The photon statistics are determined by the generalized superposition of the coherent and chaotic fields with the quasidistribution $\Phi_{\mathscr{A}}(\alpha, t)$ which is of the form (102), where the quantities $\xi(0)$, $B_{\mathscr{A}}(0)$, $C(0)$, and $K(0)$ are replaced by the following quantities:

$$\begin{aligned} \xi(t) &= \xi(0)\exp\left(-\frac{\gamma}{2}t\right) \\ B_{\mathscr{A}}(t) &= B_{\mathscr{A}}(0)\exp(-\gamma t) + (\bar{n} + 1)[1 - \exp(-\gamma t)] \\ C(t) &= C(0)\exp(-\gamma t) \\ K(t) &= B_{\mathscr{A}}^2(t) - |C(t)|^2 \end{aligned} \tag{104}$$

and are analyzed in Refs. 2 and 93. In Ref. 93 typical dependences of squeezing on the initial field are shown graphically for an initial Gaussian pure state. Some constraints for the initial field are found, under which squeezing attains its maximum. The relationship between squeezing and revivals of the coherent part of the field under study is also mentioned. The attenuation of the quantum coherence for moderate damping and moderate quantum fluctuations is also shown graphically. In Fig. 8 we can trace the time dependence of the standard squeezing variances $\langle(\Delta\hat{q})^2\rangle, \langle(\Delta\hat{p})^2\rangle$ and the principal squeezing variance λ_2 for various values of $|\xi(0)|^2$ ($\xi(0) = |\xi(0)|\exp(i\phi)$) and $|\nu|^2$ ($\nu = -|\nu|\exp(2i\theta)$) resulting in the same sum and two phases of ν. By comparison of other plots, it has been found that the symmetry of the curves is typical of real ν and the initial squeezed vacuum (Fig. 8a), and the asymmetry can be observed for $\xi(0) \neq 0$ (Fig. 8b and c). When the process starts from the squeezed vacuum, the period of the principal squeezing variance λ_2 is $\pi/(2\kappa)$ since $\langle\hat{a}(t)\rangle = 0$ in this case. This period reflects the two-photon property of the state. Time intervals shorter than this period are chosen for convenience in plotting graphs.

The squeezed and shifted Fock states $|\beta, m\rangle_g$, $\beta = \mu\alpha_0 + \nu\alpha_0^*$ [94,95] were considered at the input of the dissipative third-order nonlinear oscillator in Ref. 96. The appropriate quasidistribution for the antinormal ordering of field operators reads

$$\begin{aligned}\Phi_{\mathscr{A}}(\alpha,0) &= \frac{1}{\pi m!}\left(\frac{|\nu|}{2|\mu|}\right)^m \left|H_m\left(\frac{\alpha^* - \alpha_0^*}{\sqrt{-2\mu\nu^*}}\right)\right|^2 \\ &\quad\times \frac{1}{|\mu|}\exp\Bigg[-|\alpha|^2 - |\beta|^2 \\ &\qquad + \left(-\frac{\nu}{2\mu}\alpha^{*2} + \frac{\nu^*}{2\mu}\beta^2 + \frac{1}{\mu}\alpha^*\beta + \text{c.c.}\right)\Bigg]\end{aligned} \tag{105}$$

where $H_m(x)$ are the Hermite polynomials. In the conservative case, the quasidistribution $\Phi_{\mathscr{A}}(\alpha, l)$, where l is the length of the medium, is determined and the numerical calculation of the photon number distribution $p(n, l)$ and its factorial moments $\langle W^k(l)\rangle_{\mathscr{N}}$ is outlined. The behavior of the quasidistribution of the phase of the field amplitude was examined in the dependence on α_0. The influence of quantum fluctuations was involved approximately in $\Phi_{\mathscr{A}}(\alpha, l)$. The evolution of the displaced Fock states $|\beta, m\rangle$ in the anharmonic oscillator with quiet reservoir was studied in Ref. 97. This state arrives at some time at the superposition of two

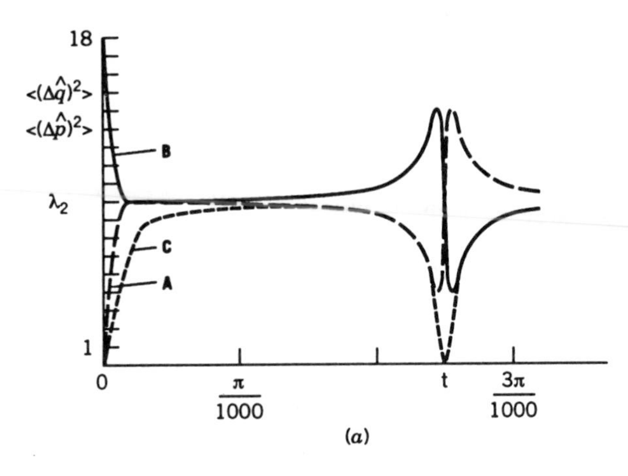

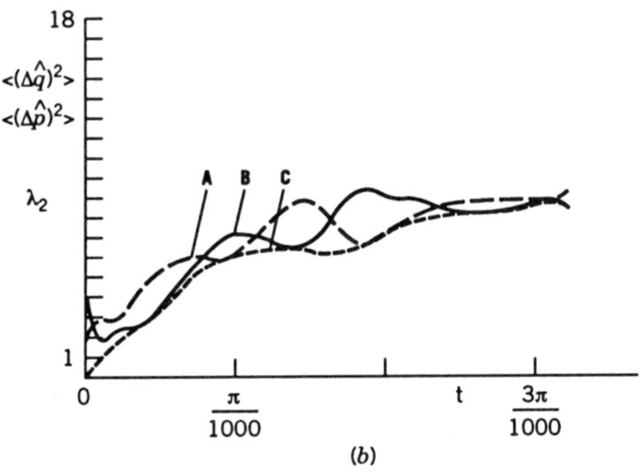

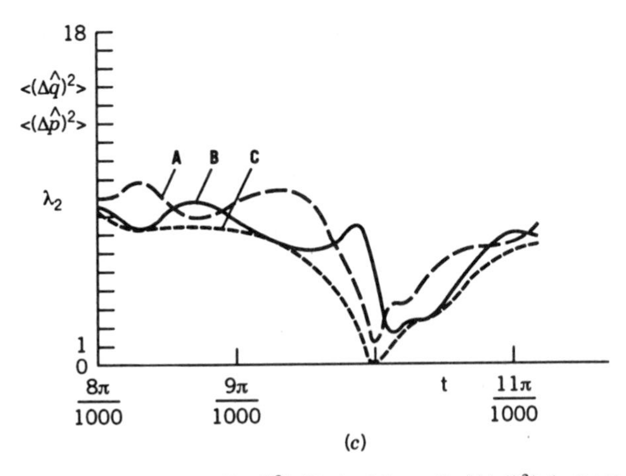

Figure 8. Time dependence of $\langle(\Delta\hat{q})^2\rangle$ (dashed line, A), $\langle(\Delta\hat{p})^2\rangle$ (solid line, B), and λ_2 (dotted line, C) for $\phi = 0$, $\kappa = 100$, $\gamma = \bar{n} = 0$: (a) $|\xi(0)|^2 = 0$, $|\nu|^2 = 4$, $\theta = \pi/2$; (b), (c) $|\xi(0)|^2 = 3$, $|\nu|^2 = 1$, $\theta = \pi/3$ (different time intervals).

displaced Fock states. Losses prevent the formation of the superposition states. The standard squeezing is also investigated.

The conservative third-order nonlinear oscillator with the SU(1, 1) coherent state, the SU(1, 1) generalized coherent state, and the k-photon coherent state at the input was analyzed in Refs. 98, 99, and 100, respectively, from the standard squeezing viewpoint.

The outline of achieved results shows great effort in the study of nonclassical properties of the Kerr medium. With the aid of the quasidistribution related to the antinormal ordering of field operators, substantial differences have been found between classical and quantum behavior. It has been shown that the initial coherent state evolves in the Kerr medium at some interaction times into the superposition of two different coherent states, which is a pure quantum effect. Conditions have been found under which the Kerr medium can produce amplitude squeezed states and number-phase squeezed states. Quantum features have been confirmed by the study of fluctuations of the photon number and phase.

The results of theoretical study motivated the use of the Kerr medium for generation of nonclassical states [17, 81]. Quantum nondemolition measurement of the number of photons based on the optical Kerr effect was described in Refs. 101–103.

2. *Symbolic Computational Method*

In this section we present an alternative method for calculating the photon statistics of multiwave mixing processes. The method is based on symbolic algebraic computations.

We implemented the standard boson operator algebra rules in our computer program [104, 105], which can calculate the normally ordered operator Taylor series of selected moments. The numerical outputs are easily obtained by insertion of the input statistics of the interacting fields. In this way, one can obtain important analytical or graphical results.

The time evolution of any operator $\hat{M}$ in the interaction picture is described by the Heisenberg equation:

$$i\hbar \frac{\mathrm{d}\hat{M}}{\mathrm{d}t} = \left[\hat{M}, \hat{H}\right]$$

One can easily find the formal solution by assuming temporal independence of the Hamiltonian $\hat{H}$:

$$\hat{M}(t) = \hat{U}^{+}(t)\hat{M}(0)\hat{U}(t) \tag{106}$$

where $\hat{U}(t) = \exp(\hat{H}t/i\hbar)$ is unitary evolution operator. Formula (106) can be expanded in Taylor series

$$\hat{M}(t) = \sum_{k=0}^{\infty} \left(\frac{t}{i\hbar}\right)^k \frac{\hat{D}_k}{k!} \tag{107}$$

where $\hat{D}_k$ is the kth order commutator defined recursively by means of

$$\hat{D}_k = \left[\hat{D}_{k-1}, \hat{H}\right] \qquad \hat{D}_0 = \hat{M}(0) \tag{108}$$

The Taylor series will be assumed to be convergent due to natural physical reasons. Some conditions for the power series convergence in the second harmonic generation can be found in Ref. 106. Good behavior of this series was also verified by the numerical results in Refs. 104 and 105.

Symbolic computations after formulas (107) and (108) will give the operator solutions for suitable moments in the form of normally ordered power series (in our cases to the 16th power in time). From these results one can gain either numerical and graphical outputs or useful analytical formulas for optimum phase conditions.

This method was applied to the second and higher harmonic generation [104, 105]. In general the process of the sth-order harmonic generation can be described by the interaction Hamiltonian:

$$\hat{H} = \frac{\hbar g}{s}\left(\hat{a}_1^s \hat{a}_2^\dagger + \hat{a}_1^{\dagger s} \hat{a}_2\right)$$

where $\hat{a}_1$ and $\hat{a}_2$ are annihilation operators of the fundamental and the s-harmonic modes, respectively. Neglecting losses one can prove the following conservation law:

$$\hat{C} = \hat{n}_1 + s\hat{n}_2 = \text{constant operator}$$

The short-time approximation gives (up to gt)

$$n_1(t) = n_1 - 2gt|\alpha_1|^s|\alpha_2|\sin(s\phi_1 - \phi_2)$$

and

$$n_2(t) = n_2 + \frac{2}{s}gt|\alpha_1|^s|\alpha_2|\sin(s\phi_1 - \phi_2)$$

Therefore, the optimum phase condition for producing the sth harmonics is $s\phi_1 - \phi_2 = \pi/2$, where we assumed the coherent input fields $|\alpha_1\rangle = \||\alpha_1|\exp(i\phi_1)\rangle$ and $|\alpha_2\rangle = \||\alpha_2|\exp(i\phi_2)\rangle$.

We will start with the stimulated processes; i.e., we assume the coherent initial signals $|\alpha_1| \neq 0$ and $|\alpha_2| \neq 0$. We show the first nonzero terms only. Variance of the fundamental mode can be obtained for an arbitrary s,

$$\langle (\Delta W_1)^2 \rangle = -2gt(s-1)|\alpha_1|^s|\alpha_2|\sin(s\phi_1 - \phi_2)$$

but variance of the sth harmonics is more complicated and can be simply given for some small s. For example,

$$\langle (\Delta W_2)^2 \rangle = -\tfrac{2}{3}(gt)^3|\alpha_1|^2|\alpha_2|^3\sin(2\phi_1 - \phi_2) \qquad s = 2$$

$$\langle (\Delta W_2)^2 \rangle = -\tfrac{4}{3}(gt)^3|\alpha_1|^3|\alpha_2|^3\left(|\alpha_1|^2 + 2\right)\sin(3\phi_1 - \phi_2) \qquad s = 3$$

$$\langle (\Delta W_2)^2 \rangle = -(gt)^3|\alpha_1|^4|\alpha_2|^3\left(2|\alpha_1|^4 + 12|\alpha_1|^2 + 17\right)\sin(4\phi_1 - \phi_2) \qquad s = 4$$

$$\langle (\Delta W_2)^2 \rangle = -(gt)^3|\alpha_1|^5|\alpha_2|^3\left(\tfrac{8}{3}|\alpha_1|^6 + 32|\alpha_1|^4 + 128|\alpha_1|^2 + 160\right) \times \sin(5\phi_1 - \phi_2) \qquad s=5$$

It is interesting that both modes are simultaneously sub-Poissonian under the phase condition $s\phi_1 - \phi_2 = \pi/2$, even if the sub-Poissonian effect in the s-order harmonic mode is weaker by two orders compared to the fundamental mode. On the other hand, in the course of s-subharmonic generation $s\phi_1 - \phi_2 = -\pi/2$, both the modes are super-Poissonian.

As an illustration of the power of the method we present Fig. 9, where one can see the temporal dependence of $\langle(\Delta W_1)^2\rangle$ for the second and the third harmonic generation for initially coherent light with $|\alpha_1| = |\alpha_2| = 1$. The optimum phase conditions were chosen to produce the nonclassical light. The full curves represent the 15th- and 16th-order approximations and the dashed curves are the quadratic approximations which could be simply reached by manual algebra manipulations. One can see that the dependences for second and third subharmonics are qualitatively similar.

Formulas obtained for principal squeezing are for the fundamental mode [108, 109]:

$$\lambda_2^{(1)} = 1 - 2(s-1)gt|\alpha_1|^{s-2}|\alpha_2|$$

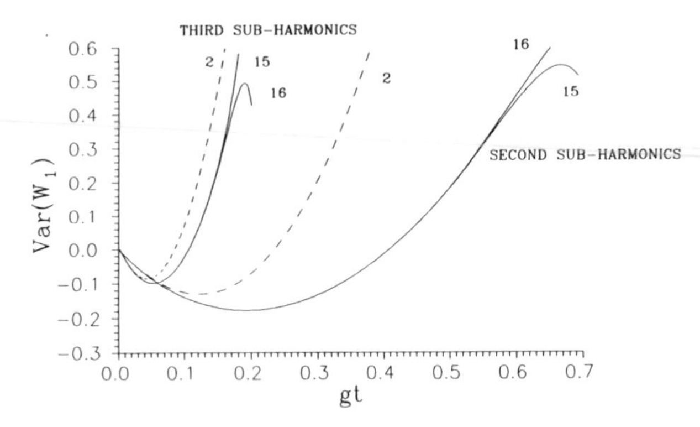

Figure 9. Time dependence of $\langle(\Delta W_1)^2\rangle$ for the second and third harmonics generation for initially coherent light with $|\alpha_1| = |\alpha_2| = 1$. The optimum phase conditions $s\phi_1 - \phi_2 = \pi/2$, $s = 2, 3$ were chosen to obtain nonclassical light. The full curves represent the 15th- and 16th-order approximations and the dashed curves are the quadratic approximations. One can see that both the dependences are qualitatively similar.

For the harmonics we can obtain

$$\lambda_2^{(2)} = 1 - \tfrac{2}{3}(gt)^3|\alpha_1|^2|\alpha_2| \qquad s = 2$$

$$\lambda_2^{(2)} = 1 - \tfrac{4}{3}(gt)^3\big(|\alpha_1|^2 + 2\big)|\alpha_1|^3|\alpha_2| \qquad s = 3$$

Thus, both the modes are principally squeezed independently of phases and the order of harmonics.

When one of the initial amplitudes is zero, we obtain the spontaneous processes and then must take higher order terms into account. We then obtain the following:

1. If the initial harmonics amplitude is zero, $|\alpha_2| = 0$ (s-harmonic generation), then for any s [112, 113]

$$\left\langle(\Delta W_1)^2\right\rangle = -\frac{s-1}{s}(gt)^2|\alpha_1|^{2s}$$

For the harmonics we give the first two formulas only:

$$\left\langle(\Delta W_2)^2\right\rangle = -\tfrac{1}{48}(gt)^6|\alpha_1|^8 \qquad s = 2$$

$$\left\langle(\Delta W_2)^2\right\rangle = -\tfrac{1}{81}(gt)^6|\alpha_1|^{12}\big(|\alpha_1|^2 + 2\big) \qquad s = 3$$

obtained in [110] and [112], respectively, which means that both the modes are sub-Poissonian independently of the values of initial phases. The effect in the harmonics is weaker by four orders in gt compared to the fundamental mode.

For principal squeezing in fundamental mode we get [114]

$$\lambda_2^{(1)} = 1 - \frac{s-1}{s}(gt)^2|\alpha_1|^{2s-2}$$

and for the harmonics

$$\lambda_2^{(2)} = 1 - \tfrac{1}{12}(gt)^4|\alpha_2|^4 \qquad s = 2$$

$$\lambda_2^{(2)} = 1 - \tfrac{1}{9}(gt)^4\left(|\alpha_1|^2 + 2\right)|\alpha_1|^6 \qquad s = 3$$

obtained in [109] and [114], respectively. Both modes are again simultaneously squeezed, and the effect in the harmonics is weaker by two orders compared to the fundamental mode.

2. If the subfrequency mode amplitude is initially zero, $|\alpha_1| = 0$ (we speak of the sth subharmonic generation process), then for $s = 2$

$$\left\langle (\Delta W_1)^2 \right\rangle = (gt)^2|\alpha_2|^2$$

and

$$\left\langle (\Delta W_2)^2 \right\rangle = \tfrac{1}{6}(gt)^4|\alpha_2|^4$$

i.e., both the modes are super-Poissonian. For principal squeezing we obtain [109]

$$\lambda_2^{(1)} = 1 - 2gt|\alpha_2|$$

$$\lambda_2^{(2)} = 1 - \tfrac{1}{6}(gt)^4|\alpha_2|^2$$

and squeezing can be observed.

Further for $s = 3$ we obtain

$$\left\langle (\Delta W_1)^2 \right\rangle = 4(gt)^2|\alpha_2|^2$$

$$\left\langle (\Delta W_2)^2 \right\rangle = \tfrac{4}{3}(gt)^4|\alpha_2|^4$$

and both modes are super-Poissonian independently of values of the initial phase. This case is connected with decreasing the harmonics, i.e., $dn_2/dt < 0$.

For principal squeezing we derived in this case

$$\lambda_2^{(1)} = 1 + 4(gt)^2|\alpha_2|^2$$

$$\lambda_2^{(2)} = 1 - \tfrac{4}{3}(gt)^4|\alpha_2|^2$$

From here we see that in the course of third harmonic generation the fundamental mode cannot be squeezed. One can easily generalize this conclusion to any $s \geq 3$.

Squeezed light was produced in a number of laboratories (see, e.g., [1, Chap. 11], [10], [115] for further discussion); therefore, it is interesting to discuss the sth harmonic generation processes with squeezed inputs. A discussion of this can be found in Ref. 105, where it is also demonstrated that the ability of nonlinear dynamics to produce nonclassical light decreases with increasing level of nonclassical behavior at the beginning.

V. CONCLUSIONS

We have summarized the basic tools for the research of quantum statistical properties of light interacting in nonlinear optical processes, such as the generalized superposition of coherent fields and quantum noise, photon-number generating function and distribution, its factorial moments, and principal squeezing variance, and we have discussed oscillations in the photon-number distribution. Nonlinear optical processes of three- and four-wave mixing, optical phase conjugation, stimulated Raman and hyper-Raman scattering, and Nth subharmonic generation have been examined, particularly from the point of view of their nonclassical behavior in the framework of Gaussian approximations, assuming either strong classical pumping or linear operator corrections to the stationary solution. Results for the anharmonic oscillator with Kerr nonlinearity have also been reviewed, including a closed non-Gaussian solution for arbitrary initial statistics and quantum-phase considerations. Finally, the symbolic computational method has been explained, providing the possibility to obtain the quantum moments to any order in the coupling constant. The method has been demonstrated using s-harmonic generation.

References

1. J. Peřina, *Quantum Statistics of Linear and Nonlinear Optical Phenomena*, Kluwer, Dordrecht, 1991.
2. V. Peřinová, J. Křepelka, J. Peřina, *Opt. Acta* **33** 1263 (1986); S. Rai, *J. Opt. Soc. Am.* **B9**, 590 (1992).
3. V. Peřinová and J. Peřina, *Opt. Acta* **28**, 769 (1981).

4. V. Peřinová, *Opt. Acta* **28**, 747 (1981).
5. M. Born and E. Wolf, *Principles of Optics*, Pergamon, Oxford, UK 1965.
6. A. Lukš, V. Peřinová, and J. Peřina, *Opt. Commun.* **67**, 149 (1988).
7. A. Lukš, V. Peřinová, and Z. Hradil, *Acta Phys. Pol. A* **74**, 713 (1988).
8. R. Loudon, *Opt. Commun.* **70** 109 (1989).
9. R. Tanaś, A. Miranowicz, and S. Kielich, *Phys. Rev. A* **43**, 4014 (1991).
10. S. Reynaud, A. Heidemann, E. Giacobino, and C. Fabre, in E. Wolf (Ed.), *Progress in Optics*, Vol. 30, Elsevier, Amsterdam, 1992, p. 1.
11. M. Kozierowski and V. I. Man'ko, *Opt. Commun.* **69**, 71 (1988) and references therein.
12. J. Peřina and J. Bajer, *Phys. Rev. A* **41**, 516 (1990).
13. W. Schleich, D. F. Walls, and J. A. Wheeler, *Phys. Rev. A* **38**, 1177 (1988) and references therein.
14. J. R. Klauder and E. C. Sudarshan, *Fundamentals of Quantum Optics*, Benjamin, New York, 1968.
15. D. F. Walls, *Nature* **306**, 141 (1983).
16. R. S. Bondurant and J. H. Shapiro, *Phys. Rev. D* **30**, 2548 (1984).
17. M. Kitagawa and Y. Yamamoto, *Phys. Rev. A* **34**, 3974 (1986).
18. Z. Hradil, *Acta Phys. Pol. A* **75**, 731 (1989); Z. Hradil, *Phys. Rev. A* **41**, 400 (1990).
19. Z. Hradil, *Phys. Rev. A* **44**, 792 (1991).
20. Z. Hradil, *Quant. Opt.* **4**, 93 (1992).
21. R. A. Fisher, *Optical Phase Conjugation*, Academic, New York, 1983.
22. G. J. Milburn, D. F. Walls, and M. D. Levenson, *J. Opt. Soc. Am. B* **3**, 390 (1984).
23. J. Bajer and J. Peřina, *Czech. J. Phys. B* **35**, 1146 (1985).
24. J. Bajer and J. Peřina, *Acta Phys. Pol. A* **71**, 149 (1987).
25. J. Bajer and J. Peřina, *Acta Phys. Pol. A* **74**, 111 (1988).
26. J. Bajer, *Czech. J. Phys. B* **40**, 646 (1990).
27. P. Kumar and J. H. Shapiro, *Phys. Rev. A* **30**, 1568 (1984).
28. J. Bajer, *Opt. Commun.* **74**, 233 (1990).
29. M. Hillery, *Opt. Commun.* **62**, 135 (1987); M. Hillery, *Phys. Rev. A* **36**, 3796 (1987).
30. C. K. Hong and L. Mandel, *Phys. Rev. Lett.* **54**, 323 (1985).
31. C. K. Hong and L. Mandel, *Phys. Rev. A* **32**, 974 (1985).
32. J. Bajer and J. Peřina, *Opt. Commun.* **85**, 261 (1991).
33. A. L. Gaeta and R. W. Boyd, *Phys. Rev. Lett.* **60**, 2618 (1988).
34. A. L. Gaeta and R. W. Boyd, in J. H. Eberly, L. Mandel, and E. Wolf (Eds.), *Coherence and Quantum Optics VI*, Plenum, New York, 1990, p. 343.
35. A. Yariv and P. Yeh, *Optical Waves in Crystals*, Wiley, New York, 1984, Chap. 13.
36. E. J. Bochove, *J. Opt. Soc. Am.* **B9**, 266 (1992); H. Arnoldus and T. F. George, *J. Mod. Opt.* **38**, 1429 (1991); *Opt. Commun.* **88**, 127 (1992).
37. M. Kárská and J. Peřina, *J. Mod. Opt.* **37**, (1990) 195 and references therein.
38. J. Peřina, M. Kárská, and J. Křepelka, *Acta Phys. Pol. A* **79**, 817 (1991).
39. J. Peřina and J. Křepelka, *J. Mod. Opt.* **38**, 2137 (1991).
40. J. Peřina and J. Křepelka, *J. Mod. Opt.* **39**, 1029 (1992).
41. Yu. P. Malakyan, *Opt. Commun.* **78**, 67 (1990).

42. Yu. P. Malakyan, *J. Mod. Opt.* **39**, 509 (1992).
43. P. S. Gupta and J. Dash, *Opt. Commun.* **88**, 273 (1992).
44. G. S. Agarwal and G. Adam, *Phys. Rev. A* **39**, 6259 (1989).
45. P. D. Drummond, K. J. McNeil, and D. F. Walls, *Opt. Acta* **27**, 321 (1980).
46. P. D. Drummond, K. J. McNeil, and D. F. Walls, *Opt. Acta* **28**, 211 (1981).
47. D. F. Walls, P. D. Drummond, and K. J. McNeil, in Ch. M. Bowden, M. Ciftan, and H. R. Robl, (Eds.), *Optical Bistability*, Plenum, New York, 1981, p. 51.
48. P. D. Drummond, K. J. McNeil, and D. F. Walls, *Opt. Commun.* **28**, 255 (1979).
49. D. F. Walls and G. J. Milburn, in P. Meystre and M. O. Scully (Eds.), Quantum Optics, Experimental Gravity and Measurement Theory, Plenum, New York, 1983, p. 209.
50. R. Vyas and S. Singh, *Phys. Rev. A* **40**, 5147 (1989).
51. R. Vyas and S. Singh, *Opt. Lett.* **14**, 1110 (1989).
52. M. D. Reid and P. D. Drummond, *Phys. Rev. A* **40**, 4493 (1989).
53. M. D. Reid and P. D. Drummond, *Phys. Rev. Lett.* **60**, 2731 (1988).
54. J. Bajer, *J. Mod. Opt.* **38**, 1085 (1991).
55. J. Peřina, *Coherence of Light*, Reidel, Dordrecht, 1985, Chap. 13.
56. H. Risken, *The Fokker-Planck Equation*, Springer, Berlin, 1984, pp. 156, 38.
57. P. D. Drummond and D. F. Walls, *J. Phys. A* **13**, 725 (1980).
58. R. Tanaś, in L. Mandel and E. Wolf (Eds.), *Coherence and Quantum Optics V*, Plenum, New York, 1984, p. 645.
59. G. J. Milburn, *Phys. Rev. A* **33**, 674 (1986).
60. F. Haake, H. Risken, C. Savage, and D. F. Walls, *Phys. Rev. A* **34**, 3969 (1986).
61. G. J. Milburn and C. A. Holmes, *Phys. Rev. Lett.* **56**, 2237 (1986).
62. V. Peřinová and A. Lukš, *J. Mod. Opt.* **35**, 1513 (1988).
63. G. J. Milburn, A. Mecozzi, and P. Tombesi, *J. Mod. Opt.* **36**, 1607 (1989).
64. D. J. Daniel and G. J. Milburn, *Phys. Rev. A* **39**, 4628 (1989).
65. V. Peřinová and A. Lukš, *Phys. Rev. A* **41**, 414 (1990).
66. S. Chaturvedi and V. Srinivasan, *J. Mod. Opt.* **38**, 777 (1991).
67. A. Lukš and V. Peřinová, *Czech. J. Phys. B* **37**, 1224 (1987).
68. D. T. Pegg and S. M. Barnett, *Europhys. Lett.* **6**, 483 (1988).
69. S. M. Barnett and D. T. Pegg, *J. Mod. Opt.* **36**, 7 (1989).
70. D. T. Pegg and S. M. Barnett, *Phys. Rev. A* **39**, 1665 (1989).
71. R. Loudon, *The Quantum Theory of Light*, Clarendon, Oxford, UK, 1973.
72. L. Susskind and J. Glogower, *Physics* **1**, 49 (1964).
73. P. Carruthers and M. M. Nieto, *Rev. Mod. Phys.* **40**, 411 (1968).
74. A. Lukš and V. Peřinová, *Czech. J. Phys.* **41**, 1205 (1991).
75. S. M. Barnett and D. T. Pegg, *J. Phys. A* **19**, 3849 (1986).
76. A. Miranowicz, R. Tanaś, and S. Kielich, *Quant. Opt.* **2**, 253 (1990).
77. V. Bužek, *Phys. Lett. A* **136**, 188 (1989).
78. C. C. Gerry and S. Rodriguez, *Phys. Rev. A* **35**, 4440 (1987).
79. R. Tanaś, *Phys. Rev. A* **38**, 1091 (1988).
80. C. C. Gerry and E. R. Vrscay, *Phys. Rev. A* **37**, 1779 (1988).

81. Y. Yamamoto, S. Machida, N. Imoto, M. Kitagawa, and G. Björk, *J. Opt. Soc. Am. B* **4**, 1645 (1987).
82. A. Lukš, V. Peřinová, and J. Křepelka, *Czech. J. Phys.* **42**, 59 (1992).
83. C. C. Gerry, *Opt. Commun.* **75**, 168 (1990).
84. Ts. Gantsog and R. Tanaś, *J. Mod. Opt.* **38**, 1021 (1991).
85. R. Tanaś, Ts. Gantsog, A. Miranowicz, and S. Kielich, *J. Opt. Soc. Am. B* **8**, 1576 (1991).
86. C. C. Gerry, *Opt. Commun.* **63**, 67 (1988).
87. R. Lynch, *Opt. Commun.* **67**, 67 (1988).
88. Ts. Gantsog and R. Tanaś, *Phys. Rev. A* **44**, 2086 (1991).
89. Z. Białynicka-Birula, *Phys. Rev.* **173**, 1207 (1968).
90. P. Tombesi and A. Mecozzi, *J. Opt. Soc. Am. B* **4**, 1700 (1987).
91. B. Yurke and D. Stoler, *Phys. Rev. Lett.* **57**, 13 (1986).
92. Ts. Gantsog and R. Tanaś, *Quant. Opt.* **3**, 33 (1991).
93. V. Peřinová, A. Lukš, and M. Kárská, *J. Mod. Opt.* **37**, 1055 (1990).
94. M. S. Kim, F. A. M. de Oliveira, and P. L. Knight, *Opt. Commun.* **72**, 99 (1989).
95. P. Král, *J. Mod. Opt.* **37**, 889 (1990).
96. P. Král, *Phys. Rev. A* **42**, 4177 (1990).
97. M. Brisudová, *J. Mod. Opt.* **38**, 2505 (1991).
98. V. Bužek, *Phys. Rev. A* **39**, 5432 (1989).
99. V. Bužek, *Acta Phys. Slov.* **39**, 344 (1989).
100. V. Bužek and I. Jex, *Acta Phys. Slov.* **39**, 351 (1989).
101. N. Imoto, H. A. Haus, and Y. Yamamoto, *Phys. Rev. A* **32**, 2287 (1985).
102. G. S. Agarwal, *Opt. Commun.* **72**, 253 (1989).
103. M. Hillery, *Phys. Rev. A* **44**, 4578 (1991).
104. J. Bajer and P. Lisoněk, *J. Mod. Opt.* **38**, 719 (1991).
105. J. Bajer and J. Peřina, *Opt. Commun.* **92**, 99 (1992).
106. V. Peřinová and J. Peřina, *Czech. J. Phys. B* **28**, 306 (1978).
107. D. Stoler, *Phys. Rev. Lett.* **33**, 1397 (1974).
108. L. Mandel, *Phys. Rev. Lett.* **49**, 136 (1982); *Opt. Commun.* **42**, 437 (1982).
109. J. Peřina, V. Peřinová, and J. Kod'ousek, *Opt. Commun.* **49**, 210 (1984).
110. M. Kozierovski and R. Tanaś, *Opt. Commun.* **21**, 229 (1977).
111. L. Mišta and J. Peřina, *Acta Phys. Pol. A* **52**, 425 (1977).
112. S. Kielich, M. Kozierowski, and R. Tanaś, in L. Mandel and E. Wolf (Eds.), *Coherence and Quantum Optics IV*, Plenum, New York, 1978, p. 511.
113. M. Hofman, *Acta Univ. Palackianae* **61**, 35 (1980).
114. M. Kozierowski, A. A. Mamedov, V. I. Man'ko, and S. M. Chumakov, *Trudy Fiz. Inst. Lebedeva* **200**, 106 (1991) (in Russian).
115. M. C. Teich and B. E. A. Saleh, in E. Wolf (Ed.), *Progress in Optics*, Vol. 26, Elsevier, Amsterdam, 1988, p. 3.

QUANTUM RESONANCE FLUORESCENCE FROM MUTUALLY CORRELATED ATOMS

Z. FICEK

Department of Physics, The University of Queensland, Brisbane, Australia

R. TANAŚ

Nonlinear Optics Division, Institute of Physics, Adam Mickiewicz University, Poznań, Poland

CONTENTS

I. INTRODUCTION

Photon antibunching and squeezing are two unique phenomena that reveal the quantum properties of the radiation field. These effects are just two examples of nonclassical light, that is, light with properties that are not predicted by the classical wave theory of light. According to quantum mechanics, electrons in atoms can occupy only certain energy levels. An electron can jump from its lowest energy level—the ground state—to a second higher energy level by absorbing light of definite frequency from a pumping beam. Next, the electron can fall back to the ground state,

Modern Nonlinear Optics, Part 1, Edited by Myron Evans and Stanisław Kielich. Advances in Chemical Physics Series, Vol. LXXXV.
ISBN 0-471-57546-1

emitting a photon. It has been predicted from the statistics of quantum theory that once the electron had returned to the ground state, there would be a delay before the electron would be re-excited by the pumping beam. This delay would result in an intermittent emission of light called photon antibunching. This intermittent emission produces a radiation in which the variance of the number of photons is less than the mean number of photons. The classical theory of electromagnetic radiation, which does not quantize energy, does not predict antibunching.

Yuen [1] has predicted the possibility of another nonclassical phenomenon, squeezed light. The Heisenberg uncertainty principle predicts that it is never possible to be absolutely precise in measuring one of two noncommuting observables. The product of the fluctuations of the two noncommuting observables must be greater than or equal to one-half of the absolute value of their commutator. For all field states that have classical analog the field quadrature variances are also greater than or equal to this commutator. For the vacuum state and coherent states the noise in the two noncommuting field quadratures is distributed symmetrically between the two quadratures and the variance of the field quadrature is equal to the commutator establishing the level of quantum noise (vacuum fluctuations). There are, however, quantum states of the field such that the variance of one of two noncommuting field observables is smaller than the vacuum fluctuations. Such a field is referred to as squeezed light. In squeezed light the quantum fluctuations in one quadrature component are reduced below their vacuum values at the expense of increased fluctuations in the other component, such that the uncertainty relation is not violated.

Photon antibunching has been predicted theoretically for the first time in resonance fluorescence of a two-level atom [2, 3]. Since then, a number of papers have appeared analyzing the possibilities of obtaining photon antibunching in various processes offered by nonlinear optics [4–11]. Significant contribution to these studies has been given by Kielich and co-workers [12–17]. The possibility of obtaining squeezed light has been extensively studied since the first theoretical papers by Walls and Zoller [18] and Mandel [19] on reduction of noise and photon statistics in resonance fluorescence of a two-level atom. Many linear and nonlinear processes have predicted a large amount of reduction of noise below the classical limit. Almost a complete reduction of noise (98% below the vacuum limit) was found by Tanaś and Kielich [20] in a self-squeezed light produced by a propagation of a coherent laser beam in a nonlinear medium.

Several experimental groups have been successful in producing nonclassical light. However, photon antibunching has been observed only in fluorescing sodium atoms [21]. This was the first experiment in which the nonclassical effect was observed in optics. A number of groups have been actively involved in the actual generation of squeezed light. Slusher et al. [22] generated for the first time a squeezed light in which a 7% noise reduction below the vacuum limit was observed. Wu et al. [23] reported more than 50% reduction of noise below the vacuum limit in an optical oscillator. Heidmann et al. [24] used a two-mode optical parametric oscillator operating above threshold to generate two highly correlated beams of light. The measured noise in the intensity difference of the two beams was 30% below the classical limit. In an improved experiment, Debuisschert et al. [25] observed a 69% noise reduction in the intensity difference. Yamamoto et al. [26] developed semiconductor lasers with intensity fluctuations reduced by 95% below the noise level of usual lasers.

The interest in the investigation and generation of squeezed light is due not only to the reduction of the natural noise of light but also to the possibilities of practical applications. For example, squeezed light may be useful in detecting gravitational waves, which would require a very sensitive detector operating at a very low noise level. Squeezed light could also be useful in optical communications, where it might be important to cut down noise, and in making sensitive spectroscopic measurements, for example, in biological samples.

As mentioned above, photon antibunching and squeezing are two nonclassical effects predicted in resonance fluorescence of two-level atoms. In this paper we review the past work and the present status of photon antibunching and squeezed light produced in resonance fluorescence. In Section II, we derive the master equation for two-level atoms interacting with a quantized electromagnetic field. In Section III, we give the definitions of photon antibunching and squeezing. Section IV deals with photon antibunching and squeezing in spontaneous emission and resonance fluorescence of a single two-level atom. In Section V, we discuss the effect of the interatomic interactions on the two nonclassical effects. In Section VI, we present new results on squeezing in two-atom spontaneous emission. These results show that the interatomic interactions can create squeezed light in spontaneous emission if the atoms were initially prepared in a linear superposition of their ground and excited states. Finally, in Section VII, we summarize our results.

II. MASTER EQUATION

We consider a collection of N identical nonoverlapping atoms, separated by distances r_{ij} $(i \neq j)$ and interacting with a quantized multimode electromagnetic field. Each atom is modeled as a two-level system with the ground state $|g_i\rangle$ $(i = 1, 2, \ldots, N)$ and the excited state $|e_i\rangle$. In the electric dipole approximation the Hamiltonian of this system has the following form:

$$H = H_0 + H_{\text{int}} \tag{1}$$

with

$$H_0 = \hbar\omega_0 \sum_{i=1}^{N} S_i^z + \hbar \sum_{\mathbf{k},s} \omega_{\mathbf{k}} a_{\mathbf{k}s}^{\dagger} a_{\mathbf{k}s} \tag{2}$$

and

$$H_{\text{int}} = \mathrm{i}\hbar \sum_{\mathbf{k},s} \sum_{i=1}^{N} \left[\boldsymbol{\mu}_i \cdot \mathbf{g}_{\mathbf{k}s}(\mathbf{r}_i) a_{\mathbf{k}s} (S_i^+ + S_i^-) - \text{h.c.} \right] \tag{3}$$

where ω_0 is the atomic transition frequency, s is the polarization index $(s = 1, 2)$, $S_i^+ = |e_i\rangle\langle g_i|$ and $S_i^- = |g_i\rangle\langle e_i|$ are operators raising and lowering the energy of ith atom, and S_i^z describes its energy. These operators fulfill the well-known commutation relations

$$\left[S_i^+, S_j^- \right] = 2S_i^z\, \delta_{ij} \qquad \left[S_i^z, S_j^{\pm} \right] = \pm S_i^{\pm}\, \delta_{ij} \tag{4}$$

In Eq. (3), $\mathbf{g}_{\mathbf{k}s}(\mathbf{r}_i)$ is the coupling constant between the quantized electromagnetic field and the electric dipole moments $\boldsymbol{\mu}_i = \langle e_i | \boldsymbol{\mu} | g_i \rangle$, and is given by

$$\mathbf{g}_{\mathbf{k}s}(\mathbf{r}_i) = \left(\frac{2\pi\omega_k}{\hbar V} \right)^{1/2} \hat{e}_{\mathbf{k}s}\, \mathrm{e}^{\mathrm{i}\mathbf{k}\cdot\mathbf{r}_i} \tag{5}$$

where $\hat{e}_{\mathbf{k}s}$ is the unit polarization vector, $\mathbf{r}_i$ is a coordinate of the ith atom.

A master equation for the reduced density operator ρ of the N-atom system interacting with the quantized electromagnetic field is derived from the Hamiltonian (1). It can be derived using any of a number of traditional techniques [27].

We apply a Born-Markov method [28] adapted to the situation of a stationary reservoir. The time evolution of the density operator $W(t)$ of the atoms-field system in the interacting picture obeys the equation

$$i\hbar \frac{\partial}{\partial t} W^{\mathrm{I}}(t) = \left[H_{\text{int}}^{\mathrm{I}}(t), W^{\mathrm{I}}(t) \right] \tag{6}$$

where H_{int} is given by Eq. (3), and the superscript I stands for operators in the interacting picture.

Formally integrating Eq. (6) gives

$$W^{\text{I}}(t) = W^{\text{I}}(0) + \frac{1}{\text{i}\hbar}\int_0^t \text{d}t'\left[H_{\text{int}}^{\text{I}}(t'), W^{\text{I}}(t')\right] \tag{7}$$

Substituting this solution into the right side of Eq. (6), and taking the trace over the reservoir states of each side of Eq. (6), we get

$$\begin{aligned}\text{i}\hbar\frac{\partial}{\partial t}\rho^{\text{I}}(t) &= \text{Tr}_{\text{R}}\left[H_{\text{int}}^{\text{I}}(t), W^{\text{I}}(0)\right]\\ &\quad + \frac{1}{\text{i}\hbar}\int_0^t \text{d}t'\,\text{Tr}_{\text{R}}\left\{\left[H_{\text{int}}^{\text{I}}(t), \left[H_{\text{int}}^{\text{I}}(t'), W^{\text{I}}(t')\right]\right]\right\}\end{aligned} \tag{8}$$

where $\rho^{\text{I}}(t) = \text{Tr}_{\text{R}}\, W^{\text{I}}(t)$ is the reduced density operator of the atomic system.

We choose an initial state with no correlations between the atomic system and the quantized electromagnetic field, i.e., $W^{\text{I}}(0) = \rho^{\text{I}}(0)\rho_{\text{R}}(0)$, where $\rho_{\text{R}}(0)$ is the density operator for the field reservoir. We also assume that the interaction Hamiltonian satisfies the condition [29, 30]

$$\text{Tr}_{\text{R}}\left[H_{\text{int}}^{\text{I}}(t), \rho_{\text{R}}(0)\right] = 0 \tag{9}$$

This can easily be arranged. The left side of Eq. (9) is a system operator. If the left side of Eq. (9) is nonzero, the system Hamiltonian can be altered to include any part in H_{int} so that when added to the left side of Eq. (9) zero occurs. On the basis of these assumptions Eq. (8) reduces to

$$\frac{\partial}{\partial t}\rho^{\text{I}}(t) + \frac{1}{\hbar^2}\int_0^t \text{d}t'\,\text{Tr}_{\text{R}}\left\{\left[H_{\text{int}}^{\text{I}}(t), \left[H_{\text{int}}^{\text{I}}(t'), W^{\text{I}}(t')\right]\right]\right\} = 0 \tag{10}$$

We now employ the Born approximation in which the atom-field interaction is supposed to be weak, and there is no effect of the atoms on the reservoir. With this approximation we can write

$$W^{\text{I}}(t') = \rho^{\text{I}}(t')\rho_{\text{R}}(0) \tag{11}$$

and after changing time variable to $t' = t - \tau$, Eq. 10 simplifies to

$$\frac{\partial}{\partial t}\rho^{\mathrm{I}}(t) + \frac{1}{\hbar^2}\int_0^t \mathrm{d}\tau \, \mathrm{Tr}_{\mathrm{R}}\{[H_{\mathrm{int}}^{\mathrm{I}}(t), [H_{\mathrm{int}}^{\mathrm{I}}(t-\tau), \rho_{\mathrm{R}}(0)\rho^{\mathrm{I}}(t-\tau)]]\} = 0 \tag{12}$$

After a Laplace transform over time t, with Eq. (3) and assuming that all modes of the quantized electromagnetic field are in a vacuum state defined by

$$\begin{aligned} &\mathrm{Tr}_{\mathrm{R}}[\rho_{\mathrm{R}}(0)a_{\mathbf{k}s}^{\dagger}a_{\mathbf{k}'s'}] = 0 \qquad \mathrm{Tr}_{\mathrm{R}}[\rho_{\mathrm{R}}(0)a_{\mathbf{k}s}a_{\mathbf{k}'s'}^{\dagger}] = \delta^3(\mathbf{k}-\mathbf{k}')\,\delta_{ss'} \\ &\mathrm{Tr}_{\mathrm{R}}[\rho_{\mathrm{R}}(0)a_{\mathbf{k}s}a_{\mathbf{k}'s'}] = \mathrm{Tr}_{\mathrm{R}}[\rho_{\mathrm{R}}(0)a_{\mathbf{k}s}^{\dagger}a_{\mathbf{k}'s'}^{\dagger}] = 0 \end{aligned} \tag{13}$$

we obtain (ignoring the superscript I)

$$\begin{aligned} \rho(0) - z\rho(z) = &-\sum_{i,j}\gamma_{ij}(z)[\rho(z)S_i^+S_j^- + S_i^+S_j^-\rho(z) - 2S_j^-\rho(z)S_i^+] \\ &- \mathrm{i}\sum_i \Omega_{ii}(z)[S_i^+S_i^-, \rho(z)] \\ &- \mathrm{i}\sum_{i\neq j}\Omega_{ij}(z)[S_i^+S_j^-, \rho(z)] \end{aligned} \tag{14}$$

Here $\rho(z)$ is the Laplace transform of $\rho(t)$, and the parameters are given by

$$\begin{aligned} \gamma_{ij}(z) &= \frac{1}{c}\sum_s \int [\boldsymbol{\mu}\cdot\mathbf{g}_{\mathbf{k}s}(\mathbf{r}_i)][\boldsymbol{\mu}^*\cdot\mathbf{g}_{\mathbf{k}s}^*(\mathbf{r}_j)]\frac{z/c}{(z/c)^2 + (k_0 - k)^2}d^3\mathbf{k}, \\ \Omega_{ii}(z) &= \frac{1}{c}\sum_s \int |\boldsymbol{\mu}\cdot\mathbf{g}_{\mathbf{k}s}(\mathbf{r}_i)|^2\left[\frac{k - k_0}{(z/c)^2 + (k-k_0)^2} - \frac{k + k_0}{(z/c)^2 + (k+k_0)^2}\right]d^3\mathbf{k}, \end{aligned} \tag{15}$$

$$\begin{aligned} \Omega_{ij}(z) = &\frac{1}{c}\sum_s \int [\boldsymbol{\mu}^*\cdot\mathbf{g}_{\mathbf{k}s}^*(\mathbf{r}_i)][\boldsymbol{\mu}\cdot\mathbf{g}_{\mathbf{k}s}(\mathbf{r}_j)] \\ &\times\left[\frac{k - k_0}{(z/c)^2 + (k-k_0)^2} + \frac{k + k_0}{(z/c)^2 + (k+k_0)^2}\right]d^3\mathbf{k}, \end{aligned}$$

where z is the complex Laplace transform parameter, and $\boldsymbol{\mu} = \boldsymbol{\mu}_1 = \boldsymbol{\mu}_2$. To obtain Eq. (15) we have used the commutation relations (4) and made the rotating-wave approximation [31]; i.e., we neglected rapidly oscillating terms with frequency $2\omega_0$ (the so-called counter rotating terms).

Now we employ the Markov approximation. This neglects retardation effects [32] and is valid in the long-time limit $t \gg \omega_0^{-1}$, providing this is short compared with the typical relaxation times of the system, and is small in comparison with the time required for appreciable changes in population of the atomic levels, i.e.,

$$(r_{ij})_{\max} \ll c\,\Delta t \tag{16}$$

With these approximations we can replace the $\gamma_{ij}(z)$, $\Omega_{ii}(z)$, and $\Omega_{ij}(z)$ parameters by their limiting values as $z \to 0^+$. After this, the inverse Laplace transform of Eq. (14) leads to the master equation

$$\begin{aligned}\frac{\partial \rho}{\partial t} = &- \sum_{i,j} \gamma_{ij}\left(\rho S_i^+ S_j^- + S_i^+ S_j^- \rho - 2 S_j^- \rho S_i^+\right) \\ &- \mathrm{i} \sum_{i} \Omega_{ii}\left[S_i^+ S_i^-, \rho\right] - \mathrm{i} \sum_{i \neq j} \Omega_{ij}\left[S_i^+ S_j^-, \rho\right]\end{aligned} \tag{17}$$

where the coefficients in the equation are

$$\begin{aligned}\gamma_{ij} &= \frac{\pi k_0^2}{c} \int \mathrm{d}\Omega_k \sum_s \left[\boldsymbol{\mu}^* \cdot \mathbf{g}_{\mathbf{k}s}^*(\mathbf{r}_i)\right]\left[\boldsymbol{\mu} \cdot \mathbf{g}_{\mathbf{k}s}(\mathbf{r}_j)\right] \\ \Omega_{ii} &= \frac{2k_0}{c} \int \mathrm{d}k \frac{k^2}{(k-k_0)^2} \int \mathrm{d}\Omega_k \sum_s \left|\boldsymbol{\mu} \cdot \mathbf{g}_{\mathbf{k}s}(\mathbf{r}_i)\right|^2 \\ \Omega_{ij} &= \frac{1}{c} \int \mathrm{d}k \frac{2k^3}{(k-k_0)^2} \int \mathrm{d}\Omega_k \sum_s \left[\boldsymbol{\mu}^* \cdot \mathbf{g}_{\mathbf{k}s}^*(\mathbf{r}_i)\right]\left[\boldsymbol{\mu} \cdot \mathbf{g}_{\mathbf{k}s}(\mathbf{r}_j)\right]\end{aligned} \tag{18}$$

and $\Omega_k = (\theta_k, \varphi_k)$ is a solid angle over which the quantized electromagnetic field is distributed.

We now examine the values of the coefficients that appear in Eq. (18). On substituting Eq. (5) and on integrating over the total solid angle 4π we get

$$\gamma_{ij} = \frac{3}{2}\gamma\left\{\left[1 - \left(\hat{\mu}\cdot\hat{r}_{ij}\right)^2\right]\frac{\sin(k_0 r_{ij})}{k_0 r_{ij}} + \left[1 - 3\left(\hat{\mu}\cdot\hat{r}_{ij}\right)^2\right]\left[\frac{\cos(k_0 r_{ij})}{\left(k_0 r_{ij}\right)^2} - \frac{\sin(k_0 r_{ij})}{\left(k_0 r_{ij}\right)^3}\right]\right\} \tag{19}$$

$$\Omega_{ij} = \frac{3}{2}\gamma\left\{-\left[1 - \left(\hat{\mu}\cdot\hat{r}_{ij}\right)^2\right]\frac{\cos(k_0 r_{ij})}{k_0 r_{ij}} + \left[1 - 3\left(\hat{\mu}\cdot\hat{r}_{ij}\right)^2\right]\left[\frac{\sin(k_0 r_{ij})}{\left(k_0 r_{ij}\right)^2} + \frac{\cos(k_0 r_{ij})}{\left(k_0 r_{ij}\right)^3}\right]\right\} \tag{20}$$

where $2\gamma = 4k_0^3\mu^2/3\hbar$ is the Einstein A coefficient for spontaneous emission, $\hat{\mu}$ and $\hat{r}_{ij}$ are unit vectors along the transition electric dipole moment and the vector $\mathbf{r}_{ij}$, respectively. Moreover, $r_{ij} = |\mathbf{r}_{ij}|$ and $k_0 = \omega_0/c = 2\pi/\lambda$, where λ is the resonant wavelength.

The evaluation of Ω_{ii} is an involved problem. The term Ω_{ii} represents the part of the Lamb shift induced by the first-order coupling in the interaction Hamiltonian (3). After performing integrations, Ω_{ii} takes the following form:

$$\Omega_{ii} = -\frac{2\gamma}{\pi}\ln\left\{\left|\frac{\omega_c}{\omega_0} - 1\right|\left(\frac{\omega_c}{\omega_0} + 1\right)\right\} \tag{21}$$

where ω_c is the cutoff frequency. It is well known that to obtain a complete calculation of the Lamb shift, it is necessary to include a second-order, multilevel Hamiltonian including electron mass renormalization [33]. If these are included, the standard nonrelativistic vacuum-Lamb-shift result is obtained.

With the parameters (19)–(21) and on transforming Eq. (17) to the Schrödinger picture, the master equation reduces to

$$\frac{\partial\rho}{\partial t} = -\mathrm{i}\omega_0\sum_i\left[S_i^z, \rho\right] - \mathrm{i}\sum_{i\neq j}\Omega_{ij}\left[S_i^+S_j^-, \rho\right] - \sum_{ij}\gamma_{ij}\left(\rho S_i^+S_j^- + S_i^+S_j^-\rho - 2S_j^-\rho S_i^+\right) \tag{22}$$

where ω_0 is the renormalized frequency which is equal to the sum of the atomic frequency ω_0 and Ω_{ii}. The above master equation has been derived assuming that the atoms are coupled to the vacuum modes of the quantized electromagnetic field with no interaction and coupling to external fields. The derivation of the master equation is easily extended to take such interactions into account. With the external coherent laser field the master equation (22) takes the form [27, 34, 35]

$$\begin{aligned}\frac{\partial \rho}{\partial t} = & -\mathrm{i}\omega_0 \sum_i [S_i^z, \rho] - \mathrm{i} \sum_{i \neq j} \Omega_{ij} [S_i^+ S_j^-, \rho] \\ & - \sum_{ij} \gamma_{ij} (\rho S_i^+ S_j^- + S_i^+ S_j^- \rho - 2 S_j^- \rho S_i^+) \\ & - \frac{\mathrm{i}}{2} \sum_i \{[\rho, S_i^+]\Omega - [S_i^-, \rho]\Omega^*\}\end{aligned} \tag{23}$$

where $\Omega = \boldsymbol{\mu} \cdot \mathscr{E}_0/\hbar$ is the Rabi frequency describing a strength of interaction between the atoms and an external coherent field $\mathscr{E}_0$. The coefficient γ_{ij} is given by Eq. (19), and Ω_{ij} is given by Eq. (20). For $i \neq j$, they depend on the interatomic separation r_{ij} and describe collective properties of the multiatom system. For large interatomic separations $k_0 r_{ij}$ goes to infinity, and then γ_{ij} and Ω_{ij} go to zero; i.e., there is no coupling between the atoms. For small interatomic separation, $k_0 r_{ij} \ll 1$, and then γ_{ij} reduces to γ, and Ω_{ij} reduces to the static dipole–dipole potential [36] which, for $k_0 r_{ij} \to 0$, tends to infinity.

Equation (23) is the final form of the master equation and will play a basic role in our calculations of photon antibunching and squeezing in interaction of the atomic systems with the quantized electromagnetic field.

III. NONCLASSICAL STATES OF LIGHT

To determine the nonclassical states of light we define the normalized second-order correlation function and variances of the electromagnetic field $\mathbf{E}(\mathbf{r}, t) = \mathbf{E}^{(+)}(\mathbf{r}, t) + \mathbf{E}^{(-)}(\mathbf{r}, t)$. The normalized second-order two-times correlation function is determined by the relation [37, 38]

$$\begin{aligned}g^{(2)}(\mathbf{R}_1, t; \mathbf{R}_2, t+\tau) \\ = \frac{G^{(2)}(\mathbf{R}_1, t; \mathbf{R}_2, t+\tau)}{G^{(1)}(\mathbf{R}_1, t) G^{(1)}(\mathbf{R}_2, t+\tau)}\end{aligned} \tag{24}$$

where

$$G^{(2)}(\mathbf{R}_1, t; \mathbf{R}_2, t+\tau) = \langle E^{(-)}(\mathbf{R}_1, t) E^{(-)}(\mathbf{R}_2, t+\tau) E^{(+)}(\mathbf{R}_2, t+\tau) E^{(+)}(\mathbf{R}_1, t)\rangle \tag{25}$$

$$G^{(1)}(\mathbf{R}, t) = \langle E^{(-)}(\mathbf{R}, t) E^{(+)}(\mathbf{R}, t)\rangle \tag{26}$$

The correlation function $G^{(1)}(\mathbf{R}, t)$ is proportional to a probability of finding one photon around the direction $\mathbf{R}$ at time t, whereas $G^{(2)}(\mathbf{R}_1, t; \mathbf{R}_2, t+\tau)$ is proportional to a joint probability of finding one photon around the direction $\mathbf{R}_1$ at time t and another photon around the direction $\mathbf{R}_2$ at the moment of time $t+\tau$. For a coherent light the probability of finding a photon around the direction $\mathbf{R}_1$ at time t is independent of the probability of finding another photon around the direction $\mathbf{R}_2$ at time $t+\tau$ and the correlation function $G^{(2)}(\mathbf{R}_1, t; \mathbf{R}_2, t+\tau)$ simply factorizes on $G^{(1)}(\mathbf{R}_1, t)G^{(1)}(\mathbf{R}_2, t+\tau)$, giving $g^{(2)}(\mathbf{R}_1, t; \mathbf{R}_2, t+\tau) = 1$ for all τ. For a chaotic field the correlation function $G^{(2)}(\mathbf{R}_1, t; \mathbf{R}_2, t+\tau)$ for $\tau = 0$ is greater than for $\tau > 0$ giving $g^{(2)}(\mathbf{R}_1, t; \mathbf{R}_2, t) > g^{(2)}(\mathbf{R}_1, t; \mathbf{R}_2, t+\tau)$. This is a manifestation of the tendency of photons to be emitted by a chaotic light source in correlated pairs, and is called photon bunching. Photon antibunching, as the name implies, is the opposite of bunching, and describes a situation in which fewer photons appear close together than further apart. The condition for photon antibunching is $g^{(2)}(\mathbf{R}_1, t; \mathbf{R}_2, t) < g^{(2)}(\mathbf{R}_1, t; \mathbf{R}_2, t+\tau)$ and implies that the probability of detecting two photons at the same time t is smaller than the probability of detecting two photons at different times t and $t+\tau$. Moreover, the fact that there is a small probability of detecting photon pairs with zero time separation indicates that the one time correlation function $g^{(2)}(\mathbf{R}_1, t; \mathbf{R}_2, t)$ is smaller than one. This effect is called photon anticorrelation. The normalized correlation function (24) for $\tau = 0$ may be written as

$$g^{(2)}(\mathbf{R}_1, t; \mathbf{R}_2, t) = 1 + \frac{\int P(\varepsilon)\{|\varepsilon|^2 - \langle|\varepsilon|^2\rangle\}^2 \, d^2\varepsilon}{\langle|\varepsilon|^2\rangle^2} \tag{27}$$

where $P(\varepsilon)$ is the Glauber P representation for the electromagnetic field with the complex amplitude ε. Hence, we see that photon antibunching has no classical analog in the sense that its diagonal coherent-state representation cannot be nonnegative.

Another nonclassical effect, which is very promising for further application in science and technology, is the squeezed state of light. To define a squeezed state of light let us introduce the quadature components $E_\theta, E_{\theta-\pi/2}$ at frequency ω, wave vector $\mathbf{k}$, and given in terms of the positive, negative frequency components $E^{(+)}(\mathbf{R}, t)$, $E^{(-)}(\mathbf{R}, t)$ of the electromagnetic field as

$$E_\theta = E^{(+)}(\mathbf{R}, t)\, \mathrm{e}^{\mathrm{i}(\omega t - \mathbf{k}\cdot\mathbf{R}+\theta)} + E^{(-)}(\mathbf{R}, t)\, \mathrm{e}^{-\mathrm{i}(\omega t - \mathbf{k}\cdot\mathbf{R}+\theta)} \tag{28}$$

$$E_{\theta-\pi/2} = -\mathrm{i}\left[E^{(+)}(\mathbf{R}, t)\, \mathrm{e}^{\mathrm{i}(\omega t - \mathbf{k}\cdot\mathbf{R}+\theta)} - E^{(-)}(\mathbf{R}, t)\, \mathrm{e}^{-\mathrm{i}(\omega t - \mathbf{k}\cdot\mathbf{R}+\theta)}\right] \tag{29}$$

and satisfying the commutation relation

$$\left[E_\theta, E_{\theta-\pi/2}\right] = 2\mathrm{i}C \tag{30}$$

where C is a positive c-number.

The fluctuations $\Delta E_\theta, \Delta E_{\theta-\pi/2}$ of the quadrature operators then satisfy an uncertainty relation

$$(\Delta E_\theta)^2 (\Delta E_{\theta-\pi/2})^2 \geq |C|^2 \tag{31}$$

Thus, a large fluctuation in one quadrature component is accompanied by a small fluctuation in the other. The situation for equality in (31) is called a minimum uncertainty state.

Introducing the fluctuation operator

$$\Delta E_\alpha = E_\alpha - \langle E_\alpha \rangle \qquad \alpha = \theta, \theta - \pi/2 \tag{32}$$

and using Eqs. (28) and (29), we can write

$$(\Delta E_\alpha)^2 = \left\langle (\Delta E_\alpha)^2 \right\rangle = \left\langle :(\Delta E_\alpha)^2: \right\rangle + C \tag{33}$$

where the form $:E_\alpha:$ is referred to as the normal ordering of E_α in which all annihilation operators are placed to the right of all certain operators.

For a coherent state of field $\left\langle :(\Delta E_\alpha)^2: \right\rangle = 0$ and from Eq. (31) we have

$$\left\langle (\Delta E_\alpha)^2 \right\rangle = C \qquad \alpha = \theta, \theta - \pi/2 \tag{34}$$

Hence, the coherent state is a minimum uncertainty state with equal fluctuations for both quadrature components. For a chaotic field, both quadrature components $\left\langle (\Delta E_\theta)^2 \right\rangle$ and $\left\langle (\Delta E_{\theta-\pi/2})^2 \right\rangle$ are greater than C and we call this a chaotic state. It is possible to generate states for which

$\langle(\Delta E_\alpha)^2\rangle$ is less than C for one of the quadrature components. These states are called squeezed states. According to (33), a squeezed state of the field is characterized by the condition that either $\langle :(\Delta E_\theta)^2: \rangle$ or $\langle :(\Delta E_{\theta-\pi/2})^2: \rangle$ is negative. This condition can be written as

$$\langle :(\Delta E_\theta)^2: \rangle = \int (\Delta \operatorname{Re} \varepsilon_\theta)^2 P(\varepsilon_\theta)\, d^2\varepsilon_\theta \tag{35}$$

Hence, we see that squeezed states, similar to photon antibunching, have no classical analog in the sense that their diagonal coherent-state representation cannot be nonnegative.

In the next sections we consider the possibility of obtaining both photon antibunching and squeezed states in resonance fluorescence of two-level atoms.

IV. SINGLE-ATOM RESONANCE FLUORESCENCE

The interaction of the electromagnetic field with the atoms leads to the phenomenon of resonance fluorescence. This phenomenon has attracted the attention of many researchers in recent years in that photon antibunching and squeezing were first discovered in resonance fluorescence.

To analyze photon antibunching and squeezed states in resonance fluorescence we use the master equation (23) and the following relation between the radiation field and atomic operators in the far-field limit [27, 34]:

$$\mathbf{E}^{(+)}(\mathbf{R}, t) = \mathbf{E}_0^{(+)}(\mathbf{R}, t) - k^2 \sum_{i=1}^{N} \frac{\hat{R} \times (\hat{R} \times \boldsymbol{\mu})}{R} S_i^-\left(t - \frac{R}{c}\right) e^{-i\mathbf{k}\cdot\mathbf{r}_i} \tag{36}$$

where $\hat{R}$ is the unit vector in the direction $\mathbf{R} = \hat{R}R$ of the observation point, $\mathbf{r}_i$ is the position vector of the ith atom, and $\mathbf{E}_0^{(+)}(\mathbf{R}, t)$ denotes the positive frequency part of the vacuum field. Insertion of (36) into (25), (26), and (33) leads to

$$\begin{aligned} G^{(2)}(\mathbf{R}_1, t; \mathbf{R}_2, t+\tau) &= \psi^2(\mathbf{R}_1)\psi^2(\mathbf{R}_2) \sum_{i,j,k,l} \langle S_i^+(t) S_j^+(t+\tau) S_k^-(t+\tau) S_l^-(t) \rangle \\ &\quad \times \exp\left[ik\left(\mathbf{r}_{il}\cdot\hat{R}_1 + \mathbf{r}_{jk}\cdot\hat{R}_2\right)\right] \end{aligned} \tag{37}$$

$$G^{(1)}(\mathbf{R}, t) = \psi^2(\mathbf{R}) \sum_{i,j} \langle S_i^+(t) S_j^-(t) \rangle \exp(ik\mathbf{r}_{ij}\cdot\mathbf{R}) \tag{38}$$

$$\langle :(\Delta E_\alpha)^2: \rangle = \psi^2(\mathbf{R})\left[\langle(\Delta R_\alpha)^2\rangle + \tfrac{1}{2}\langle R_3\rangle\right] \tag{39}$$

with $\psi^2(\mathbf{R}) = (2k^4\mu^2/R^2)\sin^2\psi_0$, where ψ_0 is the angle between the observation direction $\mathbf{R}$ and the atomic transition dipole moment $\boldsymbol{\mu}$, and $\mathbf{r}_{ij} = \mathbf{r}_j - \mathbf{r}_i$ is the distance between atoms i and j.

In Eq. (39), $R_\alpha(\alpha = \theta, \theta - \pi/2)$ and R_3 are Dicke's spin variables [39], which can be expressed in terms of the atomic operators S_i^+ and S_i^- as

$$R_\theta = \frac{1}{2}(S_\theta^+ + S_\theta^-)$$
$$R_{\theta-\pi/2} = \frac{1}{2\mathrm{i}}(S_\theta^+ - S_\theta^-) \tag{40}$$

and

$$R_3 = \frac{1}{2}[S_\theta^+, S_\theta^-]$$

where

$$S_\theta^\pm = \sum_i S_i^\pm \exp\left[\pm\mathrm{i}\left(k\hat{R}\cdot\mathbf{r}_i - \theta\right)\right] \tag{41}$$

Having available the fluorescent field correlation functions expressed by the atomic correlation functions according to Eqs. (37)–(39), we can directly apply our master equation (23) to calculate photon antibunching and squeezed states in resonance fluorescence.

First, we consider the simplest process of spontaneous emission from a single two-level atom. In this case Eqs. (37)–(39) simplify to

$$G^{(2)}(\mathbf{R}_1, t; \mathbf{R}_2, t+\tau) = \psi^2(\mathbf{R}_1)\psi^2(\mathbf{R}_2)\langle S_1^+(t)S_1^+(t+\tau)S_1^-(t+\tau)S_1^-(t)\rangle \tag{42a}$$

$$G^{(1)}(\mathbf{R}, t) = \psi^2(\mathbf{R})\langle S_1^+(t)S_1^-(t)\rangle \tag{42b}$$

$$\langle :(\Delta E_\alpha)^2: \rangle = \psi^2(\mathbf{R})\left[\tfrac{1}{2}\langle S_\theta^+(t)S_\theta^-(t)\rangle - \tfrac{1}{4}\langle S_\theta^+(t) + S_\theta^-(t)\rangle^2\right] \tag{42c}$$

where

$$S_\theta^\pm = S_1^\pm \exp\left[\pm\mathrm{i}\left(k\hat{R}\cdot\mathbf{r}_1 - \theta\right)\right] \tag{42d}$$

From the master equation (27) it is easy to find that

$$\langle S_t^+(t)S_1^+(t+\tau)S_1^-(t+\tau)S_1^-(t)\rangle = \langle S_1^+(t)S_1^+(t)S_1^-(t)S_1^-(t)\rangle\,\mathrm{e}^{-2\gamma\tau}$$
$$\langle S_1^+(t)S_1^-(t)\rangle = \langle S_1^+(0)S_1^-(0)\rangle\,\mathrm{e}^{-2\gamma t} \tag{43}$$
$$\langle S_\theta^\pm(t)\rangle = \langle S_1^\pm(0)\rangle\,\mathrm{e}^{\mp\mathrm{i}\theta}\,\mathrm{e}^{-\gamma t}$$

Since $[S_1^{\pm}(t)]^2 \equiv 0$, we have that $G^{(2)}(\mathbf{R}_1, t; \mathbf{R}_2, t+\tau) = 0$ for all times t and τ, and there is no photon antibunching in spontaneous emission from a single atom. This result has a simple physical interpretation. In a single-atom spontaneous emission we have only one photon, and a joint probability of detecting of two photons is always zero for all t and τ.

To calculate squeezed states in a single-atom spontaneous emission we have to know one-time correlation functions $\langle S_\theta^+(t) S_\theta^-(t)\rangle$ and $\langle S_\theta^{\pm}(t)\rangle$. From Eqs. (42) and (43) we have that

$$
\begin{aligned}
F_\theta(t) &= \left\langle :(\Delta E_\theta)^2: \right\rangle / \psi^2(\mathbf{R}) \\
&= \tfrac{1}{2}\left[\left\langle S_1^+(0) S_1^-(0)\right\rangle - \tfrac{1}{2}\left\langle S_1^+(0)\, \mathrm{e}^{i\theta} + S_1^-(0)\, \mathrm{e}^{-i\theta}\right\rangle^2\right] \mathrm{e}^{-2\gamma t}
\end{aligned}
\tag{44}
$$

The fluctuations in the quadrature component E_θ depend on the initial values of the expectation values $\langle S_1^+(0) S_1^-(0)\rangle$ and $\langle S_1^{\pm}(0)\rangle$. It is seen from Eq. (44) that the fluctuations $F(t)$ can be negative (squeezing) only if the single atom has the nonvanishing dipole moment $\langle S_1^{\pm}(t)\rangle$. If the atom is initially in the ground $|g_1\rangle$ or excited $|e_1\rangle$ state then $\langle S_1^{\pm}(0)\rangle = 0$ and there is no squeezing in the spontaneous emission from a single atom. To obtain negative values of $F_\theta(t)$ we have to prepare the atom in a linear superposition of its ground and excited states. Consider the initial ($t = 0$) superposition state [40]:

$$
|\psi_A\rangle = \cos\tfrac{1}{2}\theta_0 |e_1\rangle + \sin\tfrac{1}{2}\theta_0\, \mathrm{e}^{i\varphi_0} |g_1\rangle \tag{45}
$$

where $0 \le \theta_0 \le \pi$ and $0 \le \varphi_0 \le 2\pi$. In this state

$$
\begin{aligned}
\left\langle S_1^+(0) S_1^-(0)\right\rangle &= \cos^2\tfrac{1}{2}\theta_0 \\
\left\langle S_1^{\pm}(0)\right\rangle &= \cos\tfrac{1}{2}\theta_0 \sin\tfrac{1}{2}\theta_0\, \mathrm{e}^{\pm i\varphi_0}
\end{aligned}
\tag{46}
$$

and

$$
F_\theta(t) = \tfrac{1}{2}\left[\cos^2\left(\tfrac{1}{2}\theta_0\right) - \tfrac{1}{2}\sin^2(\theta_0)\cos^2(\varphi_0 - \theta)\right] \mathrm{e}^{-2\gamma t} \tag{47}
$$

Figure 1 shows the time evolution of $F_\theta(t)$, given by Eq. (47), for $(\varphi_0 - \theta) = 0$ and different θ_0. It is evident from Fig. 1 that for some values of θ_0 the atom radiates field that is squeezed. However, squeezing decreases during spontaneous emission as the atom decays toward its ground state and disappears in the steady state. The minimum value of $F_\theta(t = 0)$ corre-

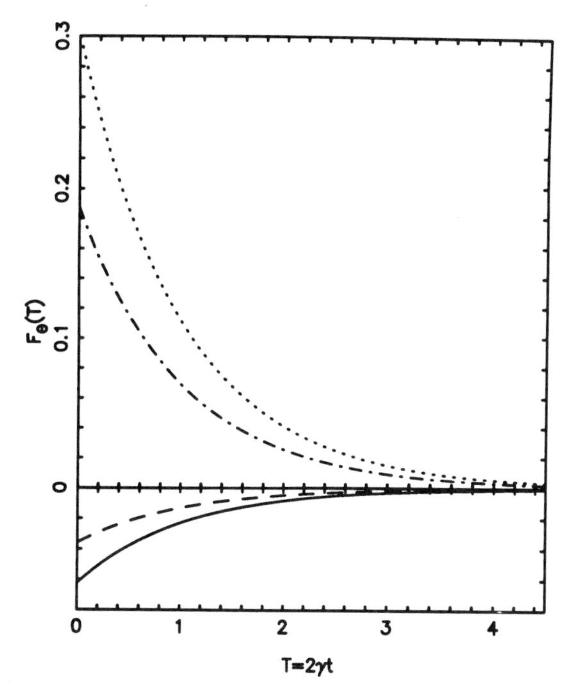

Figure 1. Time dependence of $F_\theta(T) = \langle:(\Delta E_\theta)^2:\rangle/\psi^2(\mathbf{R})$ for $\varphi_0 - \theta = 0$ and different values of θ_0: $\theta_0 = 120°$ (solid line), $\theta_0 = 100°$ (dashed line), $\theta_0 = 60°$ (dash-dotted line), $\theta = 45°$ (dotted line).

sponding to optimum squeezing occurs for $\theta_0 = 2\pi/3$ and $(\varphi_0 - \theta) = 0$ when $F(0) = -\frac{1}{16}$. The results presented in Fig. 1 have a simple physical interpretation: Squeezing is sensitive to the phase θ. Spontaneous emission does not introduce any phase information. Therefore, the essential condition of squeezing in the spontaneous emission is to prepare the initial state of the system which includes phase information.

We have shown here that the spontaneous emission of a single atom does not show photon antibunching, but shows squeezing if the atom is initially in a suitable superposition of its states. The situation is different when the atom interacts with an external coherent laser field. In resonance fluorescence the atom is re-excited by the laser field after emitting a fluorescent photon. This excitation allows the atom to re-establish a dipole moment and to radiate more fluorescent photons. In this case photon antibunching can appear and the squeezing can persist in the steady state. To show this more quantitatively we start from the master equation (23), which for the atomic correlation functions with $\Omega \neq 0$ leads to the following equation of motion [31]:

$$\begin{aligned}
\frac{\mathrm{d}}{\mathrm{d}t}\langle S_1^-(t)\rangle &= \tfrac{1}{2}\mathrm{i}\Omega - \gamma\langle S_1^-(t)\rangle - \mathrm{i}\Omega\langle S_1^+(t)S_1^-(t)\rangle \\
\frac{\mathrm{d}}{\mathrm{d}t}\langle S_1^+(t)\rangle &= -\tfrac{1}{2}\mathrm{i}\Omega - \gamma\langle S_1^+(t)\rangle + \mathrm{i}\Omega\langle S_1^+(t)S_1^-(t)\rangle \qquad (48)\\
\frac{\mathrm{d}}{\mathrm{d}t}\langle S_1^+(t)S_1^-(t)\rangle &= -2\gamma\langle S_1^+(t)S_1^-(t)\rangle - \mathrm{i}\Omega(\langle S_1^+(t)\rangle - \langle S_1^-(t)\rangle)
\end{aligned}$$

where we have assumed that the Rabi frequency is real and the laser frequency ω_L is exactly equal to the atomic transition frequency ω_0; i.e., detuning is zero.

The system of Eq. (48) can be easily solved by Laplace transform techniques. The time evolution of the atomic correlation functions for the atom initially in the ground state $|g_1\rangle$ is given by [3]

$$\begin{aligned}
\langle S_1^{\pm}(t)\rangle &= \frac{\mp 2\mathrm{i}\beta}{1+8\beta^2} \\
&\quad + \frac{2\beta}{u(1+8\beta^2)}\left[\left(\frac{1}{4} - 4\beta^2\right)\cos(ut) - u\sin(ut)\right] \qquad (49)\\
&\quad \times\exp\left(-\frac{3}{4}t\right)
\end{aligned}$$

$$\langle S_1^+(t)S_1^-(t)\rangle = \frac{4\beta^2}{1+8\beta^2}\left\{1 - \left[\cos(ut) + \frac{3}{4u}\sin(ut)\right]\exp\left(-\frac{3}{4}t\right)\right\} \qquad (50)$$

where, for simplicity, we have introduced the notation

$$t = 2\gamma t \qquad \beta = \frac{\Omega}{4\gamma} \qquad u = \left(4\beta^2 - \frac{1}{16}\right)^{1/2} \qquad (51)$$

To study the normalized second-order correlation function $g^{(2)}(\mathbf{R}_1, t; \mathbf{R}_2, t+\tau)$, we have to find the correlation function

$$\langle S_1^+(t)S_1^+(t+\tau)S_1^-(t+\tau)S_1^-(t)\rangle.$$

From the quantum regression theorem [41], it is well known that for $\tau > 0$

the two-time average

$$\langle S_1^+(t)S_1^+(t+\tau)S_1^-(t+\tau)S_1^-(t)\rangle$$

satisfies the same equation of motion as the one-time average $\langle S_1^+(t)S_1^-(t)\rangle$. By Eqs. (24), (42), and (50), we find that in the steady state ($t \to \infty$) the normalized second-order correlation function $g^{(2)}(\mathbf{R}_1, t; \mathbf{R}_2, t+\tau)$ takes the form [2, 3]

$$\begin{aligned} g^{(2)}(\tau) &= \lim_{t\to\infty} g^{(2)}(\mathbf{R}_1, t; \mathbf{R}_2, t+\tau) \\ &= 1 - \left[\cos(u\tau) + \frac{3}{4u}\sin(u\tau)\right]\exp\left(-\frac{3}{4}\tau\right) \end{aligned} \tag{52}$$

For $\tau = 0$, the correlation function $g^{(2)}(0) = 0$, showing a complete photon anticorrelation between emitted photons. As τ increases ($\tau > 0$), the correlation function $g^{(2)}(\tau)$ increases. This effect reflects the existence of photon antibunching in a single-atom resonance fluorescence.

Using the time-dependent solutions (49) and (50) we can discuss squeezing in resonance fluorescence. In the steady-state ($t \to \infty$) the fluctuations in the quadrature component E_θ are [18, 42]

$$F_\theta(\infty) = \frac{2\beta^2}{\left(1+8\beta^2\right)^2}(8\beta^2 + \cos 2\theta) \tag{53}$$

Resonance fluorescence will exhibit squeezed fluctuations if $8\beta^2 + \cos 2\theta < 0$, i.e., for a weak driving field. Maximum squeezing in steady-state resonance fluorescence occurs for $\theta = \pi/2$ and $\beta^2 = \frac{1}{24}$, when $F_\theta(\infty) = -\frac{1}{32}$. This value, compared with the $-\frac{1}{16}$ derived above, is one-half of that obtained for the spontaneous emission. The mechanism responsible for the generation of squeezing in resonance fluorescence differs from that in spontaneous emission. When the initial phase information is introduced to the single atom, then the spontaneous emission produces squeezing. Resonance fluorescence produces squeezing through phase information introduced by a coherent field. Resonance fluorescence, however, can produce optimum squeezing obtained in the spontaneous emission providing that the resonance is time dependent [42]. This is shown in Figs. 2 and 3, where we plot $F_\theta(T) = \langle :\Delta E_\theta)^2: \rangle / \psi^2(\mathbf{R})$, as given by Eq. 42c, versus the time $T = 2\gamma t$, for $\theta = \varphi_0 = 0$ and various values of the parameters β and θ_0. It is seen from Fig. 2 that as the intensity of the laser field increases, the optimum squeezing in $F_\theta(T)$ shifts to the region of shorter

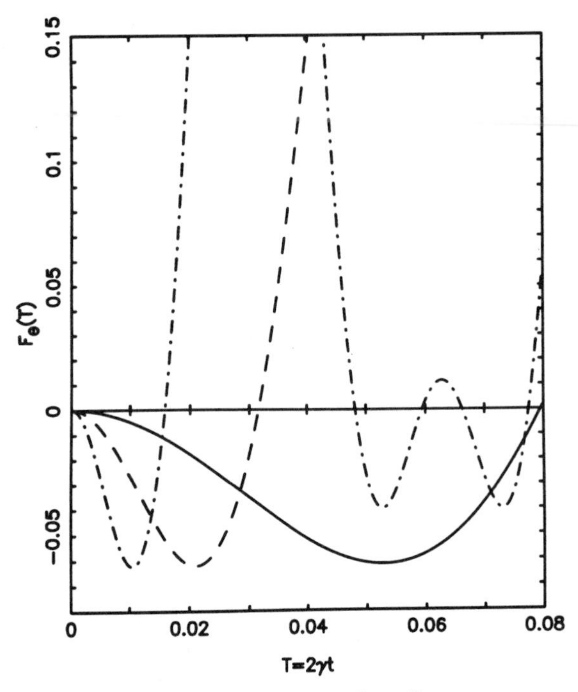

Figure 2. Time dependence of $F_\theta(T) = \langle:(\Delta E_\theta)^2:\rangle/\psi^2(\mathbf{R})$ for $\theta = \theta_0 = \varphi_0 = 0$ and for different field strengths $\beta = 10$ (solid line), $\beta = 25$ (dashed line), $\beta = 50$ (dash-dotted line).

times, and $F_\theta(T)$ itself shows an oscillatory behavior reflecting the Rabi oscillations. The optimum squeezing reaches a value of $-\frac{1}{16}$ at a very short time t and for $\beta = 50$. Figure 3 shows that the optimum squeezing can be obtained at different times t, depending on the initial state of the atom. If the atom is initially prepared in an equal superposition of its ground and excited states then the optimum squeezing appears at a time shorter than for the atom prepared initially in its ground state.

The simple model presented here provides the underlying mechanism for obtaining photon antibunching and squeezing in resonance fluorescence. In this model two-level atoms, independent of each other, interact with the electromagnetic field. Steady-state resonance fluorescence shows photon antibunching for an arbitrary intensity of the exciting field, whereas squeezing occurs only for a weak intensity of the exciting field. For a strong exciting field, squeezing occurs in the transient regime of resonance fluorescence.

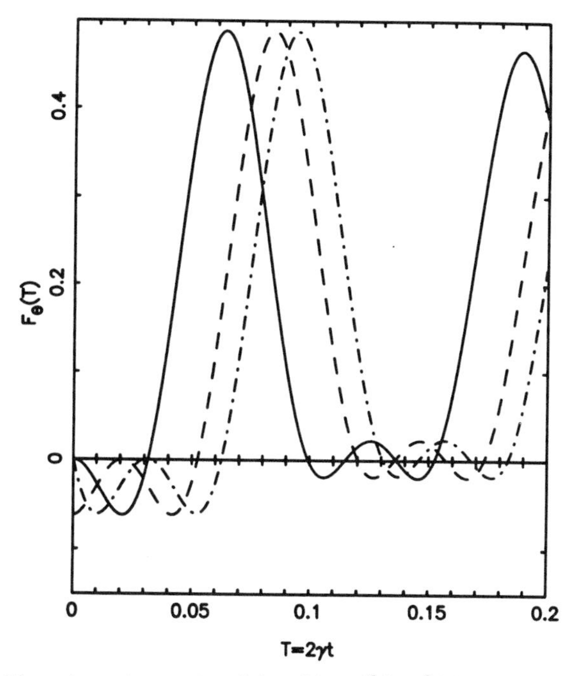

Figure 3. Time dependence of $F_\theta(T) = \langle:(\Delta E_\theta)^2:\rangle/\psi^2(\mathbf{R})$ for $\beta = 25$, $\varphi_0 = 0$ and different values of θ_0: $\theta_0 = \pi$ (solid line), $\theta_0 = \pi/2$ (dashed line), $\theta_0 = 2\pi/3$ (dash-dotted line).

V. MULTIATOM RESONANCE FLUORESCENCE

The remainder of this chapter is concerned with the interaction between two-level atoms and the quantized electromagnetic field. Our analysis so far has been concerned with a simple model in which the electromagnetic field interacts with a single two-level atom. However, photon antibunching and squeezing can be considerably modified when more atoms interact with the electromagnetic field. We are particularly interested in the role of the interatomic interactions in modifying the nonclassical effects. We focus our attention on photon antibunching and squeezing in a two-atom resonance fluorescence. Although a two-atom system is admittedly an elementary model, it offers some advantages over the multiatom problem. Because of its simplicity, one obtains detailed and almost exact dynamical solutions with a variety of initial conditions.

The simplest formulation of the problem of multiatom resonance fluorescence is associated with consideration of two atoms in the Dicke model.

In this model it is assumed that the interatomic separations are much smaller than the resonant wavelength, and level shifts associated with the presence of the dipole–dipole interaction between the atoms are ignored. With these assumptions the collective parameters γ_{ij} $(i \neq j)$, which appear in the master equation (23), reduce to $\gamma = \gamma_{ii}$, whereas the parameter Ω_{ij} is equal to zero. For the two-atom Dicke model the master equation (23) reduces to [27]

$$\frac{\partial \rho}{\partial t} = \frac{1}{2}\mathrm{i}\Omega[S^{+} + S^{-}, \rho] - \gamma(S^{+}S^{-}\rho + \rho S^{+}S^{-} - 2S^{-}\rho S^{+}) \quad (54)$$

where $S^{\pm} = S_1^{\pm} + S_2^{\pm}$ and $S^z = S_1^z + S_2^z$ are the collective atomic dipole operators. For simplicity, the laser frequency ω_{L} is assumed to be exactly equal to the atomic transition frequency ω_0.

For a strongly driven system, $\Omega \gg 2\gamma$, an approximation technique has been suggested by Agarwal et al. [43] and Kilin [44], which greatly simplifies the master equation (54). This technique transforms (54) to new collective operators $R^{\pm}$ and R^z as follows:

$$\begin{aligned} S^{\pm} &= \pm\tfrac{1}{2}\mathrm{i}(R^{+} + R^{-}) + R^{z} \\ S^{z} &= -\tfrac{1}{2}\mathrm{i}(R^{+} - R^{-}) \end{aligned} \quad (55)$$

The operators R are a rotation of the operators S. For a strong field, the $R^{\pm}$ vary with time approximately as $\exp(\pm\mathrm{i}\Omega t)$, while R^z varies slowly. Substituting $S^{\pm}$ and S^z from Eq. (55) into the master equation (54) and dropping rapidly oscillating terms such as $R^{\pm}R^{z}, R^{+}R^{+}$, we find the approximate master equation [43–45]:

$$\begin{aligned} \frac{\partial \rho}{\partial t} = \mathrm{i}\Omega[R^{z}, \rho] - \gamma\{&(R^{z}R^{z}\rho + \rho R^{z}R^{z} - 2R^{z}\rho R^{z}) \\ &+ \tfrac{1}{4}[(R^{+}R^{-}\rho + \rho R^{+}R^{-} - 2R^{-}\rho R^{+}) \\ &+ (R^{-}R^{+}\rho + \rho R^{-}R^{+} - 2R^{+}\rho R^{-})]\} \end{aligned} \quad (56)$$

Equation (56) enables us to obtain the equation of motion for the expectation value of an arbitrary operator Q as $\langle \dot{Q} \rangle = \mathrm{tr}(\dot{\rho} Q)$. In particular, the equations of motion for the transformed dipole operators are

$$
\begin{aligned}
\frac{\mathrm{d}}{\mathrm{d}t}\langle R^{z}\rangle &= -\gamma\langle R^{z}\rangle \\
\frac{\mathrm{d}}{\mathrm{d}t}\langle R^{\pm}\rangle &= -\left(\tfrac{3}{2}\gamma \pm \mathrm{i}\Omega\right)\langle R^{\pm}\rangle \\
\frac{\mathrm{d}}{\mathrm{d}t}\langle R^{+}R^{+}\rangle &= -(5\gamma + 2\mathrm{i}\Omega)\langle R^{+}R^{+}\rangle
\end{aligned}
\tag{57}
$$

Equation (57) is simple in form and can be solved exactly. Performing the inverse transformation from R to the S operators, and using the quantum regression theorem [41], we obtain from Eqs. (55) and (57) the following solutions for the normalized second-order correlation function [43, 46]:

$$
\begin{aligned}
g^{(2)}(\tau) = 1 &+ \tfrac{1}{32}\exp(-3\gamma\tau) + \tfrac{3}{32}\exp(-5\gamma\tau)\cos(2\Omega\tau) \\
&- \tfrac{3}{8}\exp\left(-\tfrac{3}{2}\gamma\tau\right)\cos(\Omega\tau)
\end{aligned}
\tag{58}
$$

And for the fluctuation in the quadrature components $E_{\theta=0}$ and $E_{\theta=\pi/2}$ of two atoms starting from their ground states [45], we obtain

$$
\begin{aligned}
&\left\langle :(\Delta E_{\theta=0})^{2}:\right\rangle \Big/ 2\psi^{2}(\mathbf{R}) \\
&\quad = \tfrac{2}{3} + \tfrac{1}{12}\exp(-3\gamma t) - \tfrac{1}{4}\exp(-5\gamma t)\cos(2\Omega t) \\
&\qquad - \exp(-3\gamma t)\sin^{2}(\Omega t) - \tfrac{1}{2}\exp\left(-\tfrac{3}{2}\gamma t\right)\cos(\Omega t)
\end{aligned}
\tag{59}
$$

and

$$
\left\langle :(\Delta E_{\theta=\pi/2})^{2}:\right\rangle \Big/ 2\psi^{2}(\mathbf{R}) = \tfrac{2}{3} - \tfrac{1}{6}\exp(-3\gamma t) - \tfrac{1}{2}\exp\left(-\tfrac{3}{2}\gamma t\right)\cos(\Omega t)
\tag{60}
$$

Figure 4 shows that the photon antibunching ($g^{(2)}(\tau) > g^{(2)}(0)$) is preserved when resonance fluorescence is from two interacting atoms. However, the photon anticorrelation effect ($g^{(2)}(0) < 1$) is reduced compared to that for a single atom. The time dependence of fluctuations in the quadrature component $E_{\theta=0}$ is shown in Fig. 5. As for a single atom, squeezing appears only in the transient regime of resonance fluorescence.

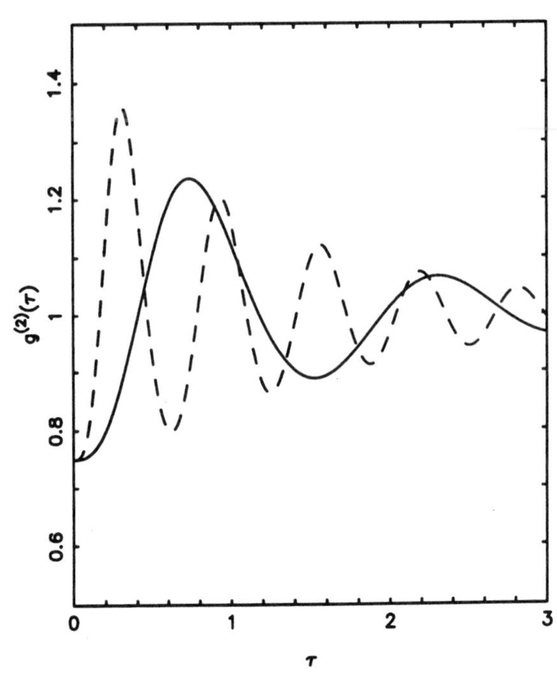

Figure 4. Time dependence of the normalized second-order correlation function $g^{(2)}(\tau)$ for two atoms and for various β: $\beta = 1$ (solid line), $\beta = 3$ (dashed line).

Its maximum value, however, is reduced compared to that for a single atom. These results indicate that cooperative effects reduce photon anticorrelations and squeezing in resonance fluorescence. Moreover, as for single atoms, there is no squeezing in the steady-state resonance fluorescence when the atoms are excited by a strong laser field.

In the Dicke model the dipole–dipole interaction between the atoms is ignored. This approximation has no justification, since for a small interatomic separation the parameter Ω_{ij}, which appears in the master equation (23), is very large and goes to infinity as the interatomic separation r_{12} goes to zero. Therefore, it seems natural to study in some detail what happens when the dipole–dipole interaction terms is included in the two-atom Dicke model. We will also assume an arbitrary separation r_{12} between the atoms and nonzero detuning between the laser frequency ω_L and the atomic transition frequency ω_0. With the dipole–dipole interaction included, the master equation (23) leads to a closed set of nine equations of motion for the atomic correlation functions. This set of equations can be solved exactly in the steady-state limit, and the solution

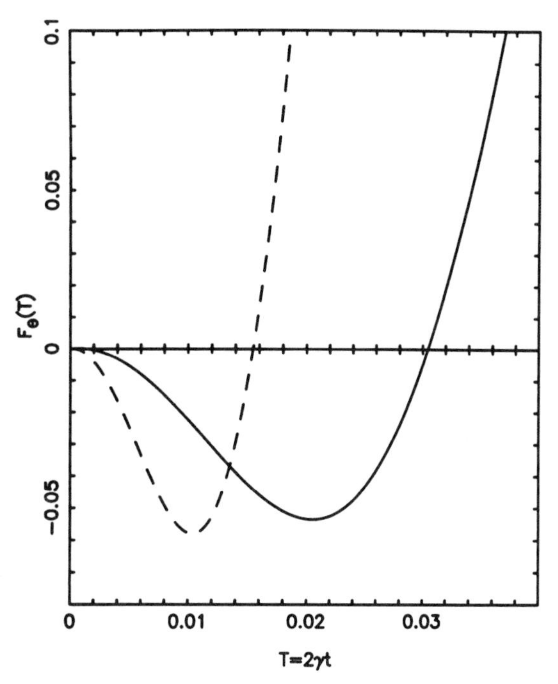

Figure 5. Time dependent of $F_\theta(T) = \langle :(\Delta E_{\theta=0})^2: \rangle / 2\psi^2(\mathbf{R})$ for two atoms and different β: $\beta = 25$ (solid line), $\beta = 50$ (dashed line).

is [47, 48]

$$\begin{aligned}
\langle X_1 \rangle &= -8\beta\left[8\beta^2 + (1+a)(1+\Delta^2)\right]/D \\
\langle X_2 \rangle &= 8\beta^2\left[8\beta^2 + (1+\Delta^2)\right]/D \\
\langle X_3 \rangle &= 8(1+\Delta^2)\beta^2/D \\
\langle X_4 \rangle &= 8\beta^2[(1+a) - \Delta(\Delta+b)]/D \\
\langle X_5 \rangle &= -32\beta^3/D \\
\langle X_6 \rangle &= 16\beta^4/D \\
\langle X_7 \rangle &= 8\beta\left[8\Delta\beta^2 + (\Delta+b)(1+\Delta^2)\right]/D \\
\langle X_8 \rangle &= 32\Delta\beta^3/D \\
\langle X_9 \rangle &= -8\beta^2[\Delta(1+a) + (\Delta+b)]/D
\end{aligned} \tag{61}$$

with

$$D = 64\beta^4 + 16(1 + \Delta^2)\beta^2 + (1 + \Delta^2)\left[(1 + a)^2 + (\Delta + b)^2\right] \quad (62)$$

where

$$\begin{aligned}
&X_1 = S_1^+ + S_2^+ + S_1^- + S_2^- \qquad X_2 = S_1^+ S_1^- + S_2^+ S_2^- \\
&X_3 = S_1^+ S_2^- + S_2^+ S_1^- \qquad X_4 = S_1^+ S_2^+ + S_1^- S_2^- \\
&X_5 = S_1^+ S_1^- S_2^- + S_1^+ S_2^+ S_1^- + S_2^+ S_1^- S_2^- + S_1^+ S_2^+ S_2^- \\
&X_6 = S_1^+ S_2^+ S_1^- S_2^- \qquad X_7 = -\mathrm{i}(S_1^- + S_2^- - S_1^+ - S_2^+) \\
&X_8 = -\mathrm{i}(S_1^+ S_1^- S_2^- - S_1^+ S_2^+ S_1^- + S_2^+ S_1^- S_2^- - S_1^+ S_2^+ S_2^-) \\
&X_9 = -\mathrm{i}(S_1^- S_2^- - S_1^+ S_2^+)
\end{aligned} \quad (63)$$

and

$$\beta = \Omega/4\gamma \qquad a = \gamma_{12}/\gamma \qquad b = \Omega_{12}/\gamma \qquad \Delta = (\omega_0 - \omega)/\gamma \quad (64)$$

The above steady-state solution, which includes the collective damping parameter a, the dipole–dipole interaction parameter b, and the detuning Δ, permits the calculation of photon anticorrelation and squeezing in the fluorescence field emitted by the two-atom system.

Having available the steady-state solution (61), we can calculate the correlation functions (37)–(39) for $\tau = 0$ and $t \to \infty$. In this limit, we have

$$G^{(2)}(\mathbf{R}_1, t; \mathbf{R}_2, t) = 4\psi^2(\mathbf{R}_1)\psi^2(\mathbf{R}_2)\langle X_6\rangle\left\{1 + \cos\left[k\mathbf{r}_{12}\cdot\left(\hat{R}_1 - \hat{R}_2\right)\right]\right\} \quad (65)$$

$$G^{(1)}(\mathbf{R}, t) = \psi^2(\mathbf{R})\left[\langle X_2\rangle + \langle X_3\rangle\cos\left(k\mathbf{r}_{12}\cdot\hat{R}\right)\right] \quad (66)$$

$$\begin{aligned}
\left\langle :(\Delta E_\alpha)^2: \right\rangle = \tfrac{1}{2}\psi^2(\mathbf{R})\Big\{&\langle X_4\rangle\cos(2\alpha) - \langle X_9\rangle\sin(2\alpha) \\
&+ \langle X_2\rangle + \langle X_3\rangle\cos\left(k\hat{R}\cdot\mathbf{r}_{12}\right) \\
&- \tfrac{1}{2}\left[\langle X_1\rangle\cos\alpha - \langle X_7\rangle\sin\alpha\right]^2\cos^2\left(\tfrac{1}{2}k\hat{R}\cdot\mathbf{r}_{12}\right)\Big\}
\end{aligned} \quad (67)$$

Since $\langle X_6\rangle$ is different from zero, the correlation function $G^{(2)}(\mathbf{R}_1, t; \mathbf{R}_2, t)$ is different from zero for $\hat{R}_1 = \hat{R}_2$ and the photon anticorrelation effect is reduced. However, for $\hat{R}_1 \neq \hat{R}_2$ and

$$\cos\theta' - \cos\theta'' = \lambda/2r_{12} \tag{68}$$

where θ' and θ'' are the angles between $\mathbf{r}_{12}$ and $\hat{R}_1(\hat{R}_2)$, respectively, we have $G^{(2)}(\mathbf{R}_1, t; \mathbf{R}_2, t) = 0$. Thus, we can obtain total anticorrelation between the photons emitted from two atoms, providing that the photons are observed in two different directions. This anticorrelation effect is due to spatial interference [49] causing

$$\left\{1 + \cos\left[k\mathbf{r}_{12} \cdot \left(\hat{R}_1 - \hat{R}_2\right)\right]\right\} = 0$$

It is interesting to note that for the two-atom resonance fluorescence the phase dependence of the variance (67) of the quadrature component

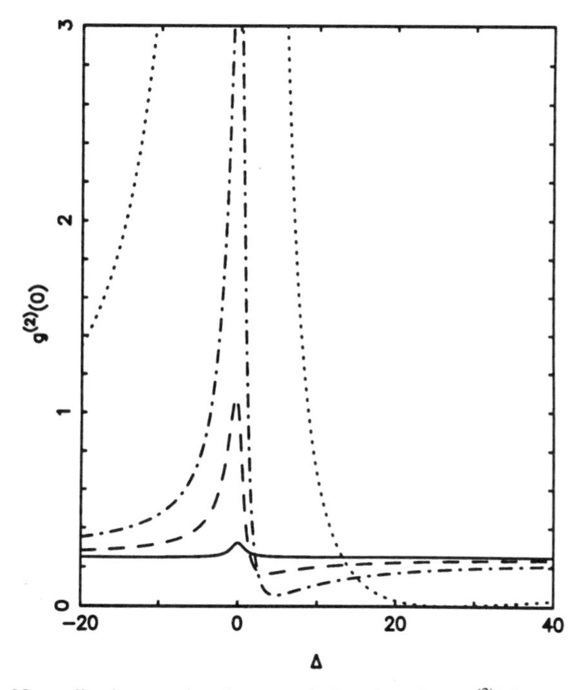

Figure 6. Normalized second-order correlation function $g^{(2)}(0)$ in function of the detuning Δ for $\hat{R}_1 = \hat{R}_2 \perp \mathbf{r}_{12}, \beta = 0.2$ and different interatomic separations $r_{12} : r_{12} = 10\lambda$ (solid line), $r_{12} = 0.25\lambda$ (dashed line), $r_{12} = 0.16\lambda$ (dash-dotted line), $r_{12} = 0.08\lambda$ (dotted line).

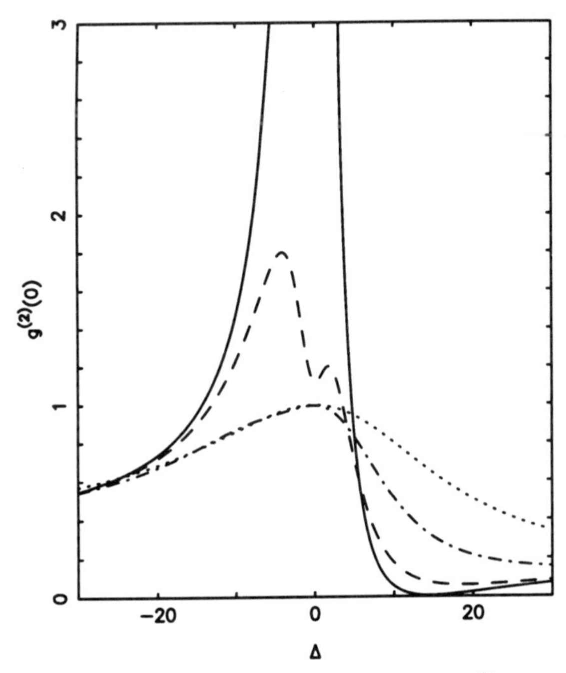

Figure 7. Normalized second-order correlation function $g^{(2)}(0)$ in function of the detuning Δ for $\hat{R}_1 = \hat{R}_2 \perp \mathbf{r}_{12}, r_{12} = 0.1\lambda$ for different field strengths β: $\beta = 0.2$ (solid line), $\beta = 2$ (dashed line), $\beta = 5$ (dash-dotted line), $\beta = 10$ (dotted line).

E_α is introduced not only through the dipole moments $\langle X_1 \rangle, \langle X_7 \rangle$ but also through the two-photon correlation functions $\langle X_4 \rangle$ and $\langle X_9 \rangle$.

The normalized second-order correlation function $g^{(2)}(0)$ is illustrated graphically in Figs. 6 and 7 as a function of the detuning Δ for $\mathbf{r}_{12} \perp \hat{R}(\hat{R} = \hat{R}_1 = \hat{R}_2)$, and for different values of the interatomic separation r_{12} and of the field strength β. These graphs show that $g^{(2)}(0)$ strongly depends on the detuning Δ, and that the total photon anticorrelation $[g^{(2)}(0) = 0]$ can be obtained for certain values of Δ. This happens for $\Delta = -b$, i.e., when the dipole–dipole interaction b and the detuning Δ cancel out mutually. In other words, this means that the laser frequency is tuned to resonance with a particular pair of energy levels of the two-atom system that are shifted by the dipole–dipole interaction. Other levels are far from resonance, and the two-atom system behaves like an individual two-level system.

The variance $F(\theta) = \langle : (\Delta E_\theta)^2 : \rangle / \psi^2(\mathbf{R})$ of the quadrature component E_θ, as given by Eq. (67), is plotted in Figs. 8 and 9 versus the detuning Δ

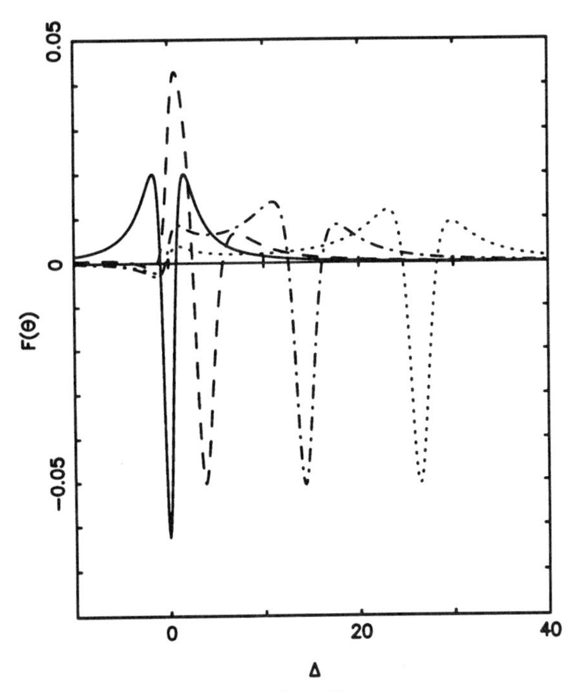

Figure 8. The variance $F(\theta) = \langle:(\Delta E_\theta)^2:\rangle/\psi^2(\mathbf{R})$ as a function of Δ for $\hat{R} \perp \mathbf{r}_{12}, \theta = 0, \beta = 0.2$, and for various interatomic separations $r_{12} : r_{12} = 10\lambda$ (solid line), $r_{12} = 0.16\lambda$ (dashed line), $r_{12} = 0.1\lambda$ (dash-dotted line), $r_{12} = 0.08\lambda$ (dotted line).

for $\mathbf{r}_{12} \perp \hat{R}$ and for various interatomic separations r_{12} at fixed θ as well as at different values of the phase θ and fixed r_{12}. It is evident from Fig. 8 that, as the interatomic distance r_{12} becomes sufficiently small and the dipole–dipole interaction between the atoms becomes considerable, the squeezing in $F(\theta)$, which for independent atoms has its maximum for $\Delta = 0$, shifts to region of finite Δ. In fact, as in the normalized second-order correlation function $g^{(2)}(0)$, the minimum in $F(\theta)$ appears for $\Delta = -b$ and can again be attributed to the change in energy-level structure of the two-atom system due to dipole–dipole interaction. Figure 9 shows that unlike in a single-atom resonance fluorescence, large squeezing can appear for a strong driving field and $\theta = \pi/2$. At $\Delta = -b$, values of squeezing in $F(\theta)$ can be obtained that are comparable to the value obtained in the transient regime of $F(\theta)$ for a single atom. Thus, stationary two-atom resonance fluorescence shows squeezing for a strong driving

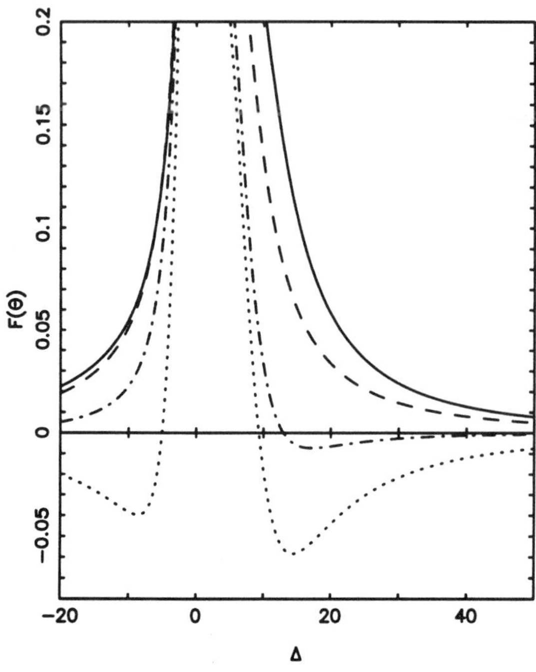

Figure 9. The variance $F(\theta)$ as a function of Δ for $\hat{R} \perp \mathbf{r}_{12}, r_{12} = \lambda/6, \beta = 2$ and for different phases θ: $\theta = 0$ (solid line), $\theta = \pi/8$ (dashed line), $\theta = \pi/4$ (dash-dotted line), $\theta = \pi/2$ (dotted line).

field contrary to the single-atom resonance fluorescence which for strong driving fields is squeezed only in the transient regime.

VI. SQUEEZING IN TWO-ATOM SPONTANEOUS EMISSION

In Section IV we have shown that only the initially different from zero dipole moment can produce squeezing in the spontaneous emission from a single atom. Here, we examine conditions for squeezing in two-atom spontaneous emission. Equation (67) shows that the variance $\langle :(\Delta E_\alpha)^2: \rangle$ depends on the phase α not only through the nonvanishing dipole moments $\langle S_i^\pm \rangle$, but also through the two-photon coherences $\langle S_1^+ S_2^+ \rangle$ and $\langle S_1^- S_2^- \rangle$. This dependence suggests that there are two different processes that can lead to squeezing in two-atom spontaneous emission. To show this, we start from Eq. (39), which for two atoms and $\hat{R} \perp \mathbf{r}_{12}$ can be written in the form

$$\begin{aligned}\langle :(\Delta E_\theta)^2: \rangle = \tfrac{1}{2}\psi^2(\mathbf{R})\Big\{ &\langle S_1^+ S_2^+ \rangle\, e^{-2i\theta} + \langle S_1^- S_2^- \rangle\, e^{2i\theta} \\ &+\langle S_1^+ S_1^- + S_2^+ S_2^- \rangle + \langle S_1^+ S_2^- + S_2^+ S_1^- \rangle \\ &- \tfrac{1}{2}\big[\langle S_1^+ + S_2^+ \rangle\, e^{-i\theta} + \langle S_1^- + S_2^- \rangle\, e^{i\theta}\big]^2 \Big\}\end{aligned} \tag{69}$$

From the master equation (23) it is easy to show that for $\Omega = 0$ the equations of motion for the atomic correlation functions, which appear in Eq. (69), are

$$\begin{aligned}
\frac{d}{dt}Y_1 &= -(\gamma + \gamma_{12} + i\Omega_{12})Y_1 + 2(\gamma_{12} + i\Omega_{12})Y_3 \\
\frac{d}{dt}Y_2 &= -(\gamma + \gamma_{12} - i\Omega_{12})Y_2 + 2(\gamma_{12} - i\Omega_{12})Y_4 \\
\frac{d}{dt}Y_3 &= -(3\gamma + \gamma_{12} - i\Omega_{12})Y_3 \\
\frac{d}{dt}Y_4 &= -(3\gamma + \gamma_{12} + i\Omega_{12})Y_4 \\
\frac{d}{dt}Y_5 &= -2\gamma Y_5 - 2\gamma_{12}Y_6 + 8\gamma_{12}Y_7 \\
\frac{d}{dt}Y_6 &= -2\gamma Y_6 - 2\gamma_{12}Y_5 \\
\frac{d}{dt}Y_7 &= -4\gamma Y_7 \\
\frac{d}{dt}Y_8 &= -2\gamma Y_8 \\
\frac{d}{dt}Y_9 &= -2\gamma Y_9
\end{aligned} \tag{70}$$

where

$$\begin{aligned}
&Y_1 = \langle S_1^- + S_2^- \rangle \quad Y_2 = \langle S_1^+ + S_2^+ \rangle \quad Y_3 = \langle S_1^+ S_1^- S_2^- + S_2^+ S_1^- S_2^- \rangle \\
&Y_4 = \langle S_1^+ S_2^+ S_1^- + S_1^+ S_2^+ S_2^- \rangle \quad Y_5 = \langle S_1^+ S_2^- + S_2^+ S_1^- \rangle \\
&Y_6 = \langle S_1^+ S_1^- + S_2^+ S_2^- \rangle \quad Y_7 = \langle S_1^+ S_2^+ S_1^- S_2^- \rangle \quad Y_8 = \langle S_1^+ S_2^+ \rangle \\
&Y_9 = \langle S_1^- S_2^- \rangle
\end{aligned} \tag{71}$$

and γ_{12} and Ω_{12} are given by Eqs. (19) and (20), respectively.

Equations (70) are simple in form and can be solved exactly. The solutions have the following form:

$$
\begin{aligned}
Y_1(t) &= Y_1(0)\exp[-(\gamma + \gamma_{12} + i\Omega_{12})t] \\
&\quad + \frac{Y_3(0)(\gamma_{12} + i\Omega_{12})}{(\gamma - i\Omega_{12})}\{1 - \exp[-2(\gamma - i\Omega_{12})t]\} \\
&\quad \times \exp[-(\gamma + \gamma_{12} + i\Omega_{12})t] \\
Y_2(t) &= Y_1^*(t) \\
Y_3(t) &= Y_4^*(t) = Y_3(0)\exp[-(3\gamma + \gamma_{12} - i\Omega_{12})t] \\
Y_5(t) &= -\frac{4\gamma\gamma_{12}Y_7(0)}{(\gamma^2 - \gamma_{12}^2)}\exp(-4\gamma t) \\
&\quad + \left\{\frac{1}{2}[Y_5(0) - Y_6(0)] + \frac{2\gamma_{12}Y_7(0)}{(\gamma + \gamma_{12})}\right\} \\
&\quad \times \exp[-(\gamma - \gamma_{12})t] \\
&\quad + \left\{\frac{1}{2}[Y_5(0) + Y_6(0)] + \frac{2\gamma_{12}Y_7(0)}{(\gamma - \gamma_{12})}\right\} \\
&\quad \times \exp[-(\gamma + \gamma_{12})t] \qquad (72) \\
Y_6(t) &= -\frac{4\gamma_{12}^2 Y_7(0)}{(\gamma^2 - \gamma_{12}^2)}\exp(-4\gamma t) \\
&\quad + \left\{\frac{1}{2}[Y_6(0) - Y_5(0)] - \frac{2\gamma_{12}Y_7(0)}{(\gamma + \gamma_{12})}\right\}\exp[-(\gamma - \gamma_{12})t] \\
&\quad + \left\{\frac{1}{2}[Y_6(0) + Y_5(0)] + \frac{2\gamma_{12}Y_7(0)}{(\gamma - \gamma_{12})}\right\}\exp[-(\gamma + \gamma_{12})t] \\
Y_7(t) &= Y_7(0)\exp(-4\gamma t) \\
Y_8(t) &= Y_8(0)\exp(-2\gamma t) \\
Y_9(t) &= Y_8^*(t)
\end{aligned}
$$

where $Y_i(0)$ $(i = 1, \ldots, 9)$ describe the initial expectation values of the atomic correlation functions. They are dependent on the initial population of the atomic states. Consider the initial $(t = 0)$ superposition state [50]

$$
|\psi_0\rangle = [\cos\tfrac{1}{2}\theta_1|e_1\rangle + \sin\tfrac{1}{2}\theta_1\, e^{i\varphi_1}|g_1\rangle][\cos\tfrac{1}{2}\theta_2|e_2\rangle + \sin\tfrac{1}{2}\theta_2\, e^{i\varphi_2}|g_2\rangle] \qquad (73)
$$

in which the atoms are in an arbitrary linear combination of their states. If the two-atom system is initially in the states $|\psi_0\rangle$, then

$$\begin{aligned}
Y_1(0) &= \tfrac{1}{2}\left(\sin\theta_1\, e^{-i\varphi_1} + \sin\theta_2\, e^{-i\varphi_2}\right)\\
Y_2(0) &= Y_1^*(0)\\
Y_3(0) &= \tfrac{1}{2}\left(\sin\theta_1 \cos^2\left(\tfrac{1}{2}\theta_2\right) e^{-i\varphi_1} + \sin\theta_2 \cos^2\left(\tfrac{1}{2}\theta_1\right) e^{-i\varphi_2}\right)\\
Y_4(0) &= Y_3^*(0)\\
Y_5(0) &= \tfrac{1}{2}\sin\theta_1 \sin\theta_2 \cos(\varphi_1 - \varphi_2)\\
Y_6(0) &= \cos^2\left(\tfrac{1}{2}\theta_1\right) + \cos^2\left(\tfrac{1}{2}\theta_2\right)\\
Y_7(0) &= \cos^2\left(\tfrac{1}{2}\theta_1\right)\cos^2\left(\tfrac{1}{2}\theta_2\right)\\
Y_8(0) &= \tfrac{1}{4}\sin\theta_1 \sin\theta_2 \exp[i(\varphi_1 + \varphi_2)]\\
Y_9(0) &= \tfrac{1}{4}\sin\theta_1 \sin\theta_2 \exp[-i(\varphi_1 + \varphi_2)]
\end{aligned} \tag{74}$$

If we choose our initial conditions so that $\theta_1 = \theta_2 = \theta_0$, $\varphi_1 = 0$, and $\varphi_2 = \pi$, then $Y_1(0) = Y_2(0) = 0$ and $Y_8(0) = Y_9(0) = -\frac{1}{4}\sin^2\theta_0$. In such a state the expectation values of the dipole moments are zero. Nevertheless, squeezing is still possible because the two-photon coherences $Y_8(0)$ and $Y_9(0)$ lead to phase sensitivity in the variance (69). At the inital time $t = 0$ the variance (69) is

$$\left\langle :(\Delta E_\theta)^2: \right\rangle = \tfrac{1}{2}\psi^2(\mathbf{R})\left(1 + \cos\theta_0 - \sin^2\theta_0 \cos^2\theta\right) \tag{75}$$

The minimum value of $\langle :(\Delta E_\theta)^2: \rangle$ corresponding to optimum squeezing occurs for $\theta = 0$ and $\theta_0 = 2\pi/3$ when

$$\left\langle :(\Delta E_\theta)^2: \right\rangle = -\tfrac{1}{8}\psi^2(\mathbf{R}) \tag{76}$$

With the parameters $\theta_0 = 2\pi/3$, $\varphi_1 = 0$ and $\varphi_2 = \pi$ the initial state $|\varphi_0\rangle$ has the form

$$\begin{aligned}
|\varphi_0\rangle = \frac{1}{4}&[3|g_1\rangle|g_2\rangle - |e_1\rangle|e_2\rangle]\\
&+ \frac{\sqrt{3}}{4}[|e_1\rangle|g_2\rangle - |e_2\rangle g_1\rangle] = |\psi_1\rangle + |\psi_2\rangle
\end{aligned} \tag{77}$$

where $|\psi_1\rangle$ is a linear combination of the atomic states in which both atoms are in their ground or excited states, whereas $|\psi_2\rangle$ is an antisymmetric combination of the excited and ground states of the atoms. In the

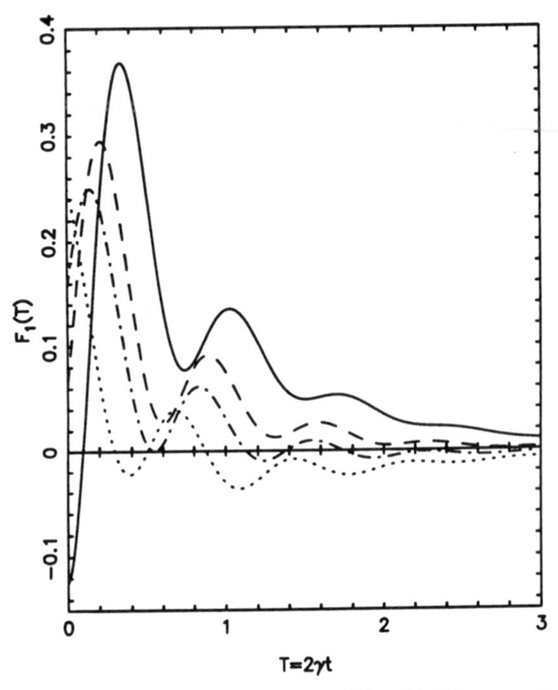

Figure 10. Time dependence of $F(T) = \langle:(\Delta E_\theta)^2:\rangle/\psi^2(\mathbf{R})$ for $r_{12} = 0.12\lambda, \theta_1 = \theta_2 = 2\pi/3, \varphi_1 = \varphi_2 = 0$, and for different phases $\theta: \theta = 0$ (solid line), $\theta = \pi/4$ (dashed line), $\theta = \pi/3$ (dash-dotted line), $\theta = \pi/2$ (dotted line).

notation of the Dicke states [39] the state $|\psi_2\rangle$ corresponds to the antisymmetric state $|0, 0\rangle$, whereas the state $|\psi_1\rangle$ corresponds to a linear combination of the Dicke states $|1, 1\rangle$ and $|1, -1\rangle$. The state $|\psi_1\rangle$, which is the linear combination of the Dicke states $|1, 1\rangle$ and $|1, -1\rangle$ and gives squeezing in the quadrature component $E_{\theta=0}$, is known in the literature as a two-atom squeezed state [51], or pairwise atomic state [52].

In Fig. 10 we plot the variance (69) as a function of time t for $r_{12} = 0.12\lambda, \theta_1 = \theta_2 = 2\pi/3, \varphi_1 = \varphi_2 = 0$ and for different values of the phase θ. These graphs show that the fluctuations in the quadrature component E_θ strongly depend on phase θ and can be squeezed at different times t. If the quadrature component E_θ were squeezed initially then it is not squeezed at later times, whereas an initially unsqueezed quadrature component can be squeezed at later times. This effect does not appear in a single atom spontaneous emission, and is due to interatomic interactions which create linear superpositions of the atomic states.

The idea of squeezing in spontaneous emission can be extended to higher order quadrature components of the electromagnetic field. Hong and Mandel [53] defined Nth-order squeezing and showed that resonance fluorescence predicts this type of squeezing. Another type of higher-order squeezing, called amplitude-squared squeezing, has been defined by Hillery [54]. It has been shown that the amplitude-squared squeezing occurs in multiatom resonance fluorescence, and does not appear in spontaneous emission and resonance fluorescence from a single two-level atom [55–57]. This is an another example that shows that the interatomic interactions can produce squeezing in spontaneous emission.

VII. SUMMARY

In this chapter we have considered two nonclassical effects: photon antibunching and squeezing in the spontaneous emission and resonance fluorescence from two-level atoms. Spontaneous emission from a single two-level atom shows squeezing if at the initial time the atom were in a suitable prepared linear combination of its excited and ground states. Resonance fluorescence from a single atom shows photon antibunching and squeezing independent of the initial preparation of the atom. However, squeezing strongly depends on the intensity of the driving laser field, and in the steady-state appears only for a weak exciting field. A strong laser field can produce squeezing only in the transient regime of resonance fluorescence. Interatomic interactions have a destructive effect on these two nonclassical effects. However, a considerable amount of photon antibunching and squeezing can be obtained in a two-atom resonance fluorescence when the detuning of the laser frequency from atomic resonance and the dipole–dipole interaction between the atoms cancel our mutually. Moreover, squeezing can appear in the steady-state resonance fluorescence even for a strong exciting laser field.

We have also discussed the possibility of obtaining squeezed states in spontaneous emission from two interacting atoms. We have shown that the two-atom system can produce squeezing even for the vanishing atomic dipole moments. The system produces squeezing through having two-photon coherences different from zero [50, 51].

In conclusion, we have demonstrated nonclassical effects exhibited by independent as well as correlated atoms. These effects are not evident when the electromagnetic field is treated classically and easily manifest the quantum nature of resonance fluorescence.

As a result of recent successful experiments, which have generated squeezed light, interest is now turning to possible applications. Gardiner [58] first pointed out that squeezed light incident upon a single two-level atom can in principle inhibit the phase decay of that atom, giving rise to line narrowing in the spectrum of resonance fluorescence [58]. Since that first paper, analyses have been extended to the treatment of atomic absorption spectra [60, 61], atomic level shifts in a squeezed vacuum [33, 62, 63], and squeezed pump lasers [64]. Multiatom and multilevels systems in a squeezed vacuum are now extensively studied [65–74], and show novel effects not observed in an ordinary vacuum.

Acknowledgments

The authors wish to thank Professor S. Kielich for the initiative for this work and valuable comments. Z. F. is indebted to the Australian Research Council for support through a QEII Fellowship.

References

1. H. P. Yuen, *Phys. Rev. A* **13**, 2226 (1976).
2. H. J. Carmichael and D. F. Walls, *J. Phys. B* **9**, 1199 (1976).
3. H. J. Kimble and L. Mandel, *Phys. Rev. A* **13**, 2123 (1976).
4. D. Stoler, *Phys. Rev. Lett.* **33**, 1397 (1974).
5. P. Chmela, *Acta Phys. Pol. A* **52**, 835 (1977).
6. H. D. Simaan and R. Loudon, *J. Phys. A* **11**, 435 (1978).
7. H. Voight, A. Bandilla, and H. H. Ritze, *Z. Phys. B* **35**, 461 (1980).
8. J. Mostowski and K. Rzazewski, *Phys. Lett* **66A**, 275 (1978).
9. P. D. Drummond, K. J. McNeil and D. F. Walls, *Opt. Commun.* **28**, 255 (1979).
10. J. Perina, in E. Wolf (Ed.), *Progress in Optics*, Vol. 18, North-Holland, Amsterdam, 1980, p. 127.
11. H. H. Ritze, *Z. Phys. B* **39**, 353 (1980).
12. M. Kozierowski and R. Tanaś, *Opt. Commun.* **21**, 229 (1977).
13. S. Kielich, M. Kozierowski and R. Tanaś, in L. Mandel and E. Wolf (Ed.), *Coherence and Quantum Optics IV*, Plenum, New York, 1978, p. 511.
14. R. Tanaś and S. Kielich, *Opt Commun.* **30**, 443 (1979).
15. P. Szlachetka, S. Kielich, J. Perina, and V. Perinova, *J. Phys. A* **12**, 1921 (1979).
16. S. Kielich, M. Kozierowski, and R. Tanaś, *Opt. Acta* **32**, 1023 (1985).
17. S. Kielich, *Nonlinear Molecular Optics*, Nauka, Moscow, 1981.
18. D. F. Walls and P. Zoller, *Phys. Rev. Lett.* **47**, 709 (1981).
19. L. Mandel, *Phys. Rev. Lett.* **49**, 136 (1982).
20. R. Tanaś and S. Kielich, *Opt. Commun.* **45**, 351 (1983); *Opt. Acta* **31**, 81 (1984).
21. H. J. Kimble, M. Dagenais, and L. Mandel, *Phys. Rev. Lett.* **39**, 691 (1977).
22. R. M. Slusher, L. W. Hollberg, B. Yurke, J. C. Mertz, and J. F. Valley, *Phys. Rev. Lett.* **55**, 2409 (1985).
23. L. A. Wu, H. J. Kimble, J. L. Hall, and H. Wu, *Phys. Rev. Lett.* **57**, 2520 (1986).
24. A. Heidmann, R. Horowicz, S. Reynaud, E. Giacobino, C. Fabre, and G. Camy, *Phys. Rev. Lett.* **59**, 2555 (1987).

25. T. Debuisschert, S. Reynaud, A. Heidmann, E. Giacobino, and C. Fabre, *Quantum Opt.* **1**, 3 (1989).
26. Y. Yamamoto, M. Imoto, and S. Machida, *Phys. Rev. A* **32**, 2287 (1986).
27. G. S. Agarwal, in G. Höhler (Ed.), *Quantum Optics*, Springer Tracts in Modern Physics, Vol. 70 Springer, Berlin, 1974, p. 25.
28. C. Cohen-Tannoudji in Frontiers in Laser Spectroscopy, R. Balian, S. Haroche, and S. Liberman (Eds.), North-Holland, Amsterdam, 1977, Vol. 1, p. 28.
29. B. J. Dalton, *Lectures in Quantum Optics*, The University of Queensland, 1990.
30. M. A. Dupertuis and S. Stenholm, *J. Opt. Soc. Am. B* **4**, 1094 (1987).
31. L. Allen and J. H. Eberly, *Optical Resonance and Two-Level Atoms*, Wiley, New York, 1975.
32. P. W. Milonni and P. L. Knight, *Phys. Rev. A* **10**, 1096 (1974); P. W. Milonni, *Phys. Rep. C* **25**, 1 (1976).
33. G. W. Ford and R. F. O'Connell, *J. Opt. Soc. Am. B* **4**, 1710 (1987).
34. R. H. Lehmberg, *Phys. Rev. A* **2**, 883 (1970).
35. Z. Ficek, R. Tanaś, and S. Kielich, *Physica A* **146**, 452 (1987).
36. M. J. Stephen, *J. Chem. Phys.* **40**, 669 (1964).
37. E. C. G. Sudarshan, *Phys. Rev. Lett.* **10**, 277 (1963).
38. R. J. Glauber, *Phys. Rev.* **130**, 2529 (1963).
39. R. H. Dicke, *Phys. Rev.* **93**, 99 (1954).
40. S. M. Barnett and P. L. Knight, *Phys. Scr.* **T21**, 5 (1988).
41. M. Lax, *Phys. Rev.* **172**, 350 (1968).
42. Z. Ficek, R. Tanaś, and S. Kielich, *J. Opt. Soc. Am. B* **1**, 882 (1984).
43. G. S. Agarwal, L. M. Narducci, D. H. Feng, and R. Gilmore, *Phys. Rev. Lett.* **42**, 1260 (1979).
44. S. Y. Kilin, *Sov. Phys. JETP* **51**, 1981 (1980).
45. Z. Ficek, R. Tanaś, and S. Kielich, *J. Phys.* **48**, 1697 (1987).
46. Z. Ficek, R. Tanaś, and S. Kielich, *Opt. Acta* **30**, 713 (1983).
47. Z. Ficek, R. Tanaś, and S. Kielich, *Phys. Rev. A.* **29**, 2004 (1984).
48. Th. Richter, *Opt. Acta* **29**, 265 (1982).
49. L. Mandel, *Phys. Scr.* **T12**, 34 (1986).
50. Z. Ficek, B. J. Dalton, and P. L. Knight, to be published.
51. S. M. Barnett and M. A. Dupertuis, *J. Opt. Soc. Am. B* **4**, 505 (1987).
52. Z. Ficek, *Phys. Rev. A* **44**, 7759 (1991).
53. C. K. Hong and L. Mandel, *Phys. Rev. A* **32**, 974 (1985).
54. M. Hillery, *Phys. Rev. A* **36**, 3796 (1987).
55. Z. Ficek, R. Tanaś, and S. Kielich, *Opt. Commun.* **69**, 20 (1988).
56. T. Quang, L. H. Lan, A. S. Shumovsky, and V. Buzek, *Opt. Commun.* **76**, 47 (1990).
57. M. H. Mahran, *Phys. Rev. A* **42**, 4199 (1990).
58. C. W. Gardiner, *Phys. Rev. Lett.* **56**, 1917 (1986).
59. H. J. Carmichael, A. S. Lane, and D. F. Walls, *Phys. Rev. Lett.* **58**, 2539 (1987); P. R. Rice and L. M. Pedvotti, *J. Opt. Soc. Am.* **B9**, 2008 (1992).
60. H. Ritsch and P. Zoller, *Phys. Rev. A* **38**, 4657 (1988).

61. S. An, M. Sargent III, and D. F. Walls, *Opt. Commun.* **67**, 373 (1988).
62. G. J. Milburn, *Phys. Rev. A* **34**, 4882 (1986).
63. G. M. Palma and P. L. Knight, *Opt. Commun.* **73**, 131 (1989).
64. M. A. Marte, H. Ritsch, and D. F. Walls, *Phys. Rev. A* **34**, 3577 (1988).
65. A. S. Shumovsky and T. Quang, *J. Phys. B* **22**, 131 (1989).
66. Z. Ficek and B. C. Sanders, *Quantum Opt.* **2**, 269 (1990).
67. G. M. Palma and P. L. Knight, *Phys. Rev. A* **39**, 1962 (1989).
68. G. S. Agarwal and R. R. Puri, *Phys. Rev. A* **41**, 3782 (1990).
69. Z. Ficek, *Opt. Commun.* **82**, 130 (1991).
70. J. Gea-Banacloche, *Phys. Rev. Lett.* **62**, 1603 (1989).
71. J. Javanainen and P. L. Gould, *Phys. Rev. A* **41**, 5088 (1990).
72. Z. Ficek, *Phys. Rev. A* **42**, 611 (1990); *Opt. Commun.* **88**, 494 (1992).
73. Z. Ficek and P. D. Drummond, *Phys. Rev. A* **43**, 6247 (1991); **43**, 6258 (1991).
74. V. Buzek, P. L. Knight, and I. K. Kudryavtsev, *Phys. Rev. A* **44**, 1931 (1991).

SQUEEZED STATES OF LIGHT IN THE SECOND AND THIRD HARMONIC GENERATED BY SELF-SQUEEZED LIGHT

S. KIELICH AND K. PIĄTEK

Nonlinear Optics Division, Institute of Physics, Adam Mickiewicz University, Poznań, Poland

CONTENTS

I. INTRODUCTION

Squeezed states in optical fields are at present a very attractive problem for theorists and experimenters (extensive accounts of the literature can be found in review articles [1–4] and in special issues of journals [5, 6]).

This work was supported by Polish Government Grant KBN 201 509 101.

Modern Nonlinear Optics, Part 1, Edited by Myron Evans and Stanisław Kielich. Advances in Chemical Physics Series, Vol. LXXXV.
ISBN 0-471-57546-1

Possibilities of their generation have been found in many nonlinear processes, such as resonance fluorescence [7–11], parametric amplification [12–19], four-wave mixing [20–31], multiphoton absorption [32, 33], the Jaynes-Cummings model [34–37], parametric down-conversion, [38–40], nonlinear propagation of light, and the harmonics generation considered in this paper.

In Section II we give a short review of the research related to light propagation and second- and third-harmonic generation in a nonlinear medium, especially from the point of view of quantum effects. Section III contains fundamental information about the squeezed states of light (also referred to in this paper as "ordinary" squeezed states). Two models of light propagation in a nonlinear medium are discussed in Sections III and IV: an anharmonic oscillator model and a model based on the effective Hamiltonian of the system. We study the squeezing effect in both approaches, analyzing the exact solutions for the quadrature variances. These results are also compared with ordinary squeezing, and the classical description of light propagation is recalled.

In Section V second-harmonic generation is studied. The first part of this section contains the classical treatment of the phenomenon. In the second part squeezing is discussed on the basis of the approximate analytical results holding in the quantum description. Section IV presents the quantum description of third-harmonic generation. Using the approximate solutions for the variances of the quadrature operators we analyze the squeezing effect.

II. HISTORY AND PERSPECTIVES

The first observation of the second harmonic of a laser beam by Franken et al. [41] has become a landmark in nonlinear optics. The classical description of the effect is due to Armstrong et al. [42]. In the next few years scientists concentrated on finding nonclassical properties of light in nonlinear media. In 1970 Walls [43] showed that the intensity of the second-harmonic beam exhibits periodic behavior in the quantum treatment, unlike the classical approach. Two years later Crosignani et al. [44] proved the impossibility of complete vanishing of the fundamental field in second-harmonic generation. Next, the second harmonic [45], higher order harmonics [46] and subharmonic [47] were studied for their quantum statistical properties. Stolarov [48], Kozierowski and Tanaś [49], and Kielich et al. [50] have shown that if the incoming beam is in a coherent state the antibunching effect (see, for example, Refs. 51–53) occurs in harmonics generation. This effect was also studied in Refs. 54 and 55.

For experiments the possibility of obtaining steady states is essential. Hence, theorists have searched for the quantum effect in an optical cavity. McNeil et al. [56] forecasted the "self-pulsing" effect in the intensities of the second-harmonic and fundamental beam in a cavity. The antibunching effect and bistability in the subharmonic and second-harmonic generation in a Fabry-Pérot cavity system were analyzed by Drummond et al. [57–59].

The search for squeezing in harmonics generation began in 1982. The first results were due to Mandel [60]. Kozierowski and Kielich [61] and Kielich et al. [62, 63] proposed a more general description. Lugiato et al. [64] predicted this nonclassical effect in the fundamental and second-harmonic beam, generated in a nonlinear crystal in a cavity. Friberg and Mandel [65] found the possibility of generation of squeezed states via a combination of parametric down-conversion and second-harmonic generation.

At the same time Tanaś [66] proposed the anharmonic oscillator model to describe laser light propagating in a nonlinear medium, which gives a squeezing effect different from ordinary squeezing. Tanaś and Kielich [67, 68] proposed the name *self-squeezing* when analyzing the effective Hamiltonian of the system. Moreover, Kielich et al. [69] applied an external magnetic field along the direction of propagation to achieve control of the self-squeezing of light. The evolution of the field in a nonlinear medium was described by Milburn [70, 71], who used the quasiprobability function $Q(\alpha, \alpha^*, t)$. He succeeded in revealing its periodic behavior and the role of dissipation in the effect and predicted squeezing in this treatment. Yurke and Stoler [72] proved that the state produced in the anharmonic oscillator model can be a superposition of a finite number of coherent states.

In 1985 Hong and Mandel [73] introduced the definition of higher-order squeezing and, in particular, studied second-harmonic generation. Kozierowski [74] searched for this effect in nth-harmonic generation. The squeezing of the square of the amplitude in second-harmonic generation was analyzed by Hillery [75], who predicted this kind of squeezing in the fundamental beam. He also found correlation between ordinary squeezing in the harmonic beam and amplitude-squared squeezing in the fundamental. Amplitude-squared squeezing was observed by Sizman et al. [76], who measured a 40% reduction of noise.

Correlation between the fundamental field and the harmonic beam was searched for in Lukš et al. [77]. The possibility of producing squeezed states in the fundamental stimulated by multiple higher-harmonic generation was proposed in Chmela et al. [78]. Kielich et al. analyzed the second harmonic [79] and third harmonic [80] generated by self-squeezed light.

Ekert and Rzążewski [81] found that the intensity of the second harmonic depends on the kind of fundamental beam. The state of the fundamental field in second-harmonic generation has been studied from the point of view of the initial phase [82]. In 1988 Pereira et al. [83] observed squeezing in the fundamental beam in second-harmonic generation. His result has been compared with the theoretical description [84].

The anharmonic oscillator model has been studied extensively at the same time. Lukš et. al. [85] defined "principal squeezing" related to the geometrical representation of the quadrature components as an ellipse. Loudon [86] proposed a different representation by Booth's elliptical lemniscate. Both representations were compared by Tanaś et al. [87]. They also proved that "crescent" squeezing, introduced by Kitagawa and Yamamoto [88] and Yamamoto et al. [89] (see also [4]), is the same as self-squeezing. Using the quasiprobability function Miranowicz et al. [90] proved the generation of superpositions of coherent states in the anharmonic oscillator model.

Recently light propagation in a nonlinear medium has been studied from the point of view of second- and fourth-order squeezing [91], the saturation effect [92], and the effect of dispersion [93]. It has been found that squeezing decreases with increasing saturation parameter and can be produced only within a limited frequency interval, determined by dispersion in the medium. Tanaś and Kielich [94] considered the role played by higher-order nonlinearity in the self-squeezing of light.

The theoretical basis for experiments has been prepared in recent years. The possibility of producing squeezed states in nth-harmonic generation in a laser resonator was proposed by Gorbaczev and Polzik [95]. Schack et al. [96] described a method of a doubly resonant cavity containing a laser medium as well as a χ^2 nonlinearity. They have predicted more than 60% squeezing in the up-converted mode. The "input–output" theory has been used by Collett and Levien [97] to generate the second harmonic. They obtained 50% squeezing. An analysis of the resonator parameters was given in Ref. 98. In 1991 You-bang Zhan [99, 100] studied in detail amplitude-cubed squeezing in second- and third-harmonic generation and amplitude-squared squeezing in nth-harmonic generation.

Quantum fluctuations in the Stokes operators of elliptically polarized light propagating in a Kerr medium were discussed by Tanaś and Kielich [101], who treated the medium as optically transparent, and by Tanaś and Gantsog [102] for a medium with dissipation. For the two-mode case, the influence of losses and noise was discussed by Horak and Peřina [103]. The influence of dissipation on the dynamics of the anharmonic oscillator, i.e., the one-mode propagation problem, was considered by Milburn and

Holmes [71], and recently the exact solutions of the master equation for the system have been discussed [104–107].

On the basis of the Pegg-Barnett formalism [108–110] the theorists have attempted to search for quantum phase properties in nonlinear processes. Tanaś et al. [111] analyzed the superposition of coherent states in the anharmonic oscillator model, using the quasiprobability function as well as the probability phase distribution $P(\theta)$. Quantum phase fluctuations in nonlinear processes are discussed in Refs. [112–114].

III. SQUEEZED STATES OF LIGHT

A. Minimum Uncertainty States and Coherent States

The Heisenberg uncertainty principle limits the possibility of measuring two observables in the same state. The variances of two observables A, B satisfy the following relation:

$$\left\langle (\Delta\hat{A})^2 \right\rangle \left\langle (\Delta\hat{B})^2 \right\rangle \geq \tfrac{1}{4} \left| \left\langle [\hat{A}, \hat{B}] \right\rangle \right|^2 \tag{1}$$

If the sign of equality holds in (1), the state is referred to as the "minimum uncertainty state."

It is well known that a single mode of an electromagnetic field in a cavity can be treated as a simple harmonic oscillator, described by the "position" and "momentum" operators. These are related to the electric and magnetic components of light. In the simplest case of a one-dimensional cavity with z axis, on the assumption of linear polarization, we can write

$$\begin{aligned} \hat{E}(z,t) &= \left(\frac{2\omega^2}{\varepsilon_0 V} \right)^{1/2} \hat{q}(t) \sin kz \\ \hat{H}(z,t) &= \left(\frac{2\varepsilon_0 c^2}{V} \right)^{1/2} \hat{p}(t) \cos kz \end{aligned} \tag{2}$$

where $\hat{E}(z,t)$ is the electric field operator, $\hat{H}(z,t)$ is the magnetic operator, $\hat{q}(t)$ is the position operator, $\hat{p}(t)$ is the momentum operator, ω is the frequency of the mode under consideration, and k is the wave vector. The commutation relation for $\hat{q}$ and $\hat{p}$ is defined as

$$[\hat{p}(t), \hat{q}(t)] = -i\hbar \tag{3}$$

Let us consider the photon number states space. When describing the electric field in the states it is helpful to introduce the annihilation $\hat{a}$ and creation $\hat{a}^+$ operators, which obey the following relations:

$$\begin{aligned} \hat{a} &= (2\hbar\omega)^{-1/2}(\omega\hat{q} + \mathrm{i}\hat{p}) \\ \hat{a}^+ &= (2\hbar\omega)^{-1/2}(\omega\hat{q} - \mathrm{i}\hat{p}) \\ [\hat{a}, \hat{a}^+] &= 1 \end{aligned} \tag{4}$$

According to Eqs. (4) the electric field can be written as

$$\hat{E}(z,t) = C[\hat{a}(t) + \hat{a}^+(t)] \tag{5}$$

where $C = (\hbar\omega/\varepsilon_0 V)^{1/2} \sin kz$, $\hat{a}(t) = \hat{a}(0)\exp(-\mathrm{i}\omega t)$. The annihilation and creation operators act on the photon number state as follows:

$$\begin{aligned} \hat{a}|n\rangle &= n^{1/2}|n-1\rangle \\ \hat{a}^+|n\rangle &= (n+1)^{1/2}|n+1\rangle \end{aligned} \tag{6}$$

Hence, the number state $|n\rangle$ can be created from the vacuum state $|0\rangle$:

$$|n\rangle = (n!)^{1/2}(\hat{a}^+)^n|0\rangle \tag{7}$$

According to Eqs. (6), the average values of the position and momentum equal zero:

$$\langle\hat{q}\rangle = \langle\hat{p}\rangle = 0 \tag{8}$$

but their variances do not:

$$\begin{aligned} \langle(\Delta\hat{q})^2\rangle &= \frac{\hbar}{2\omega}\langle(\hat{a} + \hat{a}^+)(\hat{a} + \hat{a}^+)\rangle = \frac{\hbar}{2\omega}(1+2n) \\ \langle(\Delta\hat{p})^2\rangle &= \frac{\hbar\omega}{2}(1+2n) \end{aligned} \tag{9}$$

So, using expression (3), we write the uncertainty relation (1) for $\hat{p}$ and $\hat{q}$ as follows:

$$\langle(\Delta\hat{q})^2\rangle\langle(\Delta\hat{p})^2\rangle > \frac{\hbar^2}{4} \qquad (n > 0) \tag{10}$$

and

$$\left\langle (\Delta\hat{q})_0^2 \right\rangle \left\langle (\Delta\hat{p})_0^2 \right\rangle = \frac{\hbar^2}{4} \qquad (n = 0) \tag{11}$$

These equations mean that only the vacuum state is the minimum uncertainty state among the photon number states.

In 1963 Glauber introduced coherent states, which are the eigenstates for the annihilation operator [115]:

$$\begin{aligned} \hat{a}|\alpha\rangle &= \alpha|\alpha\rangle \\ \langle\alpha|\hat{a}^+ &= \langle\alpha|\alpha^* \end{aligned} \tag{12}$$

where $\alpha = |\alpha|\exp(i\phi)$. The coherent state can be constructed from the number states:

$$|\alpha\rangle = \exp\left(-\frac{|\alpha|^2}{2}\right) \sum_{n=0}^{\infty} \frac{\alpha^n}{(n!)^{1/2}} |n\rangle \tag{13}$$

These states are characterized by the Poisson photon-number distribution:

$$|\langle n|\alpha\rangle|^2 = \exp(-|\alpha|^2)\frac{\alpha^{2n}}{n!} \tag{14}$$

The average value of the photon number takes the form

$$\langle\alpha|\hat{n}|\alpha\rangle = |\alpha|^2$$

and its variance

$$\begin{aligned} \left\langle \alpha|(\Delta\hat{n})^2|\alpha \right\rangle &= |\alpha|^2 \\ \left\langle (\Delta\hat{n})^2 \right\rangle &= \langle\hat{n}\rangle \end{aligned} \tag{15}$$

Hence, the variance is equal to the average value of the photon number in this case. Taking into account Eqs. (4) and (12), the expectation values of the position and momentum operators are given as follows:

$$\begin{aligned} \langle\alpha|\hat{q}|\alpha\rangle &= \left(\frac{\hbar}{2\omega}\right)^{1/2} (\alpha + \alpha^*) \\ \langle\alpha|\hat{p}|\alpha\rangle &= -i\left(\frac{\hbar\omega}{2}\right)^{1/2} (\alpha - \alpha^*) \end{aligned} \tag{16}$$

and for their squares we have

$$\begin{aligned} \langle\alpha|\hat{q}^2|\alpha\rangle &= \left(\frac{\hbar}{2\omega}\right)\left[\alpha^2 + (\alpha^*)^2 + 2|\alpha|^2 + 1\right] \\ \langle\alpha|\hat{p}^2|\alpha\rangle &= \left(\frac{\hbar\omega}{2}\right)\left[-\alpha^2 - (\alpha^*)^2 + 2|\alpha|^2 + 1\right] \end{aligned} \tag{17}$$

Thus, we have the following variances:

$$\begin{aligned} \left\langle(\Delta\hat{q})^2\right\rangle &= \frac{\hbar}{2\omega} \\ \left\langle(\Delta\hat{p})^2\right\rangle &= \frac{\hbar\omega}{2} \end{aligned} \tag{18}$$

It is easy to check that the left side of the uncertainty relation (1)

$$\left\langle(\Delta\hat{q})^2\right\rangle\left\langle(\Delta\hat{p})^2\right\rangle = \frac{\hbar^2}{4}$$

is equal to the right side,

$$\frac{1}{4}\left|\left\langle[\hat{p},\hat{q}]\right\rangle\right|^2 = \frac{\hbar^2}{4} \tag{19}$$

This proves that all coherent states are minimum uncertainty states.

The coherent state can be generated from the vacuum [115]

$$\begin{aligned} |\alpha\rangle &= \hat{D}(\alpha)|0\rangle \\ \hat{D}(\alpha) &= \exp(\alpha\hat{a}^+ - \alpha^*\hat{a}) \end{aligned} \tag{20}$$

where $\hat{D}(\alpha)$ is the unitary displacement operator. The above operator transforms $\hat{a}$ and $\hat{a}^+$ as follows:

$$\begin{aligned} \hat{D}^+(\alpha)\hat{a}\,\hat{D}(\alpha) &= \hat{a} + \alpha \\ \hat{D}^+(\alpha)\hat{a}^+\,\hat{D}(\alpha) &= \hat{a}^+ + \alpha^* \end{aligned} \tag{21}$$

A more detailed discussion of coherent states is to be found in the review papers (for example, [116]).

B. Quadrature Operators

The annihilation and creation operators are non-Hermitian. It is useful to break them down into Hermitian quadrature operators [2, 3]:

$$\begin{aligned} \hat{Q} &= \hat{a} + \hat{a}^{+} \\ \hat{P} &= -\mathrm{i}(\hat{a} - \hat{a}^{+}) \end{aligned} \tag{22}$$

They satisfy the commutation relation

$$\left[\hat{Q}, \hat{P}\right] = 2\mathrm{i} \tag{23}$$

The uncertainty equation for them takes the form

$$\left\langle (\Delta\hat{Q})^2 \right\rangle \left\langle (\Delta\hat{P})^2 \right\rangle \geq 1 \tag{24}$$

Considering the minimum uncertainty state (coherent state or vacuum state) we obtain the variances equal to unity:

$$\left\langle (\Delta\hat{Q})^2 \right\rangle = \left\langle (\Delta\hat{P})^2 \right\rangle = 1$$

and

$$\left\langle (\Delta\hat{Q})^2 \right\rangle \left\langle (\Delta\hat{P})^2 \right\rangle = 1 \tag{25}$$

In terms of the quadrature operators (22), the electric field can be written as

$$\hat{E}(z, t) = C\left[\hat{Q} \cos \omega t + \hat{P} \sin \omega t\right] \tag{26}$$

Hence, $\hat{Q}$ and $\hat{P}$ may be identified with the amplitudes of the two quadrature phases of the electric field [117].

The displacement operator (20) for the quadrature components,

$$\hat{D}(\alpha) = \exp\left[\mathrm{i}\left(\mathrm{Im}\, \alpha\hat{Q} - \mathrm{Re}\, \alpha\hat{P}\right)\right] \tag{27}$$

transforms them in the following way:

$$\begin{aligned} \hat{D}^{+}(\alpha)\hat{Q}\hat{D}(\alpha) &= \hat{Q} + 2\,\mathrm{Re}\, \alpha \\ \hat{D}^{+}(\alpha)\hat{P}\hat{D}(\alpha) &= \hat{P} + 2\,\mathrm{Im}\, \alpha \end{aligned} \tag{28}$$

C. Squeezed States

The variances of the quadrature operators are equal (25) for a coherent state. However, one can imagine that one of them has a value below unity but together with the second variance satisfies the uncertainty relation (25). So the following generalized definition for the quadrature operators can be introduced:

$$\begin{aligned} \hat{Q}_s &= \hat{Q} \exp(-s) \\ \hat{P}_s &= \hat{P} \exp(s) \end{aligned} \tag{29}$$

where s is called the squeezing parameter. According to Eqs. (29), the quadrature variances have the modified form:

$$\begin{aligned} \left\langle \left(\Delta \hat{Q}_s\right)^2 \right\rangle &= \exp(-2s) \\ \left\langle \left(\Delta \hat{P}_2\right)^2 \right\rangle &= \exp(2s) \end{aligned} \tag{30}$$

and the annihilation and creation operators, connected with them (22), take the new form

$$\begin{aligned} \hat{a}_s &= \hat{a} \cosh s - \hat{a}^+ \sinh s \\ \hat{a}_s^+ &= \hat{a}^+ \cosh s - \hat{a} \sinh s \end{aligned} \tag{31}$$

These generalized operators remain in commutation relations:

$$\begin{aligned} \left[\hat{Q}_s, \hat{P}_s\right] &= 2\mathrm{i} \\ \left[\hat{a}_s, \hat{a}_s^+\right] &= 1 \end{aligned} \tag{32}$$

The displacement operator has to be redefined:

$$\begin{aligned} \hat{D}_s(\alpha) &= \exp\left[\alpha \hat{a}_s^+ - \alpha^* \hat{a}_s\right] \\ \hat{D}_s(\alpha) &= \exp\left[\mathrm{i}\left(\mathrm{Im}\, \alpha \hat{Q}_s - \mathrm{Re}\, \alpha \hat{P}_s\right)\right] \end{aligned} \tag{33}$$

If $s > 0$ the exponential, in Eqs. (29), compresses the original variance of $\hat{Q}$ and expands the original variance of $\hat{P}$. The squeezing condition for one

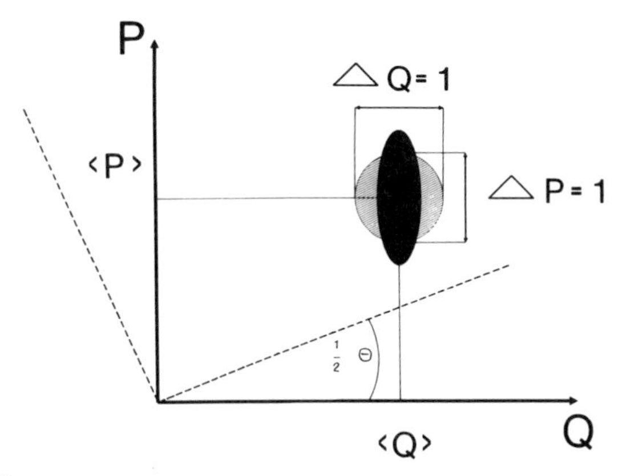

Figure 1. The error countours in the quadrature components plane for the coherent state (circle) and the squeezed state (ellipse) when $\theta = 0$. If $\theta \neq 0$ the error ellipse is inclined at angle $\theta/2$ (dashed axes). We denote $\Delta\hat{X} = \langle(\Delta\hat{X})^2\rangle^{1/2}$ for $\hat{X} = \hat{Q}, \hat{P}$.

of the quadrature components ($\hat{X} = \hat{Q}, \hat{P}$) is given as

$$\left\langle(\Delta\hat{X})^2\right\rangle < 1 \tag{34}$$

In the $\hat{Q}, \hat{P}$ plane the circular error contour for the coherent state is squeezed into an elliptical error contour (Fig. 1).

So far we have discussed the simplest case of squeezing, occurring along the $\hat{Q}$ and $\hat{P}$ component. To obtain squeezed states in general the squeeze operator has to be used [2, 3]:

$$\hat{S}(\zeta) = \exp\left[\tfrac{1}{2}\left(\zeta^*\hat{a}^2 - \zeta\hat{a}^{+2}\right)\right] \tag{35}$$

where $\zeta = s\exp(i\theta)$, $0 \leq \theta \leq 2\pi$, and $0 \leq s \leq \infty$. This operator squeezes the error contour in directions inclined at angles $\theta/2$ to the $\hat{Q}$ and $\hat{P}$ axes (dashed line in Fig. 1).

The squeeze operator (35) transforms the $\hat{a}$ and $\hat{a}^+$ operators in the following manner:

$$\begin{aligned} \hat{S}^+(\zeta)\hat{a}\hat{S}(\zeta) &= \hat{a}\cosh s - \hat{a}^+\exp(i\theta)\sinh s \\ \hat{S}^+(\zeta)\hat{a}^+\hat{S}(\zeta) &= \hat{a}^+\cosh s - \hat{a}\exp(-i\theta)\sinh s \end{aligned} \tag{36}$$

Using the squeeze operator we can define the squeezed states,

$$|\alpha, \zeta\rangle = \hat{D}(\alpha)\hat{S}(\zeta)|0\rangle \tag{37}$$

This definition (37) of squeezed states was proposed by Caves [118]. Yuen has given an alternative definition [119]:

$$|\beta, \mu, \nu\rangle = \hat{U}\hat{D}(\beta)\ |0\rangle \tag{38}$$

where $\hat{U}$ is the squeeze operator and $\hat{D}(\beta)$ the displacement operator. The two formalisms of squeezed states are equivalent [2],

$$\begin{aligned} \beta &= \mu\alpha + \nu\alpha^* \\ \mu &= \cosh s \\ \nu &= \exp(i\theta)\sinh s \end{aligned} \tag{39}$$

D. Fundamental Properties of Squeezed States

In this section some useful properties of the one-mode squeezed state are discussed.

- The average values of the annihilation and creation operators are

$$\begin{aligned} \langle\hat{a}\rangle &= \alpha \\ \langle\hat{a}^+\rangle &= \alpha^* \end{aligned} \tag{40}$$

  They do not depend on the squeezing parameter ζ.
- The average photon number is

$$\langle\hat{n}\rangle = \langle\hat{a}^+\hat{a}\rangle = |\alpha|^2 + \sinh^2 s \tag{41}$$

  The second term arises from the process of squeezing the vacuum.
- The eigenvalues of the quadrature operators are

$$\begin{aligned} \langle\hat{Q}\rangle &= \alpha + \alpha^* = 2\,\mathrm{Re}\,\alpha \\ \langle\hat{P}\rangle &= 2\,\mathrm{Im}\,\alpha \end{aligned} \tag{42}$$

  They also are independent of the squeezing parameter ζ.
- The variances of the quadrature operators are as follows:

$$\begin{aligned} \left\langle(\Delta\hat{Q})^2\right\rangle &= \exp(-2s)\cos^2\left(\tfrac{1}{2}\theta\right) + \exp(2s)\sin^2\left(\tfrac{1}{2}\theta\right) \\ \left\langle(\Delta\hat{P})^2\right\rangle &= \exp(-2s)\sin^2\left(\tfrac{1}{2}\theta\right) + \exp(2s)\cos^2\left(\tfrac{1}{2}\theta\right) \end{aligned} \tag{43}$$

  The variances are not dependent on the coherent amplitude α.

- The uncertainty relation (1) for the quadrature operators is

$$\left\langle (\Delta\hat{Q})^2 \right\rangle \left\langle (\Delta\hat{P})^2 \right\rangle = \cosh^2 2s \sin^2\theta + \cos^2\theta$$

It is obvious that the minimum quantum noise occurs for $\theta = \theta, \pi$:

$$\left\langle (\Delta\hat{Q})^2 \right\rangle \left\langle (\Delta\hat{P})^2 \right\rangle = 1$$

and the maximum for $\theta = \pi/2, 3\pi/2$:

$$\left\langle (\Delta\hat{Q})^2 \right\rangle \left\langle (\Delta\hat{P})^2 \right\rangle = \cosh^2 2s \tag{44}$$

- The condition for squeezing in the $\hat{Q}$ component is

$$\cos\theta > \tanh s$$

Therefore,

$$\left\langle (\Delta\hat{Q})^2 \right\rangle = \exp(-2s) \qquad (\text{for } \theta = 0) \tag{45}$$

- The condition for squeezing in the $\hat{P}$ component is

$$\cos\theta < -\tanh s$$

Therefore

$$\left\langle (\Delta\hat{P})^2 \right\rangle = \exp(-2s) \qquad \left(\text{for } \theta = \frac{\pi}{2}\right) \tag{46}$$

E. Two-Mode Squeezing

Obviously, it is possible to define squeezing not only for the one-mode case. If we consider light in two different frequencies ω_+ and ω_- it is useful to introduce two-mode squeezed states, which can be obtained from the vacuum state [2, 3],

$$|\alpha_+, \alpha_-, \zeta\rangle = \hat{D}_+(\alpha_+)\hat{D}_-(\alpha_-)\hat{S}(\zeta)|0\rangle$$

where the displacement operators are defined as

$$\hat{D}_\pm(\alpha_\pm) = \exp\left(\alpha_\pm \hat{a}_\pm^+ - \alpha_\pm^* \hat{a}_\pm\right)$$

and the squeeze operator is

$$\hat{S}(\zeta) = \exp\left(\zeta^* \hat{a}_+ \hat{a}_- - \zeta \hat{a}_+^+ \hat{a}_-^+\right) \tag{47}$$

This two-mode squeeze operator transforms the annihilation and creation operators [2, 117]:

$$\begin{aligned} \hat{S}^+(\zeta)\hat{a}_\pm \hat{S}(\zeta) &= \hat{a}_\pm \cosh s - \hat{a}_\mp^+ \exp(\mathrm{i}\theta)\sinh s \\ \hat{S}^+(\zeta)\hat{a}_\pm^+ \hat{S}(\zeta) &= \hat{a}_\pm^+ \cosh s - \hat{a}_\mp \exp(-\mathrm{i}\theta)\sinh s \end{aligned} \tag{48}$$

Similarly to the one-mode case, two-mode quadrature operators can be defined similarly to one-mode operators:

$$\begin{aligned} \hat{Q} &= \frac{1}{\sqrt{2}}\left(\hat{a}_+ + \hat{a}_+^+ + \hat{a}_- + \hat{a}_-^+\right) \\ \hat{P} &= \frac{-\mathrm{i}}{\sqrt{2}}\left(\hat{a}_+ - \hat{a}_+^+ + \hat{a}_- - \hat{a}_-^+\right) \end{aligned} \tag{49}$$

Because of the usefulness of two-mode squeezed states, in the next sections we given some of their more important properties:

$$\langle \hat{a}_\pm \rangle = \alpha_\pm \tag{50}$$

$$\langle \hat{n}_\pm \rangle = |\alpha_\pm|^2 + \sinh^2 s \tag{51}$$

$$\left\langle \hat{a}_\pm^+ \hat{a}_\mp \right\rangle = \alpha_\pm^* \alpha_\mp \tag{52}$$

$$\langle \hat{a}_\pm \hat{a}_\pm \rangle = \alpha_\pm^2 \tag{53}$$

$$\langle \hat{a}_\pm \hat{a}_\mp \rangle = \alpha_+ \alpha_- - \exp(\mathrm{i}\theta)\sinh s \cosh s \tag{54}$$

and for the quadrature operators:

$$\begin{aligned} \langle \hat{Q} \rangle &= 2^{1/2}(\mathrm{Re}\, \alpha_+ + \mathrm{Re}\, \alpha_-) \\ \langle \hat{P} \rangle &= 2^{1/2}(\mathrm{Im}\, \alpha_+ + \mathrm{Im}\, \alpha_-) \end{aligned} \tag{55}$$

$$\begin{aligned} \left\langle (\Delta \hat{Q})^2 \right\rangle &= \exp(-2s)\cos^2\left(\tfrac{1}{2}\theta\right) + \exp(2s)\sin^2\left(\tfrac{1}{2}\theta\right) \\ \left\langle (\Delta \hat{P})^2 \right\rangle &= \exp(-2s)\sin^2\left(\tfrac{1}{2}\theta\right) + \exp(2s)\cos^2\left(\tfrac{1}{2}\theta\right) \end{aligned} \tag{56}$$

The variances of the quadrature operators for two-mode states (56) are identical with the variances for one mode. This means that they are independent of the number of modes in the field.

In this section the theory of the squeezed states is only touched on. We have left out an account of higher-order squeezing [3, 73] and the amplitude-squared squeezing defined by Hillery [75].

Squeezed states of light are not considered only theoretically. In recent years many experimental results have been reported [29, 30, 31, 40, 76, 83, 120, 121]. To measure the variance of a quadrature component of the field a special phase-sensitive method is needed. It has been shown that homodyne and heterodyne detections are suitable. The homodyne method is used for a single quadrature measurement and the heterodyne measures both. These methods are based on the interference of squeezed light with a coherent field.

In the next sections we discuss in detail the possibilities of generating squeezed states in the propagation of light and harmonics generation in a nonlinear medium.

IV. ANHARMONIC OSCILLATOR MODEL

The anharmonic oscillator is the simplest model for the description of interaction between quantum light and a nonlinear medium. It was proposed by Tanaś [66]. In spite of its simplicity, this model gives the possibility of obtaining exact analytical results which, among other things, show the dissimilarity between the squeezing process in light propagation and the ordinary squeezing, generated by the squeeze operator (35).

It is assumed that the well-known Hamiltonian of the anharmonic oscillator can describe, for example, a single mode of the field propagating through a nonlinear medium. Then the Hamiltonian takes the form [66]

$$\hat{H} = \hbar\omega\hat{a}^{+}\hat{a} + \tfrac{1}{2}\hbar\kappa\hat{a}^{+2}\hat{a}^{2} \tag{57}$$

where $\hat{a}$, $\hat{a}^{+}$ are the annihilation and creation operators of the mode, ω is the frequency of the mode, and κ is an anharmonicity parameter (real). It is necessary to know the time evolution of $\hat{a}$ and $\hat{a}^{+}$ to obtain information about the respective quantum effects. To attain this the Heisenberg equation is constructed:

$$\frac{\mathrm{d}\hat{a}}{\mathrm{d}t} = \frac{1}{\mathrm{i}\hbar}[\hat{a}, \hat{H}] \tag{58}$$

According to the Hamiltonian (57) the equation of motion has the following form:

$$\frac{d\hat{a}}{dt} = -i(\omega + \kappa \hat{a}^{+}\hat{a})\hat{a} \tag{59}$$

Since the number-photon operator $\hat{n} = \hat{a}^{+}\hat{a}$ is a constant of motion,

$$[\hat{n}, \hat{H}] = 0 \tag{60}$$

it is possible to obtain the solution in the form

$$\hat{a}(t) = \exp\{-it[\omega + \kappa \hat{a}^{+}(0)\hat{a}(0)]\}\hat{a}(0) \tag{61}$$

where $\hat{a}(0)$, $\hat{a}^{+}(0)$ are the annihilation and creation operators at $t = 0$. The term $\exp(-i\omega t)$ is associated with the free evolution of the system, whereas the second term comes from the nonlinear interaction included in the second part of the Hamiltonian (57). This exact operator solution (61) allows us to give all the characteristics of the field at the time t, if the state of the field at $t = 0$ is known.

We assume that the field is in a coherent state $|\alpha\rangle$ initially ($t = 0$). Since the photon number is a constant of motion, the photon-number distribution retains Poissonian statistics (14). This does not mean that the field has to be in a coherent state throughout its evolution. To search for squeezing we use the quadrature operators defined in (22). As was shown in Section III this effect occurs if one of the variances of the quadrature components has a value below unity:

$$\langle(\Delta\hat{Q})^{2}\rangle < 1 \quad \text{or} \quad \langle(\Delta\hat{P})^{2}\rangle < 1 \tag{62}$$

It is convenient to introduce normal ordering of the operators. Then we can write

$$\begin{aligned} \langle:(\Delta\hat{Q})^{2}:\rangle &= \langle(\Delta\hat{Q})^{2}\rangle - 1 \\ \langle:(\Delta\hat{P})^{2}:\rangle &= \langle(\Delta\hat{P})^{2}\rangle - 1 \end{aligned} \tag{63}$$

where the colons denote normal ordering. Here, the squeezing conditions take the following form:

$$\langle:(\Delta\hat{Q})^{2}:\rangle < 0 \quad \text{or} \quad \langle:(\Delta\hat{P})^{2}:\rangle < 0 \tag{64}$$

meaning that squeezing occurs when one of the normally ordered variances takes a negative value. In terms of the annihilation and creation operators these variances can be written as

$$\begin{aligned}\left\langle :(\Delta\hat{Q})^2: \right\rangle &= \left\langle (\Delta\hat{a})^2 \right\rangle + \left\langle (\Delta\hat{a}^+)^2 \right\rangle + 2\left(\langle\hat{a}^+\hat{a}\rangle - \langle\hat{a}^+\rangle\langle\hat{a}\rangle\right) \\ \left\langle :(\Delta\hat{P})^2: \right\rangle &= -\left\langle (\Delta\hat{a})^2 \right\rangle - \left\langle (\Delta\hat{a}^+)^2 \right\rangle + 2\left(\langle\hat{a}^+\hat{a}\rangle - \langle\hat{a}^+\rangle\langle\hat{a}\rangle\right)\end{aligned} \tag{65}$$

Using solution (61) in the equations above, the following results can be derived [66]:

$$\begin{aligned}\left\langle :(\Delta\hat{Q})^2: \right\rangle &= 2\,\mathrm{Re}\left[\alpha^2 \exp\left[-i\tau + |\alpha|^2(\exp(-2i\tau) - 1)\right]\right] \\ &\quad - 2\,\mathrm{Re}\left[\alpha^2 \exp\left[2|\alpha|^2(\exp(-i\tau) - 1)\right]\right] \\ &\quad + 2|\alpha|^2\left[1 - \exp\left[2|\alpha|^2(\cos\tau - 1)\right]\right] \\ \left\langle :(\Delta\hat{P})^2: \right\rangle &= -2\,\mathrm{Re}[\,\cdots\,] + 2\,\mathrm{Re}[\,\cdots\,] + 2|\alpha|^2[\,\cdots\,]\end{aligned} \tag{66}$$

where $\tau = \kappa t$; the brackets in the second equation (66) contain the same expressions as the first equation; α is the coherent amplitude, and $|\alpha|^2$ is the average number of photons. The variances of the quadrature operators (66) are plotted against $\beta = |\alpha|^2\tau$ in Fig. 2. We assumed that $\tau = 1 \times 10^{-6}$ and chose the initial phase to have α real. Both curves oscillate between

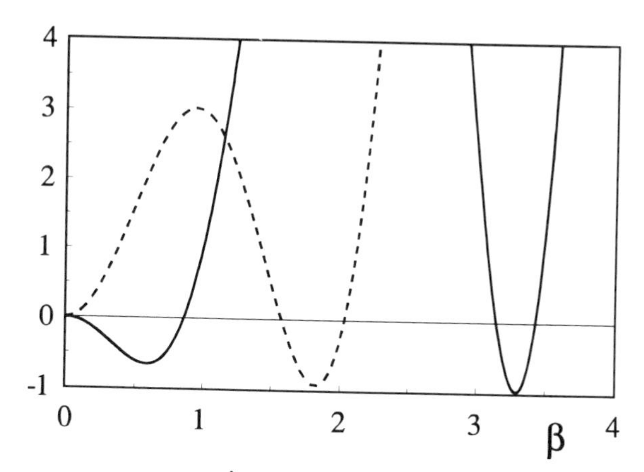

Figure 2. The variance of the $\hat{Q}$ quadrature component (solid line) and the variance of the $\hat{P}$ component (dashed line) are plotted versus $\beta = |\alpha|^2\tau$.

negative and positive values. The first minimum of the $\hat{Q}$ component (solid line) has a value of -0.66 and appears for $\beta = 0.6$. The second minimum is deeper and reaches -0.98. The first minimum of the $\hat{P}$ component (dashed line) occurs for $\beta = 1.82$ and has a value of -0.93. The next minimum is deeper and reaches 0.99. Note that if one of the variances is squeezed then the other is not.

This analysis supposes that a considerable amount of squeezing can be obtained for a large number of photons ($|\alpha|^2 \gg 1$). In this case we can assume that $|\alpha|^2\tau$ takes a value of the order of unity. Moreover, we can make the assumption that $\tau \ll 1$, because of the small value of the anharmonicity parameter κ. These assumptions allow us to expand equations (66) in power series and to retain only the leading terms. Hence we have the following approximate formulas for the quadrature variances:

$$\begin{aligned} \left\langle :(\Delta\hat{Q})^2: \right\rangle &\approx 2\beta[\beta - (\sin 2\beta + \beta\cos 2\beta)] \\ \left\langle :(\Delta\hat{P})^2: \right\rangle &\approx 2\beta[\beta + \sin 2\beta + \beta\cos 2\beta] \end{aligned} \tag{67}$$

These equations are simpler and we shall use them to compare the results obtained in the next section.

Formulas (66) and (67) mean that the states obtained in the anharmonic oscillator model do not preserve minimum uncertainty in the sense of the coherent states, i.e., fluctuations in one of the quadrature components of the field can be reduced.

We would like to emphasize that the solution (61) differs from the transformation (36) which defines ordinary squeezing.

In the next section it is shown that the result derived from the anharmonic oscillator is a particular case of a more general model.

V. SELF-SQUEEZING OF LIGHT IN NONLINEAR MEDIUM

A. Classical Treatment: Self-phase Modulation

Before giving a description of the squeezing effect in light propagation through a nonlinear medium, as proposed by Tanaś and Kielich [67, 68], it may be helpful to recall the classical treatment and some of its more interesting results. The classical approach is based on the assumption that the electric field is described by a vector E, which can be the sum of the positive and negative frequency parts at the time–space point $(\mathbf{r}, t)$:

$$E(\mathbf{r}, t) = E^{+}(\mathbf{r}, t) + E^{-}(\mathbf{r}, t) \tag{68}$$

The positive and negative parts can be written as

$$\begin{aligned} E^{+}(\mathbf{r},t) &= \sum_i E^{+}(\omega_i)\exp\left[\mathrm{i}(\mathbf{k}_{\omega_i}\cdot\mathbf{r}-\omega_i t)\right] \\ E^{-}(\mathbf{r},t) &= \sum_i E^{-}(\omega_i)\exp\left[-\mathrm{i}(\mathbf{k}_{\omega_i}\cdot\mathbf{r}-\omega_i t)\right] \end{aligned} \tag{69}$$

where ω_i is the frequency and $\mathbf{k}_{\omega_i}$ the wave vector of ith mode. On taking into account one mode only, formulas (69) take the following form:

$$\begin{aligned} E^{+}(\mathbf{r},t) &= E^{+}(\omega)\exp[\mathrm{i}(\mathbf{k}\cdot\mathbf{r}-\omega t)] \\ E^{-}(\mathbf{r},t) &= E^{-}(\omega)\exp[-\mathrm{i}(\mathbf{k}\cdot\mathbf{r}-\omega t)] \end{aligned} \tag{70}$$

We are interested in the interaction between the field and the nonlinear medium. It is contained in the time-averaged free energy [122],

$$F = -\tfrac{3}{4}\chi_{ijkl}E_i^{-}(\omega)E_j^{-}(\omega)E_k^{+}(\omega)E_l^{+}(\omega) + \text{c.c.} \tag{71}$$

where $E^{\pm}_{i,j,k,l}(\omega)$ are the components of the vector $E^{\pm}$, χ_{ijkl} is the fourth-rank tensor, describing the third-order nonlinear susceptibility, and the summation is defined by Einstein's convention. The free energy is the starting point for the calculation of the components of the nonlinear polarization vector at the frequency ω. They can be obtained from the well-known formula

$$P_i^{\pm}(\omega) = -\frac{\partial F}{\partial E_i^{\mp}} \tag{72}$$

Hence, on using Eq. 71, the vector components take the form

$$\begin{aligned} P_i^{+}(\omega) &= 3\chi_{ijkl}(-\omega,-\omega,\omega,\omega)E_j^{-}(\omega)E_k^{+}(\omega)E_l^{+}(\omega) \\ P_i^{-}(\omega) &= 3\chi_{ijkl}(-\omega,-\omega,\omega,\omega)E_j^{-}(\omega)E_k^{-}(\omega)E_l^{+}(\omega) \end{aligned} \tag{73}$$

The above formulas do not include dissipative and resonant processes. In this case the following symmetry relation is fulfilled [122]:

$$\chi^{*}_{ijkl}(-\omega,-\omega,\omega,\omega) = \chi_{klij}(-\omega,-\omega,\omega,\omega) \tag{74}$$

On the assumption that the medium is isotropic and has a center of symmetry, it is possible to write the third-order susceptibility as follows [123]:

$$\chi_{ijkl}(-\omega,-\omega,\omega,\omega) = \chi_{xxyy}\delta_{ij}\delta_{kl} + \chi_{xyxy}\delta_{ik}\delta_{jl} + \chi_{xyyx}\delta_{il}\delta_{jk} \tag{75}$$

Moreover, this tensor is symmetrical in the pairs of indices i, j and k, l. So, instead of three, we have two independent components: χ_{xxyy}, and $\chi_{xyxy} = \chi_{xyyx}$.

Considering light propagation in an isotropic medium, it is of advantage to introduce a circular basis to describe the field and the polarization of the medium. When the field propagates along the z axis, then the right- and left-polarized components take the forms

$$E_{\pm}^{+} = 2^{-1/2}\left[E_x^{+}(\omega) \mp \mathrm{i}E_y^{+}(\omega)\right] \tag{76}$$

Using formulas (74)–(76) we obtain the following equation for the average free energy (71):

$$\begin{aligned} F = -\tfrac{1}{2}\Big\{ & g_1^{\omega}\left[E_{+}^{-}(\omega)^2 E_{+}^{+}(\omega)^2 + E_{-}^{-}(\omega)^2 E_{-}^{+}(\omega)^2\right] \\ & + 4g_2^{\omega}\left[E_{+}^{-}(\omega)E_{-}^{-}(\omega)E_{+}^{+}(\omega)E_{-}^{+}(\omega)\right]\Big\} \end{aligned} \tag{77}$$

where the nonlinear coupling parameters $g_1^{\omega}, g_2^{\omega}$ are defined as follows:

$$\begin{aligned} g_1^{\omega} &= 6\chi_{xyxy}(-\omega, -\omega, \omega, \omega) \\ g_2^{\omega} &= 3\left[\chi_{xxyy}(-\omega, -\omega, \omega, \omega) + \chi_{xyxy}(-\omega, -\omega, \omega, \omega)\right] \end{aligned} \tag{78}$$

Applying Eqs. (72) and (77), we write the components of the nonlinear polarization vector in the new representation as

$$P_{\pm}^{+}(\omega) = \left[g_1^{\omega}|E_{\pm}^{-}(\omega)|^2 + 2g_2^{\omega}|E_{\mp}^{-}(\omega)|^2\right]E_{\pm}^{+}(\omega) \tag{79}$$

This expression can be inserted into the Maxwell wave equation. In the slowly varying amplitude approximation the following equation is obtained [122]:

$$\frac{\mathrm{d}E_{\pm}^{+}}{\mathrm{d}z} = \mathrm{i}\frac{2\pi\omega}{n_{\omega}c}P_{\pm}^{+} \tag{80}$$

On insertion of (79) into (80) we have

$$\frac{\mathrm{d}E_{\pm}^{+}(\omega, z)}{\mathrm{d}z} = \mathrm{i}\frac{2\pi\omega}{n_{\omega}c}\left[g_1^{\omega}|E_{\pm}^{-}(\omega)|^2 + 2g_2^{\omega}|E_{\mp}^{-}(\omega)|^2\right]E_{\pm}^{+}(\omega) \tag{81}$$

where $n_\omega = k_\omega c/\omega$ is the refractive index for the frequency ω. Since $|E_\pm^+|^2$ does not depend on z (the derivative vanishes), Eq. (81) has the simple exponential solution

$$E_\pm^+(\omega, z) = \exp(i\phi_\pm z) E_\pm^+(\omega, 0) \tag{82}$$

where

$$\phi_\pm = \frac{2\pi\omega}{n_\omega c}\left[g_1^\omega |E_\pm^-(\omega)|^2 + 2g_2^\omega |E_\mp^-(\omega)|^2 \right] \tag{83}$$

is the light intensity-dependent phase of the light. This phase is responsible for the emergency of circular birefringence. The refractive index is connected with the nonlinear polarization by the relation [122, 123]

$$n_\omega \delta n_\pm E_\pm(r,t) = 2\pi P_\pm(r,t) \tag{84}$$

According to Eq. (79) one easily finds the variations $\delta n_\pm$:

$$\delta n_\pm(\omega) = \frac{2\pi}{n_\omega}\left[g_1^\omega |E_\pm^-(\omega)|^2 + 2g_2^\omega |E_\mp^-(\omega)|^2 \right] \tag{85}$$

So, the difference between the right- and left-polarized indices is determined as

$$\begin{aligned} \delta n_+(\omega) - \delta n_-(\omega) = \frac{2\pi}{n_\omega}\Big[& 6\chi_{xxyy}(-\omega, -\omega, \omega, \omega) \\ & \times \left(|E_-^-(\omega)^2| - |E_+^-(\omega)|^2 \right) \Big] \end{aligned} \tag{86}$$

The expression above is related to the self-rotation of the polarization ellipse, described by Maker et al. [124]. This effect occurs during interaction between a classical field and a nonlinear medium. To search for squeezing it is necessary to use a quantum description.

B. Quantum Treatment: Self-squeezing

The quantum description is based on the analytical form of the Hamiltonian. Generally, the Hamiltonian can be written as

$$H = H_{\mathrm{M}} + H_{\mathrm{FREE}} + H_{\mathrm{I}} \tag{87}$$

where H_M is the Hamiltonian for the nonlinear medium and H_{FREE} is the Hamiltonian for the free field. Our interest bears on H_I, the Hamiltonian describing the interaction between the medium and the propagating light. In nonlinear optics [125], it is useful to construct an effective interaction Hamiltonian. Such a Hamiltonian can be obtained from the averaged free energy of the system,

$$H_I = \int_V F \, dV \tag{88}$$

Formally, in the quantum approach, we replace the field vectors by field boson operators, defined as

$$\hat{E}_{\pm}^{+}(\omega) = i\left(\frac{2\pi\hbar\omega}{n_{\omega}^{2}V}\right)^{1/2} \hat{a}_{\pm} \tag{89}$$

where $\hat{a}_{\pm}, \hat{a}_{\pm}^{+}$ are the annihilation and creation operators, satisfying the commutation relations

$$\begin{aligned} \left[\hat{a}_i, \hat{a}_j\right] &= \left[\hat{a}_i^{+}, \hat{a}_j^{+}\right] = 0 \\ \left[\hat{a}_i, \hat{a}_j^{+}\right] &= \delta_{ij} \end{aligned} \tag{90}$$

On insertion of the averaged free energy (77) into the formula (88) one finds [67, 68]

$$\hat{H}_I = -\frac{\hbar}{2}\left[\bar{g}_1^{\omega}\left(\hat{a}_{+}^{+2}\hat{a}_{+}^{2} + \hat{a}_{-}^{+2}\hat{a}_{-}^{2}\right) + 4\bar{g}_2^{\omega}\hat{a}_{+}^{+}\hat{a}_{-}^{+}\hat{a}_{+}\hat{a}_{-}\right] \tag{91}$$

where the nonlinear coupling parameters have been denoted by

$$\begin{aligned} \bar{g}_1^{\omega} &= \frac{V}{\hbar}\left(\frac{2\pi\hbar\omega}{n_{\omega}^{2}V}\right)^{2} g_1^{\omega} \\ \bar{g}_1^{\omega} &= \frac{V}{\hbar}\left(\frac{2\pi\hbar\omega}{n_{\omega}^{2}V}\right)^{2} g_2^{\omega} \end{aligned} \tag{92}$$

We use the Hamiltonian (91) to find the field operator time dependence, according to the Heisenberg equation. Taking into account only the effective interaction Hamiltonian, we write

$$\frac{d\hat{E}_{\pm}^{+}(\omega)}{dt} = \frac{1}{i\hbar}\left[\hat{E}_{\pm}^{+}(\omega), \hat{H}_{I}\right] \tag{93}$$

The time evolution, generally, is considered in a quantum cavity. Since the propagating field must depend on the path z traversed in the medium, we replace t by $-n_{\omega}z/c$ and obtain the Heisenberg equation in the new form

$$\frac{d\hat{a}_{\pm}(z)}{dz} = -\frac{in_{\omega}}{\hbar c}\left[\hat{a}_{\pm}, \hat{H}_{I}\right] \tag{94}$$

On insertion of (91) into (94) and using the commutation relations (90), one easily finds

$$\frac{d\hat{a}_{\pm}(z)}{dz} = i\frac{n_{\omega}}{c}\left[\bar{g}_{1}^{\omega}\hat{a}_{\pm}^{+}\hat{a}_{\pm} + 2\bar{g}_{2}^{\omega}\hat{a}_{\mp}^{+}\hat{a}_{\mp}\right]\hat{a}_{\pm} \tag{95}$$

Since the number of photons in the two circular components $\hat{a}_{+}^{+}\hat{a}_{+}, \hat{a}_{-}^{+}\hat{a}_{-}$ are constants of motion, Eq. (95) has the simple exponential solution [67, 68]

$$\hat{a}_{\pm}(z) = \exp\left[i(\varepsilon\hat{a}_{\pm}^{+}(0)\hat{a}_{\pm}(0) + \delta\hat{a}_{\mp}^{+}(0)\hat{a}_{\mp}(0))\right]\hat{a}_{\pm}(0) \tag{96}$$

with the nonlinear parameters

$$\varepsilon = \frac{n_{\omega}}{c}\bar{g}_{1}^{\omega} \qquad \delta = 2\frac{n_{\omega}}{c}\bar{g}_{2}^{\omega} \tag{97}$$

This exact operator solution (96) for the field propagating in an isotropic nonlinear medium can be used to search for quantum effects.

Note that this solution is the general two-mode case of the single-mode solution (61), calculated on the anharmonic oscillator model. If the light is circularly, say right-polarized, then the second term in the exponential vanishes and we get the result for the anharmonic oscillator.

We assume that the incoming beam is in a coherent state with amplitude consisting of two components $|\alpha\rangle = |\alpha_{+}, \alpha_{-}\rangle$. To search for squeezing we introduce the quadrature components of the field (22). On insertion of (96) into the formulas (65) we obtain the following normally ordered quadrature variances [67, 68]:

$$
\begin{aligned}
\left\langle :\left(\Delta \hat{Q}_{+}\right)^{2}: \right\rangle = 2\,\mathrm{Re}\Big[\alpha_{+}^{2} \exp[\mathrm{i}\varepsilon \\
&+ (\exp(\mathrm{i}2\varepsilon) - 1)|\alpha_{+}|^{2} + (\exp(\mathrm{i}2\delta) - 1)|\alpha_{-}|^{2}\Big] \\
&- \alpha_{+}^{2} \exp\Big[2(\exp(\mathrm{i}\varepsilon) - 1)|\alpha_{+}|^{2} \\
&+ (\exp(\mathrm{i}\delta) - 1)|\alpha_{-}|^{2}\Big]\Big] \\
&+ 2|\alpha_{+}|^{2}\Big[1 - \exp\Big[2(\cos\varepsilon - 1)|\alpha_{+}|^{2} \\
&+ 2(\cos\delta - 1)|\alpha_{-}|^{2}\Big]\Big] \\
\left\langle :\left(\Delta \hat{P}_{+}\right)^{2}: \right\rangle &= -2\,\mathrm{Re}[\cdots] + 2|\alpha_{+}|^{2}[\cdots]
\end{aligned}
\tag{98}
$$

where the expressions in brackets in the second equation are the same as in the first equation. On replacing the indices + and − we obtain the variances of the left-polarized quadrature components $\hat{Q}_{-}$, $\hat{P}_{-}$. If one of the variances has a value less than zero the field is in the squeezed state.

Because of the complexity of the exact analytical results (98) it is difficult (without numerical analysis) to determine whether they are negative or positive. In real physical processes the nonlinear parameters are very small $\varepsilon \ll 1$, $\delta \ll 1$. This means that significant changes in fluctuations appear for large numbers of photons in the components $|\alpha|^{2} \gg 1$, in other words, for strong field. This fact allows us to expand the expressions (98) in power series and to neglect all terms less than $\varepsilon z|\alpha_{\pm}\pm^{2}$ or $\delta z|\alpha_{\pm}|^{2}$. On the assumption that the phase of the incoming beam is zero, i.e., $|\alpha_{\pm}| = \alpha_{\pm}$, we obtain the following simpler formulas for the normally ordered variances:

$$
\begin{aligned}
\left\langle :\left(\Delta \hat{Q}_{\pm}\right)^{2}: \right\rangle &= 2(\beta_{\pm}^{2} + \gamma_{+}\gamma_{-}) - 2\big[\beta_{\pm} \sin 2(\beta_{\pm} + \gamma_{\mp}) \\
&\quad + (\beta_{\pm}^{2} + \gamma_{+}\gamma_{-})\cos 2(\beta_{\pm} + \gamma_{\mp})\big] \\
\left\langle :\left(\Delta \hat{P}_{\pm}\right)^{2}: \right\rangle &= 2(\cdots) + 2[\cdots]
\end{aligned}
\tag{99}
$$

where the brackets in the second equation include the same expressions as those in the first equation. The parameters are defined as

$$\beta_{\pm} = \varepsilon z |\alpha_{\pm}|^2 \quad \text{and} \quad \gamma_{\pm} = \delta z |\alpha_{\pm}|^2 \tag{100}$$

Considering only one mode of the field, for example $|\alpha_{-}|^2 = 0$, Eqs. (99) go over into formulas (67) obtained on the anharmonic oscillator model.

The numerical results based on the exact solution (98) have been discussed in detail by Tanaś and Kielich [67, 68], showing the possibility of obtaining 98% of squeezing in one of the components for a proper choice of the initial phase. We should note that the canonical nonlinear transformation (96) differs from the transformation (48) for ordinary squeezing and this is the reason why the states obtained in this model also have different properties. Tanaś and Kielich [67, 68] proposed the term "self-squeezing" for the effect, because it depends on the intensity of the mode undergoing it. In 1986, when analyzing the states created due to self-phase modulation in a nonlinear medium, Kitagawa et al. [88, 89] obtained a quasiprobability density with crescent shape. In fact, crescent squeezing is the same as self-squeezing. Tanaś et al. compared the two representations in Ref. 87.

VI. SECOND-HARMONIC GENERATION BY SELF-SQUEEZED LIGHT IN NONLINEAR MEDIUM

A. Second-Harmonic Generation: Classical Treatment

Second-harmonic generation is an important and highly useful nonlinear process. Its first observation by Franken et al. [41] has been the source of much progress in nonlinear optics. Classical effects in second-harmonic generation have been studied extensively [122, 123], and before we describe squeezing we would like to recall some of them.

In the classical approach it is assumed that the field at the space–time point $(\mathbf{r}, t)$ is the superposition of two fields with the fundamental frequency ω and the second-harmonic frequency 2ω:

$$\begin{aligned} E(\mathbf{r}, t) &= E^{+}(\omega)\exp[\mathrm{i}(\mathbf{k}_{\omega} \cdot \mathbf{r} - \omega t)] \\ &\quad + E^{+}(2\omega)\exp[i(\mathbf{k}_{2\omega} \cdot \mathbf{r} - 2\omega t)] + \text{c.c.} \end{aligned} \tag{101}$$

where $\mathbf{k}_{\omega}, \mathbf{k}_{2\omega}$ are the wave vectors of the fundamental and second-harmonic light waves. We are interested in the interaction between the field and the nonlinear medium. Following Bloembergen [122] and Kielich

[126], the time-averaged free energy can be written in the form

$$\begin{aligned} F = & -\chi_{ijk}(-2\omega, \omega, \omega) E_i^-(2\omega) E_j^+(\omega) E_k^+(\omega) \exp(i\mathbf{\Delta k}_2 \cdot \mathbf{r}) + \text{c.c.} \\ & - \tfrac{3}{4}\left[\chi_{ijkl}(-\omega, -\omega, \omega, \omega) E_i^-(\omega) E_j^-(\omega) E_k^+(\omega) E_l^+(\omega) + \text{c.c.} \right] \\ & - 3\left[\chi_{ijkl}(-\omega, -2\omega, \omega, 2\omega) E_i^-(\omega) E_j^-(2\omega) E_k^+(\omega) \right. \\ & \qquad \left. \times E_l^+(2\omega) + \text{c.c.} \right] \\ & - \tfrac{3}{4}\left[\chi_{ijkl}(-2\omega, -2\omega, 2\omega, 2\omega) E_i^-(2\omega) E_j^-(2\omega) \right. \\ & \qquad \left. \times E_k^+(2\omega) E_l^+(2\omega) + \text{c.c.} \right] \end{aligned} \tag{102}$$

where $\mathbf{\Delta k}_2 = 2\mathbf{k}_\omega - \mathbf{k}_{2\omega}$. $E^{\pm}(\omega), E^{\pm}(2\omega)$ are the components of the field vectors. Recall from the preceding section that nonlinear polarization can be obtained from the averaged free energy (72). In this case we get the following form of the polarization components:

$$\begin{aligned} P_i^+(\omega) = & \, 2\chi_{ijk}(-\omega, -\omega, 2\omega) E_j^-(\omega) E_k^+(2\omega) \exp(-i\mathbf{\Delta k}_2 \cdot \mathbf{r}) \\ & + 3\chi_{ijkl}(-\omega, -\omega, \omega, \omega) E_j^-(\omega) E_k^+(\omega) E_l^+(\omega) \\ & + 6\chi_{ijkl}(-\omega, -2\omega, \omega, 2\omega) E_j^-(2\omega) E_k^+(\omega) E_l^+(2\omega) \end{aligned} \tag{103}$$

and for the second-harmonic frequency

$$\begin{aligned} P_i^+(2\omega) = & \, \chi_{ijk}(-2\omega, \omega, \omega) E_j^+(\omega) E_k^+(\omega) \exp(i\mathbf{\Delta k}_2 \cdot \mathbf{r}) \\ & + 6\chi_{ijkl}(-2\omega, -\omega, 2\omega, \omega) E_j^-(\omega) E_k^+(2\omega) E_l^+(\omega) \\ & + 3\chi_{ijkl}(-2\omega, -2\omega, 2\omega, 2\omega) E_j^-(2\omega) \\ & \times E_k^+(2\omega) E_l^+(2\omega) \end{aligned} \tag{104}$$

In Eq. (103) the third-rank tensor $\chi_{ijk}(-\omega, -\omega, 2\omega)$ describing second-order susceptibility is related to the reconversion of part of the second harmonic back into the fundamental beam, $\chi_{ijkl}(-\omega, -\omega, \omega, \omega)$. As we showed in Section V, this is related to self-induced ellipse rotation (86) and $\chi_{ijkl}(-\omega, -2\omega, \omega, 2\omega)$ determines the optical Kerr effect at ω due to the intensity $|E^-(2\omega)|^2$. In Eq. (104) the tensor $\chi_{ijk}(-2\omega, \omega, \omega)$ is responsible for second-harmonic generation [41]; $\chi_{ijkl}(-2\omega, -\omega, 2\omega, \omega)$ determines the variation of the refractive index at 2ω, stimulated by the intensity $|E^-(\omega)|^2$; and $\chi_{ijkl}(-2\omega, -2\omega, 2\omega, 2\omega)$ is connected with the effect of self-induced intensity-dependent refractive index at 2ω.

Since Eqs. (103) and (104) concern a nonresonant, nondissipative process, it is possible to derive the following symmetry relations for the susceptibility tensors:

$$\begin{aligned}
\chi^*_{ijkl}(-\omega,-2\omega,\omega,2\omega) &= \chi_{lkji}(-2\omega,-\omega,2\omega,\omega)\\
\chi^*_{ijkl}(-\omega,-2\omega,\omega,2\omega) &= \chi_{klij}(-\omega,-2\omega,\omega,2\omega)\\
\chi^*_{ijkl}(-\omega,-\omega,\omega,\omega) &= \chi_{klij}(-\omega,-\omega,\omega,\omega)\\
\chi^*_{ijk}(-\omega,-\omega,2\omega) &= \chi_{kij}(-2\omega,\omega,\omega)
\end{aligned} \tag{105}$$

Let us consider a nonlinear isotropic medium with a center of symmetry. In this case the tensors $\chi_{ijk}(-2\omega,\omega,\omega)$, which are responsible for the generation of the second harmonic, vanish. To arouse the wave at frequency 2ω an externally dc electric field has to be applied to destroy the center of symmetry. Then the medium becomes capable of generating the second-harmonic beam. Assuming the dc electric field to act along the y axis, the third-rank tensors can be written [127] as follows:

$$\chi^{2\omega}_{ijk}(E^0) = \chi^{2\omega}_{ijk}(0) + \chi^{2\omega}_{xxyy}\delta_{ij}E^0_k + \chi^{2\omega}_{xyxy}\delta_{ik}E^0_j + \chi^{2\omega}_{yxxy}\delta_{jk}E^0_i \tag{106}$$

where E^0 is the external dc field. Since the second harmonic propagates along the z axis, like the fundamental beam, the $\chi_{ijk}(0)$ vanish [79]. Moreover, considering the isotropic medium with center of symmetry we take into account the symmetry relation (75). As was shown in Section V, it is convenient to have recourse to circular components. If the field propagates along the z axis they are defined by formulas (76). On using this basis the averaged free energy takes the form [79]

$$\begin{aligned}
F = &-\tfrac{1}{2}\Big[g^{\omega}_1\big[E^-_+(\omega)^2E^+_+(\omega)^2 + E^-_-(\omega)^2E^+_-(\omega)^2\big]\\
&\quad +4g^{\omega}_2E^-_+(\omega)E^-_-(\omega)E^+_+(\omega)E^+_-(\omega)\\
&\quad +g^{2\omega}_1\big[E^-_+(2\omega)^2E^+_+(2\omega)^2 + E^-_-(2\omega)^2E^+_-(2\omega)^2\big]\\
&\quad +4g^{2\omega}_2E^-_+(2\omega)E^-_-(2\omega)E^+_+(2\omega)E^+_-(2\omega)\Big]\\
&-\mathrm{i}\Big[g^{2\omega}_3\big[E^-_+(2\omega)E^+_-(\omega)^2 - E^-_-(2\omega)E^+_-(\omega)^2\big]\\
&-2g^{2\omega}_4\big[E^-_+(2\omega) - E^-_-(2\omega)\big]E^+_+(\omega)E^+_-(\omega)\Big]\exp(\mathrm{i}\Delta k_2 r) + \text{c.c.}\\
&-g^{2\omega}_5\big[E^-_+(2\omega)E^-_-(\omega)E^+_+(\omega)E^+_-(2\omega)\\
&\quad +E^-_-(2\omega)E^-_+(\omega)E^+_-(\omega)E^+_+(2\omega)\big]\\
&-g^{2\omega}_6\big[E^-_+(2\omega)E^-_-(\omega)E^+_-(\omega)E^+_+(2\omega)\\
&\quad +E^-_-(2\omega)E^-_+(\omega)E^+_+(\omega)E^+_-(2\omega)\big]\\
&-g^{2\omega}_7\big[E^-_+(2\omega)E^-_+(\omega)E^+_+(\omega)E^+_+(2\omega)\\
&\quad +E^-_-(2\omega)E^-_-(\omega)E^+_-(\omega)E^+_-(2\omega)\big]
\end{aligned} \tag{107}$$

The nonlinear coupling parameters are defined as ($\Omega = \omega$ or 2ω)

$$
\begin{aligned}
g_1^{\Omega} &= 6\chi_{xyxy}(-\Omega, -\Omega, \Omega, \Omega) \\
g_2^{\Omega} &= 3\left[\chi_{xxyy}(-\Omega, -\Omega, \Omega, \Omega) + \chi_{xyxy}(-\Omega, -\Omega, \Omega, \Omega)\right] \\
g_3^{2\omega} &= 2^{1/2}\chi_{xxyy}(-2\omega, \omega, \omega, 0)E_y^0 \\
g_4^{2\omega} &= 2^{-1/2}\left[\chi_{xxyy}(-2\omega, \omega, \omega, 0) + \chi_{xyxy}(-2\omega, \omega, \omega, 0)\right]E_y^0 \\
g_5^{2\omega} &= 3\left[\chi_{xxyy}(-2\omega, -\omega, \omega, 2\omega) + \chi_{xyxy}(-2\omega, -\omega, \omega, 2\omega)\right] \\
g_6^{2\omega} &= 3\left[\chi_{xxyy}(-2\omega, -\omega, \omega, 2\omega) + \chi_{xyyx}(-2\omega, -\omega, \omega, 2\omega)\right] \\
g_7^{2\omega} &= 3\left[\chi_{xyxy}(-2\omega, -\omega, \omega, 2\omega) + \chi_{xyyx}(-2\omega, -\omega, \omega, 2\omega)\right]
\end{aligned}
\tag{108}
$$

In accordance with formula (72) we obtain the following components of the nonlinear polarization:

$$
\begin{aligned}
P_{\pm}^{+}(\omega) = &\left[g_1^{\omega}|E_{\pm}^{-}(\omega)|^2 + 2g_2^{\omega}|E_{\mp}^{-}(\omega)|^2\right]E_{\pm}^{+}(\omega) \\
&-2\mathrm{i}\left[\pm g_3^{-2\omega}E_{\pm}^{+}(2\omega)E_{\pm}^{-}(\omega)\right. \\
&\left.-g_4^{-2\omega}\left[E_{+}^{+}(2\omega) - E_{-}^{+}(2\omega)\right]E_{\mp}^{-}(\omega)\right] \\
&\times\exp(-\mathrm{i}\boldsymbol{\Delta}\mathbf{k}_2\cdot\mathbf{r}) + g_5^{2\omega}E_{-}^{-}(2\omega)E_{-}^{+}(\omega)E_{+}^{+}(2\omega) \\
&+\left[g_6^{2\omega}|E_{\mp}^{-}(2\omega)|^2 + g_7^{2\omega}|E_{\pm}^{-}(2\omega)|^2\right]E_{\pm}^{+}(\omega)
\end{aligned}
\tag{109}
$$

and at 2ω

$$
\begin{aligned}
P_{\pm}^{+}(2\omega) = &\left[g_1^{2\omega}|E_{\pm}^{-}(2\omega)|^2 + 2g_2^{2\omega}|E_{\mp}^{-}(2\omega)|^2\right]E_{\pm}^{+}(2\omega) \\
&+\mathrm{i}\left[\pm g_3^{2\omega}E_{\pm}^{+}(\omega)^2 \mp g_4^{2\omega}E_{+}^{+}(\omega)E_{-}^{+}(\omega)\right] \\
&\times\exp(\mathrm{i}\boldsymbol{\Delta}\mathbf{k}_2\cdot\mathbf{r}) + g_5^{2\omega}E_{-}^{-}(\omega)E_{+}^{+}(\omega)E_{-}^{+}(2\omega) \\
&+\left[g_6^{2\omega}|E_{\mp}^{-}(\omega)|^2 + g_7^{2\omega}|E_{\pm}^{-}(\omega)|^2\right]E_{\pm}^{+}(2\omega)
\end{aligned}
\tag{110}
$$

On insertion of these expressions into the Maxwell equation (80) and neglecting terms unrelated to the self-induced intensity-dependent effect, one finds

$$
\frac{\mathrm{d}E_{\pm}^{+}(\Omega)}{\mathrm{d}z} = \mathrm{i}\frac{2\pi\Omega}{n_{\Omega}c}\left[g_1^{\Omega}|E_{\pm}^{-}(\Omega)|^2 + 2g_2^{\Omega}|E_{\mp}^{-}(\Omega)|^2\right]E_{\pm}^{+}(\Omega) \tag{111}
$$

Since $(\mathrm{d}/\mathrm{d}z)|E_{\pm}^{-}|^2 = 0$, Eq. (111) possesses the simple solution

$$
E_{\pm}^{+}(\Omega, z) = \exp(\mathrm{i}\phi_{\pm}z)E_{\pm}^{+}(\Omega, 0) \tag{112}
$$

where the phase shifts have the form:

$$\phi_{\pm} = \mathrm{i}\frac{2\pi\Omega}{n_{\Omega}c}\left[g_1^{\Omega}|E_{\pm}^{-}(\Omega)|^2 + 2g_2^{\Omega}|E_{\mp}^{-}(\Omega)|^2\right] \tag{113}$$

Equation (112) represents the general solution for the fundamental wave ($\Omega = \omega$) (obtained above in Eq. (82)) and, at the same time, for the second harmonic ($\Omega = 2\omega$).

Similar to the case of light propagation at ω alone (Section V), it is possible to analyze the birefringence effects for the fundamental and second harmonic. Applying formula (84) and inserting the nonlinear polarization of Eqs. (109) and (110), one can easily calculate the variations of the refractive indices:

$$\begin{aligned} \delta n_{\pm}(\omega) &= \frac{2\pi}{n_{\omega}}\left[g_1^{\omega}|E_{\pm}^{-}(\omega)|^2 + 2g_2^{\omega}|E_{\mp}^{-}(\omega)|^2\right. \\ &\qquad \left. +g_6^{2\omega}|E_{\mp}^{-}(2\omega)|^2 + g_7^{2\omega}|E_{\pm}^{-}(2\omega)|^2\right] \\ \delta n_{\pm}(2\omega) &= \frac{2\pi}{n_{2\omega}}\left[g_1^{2\omega}|E_{\pm}^{-}(2\omega)|^2 + 2g_2^{2\omega}|E_{\mp}^{-}(2\omega)|^2\right. \\ &\qquad \left. +g_6^{2\omega}|E_{\mp}^{-}(\omega)|^2 + g_7^{2\omega}|E_{\pm}^{-}(\omega)|^2\right] \end{aligned} \tag{114}$$

Hence, the difference between the two circular components takes the form

$$\begin{aligned} &\delta n_{+}(\omega) - \delta n_{-}(\omega) \\ &\quad = \frac{2\pi}{n_{\omega}}\Big[6\chi_{xxyy}(-\omega,-\omega,\omega,\omega)\left[|E_{-}^{-}(\omega)|^2 - |E_{+}^{-}(\omega)|^2\right] \\ &\quad +3\left[\chi_{xxyy}(-\omega,-2\omega,\omega,2\omega) - \chi_{xyyx}(-\omega,-2\omega,\omega,2\omega)\right] \\ &\quad \times\left(|E_{-}^{-}(2\omega)|^2 - |E_{+}^{-}(2\omega)|^2\right)\Big] \end{aligned} \tag{115}$$

and, at 2ω,

$$\begin{aligned} &\delta n_{+}(2\omega) - \delta n_{-}(2\omega) \\ &\quad = \frac{2\pi}{n_{2\omega}}\Big[6\chi_{xxyy}(-2\omega,-2\omega,2\omega,2\omega)\left[|E_{-}^{-}(2\omega)|^2 - |E_{+}^{-}(2\omega)|^2\right] \\ &\quad +3\left[\chi_{xxyy}(-2\omega,-\omega,2\omega,\omega) - \chi_{xyyx}(-2\omega,-\omega,2\omega,\omega)\right] \\ &\quad \times\left(|E_{-}^{-}(\omega)|^2 - |E_{+}^{-}(\omega)|^2\right)\Big] \end{aligned} \tag{116}$$

The first term of Eq. (115) was discussed in Section V. The second term is responsible for the additional anisotropy caused by the intensity of the second harmonic. The effect determined by expression (116) has not been studied experimentally.

B. Squeezing in Second-Harmonic Generation

As was done in Subsection V.B, the electric field vectors should be replaced by boson operators in the quantum description. The operator for the fundamental field was defined in (89). Similarly, the operator for the second-harmonic field can be determined as

$$\hat{E}^{+}_{\pm}(2\omega) = \mathrm{i}\left(\frac{2\pi\hbar 2\omega}{n^2_{2\omega}V}\right)^{1/2}\hat{b}_{\pm} \tag{117}$$

where $\hat{b}_{\pm}, \hat{b}^{+}_{\pm}$ are the boson annihilation and creation operators for photons at the frequency 2ω. These operators obey the boson commutation relations (90) and additionally

$$\left[\hat{a}_i, \hat{b}_j\right] = \left[\hat{a}^{+}_i, \hat{b}^{+}_j\right] = \left[\hat{a}_i, \hat{b}^{+}_j\right] = 0 \tag{118}$$

To consider the squeezing effect it is necessary to have available the form of the field propagating through the nonlinear medium. We get it from the Heisenberg equation (93), taking into account the interaction Hamiltonian (slowly varying amplitude approximation), which can be derived from formula (88). On insertion of the averaged free energy (107) into (88) the interaction Hamiltonian, in our case, takes the form

$$\begin{aligned}
\hat{H}_{\mathrm{I}} = &-\frac{\hbar}{2}\Big[\bar{g}^{\omega}_1\left(\hat{a}^{+2}_{+}\hat{a}^2_{+} + \hat{a}^{+2}_{-}\hat{a}^2_{-}\right) + 4\bar{g}^{\omega}_2\hat{a}^{+}_{+}\hat{a}^{+}_{-}\hat{a}_{+}\hat{a}_{-} \\
&+\bar{g}^{2\omega}_1\left(\hat{b}^{+2}_{+}\hat{b}^2_{+} + \hat{b}^{+2}_{-}\hat{b}^2_{-}\right) + 4\bar{g}^{2\omega}_2\hat{b}^{+}_{+}\hat{b}^{+}_{-}\hat{b}_{+}\hat{b}_{-}\Big] \\
&-\hbar\Big[\bar{g}^{2\omega}_3\left(\hat{b}^{+}_{-}\hat{a}^2_{-} - \hat{b}^{+}_{+}\hat{a}^2_{+}\right) + 2\bar{g}^{2\omega}_4\left(\hat{b}^{+}_{+} - \hat{b}^{+}_{-}\right)\hat{a}_{+}\hat{a}_{-}\Big] \\
&\times \exp(\mathrm{i}\boldsymbol{\Delta}\mathbf{k}_2\cdot\mathbf{r}) + \mathrm{h.c.} \\
&-\hbar\Big[\bar{g}^{2\omega}_5\left(\hat{b}^{+}_{+}\hat{a}^{+}_{-}\hat{a}_{+}\hat{b}_{-} + \hat{b}^{+}_{-}\hat{a}^{+}_{+}\hat{a}_{-}\hat{b}_{+}\right) \\
&+\bar{g}^{2\omega}_6\left(\hat{b}^{+}_{+}\hat{a}^{+}_{-}\hat{a}_{-}\hat{b}_{+} + \hat{b}^{+}_{-}\hat{a}^{+}_{+}\hat{a}_{+}\hat{b}_{-}\right) \\
&+\bar{g}^{2\omega}_7\left(\hat{b}^{+}_{+}\hat{a}^{+}_{+}\hat{a}_{+}\hat{b}_{+} + \hat{b}^{+}_{-}\hat{a}^{+}_{-}\hat{a}_{-}\hat{b}_{-}\right)\Big]
\end{aligned} \tag{119}$$

where the nonlinear coupling parameters (108) are redefined:

$$
\begin{aligned}
\bar{g}^{\Omega}_{1,2} &= \frac{V}{\hbar}\left(\frac{2\pi\hbar\Omega}{n_{\Omega}^{2}V}\right)^{2} g^{\Omega}_{1,2} \\
\bar{g}^{2\omega}_{3,4} &= \frac{V}{\hbar}\left(\frac{2\pi\hbar 2\omega}{n_{2\omega}^{2}V}\right)^{1/2}\left(\frac{2\pi\hbar\omega}{n_{\omega}^{2}V}\right) g^{2\omega}_{3,4} \\
\bar{g}^{2\omega}_{5,6,7} &= \frac{V}{\hbar}\left(\frac{2\pi\hbar 2\omega}{n_{2\omega}^{2}V}\right)\left(\frac{2\pi\hbar\omega}{n_{\omega}^{2}V}\right) g^{2\omega}_{5,6,7}
\end{aligned}
\tag{120}
$$

Replacing the time t by the path of propagation z, as in Section V, the Heisenberg equations become

$$
\begin{aligned}
\frac{\mathrm{d}\hat{a}_{\pm}(z)}{\mathrm{d}z} &= -\frac{\mathrm{i}n_{\omega}}{\hbar c}\left[\hat{a}_{\pm}, \hat{H}\right] \\
\frac{\mathrm{d}\hat{b}_{\pm}(z)}{\mathrm{d}z} &= -\frac{\mathrm{i}n_{2\omega}}{\hbar c}\left[\hat{b}_{\pm}, \hat{H}\right]
\end{aligned}
\tag{121}
$$

In accordance with formula (119) one obtains the general operator equations of motion for the fundamental and second-harmonic fields:

$$
\begin{aligned}
\frac{\mathrm{d}\hat{a}_{\pm}(z)}{\mathrm{d}z} = \mathrm{i}\frac{n_{\omega}}{c}\Big[&\left(\bar{g}_1^{\omega}\hat{a}_{\pm}^{+}\hat{a}_{\pm} + 2\bar{g}_2^{\omega}\hat{a}_{\mp}^{+}\hat{a}_{\mp}\right)\hat{a}_{\pm} \\
&+2\left[\mp\bar{g}_3^{-2\omega}\hat{b}_{\pm}\hat{a}_{\pm}^{+} + \bar{g}_4^{-2\omega}\left(\hat{b}_{+} - \hat{b}_{-}\right)\hat{a}_{\mp}^{+}\right]\exp(-\mathrm{i}\boldsymbol{\Delta}\mathbf{k}_2\cdot\mathbf{r}) \\
&+\left[\bar{g}_5^{2\omega}\hat{b}_{\mp}^{+}\hat{a}_{\mp}\hat{b}_{\pm} + \left(\bar{g}_6^{2\omega}\hat{b}_{\mp}^{+}\hat{b}_{\mp} + \bar{g}_7^{2\omega}\hat{b}_{\pm}^{+}\hat{b}_{\pm}\right)\hat{a}_{\pm}\right]\Big] \\
\frac{\mathrm{d}\hat{b}_{\pm}(z)}{\mathrm{d}z} = \mathrm{i}\frac{n_{2\omega}}{c}\Big[&\left(\bar{g}_1^{2\omega}\hat{b}_{\pm}^{+}\hat{b}_{\pm} + 2\bar{g}_2^{2\omega}\hat{b}_{\mp}^{+}\hat{b}_{\mp}\right)\hat{b}_{\pm} \\
&+\left(\mp\bar{g}_3^{2\omega}\hat{a}_{\pm}^{2} \pm 2\bar{g}_4^{2\omega}\hat{a}_{+}\hat{a}_{-}\right)\exp(\mathrm{i}\boldsymbol{\Delta}\mathbf{k}_2\cdot\mathbf{r}) \\
&+\left[\bar{g}_5^{2\omega}\hat{a}_{\mp}^{+}\hat{a}_{\pm}\hat{b}_{\mp} + \left(\bar{g}_6^{2\omega}\hat{a}_{\mp}^{+}\hat{a}_{\mp} + \bar{g}_7^{2\omega}\hat{a}_{\pm}^{+}\hat{a}_{\pm}\right)\hat{b}_{\pm}\right]\Big]
\end{aligned}
\tag{122}
$$

The first equation in (122) is a generalization of the expression (96). Since both equations in (122) contain interference terms, they should be solved simultaneously. This is a difficult task and some approximations are needed. To start with we assume that the dominant process resides in self-interaction of the fundamental beam that is described by the parameters $\bar{g}_1^{\omega}$ and $\bar{g}_2^{\omega}$. Hence, this assumption means that the other coupling constants are smaller and can be neglected. We next apply the solution (96) as zero-order solution in solving (122) for the second harmonic perturbatively. On formal integration, we arrive at the following equation [79]:

$$\hat{b}_{\pm}(z) = \hat{b}_{\pm}(0) \mp i\frac{n_{2\omega}}{c}\int_0^z dz' \exp(i\Delta k_2 z')\left[\bar{g}_3^{2\omega}\hat{a}_{\pm}^2(z') - 2\bar{g}_4^{2\omega}\hat{a}_{+}(z')\hat{a}_{-}(z')\right] \tag{123}$$

where terms containing the second-harmonic operators $\hat{b}_{\pm}(z)$ have been neglected. Next, we assume that the fundamental field is in a coherent state with circular polarization, for example right. Automatically the term with $\bar{g}_4^{2\omega}$ vanishes ($\hat{a}_{-}|\alpha_{+}\rangle = 0$). Moreover, the second harmonic does not exist for $z = 0$ ($\hat{b}_{\pm}|\alpha_{+}\rangle = 0$). These assumptions enable us to find simple formulas for the variances of the quadrature operators. Inserting (96) into Eq. (123) and using the definitions of the quadrature operators for the second-harmonic field we obtain their normally ordered variances,

$$\left.\begin{matrix}\left\langle :(\Delta\hat{Q}_{+})^2: \right\rangle \\ \left\langle :(\Delta\hat{P}_{+})^2: \right\rangle\end{matrix}\right\}
\begin{aligned}
&= -2\kappa_{2\omega}^2\int_0^z dz'\int_0^z dz''\Big[\pm\alpha_{+}^4 \cos\left[(\Delta k_2+\varepsilon)(z'+z'')\right.\\
&\quad \left.+4\varepsilon z'' + |\alpha_{+}|^2\sin 2\varepsilon(z'+z'')\right]\exp\left[(\cos 2\varepsilon(z'+z'')-1)|\alpha_{+}|^2\right]\\
&\quad \mp\alpha_{+}^4\cos\left[(\Delta k_2+\varepsilon)(z'+z'') + |\alpha_{+}|^2\sin 2\varepsilon z' + |\alpha_{+}|^2\sin 2\varepsilon z''\right]\\
&\quad \times\exp\left[(\cos 2\varepsilon z' + \cos 2\varepsilon z'' - 2)|\alpha_{+}|^2\right]\\
&\quad -|\alpha_{+}|^4\cos\left[(\Delta k_2+\varepsilon)(z'-z'') + |\alpha_{+}|^2\sin 2\varepsilon(z'-z'')\right]\\
&\quad \times\exp\left[(\cos 2\varepsilon(z'-z'')-1)|\alpha_{+}|^2\right]\\
&\quad +|\alpha_{+}|^2\cos\left[(\Delta k_2+\varepsilon)(z'-z'') + |\alpha_{+}|^2\sin 2\varepsilon z' - |\alpha_{+}|^2\sin 2\varepsilon z''\right]\\
&\quad \times\exp\left[(\cos 2\varepsilon z' + \cos 2\varepsilon z'' - 2)|\alpha_{+}|^2\right]
\end{aligned} \tag{124}$$

where the coupling parameter is determined as

$$\kappa_{2\omega} = \frac{n_{2\omega}}{c} \bar{g}_3^{2\omega} \tag{125}$$

The squeezing effect occurs, in the second-harmonic beam, if one of the variances in (124) takes a negative value. These equations are very complicated and difficult to analyze. From Section V, we find that it is possible to make the assumption that $\varepsilon z \ll 1$. Then the variances can be expanded in power series and we retain only terms containing $\varepsilon z |\alpha_+|^2 \approx 1$ for $|\alpha_+|^2 \gg 1$. Moreover, we assume phase matching, i.e., $\Delta k_2 = 0$, and that the phase of the incoming beam is zero, i.e., $|\alpha_+| = \alpha_+$. On these assumptions the following approximate expressions are obtained:

$$\begin{aligned}
\left\langle :(\Delta \hat{Q}_+)^2: \right\rangle &\approx 2\frac{\eta}{\beta_2^2}\Big[2\cos\beta_2 - \cos 2\beta_2 - 1 - \beta_2(\sin 2\beta_2 - \sin\beta_2) \\
&\qquad + (\cos\beta_2 - 1 + \beta_2 \sin\beta_2)^2\Big] \\
\left\langle :(\Delta \hat{P}_+)^2: \right\rangle &\approx 2\frac{\eta}{\beta_2^2}\Big[-2\cos\beta_2 + \cos 2\beta_2 + 1 + \beta_2(\sin 2\beta_2 - \sin\beta_2) \\
&\qquad + (\sin\beta_2 - \beta_2 \cos\beta_2)^2\Big]
\end{aligned} \tag{126}$$

where we introduce

$$\beta_2 = 2\varepsilon z |\alpha_+|^2 \tag{127}$$

By η we denote the part of the fundamental beam power transferred into the second harmonic,

$$\eta = \frac{2\kappa_{2\omega}^2 |\alpha_+|^4 z^2}{|\alpha_+|^2} \approx \frac{I(2\omega)}{I(\omega)} \tag{128}$$

The approximate results (126) are convenient to analyze. The normally ordered variances of the $\hat{P}$ component of the second harmonic is plotted in Fig. 3 against $\varepsilon z |\alpha_+|^2$ together with the $\hat{Q}$ component of the fundamental beam (99) showing that the squeezing effect in the second harmonic (solid line) is correlated with the self-squeezing in the fundamental beam (dashed line), because they are negative for small values of β_2. The squeezing from the $\hat{Q}$ component of the fundamental beam can be said to

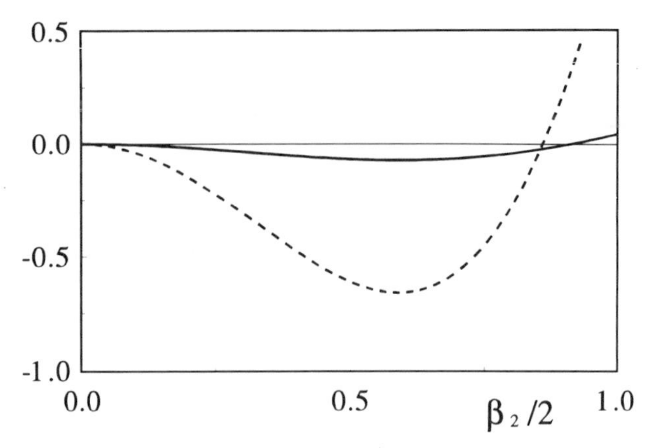

Figure 3. The approximate variance of the $\hat{Q}$ component of the fundamental field (dashed line) and the approximate variance of the $\hat{P}$ component of the second-harmonic field (solid line) are plotted versus $\beta_2/2 = \varepsilon z|\alpha_+|^2(\eta = 0.1)$.

be transferred, in some sense, into the $\hat{P}$ component of the second harmonic. However, we have to recall that these results (126) have been obtained under the approximation that there was no coupling of the second harmonic back into the fundamental. Hence, if η takes a large value this assumption breaks down.

VII. THIRD-HARMONIC GENERATION BY SELF-SQUEEZED LIGHT IN NONLINEAR MEDIUM

In a nonlinear medium that, with regard to its symmetry, admits third-harmonic (not second-harmonic) generation, two nonlinear processes occur simultaneously: first, nonlinear propagation of the fundamental field, described in Sections IV and V, leading to the self-squeezing effect, and second, third-harmonic generation. In the classical treatment the latter is a well-known phenomenon [123]. Since the self-squeezed light produces the third harmonic we can suppose that, as in the second-harmonic case (Section V), squeezing is transferred to the beam at 3ω.

To find the squeezed states in this process we have to find the equation describing the evolution of the field. As in the preceding sections, we use the Heisenberg equation (93). In quantum description, using a spherical basis, the third harmonic can be written as

$$\hat{E}^+_\pm(3\omega) = \mathrm{i}\left(\frac{2\pi\hbar 3\omega}{n_{3\omega}V}\right)^{1/2}\hat{c}_\pm \tag{129}$$

where $\hat{c}_{\pm}, \hat{c}_{\pm}^{+}$ are the annihilation and creation operators for a photon at the frequency 3ω. The interaction Hamiltonian, in terms of these operators, takes the form [80]

$$\hat{H}_{\mathrm{I}} = 2\hbar\bar{g}^{3\omega}\left(\hat{c}_{+}^{+}\hat{a}_{+}^{2}\hat{a}_{-} + \hat{c}_{-}^{+}\hat{a}_{-}^{2}\hat{a}_{+}\right)\exp(\mathrm{i}\Delta k_{3}z) + \mathrm{h.c.} \tag{130}$$

where the nonlinear coupling parameter is determined as

$$\bar{g}^{3\omega} = \frac{V}{\hbar}\left(\frac{2\pi\hbar 3\omega}{n_{3\omega}V}\right)^{1/2}\left(\frac{2\pi\hbar\omega}{n_{\omega}V}\right)^{3/2}\chi_{xxxx}(-3\omega,\omega,\omega,\omega) \tag{131}$$

The susceptibility tensor obeys relation (75). We have assumed that both beams propagate along the z axis with the linear phase mismatch $\Delta k_{3} = 3k_{\omega} - k_{3\omega}$.

Applying the Heisenberg equation (93) and replacing t by z, one easily finds the following relations [80]:

$$\begin{aligned}\frac{\mathrm{d}\hat{c}_{\pm}(z)}{\mathrm{d}z} &= 2\mathrm{i}\frac{n_{3\omega}}{c}\bar{g}^{3\omega}\hat{a}_{\pm}^{2}(z)\hat{a}_{\mp}(z)\exp(\mathrm{i}\Delta k_{3}z)\\ \frac{\mathrm{d}\hat{a}_{\pm}(z)}{\mathrm{d}z} &= 2\mathrm{i}\frac{n_{\omega}}{c}\bar{g}^{3\omega}\big[2\hat{c}_{\pm}(z)\hat{a}_{\pm}^{+}(z)\hat{a}_{\mp}^{+}(z)\\ &\qquad +\hat{c}_{\mp}(z)\hat{a}_{\mp}^{+2}(z)\big]\exp(-\mathrm{i}\Delta k_{3}z)\end{aligned} \tag{132}$$

Equations (132) show the coupling between the third harmonic and fundamental beam. On the assumption that the main process is the self-squeezing described by (95) it is possible to use the solution (96) as the zero-approximation solution solving the first equation of (132). On formal integration the following formula is obtained:

$$\hat{c}_{\pm}(z) = \hat{c}_{\pm}(0) + 2\mathrm{i}\kappa_{3\omega}\int_{0}^{z}\hat{a}_{\pm}^{2}(z')\hat{a}_{\mp}(z')\exp(\mathrm{i}\Delta k_{3}z')\,dz' \tag{133}$$

where we denote

$$\kappa_{3\omega} = \frac{n_{3\omega}}{c}\bar{g}^{3\omega} \tag{134}$$

To say whether squeezing occurs in the third-harmonic beam it is necessary to analyze the quadrature variances (65) defined for the operators $\hat{c}_{\pm}, \hat{c}_{\pm}^{+}$. We assume that the incoming field is a coherent state at $z = 0$.

Using Eq. (105) and taking into account the solution (96) the normally ordered variances are found [80]:

$$
\begin{aligned}
&\left\langle :\left(\Delta\hat{Q}_{+}\right)^{2}: \right\rangle \\
&\quad = -8\kappa_{3\omega}^{2}\int_{0}^{z}\mathrm{d}z'\int_{0}^{z}\mathrm{d}z''\Big[\operatorname{Re}\alpha_{+}^{4}\alpha_{-}^{2}\exp\big[\mathrm{i}(z'+z'')(\Delta k_{3}+\varepsilon+2\delta) \\
&\qquad +(\exp[\mathrm{i}(z'+z'')(2\varepsilon+\delta)]-1)|\alpha_{+}|^{2} \\
&\qquad +(\exp[\mathrm{i}(z'+z'')(\varepsilon+2\delta)]-1)|\alpha_{-}|^{2} \\
&\qquad +\mathrm{i}z''(5\varepsilon+2\delta)\big] \\
&\qquad -\operatorname{Re}\alpha_{+}^{4}\alpha_{-}^{2}\exp\big[\mathrm{i}(z'+z'')(\Delta k+\varepsilon+2\delta) \\
&\qquad +(\exp[\mathrm{i}z'(2\varepsilon+\delta)]+\exp[\mathrm{i}z''(2\varepsilon+\delta)]-2)|\alpha_{+}|^{2} \\
&\qquad +(\exp[\mathrm{i}z'(\varepsilon+2\delta)]+\exp[\mathrm{i}z''(\varepsilon+2\delta)]-2)|\alpha_{-}|^{2}\big] \\
&\qquad -|\alpha_{+}|^{4}|\alpha_{-}|^{2}\exp\big[-\mathrm{i}(z'-z'')(\Delta k_{3}+\varepsilon+2\delta) \qquad\qquad (135)\\
&\qquad +(\exp[-\mathrm{i}(z'-z'')(2\varepsilon+\delta)]-1)|\alpha_{+}|^{2} \\
&\qquad +(\exp[-\mathrm{i}(z'-z'')(\varepsilon+2\delta)]-)|\alpha_{-}|^{2}\big] \\
&\qquad +|\alpha_{+}|^{4}|\alpha_{-}|^{2}\exp\big[-\mathrm{i}(z'-z'')(\Delta k_{3}+\varepsilon+2\delta) \\
&\qquad +(\exp[-\mathrm{i}z'(2\varepsilon+\delta)] \\
&\qquad +\exp[\mathrm{i}z''(2\varepsilon+\delta)]-2)|\alpha_{+}|^{2}+(\exp[-\mathrm{i}z'(\varepsilon+2\delta)] \\
&\qquad +\exp[\mathrm{i}z''(\varepsilon+2\delta)]-2)|\alpha_{-}|^{2}\big]\Big]
\end{aligned}
$$

The variance of the left-polarized component is obtained by replacing all plus subscripts by minus subscripts in Eq. (135). The expressions for the quadrature operators $\hat{P}_{\pm}$ differ in the signs of their Re terms from Eq. (135). It is obvious that the variances are zero if only one of the circular components exists in the incoming beam. Hence, we take into account a beam linearly polarized along the x axis, i.e., $\alpha_{+}=\alpha_{-}=\alpha/\sqrt{2}$. Moreover, we assume that the parameters ε and δ defined in formula (97) are equal. Using the x, y, z basis, as the simplest in this case, one easily obtains the following equation:

$$\begin{aligned}\left\langle :\left(\Delta\hat{Q}_x\right)^2 : \right\rangle = & -2\kappa_{3\omega}^2|\alpha|^6 \int_0^z dz' \int_0^z dz'' \Big[\exp\left[\left(\cos 3\varepsilon(z'+z'') - 1\right)|\alpha|^2\right] \\ & \times \cos\left[(\Delta k_3 + 3\varepsilon)(z'+z'') + 9\varepsilon z'' + |\alpha|^2 \sin 3\varepsilon(z'+z'')\right] \\ & - \exp\left[\left(\cos 3\varepsilon z' + \cos 3\varepsilon z'' - 2\right)|\alpha|^2\right] \\ & \times \cos\left[(\Delta k_3 + 3\varepsilon)(z'+z'') + |\alpha|^2(\sin 3\varepsilon z' + \sin 3\varepsilon z'')\right] \\ & - \exp\left[\left(\cos 3\varepsilon(z'-z'') - 1\right)|\alpha|^2\right]\cos\left[(\Delta k_3 + 3\varepsilon)(z'-z'') \right. \\ & \left. + |\alpha|^2 \sin 3\varepsilon(z'-z'')\right] + \exp\left[\left(\cos 3\varepsilon z' + \cos 3\varepsilon z'' - 2\right)|\alpha|^2\right] \\ & \times \cos\left[(\Delta k_3 + 3\varepsilon)(z'-z'') + |\alpha|^2(\sin 3\varepsilon z' - \sin 3\varepsilon z'')\right]\Big]\end{aligned} \tag{136}$$

This equation is still too complicated. Since the parameter ε is very small in real physical situations it is possible to expand formula (136) in a power series and to neglect all terms less than $|\alpha|^2\varepsilon z \approx 1$. On the assumption of nonlinear mismatch $\Delta k_3 = 0$, the approximate equations for the quadrature variances of the third harmonic can be written in the form

$$\begin{aligned}\left\langle :\left(\Delta\hat{Q}_x\right)^2 : \right\rangle \approx & \, 2\frac{\eta}{\beta_3^2}\Big\{3\left[2\cos\beta_3 - \cos 2\beta_3 - 1 - \beta_3(\sin 2\beta_3 - \sin\beta_3)\right] \\ & + (\cos\beta_3 - 1 + \beta_3 \sin\beta_3)^2\Big\} \\ \left\langle :\left(\Delta\hat{P}_x\right)^2 : \right\rangle \approx & \, 2\frac{\eta}{\beta_3^2}\Big\{3\left[-2\cos\beta_3 + \cos 2\beta_3 + 1 + \beta_3(\sin 2\beta_3 - \sin\beta_3)\right] \\ & + (\sin\beta_3 - \beta_3\cos\beta_3)^2\Big\}\end{aligned} \tag{137}$$

where

$$\beta_3 = 3\varepsilon z|\alpha|^2 \tag{138}$$

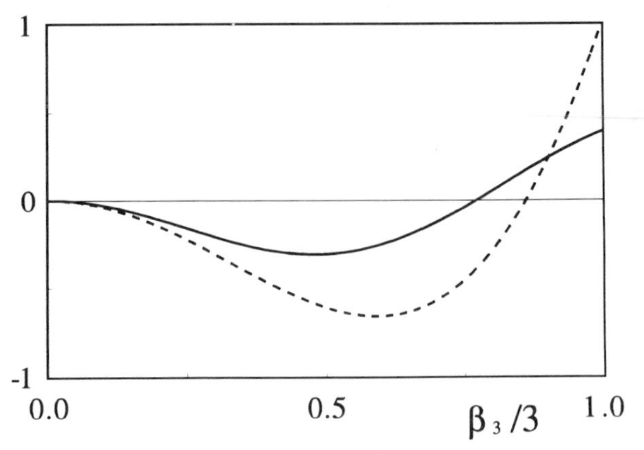

Figure 4. The approximate variance of the $\hat{Q}$ component of the fundamental field (dashed line) and the approximate variance of the $\hat{P}$ component of the third-harmonic field (solid line) are plotted versus $\beta_3/3 = \varepsilon z|\alpha|^2(\eta = 0.1)$.

and

$$\eta = \frac{3\kappa_{3\omega}^2|\alpha|^6 z^2}{|\alpha|^2} \approx \frac{I(3\omega)}{I(\omega)} \tag{139}$$

is the power-conversion ratio describing the part of the power of the fundamental that is transferred to the third harmonic.

The normally ordered variance (139) for the $\hat{P}$ component of the third-harmonic beam is plotted in Fig. 4 in comparison with the variance of the $\hat{Q}$ component of the fundamental (99). Squeezing occurs when the variances have negative values. The curves in Fig. 4 show a correlation between squeezing in the $\hat{P}$ component of the third harmonic (solid line) and self-squeezing in the $\hat{Q}$ component of the fundamental (dashed line) for small z. We can say that squeezing is transferred, in some sense, from the fundamental to the third-harmonic beam. The squeezing effect in the third harmonic depends on the conversion ratio. If η increases, then the squeezing increases too. However, we have to recall that the coupling of the third harmonic back to the fundamental beam has been ignored in our considerations. Hence, the approximation is not true for large η. Comparing Fig. 3 with Fig. 4 one can say that the correlation between the self-squeezed fundamental beam and the third harmonic is stronger than that between the fundamental and second harmonic.

VIII. CONCLUSION

In this paper we have considered the light squeezing at propagation in a nonlinear isotropic medium. The exact operator results obtained in the two quantum descriptions show a dissimilarity between the squeezing occurring in propagating light and the ordinary squeezing, briefly recalled in Section II. This effect has been named self-squeezing by Tanaś and Kielich [67, 68]. We have also analyzed the second- and third-harmonic beams, generated by self-squeezed light in an isotropic medium with a center of symmetry.

To make the medium capable of generating a wave at double frequency, an external dc field has to be applied. The classical and quantum equations describing the time evolution of the fundamental and second-harmonic beams have been derived under the assumption that the main process resided in self-interaction of the fundamental field. To discuss the squeezing in the second-harmonic beam we used the analytical form of normally ordered variances of the quadrature components. Some correlation between the squeezing in the second-harmonic and self-squeezing in the fundamental beam is found. One can say that squeezing is transferred from the fundamental into the second-harmonic beam.

In the same way, third-harmonic generation has been described (obviously without assuming an external dc electric field). The results obtained in our approach, similarly to the second-harmonic case, show correlation between squeezing in the third harmonic and self-squeezing in the fundamental. It is seen from Fig. 3 and Fig. 4 that this correlation is stronger in third-harmonic generation.

The normally ordered variances of the quadrature components of the second- and third-harmonic beams obtained in our treatment are directly proportional to the conversion ratio η. In our discussion we have taken into account that only 10% of the power of the fundamental beam is transferred into the harmonics field. For higher conversion ratio our assumption breaks down and the pairs of equations (122) and (132) have to be solved simultaneously.

We should emphasize that to obtain a considerable amount of squeezing in the second and third harmonic by the mechanism discussed in this paper, the linear mismatch should be much smaller than the intensity-dependent nonlinear mismatch.

The past few years show that interest in the optical phenomena, especially quantum phenomena, occurring in nonlinear media has been increasing steadily [128–169]. One can expect great advances in the research.

References

1. D. F. Walls, *Nature* **306**, 141 (1983).
2. R. Loudon and P. L. Knight, *J. Mod. Opt.* **34**, 709 (1987).
3. K. Zaheer and M. S. Zubairy, *Adv. At. Mol. Opt. Phys.* **28**, 143 (1990).
4. M. C. Teich and B. E. A. Saleh, *Quantum Opt.* **1**, 153 (1989).
5. Special issue of *J. Mod. Opt.* **34**, Nos. 6/7 (1987).
6. Special issue of *J. Opt. Soc. Am. B* **4**, No. 10 (1987).
7. D. F. Walls and P. Zoller, *Phys. Rev. Lett.* **47**, 709 (1981).
8. L. Mandel, *Phys. Rev. Lett.* **49**, 136 (1982).
9. Z. Ficek, R. Tanaś, and S. Kielich, *Opt. Commun.* **46**, 23 (1983).
10. W. Vogel and D. G. Welsch, *Phys. Rev. Lett.* **54**, 1802 (1985).
11. R. Short and L. Mandel, *Phys. Rev. Lett.* **51**, 384 (1983).
12. D. Stoler, *Phys. Rev. Lett.* **33**, 1397 (1974).
13. G. J. Milburn and D. F. Walls, *Opt. Commun.* **39**, 401 (1981).
14. K. Wódkiewicz and M. S. Zubairy, *Phys. Rev. A* **27**, 2003 (1983).
15. A. Lane, P. Tombesi, H. J. Carmichael, and D. F. Walls, *Opt. Commun.* **48**, 155 (1983).
16. G. Sharf and D. F. Walls, *Opt. Commun.* **50**, 245 (1984).
17. H. J. Carmichael, G. J. Milburn, and D. F. Walls, *J. Phys. A* **15**, 469 (1984).
18. M. Wolinsky and H. J. Carmichael, *Opt. Commun.* **55**, 138 (1985).
19. P. G. Fernandez, P. Colet, R. Toral, M. San Miguel, and F. J. Bermejo, *Phys. Rev. A* **43**, 4923 (1991).
20. H. P. Yuen and J. H. Shapiro, *Opt. Lett.* **4**, 334 (1979).
21. R. S. Bondurand, P. Kumar, J. H. Shapiro, and M. Maeda, *Phys. Rev. A* **30**, 343 (1984).
22. J. Peřina, V. Peřinova, C. Sibilia, and B. Bertolotti, *Opt. Commun.* **49**, 285 (1984).
23. M. D. Reid and D. F. Walls, *Opt. Commun.* **50**, 106 (1984); *Phys. Rev. A* **31**, 1622 (1985).
24. H. D. Levenson, R. M. Shelby, A. Aspect, M. Reid, and D. F. Walls, *Phys. Rev. A* **32**, 1550 (1985).
25. J. Janszky and Y. Y. Yushin, *Opt. Commun.* **60**, 92 (1986).
26. G. V. Varada, M. S. Kumar, and G. S. Agarwal, *Opt. Commun.* **62**, 328 (1987).
27. M. S. K. Razmi and J. H. Eberly, *Opt. Commun.* **76**, 265 (1990).
28. M. S. K. Razmi and J. H. Eberly, *Phys. Rev. A* **44**, 2214 (1991).
29. R. E. Slusher, L. W. Holberg, B. Yurke, J. C. Mertz, and J. F. Valley, *Phys. Rev. Lett.* **55**, 2409 (1985).
30. R. M. Shelby, M. D. Levenson, S. H. Perlmutter, R. G. Devoe, and D. F. Walls, *Phys. Rev. Lett.* **57**, 691 (1986).
31. M. W. Maeda, P. Kumar, and J. H. Shapiro, *Opt. Lett.* **12**, 161 (1987).
32. M. S. Zubairy, M. S. K. Razmi, S. Iqbal, and M. Idress, *Phys. Lett. A* **98**, 168 (1983).
33. R. Loudon, *Opt. Commun.* **49**, 67 (1984).
34. P. Meystre and M. S. Zubairy, *Phys. Lett. A* **89**, 390 (1982).
35. A. S. Shumovsky, F. L. Kien, and E. I. Aliksenderov, *Phys. Lett. A* **124**, 351 (1987).
36. S. Y. Zhu, Z. O. Lin, and X. S. Li, *Phys. Lett. A* **128**, 89 (1988).

37. J. R. Kukliński and J. L. Madajczyk, *Phys. Rev. A* **37**, 3175 (1988).
38. M. J. Gagen and G. J. Milburn, *Opt. Commun.* **76**, 253 (1990).
39. O. Aytür and P. Kumar, *Opt. Lett.* **15**, 390 (1990).
40. Ling-An Wu, H. J. Kimble, J. L. Hall, and Huifa Wu, *Phys. Rev. Lett.* **57**, 2520 (1986).
41. P. A. Franken, A. E. Hill, C. W. Peters, and G. Weinreich, *Phys. Rev. Lett.* **7**, 118 (1961).
42. J. A. Armstrong, N. Bloembergen, J. Ducuing, and P. S. Pershan, *Phys. Rev.* **127**, 1918 (1962).
43. D. F. Walls, *Phys. Lett. A* **32**, 476 (1970).
44. B. Crosignani, P. Di Porto, and S. Solimeno, *J. Phys. A* **5**, L119 (1972).
45. N. Nayak and B. K. Mahanty, *Phys. Rev. A* **15**, 1173 (1977).
46. J. Peřina, V. Peřinova, and L. Knesel, *Acta Phys. Pol. A* **51**, 725 (1977); *Czech. J. Phys. B* **27**, 487 (1977).
47. V. Peřinova and J. Peřina, *Czech. J. Phys. B* **28**, 306 (1978).
48. A. D. Stolarov, *Zh. Prikladnoy Spektroskopii* **25**, 236 (1976).
49. M. Kozierowski and R. Tanaś, *Opt. Commun.* **21**, 229 (1977).
50. S. Kielich, M. Kozierowski, and R. Tanaś, in L. Mandel and E. Wolf (Eds.), *Coherence and Quantum Optics* **4**, Plenum, New York, 1978, p. 511.
51. R. Hanbury-Brown and R. W. Twiss, *Proc. Roy. Soc. London Ser. A* **243**, 291 (1957).
52. D. F. Walls, *Nature* **280**, 451 (1979).
53. H. Paul, *Rev. Mod. Phys.* **54**, 1061 (1982).
54. V. Peřinova and J. Peřina, *Czech. J. Phys. B* **28**, 1183 (1978).
55. V. N. Gorbaczev and P. H. Zanadvorov, *Opt. Spektrosk.* **49**, 600 (1980).
56. K. J. McNeil, P. D. Drummond, and D. F. Walls, *Opt. Commun.* **27**, 292 (1978).
57. P. D. Drummond, K. J. McNeil, and D. F. Walls, *Opt. Commun.* **28**, 255 (1979).
58. P. D. Drummond, K. J. McNeil, and D. F. Walls, *Opt. Acta* **27**, 321 (1980).
59. P. D. Drummond, K. J. McNeil, and D. F. Walls, *Opt. Acta* **28**, 211 (1981).
60. L. Mandel, *Opt. Commun.* **42**, 437 (1982).
61. M. Kozierowski and S. Kielich, *Phys. Lett. A* **94**, 213 (1983).
62. S. Kielich, M. Kozierowski, and R. Tanaś, *Opt. Acta* **32**, 1023 (1985).
63. S. Kielich, R. Tanaś, and R. Zawodny, *J. Mod. Opt.* **34**, 979 (1987).
64. L. A. Lugiato, G. Strini, and F. De Martini, *Opt. Lett.* **8**, 256 (1983).
65. S. Friberg and L. Mandel, *Opt. Commun.* **48**, 439 (1984).
66. R. Tanaś, in L. Mandel and E. Wolf (Eds.), *Coherence and Quantum Optics* **5**, Plenum, New York, 1984, p. 645.
67. R. Tanaś and S. Kielich, *Opt. Commun.* **45**, 351 (1983).
68. R. Tanaś and S. Kielich, *Opt. Acta* **31**, 81 (1984).
69. S. Kielich, R. Tanaś, and R. Zawodny, *Phys. Rev. A* **36**, 5670 (1987).
70. G. J. Milburn, *Phys. Rev. A* **33**, 674 (1986).
71. G. J. Milburn and C. A. Holmes, *Phys. Rev. Lett.* **56**, 2237 (1986).
72. B. Yurke and D. Stoler, *Phys. Rev. Lett.* **57**, 13 (1986).
73. C. K. Hong and L. Mandel, *Phys. Rev. A* **32**, 974 (1985).
74. M. Kozierowski, *Phys. Rev. A* **34**, 3474 (1986).

75. M. Hillery, *Opt. Commun.* **62**, 135 (1987).
76. A. Sizman, R. J. Horowicz, G. Wagner, and G. Leuchs, *Opt. Commun.* **80**, 138 (1990).
77. A. Lukš, J. Peřina, and J. Krepelka, *Acta Phys. Pol. A* **72**, 443 (1987).
78. P. Chmela, M. Kozierowski, and S. Kielich, *Czech. J. Phys. B* **37**, 846 (1987).
79. S. Kielich, R. Tanaś, and R. Zawodny, *Appl. Phys. B* **45**, 249 (1988).
80. S. Kielich, R. Tanaś, and R. Zawodny, *J. Opt. Soc. Am. B* **4**, 1627 (1987).
81. A. Ekert and K. Rzążewski, *Opt. Commun.* **65**, 225 (1988).
82. M. Kozierowski and V. I. Man'ko, *Opt. Commun.* **69**, 71 (1988).
83. S. F. Pereira, M. Xiao, H. J. Kimble, and J. L. Hall, *Phys. Rev. A* **38**, 4931 (1988).
84. T. A. B. Kennedy, T. B. Anderson, and D. F. Walls, *Phys. Rev. A* **40**, 1385 (1989).
85. A. Lukš, V. Peřinova, and J. Peřina, *Opt. Commun.* **67**, 149 (1988).
86. R. Loudon, *Opt. Commun.* **70**, 109 (1989).
87. R. Tanaś, A. Miranowicz, and S. Kielich, *Phys. Rev. A* **43**, 4014 (1991).
88. M. Kitagawa and Y. Yamamoto, *Phys. Rev. A* **34**, 3974 (1986).
89. Y. Yamamato, S. Machida, M. Kitagawa, and G. Bjork, *J. Opt. Soc. Am. B* **4**, 1645 (1987).
90. A. Miranowicz, R. Tanaś, and S. Kielich, *Quantum Opt.* **2**, 253 (1990).
91. Lu-Bi Deng and Lian-Zhou Zang, *J. Mod. Opt.* **38**, 877 (1991).
92. D. Mihalache and D. Baboiu, *Phys. Lett. A* **159**, 303 (1991).
93. A. V. Belinski, *Kvantovaya Elektronika* **18**, 343 (1991).
94. R. Tanaś and S. Kielich, *Quantum Opt.* **2**, 23 (1990).
95. V. N. Gorbaczev and E. S. Polzik, *Opt. Commun.* **77**, 247 (1990).
96. R. Schack, A. Sizman, and A. Shenzle, *Phys. Rev. A* **43**, 6303 (1991).
97. M. J. Collett and R. B. Levien, *Phys. Rev. A* **43**, 5068 (1991).
98. V. N. Gorbaczev, *Izv. Akad. Nauk SSSR*, s. fiz. **55**, 219 (1991).
99. You-bang Zhan, *Phys. Lett. A* **160**, 498 (1991).
100. You-bang Zhan, *Phys. Lett. A* **160**, 503 (1991).
101. R. Tanaś and S. Kielich, *J. Mod. Opt.* **37**, 1935 (1990).
102. R. Tanaś and Ts. Gantsog, *J. Mod. Opt.* **39**, 749 (1992).
103. R. Horák and J. Peřina, *J. Opt. Soc. Am. B* **6**, 1239 (1989).
104. V. Peřinova and A. Lukš, *J. Mod. Opt.* **35**, 1513 (1988).
105. D. J. Daniel and G. J. Milburn, *Phys. Rev. A* **39**, 4628 (1989).
106. G. J. Milburn, A. Mecozzi, and P. Tombesi, *J. Mod. Opt.* **36**, 1607 (1989).
107. V. Peřinova and A. Lukš, *Phys. Rev. A* **41**, 414 (1990).
108. D. T. Pegg and S. M. Barnett, *Europhys. Lett.* **6**, 483 (1988).
109. S. M. Barnett and D. T. Pegg, *J. Mod. Opt.* **36**, 7 (1989).
110. D. T. Pegg and S. M. Barnett, *Phys. Rev. A* **39**, 1665 (1989).
111. R. Tanaś, Ts. Gantsog, A. Miranowicz, and S. Kielich, *J. Opt. Soc. Am. B* **8**, 1576 (1991).
112. Ts. Gantsog, R. Tanaś, and R. Zawodny, *Phys. Lett. A* **155**, 1 (1991).
113. R. Tanaś and Ts. Gantsog, *J. Opt. Soc. Am. B* **8**, 2505 (1991).
114. R. Tanaś, *J. Sov. Laser Res.* **12**, 395 (1991).

115. R. J. Glauber, *Phys. Rev.* **130**, 2529 (1963); *Phys. Rev. Lett.* **10**, 277 (1963).
116. Wei-Min Zhang, Da Hsuan Feng, and R. Gilmore, *Rev. Mod. Phys.* **62**, 867 (1990).
117. C. M. Caves and B. L. Schumaker, *Phys. Rev. A* **31**, 3068 (1985).
118. C. M. Caves, *Phys. Rev. D* **23**, 1693 (1981).
119. H. P. Yuen, *Phys. Rev. A* **13**, 2226 (1976).
120. S. Machida, Y. Yamamoto, and Y. Itaya, *Phys. Rev. Lett.* **58**, 1000 (1987).
121. T. Debuisschert, S. Reynaud, A. Heidmann, E. Giacobino, and C. Fabre, *Quantum Opt.* **1**, 3 (1989).
122. N. Bloembergen: *Nonlinear Optics*, Benjamin, Reading, MA, 1965.
123. S. Kielich: *Nonlinear Molecular Optics*, Nauka, Moscow, 1981.
124. P. D. Maker, R. W. Terhune, and C. W. Savage, *Phys. Rev. Lett.* **12**, 507 (1964).
125. D. F. Walls and R. Barakat, *Phys. Rev. A* **1**, 446 (1970).
126. S. Kielich, *Acta Phys. Pol.* **17**, 239 (1958).
127. S. Kielich, *IEEE J. Quantum Electron.* **QE-5**, 562 (1969); *J. Opto-Electron.* **2**, 5 (1970).
128. F. Kaczmarek and R. Parzyński: *Laser Physics, Part 1, Introduction to Quantum Optics*, Poznań University Press, 1990.
129. G. S. Holliday and S. Singh, *Opt. Commun.* **62**, 289 (1987).
130. D. F. Smirnov and A. S. Troshin, *Sov. Phys. USP* **153**, 233 (1987).
131. I. Abram, *Phys. Rev. A* **35**, 4661 (1987).
132. R. Lynch, *Phys. Rev. A* **36**, 4501 (1987).
133. A. V. Belinski and A. S. Chirkin, *Kvantovaya Elektronika* **16**, 889 (1989).
134. A. V. Belinski and A. S. Chirkin, *Opt. Spektrosk.* **66**, 1190 (1989).
135. G. S. Agarwal and R. P. Puri, *Phys. Rev. A* **40**, 5179 (1989).
136. M. Hillery, *Phys. Rev. A* **40**, 3147 (1989).
137. C. C. Gerry and J. B. Togeas, *Opt. Commun.* **69**, 263 (1989).
138. P. S. Gupta and J. Dash, *Opt. Commun.* **79**, 251 (1990).
139. M. Zachid and M. S. Zurbairy, *Opt. Commun.* **76**, 1 (1990).
140. Zhi-ming Zhang, Lei Xu, and Jin-lin Chai, *Phys. Lett. A* **151**, 65 (1990).
141. A. V. Belinski, *Kvantovaya Elektronika* **17**, 1182 (1990).
142. P. V. Elyutin and D. N. Klyshko, *Phys. Lett. A* **149**, 241 (1990).
143. L. Zeni, A. Cutolo, and S. Solimeno, *J. Mod. Opt.* **37**, 2085 (1990).
144. L. A. Lugiato, P. Galatola, and L. M. Narducci, *Opt. Commun.* **76**, 276 (1990).
145. Zhi-ming Zhang, Lei Xu, Jin-lin Chai, and Fu-li Li, *Phys. Lett. A* **150**, 27 (1990).
146. A. D. Wilson-Gordon, V. Bužek, and P. L. Knight, *Phys. Rev. A* **44**, 7647 (1991).
147. I. Abram and E. Cohen, *Phys. Rev. A* **44**, 500 (1991).
148. R. J. Glauber and M. Lewenstein, *Phys. Rev. A* **43**, 467 (1991).
149. M. I. Kolobov, *Phys. Rev. A* **44**, 1986 (1991).
150. Chin-lin Chai, Fu-li Li, and Zhi-ming Zhang, *J. Phys. B* **24**, 3309 (1991).
151. E. M. Wright, *Phys. Rev. A* **43**, 3836 (1991).
152. V. Bužek and I. Jex, *Phys. Rev. A* **41**, 4079 (1990).
153. M. Dance, M. J. Collett, and D. F. Walls, *Phys. Rev. Lett.* **66**, 1115 (1991).
154. R. B. Levien, M. J. Collett, and D. F. Walls, *Opt. Commun.* **82**, 171 (1991).

155. M. Hillery, *Phys. Rev. A* **44**, 4578 (1991).
156. I. H. Deutsch and J. C. Garrison, *Opt. Commun.* **86**, 311 (1991).
157. M. Rosenbluh and R. M. Shelby, *Phys. Rev. Lett.* **66**, 153 (1991).
158. N. P. Pettiaux, P. Mandel, and C. Fabre, *Phys. Rev. Lett.* **66**, 1838 (1991).
159. M. Brisudova, *J. Mod. Opt.* **38**, 2505 (1991).
160. L. Z. Zhang, L. B. Deng, and S. G. Sun, *J. Mod. Opt.* **39**, 445 (1992).
161. G. V. Varada and G. S. Agarwal, *Phys. Rev. A* **45**, 6721 (1992).
162. G. Drobný and I. Jex, *Phys. Rev. A* **45**, 1816 (1992).
163. V. Bužek, A. Vidiella-Barranco, and P. L. Knight, *Phys. Rev. A* **45**, 6570 (1992).
164. C. Cabrillo, F. J. Bermejo, P. Garcia-Fernandez, R. Toral, P. Colet, and M. San Miguel, *Phys. Rev. A* **45**, 3216 (1992).
165. Y. Qu and S. Singh, *Opt. Commun.* **90**, 111 (1992).
166. D. Yu, *Phys. Rev. A* **45**, 2121 (1992).
167. M. Hillery and D. Yu, *Phys. Rev. A* **45**, 1860 (1992).
168. A. Lukš and V. Peřinova, *Phys. Rev. A* **45**, 6710 (1992).
169. Fu-li Li, Xiao-shen Li, D. L. Lin, and T. F. George, *Phys. Rev. A* **45**, 3133 (1992).

SELF-SQUEEZING OF ELLIPTICALLY POLARIZED LIGHT PROPAGATING IN A KERR-LIKE OPTICALLY ACTIVE MEDIUM

S. KIELICH, R. TANAŚ, AND R. ZAWODNY

Nonlinear Optics Division, Institute of Physics,
Adam Mickiewicz University, Poznań, Poland

CONTENTS

I. INTRODUCTION

Quantum and stochastic properties of light fields can, in most cases, be described in terms of coherent states that are quantum field states being as close as possible to classical fields with well-defined amplitude and phase [1–3]. The well-defined diagonal Glauber-Sudarshan quasidistribution $P(\alpha)$ allows for calculations of all relevant mean values of fields that as we say "have classical counterparts." However, there are optical fields that "have no classical counterparts," that is, fields for which the quasi-distribution $P(\alpha)$ does not exist as a well-defined, positive definite dis-

This work was supported by Polish Government Grant KBN 201 509 101.

Modern Nonlinear Optics, Part 1, Edited by Myron Evans and Stanisław Kielich. Advances in Chemical Physics Series, Vol. LXXXV.
ISBN 0-471-57546-1

tribution function. Such fields have quantum properties that cannot be explained in the language of classical stochastic quantities. They require fully quantum description, and generation and detection of such field states have been the subject of numerous, both theoretical and experimental, efforts since the mid 1970s. To this day, many nonlinear processes have been analyzed as candidates for producing nonclassical states of light, which include parametric down conversion [4–19], resonance fluorescence [20–27], four-wave mixing [28–35], harmonics generation [36–48], anharmonic oscillator [49–70], light propagation in Kerr media from the point of view of photon statistics [71–73] and squeezing [74–91], and multiphoton absorption and other multiphoton processes [92–189]. The nonclassical properties of light are already the subject of review articles [190–197] and books [198–200], in which the basic information and extensive literature can be found.

Nonclassical effects, such as photon antibunching, sub-Poissonian photon statistics, and squeezing, are a result of nonlinear interaction of quantum light with a nonlinear medium; thus, the nonlinear interaction of light with matter is a crucial element in generation fields with nonclassical properties. The earliest observations of photon antibunching are due to Kimble et al. [23] in resonance fluorescence, confirming the theoretical predictions of Carmichael and Walls [20] and Kimble and Mandel [21]. Sub-Poissonian photon distribution was measured by Short and Mandel [123], and the first observation of squeezing was due to Slusher et al. [33]. Later on a number of successful experiments were performed producing light with nonclassical properties, [10, 11, 27, 32, 33, 169].

Unlike photon antibunching, squeezing is an effect that is sensitive to the phase of the field, the fluctuations of which can essentially reduce its value and even destroy it altogether. The detection of squeezing requires rather sophisticated techniques, such as balanced homodyne detection [201, 202], allowing for the elimination of the local oscillator noise. Despite the differences between photon anticorrelation and squeezing, both processes have one important common feature: Their nature is purely quantum and fields exhibiting such properties have no analogs in classical optics. The two effects can coexist in the same nonlinear process, their areas of existence can be separated, or only one of them can appear in a given process. Especially interesting, in our opinion, is the process of propagation of strong light in a nonlinear Kerr-like medium. Some time ago Tanaś and Kielich [74, 75] showed that almost perfect squeezing can be obtained in such a process, while at the same time photon statistics remain untouched. This process was referred to as self-squeezing, because the squeezing of the quantum field fluctuations is caused by the self-interaction of light via the nonlinear medium. The one-mode version of the process, which is applicable for circularly polarized light propagating in an isotropic Kerr medium, was considered by Tanaś [49] in terms of an anharmonic oscillator model. The model, which allows for exact solutions, became very popular later, and many properties of the field states generated in the model have been revealed and studied [50–70].

Classically, a strong laser field with elliptical polarization is known to rotate its polarization ellipse when propagating through a Kerr medium, an effect observed by Maker et al. [203]. To explain this effect there is no need for field quantization (see, for example, [198]). Here, however, we are interested in effects that are quantum in nature and cannot be explained with the field being classical.

To describe properly the effects associated with the propagation of elliptically polarized light in a Kerr medium, the two-mode description of the field is needed. Such a description was used in the early studies [71–75] of the quantum field effects that appear during propagation. In those studies, the Heisenberg equations of motion for the field operators were solved and their solutions used to calculate appropriate quantities revealing sub-Poissonian photon statistics or squeezing. Recently, Agarwal and Puri [82] reexamined the problem of propagation of elliptically polarized light through a Kerr medium. They discussed not only the Heisenberg equations of motion for the field operators, but also the evolution of the field states themselves. The polarization state of the field propagating in a Kerr medium can be described by the Stokes parameters, which are the expectation values of the corresponding Stokes operators when the quantum description of the field is used. Quantum fluctuations of the Stokes parameters of light propagating in a Kerr medium have recently been discussed by Tanaś and Kielich [204].

In this chapter, we consider propagation of strong light through a macroscopically isotropic, nonlinear medium, taking into account not only electric-dipole contributions to the interaction Hamiltonian, but also contributions from the electric–magnetic dipole and electric–dipole–quadrupole linear and nonlinear susceptibilities of the medium. We introduce general expressions for the effective Hamiltonians of the second and fourth order in the field strength using the circular polarization basis for the field. Such effective Hamiltonians lead to the Heisenberg equations of motion for the field operators that have exact solutions in the form of the translation operator. This means that the field propagating in the nonlinear medium undergoes a nonlinear change in phase (or self-phase modulation), which for quantum fields means essential changes of the quantum state of the field leading to self-squeezing of light. These additional contributions that we take into account mean that our results are valid for media with nonlinear optical activity. It is our aim to calculate the field expectation values describing photon antibunching and squeezing of the field propagating in such a medium.

II. THE EFFECTIVE INTERACTION HAMILTONIAN

We consider N microsystems (atoms, molecules, or elementary cells in a crystal) confined in a volume V and subjected to the electromagnetic field of a light beam with the electric field vector $\mathbf{E}(\mathbf{r}, t)$ and the magnetic field vector $\mathbf{B}(\mathbf{r}, t)$ in the point $\mathbf{r}$ at time t. The total Hamiltonian of such a system has the form

$$H = H_{\mathrm{N}} + H_{\mathrm{F}} + H_{\mathrm{I}} \tag{1}$$

where H_{N} is the Hamiltonian of the system of N microsystems, and H_{F} is the Hamiltonian of the free field.

We are interested in the explicit form of the Hamiltonian H_{I} describing the interaction of the system with the electromagnetic field. This interaction is in general nonlinear and contains all multipolar transitions both electric and magnetic [205–207]. In nonlinear optics we use, for simplicity and convenience, effective interaction Hamiltonians [208], in which it is sufficient to include terms up to the fourth order with respect to the electric and magnetic field strengths [206, 207, 209, 210].

In this chapter we take into account only contributions to the interaction Hamiltonian with even powers of the field strengths

$$H_{\mathrm{I}} = H_{\mathrm{I}}^{(2)} + H_{\mathrm{I}}^{(4)} + \cdots = \sum_{n=1}^{\infty} H_{\mathrm{I}}^{(2n)} \tag{2}$$

Restricting our considerations to the case of weak spatial dispersion (which means that we neglect higher multipoles [209, 210]), we can write for N uncorrelated molecules [211]

$$\begin{aligned} H_{\mathrm{I}}^{(2)} = -\frac{N}{2}\Big\{\alpha_{ij}E_iE_j + \tfrac{1}{3}\big[\eta_{i(jk)}E_i\nabla_k E_j + \eta_{(ik)j}(\nabla_k E_i)E_j\big] \\ + \rho_{ij}E_iB_j + \lambda_{ij}B_iE_j + \text{h.c.}\Big\} \end{aligned} \tag{3}$$

where, according to the Einstein summation convention, the summation over the repeated indices is understood in (3).

In Eq. (3) the second-rank tensor α_{ij} describes the linear electric–electric polarizability of the molecule coming from the electric-dipole–electric-dipole transitions. Similarly, the second-rank pseudotensors ρ_{ij}

and λ_{ij} denote the polarizabilities: electric–magnetic resulting from the quantum transitions electric dipole–magnetic dipole, and magnetic–electric resulting from the transitions magnetic dipole–electric dipole, respectively. The third-rank tensor $\eta_{i(jk)}$ denotes the linear electric–electric polarizability resulting from the transitions electric dipole–electric quadrupole [209, 210], while the tensor $\eta_{(ij)k}$ comes from the same transitions in reversed order.

In the same multipolar approximation the fourth-order Hamiltonian has the form [206, 207, 209, 210]

$$\begin{aligned} H_{\mathrm{I}}^{(4)} = -\frac{N}{24}\Big\{&\gamma_{ijkl}E_iE_jE_kE_l \\ &+\tfrac{1}{3}\big[\eta_{ijk(lm)}E_iE_jE_k\nabla_mE_l + \eta_{ij(km)l}E_iE_j(\nabla_mE_k)E_l \\ &\quad+\eta_{i(jm)kl}E_i(\nabla_mE_j)E_kE_l + \eta_{(im)jkl}(\nabla_mE_i)E_jE_kE_l\big] \\ &+\kappa_{ijkl}E_iE_jE_kB_l + \rho_{ijkl}E_iE_jB_kE_l \\ &+\sigma_{ijkl}E_iB_jE_kE_l + \lambda_{ijkl}B_iE_jE_kE_l + \text{h.c.}\Big\} \end{aligned} \tag{4}$$

where the fourth-rank tensor γ_{ijkl} denotes the nonlinear polarizability resulting from the four electric-dipole transitions, the fourth-rank pseudotensors κ_{ijkl}, ρ_{ijkl}, and σ_{ijkl} denote the nonlinear polarizabilities: electric–magnetic resulting from the transitions electric dipole–magnetic dipole and two electric dipoles, while λ_{ijkl} the magnetic–electric polarizability associated with the transitions magnetic dipole–electric dipole and two electric dipoles. The fifth-rank tensor $\eta_{(im)jkl}$ defines the electric quadrupole polarizability associated with the transitions electric quadrupole–electric dipole and two electric dipoles [207, 210]. The remaining tensors $\eta_{i(jm)kl}$, $\eta_{ij(km)l}$, and $\eta_{ijk(lm)}$ differ from the first by the permutation of the position of the electric–quadrupole transition, which is labeled by the indices in parentheses.

For classical fields the electric field vector can be split into two complex conjugate parts [1]:

$$\mathbf{E}(\mathbf{r},t) = \mathbf{E}^{(+)}(\mathbf{r},t) + \mathbf{E}^{(-)}(\mathbf{r},t) \tag{5}$$

where the components $\mathbf{E}^{(+)}(\mathbf{r},t)$ and $\mathbf{E}^{(-)}(\mathbf{r},t)$ are related to the time dependences $\exp(-\mathrm{i}\omega t)$ (positive frequency part) and $\exp(+\mathrm{i}\omega t)$ (negative frequency part), respectively. The transversal electric field can be expressed as a superposition of plane waves:

$$\mathbf{E}(\mathbf{r},t) = \sum_k \{\mathbf{E}^{(+)}(\mathbf{k})\exp[\mathrm{i}(\mathbf{k}\cdot\mathbf{r} - \omega_k t)] + \mathbf{E}^{(-)}(\mathbf{k})\exp[-\mathrm{i}(\mathbf{k}\cdot\mathbf{r} - \omega_k t)]\} \tag{6}$$

The same decomposition can be performed for the magnetic field vector $\mathbf{B}(\mathbf{r}, t)$, where we have the relation (in SI units)

$$B_i^{(+)} = \frac{1}{\omega}\varepsilon_{ijk}k_j E_k^{(+)} \tag{7}$$

where ε_{ijk} is the Levi-Cività antisymmetric tensor.

As usual, we assume that the light wave is propagating along the z axis of the Cartesian coordinate system $\{x, y, z\}$. It will be convenient later on to use the circular basis associated with the unit vectors of the form (the angular momentum convention is used here)

$$\mathbf{e}_{\pm} = \frac{1}{\sqrt{2}}(\mathbf{x} \pm \mathrm{i}\mathbf{y}) \tag{8}$$

where $\mathbf{e}_+$ describes right and $\mathbf{e}_-$ left polarization of the field. The vectors $\mathbf{x}$ and $\mathbf{y}$ are the unit vectors along the x and y of the Cartesian reference frame, and i is the imaginary unit ($\mathrm{i} = \sqrt{-1}$).

Assuming the microsystems to be freely oriented, the Hamiltonians (3) and (4) have to be averaged over all possible orientations. As a result of such averaging only the rotational invariants of the polarizability tensors appearing in these Hamiltonians will remain [198], which have the form

$$\begin{gathered}\langle \alpha_{ij}\rangle_\Omega = \alpha\delta_{ij}\\ \langle \eta_{i(jm)}\rangle_\Omega = \langle \eta_{(im)j}\rangle_\Omega = 0\\ \langle \rho_{ij}\rangle_\Omega = \rho\delta_{ij} \qquad \langle \lambda_{ij}\rangle_\Omega = \lambda\delta_{ij}\end{gathered} \tag{9}$$

where $\alpha = \alpha_{\alpha\alpha}/3$, $\rho = \rho_{\alpha\alpha}/3$, and $\lambda = \lambda_{\alpha\alpha}/3$ are the mean polarizabilities of the molecules.

In the nonlinear case we have [198]

$$\langle \gamma_{ijkl}\rangle_\Omega = \gamma_1\delta_{ij}\delta_{kl} + \gamma_2\delta_{ik}\delta_{jl} + \gamma_3\delta_{il}\delta_{jk} \tag{10}$$

where

$$\begin{aligned}\gamma_1 &= \tfrac{1}{30}[4\gamma_{\alpha\alpha\beta\beta} - \gamma_{\alpha\beta\alpha\beta} - \gamma_{\alpha\beta\beta\alpha}]\\ \gamma_2 &= \tfrac{1}{30}[-\gamma_{\alpha\alpha\beta\beta} + 4\gamma_{\alpha\beta\alpha\beta} - \gamma_{\alpha\beta\beta\alpha}]\\ \gamma_3 &= \tfrac{1}{30}[-\gamma_{\alpha\alpha\beta\beta} - \gamma_{\alpha\beta\alpha\beta} + 4\gamma_{\alpha\beta\beta\alpha}]\end{aligned} \tag{11}$$

and similar relations hold for the pseudotensors κ_{ijkl}, ρ_{ijkl}, σ_{ijkl}, and λ_{ijkl}.

For the nonlinear dipole–quadrupole polarizability we have [198]

$$\langle \eta_{ijk(lm)} \rangle_\Omega = \eta_1 \delta_{ij} \varepsilon_{klm} + \eta_2 \delta_{ik} \varepsilon_{jlm} + \eta_3 \delta_{il} \varepsilon_{jkm} + \eta_4 \delta_{jk} \varepsilon_{ilm} + \eta_5 \delta_{jl} \varepsilon_{ikm} + \eta_6 \delta_{kl} \varepsilon_{ijm} \tag{12}$$

where ε_{ijk} is the Levi-Cività antisymmetric tensor, while the constants $\eta_1, \eta_2, \ldots, \eta_6$ are defined by the following matrix equation:

$$\begin{pmatrix} \eta_1 \\ \eta_2 \\ \eta_3 \\ \eta_4 \\ \eta_5 \\ \eta_6 \end{pmatrix} = \frac{\eta_{\alpha\beta\gamma(\delta\phi)}}{30} \begin{pmatrix} 3 & -1 & 1 & -1 & 1 & 0 \\ -1 & 3 & -1 & -1 & 0 & 1 \\ 1 & -1 & 3 & 0 & -1 & 1 \\ -1 & -1 & 0 & 3 & -1 & -1 \\ 1 & 0 & -1 & -1 & 3 & -1 \\ 0 & 1 & 1 & -1 & -1 & 3 \end{pmatrix} \begin{pmatrix} \delta_{\alpha\beta} \varepsilon_{\gamma\delta\phi} \\ \delta_{\alpha\gamma} \varepsilon_{\beta\delta\phi} \\ \delta_{\alpha\delta} \varepsilon_{\beta\gamma\phi} \\ \delta_{\beta\gamma} \varepsilon_{\alpha\delta\phi} \\ \delta_{\beta\delta} \varepsilon_{\alpha\gamma\phi} \\ \delta_{\gamma\delta} \varepsilon_{\alpha\beta\phi} \end{pmatrix} \tag{13}$$

The Hamiltonian (6) describing the interaction of N microsystems with the electromagnetic field propagating in a definite direction can be simplified because the summation in (6) is restricted to definite k only. Having this restriction in mind and applying the circular polarization basis (8), we can write the Hamiltonian (6), up to the fourth order in the field strength, in the following form (see Appendix A):

$$\begin{aligned} H_I^{(2)} = &-\tfrac{1}{2}\chi_R^L[E_+^-E_+^+ + E_-^-E_-^+ + E_+^+E_+^- + E_-^+E_-^-] \\ &-\tfrac{i}{2}\chi_A^L[E_+^-E_+^+ - E_-^-E_-^+ + E_+^+E_+^- - E_-^+E_-^-] \end{aligned} \tag{14}$$

$$\begin{aligned} H_I^{(4)} = &-\tfrac{1}{12}\chi_R^{NL}\Big[(E_+^-)^2(E_+^+)^2 + (E_-^-)^2(E_-^+)^2 + (E_+^-E_+^+)^2 \\ &\quad + (E_-^-E_-^+)^2 + E_+^-(E_+^+)^2E_+^- + E_-^-(E_-^+)^2E_-^- \\ &\quad + \text{terms with reversed superscripts}\Big] \\ &-\tfrac{1}{24}\kappa_R^{NL}\big[4E_+^-E_-^-E_+^+E_-^+ + E_+^-E_+^+E_-^-E_-^+ + E_+^-E_-^+E_+^+E_-^- \\ &\quad + E_-^-E_-^+E_+^-E_+^+ + E_+^-E_-^+E_-^-E_+^+ + E_-^-E_+^+E_+^-E_-^+ \\ &\quad + E_+^-E_+^+E_-^+E_-^- + E_-^-E_-^+E_+^+E_+^- + E_-^-E_+^+E_-^+E_+^- \\ &\quad + \text{terms with reversed superscripts}\big] \\ &-\tfrac{i}{12}\chi_A^{NL}\Big[(E_+^-)^2(E_+^+)^2 - (E_-^-)^2(E_-^+)^2 + (E_+^-E_+^+)^2 \\ &\quad - (E_-^-E_-^+)^2 + E_+^-(E_+^+)^2E_+^- - E_-^-(E_-^+)^2E_-^- \\ &\quad + \text{terms with reversed superscripts}\Big] \end{aligned} \tag{15}$$

where we have introduced the following linear and nonlinear molecular parameters (we neglect local field corrections):

$$\chi_R^L = \frac{N}{3}\mathrm{Re}\,\alpha_{\alpha\alpha} \tag{16}$$

$$\chi_A^L = -\frac{\mathrm{i}Nk_z}{3\omega}\mathrm{Im}\,\rho_{\alpha\alpha} \tag{17}$$

$$\chi_R^{NL} = \frac{N}{15}\mathrm{Re}\left[-\gamma_{\alpha\alpha\beta\beta} + 3\gamma_{\alpha\beta\alpha\beta}\right] \tag{18}$$

$$\kappa_R^{NL} = \frac{N}{15}\mathrm{Re}\left[3\gamma_{\alpha\alpha\beta\beta} + \gamma_{\alpha\beta\alpha\beta}\right] \tag{19}$$

$$\chi_A^{NL} = -\frac{\mathrm{i}4Nk_z}{15}\left\{\frac{1}{\omega}\mathrm{Im}\left[\sigma_{\alpha\alpha\beta\beta} - 3\sigma_{\alpha\beta\alpha\beta}\right] - \frac{1}{3}\mathrm{Re}\,\eta_{\alpha(\beta\gamma)\beta\delta}\varepsilon_{\alpha\gamma\delta}\right\} \tag{20}$$

The real and imaginary parts of the linear and nonlinear polarizabilities are given in the case when the ground state of the molecule is nondegenerate by (A.38); in the case of even degeneracy by (A.42), (A.44), and (A.48); and in the case of odd degeneracy by (A.46), (A.47), and (A.49). The nonzero and independent components of the nonlinear polarizability tensors $\mathrm{Re}\,\gamma_{\alpha\beta\gamma\delta}$, $\mathrm{Im}\,\sigma_{\alpha\beta\gamma\delta}$, and $\mathrm{Re}\,\eta_{\alpha(\beta\gamma)\delta\phi}$ symmetrical with respect to the time reversal are collected in Tables I–III for 102 magnetic point groups of symmetry, whereas in Table IV the linear χ_R^L, χ_A^L and nonlinear χ_R^{NL}, κ_R^{NL}, χ_A^{NL} molecular parameters are collected for 102 magnetic point symmetry groups (Appendix B).

III. THE SOLUTION OF THE EQUATIONS OF MOTION FOR THE FIELD OPERATORS

In quantum electrodynamics the field vectors (6) and (7) become operators in the Hilbert space, and we have

$$\mathbf{E}^{(+)}(\mathbf{k}) = \mathrm{i}\sum_{\lambda} c(\omega_k)\mathbf{e}^{(\lambda)}(\mathbf{k})\hat{a}_{\mathbf{k}\lambda} \tag{21}$$

where $c(\omega_k)$ is the normalization factor, which, depending on the unit system, has the form

$$c(\omega_k) = \begin{cases} \sqrt{\dfrac{2\pi\hbar\omega_k}{V}} & \text{in CGS} \\ \sqrt{\dfrac{\hbar\omega_k}{2\varepsilon_0 V}} & \text{in SI} \end{cases} \tag{22}$$

where V is the quantization volume.

In Eq. (21) $\hat{a}_{\mathbf{k}\lambda}$ is the annihilation operator of a photon with the momentum $\hbar\mathbf{k}$ and the polarization λ defined by the unit vector $\mathbf{e}^{(\lambda)}(\mathbf{k})$. The photon annihilation and creation operators $\hat{a}_{\mathbf{k}\lambda}$ and $\hat{a}^+_{\mathbf{k}\lambda}$ satisfy the boson commutation rules

$$\begin{aligned} \left[\hat{a}_{\mathbf{k}\lambda}, \hat{a}^+_{\mathbf{k}\lambda}\right] &= \delta_{\mathbf{k}\mathbf{k}'}\delta_{\lambda\lambda'} \\ \left[\hat{a}_{\mathbf{k}\lambda}, \hat{a}_{\mathbf{k}\lambda}\right] &= \left[\hat{a}^+_{\mathbf{k}\lambda}, \hat{a}^+_{\mathbf{k}\lambda}\right] = 0 \end{aligned} \tag{23}$$

The unit vectors describing the polarization state of the field are, in general, complex quantities and satisfy the orthonormality conditions

$$e^{(\lambda)*}_{k\sigma} e^{(\lambda')}_{k\tau} = \delta_{\sigma\tau}\delta_{\lambda\lambda'} \qquad e^{(\lambda)}_{k\sigma} k_\sigma = 0 \tag{24}$$

For a quasimonochromatic wave of frequency ω propagating along the z axis of the laboratory reference frame one can discard the summation over k in Eq. (6) and, in view of (21), write

$$E^{(+)}_\sigma(z, t) = \mathrm{i}c(\omega)\exp[-\mathrm{i}(\omega t - kz)] \sum_{\lambda=1,2} e^{(\lambda)}_\sigma \hat{a}_\lambda \tag{25}$$

where $k = \omega/c$ is the value of the wave vector $\mathbf{k}$.

The field (25) represents, in fact, a two-mode field, when it is a coherent superposition of two modes with orthogonal polarizations. Usually, such two modes can be replaced by a one mode of the field with elliptical polarization

$$e_\sigma \hat{a} = e^{(1)}_\sigma \hat{a}_1 + e^{(2)}_\sigma \hat{a}_2 \tag{26}$$

where $e^{(1)}_\sigma$ and $e^{(2)}_\sigma$ denote σ components of the orthogonal unit polarization vectors $\mathbf{e}^{(1)}$ and $\mathbf{e}^{(2)}$ associated with the modes $\hat{a}_1$ and $\hat{a}_2$, and similarly e_σ denotes σ component of the polarization vector of the mode $\hat{a}$.

The transformation (26) can be interpreted as a decomposition of the initially elliptically polarized light into two orthogonal modes. Taking into account the normalization conditions (24), we get from (26)

$$\hat{a} = e_1^* \hat{a}_1 + e_2^* \hat{a}_2 \tag{27}$$

were

$$e_1^* = e_\sigma^* e_\sigma^{(1)} \qquad e_2^* = e_\sigma^* e_\sigma^{(2)} \tag{28}$$

Assuming the two modes as linearly polarized along x and y we have from (27)

$$\hat{a} = e_x^* \hat{a}_x + e_y^* \hat{a}_y \tag{29}$$

where [212]

$$\begin{aligned} e_x &= \cos\eta\cos\theta - \mathrm{i}\sin\eta\sin\theta \\ e_y &= \cos\eta\sin\theta + \mathrm{i}\sin\eta\cos\theta \end{aligned} \tag{30}$$

with θ and η denoting the azimuth and ellipticity of the polarization ellipse of the incoming light.

Analogously to the circular representation (8) of the polarization vector we can introduce, according to (26) and (29), the circular basis for the field operators:

$$\begin{aligned} \hat{a}_1 &= \hat{a}_+ = \frac{1}{\sqrt{2}}\left(\hat{a}_x - \mathrm{i}\hat{a}_y\right) \\ \hat{a}_2 &= \hat{a}_- = \frac{1}{\sqrt{2}}\left(\hat{a}_x + \mathrm{i}\hat{a}_y\right) \end{aligned} \tag{31}$$

Both representations can be used to describe the interaction of the elliptically polarized light with the medium. However, as has been shown previously [73–75] the circular representation has a clear advantage over the Cartesian representation because it allows for the simple operator solution of the equations of motion in the propagator form.

The time evolution of the field operators is described by the Heisenberg equations of motion:

$$\frac{\partial \mathbf{E}^{(\pm)}(\mathbf{r}, t)}{\partial t} = \frac{1}{\mathrm{i}\hbar}\left[\mathbf{E}^{(\pm)}(\mathbf{r}, t), H\right] \tag{32}$$

For the free field the Hamiltonian H is the free Hamiltonian H_{F}, and the solution to Eq. (32) is given by Eq. (6) describing the free (fast) evolution of the field. When the interaction of the field with the medium comes into play, the Hamiltonian H in (32) contains, beside the free part H_F, also the interaction part H_I. In this case the solution to (32) is no longer the free field (6), but an additional (slow) time dependence appears. This additional time dependence, which is due to the interaction H_I, reveals itself in the fact that the amplitudes $E^{\pm}(k)$ given by Eq. (21) become time dependent.

Usually, one considers the time evolution of a field that is confined in a cavity of volume V. In our case, we deal instead with a field propagating in a medium of a certain length z. So, instead of the time dependence, we consider the length dependence of the propagating field. However, for plane waves the transition from the cavity problem to the propagation problem can be performed by replacing the time t by z/c, where c is the speed of light [208]. Recently, Blow et al. [213] have shown that the correct treatment of the propagation processes requires the continuous-mode description, instead of the discrete-mode description used here. Blow et al. [214] have also shown that the exact solution of the quantum self-phase modulation problem can be obtained within the continuous-mode formalism. This new formalism allows us to avoid an anomalous dependence on the size of the cavity that appears when the discrete-mode formalism is applied to describe propagation effects. However, the essential features of the quantum propagation problem, such as the emergence of photon antibunching and squeezing, can be revealed with the discrete-mode formalism, and we keep using it here.

After the replacement $t \to z/c$, the Heisenberg equations of motion (32) become equations describing the dependence of the operators for the kth mode on z:

$$\frac{\partial E^{(\pm)}(k;z)}{\partial z} = \frac{1}{\mathrm{i}\hbar c}\left[E^{(\pm)}(k;z), H\right] \tag{33}$$

The next essential step is to write down the quantum form of the effective interaction Hamiltonian, which for classical fields is given by Eqs. (14) and (15). We obtain the effective interaction Hamiltonian by inserting the quantum form (25) of the field into (14) and (15) and taking the normal order of the field operators in all terms. This leads us to the following expressions for the interaction Hamiltonian:

$$H_{\mathrm{I}}^{(2)} = -\tilde{\chi}_R^L\left(\hat{a}_+^+\hat{a}_+ + \hat{a}_-^+\hat{a}_-\right) - \mathrm{i}\tilde{\chi}_A^L\left(\hat{a}_+^+\hat{a}_+ - \hat{a}_-^+\hat{a}_-\right) \tag{34}$$

$$\begin{aligned} H_{\mathrm{I}}^{(4)} = & -\tfrac{1}{2}\tilde{\chi}_R^{NL}\left(\hat{a}_+^{+2}\hat{a}_+^2 + \hat{a}_-^{+2}\hat{a}_-^2\right) - \tilde{\kappa}_R^{NL}\hat{a}_+^+\hat{a}_-^+\hat{a}_-\hat{a}_+ \\ & -\frac{\mathrm{i}}{2}\tilde{\chi}_A^{NL}\left(\hat{a}_+^{+2}\hat{a}_+^2 - \hat{a}_-^{+2}\hat{a}_-^2\right) \end{aligned} \tag{35}$$

where we have introduced the notation

$$\tilde{\chi}_R^L = c(\omega)^2 \chi_R^L \qquad \tilde{\chi}_A^L = c(\omega)^2 \chi_A^L \tag{36}$$

$$\begin{aligned} \tilde{\chi}_R^{NL} &= c(\omega)^4 \chi_R^{NL} \qquad \tilde{\kappa}_R^{NL} = c(\omega)^4 \kappa_R^{NL} \\ \tilde{\chi}_A^{NL} &= c(\omega)^4 \chi_A^{NL} \end{aligned} \tag{37}$$

According to (33)–(35) we get the equation of motion for the field annihilation operators (free evolution has been eliminated):

$$\begin{aligned} \frac{\mathrm{d}}{\mathrm{d}z}\hat{a}_{\pm}(z) = \frac{\mathrm{i}}{\hbar c}\Big\{\tilde{\chi}_R^L \pm \mathrm{i}\tilde{\chi}_A^L + \left[\tilde{\chi}_R^{NL} \pm \mathrm{i}\tilde{\chi}_A^{NL}\right]\hat{a}_{\pm}^{+}(z)\hat{a}_{\pm}(z) \\ + \tilde{\kappa}_R^{NL}\hat{a}_{\mp}^{+}(z)\hat{a}_{\mp}(z)\Big\}\hat{a}_{\pm}(z) \end{aligned} \tag{38}$$

Since $\hat{a}_{+}^{+}\hat{a}_{+}$ and $\hat{a}_{-}^{+}\hat{a}_{-}$ are constants of motion, Eq. (38) has an exact solution in the form of the translation operator

$$\begin{aligned} \hat{a}_{\pm}(z) = \exp\{\mathrm{i}[\varphi_{\pm}(z) + \varepsilon_{\pm}(z)\hat{a}_{\pm}^{+}(0)\hat{a}_{\pm}(0) \\ + \delta(z)\hat{a}_{\mp}^{+}(0)\hat{a}_{\mp}(0)]\}\hat{a}_{\pm}(0) \end{aligned} \tag{39}$$

where the notation is the following:

$$\begin{aligned} \varphi_{\pm}(z) &= \frac{z}{\hbar c}\left(\tilde{\chi}_R^L \pm \mathrm{i}\tilde{\chi}_A^L\right) \\ \varepsilon_{\pm}(z) &= \frac{z}{\hbar c}\left(\tilde{\chi}_R^{NL} \pm \mathrm{i}\tilde{\chi}_A^{NL}\right) \\ \delta(z) &= \frac{z}{\hbar c}\tilde{\kappa}_R^{NL} \end{aligned} \tag{40}$$

Taking into account the fact that the component of the dipole polarization is by definition given by [211]

$$\hat{P}^{+} = -\frac{\partial H_{\mathrm{I}}}{\partial E^{-}} \tag{41}$$

and that in the circular basis

$$\left(n_{\pm}^2 - 1\right)E_{\pm}^{+} = 4\pi P_{\pm}^{+} \tag{42}$$

we get, in view of (34) and (35), for the refractive indices of the right and left circularly polarized waves

$$n_{\pm}^2 - 1 = \frac{4\pi}{c(\omega)^2}\left\{\tilde{\chi}_R^L \pm i\tilde{\chi}_A^L + \left[\tilde{\chi}_R^{NL} \pm i\tilde{\chi}_A^{NL}\right]\hat{a}_{\pm}^{+}\hat{a}_{\pm} + \tilde{\kappa}_R^{NL}\hat{a}_{\mp}^{+}\hat{a}_{\mp}\right\} \quad (43)$$

This gives us for the circular optical birefringence in the presence of strong light the formula

$$n_{+}^2 - n_{-}^2 = \frac{4\pi}{c(\omega)^2}\Big\{2i\tilde{\chi}_A^L + \left[\tilde{\chi}_R^{NL} - \tilde{\kappa}_R^{NL}\right]\left(\hat{a}_{+}^{+}\hat{a}_{+} - \hat{a}_{-}^{+}\hat{a}_{-}\right) + i\tilde{\chi}_A^{NL}\left(\hat{a}_{+}^{+}\hat{a}_{+} + \hat{a}_{-}^{+}\hat{a}_{-}\right)\Big\} \quad (44)$$

Formula (44) is the quantum counterpart of the earlier obtained [215] classical formula in which the first term denotes the natural optical activity of the medium, the second term denotes the rotation of the polarization ellipse induced by the strong field [203], and the third term denotes the nonlinear change in optical activity caused by the strong field [198, 212, 216]. Taking into account (43) and (44), we rewrite the field (39) in the form known from ellipsometry:

$$\hat{a}_{\pm}(z) = \exp(i\alpha \pm i\phi)\hat{a}_{\pm}(0) \quad (45)$$

where we have by definition

$$\alpha = \frac{1}{2}\frac{\omega}{c}(n_{+} + n_{-})z = \alpha_0 + \delta\alpha \quad (46)$$

$$\phi = \frac{1}{2}\frac{\omega}{c}(n_{+} - n_{-}) = \phi_0 + \delta\phi \quad (47)$$

where ϕ is the angle of rotation of the polarization ellipse after the field passed the path z in the medium. In the absence of a strong field we have

$$a_0 = \frac{z}{\hbar c}\tilde{\chi}_R^L \quad (48)$$

$$\phi_0 = \frac{z}{\hbar c}\tilde{\chi}_A^L \quad (49)$$

while the changes due to the strong field are given by

$$\delta\alpha = \frac{z}{2\hbar c}\left\{\left[\tilde{\chi}_R^{NL} + \tilde{\kappa}_R^{NL}\right]\left(\hat{a}_+^+\hat{a}_+ + \hat{a}_-^+\hat{a}_-\right) + \mathrm{i}\tilde{\chi}_A^{NL}\left(\hat{a}_+^+\hat{a}_+ - \hat{a}_-^+\hat{a}_-\right)\right\} \quad (50)$$

$$\delta\phi = \frac{z}{2\hbar c}\left\{\left[\tilde{\chi}_R^{NL} - \tilde{\kappa}_R^{NL}\right]\left(\hat{a}_+^+\hat{a}_+ - \hat{a}_-^+\hat{a}_-\right) + \mathrm{i}\tilde{\chi}_A^{NL}\left(\hat{a}_+^+\hat{a}_+ + \hat{a}_-^+\hat{a}_-\right)\right\} \quad (51)$$

If in particular the light wave is linearly polarized, we have ($a_+ = a_- = a/\sqrt{2}$)

$$\delta\alpha = \frac{z}{2\hbar c}\left[\tilde{\chi}_R^{NL} + \tilde{\kappa}_R^{NL}\right]\hat{a}^+\hat{a} \quad (52)$$

$$\delta\phi = \frac{\mathrm{i}z}{2\hbar c}\tilde{\chi}_A^{NL}\hat{a}^+\hat{a} \quad (53)$$

and we see that $\delta\phi$ appears only for media with nonlinear optical activity.

A word of caution should be added here. Since the quantum counterparts to the classical formulas describing nonlinear changes of the refractive index of the medium are extracted from the operator solution for the field operators, and since the mean value of the product of two operators is not the product of their mean values, these formulas cannot be treated too seriously as the quantum expressions for the nonlinear refractive index. They simply allow for the identification of particular contributions, as in classical description, but in fact the complete solutions for the field operators enter the experimentally measurable quantities, such as light intensity, photon correlation functions, and field variances. This will become clear in the next sections.

IV. PHOTON STATISTICS

Since in the isotropic medium described by the Hamiltonian given in (34) and (35) the photon number operators $\hat{a}_+^+\hat{a}_+$ and $\hat{a}_-^+\hat{a}_-$ are constants of motion (they commute with the Hamiltonian), any function of these operators is also a constant of motion and, as a result, photon statistics of the circular components of the field do not change in the course of propagation. If the component before entering the nonlinear medium is, say, in a coherent state with the Poissonian photon distribution, the photon distribution of the beam outgoing from the medium will remain Poissonian despite the fact that, as we show in the next section, the state of the field is no longer the coherent state. So, if there are no sub-Poissonian photon statistics of the incoming beam, there will be no sub-Poissonian photon statistics in either circular component of the outgoing beam.

However, one can easily check [74, 75] that the linear polarization is not preserved during the propagation of quantum light through the nonlinear isotropic medium and this leads to the sub-Poissonian photon statistics, which is our subject in this section.

Let us take, for example, the component of the linear polarization along the x axis. The photon number operator $\hat{a}_x^+\hat{a}_x$ of this component does not commute with the interaction Hamiltonian, which means that the photon statistics of this component can change due to the interaction with the medium. Knowing the solutions (39) for the field operators $\hat{a}_+(z)$ and $\hat{a}_-(z)$, for the circular components, we can use relation (31) to write down corresponding solutions for the operators $\hat{a}_x(z)$ and $\hat{a}_y(z)$. These solutions allow us to find any characteristic for the polarization component x or y outgoing from the medium if we know the state of the field at the input. To choose one we can place a polarizer after the medium. Thus, assuming that the incoming field is in the coherent state with elliptical polarization defined by the azimuth θ and the ellipticity η, we get, for the mean number of photons with the polarization x after the path z passed by the light in the medium, the following expression:

$$\begin{aligned}\langle \hat{a}_x^+(z)\hat{a}_x(z)\rangle &= \tfrac{1}{2}\langle [\hat{a}_+^+(z) + \hat{a}_-^+(z)][\hat{a}_+(z) + \hat{a}_-(z)]\rangle \\ &= \tfrac{1}{2}\left(|\alpha_+|^2 + |\alpha_-|^2\right) + \mathrm{Re}\Big\{\alpha_+^*\alpha_- \exp\Big[-\mathrm{i}(\varphi_+ - \varphi_-) \\ &\qquad + (\mathrm{e}^{-\mathrm{i}(\varepsilon_+ - \delta)} - 1)|\alpha_+|^2 + (\mathrm{e}^{-\mathrm{i}(\varepsilon_- - \delta)} - 1)|\alpha_-|^2\Big]\Big\}\end{aligned} \tag{54}$$

where

$$\begin{aligned}\alpha_+ &= \frac{\alpha}{\sqrt{2}}(\cos\eta + \sin\eta)\,\mathrm{e}^{-\mathrm{i}\theta} \\ \alpha_- &= \frac{\alpha}{\sqrt{2}}(\cos\eta - \sin\eta)\,\mathrm{e}^{\mathrm{i}\theta}\end{aligned} \tag{55}$$

and α is the eigenvalue of the annihilation operator of the incoming light that we assume as being in the coherent state:

$$\hat{a}(0)\ |\alpha\rangle = \alpha\ |\alpha\rangle \tag{56}$$

Thus, $|\alpha|^2 = |\alpha_+|^2 + |\alpha_-|^2$ is the mean number of photons of the incoming field. The quantities $\varphi_\pm = \varphi_\pm(z)$, $\varepsilon_\pm = \varepsilon_\pm(z)$, and $\delta = \delta(z)$ are given by Eq. (40). To shorten the notation we shall omit the argument z.

In view of (55), expression (54) can be rewritten in a slightly different form:

$$\langle \hat{a}_x^+(z)\hat{a}_x(z)\rangle = \frac{|\alpha|^2}{2}[1 + \cos 2\eta \exp B \cos(2\theta + C)] \tag{57}$$

where

$$\begin{aligned} B = &\frac{|\alpha|^2}{2}(1 + \sin 2\eta)[\cos(\varepsilon_+ - \delta) - 1] \\ &+ \frac{|\alpha|^2}{2}(1 - \sin 2\eta)[\cos(\varepsilon_- - \delta) - 1] \end{aligned} \tag{58}$$

$$\begin{aligned} C = &-(\varphi_+ - \varphi_-) - \frac{|\alpha|^2}{2}(1 + \sin 2\eta)\sin(\varepsilon_+ - \delta) \\ &+ \frac{|\alpha|^2}{2}(1 - \sin 2\eta)\sin(\varepsilon_- - \delta) \end{aligned} \tag{59}$$

Formula (57) has been obtained with the effective use of the commutation relations (23); that is, to obtain it we have taken into account the quantum properties of the field. Just for reference, it is worth remembering that the corresponding formula for classical fields reads

$$\begin{aligned} &\langle \hat{a}_x^+(z)\hat{a}_x(z)\rangle_{\text{class}} \\ &\quad = \frac{1}{2}(|\alpha_+|^2 + |\alpha_-|^2) + \mathrm{Re}\Big\{\alpha_+^*\alpha_- \exp\Big[-\mathrm{i}(\varphi_+ - \varphi_-) \\ &\qquad\qquad -\mathrm{i}(\varepsilon_+ - \delta)|\alpha_+|^2 + \mathrm{i}(\varepsilon_- - \delta)|\alpha_-|^2\Big]\Big\} \\ &\quad = \frac{|\alpha|^2}{2}\left\{1 + \cos 2\eta \cos\left[2\theta - (\varphi_+ - \varphi_-)\right.\right. \\ &\qquad\qquad \left.\left. - \frac{|\alpha|^2}{2}(1 + \sin 2\eta)(\varepsilon_+ - \delta) + \frac{|\alpha|^2}{2}(1 - \sin 2\eta)(\varepsilon_- - \delta)\right]\right\} \\ &\quad = \frac{|\alpha|^2}{2}\left\{1 + \cos 2\eta \cos\left[2\theta - (\varphi_+ - \varphi_-)\right.\right. \\ &\qquad\qquad \left.\left. - \frac{|\alpha|^2}{2}\sin 2\eta(\varepsilon_+ + \varepsilon_- - 2\delta) - \frac{|\alpha|^2}{2}(\varepsilon_+ - \varepsilon_-)\right]\right\} \end{aligned} \tag{60}$$

In formula (60), as in formula (44) one can identify particular effects related to the propagation of light in the nonlinear medium. Namely,

$\varphi_+ - \varphi_-$ describes the natural optical activity, $\varepsilon_+ + \varepsilon_- - 2\delta$ describes the rotation of the polarization ellipse induced by the strong light (because of the $\sin 2\eta$ factor appearing in this term, the elliptical polarization of the field is necessary to observe this effect), and $\varepsilon_+ - \varepsilon_-$ describes the nonlinear change in optical activity.

To make the difference between the quantum formula (57) and its classical counterpart (60) more explicit, let us assume that the medium is composed of optically inactive molecules; then, we have $\varphi_+ - \varphi_- = 0$ and $\varepsilon_+ = \varepsilon_- = \varepsilon$. Moreover, assume that the incoming field is linearly polarized ($\eta = 0$) with the azimuth $\theta = \pi/2$, that is, perpendicularly to the observed polarization component. In this case the classical formula (60) gives zero, whereas the quantum formula (57) is different from zero because of the exponential function appearing in it. This means that for quantum fields during the propagation in the nonlinear isotropic medium, photons with the polarization orthogonal to the polarization of the incoming field will appear. In other words, in the nonlinear medium the linear polarization of the field is not preserved, an effect already discussed by Ritze [73]. The quantum effects in the polarization of light propagating in a Kerr medium have been recently discussed in more detail by Tanaś and Gantsog [217].

Now, we come back to the main topic of this section, that is, the problem of sub-Poissonian photon statistics. To convince ourselves whether the field outgoing from the nonlinear medium exhibits sub-Poissonian photon statistics, we have to calculate the second-order correlation function $\langle \hat{a}_x^{+2}(z)\hat{a}_x^2(z)\rangle$. Applying the solutions (39), and assuming that the incoming beam is in the coherent state $|\alpha\rangle$, we arrive at

$$
\begin{aligned}
\left\langle \hat{a}_x^{+2}(z)\hat{a}_x^2(z)\right\rangle &= \tfrac{1}{4}\left\langle \left[\hat{a}_+^{+}(z) + \hat{a}_-^{+}(z)\right]^2\left[\hat{a}_+(z) + \hat{a}_-(z)\right]^2\right\rangle \\
&= \tfrac{1}{4}\left(|\alpha_+|^4 + |\alpha_-|^4 + 4|\alpha_+|^2|\alpha_-|^2\right) \\
&\quad + \tfrac{1}{2}\mathrm{Re}\Big\{\alpha_+^{*2}\alpha_-^2 \exp\Big[-2\mathrm{i}(\varphi_+ - \varphi_-) - \mathrm{i}(\varepsilon_+ - \varepsilon_-) \\
&\qquad\qquad + (\mathrm{e}^{-2\mathrm{i}(\varepsilon_+ - \delta)} - 1)|\alpha_+|^2 + (\mathrm{e}^{2\mathrm{i}(\varepsilon_- - \delta)} - 1)|\alpha_-|^2\Big] \\
&\qquad\qquad + 2|\alpha_+|^2\alpha_+^*\alpha_- \exp\Big[-\mathrm{i}(\varphi_+ - \varphi_-) - \mathrm{i}(\varepsilon_+ - \delta) \\
&\qquad\qquad + (\mathrm{e}^{-\mathrm{i}(\varepsilon_+ - \delta)} - 1)|\alpha_+|^2 + (\mathrm{e}^{\mathrm{i}(\varepsilon_- - \delta)} - 1)|\alpha_-|^2\Big] \\
&\qquad\qquad + 2|\alpha_-|^2\alpha_-^*\alpha_+ \exp\Big[+\mathrm{i}(\varphi_+ - \varphi_-) - \mathrm{i}(\varepsilon_- - \delta) \\
&\qquad\qquad + (\mathrm{e}^{\mathrm{i}(\varepsilon_+ - \delta)} - 1)|\alpha_+|^2 + (\mathrm{e}^{-\mathrm{i}(\varepsilon_- - \delta)} - 1)|\alpha_-|^2\Big]\Big\}
\end{aligned}
\tag{61}
$$

where α_+ and α_- are given by (55).

Light is said to exhibit sub-Poissonian photon statistics if

$$\langle \hat{a}_x^{+2}(z)\hat{a}_x^2(z)\rangle - \langle \hat{a}_x^+(z)\hat{a}_x(z)\rangle^2 < 0 \tag{62}$$

Expression (61) is quite complicated and it is not easy to say without numerical analysis whether condition (62) can be satisfied. Usually, the normalized second-order correlation function is considered; it is defined by the relation

$$g^{(2)}(z) = \frac{\langle \hat{a}_x^{+2}(z)\hat{a}_x^2(z)\rangle}{\langle \hat{a}_x^+(z)\hat{a}_x(z)\rangle^2} \tag{63}$$

and condition (62) can then be written as

$$g^{(2)}(z) - 1 < 0 \tag{64}$$

Another measure of the sub-Poissonian photon statistics is the q parameter introduced by Mandel [41] and defined as

$$q = \frac{\langle (\Delta\hat{n})^2\rangle}{\langle \hat{n}\rangle} - 1 = \left[g^{(2)}(z) - 1\right]\langle \hat{a}_x^+\hat{a}_x\rangle \tag{65}$$

Negative values of the parameter q denote sub-Poissonian photon statistics, and the limit $q = -1$ is reached for number states without photon number fluctuations.

In Fig. 1 we plot both $g^{(2)}(z) - 1$ and $q(z)$ against $(\varepsilon - \delta)|\alpha|^2$ for a medium containing optically inactive molecules, and for the elliptical polarization of the incoming light with $\eta = \pi/8$ and $\theta = -\pi/4$. Both functions show oscillatory behavior with both negative and positive values. Negative values of these functions mean the sub-Poissonian photon statistics of the x component of the outgoing field. For optically nonactive molecules they depend only on the molecular parameter $\tilde{\gamma}_1(\omega)$,

$$\varepsilon(z) - \delta(z) = -\frac{2Nz}{\hbar c}\tilde{\gamma}_1(\omega) \tag{66}$$

which in Fig. 1 we have assumed as equal to 1×10^{-6}, according to the estimation made by Ritze and Bandilla [71]. To get values of $(\varepsilon - \delta)|\alpha|^2$ of the order of unity, a field is needed with the mean number of photons $|\alpha|^2 \approx 10^6$. From Fig. 1 it is seen that the values of $g^{(2)}(z) - 1$ obtained in this process are rather small—of the order $\varepsilon - \delta$. However, the q parameter reaches the value -0.63, which means considerable narrowing of the

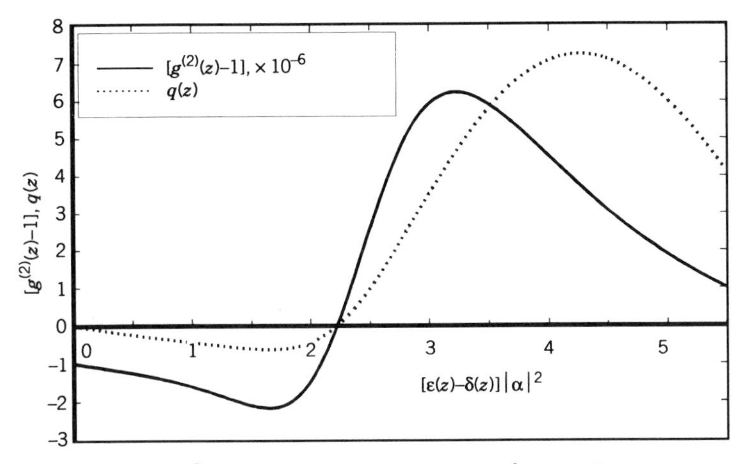

Figure 1. Plots of $g^{(2)}(z) - 1$ (scaled in units 1×10^{-6}) and $q(z)$ against the scaled intensity of light $[\varepsilon(z) - \delta(z)]|\alpha|^2$, assuming that $\varepsilon(z) - \delta(z) = 1 \times 10^{-6}$, $\theta = -\pi/4$, and $\eta = \pi/8$.

photon number distribution for strong fields. If the value $q = -1$ means 100% reduction of the photon number fluctuations, a 63% reduction can be obtained in the propagation process considered here. The possibility to get sub-Poissonian photon statistics in this process was predicted by Tanaś and Kielich [72] with the perturbative method, and confirmed by Ritze [73], who obtained exact solutions to the problem.

For optically active molecules, $\varphi_+ - \varphi_+ \neq 0$ and $\varepsilon_+ - \varepsilon_- \neq 0$, the general form of the solutions given by (57) and (61) must be used. If, however, $\tilde{\gamma}_1(\omega) = 0$, then the solutions simplify considerably, because

$$\varepsilon_+ - \delta = -(\varepsilon_- - \delta) = \mathrm{i}\frac{Nz}{\hbar c}\left[\tilde{\sigma}_2(\omega) + \tilde{\sigma}_3(\omega)\right] = \sigma \tag{67}$$

We obtain, in this case,

$$\begin{aligned}\langle \hat{a}_x^+(z)\hat{a}_x(z)\rangle = \frac{|\alpha|^2}{2}\Big\{1 &+ \cos 2\eta \exp\left[|\alpha|^2(\cos\sigma - 1)\right] \\ &\times \cos\left[2\theta - (\varphi_+ - \varphi_-) - |\alpha|^2 \sin\sigma\right]\Big\}\end{aligned} \tag{68}$$

$$\begin{aligned}\langle \hat{a}_x^{+2}(z)\hat{a}_x^2(z)\rangle = \frac{|\alpha|^4}{4}\Big\{1 &+ \frac{1}{2}\cos^2 2\eta + \frac{1}{2}\cos^2 2\eta \exp D \cos E \\ &+ 2\cos 2\eta \exp F \cos G\Big\}\end{aligned} \tag{69}$$

where

$$\begin{aligned} D &= |\alpha|^2(\cos 2\sigma - 1) \\ E &= 4\theta - 2(\varphi_+ - \varphi_-) - 2\sigma - |\alpha|^2 \sin 2\sigma \\ F &= |\alpha|^2(\cos \sigma - 1) \\ G &= 2\theta - (\varphi_+ - \varphi_-) - \sigma - |\alpha|^2 \sin \sigma \end{aligned} \tag{70}$$

Expressions (68)–(70) depend upon two molecular parameters:

$$\varphi_+ - \varphi_- = \frac{\mathrm{i}Nz}{\hbar c}\tilde{\rho} \quad \text{and} \quad \sigma = \frac{\mathrm{i}Nz}{\hbar c}\left[\tilde{\sigma}_2(\omega) + \tilde{\sigma}_3(\omega)\right]$$

describing the natural and nonlinear optical activity.

All the expressions derived here are valid for arbitrary polarization of the incoming beam defined by the parameters θ and η. It is interesting to note that for the circular polarization of entering light ($\eta = \pm\pi/4$) we have

$$\begin{aligned} \langle \hat{a}_x^+(z)\hat{a}_x(z)\rangle &= \frac{|\alpha|^2}{2} \\ \langle \hat{a}_x^{+2}(z)\hat{a}_x^2(z)\rangle &= \frac{|\alpha|^4}{4} \end{aligned} \tag{71}$$

which means $g^2(z) - 1 = 0$. This result is not surprising in view of our earlier discussion concerning the photon statistics of circular components of light propagating in a Kerr medium. In fact, it confirms our statement that the photon statistics of such light do not change during the propagation. The polarizer choosing the component x only reduces the intensity of the beam to one-half of the incoming intensity, but its statistics remain unchanged. Any deviation from the circular polarization of the incoming light will cause, as is easy to check, changes in the photon statistics of light propagating in the medium.

V. SQUEEZING

The fact that the photon statistics of elliptically polarized light do not change when the light propagates through the isotropic nonlinear medium does not mean that the state of the field does not change during such interaction. It turns out that the field can become squeezed as a result of such interaction; that is, it can be in a squeezed state, which has no classical analog and requires quantum interpretation. To show this, we introduce two Hermitian field operators $\hat{Q}_\sigma$ and $\hat{P}_\sigma$ defined as [41, 193]

$$\hat{Q}_\sigma = \hat{a}_\sigma + \hat{a}_\sigma^+ \qquad \hat{P}_\sigma = -\mathrm{i}(\hat{a}_\sigma - \hat{a}_\sigma^+) \tag{72}$$

where σ denotes $+(-)$ in the circular basis or $x(y)$ in the Cartesian basis. The operators $\hat{Q}_\sigma$ and $\hat{P}_\sigma$ satisfy the commutation rules

$$\left[\hat{Q}_\sigma, \hat{P}_{\sigma'}\right] = 2\mathrm{i}\delta_{\sigma\sigma'} \tag{73}$$

A squeezed state of the electromagnetic field is defined [195] as a state of the field in which the variance of $\hat{Q}_\sigma$ or $\hat{P}_\sigma$ is smaller than unity

$$\left\langle\left(\Delta\hat{Q}_\sigma\right)^2\right\rangle < 1 \quad \text{or} \quad \left\langle\left(\Delta\hat{P}_\sigma\right)^2\right\rangle < 1 \tag{74}$$

where $\Delta\hat{Q}_\sigma = \hat{Q}_\sigma - \langle\hat{Q}_\sigma\rangle$. On introducing the normal order of the creation and annihilation operators, definition (74) can be rewritten in the form [41]

$$\left\langle:\left(\Delta\hat{Q}_\sigma\right)^2:\right\rangle < 0 \quad \text{or} \quad \left\langle:\left(\Delta\hat{P}_\sigma\right)^2:\right\rangle < 0 \tag{75}$$

To calculate the quantities occurring in definition (75) for the process of light propagation in the nonlinear medium considered here, it suffices to insert into (75) the operator solutions (39) and next calculate the expectation value in the initial state of the field, which we assume to be the coherent state $|\alpha\rangle$ defined by (56). If one of the normally ordered variances appears to be negative, then the corresponding component of the field is in a squeezed state, which has no classical analog. Our calculations give for the normally ordered variances of the resulting field the following expressions:

$$\begin{aligned}
&\left\langle:\left[\Delta\hat{Q}_\pm(z)\right]^2:\right\rangle \\
&\quad = \left\langle:\left[\hat{a}_\pm(z) + \hat{a}_\pm^+(z)\right]^2:\right\rangle - \left\langle\hat{a}_\pm(z) + \hat{a}_\pm^+(z)\right\rangle^2 \\
&\quad = 2\,\mathrm{Re}\Big\{\alpha_\pm^2 \exp\left[2\mathrm{i}\varphi_\pm + \varepsilon_\pm + (\mathrm{e}^{2\mathrm{i}\varepsilon_\pm} - 1)|\alpha_\pm|^2 + (\mathrm{e}^{2\mathrm{i}\delta} - 1)|\alpha_\mp|^2\right. \\
&\qquad\qquad \left. - \alpha_\pm^2 \exp\left[2\mathrm{i}\varphi_\pm + 2(\mathrm{e}^{\mathrm{i}\varepsilon_\pm} - 1)|\alpha_\pm|^2 + 2(\mathrm{e}^{\mathrm{i}\delta} - 1)|\alpha_\mp|^2\right]\right\} \\
&\qquad + 2|\alpha_\pm|^2\Big\{1 - \exp\left[2(\cos\varepsilon_\pm - 1)|\alpha_\pm|^2 + 2(\cos\delta - 1)|\alpha_\mp|^2\right]\Big\}
\end{aligned} \tag{76}$$

where $\alpha_{\pm}$ are given by (55), while $\varphi_{\pm} = \varphi_{\pm}(z)$, $\varepsilon = \varepsilon(z)$, and $\delta = \delta(z)$ are given by (40). For the operators $\hat{P}_{\pm}$ we get

$$\left\langle :\left[\Delta \hat{P}_{\pm}(z)\right]^2: \right\rangle = -2\,\mathrm{Re}\{\cdots\} + 2|\alpha_{\pm}|^2\{\cdots\} \tag{77}$$

where the expressions in the braces are the same as in (76).

Especially interesting is the case of circularly polarized incoming field, because photon statistics of such a field do not change. Let us assume that the incoming beam is circularly polarized with $\eta = \pi/4$ and $\theta = 0$. Then $|\alpha_+|^2 = |\alpha|^2$, $|\alpha_-|^2 = 0$, and formula (76) takes the much simpler form

$$\begin{aligned}\left\langle :\left[\Delta \hat{Q}_{+}(z)\right]^2: \right\rangle = 2|\alpha|^2\Big\{&\exp\left[|\alpha|^2(\cos 2\varepsilon_+ - 1)\right]\cos\left(\varphi_+ + \varepsilon_+ + |\alpha|^2 \sin 2\varepsilon_+\right) \\ &- \exp\left[2|\alpha|^2(\cos\varepsilon_+ - 1)\right]\cos\left(\varphi_+ + 2|\alpha|^2\sin\varepsilon_+\right)\Big\} \\ &+ 2|\alpha|^2\left\{1 - \exp\left[2|\alpha|^2(\cos\varepsilon_+ - 1)\right]\right\}\end{aligned} \tag{78}$$

Similarly (77) goes over into

$$\left\langle :\left[\Delta \hat{P}_{+}(z)\right]^2: \right\rangle = -2|\alpha|^2\{\cdots\} + 2|\alpha|^2\{\cdots\} \tag{79}$$

with the contents of the braces the same as in (78).

Assuming the initial phase φ_0 so that $\varphi_+ + \varphi_0 = 0$, and assuming the value $\varepsilon_+(z) = 1 \times 10^{-6}$, similarly as for the photon statistics case, we have plotted in Fig. 2 expressions (78) and (79) as functions of $\varepsilon_+ |\alpha|^2$. It is seen that the normally ordered variances (78) and (79) exhibit oscillatory behavior on $\varepsilon_+ |\alpha|^2$, taking both positive and negative values. Whenever one of the variances takes negative values, the corresponding component of the field $\hat{Q}_+(z)$ or $\hat{P}_+(z)$ is said to be squeezed. This means that despite the Poissonian photon statistics, the field can be in a squeezed state. It is also worth noting that the values of squeezing possible in the propagation process are quite large. With our definition of the operators $\hat{Q}_+$ and $\hat{P}_+$ the value allowed by quantum mechanics for (78) or (79) is minus unity, which means no quantum fluctuations in the corresponding component of the field. It is seen from Fig. 2 that the first minimum of $\langle :[\Delta\hat{Q}_+(z)]^2: \rangle$ has a value of -0.66, while the second minimum already has a value of -0.97, which means 97% of the value allowed by quantum mechanics. This result can even be improved by tuning the initial phase φ_0 [75]; this means a reduction of the quantum fluctuations in the field by two orders of magnitude with respect to the fluctuations in the vacuum (or a coherent state). The first minimum of $\langle :[\Delta\hat{P}_+(z)]^2: \rangle$ has a value of -0.92, which also means a considerable reduction of quantum fluctuations. Expressions (78) and (79) after corresponding changes of the variables become identical to the results obtained for the anharmonic oscillator [49].

For the x component of the polarization of the incoming beam, we get for the normally ordered variances the following formulas:

$$\begin{aligned}
&\left\langle :\left[\Delta\hat{Q}_x(z)\right]^2:\right\rangle \\
&\quad= \left\langle :\left[\hat{a}_x(z) + \hat{a}_x^+(z)\right]^2:\right\rangle - \langle \hat{a}_x(z) + \hat{a}_x^+(z)\rangle^2 \\
&\quad= \tfrac{1}{2}\Big\{\left\langle :\left[\hat{a}_+(z) + \hat{a}_-(z) + \hat{a}_+^+(z) + \hat{a}_-^+(z)\right]^2:\right\rangle \\
&\qquad\qquad - \langle \hat{a}_+(z) + \hat{a}_-(z) + \hat{a}_+^+(z) + \hat{a}_-^+(z)\rangle^2\Big\} \\
&\quad= \mathrm{Re}\Big\{\alpha_+^2 \exp\left[2i\varphi_+ + i\varepsilon_+ + (e^{2i\varepsilon_+} - 1)|\alpha_+|^2 + (e^{2i\delta} - 1)|\alpha_-|^2\right] \\
&\qquad - \alpha_+^2 \exp\left[2i\varphi_+ + 2(e^{i\varepsilon_+} - 1)|\alpha_+|^2 + 2(e^{i\delta} - 1)|\alpha_-|^2\right] \\
&\qquad + \alpha_-^2 \exp\left[2i\varphi_- + i\varepsilon_- + (e^{2i\varepsilon_-} - 1)|\alpha_-|^2 + (e^{2i\delta} - 1)|\alpha_+|^2\right] \\
&\qquad - \alpha_-^2 \exp\left[2i\varphi_- + 2(e^{i\varepsilon_-} - 1)|\alpha_-|^2 + 2(e^{i\delta} - 1)|\alpha_+|^2\right] \\
&\qquad + 2\alpha_+\alpha_- \exp\left[i(\varphi_+ + \varphi_- + \delta) + (e^{i(\varepsilon_+ + \delta)} - 1)|\alpha_+|^2 \right.\\
&\qquad\qquad\qquad \left. + (e^{i(\varepsilon_- - \delta)} - 1)|\alpha_-|^2\right] \\
&\qquad - 2\alpha_+\alpha_- \exp\left[i(\varphi_+ + \varphi_-) + (e^{i\varepsilon_+} + e^{i\delta} - 2)|\alpha_+|^2 \right.\\
&\qquad\qquad\qquad \left. + (e^{i\varepsilon_-} + e^{i\delta} - 2)|\alpha_-|^2\right]\Big\} \\
&\quad + \Big\{|\alpha_+|^2 - |\alpha_+|^2 \exp\left[2(\cos\varepsilon_+ - 1)|\alpha_+|^2 + 2(\cos\delta - 1)|\alpha_-|^2\right] \\
&\qquad + |\alpha_-|^2 - |\alpha_-|^2 \exp\left[2(\cos\varepsilon_- - 1)|\alpha_-|^2 + 2(\cos\delta - 1)|\alpha_+|^2\right] \\
&\quad + 2\,\mathrm{Re}\Big[\alpha_+^*\alpha_- \exp\left[-i(\varphi_+ - \varphi_-) + (e^{-i(\varepsilon_+ - \delta)} - 1)|\alpha_+|^2 \right.\\
&\qquad\qquad\qquad \left. + (e^{-i(\varepsilon_- - \delta)} - 1)|\alpha_-|^2\right] \\
&\qquad - \alpha_+^*\alpha_- \exp\left[-i(\varphi_+ - \varphi_-) + (e^{-i\varepsilon_+} + e^{i\delta} - 2)|\alpha_+|^2 \right.\\
&\qquad\qquad\qquad \left. + (e^{i\varepsilon_-} + e^{-i\delta} - 2)|\alpha_-|^2\right]\Big]\Big\}
\end{aligned} \tag{80}$$

and for the other component we have

$$\left\langle :\left[\Delta\hat{P}_x(z)\right]^2:\right\rangle = -\mathrm{Re}\{\cdots\} + \{\cdots\} \tag{81}$$

with the contents of the braces the same as in (80).

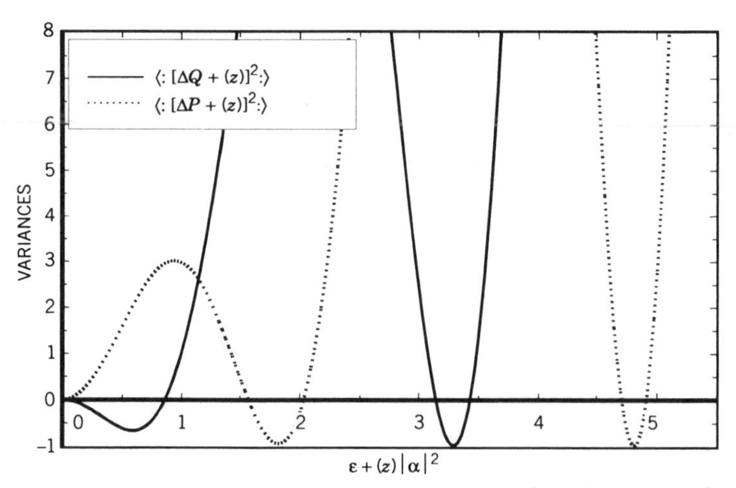

Figure 2. Plots of the normally ordered field variances $\langle:[\Delta\hat{Q}_+(z)]^2:\rangle$ and $\langle:[\Delta\hat{P}_+(z)]^2:\rangle$ against the scaled intensity of light $\varepsilon_+(z)|\alpha|^2$, assuming that $\varepsilon_+(z) = 1 \times 10^{-6}$, $\theta = 0$, $\varphi_+ + \varphi_0 = 0$, and $\eta = \pi/4$.

Expressions (80) and (81) are exact. They are, however, very complicated and only numerical analysis allows us to give a definite answer to whether the field is squeezed or not. They can be considerably simplified under certain assumptions concerning the polarization of the field and the character of the medium.

For optically inactive molecules detailed analysis of (80) and (81) has been carried out [74, 75], showing that the components $\hat{Q}_x(z)$ and $\hat{P}_x(z)$ can also become squeezed after passing through the nonlinear medium. The maximum values of squeezing for these components are the same as for $\hat{Q}_+$ and $\hat{P}_+$, although the minima can appear for different values of the field intensity. It is also interesting that for linear polarization of the incoming field perpendicular to the measured polarization ($\theta = \pi/2$), the outgoing field, which is completely quantum in nature, can also show squeezing.

For optically active molecules, expressions (80) and (81) in their extended form must be used. However, as a rule the tensors describing the nonlinear optical activity have orders of magnitude smaller values than the tensors $\gamma_{ijkl}(\omega)$, and in practice $\varepsilon_+ \approx \varepsilon_-$. Thus, in most cases one can neglect contributions from the nonlinear optical activity, which would be essential only if the result were a function of the difference $\varepsilon_+ - \varepsilon_-$.

The mechanism of producing squeezed states described in this chapter is universal in the sense that it takes place for any molecules or atoms including those with spherical symmetry.

VI. CONCLUSIONS

The subject of this chapter was the problem of producing quantum fields that have no classical analogs in the process of propagation of strong light in a nonlinear Kerr-like medium. The possibility of appearance of two nonclassical effects such as sub-Poissonian photon statistics and squeezing was discussed in detail. We showed that both effects can be produced in corresponding components of the field. Thus, the process of propagation of light in isotropic media can be a source of nonclassical fields. Despite the small values of $\varepsilon(z) - \delta(z)$ for real physical situations, the value of the q parameter measuring the sub-Poissonian character of the photon number distribution can be reduced to -0.63 for strong fields which is 63% of the limit allowed by quantum mechanics.

The process discussed turns out to be even more effective in producing squeezed states of the field. In this way one can get more than 97% of squeezing. The effect of squeezing also occurs for circularly polarized light, for which there is no change in photon statistics. This means that the squeezed states can exist with Poissonian photon statistics, which is characteristic for coherent states of the field. Our considerations show explicitly the difference between sub-Poissonian photon statistics and squeezing. The form of solutions (39) for the field operators, which is as a matter of fact nonlinear phase modulation, lead to squeezing of the field states, while the number of photons $\hat{a}_+^+\hat{a}_+(\hat{a}_-^+\hat{a}_-)$, which does not depend on the phase, does not change. To get sub-Poissonian photon statistics, a nonlinear change in the number of photons is needed. It can be achieved for the x and y components of the field for which the number of photons $\hat{a}_x^+\hat{a}_x(\hat{a}_y^+\hat{a}_y)$ does change due to the interaction.

We would like to emphasize the fact that our solutions are exact analytical solutions, which is rather exceptional for this type of problems, and this fact is worthy of attention on its own right. We have referred to the process of squeezing that occurs during the propagation of light in a nonlinear medium as self-squeezing [74, 75]. This is the field itself that causes squeezing of its own quantum fluctuations. The effect of self-squeezing accompanies to a certain degree all other nonlinear processes. In recent years experimental and theoretical studies of nonlinear optical activity have been developed [76, 162, 215, 218].

Squeezing is an effect that depends on the phase of the field, and it is interesting to study the quantum phase properties of the field. Quite recently, since Pegg and Barnett [219] introduced the Hermitian phase formalism, considerable progress has been achieved in studies of phase properties of optical fields. For fields propagating in a Kerr medium such results have been reported by Gantsog and Tanaś [220, 221]. It turned out also that the problem of propagation has exact analytical solutions even when the linear dissipation is included [222]; thus, the quantum phase properties of light propagating in a Kerr medium with dissipation have also been studied [223, 224]. The quantum phase properties of optical fields, however, are beyond the scope of this work, and require separate treatment.

APPENDIX A

Restricting our considerations to weak spatial dispersion and omitting in (6) summation over k, the linear $H_{\mathrm{I}}^{(2)}/N$ and nonlinear $H_{\mathrm{I}}^{(4)}/N$ interaction Hamiltonians of the microsystem (atom, molecule) with the electromagnetic field can be written as (in SI units)

$$H_{\mathrm{I}}^{(2)}/N = -\left[H(-\omega;\omega) + H(\omega;-\omega)\right] \tag{A.1}$$

$$\begin{aligned} H_{\mathrm{I}}^{(4)}/N = -\big[&H(-\omega;-\omega,\omega,\omega) + H(-\omega;\omega,-\omega,\omega) \\ &+H(-\omega;\omega,\omega,-\omega) + H(\omega;\omega,-\omega,-\omega) \\ &+H(\omega;-\omega,\omega,-\omega) + H(\omega;-\omega,-\omega,\omega)\big] \end{aligned} \tag{A.2}$$

where

$$\begin{aligned} H(-\omega;\omega) = \tfrac{1}{2}\Big\{&\alpha_{ij}(-\omega;\omega)E_i^- E_j^+ \\ &+\tfrac{1}{3}\left[\eta_{i(jk)}(-\omega;\omega)E_i^-\nabla_k E_j^+ + \eta_{(ik)j}(-\omega;\omega)(\nabla_k E_i^-)E_j^+\right] \\ &+\rho_{ij}(-\omega;\omega)E_i^- B_j^+ + \lambda_{ij}(-\omega;\omega)B_i^- E_j^+\Big\} + \text{h.c.} \end{aligned} \tag{A.3}$$

$$\begin{aligned} H(-\omega;-\omega,\omega,\omega) = \tfrac{1}{24}\Big\{&\gamma_{ijkl}(-\omega;-\omega,\omega,\omega)E_i^- E_j^- E_k^+ E_l^+ \\ &+\tfrac{1}{3}\big[\eta_{ijk(lm)}(-\omega;-\omega,\omega,\omega)E_i^- E_j^- E_k^+ \nabla_m E_l^+ \\ &\quad+\eta_{ij(km)l}(-\omega;-\omega,\omega,\omega)E_i^- E_j^-(\nabla_m E_k^+)E_l^+ \\ &\quad+\eta_{i(jm)kl}(-\omega;-\omega,\omega,\omega)E_i^-(\nabla_m E_j^-)E_k^+ E_l^+ \\ &\quad+\eta_{(im)jkl}(-\omega;-\omega,\omega,\omega)(\nabla_m E_i^-)E_j^- E_k^+ E_l^+\big] \\ &+\rho_{ijkl}(-\omega;-\omega,\omega,\omega)E_i^- E_j^- E_k^+ B_l^+ \\ &+\kappa_{ijkl}(-\omega;-\omega,\omega,\omega)E_i^- E_j^- B_k^+ E_l^+ \\ &+\sigma_{ijkl}(-\omega;-\omega,\omega,\omega)E_i^- B_j^- E_k^+ E_l^+ \\ &+\lambda_{ijkl}(-\omega;-\omega,\omega,\omega)B_i^- E_j^- E_k^+ E_l^+\Big\} + \text{h.c.} \end{aligned} \tag{A.4}$$

with

$$\mathbf{E}^{\pm} = \mathbf{E}^{\pm}(\mathbf{r},t) \qquad \mathbf{B}^{\pm} = \mathbf{B}^{\pm}(\mathbf{r},t) \tag{A.5}$$

and

$$\mathbf{E}^{\pm}(\mathbf{r},t) = \mathbf{E}^{\pm}(k)\exp[\pm \mathrm{i}(\mathbf{k}\cdot\mathbf{r} - \omega t)] \tag{A.6}$$

The last relation is the result of restricting the summations over k in (6) to one term with definite k; that is, we consider a one-mode field.

A characteristic feature of the linear and nonlinear response of the medium to a force oscillating with the frequency ω is its dependence on ω, which is referred to as time dispersion. Accordingly, in the case of an optical force the Hamiltonians (3) and (4) should depend on the time dispersion. This dependence enters the Hamiltonians via the coupling constants, i.e., the linear and nonlinear and electric and magnetic polarizabilities of the molecules. The linear and nonlinear and electric and magnetic polarizabilities for an individual molecule obtained according to quantum mechanical formulas can be found in Refs. 206, 207, and 225 and are given by the formulas

$$
{}^{(a)}_{A}\boldsymbol{\chi}^{(b)}_{B}(-\omega;\omega) = \frac{\rho_{\Psi\Psi}}{\hbar}\sum_{\Phi f\neq\Psi p}\left\{\frac{\langle\Psi p|\mathbf{M}^{(a)}_{A}|\Phi f\rangle\langle\Phi f|\mathbf{M}^{(b)}_{B}|\Psi p\rangle}{\omega+\omega_{\Phi\Psi}} + \frac{\langle\Psi p|\mathbf{M}^{(b)}_{B}|\Phi f\rangle\langle\Phi f|\mathbf{M}^{(a)}_{A}|\Psi p\rangle}{-\omega+\omega_{\Phi\Psi}}\right\} \tag{A.7}
$$

$$
{}^{(a)}_{A}\boldsymbol{\chi}^{(b,c,d)}_{BCD}(\omega) = \frac{S\{[\mathbf{M}^{(a)}_{A},\mathbf{M}^{(b)}_{B}],[\mathbf{M}^{(c)}_{C},\mathbf{M}^{(d)}_{D}]\}\rho_{\Psi\Psi}}{3!\hbar} \times \sum_{\Phi f,\,\Lambda l,\,\Upsilon u\neq\Psi p}\langle\Psi p|\mathbf{F}_{\Phi f\Lambda l\Upsilon u}|\Psi p\rangle
$$

$$
\begin{aligned}
\mathbf{F}_{\Phi f\Lambda l\Upsilon u} &= \frac{\mathbf{M}^{(a)}_{A}|\Phi f\rangle\langle\Phi f|\mathbf{M}^{(b)}_{B}|\Lambda l\rangle\langle\Lambda l|\mathbf{M}^{(c)}_{C}|\Upsilon u\rangle\langle\Upsilon u|\mathbf{M}^{(d)}_{D}}{(\omega+\omega_{\Phi\Psi})(2\omega+\omega_{\Lambda\Psi})(\omega+\omega_{\Upsilon\Psi})} \\
&+ \frac{\mathbf{M}^{(a)}_{A}|\Phi f\rangle\langle\Phi f|\mathbf{M}^{(c)}_{C}|\Lambda l\rangle\langle\Lambda l|\mathbf{M}^{(b)}_{B}|\Upsilon u\rangle\langle\Upsilon u|\mathbf{M}^{(d)}_{D}}{(\omega+\omega_{\Phi\Psi})\omega_{\Lambda\Psi}(\omega+\omega_{\Upsilon\Psi})} \\
&+ \frac{\mathbf{M}^{(b)}_{B}|\Phi f\rangle\langle\Phi f|\mathbf{M}^{(c)}_{C}|\Lambda l\rangle\langle\Lambda l|\mathbf{M}^{(d)}_{D}|\Upsilon u\rangle\langle\Upsilon u|\mathbf{M}^{(a)}_{A}}{(\omega+\omega_{\Phi\Psi})\omega_{\Lambda\Psi}(-\omega+\omega_{\Upsilon\Phi})} \\
&+ \frac{\mathbf{M}^{(d)}_{D}|\Phi f\rangle\langle\Phi f|\mathbf{M}^{(a)}_{A}|\Lambda l\rangle\langle\Lambda l|\mathbf{M}^{(b)}_{B}|\Upsilon u\rangle\langle\Upsilon u|\mathbf{M}^{(c)}_{C}}{(-\omega+\omega_{\Phi\Psi})\omega_{\Lambda\Psi}(\omega+\omega_{\Upsilon\Psi})}
\end{aligned} \tag{A.8}
$$

$$
\begin{aligned}
&+ \frac{\mathbf{M}^{(c)}_{C}|\Phi f\rangle\langle\Phi f|\mathbf{M}^{(b)}_{B}|\Lambda l\rangle\langle\Lambda l|\mathbf{M}^{(d)}_{D}|\Upsilon u\rangle\langle\Upsilon u|\mathbf{M}^{(a)}_{A}}{(-\omega+\omega_{\Phi\Psi})\omega_{\Lambda\Psi}(-\omega+\omega_{\Upsilon\Psi})} \\
&+ \frac{\mathbf{M}^{(c)}_{C}|\Phi f\rangle\langle\Phi f|\mathbf{M}^{(d)}_{D}|\Lambda l\rangle\langle\Lambda l|\mathbf{M}^{(a)}_{A}|\Upsilon u\rangle\langle\Upsilon u|\mathbf{M}^{(b)}_{B}}{(-\omega+\omega_{\Phi\Psi})(-2\omega+\omega_{\Lambda\Psi})(-\omega+\omega_{\Upsilon\Psi})}
\end{aligned}
$$

where $S\{[M_A^{(a)}, \mathbf{M}_B^{(b)}], [\mathbf{M}_C^{(c)}, \mathbf{M}_D^{(d)}]\}$ is the operator denoting summation over the permutations of the elements contained in the square brackets, and we have for the linear polarizabilities

$$\begin{aligned}
\alpha_{ij}(-\omega;\omega) &= {}^{(1)}_{e}\chi^{(1)}_{e\,ij}(-\omega;\omega)\\
\rho_{ij}(-\omega;\omega) &= {}^{(1)}_{e}\chi^{(1)}_{m\,ij}(-\omega;\omega)\\
\lambda_{ij}(-\omega;\omega) &= {}^{(1)}_{m}\chi^{(1)}_{e\,ij}(-\omega;\omega)\\
\eta_{i(jk)}(-\omega;\omega) &= {}^{(1)}_{e}\chi^{(2)}_{e\,i(jk)}(-\omega;\omega)\\
\eta_{(ik)j}(-\omega;\omega) &= {}^{(2)}_{e}\chi^{(1)}_{e\,(ik)j}(-\omega;\omega)
\end{aligned} \tag{A.9}$$

and for the nonlinear polarizabilities

$$\begin{aligned}
\gamma_{ijkl}(-\omega;-\omega,\omega,\omega) &= {}^{(1)}_{e}\chi^{(1,1,1)}_{eee\,ijkl}(-\omega;-\omega,\omega,\omega)\\
\rho_{ijkl}(-\omega;-\omega,\omega,\omega) &= {}^{(1)}_{e}\chi^{(1,1,1)}_{eem\,ijkl}(-\omega;-\omega,\omega,\omega)\\
\kappa_{ijkl}(-\omega;-\omega,\omega,\omega) &= {}^{(1)}_{e}\chi^{(1,1,1)}_{eme\,ijkl}(-\omega;-\omega,\omega,\omega)\\
\sigma_{ijkl}(-\omega;-\omega,\omega,\omega) &= {}^{(1)}_{e}\chi^{(1,1,1)}_{mee\,ijkl}(-\omega;-\omega,\omega,\omega)\\
\lambda_{ijkl}(-\omega;-\omega,\omega,\omega) &= {}^{(1)}_{m}\chi^{(1,1,1)}_{eee\,ijkl}(-\omega;-\omega,\omega,\omega)\\
\eta_{ijk(lm)}(-\omega;-\omega,\omega,\omega) &= {}^{(1)}_{e}\chi^{(1,1,2)}_{eee\,ijk(lm)}(-\omega;-\omega,\omega,\omega)\\
\eta_{ij(km)l}(-\omega;-\omega,\omega,\omega) &= {}^{(1)}_{e}\chi^{(1,2,1)}_{eee\,ij(km)l}(-\omega;-\omega,\omega,\omega)\\
\eta_{i(jm)kl}(-\omega;-\omega,\omega,\omega) &= {}^{(1)}_{e}\chi^{(2,1,1)}_{eee\,i(jm)kl}(-\omega;-\omega,\omega,\omega)\\
\eta_{(im)jkl}(-\omega;-\omega,\omega,\omega) &= {}^{(2)}_{e}\chi^{(1,1,1)}_{eee\,(im)jkl}(-\omega;-\omega,\omega,\omega)
\end{aligned} \tag{A.10}$$

Expressions (A.7) and (A.8) are the quantum mechanical formulas for the linear and nonlinear polarizabilities of the microsystem being in one of the stationary states of the f-fold degenerate energy level with the energy $\hbar\omega_\Psi$. This state, $|\Psi f\rangle$, is defined by the quantum numbers Ψ labeling the energy levels of the microsystem and the quantum numbers f labeling the states belonging to the level Ψ. Moreover, $\rho_{\Psi\Psi}$ denotes the expectation value of the unperturbed density matrix in the state $|\Psi\rangle$ (the probability that the microsystem is in the stationary state $|\Psi\rangle$), $\omega_{\phi\Psi} = \omega_{\phi} - \omega_{\Psi}$ is the transition frequency between the levels ϕ and Ψ, and $M_A^{(a)}$ is the operator of the multipolar electric ($A = e$) or magnetic ($A = m$) moment of the order a.

From definitions (A.7) and (A.8) one can easily show that

$$
\begin{aligned}
{}^{(a)}_{A}\boldsymbol{\chi}^{(b)}_{B}(-\omega;\omega)^{*} &= {}^{(a)}_{A}\boldsymbol{\chi}^{(b)}_{B}(\omega;-\omega) \\
{}^{(a)}_{A}\boldsymbol{\chi}^{(b,c,d)}_{BCD}(-\omega;-\omega,\omega,\omega)^{*} &= {}^{(a)}_{A}\boldsymbol{\chi}^{(b,c,d)}_{BCD}(\omega;\omega,-\omega,-\omega)
\end{aligned}
\tag{A.11}
$$

It is also easy to check that the polarizabilities are invariant with respect to the following permutations:

$$
\begin{aligned}
{}^{(a)}_{A}\boldsymbol{\chi}^{(b,c,d)}_{BCD}(-\omega;-\omega,\omega,\omega) &= {}^{(a)}_{A}\boldsymbol{\chi}^{(c,b,d)}_{CBD}(-\omega;-\omega,\omega,\omega) \\
&= {}^{(a)}_{A}\boldsymbol{\chi}^{(d,c,b)}_{DCB}(-\omega;\omega,\omega,-\omega) \\
&= {}^{(a)}_{A}\boldsymbol{\chi}^{(b,d,c)}_{BDC}(-\omega;-\omega,\omega,\omega) \\
&= {}^{(b)}_{B}\boldsymbol{\chi}^{(a,c,d)}_{ACD}(-\omega;-\omega,\omega,\omega) \\
&= {}^{(b)}_{B}\boldsymbol{\chi}^{(a,d,c)}_{ADC}(-\omega;-\omega,\omega,\omega)
\end{aligned}
\tag{A.12}
$$

which allows us to replace the polarizabilities ${}^{(a)}_{(A}\boldsymbol{\chi}^{(b,c,d)}_{BCD}(-\omega;\omega,-\omega,\omega)$ occurring in $H(-\omega,\omega,-\omega,\omega)$, and the polarizabilities ${}^{(a)}_{(A}\boldsymbol{\chi}^{(b,c,d)}_{BCD}(-\omega;\omega,\omega-\omega)$ occurring in $H(-\omega;\omega,\omega,-\omega)$ with ${}^{(a)}_{(A}\boldsymbol{\chi}^{(c,b,d)}_{CBD}(-\omega;-\omega,\omega,\omega)$ and ${}^{(a)}_{(A}\boldsymbol{\chi}^{(d,c,b)}_{DCB}(-\omega;-\omega,\omega,\omega)$, respectively. This means the same frequency dependence, that is, $(-\omega;-\omega,\omega,\omega)$, as in $H(-\omega;-\omega,\omega,\omega)$. Similarly, the polarizabilities occurring in $H(\omega;-\omega,\omega,-\omega)$ and $H(\omega;-\omega,-\omega,\omega)$ can be replaced with those occurring in $H(\omega;\omega,-\omega,-\omega)$.

The symmetry properties (A.12) allow for the reduction of the number of the coupling constants occurring in (A.2) from six to two types, and (A.11) allows us to replace the coupling constants of the type ${}^{(a)}_{(A}\boldsymbol{\chi}^{(b,c,d)}_{BCD}(\omega;\omega,-\omega,-\omega)$ with ${}^{(a)}_{(A}\boldsymbol{\chi}^{(b,c,d)}_{BCD}(-\omega;-\omega,\omega,\omega)$, which reduces the problem to one type of the coupling only.

Generally, the linear and nonlinear polarizabilities are complex quantities and can be written in the form

$$ {}^{(a)}_{A}\boldsymbol{\chi}^{(b)}_{B}(-\omega;\omega) = {}^{(a)}_{A}\boldsymbol{\chi}^{(b)\prime}_{B}(-\omega;\omega) + \mathrm{i}\,{}^{(a)}_{A}\boldsymbol{\chi}^{(b)\prime\prime}(-\omega;\omega) \tag{A.13} $$

$$ \begin{aligned} {}^{(a)}_{A}\boldsymbol{\chi}^{(b,c,d)}_{BCD}(-\omega;-\omega,\omega,\omega) &= {}^{(a)}_{A}\boldsymbol{\chi}^{(b,c,d)\prime}_{BCD}(-\omega;-\omega,\omega,\omega) \\ &\quad + \mathrm{i}\,{}^{(a)}_{A}\boldsymbol{\chi}^{(b,c,d)\prime\prime}_{BCD}(-\omega;-\omega,\omega,\omega) \end{aligned} \tag{A.14} $$

where

$$ {}^{(a)}_{A}\boldsymbol{\chi}^{(b)\prime}_{B}(-\omega;\omega) = \mathrm{Re}\,{}^{(a)}_{A}\boldsymbol{\chi}^{(b)}_{B}(-\omega;\omega) \tag{A.15} $$

$$ {}^{(a)}_{A}\boldsymbol{\chi}^{(b)\prime\prime}_{B}(-\omega;\omega) = \mathrm{Im}\,{}^{(a)}_{A}\boldsymbol{\chi}^{(b)}_{B}(-\omega;\omega) \tag{A.16} $$

and similarly for ${}^{(a)}_{(A}\boldsymbol{\chi}^{(b,c,d)\prime}_{BCD}(-\omega;-\omega,\omega,\omega)$ and ${}^{(a)}_{(A}\boldsymbol{\chi}^{(b,c,d)\prime\prime}_{BCD}(-\omega;-\omega,\omega,\omega)$.

Applying (A.7) and (A.8) we can obtain quantum mechanical expressions for the real and imaginary parts of the linear and nonlinear polarizabilities, which allows us to check the following additional permutation relations [226–228]:

$$ {}^{(a)}_{A}\chi^{(b)\prime}_{B\,i_1\cdots i_a j_1\cdots j_b}(-\omega;\omega) = {}^{(b)}_{B}\chi^{(a)\prime}_{A\,j_1\cdots j_b i_1\cdots i_a}(-\omega;\omega) \tag{A.17} $$

$$ {}^{(a)}_{A}\chi^{(b)\prime\prime}_{B\,i_1\cdots i_a j_1\cdots j_b}(-\omega;\omega) = -{}^{(b)}_{B}\chi^{(a)\prime\prime}_{A\,j_1\cdots j_b i_1\cdots i_a}(-\omega;\omega) \tag{A.18} $$

$$ \begin{aligned} &{}^{(a)}_{A}\chi^{(b,c,d)\prime}_{BCD\,i_1\cdots i_a j_1\cdots j_b k_1\cdots k_c l_1\cdots l_d}(-\omega;-\omega,\omega,\omega) \\ &\quad = {}^{(c)}_{C}\chi^{(d,a,b)\prime}_{DAB\,k_1\cdots k_c l_1\cdots l_d i_1\cdots i_a j_1\cdots j_b}(-\omega;-\omega,\omega,\omega) \end{aligned} \tag{A.19} $$

$$ \begin{aligned} &{}^{(a)}_{A}\chi^{(b,c,d)\prime\prime}_{BCD\,i_1\cdots i_a j_1\cdots j_b k_1\cdots k_c l_1\cdots l_d}(-\omega;-\omega,\omega,\omega) \\ &\quad = -{}^{(c)}_{C}\chi^{(d,a,b)\prime\prime}_{DAB\,k_1\cdots k_c l_1\cdots l_d i_1\cdots i_a j_1\cdots j_b}(-\omega;-\omega,\omega,\omega) \end{aligned} \tag{A.20} $$

Above, the indices $i_1 \cdots i_a$ label the components of the operator $\mathbf{M}^{(a)}_{A}$ of the electric ($A = e$) or magnetic ($A = m$) multipole moment of order a, which is a tensor of rank a having the following form in the Cartesian frame (z, y, z):

$$ \mathbf{M}^{(a)}_{A} = \mathrm{e}_{i_1} \cdots \mathrm{e}_{i_a} M^{(a)}_{A\,i_1\cdots i_a} \tag{A.21} $$

where the indices i_p can take the values x, y, and z for $p = 1, \ldots, a$, whereas e_{i_p} represents the unit vector along the i_p axis of the frame (z, y, z). If $A = e$ then $\mathbf{M}_e^{(a)}$ is a polar tensor, which is symmetric with respect to the $a!$ permutations of the indices $i \ldots i_a$; if $A = m$ then $\mathbf{M}_m^{(a)}$ is an axial tensor symmetric with respect to the $(a-1)!$ permutations of the indices $i_2 \ldots i_a$. The indices with respect to which the tensor is invariant are separated in (A.3) and (A.4) by the parentheses.

Acting with the time reversal operator R on the quantum mechanical expressions defining the real and imaginary parts of the linear as well as nonlinear polarizability, it is easy to check that

$$
\begin{array}{ll}
\operatorname{Re} \alpha_{ij}(-\omega; \omega) & \operatorname{Im} \rho_{ij}(-\omega; \omega) \\
\operatorname{Im} \lambda_{ij}(-\omega; \omega) & \\
\operatorname{Re} \eta_{i(jk)}(-\omega; \omega) & \operatorname{Re} \eta_{(ij)k}(-\omega; \omega) \\
\operatorname{Re} \gamma_{ijkl}(-\omega; -\omega, \omega, \omega) & \\
\operatorname{Re} \eta_{ijk(lm)}(-\omega; -\omega, \omega, \omega) & \operatorname{Re} \eta_{ij(km)l}(-\omega; -\omega, \omega, \omega) \\
\operatorname{Re} \eta_{i(jm)kl}(-\omega; -\omega, \omega, \omega) & \operatorname{Re} \eta_{(im)jkl}(-\omega; -\omega, \omega, \omega) \\
\operatorname{Im} \rho_{ijkl}(-\omega; -\omega, \omega, \omega) & \operatorname{Im} \kappa_{ijkl}(-\omega; -\omega, \omega, \omega) \\
\operatorname{Im} \sigma_{ijkl}(-\omega; -\omega, \omega, \omega) & \operatorname{Im} \lambda_{ijkl}(-\omega; -\omega, \omega, \omega)
\end{array}
\tag{A.22}
$$

are invariant with respect to the time reversal [229], and the remaining expressions change their sign under time reversal.

One of the postulates of quantum mechanics is the invariance of the Schrödinger equation with respect to the time-reversal operation R [229]. This postulate implies that if the function Ψ is a solution of the Schrödinger equation, then the function $\Psi' = R\Psi$ is also a solution of this equation with the same energy $E_\Psi = E_{R\Psi}$. A physical quantity O is invariant with respect to the time reversal if its value is the same in Ψ and $R\Psi$, and it is antisymmetrical if the two values differ in sign. Of course, the relation $R\Psi = \Psi$ rules out the quantities antisymmetrical with respect to the time reversal.

Assuming that the microsystems are freely oriented, we can average (A.1) and (A.2) over all orientations of the microsystems. Performing such averaging according to (9)–(12) and applying (7) as well as the permutation relations (A.11)–(A.12), the linear $H_{\mathrm{I}}^{(2)}/N$ and nonlinear $H_{\mathrm{I}}^{(4)}/N$ interaction Hamiltonians describing the interaction of the microsystem with the electromagnetic field take, in the circular polarization basis (8), the following forms:

$$
\begin{aligned}
\left\langle H_{\mathrm{I}}^{(2)}/N\right\rangle_{\Omega} = &-\tfrac{1}{2}\underline{\chi}_{R}^{L}\left[E_{-}^{+}E_{+}^{+}+E_{-}^{-}E_{-}^{+}+E_{+}^{+}E_{-}^{+}+E_{-}^{+}E_{-}^{-}\right] \\
&-\frac{\mathrm{i}}{2}\underline{\chi}_{A}^{L}\left[E_{+}^{-}E_{+}^{+}-E_{-}^{-}E_{-}^{+}+E_{+}^{+}E_{+}^{-}-E_{-}^{+}E_{-}^{-}\right]
\end{aligned}
\tag{A.23}
$$

$$
\begin{aligned}
\left\langle H_{\mathrm{I}}^{(4)}/N\right\rangle_{\Omega} = &-\tfrac{1}{12}\underline{\chi}_{R}^{NL}\Big[(E_{+}^{-})^{2}(E_{+}^{+})^{2}+(E_{-}^{-})^{2}(E_{-}^{+})^{2} \\
&\quad +(E_{+}^{+}E_{-}^{+})^{2}+(E_{-}^{-}E_{-}^{+})^{2} \\
&\quad +\text{terms with reversed superscripts}\Big] \\
&-\tfrac{1}{12}\underline{\tilde{\chi}}_{R}^{NL}\Big[E_{+}^{-}(E_{+}^{+})^{2}E_{+}^{-}+E_{-}^{-}(E_{-}^{+})^{2}E_{-}^{-}\Big] \\
&-\tfrac{1}{12}\underline{\tilde{\chi}}_{R}^{NL*}\Big[E_{+}^{+}(E_{+}^{-})^{2}E_{+}^{+}+E_{-}^{+}(E_{-}^{-})^{2}E_{-}^{+}\Big] \\
&-\tfrac{1}{6}\underline{\kappa}_{R}^{NL}\Big[E_{+}^{-}E_{-}^{-}E_{+}^{+}E_{-}^{+}+\tfrac{1}{4}\big(E_{+}^{-}E_{+}^{+}E_{-}^{+}E_{-}^{+}+E_{-}^{-}E_{-}^{+}E_{+}^{-}E_{+}^{+} \\
&\quad +E_{+}^{-}E_{-}^{+}E_{-}^{-}E_{+}^{+}+E_{-}^{-}E_{+}^{+}E_{+}^{-}E_{-}^{+}\big) \\
&\quad +\text{terms with reversed superscripts}\Big] \\
&-\tfrac{1}{24}\underline{\tilde{\kappa}}_{R}^{NL}\left[E_{+}^{-}E_{+}^{+}E_{-}^{+}E_{-}^{-}+E_{-}^{-}E_{-}^{+}E_{+}^{+}E_{+}^{-}\right. \\
&\quad \left.+E_{+}^{-}E_{-}^{+}E_{+}^{+}E_{-}^{-}+E_{-}^{-}E_{+}^{+}E_{-}^{+}E_{+}^{-}\right] \\
&-\tfrac{1}{24}\underline{\tilde{\kappa}}_{R}^{NL*}\left[E_{+}^{+}E_{+}^{-}E_{-}^{-}E_{-}^{+}+E_{-}^{+}E_{-}^{-}E_{+}^{-}E_{+}^{+}\right. \\
&\quad \left.+E_{+}^{+}E_{-}^{-}E_{+}^{-}E_{-}^{+}+E_{-}^{+}E_{+}^{-}E_{-}^{-}E_{+}^{+}\right] \\
&-\tfrac{1}{12}\underline{\chi}_{R}^{NL}\Big[(E_{+}^{-})^{2}(E_{+}^{+})^{2}-(E_{-}^{-})^{2}(E_{-}^{+})^{2} \\
&\quad +(E_{-}^{-}E_{+}^{+})^{2}-(E_{-}^{-}E_{-}^{+})^{2} \\
&\quad +E_{+}^{-}(E_{+}^{+})^{2}E_{+}^{-}-E_{-}^{-}(E_{-}^{+})^{2}E_{-}^{-} \\
&\quad +\text{terms with reversed superscripts}\Big]
\end{aligned}
\tag{A.24}
$$

where

$$\underline{\chi}_R^L = \tfrac{1}{3}\mathrm{Re}\,\alpha_{\alpha\alpha}(-\omega;\omega) \tag{A.25}$$

$$\bar{\chi}_A^L = -\frac{\mathrm{i}k_z}{3\omega}\mathrm{Im}\,\rho_{\alpha\alpha}(-\omega;\omega) \tag{A.26}$$

$$\underline{\chi}_R^{NL} = \tfrac{1}{15}\mathrm{Re}\big[-\gamma_{\alpha\alpha\beta\beta}(-\omega;-\omega,\omega,\omega) + 3\gamma_{\alpha\beta\alpha\beta}(-\omega;-\omega,\omega,\omega)\big] \tag{A.27}$$

$$\tilde{\underline{\chi}}_R^{NL} = \tfrac{1}{15}\big[-\gamma_{\alpha\alpha\beta\beta}(-\omega;-\omega,\omega,\omega) + 3\gamma_{\alpha\beta\alpha\beta}(-\omega;-\omega,\omega,\omega)\big] \tag{A.28}$$

$$\underline{\kappa}_R^{NL} = \tfrac{1}{15}\mathrm{Re}\big[3\gamma_{\alpha\alpha\beta\beta}(-\omega;-\omega,\omega,\omega) + \gamma_{\alpha\beta\alpha\beta}(-\omega;-\omega,\omega,\omega)\big] \tag{A.29}$$

$$\tilde{\underline{\kappa}}_R^{NL} = \tfrac{1}{15}\big[3\gamma_{\alpha\alpha\beta\beta}(-\omega;-\omega,\omega,\omega) + \gamma_{\alpha\beta\alpha\beta}(-\omega;-\omega,\omega,\omega)\big] \tag{A.30}$$

$$\underline{\chi}_A^{NL} = -\frac{\mathrm{i}4k_z}{15}\left\{\frac{1}{\omega}\mathrm{Im}\big[\sigma_{\alpha\alpha\beta\beta}(-\omega;-\omega,\omega,\omega) - 3\sigma_{\alpha\beta\alpha\beta}(-\omega;-\omega,\omega,\omega)\big] - \tfrac{1}{3}\mathrm{Re}\,\eta_{\alpha(\beta\gamma)\beta\delta}(-\omega;-\omega,\omega,\omega)\varepsilon_{\alpha\gamma\delta}\right\} \tag{A.31}$$

and

$$E = E(k) \tag{A.32}$$

The total interaction Hamiltonian describing the interaction of the optical field with an ensemble of N microsystems confined in the unit volume is, according to (1) and (2), equal to

$$H_\mathrm{I} = H_\mathrm{I}^{(2)} + H_\mathrm{I}^{(4)} + \cdots = \sum_{n=1}^{\infty} H_\mathrm{I}^{(2n)} \tag{A.33}$$

where

$$H_\mathrm{I}^{(2n)} = \sum_{p=1}^{N} \langle H_\mathrm{I}^{(2n)}/N\rangle_\Omega \tag{A.34}$$

and the summation runs over all microsystems.

Let us assume that the microsystems are identical. In this case the summation over p simplifies considerably, and it becomes trivial if the ground state of the microsystem is nondegenerate. For the nondegenerate case, we have

$$R\Psi = \Psi \tag{A.35}$$

which implies vanishing of the polarizabilities antisymmetrical with respect to the time reversal, and we have

$$\tilde{\underline{\chi}}_R^{NL} = \tilde{\underline{\chi}}_R^{NL*} = \underline{\chi}_R^{NL} \tag{A.36}$$

$$\tilde{\underline{\kappa}}_R^{NL} = \tilde{\underline{\kappa}}_R^{NL*} = \underline{\kappa}_R^{NL} \tag{A.37}$$

where the summation over p reduces to the multiplication of the parameters (A.25)–(A.31) over N, which leads to (14)–(20) with the molecular constants:

$$\begin{aligned} \operatorname{Re}\alpha_{\alpha\alpha} &= \operatorname{Re}\alpha_{\alpha\alpha}(-\omega;\omega) \\ \operatorname{Im}\rho_{\alpha\alpha} &= \operatorname{Im}\rho_{\alpha\alpha}(-\omega;\omega) \\ \operatorname{Re}\gamma_{\alpha\beta\gamma\delta} &= \operatorname{Re}\gamma_{\alpha\beta\gamma\delta}(-\omega;-\omega,\omega,\omega) \\ \operatorname{Im}\sigma_{\alpha\beta\gamma\delta} &= \operatorname{Im}\sigma_{\alpha\beta\gamma\delta}(-\omega;-\omega,\omega,\omega) \\ \operatorname{Re}\eta_{\alpha(\beta\gamma)\beta\delta} &= \operatorname{Re}\eta_{\alpha(\beta\gamma)\beta\delta}(-\omega;-\omega,\omega,\omega) \end{aligned} \tag{A.38}$$

If the ground state Ψ of the microsystem is $2u$-fold degenerate and the states are labeled by the indices $f = 1, 2, \ldots, u$ associated with the particular wave functions in such a way that

$$R\Psi f = \Psi(2f) \tag{A.39}$$

then, according to quantum mechanics, the probability of finding a microsystem in any state of the s-fold degenerate level Ψ is the same for all states and equal to $\rho_{\Psi\Psi}$. This means that for a large number N of identical microsystems in the unit volume, in each state of the s-fold degenerate level Ψ will be the same number of microsystems, equal to $(N/s)\rho_{\Psi\Psi}$. In our case this number will be equal to $N/2u$ because we have already incorporated $\rho_{\Psi\Psi}$ into (A.7 and (A.8) defining the linear and nonlinear polarizabilities. As a result, the summation over p can be replaced by the summation over the states of the $2u$-fold degenerate level Ψ

$$\begin{aligned} &\sum_{p=1}^{N} \gamma_{\alpha\beta\gamma\delta}(-\omega;-\omega,\omega,\omega) \\ &\quad = \frac{N}{2u}\sum_{f=1}^{u}\left[\gamma_{\alpha\beta\gamma\delta}(-\omega;-\omega,\omega,\omega)_{\Psi f} + \gamma_{\alpha\beta\gamma\delta}(-\omega;-\omega,\omega,\omega)_{\Psi(2f)}\right] \end{aligned} \tag{A.40}$$

where the first term is the nonlinear polarizability in the state Ψf, and the second term is in the state $\Psi(2f)$; the polarizabilities are complex quantities. Keeping in mind the fact that the polarizabilities symmetrical with respect to the time reversal are the same in the states Ψf and $\Psi(2f) = R\Psi f$, while antisymmetrical differ in sign, in view of (A.40), (A.10), (A.14), and (A.22), we have

$$\sum_{p=1}^{N} \gamma_{\alpha\beta\gamma\delta}(-\omega; -\omega, \omega, \omega) = N \operatorname{Re} \gamma_{\alpha\beta\gamma\delta} \tag{A.41}$$

where the mean polarizability is given by

$$\operatorname{Re} \gamma_{\alpha\beta\gamma\delta} = \frac{1}{u} \sum_{f=1}^{u} \operatorname{Re} \gamma_{\alpha\beta\gamma\delta}(-\omega; -\omega, \omega, \omega)_{\Psi f} \tag{A.42}$$

Similarly, we get

$$\sum_{p=1}^{N} \sigma_{\alpha\beta\gamma\delta}(-\omega; -\omega, \omega, \omega) = \mathrm{i}\, N \operatorname{Im} \sigma_{\alpha\beta\gamma\delta} \tag{A.43}$$

where

$$\operatorname{Im} \sigma_{\alpha\beta\gamma\delta} = \frac{1}{u} \sum_{f=1}^{u} \operatorname{Im} \sigma_{\alpha\beta\gamma\delta}(-\omega; -\omega, \omega, \omega)_{\Psi f} \tag{A.44}$$

In the case of odd, $(2u + 1)$-fold, degeneracy of the level Ψ, we assume additionally that for at least one state, say, $\Psi 0$, the following relation is satisfied:

$$R\Psi 0 = \Psi 0 \tag{A.45}$$

and for the remaining states (A.39) holds. As a result of (A.45) the polarizability antisymmetrical with respect to the time reversal vanishes in the state Ψ. In a similar way as before, in place of (A.42) and (A.44), we get

$$\operatorname{Re}\gamma_{\alpha\beta\gamma\delta} = \frac{1}{2u+1}\left[\operatorname{Re}\gamma_{\alpha\beta\gamma\delta}(-\omega; -\omega, \omega, \omega)_{\Psi 0} + 2\sum_{f=1}^{u}\operatorname{Re}\gamma_{\alpha\beta\gamma\delta}(-\omega; -\omega, \omega, \omega)_{\Psi f}\right] \tag{A.46}$$

$$\operatorname{Im}\sigma_{\alpha\beta\gamma\delta} = \frac{1}{2u+1}\left[\operatorname{Im}\sigma_{\alpha\beta\gamma\delta}(-\omega; -\omega, \omega, \omega)_{\Psi 0} + 2\sum_{f=1}^{u}\operatorname{Im}\sigma_{\alpha\beta\gamma\delta}(-\omega; -\omega, \omega, gu)_{\Psi f}\right] \tag{A.47}$$

Using (A.41)–(A.47), we can perform the summation over p in (A.34). For both even and odd degeneracies, $H_{\mathrm{I}}^{(2)}$ and $H_{\mathrm{I}}^{(4)}$ are still described by (14) and (15) with the molecular parameters (16)–(20), which for even degeneracies are given by

$$\begin{aligned}
\operatorname{Re}\alpha_{\alpha\alpha} &= \frac{1}{u}\sum_{f=1}^{u}\operatorname{Re}\alpha_{\alpha\alpha}(-\omega; \omega)_{\Psi f} \\
\operatorname{Im}\rho_{\alpha\alpha} &= \frac{1}{u}\sum_{f=1}^{u}\operatorname{Im}\rho_{\alpha\alpha}(-\omega; \omega\omega)_{\Psi f} \\
\operatorname{Re}\eta_{\alpha(\beta\gamma)\beta\delta} &= \frac{1}{u}\sum_{f=1}^{u}\operatorname{Re}\eta_{\alpha(\beta\gamma)\beta\delta}(-\omega; -\omega, \omega, \omega)_{\Psi f}
\end{aligned} \tag{A.48}$$

whereas $\operatorname{Re}\gamma_{\alpha\alpha\beta\beta}$ and $\operatorname{Re}\gamma_{\alpha\beta\alpha\beta}$ are defined by (A.42), and $\operatorname{Im}\sigma_{\alpha\alpha\beta\beta}$ and $\operatorname{Im}\sigma_{\alpha\beta\alpha\beta}$ are defined by (A.44). For odd degeneracies the parameters are given by

$$\operatorname{Re}\alpha_{\alpha\alpha} = \frac{1}{2u+1}\left[\operatorname{Re}\alpha_{\alpha\alpha}(-\omega; \omega)_{\Psi 0} + 2\sum_{f=1}^{u}\operatorname{Re}\alpha_{\alpha\alpha}(-\omega; \omega)_{\Psi f}\right]$$

$$\operatorname{Im} \rho_{\alpha\alpha} = \frac{1}{2u+1}\left[\operatorname{Im} \rho_{\alpha\alpha}(-\omega;\omega)_{\Psi 0} + 2\sum_{f=1}^{u} \operatorname{Im} \rho_{\alpha\alpha}(-\omega;\omega)_{\Psi f}\right] \tag{A.49}$$

$$\operatorname{Re} \eta_{\alpha(\beta\gamma)\beta\delta} = \frac{1}{2u+1}\left[\operatorname{Re} \eta_{\alpha(\beta\gamma)\beta\delta}(-\omega;-\omega,\omega,\omega)_{\Psi 0} + 2\sum_{f=1}^{u} \operatorname{Re} \eta_{\alpha(\beta\gamma)\beta\delta}(-\omega;-\omega,\omega,\omega)_{\Psi f}\right]$$

while $\operatorname{Re} \gamma_{\alpha\alpha\beta\beta}$ and $\operatorname{Re} \gamma_{\alpha\beta\alpha\beta}$ are defined by (A.46), and $\operatorname{Im} \sigma_{\alpha\alpha\beta\beta}$ and $\operatorname{Im} \sigma_{\alpha\beta\alpha\beta}$ are define by (A.47).

Taking into account (A.38), (A.42), (A.44), and (A.46)–(A.49) defining the nonlinear molecular polarizabilities for both the nondegenerate and degenerate electronic states and (A.10) and (A.14)–(A.16), one can check that $\operatorname{Re} \gamma_{\alpha\beta\gamma\delta}$, $\operatorname{Im} \sigma_{\alpha\beta\gamma\delta}$, and $\operatorname{Re} \eta_{\alpha(\beta\gamma)\delta\Phi}$ have the same permutation symmetry as ${}^{(1)}_{e}\chi^{(1,1,1)\prime}_{eee\,\alpha\beta\gamma\delta}(-\omega;-\omega,\omega,\omega)$, ${}^{(1)}_{e}\chi^{(1,1,1)\prime\prime}_{mee\,\alpha\beta\gamma\delta}(-\omega;-\omega,\omega,\omega)$, and ${}^{(1)}_{e}\chi^{(2,1,1)\prime}_{eee\,\alpha(\beta\gamma)\delta\Phi}(-\omega;-\omega,\omega,\omega)$, respectively. This symmetry can be found from (A.12), (A.19), and (A.20).

APPENDIX B

In Appendix A we showed that the tensor $\operatorname{Re} \gamma_{\alpha\beta\gamma\delta}$ is invariant with respect to the permutations of the indices α and β as well as γ and δ, and also with respect to the permutations of the pairs of indices $\alpha\beta$ and $\gamma\delta$. The tensor $\operatorname{Im} \sigma_{\alpha\beta\gamma\delta}$ defining the nonlinear molecular parameter χ_A^{NL} is invariant under the permutation γ and δ, and the tensor $\operatorname{Re} \eta_{\alpha(\beta\gamma)\delta\phi}$ is symmetric with respect to the permutations δ and ϕ as well as β and γ (β and γ are associated with the electric quadrupole operator).

To find the molecular parameters χ_R^L, χ_A^L, χ_R^{NL}, κ_R^{NL}, and χ_A^{NL} for the molecules with a definite molecular symmetry one has to know the explicit form of the polarizability tensors $\operatorname{Re} \gamma_{\alpha\beta\gamma\delta}$, $\operatorname{Im} \sigma_{\alpha\beta\gamma\delta}$, and $\operatorname{Re} \eta_{\alpha(\beta\gamma)\delta\phi}$. Applying the group theory methods [230, 228] the components of the tensors have been found for 102 magnetic point groups, and the results are presented in Tables I–III. The molecular parameters χ_R^L, χ_A^L, χ_R^{NL}, κ_R^{NL}, and χ_A^{NL} for all these symmetry groups are collected in Table IV.

TABLE I
Fourth-Rank Polar i-tensor Re $\gamma_{\alpha\beta\gamma\delta}$ for 102 Magnetic Point Groups

Magnetic Point Group	N	I	Form of the i-tensor Re $\gamma_{\alpha\beta\gamma\delta}$
$1, \bar{1}, \underline{\bar{1}}$	81	21	$a_1 \equiv$ 1111, 2222, 3333, 1122 = 2211, 1133 = 3311, 2233 = 3322, 1212 = 1221 = 2121 = 2112,1313 = 1331 = 3131 = 3113, 2323 = 2332 = 3232 = 3223 $b_1 \equiv$ 1112 = 1121 = 1211 = 2111, 2221 = 2212 = 2122 = 1222, 1233 = 2133 = 3312 = 3321, 1323 = 1332 = 3123 = 3132 = 2313 = 2331 = 3213 = 323 $c_1 \equiv$ 1113 = 1131 = 1311 = 3111, 3331 = 3313 = 3133 = 1333, 1322 = 3122 = 2213 = 2231, 1232 = 1223 = 2132 = 2123 = 3212 = 2312 = 3221 = 232 2223 = 2232 = 2322 = 3222, 3332 = 3323 = 3233 = 2333, 2311 = 3211 = 1123 = 1132, 2131 = 2113 = 1231 = 1213 = 3121 = 1321 = 3112 = 131
$2, \underline{2}, m, \underline{m}, 2/m, \underline{2}/m,$ $2/\underline{m}, \underline{2}/\underline{m}$	41	13	a_1, b_1
$222, \underline{22}2, mm2, \underline{mm}2,$ $\underline{2m}m, mmm, \underline{mm}m,$ $\underline{mmm}, mm\underline{m}$	21	9	a_1
$4, \underline{4}, \bar{4}, \underline{\bar{4}}, 4/m, \underline{4}/m,$ $4/\underline{m}, \underline{4}/\underline{m}$	29	7	$d_1 \equiv$ 1111 = 2222, 3333, 1122 = 2211, 1133 = 3311 = 2233 = 3322, 1212 = 1221 = 2112 = 2121, 2323 = 2332 = 3223 = 3232 = 1313 = 1331 = 3113 = 313 $e_1 \equiv$ 1112 = −2221 = 1121 = −2212 = 1211 = −2122 = 2111 = −1222
$422, \underline{4}2\underline{2}, 4\underline{22}, 4mm,$ $\underline{4}m\underline{m}, 4\underline{mm}, \bar{4}2m, \underline{\bar{4}}2\underline{m},$ $\underline{\bar{4}}m\underline{2}, \bar{4}\underline{2m}, 4/mmm,$ $\underline{4}/mm\underline{m}, 4/m\underline{mm},$ $4/\underline{mmm}, 4/\underline{m}mm,$ $\underline{4}/\underline{mm}m$	21	6	d_1

TABLE I *(Continued)*

Magnetic Point Group	N	I	Form of the i - tensor Re $g_{\alpha\beta\gamma\delta}$
$3, \bar{3}, \underline{\bar{3}}$	53	7	$h_1 \equiv 1111 = 2222 = 1122 + 2(1212), 3333, 1122 = 2211,$ $1212 = 1221 = 2112 = 2121, 1133 = 3311 = 2233 = 3322,$ $1313 = 1331 = 3113 = 3131 = 2323 = 2332 = 3223 = 3232$ $j_1 \equiv 1113 = -1232 = -1223 = -2132 =$ $1131 = -2123 = -3221 = -2312 =$ $1311 = -3212 = -2321 = -1322 =$ $3111 = -3122 = -2213 = -2331,$ $k_1 \equiv 2223 = -2131 = -2113 = -1231 =$ $2232 = -1213 = -3112 = -1321 =$ $2322 = -3121 = -1312 = -2311 =$ $3222 = -3211 = -1123 = -1132$
$32, 3\underline{2}, 3m, \bar{3}m, 3\underline{m},$ $\bar{3}\underline{m}, \underline{\bar{3}m}, \underline{\bar{3}}m$	37	6	h_1, j_1
$6, \underline{6}, \bar{6}, \underline{\bar{6}}, 6/m, \underline{6}/m,$ $6/\underline{m}, \underline{6}/\underline{m}, 622, \underline{6}2\underline{2},$ $6\underline{22}, 6mm, \underline{6}m\underline{m}, 6\underline{mm},$ $\bar{6}m2, \underline{\bar{6}}m\underline{2}, \underline{\bar{6}}2\underline{m}, \bar{6}\underline{m2},$ $6/mmm, \underline{6}/\underline{m}m\underline{m},$ $6/m\underline{mm}, 6/\underline{mmm},$ $6/\underline{m}mm, \underline{6}/mm\underline{m},$ $\infty, \infty/m, \infty m, \infty/\underline{m},$ $\infty\underline{m}, \infty/mm, \infty/\underline{m}m,$ $\infty/m\underline{m}$	21	5	h_1
$23, m\underline{3}, \underline{m3}, 432, \underline{4}3\underline{2},$ $\bar{4}3m, \underline{\bar{4}}3\underline{m}, m3m,$ $\underline{m}3m, m3\underline{m}, \underline{m}3\underline{m},$	21	3	$m_1 \equiv 1111 = 2222 = 3333,$ $1122 = 2233 = 3311 = 2211 = 3322 = 1133,$ $1212 = 2323 = 3131 = 1221 = 2332 = 3113 =$ $2121 = 3232 = 1313 = 1331 = 2112 = 3223 = 1331$
Y, Y_h, K, K_h	21	2	m_1 and $1111 = 2222 = 3333 = 1122 + 2(1212)$

Note. The components of the nonlinear polarizability tensor Re $\gamma_{\alpha\beta\gamma\delta}$ are denoted by the subscripts $\alpha\beta\gamma\delta$, taking values 1, 2, 3 in the molecular reference frame. N and I denote the number of nonzero and independent components, respectively. Sets of components recurring in various point groups are denoted by lowercase letters.

TABLE II
Fourth-Rank Axial i-tensor Im $\sigma_{\alpha\beta\gamma\delta}$ for 102 Magnetic Point Groups

Magnetic Point Group	N	I	Form of the i-tensor Im $\sigma_{\alpha\beta\gamma\delta}$
1	81	54	$a_2 \equiv$ 1111, 2222, 3333, 1122, 2211, 1133, 3311, 2233, 3322, 1212 = 1221, 1313 = 1331, 2323 = 2332, 2121 = 2112, 3131 = 3113, 3232 = 3223 $b_2 \equiv$ 1112 = 1121, 1211, 2111, 2221 = 2212, 1222, 2122, 1233, 2133, 3312 = 3321, 1323 = 1332, 3123 = 3132, 2313 = 2331, 3213 = 3231 $c_2 \equiv$ 1113 = 1131, 1311, 3111, 3331 = 3313, 1333, 3133, 1322, 3122, 2213 = 2231, 1232 = 1223, 2132 = 2123, 3212 = 3221, 2312 = 2321, 2223 = 2232, 2322, 3222, 3332 = 3323, 2333, 3233, 2311, 3211, 1123 = 1132, 2131 = 2113, 1231 = 1213, 3121 = 3112, 1321 = 1312
$2, \underline{2}$	41	28	a_2, b_2
$m, \underline{m}$	40	26	c_2
$222, \underline{22}2$	21	15	a_2
$mm2, \underline{mm}2, 2\underline{mm}$	20	13	b_2
$4, \underline{4}$	39	14	$d_2 \equiv$ 1111 = 2222, 3333, 1122 = 2211, 1212 = 1221 = 2121 = 2112, 1133 = 2233, 1313 = 1331 = 2323 = 2332, 3311 = 3322, 3131 = 3113 = 3232 = 3223 $e_2 \equiv$ 1112 = −2221 = 1121 = −2212, 1211 = −2122, 2111 = −1222, 1233 = −2133, 1323 = −2313 = 1332 = −2331, 3123 = −3213 = 3132 = −3231
$\bar{4}, \underline{\bar{4}}$	40	14	$f_2 \equiv$ 1111 = −2222, 1122 = −2211, 1212 = 1221 = −2121 = −2112, 1133 = −2233, 1313 = 1331 = −2323 = −2332, 3311 = −3322, 3131 = 3113 = −3232 = −3223 $g_2 \equiv$ 1112 = 2221 = 1121 = 2212, 1211 = 2122, 2111 = 1222, 1233 = 2133, 1323 = 2313 = 1332 = 2331, 3312 = 3321, 3123 = 3213 = 3132 = 3231
$422, \underline{4}2\underline{2}, 4\underline{22}$	21	8	d_2
$4mm, \underline{4}m\underline{m}, 4\underline{mm}$	18	6	e_2
$\bar{4}2m, \underline{\bar{4}}2\underline{m}, \underline{\bar{4}}m\underline{2}, \bar{4}\underline{2m}$	20	7	f_2
3	71	18	$h_2 \equiv$ 1111 = 2222 = 1122 + 2(1212), 3333, 1212 = 1221 = 2112 = 2121, 1122 = 2211, 1313 = 1331 = 2323 = 2332, 1133 = 2233, 3131 = 3113 = 3232 = 3223, 3311 = 3322 $i_2 \equiv$ 1112 = −2221 = 1121 = −2212 = $-\frac{1}{2}$(1211 + 211 2111 = −1222, 1211 = −2122, 1233 = −2133, 1323 = −2313 = 1332 = −2331, 3123 = −3213 = 3132 = −3231

TABLE II *(Continued)*

Magnetic Point Group	N	I	Form of the i - tensor Re $g_{\alpha\beta\gamma\delta}$
			$j_2 \equiv 1113 = -1223 = -1232 = -2123 =$ $1131 = -2132 = -2213 = -2231,$ $1311 = -1322 = -2312 = -2321,$ $3111 = -3122 = -3212 = -3221$ $k_2 \equiv 2223 = -2113 = -2131 = -1213 =$ $2232 = -1231 = -1123 = -1132,$ $2322 = -2311 = -1321 = -1312,$ $3222 = -3211 = -3121 = -3112$
$32, 3\underline{2}$	37	10	h_2, j_2
$3m, 3\underline{m}$	34	8	i_2, k_2
$6, \underline{6}, \infty$	39	12	h_2, i_2
$\bar{6}, \underline{\bar{6}}$	32	6	j_2, k_2
$622, \underline{6}\underline{2}2, 6\underline{22}$	21	7	h_2
$6mm, \underline{6}m\underline{m}, 6\underline{mm}, \infty m,$ $\infty\underline{m}$	18	5	i_2
$\bar{6}m2, \underline{\bar{6}}m\underline{2}, \underline{\bar{6}}2\underline{m}, \bar{6}\underline{m2}$	16	3	k_2
23	21	5	$l \equiv 1111 = 2222 = 3333.$ $1122 = 2233 = 3311, 2211 = 1133 = 3322,$ $1212 = 2323 = 3131 = 1221 = 2332 = 3113,$ $2121 = 3232 = 1313 = 2112 = 3223 = 1331$
$432, \underline{4}3\underline{2}$	21	3	$m_2 \equiv 1111 = 2222 = 3333,$ $1122 = 2233 = 3311 = 2211 = 1133 = 3322,$ $1212 = 2323 = 3131 = 1221 = 2332 = 3113 =$ $2121 = 3232 = 1313 = 2112 = 3223 = 1331$
$\bar{4}3m, \underline{\bar{4}}3\underline{m}$	18	2	$0_2 \equiv 1122 = 2233 = 3311 = -2211 = -1133 = -3322,$ $1212 = 2323 = 3131 = -2121 = -3232 = -1313 =$ $1221 = 2332 = 3113 = -2112 = -3223 = -1331$
Y, K	21	2	m_2 and $1111 = 2222 = 3333 = 1122 + 2(1212)$

In the remaining groups:

$\bar{1}, \underline{\bar{1}}, 2/m, \underline{2}/m, 2/\underline{m}, \underline{2}/\underline{m}, mmm, \underline{mm}m, \underline{mmm}, mm\underline{m}, 4/m, \underline{4}/m, 4/\underline{m}, \underline{4}/\underline{m}, 4/mmm,$
$\underline{4}/mm\underline{m}, 4/m\underline{mm}, 4/\underline{mmm}, 4/\underline{m}mm, \underline{4}/\underline{mm}m, \bar{3}, \underline{\bar{3}}, \bar{3}m, \bar{3}\underline{m}, \underline{\bar{3}m}, \bar{3}m, 6/m, \underline{6}/m, 6/\underline{m}, \underline{6}/\underline{m},$
$6/mmm, \underline{6}/\underline{m}m\underline{m}, 6/m\underline{mm}, 6/\underline{mmm}, 6/m\underline{m}m, \underline{6}/m\underline{m}m, \infty/m, \infty/\underline{m}, \infty/mm, \infty/\underline{m}m, \infty/m\underline{m},$
$m3, \underline{m}3, m3m, \underline{m}3m, m3\underline{m}, \underline{m}3\underline{m}, Y_h$ and K_h
all components vanish

Note. The components of the nonlinear polarizability tensor Im $\sigma_{\alpha\beta\gamma\delta}$ are denoted by the subscripts $\alpha\beta\gamma\delta$, taking values 1, 2, 3 in the molecular reference frame. N and I denote the number of nonzero and independent components, respectively. Sets of components recurring in various point groups are denoted by lowercase letters.

TABLE III
Fifth-Rank Polar i-tensor Re $\eta_{\alpha(\beta\gamma)\delta\phi}$ for 102 Magnetic Point Groups

Magnetic Point Group	N	I	Form of the i-tensor Re $\eta_{\alpha(\beta\gamma)\delta\phi}$
1 $4m2$,	243	108	$a_3 \equiv$ 11123 = 11132, 11312 = 11321 = 13112 = 13121, 31112 = 31121, 11213 = 11231 = 12113 = 12131, 12311 = 13211, 31211 = 32111, 21113 = 21131, 21311 = 23111, 22213 = 22231, 22321 = 22312 = 23221 = 23212, 32221 = 32212, 22123 = 22132 = 21223 = 21232, 21322 = 23122, 32122 = 31222, 12223 = 12232, 12322 = 13222, 33312 = 33321, 33123 = 33132 = 31332 = 31323, 31233 = 32133, 32331 = 32313 = 33231 = 33213, 13332 = 13323, 13233 = 12333, 23331 = 23313, 23133 = 21333 $b_3 \equiv$ 33333, 11113 = 11131, 11311 = 13111, 31111, 22223 = 22232, 22322 = 23222, 32222, 11223 = 11232 = 12123 = 12132, 11322 = 13122, 31122, 32211, 22113 = 22131 = 21213 = 21231, 22311 = 23211, 12312 = 12321 = 13212 = 13221, 12213 = 12231, 31212 = 31221 = 32112 = 32121, 21123 = 21132, 23121 = 23112 = 21321 = 21312, 11333 = 13133, 13313 = 13331, 33311, 31133, 31313 = 31331 = 33113 = 33131, 22333 = 23233, 23323 = 23332, 33322, 32233, 32323 = 32332 = 33223 = 33232 $c_3 \equiv$ 11111, 22222, 11112 = 11121, 11211 = 12111, 21111, 22221 = 22212, 22122 = 21222, 12222, 33331 = 33313, 33133 = 31333, 13333, 33332 = 33323, 33233 = 32333, 23333, 11332 = 11323 = 13132 = 13123, 11233 = 12133, 13213 = 13231 = 12313 = 12331, 32311 = 33211, 21313 = 21331 = 23113 = 23131, 21133, 23311, 31312 = 31321 = 33112 = 33121, 31132 = 31123, 31231 = 31213 = 32131 = 32113, 13312 = 13321, 22331 = 22313 = 23231 = 23213, 22133 = 21233, 23123 = 23132 = 21323 = 21332, 31322 = 33122, 12323 = 12332 = 13223 = 13232, 12233, 13322, 32321 = 32312 = 33221 = 33212, 32231 = 32213, 32132 = 32123 = 31232 = 31223, 23321 = 23312, 11222 = 12122, 12212 = 12221, 21122, 21221 = 21212 = 22121 = 22112, 22211, 22111 = 21211, 21121 = 21112, 12211, 12112 = 12121 = 11212 = 11221, 11122, 33111 = 31311, 31131 = 31113, 13311, 13113 = 13131 = 11313 = 11331, 11133, 33222 = 32322, 32232 = 32223, 23322, 23223 = 23232 = 22323 = 22332, 22233

TABLE III *(Continued)*

Magnetic Point Group	N	I	Form of the i-tensor Re $\eta_{\alpha(\beta\gamma)\delta\phi}$
$2, \underline{\bar{2}}$	121	52	a_3, b_3
$m, \underline{m}$	122	56	c_3
$222, \underline{222}$	60	24	a_3
$mm2, \underline{mm}2, \underline{2mm}$	61	28	b_3
$4, \underline{4}$	117	26	$d_3 \equiv$ 11123 = 11132 = −22213 = −22231, 31112 = 31121 = −32221 = −32212, 11312 = 11321 = −22321 = −22312 = 13112 = 13121 = −23221 = −23212, 11213 = 11231 = −22123 = −22132 = 12113 = 12131 = −21223 = −21232, 12311 = 13211 = −21322 = −23122, 31211 = 32111 = −32122 = −31222, 21113 = 21131 = −12223 = −12232, 21311 = 23111 = −12322 = −13222, 12333 = 13233 = −21333 = −23133, 13323 = 13332 = −23313 = −23331, 31323 = 31332 = −32313 = −32331 = 33123 = 33132 = −33213 = −33231 $e_3 \equiv$ 33333, 11113 = 11131 = 22223 = 22232, 11311 = 13111 = 22322 = 23222, 31111 = 32222, 11322 = 13122 = 22311 = 23211, 12123 = 12132 = 11223 = 11232 = 21213 = 21231 = 22113 = 22131, 12312 = 12321 = 13212 = 13221 = 21321 = 21312 = 23121 = 23112, 31212 = 31221 = 32112 = 32121, 12213 = 12231 = 21123 = 21132, 31122 = 32211, 11333 = 13133 = 22333 = 23233, 31133 = 32233, 13313 = 13331 = 23323 = 23332, 33311 = 33322, 31313 = 31331 = 33113 = 33131 = 32323 = 32332 = 33223 = 33232
$\bar{4}, \underline{\bar{4}}$	116	26	$f_3 \equiv$ 11123 = 11132 = 22213 = 22231, 31112 = 31121 = 32221 = 32212, 11312 = 11321 = 22321 = 22312 = 13112 = 13121 = 23221 = 23212, 11213 = 11231 = 22123 = 22132 = 12113 = 12131 = 21223 = 21232, 12311 = 13211 = 21322 = 23122, 31211 = 32111 = 32122 = 31222, 21113 = 21131 = 12223 = 12232, 21311 = 23111 = 12322 = 13222, 12333 = 13233 = 21333 = 23133, 13323 = 13332 = 23313 = 23331, 31233 = 32133, 31323 = 31332 = 32313 = 32331 = 33123 = 33132 = 33213 = 33231, 33312 = 33321 $g_3 \equiv$ 11113 = 11131 = −22223 = −22232, 11311 = 13111 = −22322 = −23222, 31111 = −32222,

TABLE III *(Continued)*

Magnetic Point Group	N	I	Form of the i-tensor Re $\eta_{\alpha(\beta\gamma)\delta\phi}$
			11322 = 13122 = −22311 = −23211, 12123 = 12132 = −21213 = −21231 = 11223 = 11232 = −22113 = −22131, 12312 = 12321 = −21321 = −21312 = 13212 = 13221 = −23121 = 23112, 12213 = 12231 = −21123 = −21132, 31122 = −32211, 11333 = 13133 = −22333 = −23233, 31133 = −32233, 13313 = 13331 = −23323 = −23332, 33311 = −33322, 31313 = 31331 = −32323 = −32332 = 33113 = 33131 = −33223 = −33232
$422, \underline{4}2\underline{2}, 4\underline{22}$,	56	11	d_3
$4mm, \underline{4}m\underline{m}, 4\underline{mm}$	61	15	e_3
$\bar{4}2m, \underline{\bar{4}}2\underline{m}, \underline{\bar{4}}m\underline{2}, \bar{4}\underline{2m}$	60	13	f_3
3	229	36	$h_3 \equiv$ 11123 = 11132 = −22213 = −22231 = −2(11213) − 21113, 31112 = 31121 = −32221 = −32212 = 32122 = 31222 = −31211 = −32111, 11312 = 11321 = −22321 = −22312 = 13112 = 13121 = −23221 = −23212 = −(1/2)[12311 + 21311], 11213 = 11231 = −22123 = −22132 = 12113 = 12131 = −21223 = −21232, 12311 = 13211 = −21322 = −23122, 21113 = 21131 = −12223 = −12232, 21311 = 23111 = −12322 = −13222, 12333 = 13233 = −21333 = −23133, 13323 = 13332 = −23313 = −23331, 31323 = 31332 = −32313 = −32331 = 33123 = 33132 = −33213 = −33231 $i_3 \equiv$ 11133 = −12233 = −21233 = −22133, 13311 = −13322 = −23312 = −23321, 11313 = −12323 = −21323 = −22313 = 11331 = −12332 = −21332 = −22331 = 13113 = −13223 = −23123 = −23213 = 13131 = −13232 = −23123 = −23231, 31131 = −31232 = −32132 = −32231 = 31113 = −32123 = −32213 = −31223, 31311 = −31322 = −32312 = −32321 = 33111 = −33122 = −33212 = −33221, 11111 = −(2/3)[22221 + 22122 + (1/2)12222], 11122 = (2/3)[2(22221) − 22122 − (1/2)12222], 11212 = 11221 = 12112 = 12121 = (1/3)[22221 + 22122 − 12222], 12211 = (1/3)[−2(22221) + 4(22122) − 12222], 21112 = 21121 = (2/3)[(1/2)22221 − 22122 + 12222], 21211 = 22111 = (2/3)[−22221 + (1/2)22122 + 12222], 12222, 22221 = 22212, 22122 = 21222

TABLE III *(Continued)*

Magnetic Point Group	N	I	Form of the i-tensor Re $\eta_{\alpha(\beta\gamma)\delta\phi}$
			$j_3 \equiv$ 33333, 11333 = 13133 = 22333 = 23233, 31133 = 32233, 13313 = 13331 = 23323 = 23332, 33311 = 33322, 31313 = 31331 = 33113 = 33131 = 32323 = 32332 = 33223 = 33232, 11113 = 11131 = 22223 = 22232 = 2(11223) + 12213, 11311 = 13111 = 22322 = 23222 = 11322 + 2(12312), 31111 = 32222 = 31122 + 2(31212), 11223 = 11232 = 12123 = 12132 = 22113 = 22131 = 21213 = 21231, 12213 = 12231 = 21123 = 21132, 11322 = 13122 = 22311 = 23211, 12312 = 12321 = 13212 = 13221 = 21321 = 21312 = 23121 = 23112, 31212 = 31221 = 32112 = 32121, 31122 = 32211 $k_3 \equiv$ 22233 = − 21133 = − 12133 = − 11233, 23322 = − 23311 = − 13321 = − 13312, 22332 = − 21331 = − 12331 = − 11332 = 22323 = − 21313 = − 12313 = − 11323 = 23232 = − 23131 = − 13231 = − 13132 = 23223 = − 23113 = − 13213 = − 13123, 32232 = − 32131 = − 31231 = − 31132 = 32223 = − 32113 = − 31213 = − 31123, 33222 = − 33121 = − 33112 = − 32311 = 32322 = − 31321 = − 31312 = − 33211, 22222 = − (2/3)[11112 + 11211 + (1/2)21111], 22211 = (1/3)[4(11112) − 2(11211) − 21111], 22121 = 22112 = 21221 = 21212 = (1/3)[11112 + 11211 − 21111], 21122 = (2/3)[− 11112 + 2(11211) − (1/2)21111], 12221 = 12212 = (1/3)[11112 − 2(11211) − 2(21111)], 12122 = 11222 = (2/3)[− 11112 + (1/2)11211 + 21111], 11112 = 11121, 11211 = 12111, 21111
$32, 3\underline{2}$	112	16	h_3, i_3
$3m, 3\underline{m}$	117	20	j_3, k_3
$6, \infty, \underline{6}$	117	20	h_3, j_3
$\bar{6}, \underline{\bar{6}}$	112	16	i_3, k_3
$622, \underline{6}\underline{2}2, 6\underline{22}$	56	8	h_3
$6mm, \underline{6}m\underline{m}, 6\underline{mm}$, $\infty m, \infty\underline{m}$	61	12	j_3
$\bar{6}m2, \underline{\bar{6}}\underline{2m}, \underline{\bar{6}}m\underline{2}, \bar{6}\underline{m2}$	56	8	i_3
23	60	8	$l_3 \equiv$ 11123 = 11132 = 22231 = 33312 = 22213 = 33321 11312 = 22123 = 33231 = 11321 = 22132 = 33213 = 13112 = 21223 = 32331 = 13121 = 21232 = 32313, 11213 = 22321 = 33132 = 11231 = 22312 = 33123 =

TABLE III *(Continued)*

Magnetic Point Group	N	I	Form of the i-tensor Re $\eta_{\alpha(\beta\gamma)\delta\phi}$
			12113 = 23221 = 31332 = 12131 = 23212 = 31323, 21113 = 32221 = 13332 = 21131 = 32212 = 13323, 21311 = 32122 = 13233 = 23111 = 31222 = 12333, 31112 = 12223 = 23331 = 31121 = 12232 = 23313, 31211 = 12322 = 23133 = 32111 = 13222 = 21333, 13211 = 21322 = 32133 = 12311 = 23122 = 31233
$432, \underline{4}3\underline{2}$	48	3	m_3 ≡11312 = 22123 = 33231 = 11321 = 22132 = 33213 = 13112 = 21223 = 32331 = 13121 = 21232 = 32313 = −11213 = −22321 = −33132 = −11231 = −22312 = −33123 = −12113 = −23221 = −31332 = −12131 = −23212 = −31323, 21113 = 32221 = 13332 = 21131 = 32212 = 13323 = −31112 = −12223 = −23331 = −31121 = −12232 = −23313, 21311 = 32122 = 13233 = 23111 = 31222 = 12333 = −31211 = −12322 = −23133 = −32111 = −13222 = −21333
$\bar{4}3m, \underline{\bar{4}}3\underline{m}$	60	5	0_3 ≡ 11123 = 22231 = 33312 = 11132 = 22213 = 33321, 13211 = 21322 = 32133 = 12311 = 23122 = 31233, 11312 = 22123 = 33231 = 11321 = 22132 = 33213 = 13112 = 21223 = 32331 = 13121 = 21232 = 32313 = 11213 = 22321 = 33132 = 11231 = 22312 = 33123 = 12113 = 23221 = 31332 = 12131 = 23212 = 31323, 21113 = 32221 = 13332 = 21131 = 32212 = 13323 = 31112 = 12223 = 23331 = 31121 = 12232 = 23313, 21311 = 32122 = 13233 = 23111 = 31222 = 12333 = 31211 = 12322 = 23133 = 13222 = 21333 = 32111
Y, K	48	1	11213 = 22321 = 33132 = 11231 = 22312 = 33123 = 12113 = 23221 = 31332 = 12131 = 23212 = 31323 = −11312 = −22123 = −33231 = −11321 = −22132 = −33213 = −13112 = −21223 = −32331 = −13121 = −21232 = −32313 = (1/2)21311, 21311 = 32122 = 13233 = 23111 = 31222 = 12333 = 31112 = 12223 = 23331 = 31121 = 12232 = 23313 = −31211 = −12322 = −23133 = −32111 = −13222 = −21333 = −21113 = −32221 = −13332 = −21131 = −32212 = −13323

In the remaining magnetic point groups:
$\bar{1}$, $\underline{\bar{1}}$, $2/m$, $\underline{2}/m$, $2/\underline{m}$, $\underline{2}/\underline{m}$, mmm, $\underline{mm}m$, $\underline{mmm}$, $mm\underline{m}$, $4/m$, $\underline{4}/m$, $4/\underline{m}$, $\underline{4}/\underline{m}$, $4/mmm$, $\underline{4}/mm\underline{m}$, $4/m\underline{mm}$, $4/\underline{mmm}$, $4/\underline{m}mm$, $\underline{4}/\underline{mm}m$, $\bar{3}$, $\underline{\bar{3}}$, $\bar{3}m$, $\bar{3}\underline{m}$, $\underline{\bar{3}m}$, $\underline{\bar{3}}m$, $6/m$, $\underline{6}/m$, $6/\underline{m}$, $\underline{6}/\underline{m}$, $6/mmm$, $\underline{6}/\underline{m}m\underline{m}$, $6/m\underline{mm}$, $6/\underline{mmm}$, $6/\underline{m}mm$, $\underline{6}/m\underline{m}m$, ∞/m, $\infty/\underline{m}$, ∞/mm, $\infty/\underline{m}m$, $\infty/m\underline{m}$, $m3$, $\underline{m}3$, $m3m$, $\underline{m}3m$, $m3\underline{m}$, $\underline{m}3\underline{m}$, Y_h and K_h all components vanish

Note. The components of the nonlinear polarizability tensor Re $\eta_{\alpha(\beta\gamma)\delta\phi}$ are denoted by the subscripts $\alpha\beta\gamma\delta\phi$, taking values 1, 2, 3 in the molecular reference frame. N and I denote the number of nonzero and independent components, respectively. Sets of components recurring in various point groups are denoted by lowercase letters.

TABLE IV
Linear χ_R^L, χ_A^L and Nonlinear $\chi_R^{NL}, \kappa_R^{NL}, \chi_A^{NL}$ Molecular Parameters for 102 Magnetic Point Groups

Magnetic Point Group	Hamiltonian H_{I}: $H_{\text{I}}^{(2)}$ χ_R^L	$H_{\text{I}}^{(2)}$ χ_A^L	$H_{\text{I}}^{(4)}$ χ_R^{NL}	$H_{\text{I}}^{(4)}$ κ_R^{NL}	$H_{\text{I}}^{(4)}$ χ_A^{NL}
$1, 2, \underline{2}, 222, \underline{22}2$	χ_R^L	χ_A^L	χ_R^{NL}	κ_R^{NL}	χ_A^{NL}
$\bar{1}, \underline{\bar{1}}, m, \underline{m}, 2/m, \underline{2}/m, 2/\underline{m}, \underline{2}/\underline{m}, mm2, \underline{mm}2, \underline{2m}m, mmm, \underline{mm}m, \underline{mmm}, mm\underline{m}$	χ_R^L	0	χ_R^{NL}	κ_R^{NL}	0
$4, \underline{4}, 422, \underline{4}2\underline{2}, 4\underline{22}$	a_1	b_1	c_1	d_1	e_1
$\bar{4}, \underline{\bar{4}}, 4/m, 4/\underline{m}, \underline{4}/\underline{m}, 4mm, \underline{4}m\underline{m}, 4/\underline{mm}, \bar{4}2m, \underline{\bar{4}}2\underline{m}, \underline{\bar{4}}m\underline{2}, 4\underline{2m}, 4/mmm, \underline{4}/mm\underline{m}, 4/m\underline{mm}, 4/\underline{mmm}, 4/\underline{m}mm, \underline{4}/\underline{mmm}$	a_1	0	c_1	d_1	0
$3, 32, 3\underline{2}, 6, \underline{6}, \infty, 622, \underline{6}2\underline{2}, 6\underline{22}$	a_1	b_1	c_2	d_2	e_2
$\bar{3}, \underline{\bar{3}}, 3m, 3\underline{m}, \bar{3}m, \bar{3}\underline{m}, \underline{\bar{3}m}, \underline{\bar{3}}m, \bar{6}, \underline{\bar{6}}, 6/m, \underline{6}/m, 6/\underline{m}, \underline{6}/\underline{m}, \infty/m, \infty/\underline{m}, 6mm, \underline{6}m\underline{m}, 6/\underline{mm}, \infty m, \infty\underline{m}, \bar{6}m2, \underline{\bar{6}}2\underline{m}, \underline{\bar{6}}m\underline{2}, \bar{6}\underline{m2}, 6/mmm, \underline{6}/\underline{m}m\underline{m}, \underline{6}/m\underline{mm}, 6/\underline{mmm}, 6/\underline{m}mm, \underline{6}/m\underline{m}m, \infty/mm, \infty/\underline{m}m, \infty/m\underline{m}$	a_1	0	c_2	d_2	0
23	a_2	b_2	c_3	d_3	e_3
$432, \underline{4}3\underline{2}$	a_2	b_2	c_3	d_3	e_4
$m3, \underline{m}3, \bar{4}3m, \underline{\bar{4}}3\underline{m}, m3m, \underline{m}3m, m3\underline{m}, \underline{m}3\underline{m}$	a_2	0	c_3	d_3	0
Y, K	a_2	b_2	c_4	d_4	e_5
Y_h, K_h	a_2	0	c_4	d_4	0

Note. where we used the notation

$$\chi_R^L = \frac{N}{3}\text{Re}(\alpha_{11} + \alpha_{22} + \alpha_{33}) \qquad \chi_A^L = -\frac{\text{i}Nk_z}{3\omega}\text{Im}(\rho_{11} + \rho_{22} + \rho_{33})$$

$$a_1 = \frac{N}{3}\text{Re}(2\alpha_{11} + \alpha_{33}) \qquad b_1 = -\frac{\text{i}Nk_z}{3\omega}\text{Im}(2\rho_{11} + \rho_{33})$$

$$a_2 = N\,\text{Re}\,\alpha_{11} \qquad b_2 = -\frac{\text{i}Nk_z}{\omega}\text{Im}\,\rho_{11}$$

$$\chi_R^{NL} = \frac{2N}{15}\text{Re}\{\gamma_{1111} + \gamma_{2222} + \gamma_{3333} - \gamma_{1122} - \gamma_{1133} - \gamma_{2233} + 3[\gamma_{1212} + \gamma_{1313} + \gamma_{2323}]\}$$

$$c_1 = \frac{2N}{15}\text{Re}\{2\gamma_{1111} + \gamma_{3333} - \gamma_{1122} - 2\gamma_{1133} + 3[\gamma_{1212} + 2\gamma_{1313}]\}$$

$$c_2 = \frac{2N}{15}\text{Re}\{\gamma_{3333} + \gamma_{1122} - 2\gamma_{1133} + 7\gamma_{1212} + 6\gamma_{1313}\}$$

$$c_3 = \frac{2N}{5}\text{Re}\{\gamma_{1111} - \gamma_{1122} + 3\gamma_{1212}\} \quad c_4 = 2N\,\text{Re}\,\gamma_{1212}$$

TABLE IV *(Continued)*

$$\kappa_R^{NL} = \frac{2N}{15}\mathrm{Re}\{2(\gamma_{1111}+\gamma_{2222}+\gamma_{3333})+3(\gamma_{1122}+\gamma_{1133}+\gamma_{2233}) +\gamma_{1212}+\gamma_{1313}+\gamma_{2323}\}$$

$$d_1 = \frac{2N}{15}\mathrm{Re}\{2(2\gamma_{1111}+\gamma_{3333})+3(\gamma_{1122}+2\gamma_{1133})+\gamma_{1212}+2\gamma_{1313}\}$$

$$d_2 = \frac{2N}{15}\mathrm{Re}\{2\gamma_{3333}+7\gamma_{1122}+6\gamma_{1133}+9\gamma_{1212}+2\gamma_{1313}\}$$

$$d_3 = \frac{2N}{5}\mathrm{Re}\{2\gamma_{1111}+3\gamma_{1122}+\gamma_{1212}\} \quad d_4 = 2N\,\mathrm{Re}\{\gamma_{1122}+\gamma_{1212}\}$$

$$\chi_A^{NL} = -\frac{i4Nk_z}{15}\left\{\frac{1}{\omega}\mathrm{Im}[-2(\sigma_{1111}+\sigma_{2222}+\sigma_{3333})+\sigma_{1122}+\sigma_{1133} +\sigma_{2233}+\sigma_{2211}+\sigma_{3311}+\sigma_{3322}-3(\sigma_{1212} +\sigma_{1313}+\sigma_{2323}+\sigma_{2121}+\sigma_{3131}+\sigma_{3232})] -\frac{1}{3}\mathrm{Re}[\eta_{1(23)33}-\eta_{1(32)22}+\eta_{2(31)11}-\eta_{2(13)33}+\eta_{3(12)22}-\eta_{3(21)11} +\eta_{3(31)23}-\eta_{2(21)23}+\eta_{1(12)31}-\eta_{3(32)31}+\eta_{2(23)12}-\eta_{1(13)12} -\eta_{1(13)32}+\eta_{1(22)23}-\eta_{2(11)13}+\eta_{2(33)31}-\eta_{3(22)21}+\eta_{3(11)12}]\right\}$$

$$e_1 = -\frac{i8Nk_z}{15}\left\{\frac{1}{\omega}\mathrm{Im}[-2\sigma_{1111}-\sigma_{3333}+\sigma_{1122}+\sigma_{1133}+\sigma_{3311} -3(\sigma_{1212}+\sigma_{1313}+\sigma_{3131})] -\frac{1}{3}\mathrm{Re}[\eta_{1(23)33}+\eta_{2(31)11}+\eta_{3(12)22}+\eta_{3(31)23} +\eta_{1(12)31)}+\eta_{2(23)12}+\eta_{1(22)23}+\eta_{2(33)31}+\eta_{3(11)12}]\right\}$$

$$e_2 = -\frac{i8Nk_z}{15}\left\{\frac{1}{\omega}\mathrm{Im}[-\sigma_{3333}-\sigma_{1122}+\sigma_{1133}+\sigma_{3311}-7\sigma_{1212}-3(\sigma_{1313}+\sigma_{3131})] -\frac{1}{3}\mathrm{Re}[\eta_{1(23)33}+\eta_{2(31)11}+2\eta_{3(12)22}+\eta_{3(31)23} +\eta_{1(12)31}+\eta_{2(23)12}+\eta_{1(22)23}+\eta_{2(33)31}]\right\}$$

$$e_3 = -\frac{i4Nk_z}{5}\left\{\frac{1}{\omega}\mathrm{Im}[-2\sigma_{1111}+\sigma_{1122}+\sigma_{2211}-3(\sigma_{1212}+\sigma_{2121})] -\frac{1}{3}\mathrm{Re}[\eta_{1(23)33}-\eta_{1(32)22}+\eta_{3(31)23}-\eta_{2(21)23}-\eta_{1(33)32}+\eta_{1(22)23}]\right\}$$

$$e_4 = -\frac{i8Nk_z}{5}\left\{\frac{1}{\omega}\mathrm{Im}[-\sigma_{1111}+\sigma_{1122}-3\sigma_{1212}]-\frac{1}{3}\mathrm{Re}[\eta_{1(23)33}-\eta_{3(31)23}+\eta_{1(22)23}]\right\}$$

$$e_5 = i8Nk_z\left\{\frac{1}{\omega}\mathrm{Im}\,\sigma_{1212}+\frac{1}{6}\mathrm{Re}\,\eta_{1(23)33}\right\}$$

References

1. R. J. Glauber, in C. D. Witt, A. Blandin, and C. Cohen-Tannoudji (Eds.), *Quantum Optics and Electronics*, Gordon & Breach, New York, 1965, p. 63.
2. L. Mandel and E. Wolf, *Rev. Mod. Phys.* **37**, 231 (1965).
3. J. R. Klauder and E. C. G. Sudarshan, *Fundamentals of Quantum Optics*, Benjamin, New York, 1968.
4. D. Stoler, *Phys. Rev. Lett.* **33**, 1397 (1974).
5. P. Chmela, *Acta Phys. Pol. A* **55**, 945 (1979).
6. K. Wódkiewicz and M. S. Zubairy, *Phys. Rev. A* **27**, 2003 (1983).
7. C. M. Caves and B. L. Schumaker, *Phys. Rev. A* **31**, 3068 (1985).
8. S. Friberg, C. K. Hong, and L. Mandel, *Opt. Commun.* **54**, 311 (1985).
9. W. Becker, S. A. Shakir, and M. S. Zubairy, *Opt. Commun.* **59**, 395 (1986).
10. L. A. Wu, H. J. Kimble, J. L. Hall, and H. Wu, *Phys. Rev. Lett.* **57**, 2520 (1986).
11. A. Heidmann, R. J. Horowicz, S. Reynaud, E. Giacobino, C. Fabre, and G. Camy, *Phys. Rev. Lett.* **59**, 2555 (1987).
12. J. G. Rarity, P. R. Tapster, and E. Jakeman, *Opt. Commun.* **62**, 201 (1987).
13. D. Yao and Y. Ni, *Phys. Lett. A* **120**, 134 (1987).
14. M. S. Abdalla, R. K. Colegrave, and A. A. Selim, *Physica A* **151**, 467 (1988).
15. P. V. Elyutin and D. N. Klyshko, *Phys. Lett. A* **149**, 241 (1990).
16. A. S. Akhmanov, A. V. Belinskii, and A. S. Chirkin, *Kvant. Elektr.* **15**, 873 (1988).
17. A. V. Belinskii and A. S. Chirkin, *Opt. Spektrosk.* **66**, 1190 (1989).
18. A. V. Belinskii, *Kvant. Elektr.* **17**, 1182 (1990).
19. A. V. Belinskii and A. S. Chirkin, *Opt. Spektrosk.* **69**, 393 (1990).
20. H. J. Carmichael and D. F. Walls, *J. Phys. B* **9**, 1199 (1976).
21. H. J. Kimble and L. Mandel, *Phys. Rev. A* **13**, 2123 (1976).
22. C. Cohen-Tannoudji and S. Reynaud, *J. Phys. B* **10**, 345 (1977).
23. H. J. Kimble, M. Dagenais, and L. Mandel, *Phys. Rev. Lett.* **39**, 691 (1977).
24. M. Dagenais and L. Mandel, *Phys. Rev. A* **18**, 2217 (1978).
25. Z. Ficek, R. Tanaś, and S. Kielich, *Acta Phys. Pol. A* **29**, 2004 (1984).
26. Z. Ficek, R. Tanaś, and S. Kielich, *Acta Phys. Pol. A* **67**, 583 (1985).
27. S. F. Pereira, M. Xiao, H. J. Kimble, and J. L. Hall, *Phys. Rev. A* **38**, 4931 (1988).
28. R. S. Bondurant, P. Kumar, H. J. Shapiro, and M. Maeda, *Phys. Rev. A* **30**, 343 (1984).
29. J. Janszky and Y. Y. Yushin, *Opt. Commun.* **49**, 290 (1984).
30. P. Kumar and J. H. Shapiro, *Phys. Rev. A* **30**, 1568 (1984).
31. P. Kumar, J. H. Shapiro, and R. S. Bondurant, *Opt. Commun.* **50**, 183 (1984).
32. M. D. Levenson, R. M. Shelby, A. Aspect, M. D. Reid, and D. F. Walls, *Phys. Rev. A* **32**, 1550 (1985).
33. R. E. Slusher, L. W. Hollberg, B. Yurke, J. C. Mertz, and J. F. Valley, *Phys. Rev. Lett.* **55**, 2409 (1985).
34. M. D. Reid and D. F. Walls, *Phys. Rev. A* **34**, 4929 (1986).
35. N. A. Ansari and M. S. Zubairy, *Phys. Rev. A* **44**, 2214 (1991).
36. M. Kozierowski and R. Tanaś, *Opt. Commun.* **21**, 229 (1977).

37. J. Mostowski and K. Rzążewski, *Phys. Lett. A* **66**, 275 (1978).
38. R. Neumann and H. Haug, *Opt. Commun.* **31**, 267 (1979).
39. J. Wagner, P. Kurowski, and W. Martiensen, *Z. Phys. B* **33**, 391 (1979).
40. P. Chmela, *Opt. Commun.* **42**, 201 (1982).
41. L. Mandel, *Opt. Commun.* **42**, 437 (1982).
42. M. Kozierowski and S. Kielich, *Phys. Lett. A* **94**, 213 (1983).
43. L. A. Lugiato, G. Strini, and F. D. Martini, *Opt. Lett.* **8**, 256 (1983).
44. S. Friberg and L. Mandel, *Opt. Commun.* **48**, 439 (1984).
45. C. K. Hong and L. Mandel, *Phys. Rev. Lett.* **54**, 323 (1985).
46. S. Kielich, R. Tanaś, and R. Zawodny, *J. Mod. Opt.* **34**, 979 (1987).
47. S. Kielich, R. Tanaś, and R. Zawodny, *J. Opt. Soc. Am. B* **4**, 1627 (1987).
48. A. V. Belinskii and A. S. Chirkin, *Kvant. Elektr.* **16**, 889 (1989).
49. R. Tanaś, in L. Mandel and E. Wolf (Eds.), *Coherence and Quantum Optics V*, Plenum, New York, 1984, p. 645.
50. G. J. Milburn, *Phys. Rev. A* **33**, 674 (1986).
51. G. J. Milburn and C. A. Holmes, *Phys. Rev. Lett.* **56**, 2237 (1986).
52. B. Yurke and D. Stoler, *Phys. Rev. Lett.* **57**, 13 (1986).
53. C. C. Gerry, *Phys. Rev. A* **35**, 2146 (1987).
54. P. Tombesi and A. Mecozzi, *J. Opt. Soc. Am. B* **4**, 1700 (1987).
55. C. C. Gerry and S. Rodrigues, *Phys. Rev. A* **36**, 5444 (1987).
56. G. S. Agarwal, *Opt. Commun.* **62**, 190 (1987).
57. C. C. Gerry and E. R. Vrscay, *Phys. Rev. A* **37**, 4265 (1988).
58. R. Lynch, *Opt. Commun.* **67**, 67 (1988).
59. V. Peřinova and A. Lukš, *J. Mod. Opt.* **35**, 1513 (1988).
60. R. Tanaś, *Phys. Rev. A* **38**, 1091 (1988).
61. G. J. Milburn, A. Mecozzi, and P. Tombesi, *J. Mod. Opt.* **36**, 1607 (1989).
62. D. J. Daniel and G. J. Milburn, *Phys. Rev. A* **39**, 4628 (1989).
63. R. Tanaś, *Phys. Lett. A* **141**, 217 (1989).
64. V. Bužek and I. Jex, *Int. J. Mod. Phys. B* **4**, 659 (1990).
65. C. C. Gerry, *Phys. Lett. A* **146**, 363 (1990).
66. C. C. Gerry, *J. Mod. Opt.* **38**, 1773 (1991).
67. A. Joshi and R. R. Puri, *J. Mod. Opt.* **38**, 473 (1991).
68. A. Miranowicz, R. Tanaś, and S. Kielich, *Quantum Opt.* **2**, 253 (1990).
69. V. Peřinova and A. Lukš, *Phys. Rev. A* **41**, 414 (1990).
70. R. Tanaś, A. Miranowicz, and S. Kielich, *Phys. Rev. A* **43**, 4014 (1991).
71. H. H. Ritze and A. Bandilla, *Opt. Commun.* **30**, 125 (1979).
72. R. Tanaś and S. Kielich, *Opt. Commun.* **30**, 443 (1979).
73. H. H. Ritze, *Z. Phys. B* **39**, 353 (1980).
74. R. Tanaś and S. Kielich, *Opt. Commun.* **45**, 351 (1983).
75. R. Tanaś and S. Kielich, *Opt. Acta* **31**, 81 (1984).
76. S. Kielich and R. Tanaś, *Izv. Akad. Nauk SSSR, Ser. Fiz.* **48**, 518 (1984).
77. N. Imoto, H. A. Haus, and Y. Yamamoto, *Phys. Rev. A* **32**, 2287 (1985).

78 M. Kitagawa and Y. Yamamoto, *Phys. Rev. A* **34**, 3974 (1986).
79. S. Kielich, R. Tanaś, and R. Zawodny, *Phys. Rev. A* **36**, 5670 (1987).
80. T. A. B. Kennedy and P. D. Drummond, *Phys. Rev. A* **38**, 1319 (1988).
81. G. S. Agarwal, *Opt. Commun.* **72**, 253 (1989).
82. G. S. Agarwal and R. R. Puri, *Phys. Rev. A* **40**, 5179 (1989).
83. R. Horák, *Opt. Commun.* **72**, 239 (1989).
84. R. Tanaś and S. Kielich, *Quantum Opt.* **2**, 23 (1990).
85. E. M. Wright, *J. Opt. Soc. Am. B* **7**, 1142 (1990).
86. A. D. Wilson-Gordon, V. Bužek, and P. L. Knight, *Phys. Rev. A* **44**, 7647 (1991).
87. E. M. Wright, *Phys. Rev. A* **43**, 3836 (1991).
88. C. Brosseau, R. Barakat, and E. Rockower, *Opt. Commun.* **82**, 204 (1991).
89. D. Mihalache and D. Baboiu, *Phys. Lett. A* **159**, 303 (1991).
90. M. J. Werner and H. Risken, *Phys. Rev. A* **44**, 4623 (1991).
91. B. C. Sanders and G. J. Milburn, *Phys. Rev. A* **45**, 1919 (1992).
92. N. Tornau and A. Bach, *Opt. Commun.* **11**, 46 (1974).
93. H. D. Simaan and R. Loudon, *J. Phys. A* **8**, 1140 (1975).
94. H. Paul, U. Mohr, and W. Brunner, *Opt. Commun.* **17**, 145 (1976).
95. M. L. Berre-Rousseau, E. Ressayre, and A. Tallet, *Phys. Rev. Lett.* **43**, 1314 (1979).
96. P. Chmela, *Czech. J. Phys. B* **29**, 129 (1979).
97. P. Chmela, *Opt. Quant. Electron.* **11**, 103 (1979).
98. J. Peřina, *Opt. Acta* **26**, 821 (1979).
99. H. P. Yuen and J. H. Shapiro, *Opt. Lett.* **4**, 334 (1979).
100. A. Bandilla and H. H. Ritze, *Opt. Commun.* **32**, 195 (1980).
101. S. Carusotto, *Opt. Acta* **27**, 1567 (1980).
102. P. D. Drummond and C. W. Gardiner, *J. Phys. A* **13**, 2353 (1980).
103. G. P. Hildred, *Opt. Acta* **27**, 1621 (1980).
104. G. Oliver and C. Bendjaballah, *Phy. Rev. A* **22**, 630 (1980).
105. H. Paul and W. Brunner, *Opt. Acta* **27**, 263 (1980).
106. H. Voigt, A. Bandilla, and H. H. Ritze, *Z. Phys. B* **36**, 295 (1980).
107. M. S. Zubairy and J. J. Yeh, *Phys. Rev. A* **21**, 1624 (1980).
108. S. Carusotto, *Physica A* **107**, 509 (1981).
109. P. Chmela, R. Horák, and J. Peřina, *Opt. Acta* **28**, 1209 (1981).
110. P. D. Drummond, K. J. McNeil, and D. F. Walls, *Opt. Acta* **28**, 211 (1981).
111. G. J. Milburn and D. F. Walls, *Opt. Commun.* **39**, 401 (1981).
112. H. Paul and W. Brunner, *Ann. Phys.* **7**, 89 (1981).
113. J. Peřina, *Opt. Acta* **28**, 1529 (1981).
114. P. Chmela, *Opt. Quant. Electron.* **14**, 333, 425 (1982).
115. J. D. Cresser, J. Häger, G. Leuchs, M. Rateike, and H. Walther, in *Topics in Current Physics*, Vol. 27, Springer, Berlin, 1982.
116. D. N. Klyshko, *Zh. Eksp. Teor. Fiz.* **83**, 1313 (1982).
117. L. A. Lugiato and G. Strini, *Opt. Commun.* **41**, 67 (1982).
118. D. F. Walls, G. J. Milburn, and H. J. Carmichael, *Opt. Acta* **29**, 1179 (1982).

119. S. Kielich and R. Tanaś, in *Proc. European Optical Conference*, Rydzyna, Poland, 1983, pp. 5–16.
120. A. Lane, P. Tombesi, H. J. Carmichael, and D. F. Walls, *Opt. Commun.* **48**, 155 (1983).
121. G. J. Milburn and D. F. Walls, *Phys. Rev. A* **27**, 392 (1983).
122. J. Peřina and V. Peřinova, *Opt. Acta* **30**, 955 (1983).
123. R. Short and L. Mandel, *Phys. Rev. Lett.* **51**, 384 (1983).
124. M. S. Zubairy, M. S. K. Razmi, S. Iqbal, and M. Idress, *Phys. Lett. A* **98**, 168 (1983).
125. H. J. Carmichael, G. J. Milburn, and D. F. Walls, *J. Phys. A* **17**, 469 (1984).
126. P. Chmela, *Opt. Quant. Electron.* **16**, 445, 495 (1984).
127. M. J. Collett and C. W. Gardiner, *Phys. Rev. A* **30**, 1386 (1984).
128. C. W. Gardiner and C. M. Savage, *Opt. Commun.* **50**, 173 (1984).
129. M. Hillery, R. F. O'Connell, M. O. Scully, and E. P. Wigner, *Phys. Rep.* **106**, 121 (1984).
130. M. Hillery, M. S. Zubairy, and K. Wódkiewicz, *Phys. Lett. A* **103**, 259 (1984).
131. M. Kozierowski, S. Kielich, and R. Tanaś, in L. Mandel and E. Wolf (Eds.), *Coherence and Quantum Optics V*, Plenum, New York, 1984, p. 71.
132. P. A. Lakshmi and G. S. Agarwal, *Phys. Rev. A* **29**, 2260 (1984).
133. M. D. Levenson, *J. Opt. Soc. Am. B* **1**, 525 (1984).
134. R. Loudon, *Opt. Commun.* **49**, 67 (1984).
135. R. Loudon and T. J. Shepherd, *Opt. Acta* **31**, 1243 (1984).
136. G. J. Milburn, *Opt. Acta* **31**, 671 (1984).
137. G. J. Milburn, D. F. Walls, and M. D. Levenson, *J. Opt. Soc. Am. B* **1**, 390 (1984).
138. A. D. Petrenko and N. I. Zheludev, *Opt. Acta* **31**, 1177 (1984).
139. J. Peřina, V. Peřinova, and J. Kodoušek, *Opt. Commun.* **49**, 210 (1984).
140. J. Peřina, V. Peřinova, C. Sibilia, and M. Bertolotti, *Opt. Commun.* **49**, 285 (1984).
141. M. D. Reid and D. F. Walls, *Opt. Commun.* **50**, 106 (1984).
142. S. Reynaud and A. Heidmann, *Opt. Commun.* **50**, 271 (1984).
143. T. S. Santhanam and M. V. Satyanarayana, *Phys. Rev. D* **30**, 2251 (1984).
144. G. Scharf and D. F. Walls, *Opt. Commun.* **50**, 245 (1984).
145. M. Schubert, W. Vogel, and D. G. Welsch, *Opt. Commun.* **52**, 247 (1984).
146. B. L. Schumaker and C. M. Caves, *J. Opt. Soc. Am. B* **1**, 524 (1984).
147. H. J. Shapiro, P. Kumar, and M. W. Maeda, *J. Opt. Soc. Am. B* **1**, 517 (1984).
148. M. C. Teich, B. E. A. Saleh, and J. Peřina, *J. Opt. Soc. Am. B* **1**, 366 (1984).
149. B. Yurke and J. S. Denker, *Phys. Rev. A* **29**, 1419 (1984).
150. C. W. Gardiner and M. J. Collett, *Phys. Rev. A* **31**, 3761 (1985).
151. A. Heidmann, J. M. Raimond, and S. Reynaud, *Phys. Rev. Lett.* **54**, 326 (1985).
152. M. Hillery, *Phys. Rev. A* **31**, 338 (1985).
153. C. K. Hong and L. Mandel, *Phys. Rev. A* **32**, 974 (1985).
154. E. Jakeman and J. G. Walker, *Opt. Commun.* **55**, 219 (1985).
155. P. Kask, P. Piksarv, and U. Mets, *EuroBiophys. J.* **12**, 163 (1985).
156. S. Kielich, M. Kozierowski, and R. Tanaś, *Opt. Acta* **32**, 1023 (1985).
157. P. A. Lakshmi and G. S. Agarwal, *Phys. Rev. A* **32**, 1643 (1985).
158. M. W. Maeda, P. Kumar, and J. H. Shapiro, *Phys. Rev. A* **32**, 3803 (1985).

159. K. E. Süsse, W. Vogel, D. G. Welsch, and D. Kühlke, *Phys. Rev. A* **31**, 2435 (1985).
160. J. G. Walker and E. Jakeman, *Opt. Acta* **32**, 1303 (1985).
161. M. Wolinsky and H. J. Carmichael, *Opt. Commun.* **55**, 138 (1985).
162. S. A. Akhmanov, N. I. Zheludev, and R. S. Zadoyan, *Zh. Eksp. Theor. Fiz* **91**, 984 (1986).
163. P. Garcia-Fernandez, L. Sainz de Los Terreros, F. J. Bermejo, and J. Santor, *Phys. Lett. A* **118**, 400 (1986).
164. H. A. Haus and Y. Yamamoto, *Phys. Rev. A* **34**, 270 (1986).
165. Y. Yamamoto and H. A. Haus, *Rev. Mod. Phys.* **58**, 1001 (1986).
166. P. Chmela, *Czech. J. Phys. B* **37**, 1130 (1987).
167. J. Janszky and Y. Yushin, *Phys. Rev. A* **36**, 1288 (1987).
168. R. Lynch, *J. Opt. Soc. Am. B* **4**, 1723 (1987).
169. S. Machida, Y. Yamamoto, and Y. Itaya, *Phys. Rev. Lett.* **58**, 1000 (1987).
170. V. I. Zakharov and V. G. Tyuterev, *Lasers and Particle Beams* **5**, 27 (1987).
171. T. A. B. Kennedy and E. M. Wright, *Phys. Rev. A* **38**, 212 (1988).
172. P. Tombesi and A. Mecozzi, *Phys. Rev. A* **37**, 4778 (1988).
173. C. C. Gerry and C. Johnson, *Phys. Rev. A* **40**, 2781 (1989).
174. P. Tombesi, *Phys. Rev. A* **39**, 4288 (1989).
175. E. K. Bashkirov and A. S. Shumovsky, *Int. J. Mod. Phys. B* **4**, 1579 (1990).
176. D. N. Klyshko, *Phys. Lett. A* **146**, 93 (1990).
177. P. Kumar, O. Aytür, and J. Huang, *Phys. Rev. Lett.* **64**, 1015 (1990).
178. J. Peřina and J. Bajer, *Phys. Rev. A* **41**, 516 (1990).
179. Z.-M. Zhang, L. Xu, and J.-L. Chai, *Phys. Lett. A* **151**, 65 (1990).
180. B. A. Zon and H. A. Kuznietsova, *Opt. Spektrosk* **69**, 192 (1990).
181. L.-B. Deng and L.-Z. Zhang, *J. Mod. Opt.* **38**, 877 (1991).
182. C. C. Gerry, R. Grobe, and E. R. Vrscay, *Phys. Rev. A* **43**, 361 (1991).
183. L. Hardy, *Europhys. Lett.* **15**, 591 (1991).
184. M. Hillery, *Phys. Rev. A* **44**, 4578 (1991).
185. M. J. Holland, D. F. Walls, and P. Zoller, *Phys. Rev. Lett.* **67**, 1716 (1991).
186. A. Kumar, *Phys. Rev. A* **44**, 2130 (1991).
187. C. T. Lee, *Phys. Rev. A* **44**, 2775 (1991).
188. M. J. Werner and H. Risken, *Quantum Opt.* **3**, 185 (1991).
189. M. Zahler and Y. B. Aryeh, *Phys. Rev. A* **43**, 6368 (1991).
190. D. F. Walls, *Nature* **280**, 451 (1979).
191. R. Loudon, *Rep. Progress Phys.* **43**, 913 (1980).
192. H. Paul, *Rev. Mod. Phys.* **54**, 1061 (1982).
193. D. F. Walls, *Nature* **306**, 141 (1983).
194. G. Leuchs, in G. T. Moore and M. O. Scully (Eds), *Frontiers of Nonequilibrium Statistical Physics*, Plenum, New York, 1986.
195. R. Loudon and P. L. Knight, *J. Mod. Opt.* **34**, 709 (1987).
196. M. C. Teich and B. E. A. Saleh, *Progress in Optics* **26**, 1 (1988).
197. K. Zaheer and M. S. Zubairy, *Adv. Atom. Mol. Opt. Phys.* **28**, 143 (1990).

198. S. Kielich, *Nonlinear Molecular Optics*, Nauka, Moscow, 1981.
199. J. Peřina, *Quantum Statistics of Linear and Nonlinear Optical Phenomena*, Reidel, Dordrecht, 1984.
200. M. Schubert and B. Wilhelmi, *Nonlinear Optics and Quantum Electronics*, Wiley, New York, 1986.
201. H. P. Yuen and V. W. S. Chan, *Opt. Lett.* **8**, 177 (1983).
202. B. L. Schumaker, *Opt. Lett.* **9**, 189 (1984).
203. P. D. Maker, R. W. Terhune, and C. M. Savage, *Phys. Rev. Lett.* **12**, 57 (1964).
204. R. Tanaś and S. Kielich, *J. Mod. Opt.* **37**, 1935 (1990).
205. J. Fiutak, *Can. J. Phys.* **41**, 12 (1963).
206. S. Kielich, *Proc. Phys. Soc.* **86**, 709 (1965).
207. S. Kielich, *Physica* **32**, 385 (1966).
208. Y. R. Shen, *Phys. Rev.* **155**, 921 (1967).
209. S. Kielich, *Acta Phys. Pol.* **28**, 459 (1965).
210. S. Kielich, *Acta Phys. Pol.* **30**, 851 (1966).
211. P. S. Pershan, *Phys. Rev.* **130**, 919 (1963).
212. P. W. Atkins and A. D. Wilson, *Mol. Phys.* **24**, 33 (1972).
213. K. J. Blow, R. Loudon, S. J. D. Phoenix, and T. J. Shepherd, *Phys. Rev. A* **42**, 4102 (1990).
214. K. J. Blow, R. Loudon, and S. J. D. Phoenix, *J. Opt. Soc. Am. B* **8**, 1750 (1991).
215. S. Kielich and R. Zawodny, *Opt. Commun.* **15**, 267 (1975).
216. S. Kielich, *Opto-Electronics* **1**, 75 (1969).
217. R. Tanaś and T. Gantsog, *Opt. Commun.* **87**, 369 (1992).
218. S. M. Arakelian and J. S. Chilingarian, *Nonlinear Optics of Liquid Crystals*, Nauka, Moscow, 1984.
219. D. T. Pegg and S. M. Barnett, *Phys. Rev. A* **39**, 1665 (1989).
220. T. Gantsog and R. Tanaś, *J. Mod. Opt.* **38**, 1021 (1991).
221. T. Gantsog and R. Tanaś, *J. Mod. Opt.* **38**, 1537 (1991).
222. S. Chaturvedi and V. Srinivasan, *Phys. Rev. A* **43**, 4054 (1991).
223. T. Gantsog and R. Tanaś, *Phys. Rev. A* **44**, 2086 (1991).
224. R. Tanaś and T. Gantsog, *J. Opt. Soc. Am. B* **8**, 2505 (1991).
225. S. Kielich, *Progress in Optics* **20**, 155 (1983).
226. S. Kielich, *Acta Phys. Pol.* **29**, 875 (1966).
227. R. Zawodny and H. Drozdowicz, in S. Kielich (Ed.), *Selected Problems in Nonlinear Optics Ser. Fizyka No.27*, Poznań University Press, 1978, p. 119.
228. R. Zawodny, thesis, Poznań, 1977.
229. E. P. Wigner, *Group Theory and Application to Quantum Mechanics of Atomic Spectra*, Academic, New York, 1959.
230. R. R. Birss, *Symmetry and Magnetism*, North-Holland, Amsterdam, 1965.

AUTHOR INDEX

Numbers in parentheses are reference numbers and indicate that the author's work is referred to although his name is not mentioned in the text. Numbers in *italic* show the pages on which the complete references are listed.

SUBJECT INDEX